Enzyme Systems that Metabolise Drugs and Other Xenobiotics

Current Toxicology Series

Series Editors

Diana Anderson
Department of
Biomedical Sciences
University of Bradford, UK

Michael D Waters
Consultant
Chapel Hill
NC, USA

Timothy C Marrs
Food Standards Agency
London, UK

Toxicology is now considered to be a topic worthy of study and research in its own right, having originally arisen as a subsection of pharmacology. This rapid growth in the significance of toxicology necessitates specialised yet comprehensive information that is easily accessible both to professionals and to the increasing number of students with an interest in the subject area.

Comprising professional and reference books, primarily aimed at an academic/industrial/professional audience, the *Current Toxicology Series* will cover a wide variety of 'core' toxicology topics, thus building into a comprehensive range of books suitable for use both as an updating tool and as a reference source.

Published titles

Nutrition and Chemical Toxicity
Edited by C Ioannides (0 471 974536)

Toxicology of Contact Dermatitis: Allergy, Irritancy and Urticaria
D Basketter, F Gerberick, I Kimber and C Willis (0 471 97201 0)

Food Borne Carcinogens: Heterocyclic Amines
Edited by M. Nagao and T. Sugimura (0 471 98399 3)

Enzyme Systems that Metabolise Drugs and Other Xenobiotics
Edited by C Ioannides (0 471 89466 4)

Enzyme Systems that Metabolise Drugs and Other Xenobiotics

Edited by

Costas Ioannides
School of Biomedical and Life Sciences, University of Surrey, Guildford, UK

JOHN WILEY & SONS, LTD

Baffins Lane, Chichester,
West Sussex PO19 1UD, UK

National 01243 779777
International (+44) 1243 779777
e-mail (for orders and customer service enquiries): cs-books@wiley.co.uk
Visit our Home Page on: http://www.wiley.co.uk or http://www.wiley.com

Other Wiley Editorial Offices

John Wiley & Sons, Inc., 605 Third Avenue,
New York, NY 10158-0012, USA

WILEY-VCH Verlag GmbH, Pappelallee 3,
D-69469 Weinhaim, Germany

John Wiley & Sons Australia, Ltd., 33 Park Road, Milton,
Queensland 4064, Australia

John Wiley & Sons (Asia) Pte, Ltd., 2 Clementi Loop #02-01,
Jin Xing Distripark, Singapore 129809

John Wiley & Sons (Canada), Ltd., 22 Worcester Road,
Rexdale, Ontario M9W 1L1, Canada

Library of Congress Cataloging-in-Publication Data

Enzyme systems that metabolise drugs and other xenobiotics / edited by Costas Ioannides.
p. ; cm. – (Current toxicology series)
Includes bibliographical references and index.
ISBN 0-471-89466-4 (alk. paper)
1. Drugs–Metabolism. 2. Xenobiotics–Metabolism. 3. Enzymes–Mechanism of action. I. Ioannides, Costas. II. Series.
[DNLM: 1. Pharmaceutical Preparations–metabolism. 2. Enzymes–physiology. 3. Metabolic Detoxication, Drug–physiology. 4. Xenobiotics–metabolism. QV 38 E61 2001]
RM301.55.E56 2001
615′.7–dc21

2001039011

British Library Cataloguing in Publication Data

A catalogue record for this book is available from the British Library

ISBN 0 471 89466 4

Typeset in 10/12pt Times from the authors' disks by Keytec Typesetting Ltd., Bridport, Dorset, England
Printed and bound in Great Britain by Antony Rowe, Chippenham, Wilts
This book is printed on acid-free paper responsibly manufactured from sustainable forestry, in which at least two trees are planted for each one used for paper production.

Contents

Contributors   vii
Forward   ix

1 Xenobiotic Metabolism: An Overview   1
   C. Ioannides

2 Cytochrome P450   33
   F.P. Guengerich

3 Flavin Monooxygenases   67
   J.R. Cashman

4 Amine Oxidases and the Metabolism of Xenobiotics   95
   K.F. Tipton and M. Strolin Benedetti

5 Molybdenum Hydroxylases   147
   C. Beedham

6 Prostaglandin Synthases   189
   G.H. Degen, C. Vogel and J. Abel

7 Lipoxygenases   231
   A.P. Kulkarni

8 UDP-Glucuronosyltransferases   281
   K.W. Bock

9 Glutathione *S*-transferases   319
   P.J. Sherratt and J.D. Hayes

10 Sulphotransferases   353
   H. Glatt

11 Arylamine Acetyltransferases   441
   G.N. Levy and W.W. Weber

12 Mammalian Xenobiotic Epoxide Hydrolases   459
   M. Arand and F. Oesch

13 Methyltransferases   485
   C.R. Creveling

14 The Amino Acid Conjugations   501
     G.B. Steventon and A.J. Hutt

15 Deconjugating Enzymes; Sulphatases and Glucuronidases   521
     C. Kunert-Keil, C.A. Ritter, H.K. Kroemer, and B. Sperker

16 Nitroreductases and Azoreductases   555
     S. Zbaida

Index   567

Contributors

Josef Abel, Institut für Arbeitsphysiologie an der Universität Dortmund, Ardeystrasse 67, D-44139 Dortmund, Germany

Michael Arand, Institute of Toxicology, University of Mainz, D-55131 Mainz, Germany

Christine Beedham, School of Pharmacy, University of Bradford, Bradford BD7 1DP, UK

Margherita Strolin Benedetti, UCB Pharma, BP 314, 21 rue de Neuilly, 92003 Nanterre Cedex, France

K.W. Bock, Institute of Toxicology, University of Tübingen, Wihelmstrasse 56, D-72074 Tübingen, Germany

John R. Cashman, Human BioMolecular Research Institute, 5310 Eastgate Mall, San Diego, CA 92121, USA

C.R. Creveling, NIDDK, NIH, Bethedsa, MD 20892, USA

Gisela H. Degen, Institut für Arbeitsphysiologie an der Universität Dortmund, Ardeystrasse 67, D-44139 Dortmun, Germany

Hansruedi Glatt, German Institute of Human Nutrition (DlfE), Department of Toxicology, Arthur-Scheunert-Allee 114-116, 14558 Potsdam-Rehbrucke, Germany

F. Peter Guengerich, Department of Biochemistry and Center in Molecular Toxicology, Vanderbilt University School of Medicine, 638 Medical Research Building 1, 23rd and Pierce Avenues, Nashville, TN 37232-0146, USA

John D. Hayes, Biomedical Research Centre, Ninewells Hospital and Medical School, University of Dundee, Dundee DD1 9SY, UK

A.J. Hutt, Department of Pharmacy, School of Health and Life Sciences, King's College London, Franklin-Wilkins Building, 150 Stamford Street, London SE1 9NN, UK

Costas Ioannides, School of Biomedical and Life Sciences, University of Surrey, Guildford, Surrey GU2 7XH, UK

Heyo K. Kroemer, Institut für Pharmakologie, Ernst Moritz Arndt Universität Greifswald, Greifswald, Germany

Arun P. Kulkarni, Florida Toxicology Research Center, Department of Environmental and Occupational Health, College of Public Health, MDC-056 University of Florida, 13201 Bruce B Down Boulevard, Tampa, FL 33612-3085, USA

Christiane Kunert-Keil, Institut für Pharmakologie, Ernst Moritz Arndt Universität Greifswald, Greifswald, Germany

Gerald N. Levy, Department of Pharmacology, 1150 W Medical Center Drive, University of Michigan, Ann Arbor, MI 48109-0632, USA

Franz Oesch, Institute of Toxicology, University of Mainz, D-55131 Mainz, Germany

Christoph A. Ritter, Institut für Pharmakologie, Ernst Moritz Arndt Universität Greifswald, Greifswald, Germany

Philip J. Sherratt, Biomedical Research Centre, Ninewells Hospital and Medical School, University of Dundee, Dundee DD1 9SY, UK

Bernhard Sperker, Institut für Pharmakologie, Ernst Moritz Arndt Universität Greifswald, Greifswald, Germany

G.B. Steventon, Department of Pharmacy, School of Health and Life Sciences, King's College London, Franklin-Wilkins Building, 150 Stamford Street, London SE1 9NN, UK

Keith F. Tipton, Department of Biochemistry, Trinity College, Dublin 2, Ireland

Christoph Vogel, Institut für Arbeitsphysiologie an der Universität Dortmund, Ardeystrasse 67, D-44139 Dortmund, Germany

Wendell W. Weber, Department of Pharmacology, 1150 W Medical Center Drive, University of Michigan, Ann Arbor, MI 48109-0632, USA

Shmuel Zbaida, Department of Drug Matabolism and Pharmacokinetics, Mail Stop: K-15-3; 3700, Schering-Plough Research Institute, 2015 Galloping Hill Road, Kenilworth, NJ 07033-1300, USA

Foreword

It is now generally recognised that an appreciation of xenobiochemistry is essential for the safe and effective usage of the numerous chemicals exploited by today's Society as medicinal drugs, food additives and agrochemicals. Moreover, such knowledge enhances our understanding of the biohandling of the plethora of chemicals, both synthetic and naturally-occurring, to which we are continuously, and inadvertently exposed, for example through our daily diet, exposure to tobacco products and inhalation of vehicle exhaust fumes.

The last fifty years have witnessed tremendous and unprecedented advances in our understanding of the chemistry and biochemistry governing the biotransformation of xenobiotics, whereby the body rids itself of unwanted, and potentially deleterious, lipophilic xenobiotics by metabolically converting them to more readily excretable hydrophilic metabolites, a process known as detoxication. However, it is now evident that such metabolism is not always beneficial to the living organism, as in some cases the generated metabolites are highly reactive, capable of covalent interaction with critical cellular macromolecules leading to various forms of toxicity, a process known as bioactivation.

The present volume consolidates our current knowledge of the xenobiotic-metabolising enzyme systems. Previous books on the topic were usually restricted to a single enzyme system involved in phase 1 or phase 2 metabolism, whereas this volume, for the first time, brings together under a single cover the majority of enzyme systems utilised in both phases of metabolism. Each chapter discusses critically and in detail one specific enzyme system, dealing with its substrate specificity, underlying mechanism of action emphasising its role in the activation and detoxication of xenobiotics, tissue and species distribution, age-dependent development, sex differences, and the endogenous and exogenous factors that regulate its activity. The authors of each chapter are internationally recognised authorities in their respective fields, having made original contributions to the understanding of their chosen topic.

This book will be invaluable to those in the Chemicals Industry, particularly Pharmaceutical Industry, who investigate the metabolism and pharmacokinetics of new chemicals or investigate possible drug interactions, and those concerned with their safety and risk assessment. The enthusiastic neophyte will also benefit from the wealth of information to be found in this book as well as the postgraduate student pursuing an advanced course in this area.

Professor John W. Gorrod
Toxicology Unit
John Tabor Laboratories
University of Essex

1 Xenobiotic Metabolism: An Overview

Costas Ioannides

University of Surrey, UK

The meteoric advances in analytical technology that occurred during the last two decades have made possible the analysis of complex mixtures and the determination of chemicals present in minute quantities, thus allowing us to appreciate for the first time the number and diversity of chemicals to which we are exposed, from conception to death. To a surprisingly large number of people, the word 'chemicals' has become synonymous with manmade chemicals that have been synthesised by the chemist, and scant attention is usually paid to the myriad chemicals that occur in nature. Of the chemicals that humans ingest, 99.9% are natural, largely of plant origin (Ames and Gold 1998; Ames *et al.* 1990). These are chemicals that the plants generate to defend themselves and consequently are biologically active. Ames has estimated that North Americans consume some 5000–10 000 different chemicals and their breakdown products at a dose of 1.5 g per person per day. To this, one has to add the chemicals that are generated during cooking, 2 g of burnt material per person per day which also contains numerous chemicals, including many established chemical carcinogens such as heterocyclic amines and polycyclic aromatic hydrocarbons (Skog and Jägerstad 1998). Indeed, more than 1000 chemicals have been detected in coffee. Exposure to chemicals is, thus, continuous, inevitable and unavoidable.

Naturally occurring chemicals, similar to their anthropogenic counterparts, are biologically active, and thus capable of modulating physiological processes in humans. Undoubtedly the major source of human exposure to chemicals is diet. Many chemicals present in our everyday diet have been shown to display carcinogenicity in humans. Most of these derive from plants, such as hydrazines in the edible mushroom *agaricus bisporus* (Toth 1995), and safrole and estragole in spices (Mori *et al.* 1998). Furthermore, food may be contaminated by chemicals emanating from packages, such as the phthalate esters, or produced during storage, such as mycotoxins which are generated by contaminating fungi (Wang *et al.* 1998). Finally, potent chemical carcinogens, such as heterocyclic amines and polycyclic aromatic hydrocarbons, are

Enzyme Systems that Metabolise Drugs and Other Xenobiotics. Edited by C. Ioannides.
© 2002 John Wiley & Sons Ltd

generated during the normal cooking of food (Skog and Jägerstadt 1998). The biological activity of naturally occurring chemicals, however, is not always detrimental to the living organism, and it is increasingly being recognised that plant constituents in food, such as flavanols in tea and organosulphates in garlic, may possess anticancer activity (Mori and Nishikawa 1996; Ahmad *et al.* 1998), and this realisation has led to enormous current research effort, aimed at better defining these dietary anticancer chemicals and elucidating their mechanism of action (American Institute for Cancer Research 1997).

The body cannot exploit the chemicals to which it is exposed to generate energy, build new tissues, as chemical messengers or as cofactors, and consequently these chemicals are frequently referred to as xenobiotics (*Greek*: foreign to life). It recognises these as being foreign, and potentially detrimental, and its first line of defence is to eliminate them. Chemicals, however, that find their way into the body are lipophilic as they need to traverse lipoid membranes in order to reach the systemic circulation, through which they are distributed to the body. The capacity of the body to excrete lipophilic compounds is poor since, for example, lipophilic compounds that are excreted by glomerular filtration and active secretion in the kidneys are extensively reabsorbed because of their lipophilic character.

The body has, therefore, developed a number of efficient enzyme systems adept at metabolically converting lipophilic xenobiotics to hydrophilic metabolites, thus facilitating their elimination and minimising its exposure to them. The biological half-life of the chemical is shortened, thus reducing exposure and, moreover, the risk of accumulation on repeated or continuous exposure is decreased. In most cases, such metabolism also abolishes the biological activity of the xenobiotic, by, on the one hand, preventing its distribution to its biological receptor and, on the other, by markedly diminishing its affinity for the receptor, in comparison with the parent compound. Thus metabolism has a very profound effect on the biological activity of a chemical, regulating its nature, i.e. intensity and duration. It is unlikely that the human body could withstand the constant onslaught of chemicals and survive in such a hostile chemical environment if such effective defence mechanisms were not operative.

Metabolism of xenobiotics

The process of transforming lipophilic chemicals to polar entities occurs in two distinct metabolic phases. During Phase I, functionalisation, an atom of oxygen is incorporated into the chemical, and functional groups such as –OH, –COOH etc. are generated, as in the hydroxylation of diazepam (Figure 1.1); alternatively, such functional groups may be also formed following reduction, as in the case of the antitumour agent misonidazole (Figure 1.2), or unmasked, as in the deethylation of ethoxycoumarin (Figure 1.1). Such metabolites not only are more polar than the parent compound but, furthermore, are capable of undergoing Phase II metabolism, conjugation, where endogenous substrates, e.g. glucuronic acid and sulphate, are added to them to form highly hydrophilic molecules, ensuring in this way their elimination (Figure 1.1). Xenobiotics that already possess a functional group can bypass Phase I metabolism and directly participate in conjugation reactions; the mild analgesic

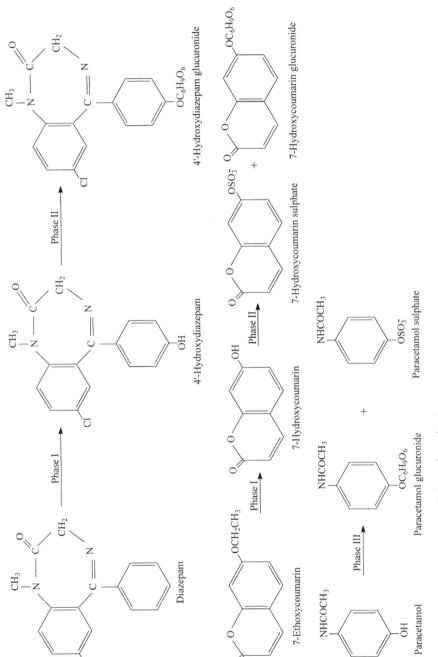

Figure 1.1 Phase I and Phase II metabolism of xenobiotics.

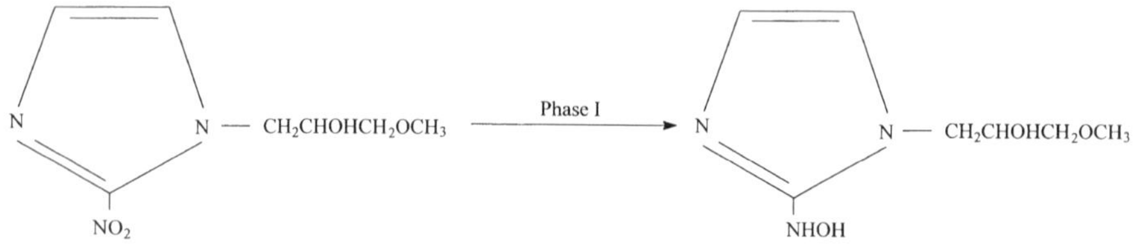

Misonidazole

N-Hydroxymisonidazole

Figure 1.2 Reductive metabolism of xenobiotics.

paracetamol (acetaminophen) is metabolised primarily by sulphate and glucuronide conjugation (Figure 1.1).

It is also possible, but rare, that metabolism may lead to the formation of a more lipophilic metabolite. An important pathway in the metabolism of sulphohamides is *N*-acetylation, producing the less water-soluble acetylated metabolites (Figure 1.3). Such metabolites, as a result of their poor water solubility, may crystallise out in the kidney tubule, causing tissue necrosis and giving rise to a condition known as 'crystalluria'.

A number of enzyme systems contribute to xenobiotic metabolism, both Phase I and Phase II, the majority of which are localised in the endoplasmic reticulum and the cytosolic fraction of the cell. In mammals, they are encountered in every tissue, but particularly in the liver, which consequently functions as the principal site of xenobiotic metabolism. A Phase III metabolism is sometimes referred to and involves the further metabolism of products emanating from Phase II metabolism, the reactions being catalysed by enzymes that participate in the other two phases of metabolism. For example, conjugation with glutathione is an important route of metabolism, through which the cell detoxicates reactive metabolites and protects itself from their detrimental effects (see below). Glutathione conjugates may be further metabolised to mercapturates that are readily excreted in the urine and bile. This involves a sequential loss of glutamate and glycine to yield the cysteinyl derivative that is *N*-acetylated to generate the mercapturate. The cysteinyl derivative may be also metabolised by a pyridoxal phosphate-dependent β-lyase to form a thiol, releasing glutamate and ammonia (Figure 1.4). The thiol may give rise to toxicity or may undergo methylation (Anders and Dekant 1998).

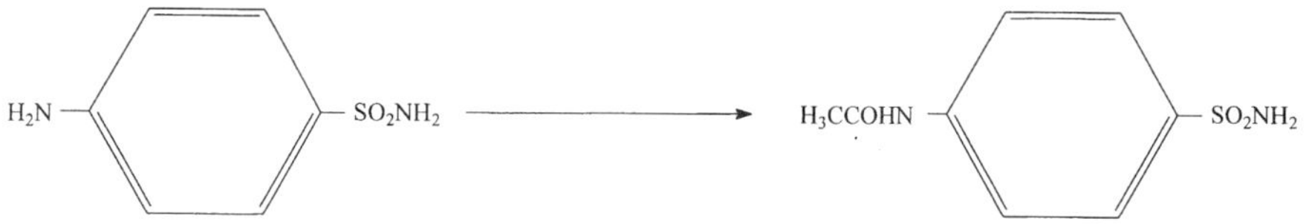

Sulphanilamide

Acetylsulphanilamide

Figure 1.3 N-Acetylation of sulphanilamide.

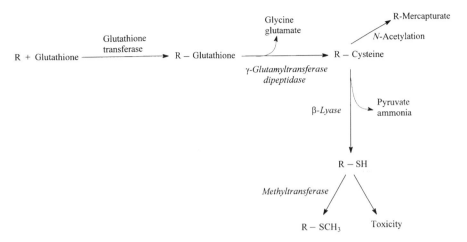

Figure 1.4 Phase III metabolism of glutathione conjugates.

Xenobiotic metabolism by gut microflora

Xenobiotics taken orally as well as those that are excreted into the bile may be subject to metabolism effected by gastrointestinal microorganisms (Rowland 1988). These microorganisms are particularly adept at carrying out reductive metabolism, e.g. nitro- and azo-reduction, and are also effective in catalysing hydrolytic reactions (see below). Metabolism by gut microflora of biliary excreted conjugates may result in cleavage of the conjugate releasing the less polar precursor which, as a result of the increased lipophilicity, may be reabsorbed through the intestine, and this cycling phenomenon is known as 'enterohepatic circulation'. The consequence is that the chemical burden of the living organism is increased since it is continuously re-exposed to the same chemical, until elimination is achieved by renal excretion and loss in the intestine.

Phase I metabolic pathways

AROMATIC HYDROXYLATION

This is one of the commonest pathways of metabolism leading to the generation of phenolic products which are then subject to metabolism by Phase II conjugation. Almost every compound containing a benzene ring undergoes aromatic hydroxylation. For example, the industrial chemical biphenyl is hydroxylated at two different sites, producing 2- and 4-hydroxybiphenyl (Figure 1.5). Aromatic hydroxylation may proceed via the formation of an epoxide which rearranges to a phenol (see below).

EPOXIDATION

This is a frequent pathway in the metabolism of many carcinogenic compounds such as aromatic hydrocarbons, mycotoxins, e.g. aflatoxin B_1 and vinyl chloride

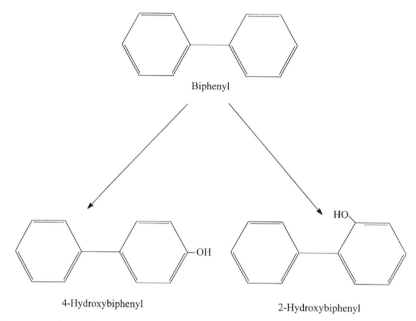

Biphenyl

4-Hydroxybiphenyl   2-Hydroxybiphenyl

Figure 1.5 Biphenyl hydroxylation.

(Figure 1.6); an atom of oxygen is added across olefinic and acetylenic bonds. Such epoxides are frequently very reactive and have been implicated in the toxicity and carcinogenicity of many chemicals.

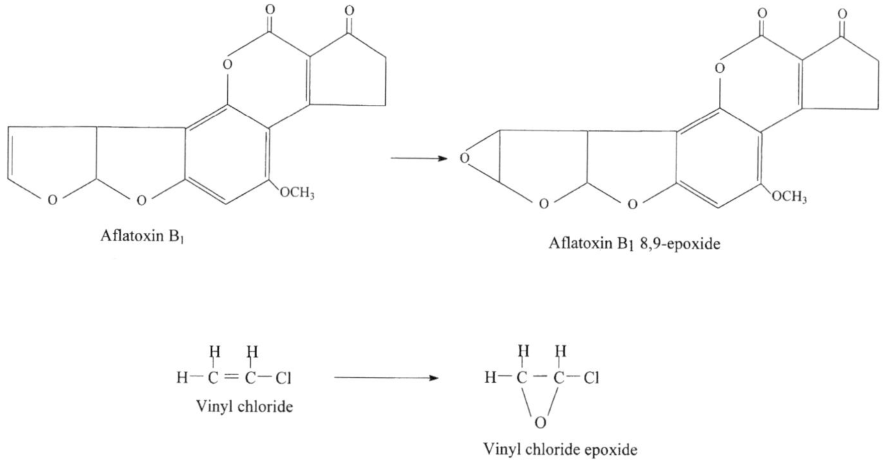

Aflatoxin B$_1$ Aflatoxin B$_1$ 8,9-epoxide

Vinyl chloride   Vinyl chloride epoxide

Figure 1.6 Epoxidation of aflatoxin B$_1$ and vinyl chloride.

ALIPHATIC HYDROXYLATION

Many medicinal drugs are metabolised and deactivated through aliphatic hydroxylation, including the barbiturate pentobarbitone (Figure 1.7). Further oxidation of the alcohol may occur, to yield the corresponding acid.

DEALKYLATION REACTIONS

This is a frequently observed reaction where alkyl groups attached to oxygen, nitrogen or sulphur may be dealkylated as in the case of ethoxycoumarin (Figure 1.1). These dealkylation reactions proceed through oxidation of the alkyl group and rearrangement resulting in loss of the alkyl group as the aldehyde (Figure 1.8).

NITROGEN AND SULPHUR OXIDATION

Nitrogen may be oxidised to form oxides and hydroxylamines, the latter being a pathway of metabolism of compounds having an exocyclic amino group, including many carcinogenic aromatic amines, and generally leads to the expression of toxicity and carcinogenicity (Figure 1.9).

OXIDATIVE DEAMINATION

This results in the release of ammonia and, as in the dealkylation reactions, it involves an initial carbon oxidation (see above) to generate an unstable intermediate (Figure 1.10).

OXIDATIVE DEHALOGENATION

Oxidative dehalogenation is an important pathway in the metabolism of halogenated hydrocarbons such as the anaesthetic halothane, which may also be metabolised by reductive dehalogenation (Figure 1.11).

NITROREDUCTION

Nitroreduction is catalysed not only by mammalian enzymes but also by microbial enzymes which thus contribute to the presystemic metabolism of xenobiotics. For example, the antibacterial drug chloramphenicol undergoes reduction, mediated largely by the gut flora (Figure 1.12).

AZOREDUCTION

Both microbial and mammalian enzymes can catalyse the azoreduction of chemicals. For example, the active form of the anti-inflammatory drug sulphasalazine is 5-aminosalicylic acid, which is released into the intestine following azoreduction catalysed by the gut flora (Figure 1.13).

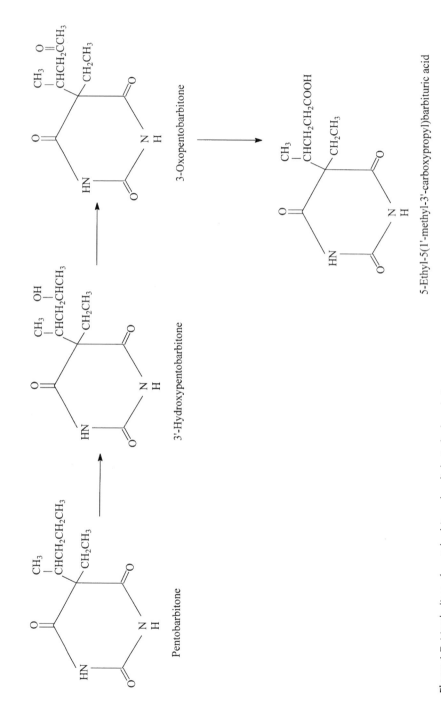

Figure 1.7 Metabolism of pentobarbitone by aliphatic hydroxylation.

$$R - CH_3 \longrightarrow R - CH_2OH \longrightarrow RH + HCHO$$

Figure 1.8 Mechanism of demethylation reactions.

REDUCTIVE DEHALOGENATION

This is an important pathway in the metabolism of halogenated hydrocarbons such as the carcinogen carbon tetrachloride and the anaesthetic halothane (Figure 1.11). Haloalkane-free radicals are initially formed as a result of the homolytic cleavage of the carbon–halogen bond.

HYDROLYSIS

A number of esters, amides and hydrazides are hydrolysed by esterases and amidases. The butyl ester of chlorambucil is hydrolysed to the free drug, the herbicide propanil is principally metabolised by amide hydrolysis, and the hydrazide isoniazid, an anti-tubercular drug, is hydrolysed to isonicotinic acid (Figure 1.14). These hydrolytic enzymes are present in many tissues, usually residing in the cytosol but also encountered in the endoplasmic reticulum. The plasma esterases also make a major contribution in the hydrolysis of many chemicals.

Phase II metabolic pathways

Conjugation reactions with endogenous substrates yield highly hydrophilic, and thus readily excretable, metabolites. However, functional groups may be also methylated or acetylated to produce less hydrophilic compounds.

GLUCURONIDE CONJUGATION

This is one of the most frequently utilised routes of conjugation where glucuronic acid, made available in the form of uridine diphosphate glucuronic acid (UDPGA), is added to the molecule thus conferring to it a high degree of hydrophilicity leading to its excretion in the bile and urine (Figure 1.1). The most readily conjugated functional groups are phenols and alcohols, yielding ester glucuronides, and carboxylic acids forming ether glucuronides. N-Glucuronidation is an important step in the metabolism of aromatic amines, many of which are carcinogenic, leading to their deactivation and excretion.

SULPHATE CONJUGATION

Conjugation with sulphate, facilitated by cytosolic sulphotransferases, is also a major route of Phase II metabolism, where inorganic sulphate, made available in the form of 3'-phosphoadenosine-5'-phosphosulphate (PAPS), is added to the molecule (Figure 1.1). This is the most important pathway in the metabolism of phenols and is a very

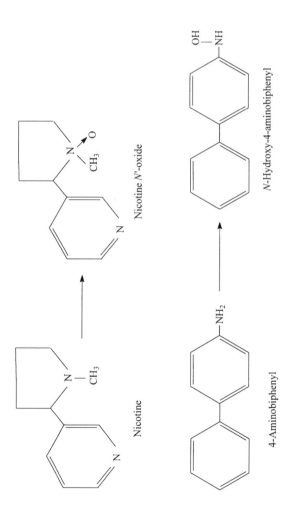

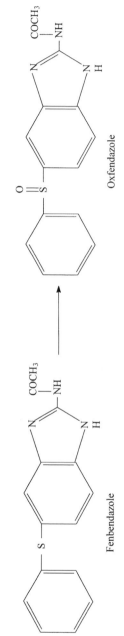

Figure 1.9 N- and S-oxidation of xenobiotics.

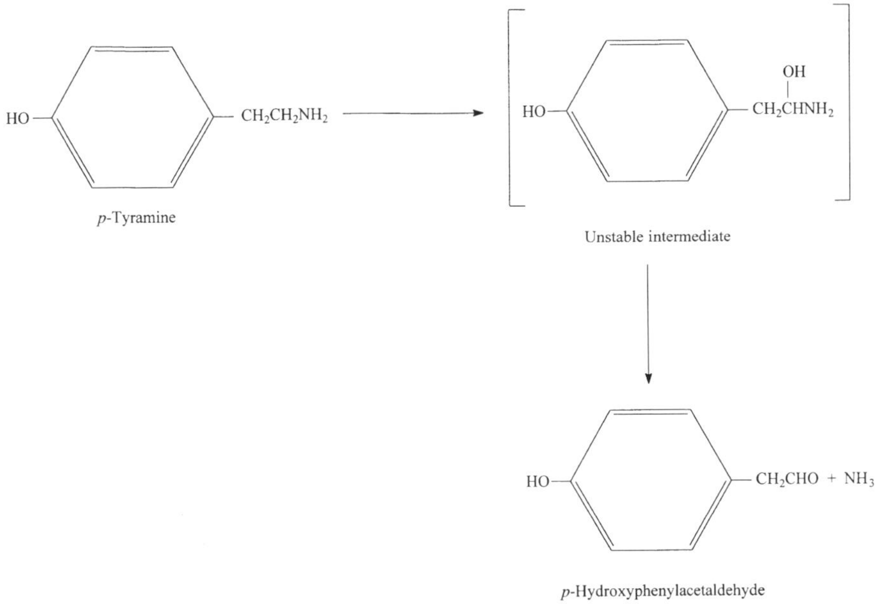

p-Tyramine

Unstable intermediate

p-Hydroxyphenylacetaldehyde

Figure 1.10 Oxidative deamination of *p*-tyramine.

efficient conjugating system as long as inorganic sulphate is available. Amino groups can also be sulphated to yield readily excretable sulphamates.

GLUTATHIONE CONJUGATION

This is one of the most important pathways of metabolism that allows the cell to defend itself from chemical insult. It utilises the nucleophilic tripeptide glutathione, possessing a nucleophilic sulphur atom, whose cellular concentrations are high (Vermeulen 1996), to detoxify reactive electrophiles. This reaction is catalysed by the glutathione *S*-transferases but may also occur chemically. These enzymes are localised primarily in the cytosol, and to a much lesser extent in the endoplasmic reticulum. The glutathione conjugate is excreted either unchanged or following further processing to produce the mercapturate (see above). Paracetamol, in addition to the deactivation reactions depicted in Figure 1.1, to a small extent undergoes Phase I oxidation to form an electrophilic quinoneimine, believed to be responsible for its hepatotoxicity during overdose. The cell neutralises this reactive intermediate by forming a glutathione conjugate, which is eventually excreted as the mercapturate, following Phase III metabolism. (Figure 1.15). The mercapturate is formed by sequential loss of the glutamyl and glycinyl moieties of glutathione. The resulting cysteinyl conjugate is *N*-acetylated to form the mercapturate (N-acetylcysteine derivative). The

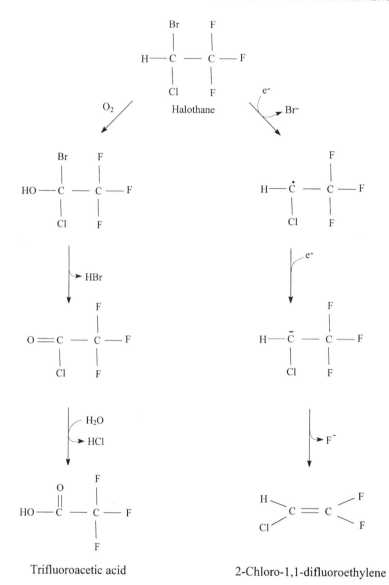

Figure 1.11 Oxidative and reductive dehalogenation of halothane.

glutathione conjugate is rarely excreted intact in the urine since its high molecular weight facilitates its excretion in the bile. The cysteine conjugate may be also further metabolised by the enzyme β-lyase (see above) to generate a thiol and release ammonia and pyruvate (Figure 1.15).

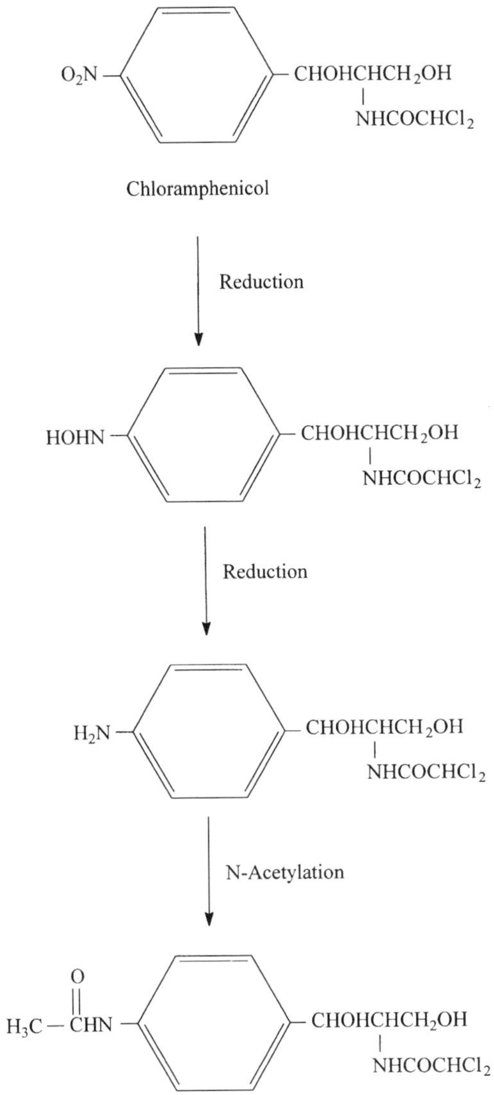

Figure 1.12 Nitroreduction of chloramphenicol.

AMINOACID CONJUGATION

The carboxylic group of organic acids may conjugate with aminoacids, glycine being the most common; other aminoacids found conjugated with xenobiotics are glutamine, taurine and ornithine. The carboxyl group of the xenobiotic forms a peptide

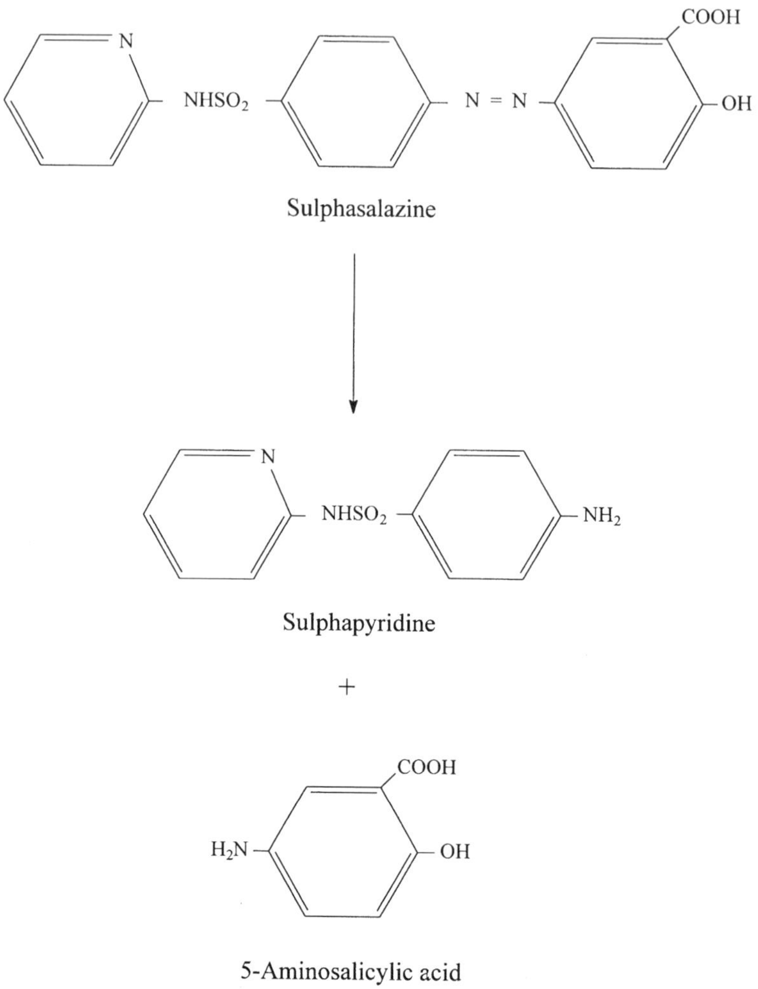

Sulphasalazine

Sulphapyridine

+

5-Aminosalicylic acid

Figure 1.13 Azoreduction of sulphasalazine.

bond with the α-amino group of the aminoacid (Figure 1.16). Initially the carboxylic group reacts with CoA to form a derivative, which then interacts with the aminoacid.

HYDRATION

This involves addition of water to epoxides to form dihydrodiols. As epoxides are generally toxic entities, this is a very important route for their detoxification. Microsomal epoxide hydrolases catalyse the conversion of epoxides of polycyclic aromatic

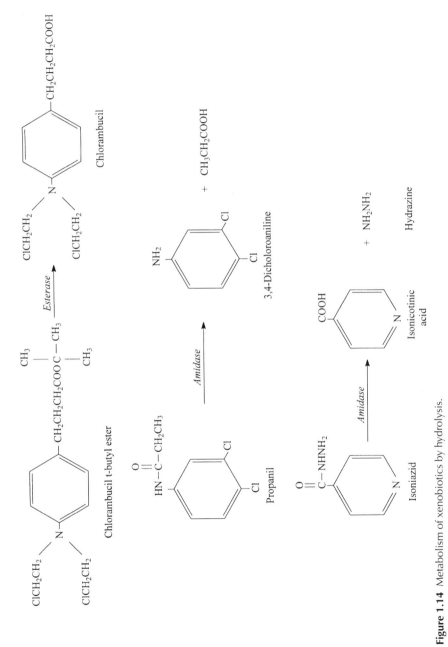

Figure 1.14 Metabolism of xenobiotics by hydrolysis.

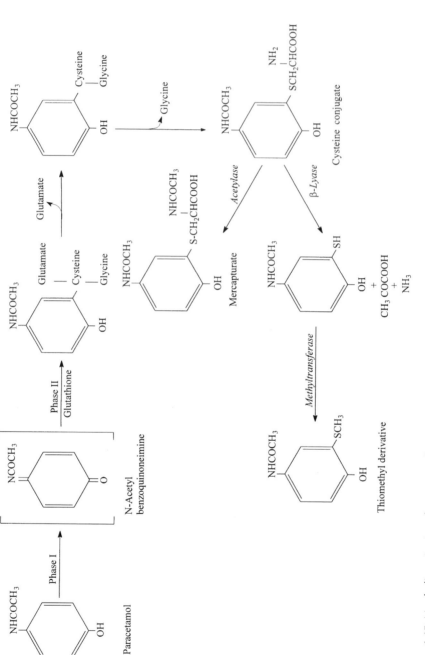

Figure 1.15 Metabolic activation of paracetamol.

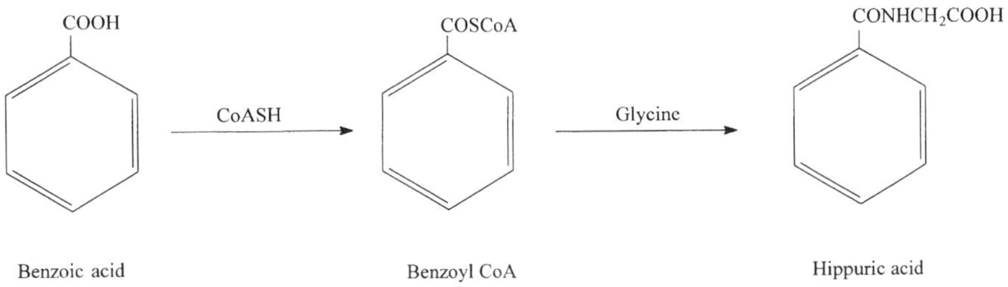

Figure 1.16 Conjugation of benzoic acid with glycine.

hydrocarbons to the corresponding dihydrodiols, as in the case of naphthalene (Figure 1.17).

METHYLATION

Hydroxyl as well as amino and thiol groups may be metabolised through methylation, the methyl donor being S-adenosyl methionine. The methyltransferases are primarily cytosolic enzymes, but are also present in the endoplasmic reticulum. Methylation is usually a minor metabolic route in xenobiotic metabolism, but plays a major role in the metabolism of endogenous substrates such as noradrenaline. Naturally-occurring polyphenolics, such as the tea flavonoid (-)-epicatechin, are metabolised through methylation (Figure 1.18). Thiols, emanating from the metabolism of glutathione conjugates, may be also subject to methylation (Figure 1.4).

ACETYLATION

This is an important metabolic route for aromatic and heterocyclic amines, hydrazines and sulphonamides. An amide bond is formed between the amino group of the chemical and acetate (Figure 1.3).

Bioactivation of xenobiotics

A well-documented paradox of xenobiotic metabolism is that, with certain chemicals, metabolism, both Phase I and Phase II, may generate highly reactive electrophilic

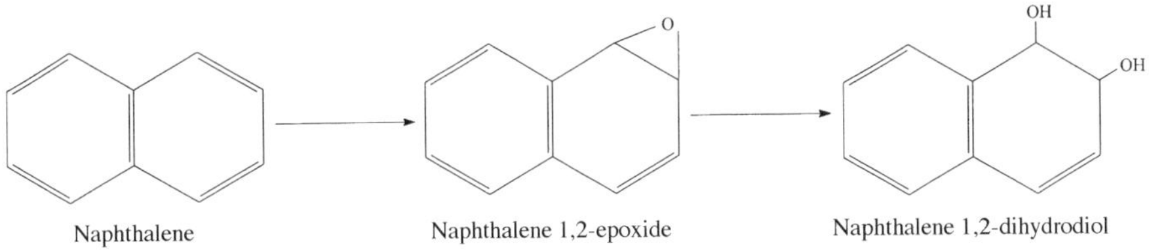

Figure 1.17 Metabolism of epoxides to dihydrodiols.

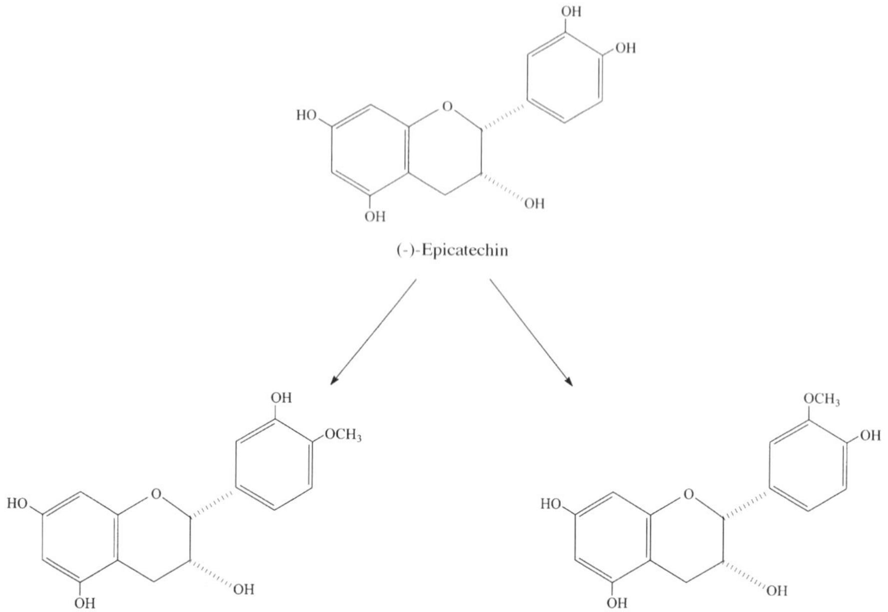

Figure 1.18 Methylation of flavonoids.

metabolites. Since these reactive species are generated intracellularly, they can readily and irreversibly interact with vital cellular macromolecules, such as DNA, RNA and proteins, to provoke various types of toxicity; thus, in this case, metabolism confers to the chemical adverse biological activity. The process through which inert chemicals are biotransformed to reactive intermediates capable of causing cellular damage is known as 'metabolic activation' or 'bioactivation' (Hinson *et al.* 1994). For these chemicals, toxicity/carcinogenicity is inextricably linked to their metabolism. Such metabolically derived reactive metabolites are epoxides, radicals, carbonium ions and nitrenium ions (Figure 1.19). The generated reactive intermediates may interact with DNA to form adducts which, if they escape the repair mechanisms of the cell, may be fixed and passed to the progeny, thus giving rise to a mutation (Figure 1.20). Reactive intermediates of chemicals may also induce DNA damage through an alternative mechanism that involves interaction with molecular oxygen to produce superoxide anions which, in the presence of traces of iron salts, can be transformed to the highly reactive hydroxyl radical (OH·), a powerful oxidant. It possesses an unpaired electron and so tends to form bonds with other species in order for the unpaired electrons to become paired. The hydroxyl radical, as well as other reactive oxygen species, can cause cellular damage similar to that resulting from the covalent interaction of the reactive species of chemicals with cellular components; they oxidise DNA to induce mutations, oxidise lipids to form lipid peroxides which appear to play an important

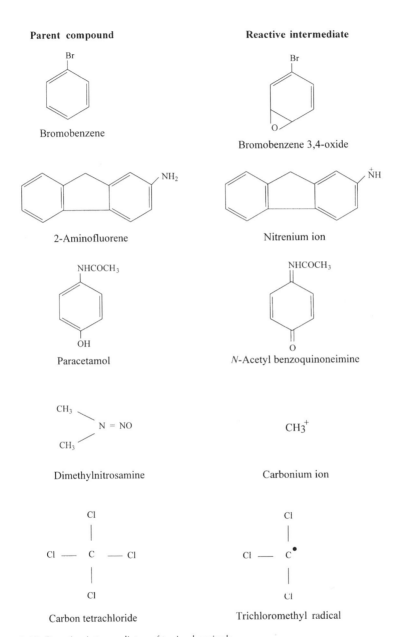

Figure 1.19 Reactive intermediates of toxic chemicals.

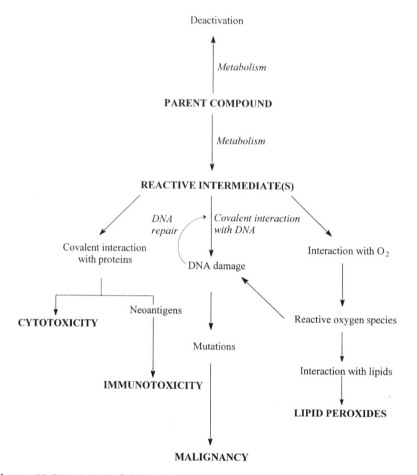

Figure 1.20 Bioactivation of chemicals.

role in the promotion and progression stages of chemical carcinogenesis, and also oxidise proteins (Wiseman and Halliwell 1996).

The reactive intermediates of chemicals may also interact covalently with proteins, disturbing physiological homeostasis, leading to cell death. In the last decade it has become apparent that reactive intermediates can also function as haptens, conferring on proteins antigenic potential and eliciting immunotoxicity (Pirmohamed and Park 1996; Dansette *et al.* 1998). Drugs such as tienilic acid, dihydralazine and halothane are metabolically converted to metabolites that bind covalently to proteins to generate neoantigens resulting in the production of autoantibodies. Subsequent exposure to these drugs may provoke an autoimmune response leading to hepatitis.

Not only oxidations, but also reductions can result in the bioactivation of chemicals, as in the case of nitropolycyclic aromatic hydrocarbons, a ubiquitous class of carcinogenic environmental contaminants (Fu 1990). Reduction of the nitro group is the first step in their bioactivation, leading eventually to the production of the genotoxic nitrenium ion (Figure 1.21).

Gut microflora can also participate in the bioactivation of chemicals, the metabolic pathways including nitro- and azo-reduction (Chadwick *et al.* 1992). Intestinal metabolism is also responsible for the cleavage of glycosidic bonds releasing carcinogenic products, as in the case of cycasin (methylazoxymethanol-β-D-glucoside) which is hydrolysed to the aglycone, methylazoxymethanol (Figure 1.22).

The most effective protective mechanism against chemical reactive intermediates is their detoxication through conjugation with endogenous nucleophilic substrates such as the tripeptide glutathione; this process also renders it sufficiently polar for its facile excretion. In this way the cell hinders the interaction of the reactive intermediates with cellular components. Glutathione conjugates are excreted into the urine and bile usually following further processing to form mercapturates (Figure 1.15). It is, thus, not surprising that the cellular concentrations of glutathione are rather high (about 10 mM); the cytosolic enzyme glutathione reductase maintains glutathione in the reduced form. Unavailability of glutathione, as, for example, following depletion consequent to exposure to megadoses of chemicals or following inadequate nutrition, renders the cell vulnerable to the toxicity of xenobiotics. For example, bromobenzene forms an epoxide which appears to mediate its toxicity, following covalent binding to proteins. The epoxide may rearrange to form the 3- and 4-bromophenols or hydrated

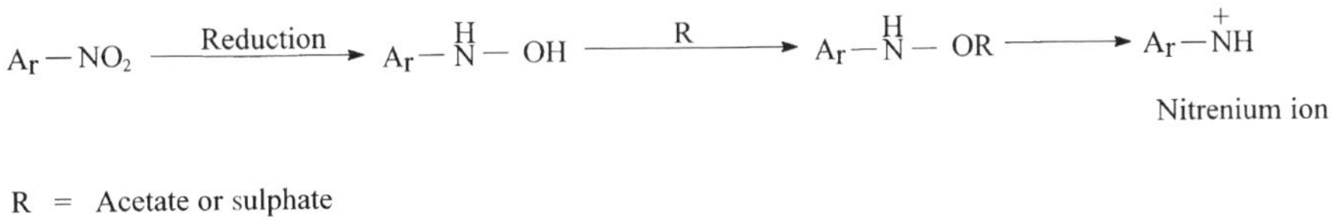

$$Ar-NO_2 \xrightarrow{\text{Reduction}} Ar-\overset{H}{N}-OH \xrightarrow{\quad R \quad} Ar-\overset{H}{N}-OR \longrightarrow Ar-\overset{+}{N}H$$

Nitrenium ion

R = Acetate or sulphate

Figure 1.21 Bioactivation of chemicals through reductive metabolism.

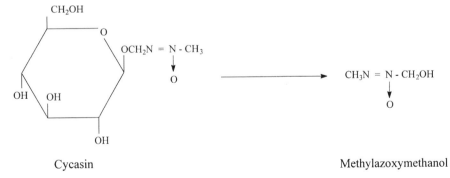

Cycasin   Methylazoxymethanol

Figure 1.22 Bioactivation of chemicals catalysed by microbial enzymes.

by epoxide hydrolase to the 3,4-diol, all these being deactivation pathways. The epoxide is also effectively detoxicated through conjugation with glutathione (Figure 1.23). Bromobenzene is markedly more toxic to animals that have been starved, since this treatment leads to depletion of glutathione. Glutathione synthesis is impaired as a result of deficiency of its constituent aminoacids, particularly cysteine (Pessayre *et al.* 1979).

Balance of activation/deactivation

Clearly, a chemical is subject to a number of metabolic pathways, the majority of which will bring about its deactivation and facilitate its excretion. However, some routes of metabolism will transform the chemical to a metabolite capable of inducing toxicity and carcinogenicity. Obviously, the amount of reactive intermediate produced, and hence incidence and degree of toxicity, will be largely dependent on the

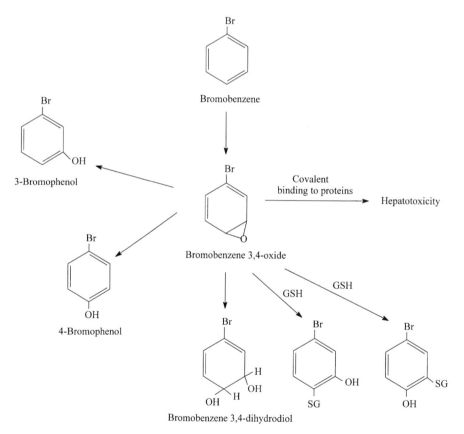

Figure 1.23 Metabolic activation and deactivation of bromobenzene.

competing pathways of activation and deactivation. The susceptibility of an animal species to the toxicity and carcinogenicity of a given chemical carcinogen is largely dependent on its complement of enzymes at the time of exposure, which determines whether activation or deactivation pathways are favoured. For example, the guinea pig is unable to carry out the N-hydroxylation of the carcinogen 2-acetylaminofluorene, the initial and rate limiting step in its bioactivation, and is consequently resistant to its carcinogenicity (Kawajiri *et al.* 1978). Similarly, the cynomolgous monkey lacks the enzyme—a cytochrome P450 protein, namely CYP1A2—that is responsible for the activation of the heterocyclic amine MeIQx (2-amino-3,8-dimethylimidazo[4,5-*f*]quinoxaline), a food carcinogen, and is consequently refractive to its carcinogenicity (Ogawa *et al.* 1999), whereas rodents such as mice and rats, which catalyse the activation of this carcinogen, are also susceptible to its carcinogenicity (Kato *et al.* 1988; Snyderwine *et al.* 1997).

In most cases, under physiological conditions, the activation pathways represent minor routes of metabolism so that the generation of reactive intermediates is minimal, the low levels formed are effectively deactivated by the defensive mechanisms, and no toxicity is apparent. However, in certain situations, the activation pathways may assume a greater role, leading to enhanced production of reactive intermediates, overwhelming the deactivation pathways, resulting in their accumulation in the body, thus increasing the likelihood of an interaction with cellular components with ensuing toxicity. Such situations may arise when:

(a) Deactivation pathways are saturated or impaired. This can occur during exposure to large quantities of a chemical, rather than chronic exposure to low doses, resulting in depletion of the body of the conjugating substrates. As already discussed, paracetamol (acetaminophen) is effectively deactivated by being conjugated with sulphate and glucuronide, these being its principal pathways of metabolism (Figure 1.1). A small fraction of the dose is oxidised to form the N-acetylbenzoquinoneimine, a reactive intermediate that is readily deactivated through conjugation with glutathione (Figure 1.15). Intake of large doses, as in suicide cases, alters the quantitative metabolic profile of the drug. Sulphate and glucuronic acid conjugation pathways become saturated as a result of unavailability of the activated forms of sulphate (PAPS) and glucuronic acid (UDPGA) respectively; the rate of their utilisation exceeds the rate of supply. As a consequence, more of the metabolism is directed towards the oxidation pathway that now assumes greater importance. Initially conjugation with glutathione deactivates the quinoneimine, but eventually the extensive use of glutathione results in its depletion, thus allowing the covalent interaction of this reactive intermediate with the –SH groups of proteins, leading to its hepatotoxicity (Holtzman 1995; Cohen and Khairallah 1997). Indeed, the preferred treatment of paracetamol intoxication involves administration of N-acetylcysteine, which deacetylates to release cysteine, the rate-limiting aminoacid in the synthesis of glutathione, thus preventing its depletion. The toxicity of paracetamol is also markedly enhanced when glutathione levels are low, as, for example, when animals are starved (Pessayre *et al.* 1979). Similarly, genetic deficiencies in conjugating systems may lead to increased sensitivity to the toxicity of chemicals that rely heavily on these enzymes for their deactivation. For example, in humans, glucuronyl transferase activity may be totally lacking in individuals with the Crigler–Najjar syndrome type I, a severe and fatal disease. These patients, being unable to eliminate bilirubin through glucuronidation, develop jaundice (de Wildt *et al.* 1999). Such patients may be sensitive to drugs whose principal pathway of metabolism is through glucuronidation. A milder condition is Gilbert's syndrome where the patient experiences only intermittent jaundice (de Wildt *et al.* 1999). These patients display low glucuronidation capacity, and when taking paracetamol they excrete less in the form of glucuronides, with more of the metabolism being directed towards oxidation forming the hepatotoxic *N*-acetylbenzoquinoneimine (Esteban and Perez-Mateo 1999). The anticancer drug iminotecan provokes severe toxicity in patients with Gilbert's syndrome as a result of suppressed glucuronidation (Wasserman *et al.* 1997). Moreover, intake of drugs metabolised by glucuronidation may induce jaundice, as the drug competes with bilirubin for glucuronidation (Burchell *et al.* 2000).

(b) The enzyme systems catalysing the activation pathways are selectively induced as a result of prior exposure to other chemicals (Okey 1990). One of the cytochrome P450 enzymes catalysing the bioactivation of paracetamol, namely CYP2E1, is stimulated by alcohol exposure and consequently chronic alcoholics are characterised by high levels of activity. As a result they are vulnerable to the hepatotoxicity of paracetamol (Seeff *et al* 1986; Zimmerman and Maddrey 1995). It is evident that any factor that disturbs the delicate balance of activation/deactivation will also influence the fate and toxicity of chemicals.

Although the outcome of Phase II conjugation reactions in the vast majority of cases is deactivation, these pathways can also generate reactive intermediates, and thus contribute in the bioactivation of chemicals (Kato and Yamazoe 1994; Glatt 1997; Glatt *et al.* 1998). For example, *O*-acetylation and *O*-sulphation of aromatic hydroxylamines generates unstable acetoxy and sulphatoxy esters which break down spontaneously to release a nitrenium ion, the species believed to bind to the DNA ultimate (Figure 1.21). Glutathione conjugates, as already discussed, can be converted to toxic thiols through the involvement of β-lyase (Koob and Dekant 1991) (Figure 1.4).

Fate of reactive intermediates

Liver is the principal site of the bioactivation of chemicals since most xenobiotic-metabolising enzyme systems are expressed in this tissue at high concentrations. Extrahepatic tissues contain a more restricted number of enzyme systems, generally being present at much lower concentrations compared to the liver. However, there are a few exceptions, for example prostaglandin synthetases, γ-glutamyltranspeptidase, the enzyme system involved in the metabolic conversion of glutathione conjugates to mercapturates, and some cytochrome P450 proteins are not appreciably expressed in the liver.

Since the liver is the centre of xenobiotic activation, it would be logical to expect that it would also be the major site for toxic and carcinogenic manifestations. In reality, however, the breast, lung and colon are far more frequent sites of tumorigenesis, despite their limited metabolic competence. It is becoming increasing apparent that the liver may function as the centre of production of reactive intermediates, but these may be exported systemically to other tissues where they can exert their deleterious effects.

Aromatic amines are potent urinary bladder carcinogens. In the liver they undergo N-hydroxylation, catalysed by cytochrome P450 enzymes and the flavin monooxygenase system, and the unstable hydroxylamines are trapped in the form of the more stable N-glucuronides. In the form of these glucuronides, the hydroxylamines are transported to the bladder where, under the acidic conditions prevailing in this tissue, the hydroxylamine is released and converted to the nitrenium ion, the ultimate carcinogen (Kadlubar *et al.* 1981). The nitrenium ion can also be formed *in situ* since the bladder has low levels of cytochrome P450 activity (Imaoka *et al.* 1997) and substantial levels of prostaglandin synthetase activity (Eling and Curtis 1992) that may contribute to the availability of the nitrenium ion (Figure 1.24). Similarly, the breast is a

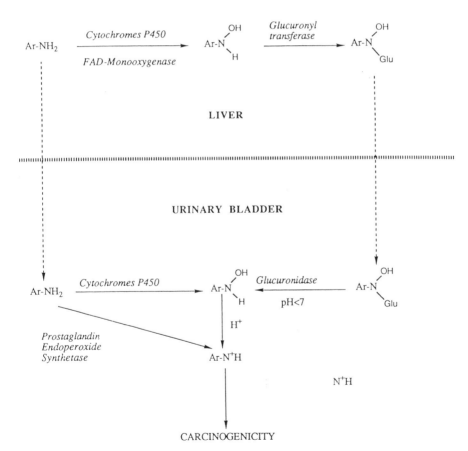

Figure 1.24 Transport of chemical reactive intermediates.

frequent site of tumorigenesis despite the fact that its capacity to metabolise and bioactivate chemicals through oxidation appears to be minimal (Davis *et al.* 1994). It has been proposed that heterocyclic carcinogenic amines, such as PhIP (2-amino-1-methyl-6-phenylimidazo[4,5-*b*]pyridine), are N-oxidised in the liver and, in the form of either the hydroxylamine or following esterification to the acetoxyester, are then transported to extrahepatic tissues such as the colon to yield DNA adducts (Kaderlik *et al.* 1994). Extrahepatic tissues appear capable of catalysing the esterification of hydroxylamines but poor in catalysing the initial oxidation of the heterocyclic amine as a result of the very low levels of cytochrome P450 (Stone *et al.* 1998). In an elegant study, the liver from rats pretreated with benzo[a]pyrene was transplanted in untreated animals. The extent of DNA binding in the lung, liver and kidney was the same in both, those animals exposed directly to the carcinogen and those who received the liver transplants, indicating clearly that the liver was the source of the reactive intermediates that interacted with DNA, not only for the liver, but also for the lung and kidney (Wall *et al.* 1991).

Reactive oxygen species and chemical toxicity and carcinogenicity

Electrophilic intermediates can induce toxicity and carcinogenicity through an additional mechanism, involving interaction with molecular oxygen to yield short-lived highly reactive oxygen species capable of oxidising lipids and proteins and eliciting DNA damage. Indeed, reactive oxygen species are currently being implicated in the aetiology and progression of a number of major chronic diseases including atherosclerosis, cancer, cardiovascular disease, diabetes, rheumatoid arthritis, reperfusion injury and ischaemia (Parke *et al.* 1991; Martínez-Cayuela 1995; Oberley and Oberley 1995; Baynes 1995).

The most reactive oxygen species is the hydroxyl radical (OH·); other biologically relevant reactive oxygen species are the superoxy anion, another radical ($O_2^{-\cdot}$), and hydrogen peroxide (H_2O_2). In the presence of iron, the OH· can be generated from either $O_2^{-\cdot}$ (Haber–Weiss reaction) or H_2O_2 (Fenton reaction).

Quinones, oxidation products of aromatic hydrocarbons, can undergo one-electron reductions to form the semiquinone radical (Figure 1.25), which may directly attack DNA. It causes DNA damage also indirectly, through reactive oxygen species produced as a consequence of their interaction with molecular oxygen (Aust *et al.* 1993).

However, the living organism is also endowed by defensive enzyme systems which effectively detoxify these reactive oxygen species, and prevent a state of 'oxidative stress', where the cell is unable to cope with the generation of reactive oxygen species. Such a state may ensue as a result of overproduction of reactive oxygen species and/or a decreased ability to deactivate them. Hydrogen peroxide is broken down enzymically by glutathione peroxidase, an enzyme present in the cytosol and mitochondria, and catalase, a peroxisomal enzyme (Figure 1.26). Superoxide dismutase, an enzyme localised in the mitochondria and cytosol, is an effective defence against the superoxy anion, and NAD(P)H-quinone reductase (DT-diaphorase), an enzyme found primarily in the cytosol, protects against quinone-derived oxygen radicals by converting the quinone to the hydroquinone through a two-electron reduction. An increasing number

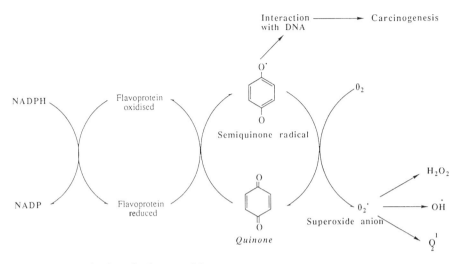

Figure 1.25 Mechanism of quinone toxicity.

$$2GSH + H_2O_2 \xrightarrow{\textit{Glutathione peroxidase}} GSSG + 2H_2O$$

$$2H_2O_2 \xrightarrow{\textit{Catalase}} 2H_2O + O_2$$

$$O_2^{\cdot -} + O_2^{\cdot -} + 2H \xrightarrow{\textit{Superoxide dismutase}} H_2O_2 + O_2$$

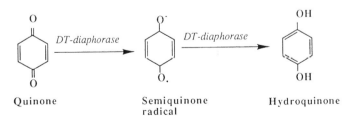

Figure 1.26 Enzyme systems protecting against reactive oxygen species.

of chemicals can function as antioxidants, scavenging oxygen radicals, and in this way afford protection against oxidative damage. Such antioxidants include endogenous chemicals, e.g. uric acid, vitamin E, melatonin etc., and numerous naturally occurring, plant-derived compounds that are ingested with food in substantial amounts, e.g. plant polyphenolics (Williamson 1998).

Conclusions

There can be no doubt that both the ever-increasing life expectancy and the high quality of life humans enjoy today could not have been achieved in the absence of the novel, not pre-existing in nature, chemicals that the chemist has synthesised. These possess biological activity that has been exploited to treat effectively, and in many instances cure, human disease and to provide a more hygienic and safe environment. It is not coincidental that the sharpest rise in life expectancy occurred at the same time as the birth of the pharmaceuticals industry. However, it must be also recognised that chemicals, both anthropogenic and naturally occurring, are also capable of inducing toxicity in humans. Manifestation of chemical toxicity is largely a consequence of the irreversible interaction of chemicals with cellular macromolecules such as DNA, RNA, lipids and proteins. However, most compounds lack the necessary chemical reactivity to participate in such interactions but can acquire it through metabolism. The reactive, toxic intermediates are generated intracellularly which allows them access to the vital components of the cell. Thus the view that metabolism is strictly a defensive mechanism against chemical injury is anachronistic. Clearly for a chemical to provoke toxicity at least two prerequisites must be fulfilled: (a) the chemical must be, or must have the propensity to be metabolically converted to a reactive inter-mediate(s), and (b) the living organism, at the time of exposure, must possess the necessary enzyme(s) required for the activation of the chemical. Both 2- and 4-aminobiphenyl can form genotoxic metabolites, the corresponding hydroxylamines, but in the case of the 2-isomer no enzyme appears to be capable of catalysing this N-oxidation, because it is situated in a conformationally hindered position (Ioannides et al. 1989). For this reason, the 4-aminobiphenyl is a potent human and animal carcinogen whereas the 2-isomer appears to be devoid of carcinogenicity.

It is apparent that toxicity is not simply a consequence of the intrinsic molecular structure of the chemical, but is also determined by the nature of the enzymes present at the time of exposure; these enzyme systems are in turn regulated genetically (Raunio and Pelkonen 1995; Cascorbi et al. 1995; van der Weide and Steijns 1999; Wormhoudt et al. 1999) but are also modulated by environmental factors such as diet (Ioannides 1999) and previous exposure to chemicals (Okey 1990; Tanaka 1998; Lin and Lu 1998) and the presence of disease (Ioannides et al. 1996; Iber et al. 1999). Toxicologists have the habit of referring to chemicals as being toxic or non-toxic, while in the strictest sense the vast majority of chemicals are innocuous, and it is the living organism that renders them toxic through metabolism. As we are becoming more competent in phenotyping humans for xenobiotic-metabolising activity, it may well become more appropriate to talk of 'toxicophilic' and 'toxicophobic' individuals rather than of toxic and non-toxic chemicals, depending on their propensity to bioactivate or detoxify chemicals. At present humans can be phenotyped for a number of cytochrome P450 proteins, for N-acetylase, xanthine oxidase and flavin monooxygenase activities (Lucas *et al.* 1999; Chung *et al.* 2000), and these studies are bound to be extended to other enzyme systems, once appropriate chemical probes are identified. It is now possible, for example, to assess the effect of environmental factors such as diet on the expression of cytochrome P450 proteins in humans. For example, the cytochrome P450 enzyme, CYP1A2, was induced following the consumption of diets supplemented with Brassica vegetables but, in contrast, was suppressed when diets supplemented with apiaceous vegetables were consumed (Kall *et al.* 1996; Lampe *et al.* 2000). Such knowledge of a person's metabolic capacity will also enable us to identify individuals vulnerable to the toxicity of a certain chemical(s), and to develop drug dose regimens tailor-made to the needs of individuals, thus increasing efficacy and minimising the risk of adverse effects.

Acknowledgements

The author would like to thank Professor L.J. King for his valuable comments on the mechanisms of the metabolic pathways, and Miss Sheila Evans for the preparation of all the figures.

References

Ahmad N, Katiyar SK and Mukhtar H (1998) Cancer chemoprevention by tea polyphenols. In *Nutrition and Chemical Toxicity*, Ioannides C (ed.), John Wiley, Chichester, pp. 301–343.

American Institute for Cancer Research (1997) *Food, Nutrition and the Prevention of Cancer: A global perspective.* Washington.

Ames BN and Gold LS (1998) The prevention of cancer. *Drug Metabolism Reviews*, **30**, 201–223.

Ames BN, Profet M and Gold LS (1990) Dietary pesticides (99.99% all natural). *Proceedings of the National Academy of Sciences, USA*, **87**, 7777–7781.

Anders MW and Dekant W (1998) Glutathione-dependent bioactivation of haloalkenes. *Annual Reviews of Pharmacology and Toxicology*, **38**, 501–537.

Aust SD, Chignell CF, Bray TM, Kalyanaraman B and Mason RP (1993) Free radicals in toxicology. *Toxicology and Applied Pharmacology*, **120**, 168–178.

Baynes JW (1995) Reactive oxygen in the aetiology and complications of diabetes, In *Drugs, Diet and Disease. Volume 2, Mechanistic approaches to diabetes*, Ioannides C and Flatt PR (eds), Ellis Horwood, London, pp. 201–240.

Burchell B, Soars M, Monaghan G, Cassidy A, Smith D and Ethel B (2000) Drug-mediated toxicity caused by genetic deficiency of UDP-glucuronyl transferases. *Toxicology Letters*, **112–113**, 333–340.

Cascorbi I, Brockmöller J, Mrozikiewicz PM, Müller A and Roots I (1999) Arylamine *N*-acetyltransferase activity in man. *Drug Metabolism Reviews*, **31**, 489–502.

Chadwick RW, George SE and Claxton LD (1992) Role of the gastrointestinal mucosa and microflora in the bioactivation of dietary and environmental mutagens or carcinogens. *Drug Metabolism Reviews*, **24**, 425–492.

Cohen SD and Khairallah EA (1997) Selective protein arylation and acetaminophen-induced hepatotoxicity. *Drug Metabolism Reviews*, **29**, 59–77.

Chung W-G, Kang J-H, Park C-S, Cho M-H and Cha Y-N (2000) Effect of age and smoking on *In vivo* CYP1A2, flavin-containing monooxygenase, and xanthine oxidase activities in Koreans: Determination by caffeine metabolism. *Clinical Pharmacology and Therapeutics*, **67**, 258–266.

Dansette PM, Bonierbale E, Minoletti C, Beaune PH, Pessayre D and Mansuy D (1998) Drug-induced immunotoxicity. *European Journal of Drug Metabolism and Pharmacokinetics*, **23**, 443–451.

Davis CD, Ghoshal A, Schut HAJ and Snyderwine EG. (1994) Metabolic activation of heterocyclic amine food mutagens in the mammary gland of lactating Fischer 344 rats. *Cancer Letters*, **84**, 67–73.

De Wildt SN, Kearns GL, Leeder JS and van den Anker JN (1999) Glucuronidation in humans. Pharmacogenetic and developmental aspects. *Clinical Pharmacokinetics*, **36**, 439–452.

Eling TE and Curtis FJ (1992) Xenobiotic metabolism by prostaglandin H synthase. *Pharmacology and Therapeutics*, **53**, 261–273.

Esteban A and Perez-Mateo M (1999) Heterogeneity of paracetamol metabolism in Gilbert's syndrome. *European Journal of Drug Metabolism and Pharmacokinetics*, **24**, 9–13.

Fu PP (1990) Metabolism of nitro-polycyclic aromatic hydrocarbons. *Drug Metabolism Reviews*, **22**, 209–268.

Glatt H (1997) Bioactivation of mutagens via sulphation. *FASEB Journal*, **11**, 314–321.

Glatt H, Bartsch I, Christoph S, Coughtrie MWH, Falany CN, Hagen M, Landsdiel R, Pabel R, Phillips DH, Seidel A and Yamazoe Y (1998) Sulfotransferase-mediated activation of mutagens studied using heterologous expression systems. *Chemico-Biological Interactions*, **109**, 195–219.

Hinson JA, Pumford NR and Nelson SD (1994) The role of metabolic activation in drug toxicity. *Drug Metabolic Reviews*, **26**, 395–412.

Holtzman JL (1995) The role of covalent binding to microsomal proteins in the hepatotoxicity of acetaminophen. *Drug Metabolism Reviews*, **27**, 277–297.

Iber H, Sewer MB, Barclay TB, Mitchell SR, Li T and Morgan ET (1999) Modulation of drug metabolism in infectious and inflammatory diseases. *Drug Metabolism Reviews*, **31**, 29–41.

Imaoka S, Yoneda Y, Matsuda T, Degawa M, Fukushima S and Funae Y (1997) Mutagenic activation of urinary bladder carcinogens by CYP4B1 and the presence of CYP4B1 in bladder mucosa. *Biochemical Pharmacology*, **54**, 677–683.

Ioannides C (1999) Effect of diet and nutrition on the expression of cytochromes P450. *Xenobiotica*, **29**, 109–154.

Ioannides C, Lewis DFV, Trinick J, Neville S, Sertkaya NN, Kajbaf M and Gorrod JW (1989) A rationale for the non-mutagenicity of 2- and 3-aminobiphenyls. *Carcinogenesis*, **10**, 1403–1407.

Ioannides C, Barnett CR, Irizar A and Flatt PR (1996) Expression of cytochrome P450 proteins in disease. In *Cytochromes P450: Metabolic and Toxicological Aspects*, Ioannides C (ed.), CRC Press Inc., Boca Raton, FL, pp. 301–327.

Kaderlik KR. Minchin RF, Mulder GJ, Ilett KF, Daugaard-Jenson M, Teitel CH and Kadlubar FF (1994) Metabolic activation pathway for the formation of DNA adducts of the carcinogen 2-amino-1-methyl-6-phenylimidazo[4,5-*b*]pyridine (PhIP) in rat extrahepatic tissues. *Carcinogenesis*, **15**, 1703–1709.

Kadlubar FF, Unruh LE, Flammang TJ, Sparks D, Mitchum RK and Mulder GJ (1981). Alteration of urinary levels of the carcinogen N-hydroxy-2-naphthylamine, and its *N*-glucuronide in the rat by control of urinary pH, inhibition of metabolic sulfation and changes in biliary excretion. *Chemico-Biological Interactions*, **33**, 129–147.

Kall MA, Vang O and Clausen J (1996) Effects of dietary broccoli on human *in vivo* drug metabolizing enzymes: evaluation of caffeine, oestrone and chlorzoxazone metabolism. *Carcinogenesis*, **17**, 793–799.

Kato R and Yamazoe Y (1994) Metabolic activation of N-hydroxylated metabolites of carcinogenic and mutagenic arylamines and arylamides by esterification. *Drug Metabolism Reviews*, **26**, 413–430.

Kato T, Ohgaki H, Hasegawa H, Sato S, Takayama S and Sugimura T (1988) Carcinogenicity in rats of a mutagenic compound 2-amino-3,8-dimethylimidazo[4,5-*f*]quinoxaline. *Carcinogenesis*, **9**, 71–73.

Kawajiri K, Yonekawa H, Hara E and Tagashira Y (1978) Biochemical basis for the resistance of guinea pigs to carcinogenesis by 2-acetylaminofluorene. *Biochemical and Biophysical Research Communications*, **85**, 959–965.

Koob M and Dekant W (1991) Bioactivation of xenobiotics by formation of toxic glutathione conjugates. *Chemico-Biological Interactions*, **77**, 107–136.

Lampe JW, King IB, Li S, Grate MT, Barale KV, Chen C, Feng Z and Potter JD (2000) Brassica vegetables increase and apiaceous vegetables decrease cytochrome P450 1A2 activity in humans: changes in caffeine metabolite ratios in response to controlled vegetable diets. *Carcinogenesis*, **21**, 1157–1162.

Lin JH and Lu AYH (1998) Inhibition and induction of cytochrome P450 and the clinical implications. *Clinical Pharmacokinetics*, **35**, 361–390.

Lucas D, Ferrara R, Gonzalez E, Bodenez P, Albores A, Manno M and Berthou F (1999) Chlorzoxazone, a selective probe for phenotyping CYP2E1 in humans. *Pharmacogenetics*, **9**, 377–388.

Martínez-Cayuela M (1995) Oxygen free radicals and human disease. *Biochimie*, **77**, 147–161.

Mori H and Nishikawa A (1996) Naturally occurring organosulphur compounds as potential anticarcinogens. In *Nutrition and Chemical Toxicity*, Ioannides C (ed.), John Wiley, Chichester, pp. 285–299.

Mori H, Tanaka T and Hirono I (1998) Toxicants in food: Naturally occurring. In *Nutrition and Chemical Toxicity*, Ioannides C (ed.), John Wiley, Chichester, pp. 1–27.

Oberley LW and Oberley TD (1995) Reactive oxygen species in the aetiology of cancer. In *Drugs, Diet and Disease. Volume 1, Mechanistic approaches to cancer*, Ioannides C and Lewis DFV (eds), Ellis Horwood, London, pp. 47–63.

Ogawa K, Tsuda H, Shirai T, Ogiso T, Wakabayashi K, Dalgard DW, Thorgeirsson UP, Adamson RH and Sugimura T (1999) Lack of carcinogenicity of 2-amino-3,8-dimethylimidazo[4,5-*f*]quinoxaline (MeIQx) in cynomolgus monkeys. *Japanese Journal of Cancer Research*, **90**, 622–628.

Okey AB (1990) Enzyme induction in the P-450 system. *Pharmacology and Therapeutics*, **45**, 241–298.

Parke AL, Ioannides C, Lewis DFV and Parke DV (1991) Molecular pathology of drug-disease interactions in chronic autoimmune inflammatory diseases, *Inflammopharmacology*, **1**, 3–36.

Pessayre D, Dolder A, Artigou JY, Wandscheer JC, Descatoire V, Degott C and Benhamou JP (1979) Effect of fasting on metabolite-mediated hepatotoxicity in the rat. *Gastroenterology*, **77**, 264–271.

Pirmohamed M and Park BK (1996) Cytochromes P450 and immunotoxicity. In *Cytochromes P450: Metabolic and Toxicological Aspects*, Ioannides C (ed.) CRC Press Inc., Boca Raton, FL, pp. 329–354.

Raunio H and Pelkonen O (1995) Cancer genetics: Genetic factors in the activation and inactivation of chemical carcinogens. In *Drugs, Diet and Disease. Volume 1, Mechanistic approaches to cancer*, Ioannides C and Lewis DFV (eds), Ellis Horwood, London, pp. 229–258.

Rowland IR (1988) Factors affecting metabolic activity of intestinal microflora. *Drug Metabolism Reviews*, **19**, 243–261.

Seeff LB, Cuccherini BA, Zimmerman HJ, Adler E and Benjamin SB (1986) Acetaminophen toxicity in the alcoholic. A therapeutic misadventure. *Annals of Internal Medicine*, **104**, 399–404.

Skog KI and Jägerstad M (1998) Toxicants in food: generated during cooking. In *Nutrition and Chemical Toxicity*, Ioannides C (ed.), John Wiley, Chichester, pp. 59–79.

Snyderwine EG, Turesky RJ, Turteltaub KW, Davis CD, Sadrieh N, Schut HAJ, Nagao M, Sugimura T, Thorgeirsson UP, Adamson RH and Thorgeirsson SS (1997) Metabolism of food-derived heterocyclic amines in nonhuman primates. *Mutation Research*, **376**, 203–210.

Stone EM, Williams JA, Grover PL, Gusterson BA and Phillips DH (1998) Interindividual variation in the metabolic activation of heterocyclic amines and their N-hydroxyderivatives in primary cultures of human mammary epithelial cells. *Carcinogenesis*, **19**, 873–879.

Tanaka E (1998) Clinically important pharmacokinetic drug–drug interactions: role of cytochrome P450 enzymes. *Journal of Clinical Pharmacy and Therapeutics*, **23**, 413–416.

Toth B (1995) Mushroom toxins and cancer (Review). *International Journal of Oncology*, **6**, 137–145.

van der Weide J and Steijns LS (1999) Cytochrome P450 enzyme system: genetic polymorphisms and impact on clinical pharmacology. *Annals of Clinical Biochemistry*, **36**, 722–729.

Vermeulen NPE (1996) Role of metabolism in chemical toxicity. In *Cytochromes P450: Metabolic and Toxicological Aspects*, Ioannides C (ed.) CRC Press Inc., Boca Raton, FL, pp. 29–53.

Wall KL, Gao W, TeKoppele JM, Kwei GY, Kauffman FC and Thurman RG (1991) The liver plays a central role in the mechanism of chemical carcinogenesis due to polycyclic aromatic hydrocarbons. *Carcinogenesis*, **12**, 783–786.

Wang J-S, Kensler TW and Groopman JD (1998) Toxicants in food: Fungal contaminants. In *Nutrition and Chemical Toxicity*, Ioannides C (ed.), John Wiley, Chichester, pp. 29–57.

Wasserman E, Myara A, Lokiec F, Goldwasser F, Trivin F, Mahjoubi M, Misset JL and Cvitkovic E (1997) Severe CPT-11 toxicity in patients with Gilbert's syndrome: two case reports. *Annals of Oncology*, **8**, 1049–1051.

Williamson G (1998) Expression of chemical toxicity in vitamin deficiency and supplementation. In *Nutrition and Chemical Toxicity*, Ioannides C (ed.), John Wiley, Chichester, pp. 237–256.

Wiseman H and Halliwell B (1996) Damage to DNA by reactive oxygen and nitrogen species: role in inflammatory disease and progression to cancer. *Biochemistry Journal*, **313**, 17–29.

Wormhoudt LW, Commandeur JNM and Vermeulen NPE (1999) Genetic polymorphisms of human N-acetyltransferase, cytochrome P450, glutathione-S-transferase, and epoxide hydrolase enzymes: relevance to xenobiotic metabolism and toxicity. *Critical Reviews in Toxicology*, **29**, 59–124.

Zimmerman HJ and Maddrey WC (1995) Acetaminophen (paracetamol) hepatotoxicity with regular intake of alcohol: Analysis of instances of therapeutic misadventure. *Hepatology*, **22**, 767–773.

2 Cytochrome P450

F. Peter Guengerich

Vanderbilt University School of Medicine, USA

Introduction

Cytochrome P450 (P450) enzymes constitute the major group of enzymes involved in the oxidation of drugs and other xenobiotic chemicals (Wislocki *et al.* 1980; Ortiz de Montellano 1995a; Guengerich 1997a). The functions of the enzymes are a major component of most problems in drug development and chemical toxicity. The P450 enzymes are haemoproteins that generally function in mixed-function oxidase reactions, using the stoichiometry:

$$NADPH + H^+ + O_2 + R \longrightarrow NADP^+ + H_2O + RO \qquad (2.1)$$

As discussed later, the identity of the product RO may be obscured by rearrangements, and other variations from the above stoichiometry are possible.

P450s have been known, at least through their actions, for over 50 years. The roots of P450 research are found in studies on the metabolism of steroids (Estabrook *et al.* 1963), drugs (Axelrod 1955; Brodie *et al.* 1958) and carcinogens (Mueller and Miller 1948, 1953). Today these enzymes are studied not only by biochemists but also by inorganic chemists, development biologists, pesticide and plant scientists and physicians. More attention has been probably given to the P450s than to any other single group of enzymes discussed in this book.

Roles of P450s in the metabolism of chemicals

P450s are involved in the metabolism of a wide variety of xenobiotic chemicals. One major role is the clearance of drugs, to which P450s collectively contribute more than any other group of enzymes. Many drugs are rendered more hydrophilic by hydroxylation so that they are excreted from the body, particularly if an oxygen atom is added to be conjugated with a hydrophilic moiety (e.g. glucuronide). The adjustment of rates of metabolism to yield optimal pharmacokinetic profiles ia a major focus in the pharmaceutical industry. P450 oxidation of drugs usually results in their deactivation,

Enzyme Systems that Metabolise Drugs and Other Xenobiotics. Edited by C. Ioannides.
© 2002 John Wiley & Sons Ltd

although not always. In some cases drugs are rendered more active by the process of metabolism and the parent drug is considered a 'pro-drug'.

P450s are also involved in the oxidation of 'endogenous' chemicals, those normally found in the body. Vitamins are usually considered in this group because of their essential nature. Deficiencies in the P450s involved in steroid oxidations are often very detrimental. However, the levels of the microsomal P450s, the primary concern of this chapter, can vary considerably in the human population with apparently no problems. For instance, 10- to 40-fold variation in the levels of CYP1A2 and 3A4 are documented, and CYP2D6 is absent in ~7% of Caucasian populations (Guengerich 1995). Although some of the 'xenobiotic-metabolising' P450s can oxidise 'endogenous' substrates (e.g. testosterone, arachidonate), it is not clear that these are critical functions, in the context of the wide variability of levels. Moreover, all the P450-knockout mice developed to date (1A1, 1A2, 1B1, 2E1) show relatively normal phenotypes in the absence of stress from drugs or deleterious chemicals (Lee *et al.* 1996b; Liang *et al.* 1996; Buters *et al.* 1999).

P450s have been of interest in chemical carcinogenesis and toxicology because of their oxidations of protoxicants to reactive electrophilic forms that cause biological damage (Conney 1982; Guengerich 1988a; Guengerich and Shimada 1991). Lists of such reactions have been published (Guengerich and Shimada 1991), and reactive products can covalently modify proteins and DNA (Nelson 1994). Considerable evidence for critical roles of P450 modulation in toxicity and cancer has been developed in experimental animal models (Guengerich 1988a; Nebert 1989), but human epidemiology is not so clear (d'Errico *et al.* 1996). Nevertheless, the role of P450s in influencing the effectiveness of toxicity of drugs in people has now been well established and some of the approaches can also be used with issues of 'environmental' chemicals (Guengerich 1998).

Nomenclature

At the time of the discovery of P450 (Klingenberg 1958) and the demonstration of its role as the terminal oxidase in microsomal mixed-function oxidase systems (Estabrook *et al.* 1963) little evidence existed for the presence of multiple forms. The name 'P450' was developed to indicate a pigment ('P') with a haem Fe^{2+}-carbon monoxide complex having an absorption spectral band at 450 nm (Omura and Sato 1962, 1964). Several lines of investigation led to the view that two or more distinct forms of P450 might exist within a single species (Alvares *et al.* 1967; Hildebrandt *et al.* 1968; Sladek and Mannering 1969). The components of the liver microsomal system—P450, NADPH-P450 reductase, and phospholipid—were separated and then reconstituted in 1968 (Lu and Coon 1968).

In the 1970s and early 1980s a large number of P450s were purified from rat and rabbit liver (Haugen *et al.* 1975; Johnson and Schwab 1984; Ryan *et al.* 1978; Guengerich 1987a) and later human liver (Distlerath and Guengerich 1987; Guengerich 1995). Complex naming systems developed, usually based upon designations of column fractions by investigators. With the growing application of recombinant DNA technology to P450 research in the 1980s, cDNA clones were characterised (Fujii-Kuriyama *et al.* 1982; Kimura *et al.* 1984). A systematic nomenclature system was developed by Nebert (Nebert *et al.* 1987, 1989; Nelson *et al.* 1993) based upon the coding sequences. Updated versions of the nomenclature can be found on the website <http://drnelson.utmem.edu/CytochromeP450.html>. For comparisons of older nomenclature to the current, see Nelson *et al.* (1993). The nomenclature system follows these basic rules: All P450s are members of the 'superfamily' (some 'functionally' related proteins such as fungal chloroperoxidase and the nitric oxide synthases are not included). The most defining region of the sequence is that surrounding the cysteine that serves as an axial thiolate ligand to the haem ion. Sequences with ≥40% sequence identity are grouped in the same 'family', specified by a numeral. Sequences with ≥60% identity fall in the same 'subfamily', denoted with a letter. The last number denotes the individual P450. Families 1–49 are for mammalian and insect genes, 50–99 for plants and fungi, and 100+ for bacteria. There are some exceptions to the rules, including the exact cutoff percentage for classifications and the assignment of families. P450s that seem to be conserved across species (e.g. 1A1, 1A2, 2E1) are given the same name in all species. Other P450s that seem to vary more are given individual names in different species, i.e. 3A4, 3A5, and 3A7 in humans and 3A6 in rabbits. Obviously there may be future problems in fitting all species and new P450s into the existing nomenclature system (see <http://drnelson.utmem.edu/CytochromeP450. html>), but the system has been useful. The terms 'P450' or 'CYP' may be used, but searches with CYP will also identify a yeast transcription factor with the same acronym (Naït-Kaoudjt *et al.* 1997).

Distribution

P450s are rather ubiquitous. In nature they appear almost everywhere from Archebacteria to humans, with the apparent exception of enterobacteria (e.g. *Escherichia coli*). The availability of genomic databases and generally accepted trademark sequences ('cysteine peptide') has made it possible to estimate the number of P450 genes in some organisms. That number currently stands at 53 in humans at the time of this writing (Spring 2000), 70 in *Caenorhabditis elegans*, 90 in Drosophila, and 330 in *Arabidopsis thaliana*. These numbers are probably approaching the final count now with the completion of sequencing of genomes.

This chapter will not deal with prokaryotic P450s except to the extent that they serve as models for the mammalian P450s of interest. The mammalian enzymes discussed in this chapter are found in the endoplasmic reticulum, associated with the auxiliary flavoprotein NADPH-P450 reductase, which delivers electrons from NADPH to the iron atom in the haem of P450. The family 11 mitochondrial P450s function in the hydroxylation of steroids and vitamin D, and accept electrons from the iron-sulphur protein adrenodoxin (Kagawa and Waterman 1995). Family 24 and 27 P450s—total of 6 including the family 11 P450s—are also targeted solely to the mitochondria. It should be pointed out that some microsomal P450s, because of particular processing patterns, are also delivered into the mitochondria (Addya *et al.* 1997; Neve and Ingelman-Sundberg 1999) and the plasma membrane (Loeper *et al.* 1998). In both cases there appears to be functional capability, although the details are not completely clear yet. The fraction of microsomal P450 targeted to the plasma membrane is very low but may be involved in the initiation of auto-immune responses involving P450s (Loeper *et al.* 1993). The fraction of some P450s localised in mito-chondria can be substantial and they may be significant contributors to some reactions (Anandatheerthavarada *et al.* 1997).

The xenobiotic-oxidising P450s are localised principally in the liver, and some are expressed essentially exclusively in the liver (e.g. CYP1A2). Others are expressed in the liver and in a variety of extrahepatic tissues, often with differential mechanisms of regulation (e.g. CYP3A4. Some P450s are expressed in several extrahepatic tissues but almost not at all in the liver (e.g. CYP1A1 and 1B1). The liver is a major site for oxidation of drugs and other xenobiotic chemicals, but P450s in extrahepatic tissues, even if expressed at much lower levels there than in the liver, can have very important roles in at least three situations:

(1) The catalytic specificity for a reaction is so absolute that the substrate will only be oxidised when it encounters this P450 (e.g. some steroid hydroxylations).
(2) Extensive oxidation occurs when the chemical first encounters this P450 because of the route of administration. For instance, there is often a major 'first-pass' effect in the small intestine for drugs ingested following oral administration (esp. CYP3A4 (Kolars *et al.* 1991; Guengerich 1999b)).
(3) Oxidation of a chemical by a P450 yields a reactive product with very limited stability, and only oxidation in the 'target' cells for evoking the toxic response is important in considering bioactivation.

The point should also be emphasised that localisation of P450s within a tissue occurs, and can be important in determining responses to chemicals. In many tissues a number of different cell types exist (e.g. ~40 in the lung) and P450s are usually found in only a few of these (e.g. within the lung, kidney, brain (Dees *et al.* 1982; Ishii-Ohba *et al.* 1984)). In the liver, the P450s are most concentrated in the parenchymal cells (hepatocytes), but different gradations through the regions of the liver (centrilobular, midzonal, periportal) are observed with individual P450s (Baron *et al.* 1981).

Regulation

This area really began with observations in the 1950s that metabolism was inducible in experimental animals (Richardson *et al.* 1952; Conney *et al.* 1956) and humans (Remmer 1957). These phenomena helped lead to the discovery of P450 multiplicity. For many years the study of regulation of P450s was rather descriptive, and until 1980 there was little real evidence that P450 protein concentrations were really increased (immunochemical methods were critical in this work). The field of P450 regulation really developed with the advent of recombinant DNA technology and molecular biology. Although many details are still unknown, what has emerged today is a gen-eralised paradigm of primarily transcriptional regulation, controlled by *trans*-acting factors; although other factors contribute as well.

In the generalised model of transcriptional regulation of P450s, a receptor interacts with a low M_r ligand in the cytosol and somehow changes its conformation, moves to the nucleus, binds to a specific region in the P450 upstream of the start site, and changes the DNA in such a way to make it more accessible to other proteins involved in transcription (Hankinson 1995; Waterman and Guengerich 1997). All the details have not been elucidated in any case, but several major systems are outlined and fit into the general context of the scheme shown in Figure 2.1.

The 'Ah (aryl hydrocarbon) locus' is involved in the regulation of CYP1A1, 1A2, and 1B1. In this system, the receptor protein binds to a polycyclic hydrocarbon. As a result, the Ah receptor no longer binds to the heat shock protein (hsp)90 but now forms a heterodimer with the ARNT (Ah receptor nuclear transporter) protein and moves to the DNA, binding to XRE (xenobiotic response element) sequences (Hankinson 1995). Other events may be involved, but the result is that transcription of the P450 1A1 or other structural gene is considerably enhanced. Similar systems are involved with PXR (pregnane X receptor) (regulation of *CYP3A* genes) (Kliewer *et al.* 1998), CAR (constitutively active receptor) (regulation of *CYP2B* genes) (Honkakosski *et al.* 1998), and PPAR (peroxisome proliferator activator receptor) (regulation of P450 4A genes and other genes involved in peroxisome proliferation) (Muerhoff *et al.* 1992). PXR, CAR, and PPAR are all in the 'steroid orphan receptor' family and, when bound to their ligands, form heterodimers with RXR (retinoid x receptor)–retinoid complexes. The interactions can be complex in that multiple forms of some of these proteins exist and the properties of the heterodimers are also influenced by the ligands that are bound.

Obviously this is an over-simplification of a complex process. Evidence exists that other proteins interact with the proteins in the heterodimers (e.g. HIF (hypoxia inducible factor) 1α with ARNT, RARs (retinoic acid receptors) with RXR). P450 genes usually contain binding sites for multiple factors and are not controlled only by a single response system. Such elements are usually responsible for the tissue-specific expression of P450s (e.g. HNFs (hepatic nuclear factors) for some P450s in liver).

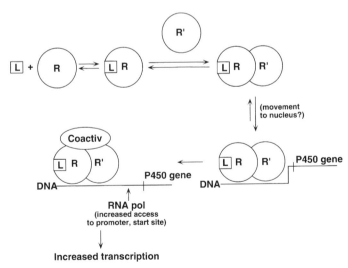

Figure 2.1 Generalised model for regulation of P450 genes by induction. L = ligand, R = receptor, R' = partner protein for heterodimer of R, Coactiv = coactivator, RNA pol = RNA polymerase.

Some of these elements are involved in basal responses; some contribute to inducibility. With some P450s, GRFs (glucocoticoid receptor elements) within the introns can enhance responses.

Although transcriptional systems have received the most attention and can be considered to be the predominant aspect of induction, post-transcriptional regulation is also known. Evidence for mRNA stabilisation is more difficult to obtain (than for enhanced mRNA synthesis), but some systems appear to involve such a mechanism (Zangar and Novak 1998; Dong et al. 1988). Mechanisms are less clear than with transcriptional control. With some P450s there is evidence that certain proteins selectively bind the mRNA and stabilise it (Geneste et al. 1996).

Another means of regulating P450 levels is post-translational, and there is evidence that such a phenomenon occurs in experimental animals. The sensitivty of P450 to proteolytic degradation can be attenuated by ligand binding in some cases (Steward et al. 1985; Eliasson et al. 1988) and enhanced by protein alkylation in others (Tephly et al. 1986; Correia et al. 1992b). Incorporation of haem is a post-translational event. This process is generally considered to occur during translation, i.e. the protein is wrapped around the haem. Haem is ultimately degraded by haem oxygenase after it leaves P450 (Schmid 1973). The question of whether haem equilibrates among P450s is an interesting one, because the general consensus is that the haem site is not very exposed to the protein surface (Guengerich 1978). Some evidence for haem exchange comes from the slightly shorter half-life for P450 haem turnover in rat liver (Sadano and Omura 1983a; Shiraki and Guengerich 1984) and from the immunoprecipitation experiments of Omura (Sadano and Omura 1983b, 1985). The only other postranslational event of possible consequence with P450s is phosphorylation. Evidence for destabilisation of P450s by serine phosphorylation has been presented for rat CYP2B1 and 2E1 (Johansson et al. 1991; Bartlomowicz et al. 1989). Presumably any post-translational modification is not critical to normal P450 function, because most P450s can be readily expressed in bacterial systems (Waterman et al. 1995; Guengerich et al. 1996). The degradation of at least some P450s does involve ubiquination (Correia et al. 1992a; Tierney et al. 1992).

Some other aspects of regulation of P450s should also be mentioned. In rodents some P450s show dramatic gender-specific expression (e.g. CYP2C11 in male rats (Kato and Kamataki 1982; Guengerich et al. 1982)). Many studies have been carried out that document the roles of steroid hormone levels (especially at the neonatal stage) and pulsatile growth hormone levels in these responses (Waxman et al. 1985; Gustafsson et al. 1983). The underlying basis appears to be in factors such as STAT (Ram et al. 1996). The relevance of this phenomenon to humans is not clear; in humans the gender effects in drug metabolism are rather small. Another interesting phenomenon is P450 down-regulation. This has been observed with some P450s, with the same chemicals that induce other P450s (e.g. 2C11 and polycyclic hydrocarbons). The mechanism involves lowered mRNA levels but is not well understood beyond this level. P450 levels are also attenuated in rodents or humans treated with interferons or in disease states where levels of interferons or interleukins are elevated (Renton et al. 1979). This phenomenon has direct relevance in humans, and individuals with colds (or following flu shots) have generally impaired drug metabolism (Renton and Knickle 1990).

In animals the toxicity of a chemical may be either enhanced or attenuated due to induction of P450s, and the effect may be tissue-specific (Nebert 1989). Exactly how much the toxicity of chemicals is modulated by P450 induction is less clear in humans; differences in lung cancer (among smokers) are associated with variable CYP1A1 inducibility (Kouri *et al.* 1982). A goal of the US National Institute of Environmental Health Sciences Environmental Genome Project is to determine if better associations can be developed between gene responses and disease outcome (Guengerich 1998).

P450 induction has practical significance in drug development. Induction may not be a fatal flaw for a new drug candidate but will raise two important issues. The first is the prospect of drug–drug interactions. Induction of a P450 involved in metabolism of a drug will lower the concentration of that drug in the body. One rather classic example is the induction of CYP3A4 by rifampicin or barbiturates, which leads to more rapid oxidation of the oral contraceptive 2-ethynyloestradiol and its loss of effectiveness (Bolt *et al.* 1977; Guengerich 1988b). Induction of a P450 by a drug which is a substrate leads to a lower AUC (area-under-the-curve) as a function of time. In practice, the clinical consequences of P450 induction are not as common as for inhibition. One of the major concerns about P450 induction by drugs in development is that some general correlations exist between induction of certain P450s in rodents and the potential for tumour promotion (Nims *et al.* 1994). For instance, 2,3,7,8-tetrachlorodibenzo-*p*-dioxin (inducer of CYP1A1, 1A2, 1B1), barbiturates (CYP2B) and some hypolipidaemic agents (CYP4A) are all tumour promoters in rodents (Pitot 1995). These phenomena may be largely unimportant for humans (e.g. epileptic patients on lifetime phenobarbital do not develop tumours (Olsen *et al.* 1989)) but are of concern regarding review of rodent bioassays by regulatory agencies.

Structure

The first three-dimensional structure of a P450 was reported by Poulos in 1987 (Poulos *et al.* 1987). At this time the structure of at least eight different P450s is known; six have been published (Poulos *et al.* 1987; Ravichandran *et al.* 1993; Hasemann *et al.* 1994; Cupp-Vickery *et al.* 1996; Park *et al.* 1997; Williams *et al.* 2000). All but one are of soluble enzymes, and all but two are of bacterial origin. One is from a Fusarium mold (Park *et al.* 1997) and one is a slightly modified form of a rabbit enzyme (Williams *et al.* 2000). Because of restrictions on the length of this chapter, these structures will be discussed only briefly and the reader is referred to the original papers (see above) for reference to the actual structures and to discussion of exactly what one can conclude from the results (particularly see Poulos 1991 and Hasemann *et al.* 1995).

All the P450 structures have common elements and look similar at a gross level. The proteins are arranged into a series of helices and folds that are rather similar. The helices are denoted A through L and are well conserved. The I and L helices make contact with the haem. Residues in the B and I helices contact the substrate, and mutations in these regions have had some of the most dramatic effects in site-directed mutagenesis experiments (Kronbach *et al.* 1991; Hasler *et al.* 1994; Halpert and He 1993; Guengerich 1997b). Another region that contacts the substrate is the F helix, where mutations have also influenced catalytic selectivity (Lindberg and Negishi 1989; Ibeanu *et al.* 1996; Harlow and Halpert 1997). Other structural, homology modelling, and site-directed mutagenesis work indicates that the six 'substrate recognition sequences' (SRS, three already cited in the discussion of the B, I, and F helices, see above) are generally conserved among most P450s (Gotoh 1992).

The most conserved part of the P450 sequence is in the region containing the cysteine that acts as the thiolate ligand to the haem iron. This sequence is used as the 'landmark' to identify P450s in genebanks. Another highly conserved residue is a threonine corresponding to threonine 252 in bacterial P450 101, which has been postulated to have various functions. It may donate a proton to the $Fe^{2+}O_2^-$ complex to facilitate (heterolytic) cleavage of the O–O bond. Arguments against the general need for such an activity include the high pK_a of threonine, the absence of this threonine in some P450s, and the functionality of CYP101 in which a threonine O-methyl ether was substituted using mutagenesis (Kimata *et al.* 1995). The dominant view today is that the threonine and its neighbouring glutamic acid are part of a proton relay system that is useful but not absolutely necessary. In CYP108 (P450eryF), the substrate binds and then mobilises H_2O to serve the same function (Cupp-Vickery *et al.* 1996).

Crystal structures in themselves do not necessarily provide information about function. Indeed, some of the P450s for which structures have been obtained are not well characterised in the sense that in some cases the physiological electron donor has not been identified (Cupp-Vickery *et al.* 1996). CYP101 (P450$_{cam}$) was the first P450 crystallised, and this structure has provided the greatest amount of information, in part because of the wealth of knowledge that exists about protein interactions, substrate specificity and kinetics (Mueller *et al.* 1995). Recently time-resolved crystallographic approaches have been used to obtain the structure of distinct redox forms of P450 101 during the catalytic cycle (Schlichting *et al.* 2000). Although the structure of the latter intermediates is not as clear and the molecular identity is not as certain, the work provides insight into the changes that occur in the course of P450 reaction. Of particular interest is the observation that changes in the structure near the active site occur at every step of the reaction. To some extent then, each of the redox forms may be considered as a slightly different enzyme. In P450 102 there is a long, open substrate channel and only relatively small conformational changes are required (Ravichandran *et al.* 1993; Hasemann *et al.* 1995). However, with several of the other P450s, the substrate is completely enclosed by the protein (e.g. CYP101, 107, 108, 2C5 (Hasemann *et al.* 1995; Williams *et al.* 2000)) and obviously a major rearrangement of the protein is necessary. This rearrangement is not well understood, even in P450 101, and has only been approached from molecular dynamics simulations (Paulsen and Ornstein 1993; Helms and Wade 1998). A key question is how fast the protein motion is and exactly why the substrate is bound. The structural change is also required for the product to leave, and similar questions can be raised. Also, the question arises as whether the product and substrate follow the same paths in entering and leaving.

The availability of a structure of a mammalian P450 is an important development. Rabbit CYP2C5 was subjected to some modification to increase the level of heterologous expression and decrease aggregation (Cosme and Johnson 2000). These approaches yielded a protein that could be crystallised in the absence of detergents (Williams *et al.* 2000). However, molecular anomalous dispersion phasing techniques and synchotron radiation were required to solve the structure. The structure is of considerable interest in terms of modelling of other mammalian P450s. In this regard, the structure has been compared to that of CYP102, which has already been employed extensively and shows the greatest similarity (Williams *et al.* 2000). The spatial arrangement of the major elements diverges from the other structures, with SRS-4 being the most conserved compared to CYP102. SRS-5 diverges significantly, with 3.3 Å root mean square (rms deviation) and the shape of the base of the active site is significantly different, affecting the orientation and position of substrates as well as selectivity for different substrates. This is not surprising, in that the substrates of CYP102 and 2C5 are fatty acids and steroids, respectively (Williams *et al.* 2000; Fulco 1991; Johnson *et al.* 1983). The spatial organisation of SRS regions 1, 2, 3, and 6 also diverges from CYP102 (4.5–6.0 Å). The high temperature factors of the CYP2C5 B-C loop might be indicative of dynamic fluctuations related to passage of substrates through this area (Williams *et al.* 2000).

The CYP2C5 structure seems to indicate a bound ligand, although no substrate was added. It is possible that this is a reagent utilised in the work (e.g. dithiothreitol). The substrate binding cavity is only slightly larger than that for camphor in CYP 101 (360 $Å^3$) (Williams *et al.* 2000). Progesterone was readily docked into this site. Many P450 modelling efforts have been published, based first on P450 101 and subsequently on CYP102 or aggregate models of the known structures. Undoubtedly structure modelling will now also be based upon CYP2C5. The general features will apply; the usefulness of modelling details of ligand interaction remains to be demonstrated.

Mechanisms of catalysis

A discussion of the chemistry is in order because of relevance to considerations of catalytic selectivity (see below). For earlier and more extensive discussion of the chemistry of catalysis, the reader is directed to several articles (Ortiz de Montellano 1986, 1995b; Dawson 1988; Guengerich and Macdonald 1984, 1990, Guengerich 1996).

General features

The general paradigm for P450 catalysis is shown in Figure 2.2. Briefly, ferric P450 binds substrate in step 1. In the bacterial CYP101 (P450$_{cam}$), this step is associated with a shift from low- to high-spin iron, an increased redox potential ($E_{1/2}$) and faster reduction (Fisher and Sligar 1985; Mueller *et al.* 1995). However, with the microsomal P450s these events may or may not occur and are not coupled (Guengerich 1983; Bäckström *et al.* 1983; Guengerich and Johnson 1997). In step 2, one electron is transferred from the reductase. This is a complex process in that the two flavins (FAD, FMN) are used and a number of redox state combinations are possible. The accepted view is that the step FADH/FMNH$_2$ $\rightarrow$ FADH/FMNH is associated with step 2 (Vermilion and Coon 1978; Vermilion *et al.* 1981). Step 3 involves binding of O$_2$ to

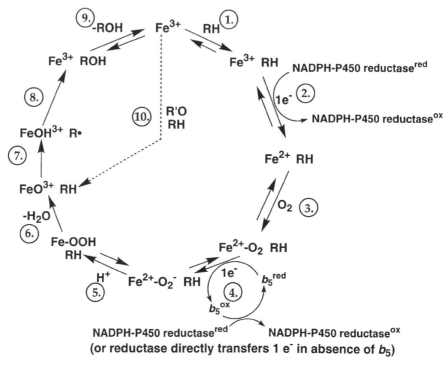

Figure 2.2 Generalised catalytic P450 mechanism.

ferrous P450 (ferric haem does not bind O_2). The 'second' electron is delivered, to this ferrous·O_2 species, in step 4. It is generally thought that the redox potential for this step is much higher (0 mV?) than for step 2 (-300 mV), although estimates are very crude (Guengerich 1983; Pompon and Coon 1984). The reductase step involved in step 4 is FAD/FMNH$_2$ → FAD/FMNH· (which then is recycled in the reductase: FAD/FMNH·+NADPH + H$^+$ → FADH$_2$/FMNH· → FADH·/FMNH$_2$) Vermilion *et al.* 1981). However, there is evidence that in some cases this electron can be provided from ferrous cytochrome b_5 (which has a redox potential of 0 mV (Rivera *et al.* 1998)).

Following the Fe^{2+}·O_2 complex (between steps 3 and 4), only limited spectral characterisation has been obtained and some of the evidence is based on biomimetic models (Ortiz de Montellano 1995b). Step 5 involves the protonation of the formal Fe^{2+}O$_2$-complex, which appears to be mediated by a Thr/Asp/H$_2$O 'relay' system in CYP101 and some other P450s for which three-dimensional structures are available (Martinis *et al.* 1989; Cupp-Vickery *et al.* 1996). This relay system is is also believed to facilitate the heterolytic cleavage of the O–O band to generate a formal FeO^{3+} complex in step 6. This complex can formally be written Fe^V = O but the prevailing view is that a better electronic representation is as Fe^{4+}O/porphyrin radical, analogous to peroxidase Compound I (Dawson 1988). Step 7 involves abstraction of a hydrogen atom (C–H → C·+ FeOH), or possibly a non-bonded or π electron, followed by rapid 'oxygen rebound' in step 8 to yield the product. The product is released in step 9, to regenerate the starting ferric, unliganded protein.

Types of oxidations

So many P450 substrates exist that can only be considered in major groups, on a mechanistic basis. Some minor paths are not presented, unless they have a major bearing on general mechanisms, but the reader is referred to discussions presented elsewhere (Ortiz de Montellano 1995b; Guengerich and Macdonald 1990).

CARBON HYDROXYLATION

This process results in the formation of alcohols and is common in the metabolism of sterols, alkanes, etc. The most generally accepted mechanism involves hydrogen abstraction followed by oxygen rebound:

$$FeO^{3+} + \ -\overset{|}{\underset{|}{C}}H \longrightarrow FeOH^{3+} + \ -\overset{|}{\underset{|}{C}}\bullet \longrightarrow Fe^{3+} + \ -\overset{|}{\underset{|}{C}}-OH \qquad (2.2)$$

Major evidence for this pathway comes from the high kinetic hydrogen effects and the scrambling of putative methylene and allylic radicals (Ortiz de Montellano 1995b; Groves and Subramanian 1984; Ortiz de Montellano 1986).

DEHYDROGENATION

Several reactions are included here:

$$(2.3)$$

The 2-electron oxidation of an alcohol can be readily explained by the carbon hydroxylation mechanism. The oxidation of an aldehyde to a carboxylic acid can also be explained in this way, although an alternate mechanism is possible (see below).

Desaturation of alkanes is not uncommon. It usually accompanies carbon hydroxylation and is postulated to result from bifurcation of the putative radical intermediate:

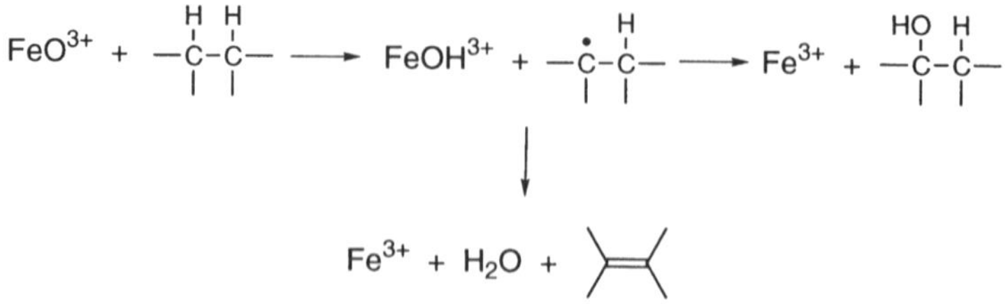

$$\text{FeO}^{3+} + \underset{\underset{H}{|}}{\overset{\overset{H}{|}}{-C}}-\underset{\underset{H}{|}}{\overset{\overset{H}{|}}{C}}- \longrightarrow \text{FeOH}^{3+} + \overset{\overset{H}{|}}{-\overset{\bullet}{C}}-\underset{\underset{H}{|}}{\overset{\overset{H}{|}}{C}}- \longrightarrow \text{Fe}^{3+} + \underset{\underset{H}{|}}{\overset{\overset{HO}{|}}{-C}}-\underset{\underset{H}{|}}{\overset{\overset{H}{|}}{C}}-$$

$$\downarrow$$

$$\text{Fe}^{3+} + \text{H}_2\text{O} + \hspace{0.5cm} \rangle\!=\!\langle$$

(2.4)

1e⁻ REDUCTION

When the oxygen tension is low, ferrous P450 may transfer an electron to a substrate before binding O_2 (Wislocki *et al.* 1980). Some reductions catalysed by P450s include those of nitroaromatics and CCl_4:

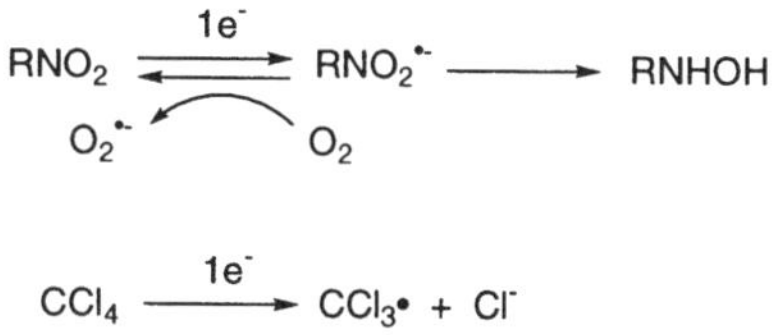

$$\text{RNO}_2 \underset{O_2^{\bullet-}}{\overset{1e^-}{\rightleftarrows}} \text{RNO}_2^{\bullet-} \underset{O_2}{\longrightarrow} \text{RNHOH}$$

(2.5)

$$\text{CCl}_4 \overset{1e^-}{\longrightarrow} \text{CCl}_3\bullet + \text{Cl}^-$$

Other reductions by P450 (e.g. epoxides) are not as well characterised (Wislocki *et al.* 1980; Yamazoe *et al.* 1978).

1e⁻ OXIDATION

This process has been postulated in a number of reactions where low redox potentials are involved (see below). The concept has relevance in the discussion of 1e⁻ oxidation of polycyclic aromatic hydrocarbons (Cavalieri and Rogan 1995), for which direct evidence is not available. Recently this laboratory has presented evidence for the 1e⁻ oxidation of 1,2,4,5-tetramethoxybenzene ($E_{1/2}$ 0.94 versus SCE), which can be observed in the steady state (Sato and Guengerich 2000):

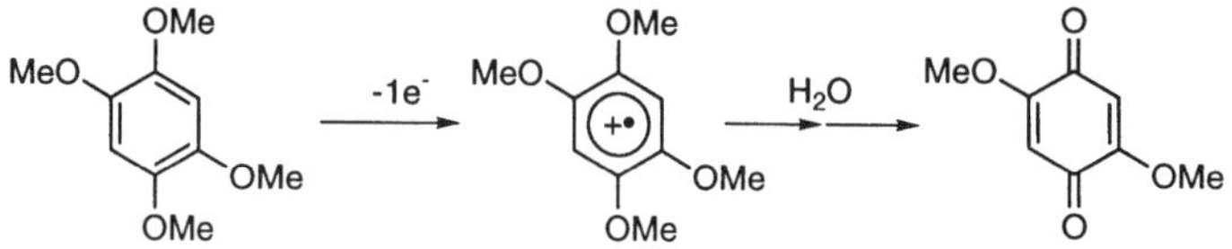

(2.6)

The rearrangement of strained cycloalkanes (e.g. quadricylane) with low potentials has also been rationalised in the context of 1e⁻ oxidation (Stearns and Ortiz de Montellano 1985):

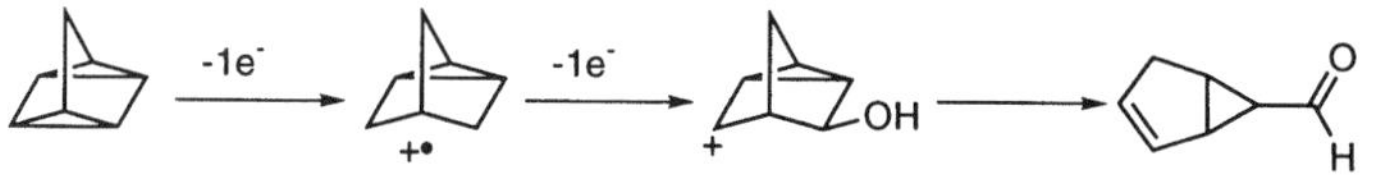

(2.7)

HETEROATOM OXYGENATION

This reaction is very common for the enzyme flavin-containing monooxygenase but it is also catalysed by P450s, even in situations where α-hydrogen atoms are available. The reaction is commonly seen with amines and sulphides and has also been implicated in a reaction with an iodide (Watanabe *et al.* 1982; Seto and Guengerich 1993; Guengerich 1989).

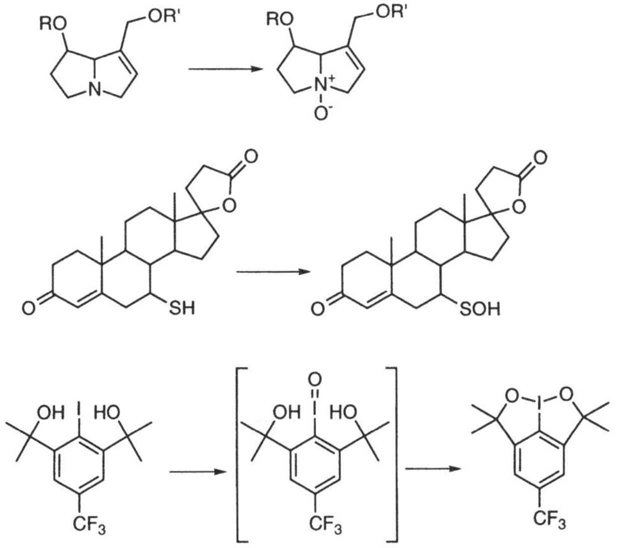

(2.8)

HETEROATOM DEALKYLATION

The cleavage of a substrate at a heteroatom is a common P450 reaction. In general, the cleavage of ethers and carboxylic acid esters and amides (Guengerich 1987; Guengerich *et al.* 1988a; Peng *et al.* 1995) is generally thought to involve the carbon hydroxylation mechanism presented above.

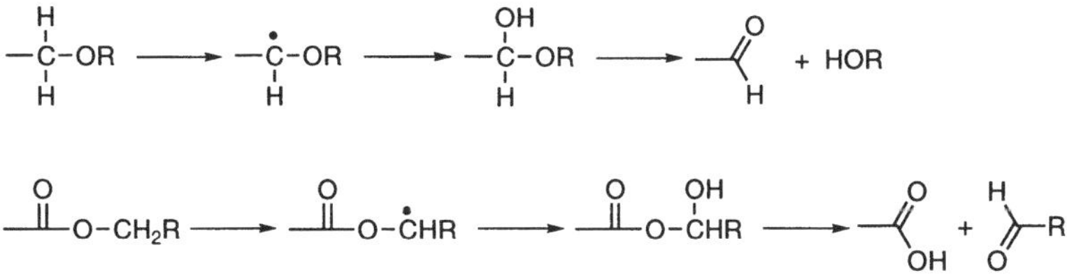

(2.9)

In situations where $E_{1/2}$ is low ($\sim$1 V versus SCE (Guengerich and Macdonald 1984; Guengerich and Macdonald 1990, 1993)), the mechanism is postulated to involve initial 1e$^-$ oxidation (if spatially accessible) followed by base-catalysed rearrangement involving the $FeO_2{}^{2+}$ entity (Okazaki and Guengerich 1993; Ortiz de Montellano 1987):

$$\text{FeO}^{3+} + \overset{\bullet\bullet}{\underset{\underset{R''}{|}}{R\overset{}{N}}}-CH_2R' \longrightarrow \text{FeO}^{2+} + \overset{+\bullet}{\underset{\underset{R''}{|}}{R\overset{}{N}}}-CH_2R' \longrightarrow \text{FeOH}^{3+} + \overset{\bullet\bullet}{\underset{\underset{R''}{|}}{R\overset{}{N}}}-\overset{\bullet}{C}HR' \tag{2.10}$$

$$\text{Fe}^{3+} + \overset{\bullet\bullet}{\underset{\underset{R''}{|}}{R\overset{}{N}H}} + \underset{H}{\overset{O}{\|}}{-}R' \longleftarrow \text{Fe}^{3+} + \overset{\bullet\bullet}{\underset{\underset{R''}{|}}{R\overset{}{N}}}-\overset{\underset{}{OH}}{C}HR'$$

Similar mechanisms can be involved for the dealkylation of sulphides (Guengerich and Macdonald 1984) and alkyl phenylethers with low $E_{1/2}$ (Yun *et al.* 2000).

$1e^-$ oxidation may be a common starting point for amine/sulphide oxygenation and dealkylation, because the two processes are usually associated with each other (Guengerich and Macdonald 1984):

$$\text{FeO}^{3+} + \overset{\bullet\bullet}{\underset{\underset{R''}{|}}{R\overset{}{N}}}-CH_2R' \longrightarrow \overset{+\bullet}{\underset{\underset{R''}{|}}{R\overset{}{N}}}-CH_2R' \left\langle \begin{array}{l} \overset{O^-}{\underset{\underset{R''}{|}}{R\overset{}{N}{}^+}}-CH_2R' \\[2em] \overset{\bullet\bullet}{\underset{\underset{R''}{|}}{R\overset{}{N}H}} + \underset{H}{\overset{O}{\|}}{-}R' \end{array} \right. \tag{2.11}$$

However, the lack of a linear free energy relationship (Seto and Guengerich 1993) and certain molecular orbital calculations (Hammons *et al.* 1985) suggest that N-oxygenation may not be so simple and might possibly involve a further electron transfer (Hammons *et al.* 1985):

$$\text{FeO}^{3+} + \overset{\bullet\bullet}{\underset{|}{R\overset{}{N}}}- \longrightarrow \text{FeO}^{2+} + \overset{+\bullet}{\underset{|}{R\overset{}{N}}}- \longrightarrow \text{FeO}^+ + \overset{+}{\underset{|}{R\overset{}{N}}}- \longrightarrow \text{Fe}^{3+} + \overset{O^-}{\underset{|}{R\overset{}{N}{}^+}}- \tag{2.12}$$

OXIDATION OF OLEFINS AND ACETYLENES

A very common product of olefin oxidation is an epoxide (Daly *et al.* 1972). However, the oxidation of olefins is probably best understood in the context of a stepwise mechanism with ionic intermediates, which may or may not follow radical intermediates that could undergo collapse first:

$$(2.13)$$

OXIDATION OF AROMATIC RINGS

Hydroxylation of aromatic rings is a reaction observed not only with P450s but also with numerous other oxygenases, including non-haem iron, copper, and flavin proteins (Walsh 1979; Guengerich 1990). Studies with the non-haem iron phenylalanine hydroxylase showed that a hydrogen label in the *para* position of the substrate could undergo a 1,2-shift, which was rationalised in the context of a hydrogen isotope effect and an epoxide intermediate:

$$(2.14)$$

This migration, commonly termed the 'NIH Shift' (Daly *et al.* 1972; Guroff *et al.* 1967), is consistent with the formation of an epoxide intermediate but does not in itself constitute proof of an epoxide. Epoxides are often intermediates in aromatic oxidations, or demonstrated by their isolation (or dihydrodiols in the presence of epoxide hydrolase) (Daly *et al.* 1972). A more generalised mechanism of aromatic oxidation includes the existence of Fe–O-substrate intermediates, with the 1,2-anion shifts well-rationalised (Guengerich and Macdonald 1990; Ortiz de Montellano 1995b).

$$FeO^{3+} + \quad \underset{*H}{\overset{R}{\bigcirc}} \quad \xrightarrow{[O]} \quad \cdots \quad \xrightarrow{D_k} \quad \underset{OH}{\overset{R}{\bigcirc}} \quad *H \tag{2.15}$$

As pointed out earlier, evidence has been for $1e^-$ oxidation of low $E_{1/2}$ aromatic systems but no clear evidence for the coupling of stepwise electron transfer and epoxidation/hydroxylation has been obtained (Anzenbacher *et al.* 1996).

Nature of reactive oxidant

Early work in the P450 field was dominated by a search for 'mobile' oxidants, e.g. O_2^-, 'oxene' (Ullrich *et al.* 1982; Strobel and Coon 1971; Ullrich 1972). However, the general view today is that catalysis is the result of interaction of substrates with Fe–oxygen complexes. However, several possibilities are still viable.

The dominant view is that of a FeO^{3+} entity analogous to peroxidase Compound I (McMurry and Groves 1986; Ortiz de Montellano 1995b; Guengerich 1991). Most of the mechanisms of oxidation are reasonably rationalised with such an intermediate (see above). Moreover, most of the P450 reactions can be reproduced in systems in which the reductase and NADPH replaced by the 'oxygen surrogate' iodosylbenzene (step 10 in Figure 2.2), which is only capable of forming a mono-oxygen Fe complex. Reactions supported by this oxygen surrogate are usually faster than the 'normal' P450 reactions but linearity suffers from the harshness of the iodosylbenzene in destroying P450 haem (Lichtenberger *et al.* 1976; Gustafsson *et al.* 1979). If P450 and per-oxidases have a similar set of reaction intermediates, why doesn't horseradish peroxidase catalyse all P450 oxidations? Apparently the steric restriction of the haem FeO entity causes electron transfer through the porphyrin edge but precludes base catalysis and oxygen transfer (Ortiz de Montellano 1987; Okazaki and Guengerich 1993).

An alternate view to the hydrogen atom abstraction pathway (Groves *et al.* 1978; McMurry and Groves 1986; Ortiz de Montellano 1995a) has been proposed by Newcomb (Newcomb *et al.* 1995a, 2000). The issue is the limited rearrangement of strained alkane 'radical clocks'. The lack of rearrangement has been interpreted as evidence that a discrete carbon-centred radical would have a lifetime too short to qualify as a bonafide intermediate (Newcomb *et al.* 1995b; 2000). An alternative to the hydrogen atom abstraction mechanism (which actually should be considered a caged system) is a rather concerted mechanism. However, such a mechanism does not explain the degree of rearrangement seen in some reactions (Newcomb *et al.* 2000; White *et al.* 1986) and may not be considered general. Another issue regarding the use of radical clocks for more than qualitative studies is their reliability inside enzymes (Jin and Lipscomb 1999; Frey 1997).

Collman has raised the hypothesis of 'agostic' mechanisms, involving Fe–C bonds:

$$\underset{Fe^V}{\overset{O}{\parallel}}\overset{\overset{H}{\mid}}{R} \ \rightleftharpoons \ \underset{Fe}{\overset{O}{\parallel}} \overset{H}{\underset{R}{<}} \ \dashrightarrow \ \underset{Fe \cdot R}{\overset{O-H}{\mid}} \longrightarrow Fe^{3+} + ROH \qquad (2.16)$$

This proposal is based largely on bioorganic models (Collman *et al.* 1998); evidence in support of this has not been presented with P450 enzyme systems.

One 'alternative' mechanism (to the FeO^{3+} odd-electron abstraction/oxygen rebound system) that does have reasonable support is the iron-peroxide mechanism, originally proposed by Akhtar *et al.* (1982) to explain the third reaction in the sterol demethylation sequences.

$$(2.17)$$

Further studies have provided evidence that this nucleophilic mechanism can be implicated in the oxidation of simple aldehydes and formation of epoxides (Vaz *et al.* 1998):

$$(2.18)$$

A mechanism involving FeOOH has also been proposed for the formation of alcohols (Newcomb *et al.* 2000).

Rate-limiting steps

The issue of rate-limiting steps in P450 reactions has been considered since the early days of research in the field (Diehl *et al.* 1970; White and Coon 1980). Early work was focussed on rates of reduction of ferric P450 (Diehl *et al.* 1970) and on intermolecular kinetic isotope effects, which were generally very small (Ortiz de Montellano 1986). Unfortunately, knowledge in this area is still very limited and restricted to certain P450/substrate pairs. Many investigators attach great significance to the steady-state kinetic parameters $k_{cat}(V_{max})$ and K_m with no interpretation of their meaning. For a more general consideration of this issues see Northrop (1998). Possibilities regarding each step in Figure 7.2 will be considered briefly.

Step 1 is generally considered to be fast, and the binding itself probably nearly diffusion-limited ($k_{on} \sim 10^8 \text{ M}^{-1} \text{s}^{-1}$), as judged by rates of change in the Soret spectrum (White and Coon 1980). It is possible (and probable; Poulos *et al.* 1995; Schlichting *et al.* 2000) that a conformational change occurs upon binding, but no rate has been measured.

Step 2, the reduction of ferric P450, has been studied extensively (White and Coon 1980; Peterson and Prough 1986). With some P450s the rate does appear to limit the overall reaction, especially in liver microsomes, where the concentration of reductase is far less than that of P450. The extent of stimulation of reduction rates by the P450 substrate can vary considerably (Guengerich and Johnson 1997). However, many P450-catalysed oxidations are very slow ($\sim 1 \text{ min}^{-1}$) and not really limited by the rate of reduction. Thus, rate-limiting reduction is probably limited to the faster substrate oxidations.

Step 3, the binding of O_2 to ferrous iron, is probably fast. The rate of binding has been measured to be $7.7 \times 10^5 \text{ M}^{-1} \text{s}^{-1}$ for CYP101 (Griffin and Peterson 1972); thus at 200 µM O_2 the rate would be $9 \times 10^3 \text{ min}^{-1}$.

Step 4, the transfer of the second electron, is considered to be the rate-limiting step in the oxidation of bacterial CYP101 (Mueller *et al.* 1995). This is probably a rate-limiting step in some reactions catalysed by microsomal P450s, in that cytochrome b_5 can stimulate some reactions by supplying the electron at this step (Peterson and Prough 1986). However, it should be noted that some cytochrome b_5 enhancements cannot be explained by electron transfer (Yamazaki *et al.* 1996; Aucus *et al.* 1998).

Steps 5 and 6 may be rate-limiting, because disruption of the threonine/glutamic acid/H_2O relay system can attenuate substrate oxidation and increase abortive O_2 reduction (Imai *et al.* 1989; Martinis *et al.* 1989; Kimata *et al.* 1995).

Step 7 involves C–H bond breaking in many cases and is subject to study using isotopic hydrogen substitution. Intramolecular isotope studies provide an estimate of the intrinsic isotope effect but do not in themselves yield information as to whether C–H bond breakage is a rate-limiting step (Ortiz de Montellano 1986). Non-competitive intermolecular hydrogen isotope effect studies can. In many cases with P450 reactions, no significant isotope effects could be measured (Ortiz de Montellano 1986; Ullrich 1969). However, in several cases these isotope effects are large (>10) and clearly implicate rate-limiting C–H cleavage (Guengerich *et al.* 1988; Yun *et al.* 2000).

Step 9, product release, has generally been considered to be rapid. However, in our own work on the interpretation of a kinetic deuterium isotope effect on K_m but not k_{cat} for the oxidation of ethanol to acetaldehyde and acetic acid (by CYP2E1), the pre-steady-state burst kinetics clearly implicated a rate-limiting step *following* product formation (Bell and Guengerich 1997; Bell-Parikh and Guengerich 1999). The best explanation, based on simulated fitting of all data for these reactions, is a scheme in which actual substrate oxidation is very fast (500–1000 min^{-1}), as clearly shown experimentally by the burst rates, and the k_{cat} approximates a putative conformational relaxation step that occurs following product formation (Bell-Parikh and Guengerich 1999). This paradigm may explain why several CYP2E1 reactions have similar rates (Guengerich *et al.* 1991). However, it should be pointed out that other CYP2E1 reactions can have slower rates, show different isotope effects, and not show burst kinetics because they are limited by other steps (Bell and Guengerich 1997; Bell-Parikh and Guengerich 1999).

These considerations should be kept in mind as one considers static models of P450 substrate interactions and their meaning. In many cases, the rates of P450 reactions may be much slower than the rate of 'abortive' P450 turnover, so the 'fit' within the active site/transition state may effectively govern reaction selectivity. However, the significance of steady-state kinetic parameters is usually not clear. In some cases the meaning of k_{cat} may be developed through the appropriate experiments. The meaning of K_m in complex systems such as P450-catalysed reactions is usually unknown, as demonstrated in our own work on K_m isotope effects with CYP2E1 (Bell and Guengerich 1997; Bell-Parikh and Guengerich 1999). Investigators should not equate K_m with K_d, the substrate dissociation constant, unless they have specific evidence to justify such a conclusion. This is simply a recapitulation of the advice of basic kinetic texts (Fersht 1999; Northrop 1998).

Catalytic specificity

BASIC CONSIDERATIONS

Catalytic selectivity has been one of the most practical issues involving P450 enzymes, particularly in the context of reactions with new substrates such as drugs, pesticides, etc. What really determines catalytic specificity is will a compound be a substrate and, if so, at which site it will be oxidized and how fast? These are questions that build on two points already discussed, structures and catalytic mechanisms.

One view of catalytic selectivity is that the protein structure is the only determinant and serves only to position substrates near the chemically reactive FeO centre. Reactivity and rates of oxidation are functions of the distance of each substrate atom to the FeO entity. Substrates that do not fit tightly into the active site can 'rattle', and regioselectivity of oxidation is largely a function of the ease of chemical oxidation at each substrate atom. There is a certain amount of credibility to this view. For instance, with relatively small substrates it has been possible to use series of chemicals to develop linear free energy relationships (Burka *et al.* 1985; Macdonald *et al.* 1989). With the growing popularity of protein models, many docking efforts have been made on the basis of the ability of substrates to fit into models with the lowest energy barriers (Jones *et al.* 1996; de Groot and Vermeulen 1997; Lewis and Lake 1995).

No one would exclude the influence of protein structure in selectivity considerations, but some mechanistic differences are also operative. For instance, if electron abstraction is involved in a mechanism, it should be feasible over a longer distance than 'direct' oxygenation or hydrogen abstraction. The possibility that either Fe^{2+}OOH (or $Fe^{2+}O_2{}^-$) or FeO^{3+} can be the oxidant will also influence selectivity (Sahali-Sahly et al. 1996). Rate-limiting steps influence rates and therefore selectivity. For instance, for some P450 2E1 reactions we showed that the rate-limiting step follows product formation but that the K_m for the reaction is isotopically sensitive (because the bond-breaking step is attenuated) (Bell and Guengerich 1997; Bell-Parikh and Guengerich 1999). In other reactions the rate of P450 reduction may be the issue (Guengerich and Johnson 1997).

Thus, the balance of the individual rate-limiting steps is a part of the catalytic selectivity. Having said this, it should be emphasised that, to a large extent, the rate-limiting steps are also a function of the steric fit of the substrate into the protein. For instance, a poor fit may not facilitate efficient reduction for some P450 substrates (Guengerich and Johnson 1997). With all these complexities, predicting substrate and regioselectivity has been very challenging and few if any good prospective examples are available. Predicting rates has been nearly impossible except within sets of small molecules where linear free energy relationships are operative (Burka et al. 1985; Macdonald et al. 1989).

PRACTICAL ASPECTS

The difficulty of predicting catalytic specificity with potential substrates has been discussed above. However, when a reaction is established to occur, it is relatively straightforward to determine which P450 is involved. General approaches have been presented elsewhere (Guengerich and Shimada 1991).

In vitro experiments can be done with microsomes (usually liver) or (less often) with isolated hepatocytes. With rat liver microsomes, considerable information can be obtained from comparisons of gender effects and treatments of the rats with various inducers. For instance, very high inducibility with phenobarbital would first suggest CYP2B1 or 2B2; a male-specific activity that is attenuated in rats treated with poly-cyclic hydrocarbons would suggest CYP2C11, etc. (Guengerich 1987a). Another means of assessing selectivity is to simply compare isolated enzymes for their parameters. Today most preparations are from heterologous expression and not derived from tissues; some are commercially available. The enzymes can be utilised in purified form (along with the reductase, phospholipids, and cytochrome b_5, when appropriate) or within the context of cruder preparations derived from expression systems, e.g. membranes (Guengerich et al. 1996; Gonzalez et al. 1991). Another approach is to use crude animal or human systems (e.g. microsomes) as the source of enzymes and then determine the effects of selective inhibitors of individual P450s (Correia 1995; Halpert and Guengerich 1997). In principle, the fraction of inhibition is indicative of the extent to which that P450 contributes. The same approach can be applied with inhibitory antibodies (Thomas et al. 1977; Guengerich 1987a; Krausz et al. 1997). Finally, if one P450 is dominant in catalysing a particular reaction, then a high correlation should be seen in different human liver samples between the activity and that form of P450 (Beaune et al. 1986). To a first approximation, the correlation coefficient r^2 indicates the fraction of the relationship (activity versus parameter of a particular P450) than can be accounted for by that P450.

In vivo experiments can also be done to establish roles of individual P450s in reactions, both in experimental animals and in humans. Induction experiments can be done and the effects on *in vivo* parameters measured. Several inhibitors can be used safely in humans (e.g. quinidine for CYP2D6, ketoconazole for CYP3A4). Genetic systems can also be of value. With humans, genotypic 'poor metabolisers' (with regard to a particular P450) are expected to also slow oxidation of a drug if that P450 is dominant in metabolism. With animals, several P450-knockout mouse lines are now available (Medinsky *et al.* 1997; Buters *et al.* 1999) (the applicability to humans must be evaluated, however).

Progress has been made in using *in vitro* information (from human liver microsomes and hepatocytes) to make *in vivo* human predications. All approaches have some deficiencies, particularly when issues such as transport are problems. Nevertheless, the general approach has promise (Iwatsubo *et al.* 1997).

Inhibition

Inhibition of P450s is a very practical issue today in the pharmaceutical industry. Before discussing the practical considerations, the basic principles will be reviewed briefly. More extensive treatments of enzyme inhibition in general (Kuby 1991; Segel 1975) and P450 inhibition (Ortiz de Montellano and Correia 1983; Ortiz de Montellano and Reich 1986; Correia 1995; Halpert and Guengerich 1997) have been published and the reader is referred to these articles.

Competitive inhibition is relatively straightforward and, in its simplest form, explained by direct competition of two ligands in the substrate binding site, yielding an increase in the apparent K_m but no change in k_{cat}. However, for discussion of the potential complexity see Segel (1975).

Mechanism-based inactivation utilises the oxidation of a substrate to an inhibitory form. Such oxidation is time-dependent and requires the usual cofactors (NADPH, O_2). In the strictest sense, mechanism-based inactivation implies the formation of an intermediate in the reaction that can either (1) react with the enzyme to destroy it or (2) go on to yield a stable product. The ratio of these two reactions (2/1) is termed the partition coefficient (which can be viewed as the average number of times the enzyme turns over before inactivation). This behaviour has been seen with many drugs and is even a strategy in the design of some drugs intended to inactivate targets, e.g. the CYP19-catalysed oxidation of androgens to oestrogens in hormone-dependent cancers (Vanden Bossche *et al.* 1994). Three major types of inactivation are seen. The first involves destruction of the P450 haem, by alkylation of a pyrrole nitrogen (Ortiz de Montellano *et al.* 1979). Following this process, the iron is lost and the porphyrin derivative is removed from the P450 protein. The second process involves the interaction of a reaction intermediate with the P450 protein. A number of products have been characterised (Roberts *et al.* 1997). The third type of P450 inactivation involves the degradation of haem and crosslinking of the fragments to the protein, which occurs with oxidants and certain chemicals (Osawa and Pohl 1989).

Some other types of inhibition often presented in introductory biochemistry texts are non-competitive and uncompetitive inhibition. In principle, the former involves a decreased k_{cat} and change in K_m. The latter involves interaction of an inhibitor with only the substrate-bound form of the enzyme. Although these cases are often presented in graph example, they are seldom observed in practice. Another type of inhibition is 'slow-binding tight inhibition', in which a time-dependent development of inhibition occurs without transformation of the compound. For instance, a conformational change is thought to be involved in such a process when prostaglandin synthase is inhibited by non-steroidal anti-inflammatory drugs (Kalgutkar and Marnett 1994). Such a process has not been documented for any P450s (Shimada *et al.* 1998). Another type of inhibition occurs when a reaction product is a reactive electrophile and binds to the P450 protein. Distinguishing such reactions from true mechanism-based inactivation may be difficult (Silverman 1995), and often this type of inhibition is grouped under the same heading.

In the course of our own work, we found that some peptides were very effective in blocking the binding of substrates to CYP3A4 but were more than an order of magnitude less efficient in blocking catalytic activity (Hosea *et al.* 2000). The basis appears to be the selective binding of the peptides (which were shown not to be substrates for oxidation) to the ferric P450 (compared to ferrous P450). This situation is of conceptual interest in that different redox states of P450 are considered as distinct enzymes in the analysis of the catalytic cycle (Hosea *et al.* 2000; Segel 1975).

A complex issue that will be mentioned briefly is that of P450 stimulation by chemicals. The phenomenon is not new (Cinti 1978) but remains poorly understood (Halpert and Guengerich 1997). Two modes are seen: (1) heterotropic cooperativity, in which an added chemical other than the substrate can enhance oxidation of the substrate (whether or not the added chemical is also a substrate), and (2) homotropic cooperativity, where plots of reaction velocity versus substrate are not hyperbolic. Several models have been proposed, including those with a large site capable of binding ≥ 2 substrates (Johnson *et al.* 1988; Huang *et al.* 1981; Hosea *et al.* 2000; Shou *et al.* 1999). These models have not been distinguished by biophysical experiments, and several possibilities must still be considered possible to account for all the observations. The cooperativity is probably not simply a biochemical curiosity. Some sigmoidal kinetic behaviour has been observed in kinetic assays with hepatocyte suspensions (Witherow and Houston 1998), and heterotropic stimulation of animal P450s has been demonstrated *in vivo* (Lasker *et al.* 1982; Lee *et al.* 1996a).

As mentioned earlier, P450 inhibition is a practical issue and has been implicated in adverse drug–drug interactions (Kivistö *et al.* 1994; Guengerich 1999a). Most pharmaceutical companies do assays early in the screening/development phase. Screens for competitive inhibition can be readily done in relatively high-throughput modes.

Several issues can be considered. One of the major decisions to make is how much information to acquire. More data may be useful in reaching decision points but more time and resources are needed. Assays for mechanism-based inactivation are much more laborious than for simple competitive inhibition. One problem, even with competitive inhibition, is the generality of results obtained with one particular reaction, and this problem is probably greatest with CYP3A4. For instance, some of the different fluorescence-based high-throughput reactions show varying effects (stimulation or inhibition) when the same compound is added (Miller *et al.* 1999). A potentially valuable new fluorescence assay was developed by Chauret *et al.* (1999) and appears to be congruent with the prototype testosterone 6β-hydroxylation. Another concern is how much organic solvent to use to dissolve the substrate. Many drugs are notoriously insoluble in H_2O and solvents are needed. In this laboratory we have generally operated with $\leqslant 1\%$ (v/v), but even this concentration can be an issue (Yoo *et al.* 1987; Chauret *et al.* 1998).

A major issue is exactly how much *in vitro* inhibition is serious *in vivo* (Iwatsubo *et al.* 1997). Calculations can be made by simply doing substitutions in the standard Michaelis–Menten formulae, letting [S] = plasma concentration of drug and [I] = plasma concentration of inhibitor, although protein binding and transport phenomena can influence the extrapolations. Mechanism-based inactivation has been more difficult to model because of its very nature. As the databases of drugs with both *in vivo* history and valid inactivation parameters ($k_{inactivation}$, K_i) grow, better prediction of *in vivo* problems should be possible.

References

Addya S, Anandatheerthavarada HK, Biswas G, Bhagwat SV, Mullick J and Avadhani NG (1997) Targeting of NH_2-terminal-processed microsomal protein to mitochondria: a novel pathway for the biogenesis of hepatic mitochrondrial P450MT2. *Journal of Cell Biology*, **139**, 589–599.

Akhtar M, Calder MR, Corina DL and Wright JN (1982) Mechanistic studies on C-19 demethylation in oestrogen biosynthesis. *Biochemical Journal*, **201**, 569–580.

Alvares AP, Schilling G, Levin W and Kuntzman R (1967) Studies on the induction of CO-binding pigments in liver microsomes by phenobarbital and 3-methylcholanthrene. *Biochemical and Biophysical Research Communications*, **29**, 521–526.

Anandatheerthavarada HK, Addya S, Dwivedi RS, Biswas G, Mullick J and Avadhani NG (1997) Localization of multiple forms of inducible cytochromes P450 in rat liver mitochrondria: immunological characteristics and patterns of xenobiotic substrate metabolism. *Archives of Biochemistry and Biophysics*, **339**, 136–150.

Anzenbacher P, Niwa T, Tolbert LM, Sirimanne SS and Guengerich FP (1996) Oxidation of 9-alkyl anthracenes by cytochrome P450 2B1, horseradish peroxidase, and iron tetraphenyl-porphin/iodosylbenzene systems. Anaerobic and aerobic mechanisms. *Biochemistry*, **35**, 2512–2520.

Aucus RJ, Lee TC and Miller WL (1998) Cytochrome b_5 augments the 17,20-lyase activity of human P450c17 without direct electron transfer. *The Journal of Biological Chemistry*, **273**, 3158–3165.

Axelrod J (1955) The enzymatic deamination of amphetamine. *The Journal of Biological Chemistry*, **214**, 753–763.

Bäckström D, Ingelman-Sundberg M and Ehrenberg A (1983) Oxidation-reduction potential of soluble and membrane-bound rabbit liver microsomal cytochrome P-450 LM2. *Acta Chemica Scandinavica*, **37**, 891–894.

Baron J, Redick JA and Guengerich FP (1981) An immunohistochemical study on the localizations and distributions of phenobarbital- and 3-methylcholanthrene-inducible cytochromes P-450 within the livers of untreated rats. *The Journal of Biological Chemistry*, **256**, 5931–5937.

Bartlomowicz B, Friedberg T, Utesch D, Molitor E, Platt K and Oesch F (1989) Regio- and stereoselective regulation of monooxygenase activities by isoenzyme-selective phosphorylation of cytochrome P450. *Biochemical and Biophysical Research Communications*, **160**, 46–52.

Beaune P, Kremers PG, Kaminsky LS, de Graeve J and Guengerich FP (1986) Comparison of monooxygenase activities and cytochrome P-450 isozyme concentrations in human liver microsomes. *Drug Metabolism and Disposition*, **14**, 437–442.

Bell LC and Guengerich FP (1997) Oxidation kinetics of ethanol by human cytochrome P450 2E1. Rate-limiting product release accounts for effects of isotopic hydrogen substitution and cytochrome b_5 on steady-state kinetics. *The Journal of Biological Chemistry*, **272**, 29643–29651.

Bell-Parikh LC and Guengerich FP (1999) Kinetics of cytochrome P450 2E1-catalyzed oxidation of ethanol to acetic acid via acetaldehyde. *The Journal of Biological Chemistry*, **274**, 23833–23840.

Bolt HM, Bolt M and Kappus H (1977) Interaction of rifampicin treatment with pharmacokinetics and metabolism of ethinyloestradiol in man. *Acta Endocrinologica*, **85**, 189–197.

Brodie BB, Gillette JR and LaDu BN (1958) Enzymatic metabolism of drugs and other foreign compounds. *Annual Review of Biochemistry*, **27**, 427–454.

Burka LT, Guengerich FP, Willard RJ and Macdonald TL (1985) Mechanism of cytochrome P-450 catalysis. Mechanism of N-dealkylation and amine oxide deoxygenation. *Journal of the American Chemical Society*, **107**, 2549–2551.

Buters JTM, Sakai S, Richter T, Pineau T, Alexander DL, Savas U, Doehmer J, Ward JM, Jefcoate CR and Gonzalez FJ (1999) Cytochrome P450 CYP1B1 determines susceptibility to 7,12-dimethylbenz[a]anthracene-induced lymphomas. *Proceedings of the National Academy of Sciences, USA*, **96**, 1977–1982.

Cavalieri EL and Rogan EG (1995) Central role of radical cations in metabolic activation of polycyclic aromatic hydrocarbons. *Xenobiotica*, **25**, 677–688.

Chauret N, Gauthier A and Nicoll-Griffith DA (1998) Effect of common organic solvents on in vitro cytochrome P450-mediated metabolic activities in human liver microsomes. *Drug Metabolism and Disposition*, **26**, 1–4.

Chauret N, Tremblay N, Lackman RL, Gauthier J-Y, Silva JM, Marois J, Yergey JA and Nicoll-Griffith DA (1999) Description of a 96-well plate assay to measure cytochrome P4503A inhibition in human liver microsomes using a selective fluorescent probe. *Analytical Biochemistry*, **276**, 215–226.

Cinti DL (1978) Agents activating the liver microsomal mixed function oxidase system. *Pharmacology and Therapeutics*, **2**, 727–749.

Collman JP, Chien AS, Eberspacher TA and Brauman J,I. (1998) An agostic alternative to the P-450 rebound mechanism. *Journal of the American Chemical Society*, **120**, 425–426.

Conney AH (1982) Induction of microsomal enzymes by foreign chemicals and carcinogenesis by polycyclic aromatic hydrocarbons: G. H. A. Clowes memorial lecture. *Cancer Research*, **42**, 4875–4917.

Conney AH, Miller EC and Miller JA (1956) The metabolism of methylated aminoazo dyes. V. Evidence for induction of enzyme synthesis in the rat by 3-methylcholanthrene. *Cancer Research*, **16**, 450–459.

Correia MA (1995) Rat and human liver cytochromes P450. Substrate and inhibitor specificities and functional markers. In *Cytochrome P450: Structure, Mechanism, and Biochemistry*, 2nd edition, Ortiz de Montellano PR (ed.), Plenum Press, New York, pp. 607–630.

Correia MA, Davoll SH, Wrighton SA and Thomas PE (1992a) Degradation of rat liver cytochromes P450 3A after their inactivation by 3,5-dicarbethoxy-2,6-dimethyl-4-ethyl-1,4-dihydropyridine: characterization of the proteolytic system. *Archives of Biochemistry and Biophysics*, **297**, 228–238.

Correia MA, Yao K, Wrighton SA, Waxman DJ and Rettie AE (1992b) Differential apoprotein loss of rat liver cytochromes P450 after their inactivation by 3,5-dicarbethoxy-2,6-dimethyl-4-ethyl-1,4-dihydropyridine: a case for distinct proteolytic mechanisms? *Archives of Biochemistry and Biophysics*, **294**, 493–503.

Cosme J and Johnson EF (2000) Engineering microsomal cytochrome P450 2C5 to be a soluble, monomeric enzyme. Mutations that alter aggregation, phospholipid dependence of catalysis, and membrane binding. *The Journal of Biological Chemistry*, **275**, 2545–2553.

Cupp-Vickery JR, Han O, Hutchinson CR and Poulos TL (1996) Substrate-assisted catalysis in cytochrome P450eryF. *Nature Structural Biology*, **3**, 632–637.

Cupp-Vickery J, Anderson R and Hatziris Z (2000) Crystal structures of ligand complexes of P450eryF exhibiting homotropic cooperativity. *Proceedings of the National Academy of Sciences, USA*, **97**, 3050–3055.

d'Errico A, Taioli E, Chen X and Vineis P (1996) Genetic metabolic polymorphisms and the risk of cancer: a review of the literature. *Biomarkers*, **1**, 149–173.

Daly JW, Jerina DM and Witkop B (1972) Arene oxides and the NIH shift: the metabolism, toxicity and carcinogenicity of aromatic compounds. *Experientia*, **28**, 1129–1264.

Dawson JH (1988) Probing structure–function relations in heme-containing oxygenases and peroxidases. *Science*, **240**, 433–439.

de Groot MJ and Vermeulen NPE (1997) Modeling the active sites of cytochrome P450s and glutathione *S*-transferases, two of the most important biotransformation enzymes. *Drug Metabolism Reviews*, **29**, 747–799.

Dees JH, Masters BSS, Muller-Eberhard U and Johnson EF (1982) Effect of 2,3,7,8-tetrachloro-dibenzo-*p*-dioxin and phenobarbital on the occurrence and distribution of four cytochrome P-450 isozymes in rabbit kidney, lung, and liver. *Cancer Research*, **42**, 1423–1432.

Diehl H, Schädelin J and Ullrich V (1970) Studies on the kinetics of cytochrome P-450 reduction in rat liver microsomes. *Hoppe–Seyler's Zeischrift fur Physioligische Chemie*, **351**, 1359–1371.

Distlerath LM and Guengerich FP (1987) Enzymology of human liver cytochromes P-450. In *Mammalian Cytochromes P-450, Volume 1*, Guengerich FP (ed.), CRC Press, Boca Raton, FL, pp. 133–198.

Dong Z, Hong J, Ma Q, Li D, Bullock J, Gonzalez FJ, Park SS, Gelboin HV and Yang CS (1988) Mechanism of induction of cytochrome P-450ac (P-450j) in chemically induced and spontaneously diabetic rats. *Archives of Biochemistry and Biophysics*, **263**, 29–35.

Eliasson E, Johansson I and Ingelman-Sundberg M (1988) Ligand-dependent maintenance of ethanol-inducible cytochrome P-450 in primary rat hepatocyte cell cultures. *Biochemical and Biophysical Research Communications*, **150**, 436–443.

Estabrook RW, Cooper DY and Rosenthal O (1963) The light reversible carbon monoxide inhibition of the steroid C21-hydroxylase system of the adrenal cortex. *Biochemische Zeitschrift*, **338**, 741–755.

Fersht A (1999) *Structure and Mechanism in Protein Science*, Freeman, New York, p. 110.

Fisher MT and Sligar SG (1985) Control of heme protein redox potential and reduction rate: linear free energy relation between potential and ferric spin state equilibrium. *Journal of the American Chemical Society*, **107**, 5018–5019.

Frey PA (1997) Radicals in enzymatic reactions. *Current Opinion in Chemical Biology*, **1**, 347–356.

Fujii-Kuriyama Y, Mizukami Y, Kawajiri K, Sogawa K and Muramatsu M (1982) Primary structure of a cytochrome P-450: coding nucleotide sequence of phenobarbital-inducible cytochrome P-450 cDNA from rat liver. *Proceedings of the National Academy of Sciences, USA*, **79**, 2793–2797.

Fulco AJ (1991) P450$_{BM-3}$ and other inducible bacterial P450 cytochromes: biochemistry and regulation. *Annual Review of Pharmacology and Toxicology*, **31**, 177–203.

Geneste O, Raffalli F and Lang MA (1996) Identification and characterization of a 44 kDa protein that binds specifically to the 3′-untranslated region of CYP2a5 mRNA: inducibility, subcellular distribution and possible role in mRNA stabilization. *Biochemical Journal*, **313**, 1029–1037.

Gonzalez FJ, Crespi CL and Gelboin HV (1991) cDNA-expressed human cytochrome P450s: a new age of molecular toxicology and human risk assessment. *Mutation Research*, **247**, 113–127.

Gotoh O (1992) Substrate recognition sites in cytochrome P450 family 2 (CYP2) proteins inferred from comparative analysis of amino acid and coding nucleotide sequences. *The Journal of Biological Chemistry*, **267**, 83–90.

Griffin BW and Peterson JA (1972) Camphor binding by *Pseudomonas putida* cytochrome P-450. Kinetics and thermodynamics of the reaction. *Biochemistry*, **11**, 4740–4746.

Groves JT and Subramanian DV (1984) Hydroxylation by cytochrome P-450 and metalloporphyrin models. Evidence for allylic rearrangement. *Journal of the American Chemical Society*, **106**, 2177–2181.

Groves JT, McClusky GA, White RE and Coon MJ (1978) Aliphatic hydroxylation by highly purified liver microsomal cytochrome P-450: Evidence for a carbon radical intermediate. *Biochemical and Biophysical Research Communications*, **81**, 154–160.

Guengerich FP (1978) Destruction of heme and hemoproteins mediated by liver microsomal reduced nicotinamide adenine dinucleotide phosphate-cytochrome P-450 reductase. *Biochemistry*, **17**, 3633–3639.

Guengerich FP (1983) Oxidation-reduction properties of rat liver cytochrome P-450 and

NADPH-cytochrome P-450 reductase related to catalysis in reconstituted systems. *Biochemistry*, **22**, 2811–2820.

Guengerich FP (1987a) Enzymology of rat liver cytochromes P-450. In *Mammalian Cytochromes P-450, Volume 1*, Guengerich FP (ed.) CRC Press, Boca Raton, FL, pp. 1–54.

Guengerich FP (1987b) Oxidative cleavage of carboxylic esters by cytochrome P-450. *The Journal of Biological Chemistry*, **262**, 8459–8462.

Guengerich FP (1988a) Roles of cytochrome P-450 enzymes in chemical carcinogenesis and cancer chemotherapy. *Cancer Research*, **48**, 2946–2954.

Guengerich FP (1988b) Oxidation of 17α-ethynylestradiol by human liver cytochrome P-450. *Molecular Pharmacology*, **33**, 500–508.

Guengerich FP (1989) Oxidation of halogenated compounds by cytochrome P-450, peroxidases, and model metalloporphyrins. *The Journal of Biological Chemistry*, **264**, 17198–17205.

Guengerich FP (1990) Enzymatic oxidation of xenobiotic chemicals. *Critical Reviews in Biochemistry and Molecular Biology*, **25**, 97–153.

Guengerich FP (1991) Reactions and significance of cytochrome P-450 enzymes. *The Journal of Biological Chemistry*, **266**, 10019–10022.

Guengerich FP (1995) Human cytochrome P450 enzymes. In *Cytochrome P450*, 2nd edition, Ortiz de Montellano PR (ed.), Plenum Press, New York, pp. 473–535.

Guengerich FP (1996) The chemistry of cytochrome P450 reactions. In *Cytochromes P450: Metabolic and Toxicological Aspects*, Ioannides C (ed.), CRC Press, Boca Raton, FL, pp. 55–74.

Guengerich FP (1997a) Cytochrome P450 enzymes. In *Biotransformation, Volume 3, Comprehensive Toxicology*, Guengerich FP (ed.), Elsevier Science, Oxford, pp. 37–68.

Guengerich FP (1997b) Comparisons of catalytic selectivity of cytochrome P450 subfamily members from different species. *Chemico-Biological Interactions*, **106**, 161–182.

Guengerich FP (1998) The environmental genome project: functional analysis of polymorphisms. *Environmental Health Perspectives*, **106**, 365–368.

Guengerich FP (1999a) Inhibition of drug metabolizing enzymes: molecular and biochemical aspects. In *Handbook of Drug Metabolism*, Woolf TF (ed.), Marcel Dekker, New York, pp. 203–227.

Guengerich FP (1999b) Human cytochrome P-450 3A4: regulation and role in drug metabolism. *Annual Review of Pharmacology and Toxicology*, **39**, 1–17.

Guengerich FP and Johnson WW (1997) Kinetics of ferric cytochrome P450 reduction by NADPH-cytochrome P450 reductase. Rapid reduction in absence of substrate and variation among cytochrome P450 systems. *Biochemistry*, **36**, 14741–14750.

Guengerich FP and Macdonald TL (1984) Chemical mechanisms of catalysis by cytochromes P-450: a unified view. *Accounts of Chemical Research*, **17**, 9–16.

Guengerich FP and Macdonald TL (1990) Mechanisms of cytochrome P-450 catalysis. *FASEB Journal*, **4**, 2453–2459.

Guengerich FP and Macdonald TL (1993) Sequential electron transfer reactions catalyzed by cytochrome P-450 enzymes. In *Advances in Electron Transfer Chemistry, Volume 3*, Mariano PS (ed.), JAI Press, Greenwich, CT, pp. 191–241.

Guengerich FP and Shimada T (1991) Oxidation of toxic and carcinogenic chemicals by human cytochrome P-450 enzymes. *Chemical Research in Toxicology*, **4**, 391–407.

Guengerich FP, Dannan GA, Wright ST, Martin MV and Kaminsky LS (1982) Purification and characterization of liver microsomal cytochromes P-450: electrophoretic, spectral, catalytic, and immunochemical properties and inducibility of eight isozymes isolated from rats treated with phenobarbital or β-naphthoflavone. *Biochemistry*, **21**, 6019–6030.

Guengerich FP, Peterson LA and Böcker RH (1988) Cytochrome P-450-catalyzed hydroxylation and carboxylic acid ester cleavage of Hantzsch pyridine esters. *The Journal of Biological Chemistry*, **263**, 8176–8183.

Guengerich FP, Kim D-H and Iwasaki M (1991) Role of human cytochrome P-450 IIE1 in the oxidation of many low molecular weight cancer suspects. *Chemical Research in Toxicology*, **4**, 168–179.

Guengerich FP, Gillam EMJ and Shimada T (1996) New applications of bacterial systems to problems in toxicology. *CRC Critical Reviews in Toxicology*, **26**, 551–583.

Guroff G, Daly JW, Jerina DM, Renson J, Witkop B and Udenfriend S (1967) Hydroxylation-induced migration: the NIH shift. *Science*, **157**, 1524–1530.

Gustafsson JÅ, Rondahl L and Bergman J (1979) Iodosylbenzene derivatives as oxygen donors in cytochrome P-450 catalyzed steroid hydroxylations. *Biochemistry*, **18**, 865–870.

Gustafsson J-Å, Mode A, Norstedt G and Skett P (1983) Sex steroid induced changes in hepatic enzymes. *Annual Review of Physiology*, **45**, 51–60.

Halpert JR and Guengerich FP (1997) Enzyme inhibition and stimulation. In *Biotransformation, Volume 3, Comprehensive Toxicology*, Guengerich FP (ed.), Elsevier Science, Oxford, pp. 21–35.

Halpert JR and He Y (1993) Engineering of cytochrome P450 2B1 specificity: conversion of an androgen 16β-hydroxylase to a 15α-hydroxylase. *The Journal of Biological Chemistry*, **268**, 4453–4457.

Hammons GJ, Guengerich FP, Weis CC, Beland FA and Kadlubar FF (1985) Metabolic oxidation of carcinogenic arylamines by rat, dog, and human hepatic microsomes and by purified flavin-containing and cytochrome P-450 monooxygenases. *Cancer Research*, **45**, 3578–3585.

Hankinson O (1995) The aryl hydrocarbon receptor complex. *Annual Review of Pharmacology and Toxicology*, **35**, 307.

Harlow GR and Halpert JR (1997) Alanine-scanning mutagenesis of a putative substrate recognition site in human cytochrome P450 3A4: role of residues 210 and 211 in flavonoid activation and substrate specificity. *The Journal of Biological Chemistry*, **272**, 5396–5402.

Harlow GR and Halpert JR (1998) Analysis of human cytochrome P450 3A4 cooperativity: construction and characterization of a site-directed mutant that displays hyperbolic steroid hydroxylation kinetics. *Proceedings of the National Academy of Sciences, USA*, **95**, 6636–6641.

Hasemann CA, Ravichandran KG, Peterson JA and Deisenhofer J (1994) Crystal structure and refinement of cytochrome P450$_{terp}$ at 2.3 Å resolution. *Journal of Molecular Biology*, **236**, 1169–1185.

Hasemann CA, Kurumbail RG, Boddupalli SS, Peterson JA and Deisenhofer J (1995) Structure and function of cytochromes P450: a comparative analysis of three crystal structures. *Structure*, **2**, 41–62.

Hasler JA, Harlow GR, Szlarz GD, John GH, Kedzie KM, Burnett VL, He YA, Kaminsky LS and Halpert JR (1994) Site-directed mutagenesis of putative substrate recognition sites in cytochrome P450 2B11: importance of amino acid residues 114, 290, and 363 for substrate specificity. *Molecular Pharmacology*, **46**, 338–345.

Haugen DA, van der Hoeven TA and Coon MJ (1975) Purified liver microsomal cytochrome P-450: separation and characterization of multiple forms. *The Journal of Biological Chemistry*, **250**, 3567–3570.

Helms V and Wade RC (1998) Hydration energy landscape of the active cavity in cytochrome P450$_{cam}$. *Proteins*, **32**, 381–396.

Hildebrandt A, Remmer H and Estabrook RW (1968) Cytochrome P-450 of liver microsomes: one pigment or many. *Biochemical and Biophysical Research Communications*, **30**, 607–612.

Honkakosski P, Zelko I, Sueyoshi T, Negishi M (1998) The nuclear orphan receptor CAR-RXR heterodimer activates the phenobarbital-responsive enhancer module of the CYP2B gene. *Molecular and Cellular Biology*, **18**, 1–7.

Hosea NA, Miller GP and Guengerich FP (2000) Elucidation of distinct binding sites for cytochrome P450 3A4. *Biochemistry*, **39**, 5929–5939.

Huang MT, Chang RL, Fortner JG and Conney AH (1981) Studies on the mechanism of activation of microsomal benzo[a]pyrene hydroxylation by flavonoids. *The Journal of Biological Chemistry*, **256**, 6829–6836.

Ibeanu GC, Ghanayem BI, Linko P, Li L, Pedersen LG and Goldstein JA (1996) Identification of residues 99, 220, and 221 of human cytochrome P450 2C19 as key determinants of omeprazole hydroxylase activity. *The Journal of Biological Chemistry*, **271**, 12496–12501.

Imai M, Shimada H, Watanabe Y, Matsushima-Hibiya Y, Makino R, Koga H, Horiuchi T and Ishimura Y (1989) Uncoupling of the cytochrome P-450cam monooxygenase reaction by a single mutation, threonine-252 to alanine or valine: a possible role of the hydroxy amino acid in oxygen activation. *Proceedings of the National Academy of Sciences, USA*, **86**, 7823–7827.

Ishii-Ohba H, Guengerich FP and Baron J (1984) Localization of epoxide-metabolizing enzymes in rat testis. *Biochimica et Biophysica Acta*, **802**, 326–334.

Iwatsubo T, Hirota N, Ooie T, Suzuki H, Shimada N, Chiba K, Ishizaki T, Green CE, Tyson CA and Sugiyama Y (1997) Prediction of *in vivo* drug metabolism in the human liver from *in vitro* metabolism data. *Pharmacology and Therapeutics*, **73**, 147–171.

Jin Y and Lipscomb JD (1999) Probing the mechanism of C-H activation: oxidation of methyl-cubane by soluble methane monooxygenase from *Methylosinus trichosporium* OB3b. *Biochemistry*, **38**, 6178–6186.

Johansson I, Eliasson E and Ingelman-Sundberg M (1991) Hormone controlled phosphorylation and degradation of CYP2B1 and CYP2E1 in isolated rat hepatocytes. *Biochemical and Biophysical Research Communications*, **174**, 37–42.

Johnson EF, Schwab GE and Dieter HH (1983) Allosteric regulation of the 16α-hydroxylation of progesterone as catalyzed by rabbit microsomal cytochrome P-450 3b. *The Journal of Biological Chemistry*, **258**, 2785–2788.

Johnson EF and Schwab GE (1984) Constitutive forms of rabbit-liver microsomal cytochrome P-450: enzymatic diversity, polymorphism and allosteric regulation. *Xenobiotica*, **14**, 3–18.

Johnson EF, Schwab GE and Vickery LE (1988) Positive effectors of the binding of an active site-directed amino steroid to rabbit cytochrome P-450 3c. *The Journal of Biological Chemistry*, **263**, 17672–17677.

Jones BC, Hawksworth G, Horne VA, Newlands A, Morsman J, Tute MS and Smith DA (1996) Putative active site template model for cytochrome P4502C9 (tolbutamide hydroxylase). *Drug Metabolism and Disposition*, **24**, 260–266.

Kagawa N and Waterman MR (1995) Regulation of steroidogenic and related P450s. In *Cytochrome P450*, 2nd edition, Ortiz de Montellano PR (ed.), Plenum Press, New York, pp. 419–442.

Kalgutkar AS and Marnett LJ (1994) Rapid inactivation of prostaglandin endoperoxide synthases by *N*-(carboxyalkyl)maleimides. *Biochemistry*, **33**, 8625–8628.

Kato R and Kamataki T (1982) Cytochrome P-450 as a determinant of sex difference of drug metabolism in the rat. *Xenobiotica*, **12**, 787–800.

Kimata Y, Shimada H, Hirose T and Ishimura Y (1995) Role of Thr-252 in cytochrome P450$_{CAM}$: a study with unnatural amino acid mutagenesis. *Biochemical and Biophysical Research Communications*, **208**, 96–102.

Kimura S, Gonzalez FJ and Nebert DW (1984) Mouse cytochrome P3-450: complete cDNA and amino acid sequence. *Nucleic Acids Research*, **12**, 2917–2927.

Kivistö KT, Neuvonen PJ and Klotz U (1994) Inhibition of terfenadine metabolism: pharmacokinetic and pharmacodynamic consequences. *Clinical Pharmacokinetics*, **27**, 1–5.

Kliewer SA, Moore JT, Wade L, Staudinger JL, Watson MA, Jones SA, McKee DD, Oliver BB, Wilson TM, Zetterströrm RH, Perlmann T and Lehmann JM (1998) An orphan nuclear receptor activated by pregnanes defines a novel steroid signaling pathway. *Cell*, **92**, 73–82.

Lee CA, Lillibridge JH, Nelson SD and Slattery JT (1996a) Effects of caffeine and theophylline on acetaminophen pharmacokinetics: P450 inhibition and activation. *Journal of Pharmacology and Experimental Therapeutics*, **277**, 287–291.

Lee SST, Buters JTM, Pineau T, Fernandez-Salguero P and Gonzalez FJ (1996b) Role of CYP2E1 in the hepatotoxicity of acetaminophen. *The Journal of Biological Chemistry*, **271**, 12063–12067.

Lewis DFV and Lake BG (1995) Molecular modelling of members of the P4502A subfamily: application to studies of enzyme specificity. *Xenobiotica*, **25**, 585–598.

Liang HCL, Li H, McKinnon RA, Duffy JJ, Potter SS, Puga A and Nebert DW (1996) *Cyp1a2*(-/-) null mutant mice develop normally but show deficient drug metabolism. *Proceedings of the National Academy of Sciences, USA*, **93**, 1671–1676.

Lichtenberger F, Nastainczyk W and Ullrich V (1976) Cytochrome P-450 as an oxene transferase. *Biochemical and Biophysical Research Communications*, **70**, 939–946.

Lindberg RLP and Negishi M (1989) Alteration of mouse cytochrome P450$_{coh}$ substrate specificity by mutation of a single amino-acid residue. *Nature*, **339**, 632–634.

Loeper J, Descatoire V, Maurice M, Beaune P, Belghiti J, Houssin D, Ballet F, Feldman G, Guengerich FP and Pessayre D (1993) Cytochromes P-450 in human hepatocyte plasma membrane: recognition by several autoantibodies. *Gastroenterology*, **104**, 203–216.

Loeper J, Le Berre A and Pompon D (1998) Topology inversion of CYP2D6 in the endoplasmic reticulum is not required for plasma membrane transport. *Molecular Pharmacology*, **53**, 408–414.

Lu AYH and Coon MJ (1968) Role of hemoprotein P-450 in fatty acid ω-hydroxylation in a soluble enzyme system from liver microsomes. *The Journal of Biological Chemistry*, **243**, 1331–1332.

Macdonald TL, Gutheim WG, Martin RB and Guengerich FP (1989) Oxidation of substituted *N,N*-dimethylanilines by cytochrome P-450: estimation of the effective oxidation-reduction potential of cytochrome P-450. *Biochemistry*, **28**, 2071–2077.

Martinis SA, Atkins WM, Stayton PS and Sligar SG (1989) A conserved residue of cytochrome P-450 is involved in heme-oxygen stability and activation. *Journal of the American Chemical Society*, **111**, 9252–9253.

McMurry TJ and Groves JT (1986) Metalloporphyrin models for cytochrome P-450. In *Cytochrome P-450*, Ortiz de Montellano PR (ed.), Plenum Press, New York, pp. 1–28.

Medinsky MA, Asgharian B, Farris GM, Gonzalez FJ, Schlosser PM, Seaton MJ and Valentine JL (1997) Benzene risk assessment: transgenic animals and human variability. *Chemical Industry Institute of Toxicology Activities*, **17**, 1–5.

Miller VP, Crespi CL, Ackermann JM, Stresser DM and Busby WF, Jr. (1999) Novel high throughput fluorescent cytochrome P450 assays. *Proceedings, 7th European Meeting, International Society for the Study of Xenobiotics, Budapest, 22–26 August*, **14**, 63.

Mueller EJ, Loida PJ and Sligar SG (1995) Twenty-five years of P450$_{cam}$ research: mechanistic insights into oxygenase catalysis. In *Cytochrome P450*, 2nd edition, Ortiz de Montellano PR (ed.), Plenum Press, New York, pp. 83–124.

Mueller GC and Miller JA (1948) The metabolism of 4-dimethylaminoazobenzene by rat liver homogenates. *The Journal of Biological Chemistry*, **176**, 535–544.

Mueller GC and Miller JA (1953) The metabolism of methylated aminoazo dyes. II. Oxidative demethylation by rat liver homogenates. *The Journal of Biological Chemistry*, **202**, 579–587.

Muerhoff AS, Griffin KJ and Johnson EF (1992) The peroxisome proliferator-activated receptor mediates the induction of CYP4A6, a cytochrome P450 fatty acid ω-hydroxylase, by clofibric acid. *The Journal of Biological Chemistry*, **267**, 19051–19053.

Naït-Kaoudjt R, Williams R, Guiard B and Gervais M (1997) Some DNA targets of the yeast CYP1 transcriptional activator are functionly asymmetric: evidence of two half-sites with different affinities. *European Journal of Biochemistry*, **244**, 301–309.

Nebert DW (1989) The *Ah* locus: genetic differences in toxicity, cancer, mutation, and birth defects. *CRC Critical Reviews in Toxicology*, **20**, 153–174.

Nebert DW, Adesnik M, Coon MJ, Estabrook RW, Gonzalez FJ, Guengerich FP, Gunsalus IC, Johnson EF, Kemper B, Levin W, Phillips IR, Sato R and Waterman MR (1987) The P450 gene superfamily: recommended nomenclature. *DNA*, **6**, 1–11.

Nebert DW, Nelson DR, Adesnik M, Coon MJ, Estabrook RW, Gonzalez FJ, Guengerich FP, Gunsalus IC, Johnson EF, Kemper B, Levin W, Phillips I, Sato R and Waterman M (1989) The P450 superfamily: updated listing of all genes and recommended nomenclature for the chromosomal loci. *DNA*, **8**, 1–13.

Nelson DR, Kamataki T, Waxman DJ, Guengerich FP, Estabrook RW, Feyereisen R, Gonzalez FJ, Coon MJ, Gunsalus IC, Gotoh O, Okuda K and Nebert DW (1993) The P450 superfamily: update on new sequences, gene mapping, accession numbers, early trivial names of enzymes, and nomenclature. *DNA and Cell Biology*, **12**, 1–51.

Nelson SD (1994) Covalent binding to proteins. In *Methods in Toxicology, Volume 1B*, Academic Press, San Diego, pp. 340–348.

Neve EPA and Ingelman-Sundberg M (1999) A soluble NH$_2$-terminally truncated catalytically active form of cytochrome P450 2E1 targeted to liver mitochrondria. *FEBS Letters*, **460**, 309–314.

Newcomb M, Le Tadic-Biadatti MH, Chestney DL, Roberts ES and Hollenberg PF (1995a) A nonsynchronous concerted mechanism for cytochrome P-450 catalyzed hydroxylation. *Journal of the American Chemical Society*, **117**, 12085–12091.

Newcomb M, Letadic MH, Putt DA and Hollenberg PF (1995b) An incredibly fast apparent oxygen rebound rate constant for hydrocarbon hydroxylation by cytochrome P-450 enzymes. *Journal of the American Chemical Society*, **117**, 3312–3313.

Newcomb M, Shen R, Choi SY, Toy PH, Hollenberg PF, Vaz ADN and Coon MJ (2000) Cytochrome P450-catalyzed hydroxylation of mechanistic probes that distinguish between radicals and cations. Evidence for cationic but not for radical intermediates. *Journal of the American Chemical Society*, **122**, 2677–2686.

Nims RW, McClain RM, Manchand PS, Belica PS, Thomas PE, Mellini DW, Utermahlen WE Jr and Lubet RA (1994) Comparative pharmacodynamics of hepatic cytochrome P450 2B induction by 5,5-diphenyl- and 5,5-diethyl-substituted barbiturates and hydantoins in the male F344/NCr rat. *Journal of Pharmacology and Experimental Therapeutics*, **270**, 348–355.

Northrop DB (1998) On the meaning of K_m and V/K in enzyme kinetics. *Journal of Chemical Education*, **75**, 1153–1157.

Okazaki O and Guengerich FP (1993) Evidence for specific base catalysis in *N*-dealkylation reactions catalyzed by cytochrome P450 and chloroperoxidase: differences in rates of deprotonation of aminium radicals as an explanation for high kinetic hydrogen isotope effects observed with peroxidases. *The Journal of Biological Chemistry*, **268**, 1546–1552.

Olsen JH, Boice JD Jr Jensen JPA and Fraumeni JF Jr (1989) Cancer among epileptic patients exposed to anticonvulsant drugs. *Journal of the National Cancer Institute*, **81**, 803–808.

Omura T and Sato R (1962) A new cytochrome in liver microsomes. *The Journal of Biological Chemistry*, **237**, 1375–1376.

Omura T and Sato R (1964) The carbon monoxide-binding pigment of liver microsomes. I. Evidence for its hemoprotein nature. *The Journal of Biological Chemistry*, **239**, 2370–2378.

Ortiz de Montellano PR (1986) Oxygen activation and transfer. In *Cytochrome P-450*, Ortiz de Montellano PR (ed.), Plenum Press, New York, pp. 217–271.

Ortiz de Montellano PR (1987) Control of the catalytic activity of prosthetic heme by the structure of hemoproteins. *Accounts of Chemical Research*, **20**, 289–294.

Ortiz de Montellano PR (1995a) *Cytochrome P450: Structure, Mechanism, and Biochemistry*, Plenum Press, New York, Ed. 2.

Ortiz de Montellano PR (1995b) Oxygen activation and reactivity. In *Cytochrome P450*, 2nd edition, Ortiz de Montellano PR (ed.), Plenum Press, New York, pp. 245–303.

Ortiz de Montellano PR and Correia MA (1983) Suicidal destruction of cytochrome P-450 during oxidative drug metabolism. *Annual Review of Pharmacology and Toxicology*, **23**, 481–503.

Ortiz de Montellano PR and Reich NO (1986) Inhibition of cytochrome P-450 enzymes. In *Cytochrome P-450*, Ortiz de Montellano PR (ed.), Plenum Press, New York, pp. 273–314.

Ortiz de Montellano PR, Kunze KL, Yost GS and Mico BA (1979) Self-catalyzed destruction of cytochrome P-450: covalent binding of ethynyl sterols to prosthetic heme. *Proceedings of the National Academy of Sciences, USA*, **76**, 746–749.

Osawa Y and Pohl LR (1989) Covalent bonding of the prosthetic heme to protein: a potential mechanism for the suicide inactivation or activation of hemoproteins. *Chemical Research in Toxicology*, **2**, 131–141.

Park S, Shimizu H, Adachi S, Shiro Y, Iizuka T, Nakagawa A, Tanaka I, Shoun H and Hori H (1997) Crystallization, preliminary diffraction and electron paramagnetic resonance studies of a single crystal of cytochrome P450nor. *FEBS Letters*, **412**, 346–350.

Paulsen MD and Ornstein RL (1993) Substrate mobility in thiocamphor-bound cytochrome P450$_{cam}$: an explanation of the conflict between the observed product profile and the X-ray structure. *Protein Engineering*, **6**, 359–365.

Peng HM, Raner GM, Vaz ADN and Coon MJ (1995) Oxidative cleavage of esters and amides to carbonyl products by cytochrome P450. *Archives of Biochemistry and Biophysics*, **318**, 333–339.

Peterson JA and Prough RA (1986) Cytochrome P-450 reductase and cytochrome b_5 in cytochrome P-450 catalysis. In *Cytochrome P-450*, Ortiz de Montellano PR (ed.), Plenum Press, New York, pp. 89–117.

Pitot HC (1995) The role of receptors in multistage carcinogenesis. *Mutation Research*, **333**, 3–14.

Pompon D and Coon MJ (1984) On the mechanism of action of cytochrome P-450: oxidation and reduction of the ferrous dioxygen complex of liver microsomal cytochrome P-450 by cytochrome b_5. *The Journal of Biological Chemistry*, **259**, 15377–15385.

Poulos TL (1991) Modeling of mammalian P450s on the basis of $P450_{cam}$ X-ray structure. *Methods in Enzymology*, **206**, 11–30.

Poulos TL, Finzel BC and Howard AJ (1987) High-resolution crystal structure of cytochrome $P450_{cam}$. *Journal of Molecular Biology*, **195**, 687–700.

Poulos TL, Finzel BC and Howard AJ (1995) Crystal structure of substrate-free *Pseudomonas putida* cytochrome P-450. *Biochemistry*, **25**, 5314–5322.

Ram PA, Park SH, Choi HK and Waxman DJ (1996) Growth hormone activation of Stat 1, Stat 3, and Stat 5 in rat liver: differential kinetics of hormone desensitization and growth hormone stimulation of both tyrosine phosphorylation and serine/threonine phosphorylation. *The Journal of Biological Chemistry*, **271**, 5929–5940.

Ravichandran KG, Boddupalli SS, Hasemann CA, Peterson JA and Deisenhofer J (1993) Crystal structure of hemoprotein domain of P450 BM-3, a prototype for microsomal P450's. *Science*, **261**, 731–736.

Remmer H (1957) The acceleration of evipan oxidation and the demethylation of methylamino-pyrine by barbiturates. *Naunyn–Schmiedeberg's Archive für experimentiel Pathologie und Pharmakologie*, **237**, 296–307.

Renton KW and Knickle LC (1990) Regulation of hepatic cytochrome P-450 during infectious disease. *Canadian Journal of Physiology and Pharmacology*, **68**, 777–781.

Renton KW, Keyler DE and Mannering GJ (1979) Suppression of the inductive effects of phenobarbital and 3-methylcholanthrene on ascorbic acid synthesis and hepatic cytochrome P-450-linked monooxygenase systems by the interferon inducers, poly rI-rC and tilorone. *Biochemical and Biophysical Research Communications*, **88**, 1017–1023.

Richardson HL, Stier AR and Borsos-Nachtnebel E (1952) Liver tumor inhibition and adrenal histologic responses in rats to which 3′-methyl-4-dimethylaminoazobenzene and 20-methyl-cholanthrene were simultaneously administered. *Cancer Research*, **12**, 356–361.

Rivera M, Seetharaman R, Girdhar D, Wirtz M, Zhang XJ, Wang XQ and White S (1998) The reduction potential of cytochrome b_5 is modulated by its exposed heme edge. *Biochemistry*, **37**, 1485–1494.

Roberts ES, Hopkins NE, Foroozesh M, Alworth WL, Halpert JR and Hollenberg PF (1997) Inactivation of cytochrome P450s 2B1, 2B4, 2B6, and 2B11 by arylalkynes. *Drug Metabolism and Disposition*, **25**, 1242–1248.

Ryan D, Lu AYH and Levin W (1978) Purification of cytochrome P-450 and P-448 from rat liver microsomes. *Methods in Enzymology*, **52**, 117–123.

Sadano H and Omura T (1983a) Turnover of two drug-inducible forms of microsomal cyto-chrome P-450 in rat liver. *Journal of Biochemistry*, **93**, 1375–1383.

Sadano H and Omura T (1983b) Reversible transfer of heme between different molecular species of microsome-bound cytochrome P-450 in rat liver. *Biochemical and Biophysical Research Communications*, **116**, 1013–1019.

Sadano H and Omura T (1985) Incorporation of heme to microsomal cytochrome P-450 in the absence of protein biosynthesis. *Journal of Biochemistry*, **98**, 1321–1331.

Sahali-Sahly Y, Balani SK, Lin JH and Baillie TA (1996) In vitro studies on the metabolic activation of the furanopyridine L-754,394, a highly potent and selective mechanism-base inhibitor of cytochrome P450 3A4. *Chemical Research in Toxicology*, **9**, 1007–1012.

Sato H and Guengerich FP (2000) Oxidation of 1,2,4,5-tetramethoxybenzene to a radical cation by cytochrome P450. *Journal of the American Chemical Society*, submitted.

Schlichting I, Berendzen J, Chu K, Stock AM, Maves SA, Benson DE, Sweet BM, Ringe D, Petsko GA and Sligar SG (2000) The catalytic pathway of cytochrome $P450_{cam}$ at atomic resolution. *Science*, **287**, 1615–1622.

Schmid R (1973) Synthesis and degradation of microsomal hemoproteins. *Drug Metabolism and Disposition*, **1**, 256–258.

Segel IH (1975) *Enzyme Kinetics*, Wiley, New York.

Seto Y and Guengerich FP (1993) Partitioning between N-dealkylation and N-oxygenation in the oxidation of N,N-dialkylarylamines catalyzed by cytochrome P450 2B1. *The Journal of Biological Chemistry*, **268**, 9986–9997.

Shimada T, Yamazaki H, Foroozesch M, Hopkins NE, Alworth WL and Guengerich FP (1998) Selectivity of polycyclic inhibitors for human cytochromes P450 1A1, 1A2, and 1B1. *Chemical Research in Toxicology*, **11**, 1048–1056.

Shiraki H and Guengerich FP (1984) Turnover of membrane proteins: kinetics of induction and degradation of seven forms of rat liver microsomal cytochrome P-450, NADPH-cytochrome P-450 reductase, and epoxide hydrolase. *Archives of Biochemistry and Biophysics*, **235**, 86–96.

Shou M, Mei Q,ettore MW, Jr., Dai R, Baillie TA and Rushmore TH (1999) Sigmoidal kinetic model for two co-operative substrate-binding sites in a cytochrome P450 3A4 active site: an example of the metabolism of diazepam and its derivatives. *Biochemical Journal*, **340**, 845–853.

Silverman RB (1995) Mechanism-based enzyme inactivators. *Methods in Enzymology*, **249**, 240–283.

Sladek NE and Mannering GJ (1969) Induction of drug metabolism. II. Qualitative differences in the microsomal N-demethylating systems stimulated by polycyclic hydrocarbons and by phenobarbital. *Molecular Pharmacology*, **5**, 186–199.

Stearns RA and Ortiz de Montellano PR (1985) Cytochrome P-450 catalyzed oxidation of quadricyclane. Evidence for a radical cation intermediate. *Journal of the American Chemical Society*, **107**, 4081–4082.

Steward AR, Wrighton SA, Pasco DS, Fagan JB, Li D and Guzelian PS (1985) Synthesis and degradation of 3-methylcholanthrene-inducible cytochromes P-450 and their mRNAs in primary monolayer cultures of adult rat hepatocytes. *Archives of Biochemistry and Biophysics*, **241**, 494–508.

Strobel HW and Coon MJ (1971) Effect of superoxide generation and dismutation on hydroxylation reactions catalyzed by liver microsomal cytochrome P-450. *The Journal of Biological Chemistry*, **246**, 7826–7829.

Tephly TR, Black KA, Green MD, Coffman BL, Dannan GA and Guengerich FP (1986) Effect of the suicide substrate 3,5-diethoxycarbonyl-2,6-dimethyl-4-ethyl-1,4-dihydropyridine on the metabolism of xenobiotics and on cytochrome P-450 apoproteins. *Molecular Pharmacology*, **29**, 81–87.

Thomas PE, Lu AYH, West SB, Ryan D, Miwa GT and Levin W (1977) Accessibility of cytochrome P450 in microsomal membranes: inhibition of metabolism by antibodies to cytochrome P450. *Molecular Pharmacology*, **13**, 819–831.

Tierney DJ, Haas AL and Koop DR (1992) Degradation of cytochrome P450 2E1: selective loss after labilization of the enzyme. *Archives of Biochemistry and Biophysics*, **293**, 9–16.

Ullrich V (1969) On the hydroxylation of cyclohexane in rat liver microsomes. *Hoppe–Seyler's Zeitschrift für Physiologische Chemie*, **350**, 357–365.

Ullrich V (1972) Enzymatic hydroxylations with molecular oxygen. *Angewandte Chemie— International Edition in English*, **11**, 701–712.

Ullrich V, Castle L and Haurand M (1982) Cytochrome P-450 as an oxene transferase. In *Oxygenases and Oxygen Metabolism*, Academic Press, New York, pp. 497–509.

Vanden Bossche H, Moereels H and Koymans LMH (1994) Aromatase inhibitors—mechanisms for non-steroidal inhibitors. *Breast Cancer Research and Treatment*, **30**, 43–55.

Vaz ADN, McGinnity DF and Coon MJ (1998) Epoxidation of olefins by cytochrome P450: evidence from site-specific mutagenesis by hydroperoxo-iron as an electrophilic oxidant. *Proceedings of the National Academy of Sciences USA*, **95**, 3555–3560.

Vermilion JL and Coon MJ (1978) Identification of the high and low potential flavins of liver microsomal NADPH-cytochrome P-450 reductase. *The Journal of Biological Chemistry*, **253**, 8812–8819.

Vermilion JL, Ballou DP, Massey V and Coon MJ (1981) Separate roles for FMN and FAD in catalysis by liver microsomal NADPH-cytochrome P-450 reductase. *The Journal of Biological Chemistry*, **256**, 266–277.

Walsh C (1979) *Enzymatic Reaction Mechanisms*, W. H. Freeman, San Francisco, pp. 34–35.

Watanabe Y, Oae S and Iyanagi T (1982) Mechanisms of enzymatic *S*-oxygenation of thioanisole derivatives and *O*-demethylation of anisole derivatives promoted by both microsomes and a reconstituted system with purified cytochrome P-450. *Bulletin of the Chemical Society of Japan*, **55**, 188–195.

Waterman MR and Guengerich FP (1997) Enzyme regulation. In *Biotransformation, Volume 3, Comprehensive Toxicology*, Guengerich FP (ed.), Elsevier Science, Oxford, pp. 7–14.

Waterman MR, Jenkins CM and Pikuleva I (1995) Genetically engineered bacterial cells and applications. *Toxicology Letters*, **82/83**, 807–813.

Waxman DJ, Dannan GA and Guengerich FP (1985) Regulation of rat hepatic cytochrome P-450: age-dependent expression, hormonal imprinting, and xenobiotic inducibility of sex-specific isoenzymes. *Biochemistry*, **24**, 4409–4417.

White RE and Coon MJ (1980) Oxygen activation by cytochrome P-450. *Annual Review of Biochemistry*, **49**, 315–356.

White RE, Miller JP, Favreau LV and Bhattacharyya A (1986) Stereochemical dynamics of aliphatic hydroxylation by cytochrome P-450. *Journal of the American Chemical Society*, **108**, 6024–6031.

Williams PA, Cosme J, Sridhar V, Johnson EF and MeRee DE (2000) Mammalian microsomal cytochrome P450 monooxygenase: structural adaptations for membrane binding and functional diversity. *Molecular Cell*, **5**, 121–131.

Wislocki PG, Miwa GT and Lu AYH (1980) Reactions catalyzed by the cytochrome P-450 system. In *Enzymatic Basis of Detoxication, Volume 1*, Jakoby WB (ed.), Academic Press, New York, pp. 135–182.

Witherow LE and Houston JB (1998) Sigmoidal kinetics of CYP3A4 substrates: an approach for scaling dextromethorphan metabolism in rat microsomes and isolated hepatocytes to predict *in vivo* clearance. *Proceedings of the International Society for the Study of Xenobiotics*, **12**, 26.

Yamazaki H, Johnson WW, Ueng Y-F, Shimada T and Guengerich FP (1996) Lack of electron transfer from cytochrome b_5 in stimulation of catalytic activities of cytochrome P450 3A4: characterization of a reconstituted cytochrome P450 3A4/NADPH-cytochrome P450 reductase system and studies with apo-cytochrome b_5. *The Journal of Biological Chemistry*, **271**, 27438–27444.

Yamazoe Y, Sugiura M, Kamataki T and Kato R (1978) Reconstitution of benzo[a]pyrene 4,5-oxide reductase activity by purified cytochrome P-450. *FEBS Letters*, **88**, 337–340.

Yoo JSH, Cheung RJ, Patten CJ, Wade D and Yang CS (1987) Nature of *N*-nitrosodimethylamine demethylase and its inhibitors. *Cancer Research*, **47**, 3378–3383.

Yun C-H, Miller GP and Guengerich FP (2000) Rate-determining steps in phenacetin oxidations by human cytochrome P450 1A2 and selected mutants. *Biochemistry*, submitted.

Zangar RC and Novak RF (1998) Posttranslational elevation of cytochrome P450 3A levels and activity by dimethyl sulfoxide. *Archives of Biochemistry and Biophysics*, **353**, 1–9.

3 Flavin Monooxygenases

John R. Cashman

Human BioMolecular Research Institute, USA

Introduction

For many years it was assumed that the metabolism of nitrogen- and sulphur-containing chemicals, drugs and xenobiotics to their corresponding *N*-oxides and *S*-oxides was an exclusive property of cytochrome P-450 (CYP) (Cashman 1995). However, after the isolation and purification of the flavin-containing monooxygenase (FMO) from pig liver in the mid-1960s, it was apparent that FMO could catalyse the oxygenation of many nitrogen-, sulphur-, phosphorous-, selenium and other nucleophilic heteroatom-containing chemicals (Ziegler 1980). Today, it is recognised that FMO catalyses the oxygenation of numerous heteroatom-containing drugs (Cashman 1997, 2000), chemicals (Ziegler 1993) and agricultural agents (Hodgson *et al.* 1998; Hodgson and Levi 1992). In this chapter the term 'oxygenation' is used to signify a one-step two-electron substrate oxygenation by FMO as opposed to two sequential one-electron oxidations by CYP. Despite the pioneering studies of Ziegler, Hlavica (Heinze *et al.* 1970; Ziegler 1988) and others, the involvement of the FMO in drug and xenobiotic metabolism has historically been underestimated probably due to a fundamental biochemical property of the enzyme. FMO is considerably more thermally labile than most CYPs and investigations of drug metabolism that did not account for this fact invariably led to the conclusion that FMO was not important in the oxygenation of the chemical studied. Although thermal instability confounds studies with FMO, it also points to a key property of FMO that can be used to distinguish the involvement of FMO in drug metabolism.

In the 1970s, important studies describing the mechanism of molecular oxygen addition to dihydroisoalloxazines and related flavin models by Bruice (Kemal *et al.* 1977), and Balou and Massey (Ballou *et al.* 1969) and others shed considerable light on the chemical basis for FMO action. Studies of the chemistry of 4a-hydroperoxy flavins and the kinetics of reaction with nucleophiles provided insight as to how FMO could accomplish many of the same chemical oxygenations (Ball and Bruice 1983; Kemal and Bruice 1976; Doerge and Corbett 1984; Miller *et al.* 1986). Much of the

Enzyme Systems that Metabolise Drugs and Other Xenobiotics. Edited by C. Ioannides.
© 2002 John Wiley & Sons Ltd

current understanding of the mechanism of flavoprotein and FMO catalysis was developed in the late 1970s and early 1980s.

In the mid-1980s several investigators provided evidence that multiple forms of FMO could be present in an animal (Williams *et al.* 1984; Tynes *et al.* 1985; Hlavica and Golly 1991). Thus a 'pulmonary' FMO with properties distinct from that of the 'hepatic' FMO was identified, purified and characterised. Today, we recognise that there are six forms of mammalian FMO and some can be present in multiple tissues of the same organism. The description of multiple forms of FMO was advanced by elucidation of the primary sequences by amino acid and nucleotide analysis (Hines *et al.* 1994; Lawton *et al.* 1994). While the significance of human FMOs had been recognised since the 1960s, it was not until the late 1980s and 1990s that the FMOs were characterised by purification and cDNA cloning (Lomri *et al.* 1992; Lawton and Philpot 1993; Dolphin *et al.* 1992; Phillips *et al.* 1995; Philpot *et al.* 1996).

Today, the number of human FMOs being described in the literature is expected to increase because of the ease of obtaining new sequences with the polymerase chain reaction (PCR) and the availability of the human genome sequence. Numerous allelic variants have been reported and some clinical significance has been associated with the FMO variants. The number of flavoproteins related to FMOs is also likely to increase as investigators become interested in studying FMO in other species including plants, insects and other organisms.

The physiological role of FMO is unknown. However, FMO has been suggested to have evolved to detoxicate nucleophilic heteroatom-containing chemicals and xenobiotics found in foodstuffs by converting them to polar, readily excreted, water-soluble metabolites (Ziegler 1990). FMOs have very broad substrate specificity and, barring steric limitations, accept most nucleophilic heteroatom-containing substrates for oxygenation (Cashman 1995). Recent studies have shown that FMO is capable of oxygenating several endogenous and dietary compounds with significant physiological activities such as biogenic amines (Cashman 2000). It is likely that, as more species are investigated the physiological role of FMO in cellular homeostasis will become clearer.

Nomenclature

FMO is a general definition that may include a number of flavoproteins. For the purposes of describing mammalian FMOs in this chapter, the term FMO applies to flavoproteins among different families that possess a consensus amino acid sequence equal to or greater than 50% identity, and identities among orthologous forms greater or equal to 82% Thus, numerous flavoproteins with similar functional properties (i.e. cyclohexanone monooxygenase (Ryerson *et al.* 1982), yeast FMO (Suh *et al.* 1996) that meet the criteria as a multi-substrate flavoprotein monooxygenase are excluded as an FMO because of low sequence similarity. The prefix '*FMO*' is used to designate the gene or cDNA of an FMO.

As FMOs were discovered and characterised, the common or trivial names assigned to enzymes were formalised and a system of nomenclature was adopted. The nomenclature was developed on the basis of primary amino acid sequence identity. If an FMO has an amino acid sequence identity with ⩾ 82% identity it is grouped within a family, and the family is indicated by the first numeral of the designation (i.e. 1, 2, 3 . . . .). The order of naming follows the chronology of publication of full-length sequences for each member of the family (Table 3.1). The nomenclature conforms to that approved by the Human Gene Mapping Nomenclature Committee (Dolphin *et al.* 1991). Compared with the CYP gene families, the FMO gene family is relatively simple. Allelic variants have been observed for FMO that usually possess only single base changes. Allelic variation can occur as a function of the population and possibly age and gender and can result in an FMO with altered activity. However, there are other missense, nonsense and deletion or truncation mutants of FMO that can significantly affect enzyme function and these will be discussed below in greater detail.

Gene Organization

The *FMO* genes are localised on chromosome 1q and the human *FMO* gene family may exist as a gene cluster (McCombie *et al.* 1996; Dolphin *et al.* 1997, Gelb *et al.* 1997). The general pattern of intron/exon organisation for *FMO* is assumed to be similar in various animals and humans although this has not been exhaustively examined. Evidence for multiple gene promoters and other regulatory elements have been reported for rabbit *FMO1* (Luo and Hines 1996, 1997) and rabbit *FMO2* (Shehin-Johnson *et al.* 1996). *FMO* cDNA primers can be selected to amplify certain introns on the basis that the junctional sites are conserved across gene family and across species lines. For example, the intron/exon boundaries determined for human *FMO3* (Treacy *et al.* 1998) relied on the gene structure of rabbit *FMO2* (Wyatt *et al.* 1996). After a PCR fragment was obtained and verified by sequencing, the strategy enabled amplification of human *FMO3* introns 1 and 4–8. Introns that are hard to amplify by the above approach can be derived from a consideration of the sequence available in GenBank. For human *FMO3*, sequence analysis indicated that *FMO3* had nine exons ranging in size from 80 to 705 bp. The similarity in exon/intron organisation for the *FMO* genes may suggest that the FMO family members arose from gene

Table 3.1 Summary of mammalian flavin-containing monooxygenases[a]

Designation	Trivial name	Species	Accession number
FMO1	1A1	Rabbit	M32030
FMO1	Ziegler's enzyme	Pig	M32031
FMO1	FMO-1	Human	M64082
FMO2	1B1	Rabbit	M32029
FMO2	Lung enzyme	Rabbit	—
FMO3	1D1	Rabbit	L10037
FMO3	HLFMO	Human	M83772
FMO4	1E1	Rabbit	L10392
FMO4	FMO2	Human	Z11737
FMO5	1C1	Rabbit	L08449
FMO6	—	Human	AL021026

[a] Adapted from Cashman (1995).

duplication and further mutagenesis. Diversification of the *FMO* gene presumably led to selective advantages and new function. Because FMO has been suggested to play a role in detoxicating nucleophilic heteroatom-containing foodstuffs, it is possible that FMO played a role in certain populations to process some biological natural products and protect that population. Further allelic variation of *FMO* (as described below for human *FMO3*) altering the catalytic activity and or substrate specificity could render certain individuals or populations more or less susceptible to the effects of environmental xenobiotics. Human *FMO3*, for example may be another example of an 'environmental gene'. The large allelic variation of codon 158 of human *FMO3* that approaches 50% may represent an example of a protective mechanism of 'animal–plant warfare'. It is possible that evolutionarily conserved allelic variation of human *FMO3* prevalent in certain geographical locations possessing certain plant toxins helps protect humans from plant toxin exposure (Gonzalez and Nebert 1990).

Structural aspects

The primary amino acid sequences of perhaps two to three dozen mammalian FMOs and variants are now known but the three-dimensional structure of FMO is not known. The lack of an X-ray structure probably comes from the fact that mammalian FMOs are highly lipophilic enzymes that are associated with the membrane and are hard to crystallise. Despite the difficulties of working with a membrane-associated enzyme, considerable structural information is known. Most of the sequence information has been deduced from oligonucleotide sequencing (Gasser *et al.* 1990; Lomri *et al.* 1993a; Lawton *et al.* 1993, 1994). For some FMOs, automated Edman degradation sequence and, to a lesser extent, mass spectral sequence analysis has provided substantial sequence information especially for FMO1. Ozols has provided extensive amino acid sequence information of rabbit FMOs (Ozols 1991, 1994). Combined with the sequence deduced from the cDNA data, the amino acid sequence data has provided insight into the cofactor binding domains and the general structural motifs of the protein and has provided some evidence for microheterogeneity (Ozols 1994). It was soon clear that widely studied FMOs such as pig FMO1 and rabbit FMO2 were N-terminal blocked (Guan *et al.* 1990). While there were methods available to deacetylate proteins, the most straightforward method to identify post-translational modifications of FMO was by direct peptide sequencing using mass spectrometry. In addition, the cDNA sequence data could not by itself provide information about post-translational modifications, and the cDNA and peptide data suggested that pig *FMO1* had consensus sequences for N-glycosylation (Guan *et al.* 1991). By using a combination of biochemical methods and mass spectrometry (i.e. gas chromatography mass spectrometry, HPLC mass spectrometry, electrospray mass spectrometry and matrix-assisted laser desorption mass spectrometry) the site of FMO N-glycosylation was identified (Korsmeyer *et al.* 1998). For pig FMO1, the only residue that was N-glycosylated was Asn 120. Determination of the site of N-glycosylation helped to support construction of molecular models of FMO1. Information about the site of N-glycosylation also potentially revealed information as to how pig FMO1 associated with the membrane. However, N-glycosylation of FMO1 probably is not essential for enzyme activity because cDNA-expression in bacteria (that lacks the ability to N-glycosylate) nevertheless provides active FMO enzyme.

The amino acid composition of a number of FMOs has been reported (Lawton et al. 1994). Although molecular models of FMO based on the crystal structure of other flavoproteins have been proposed, the level of suitability and resolution are not really sufficient to make firm conclusions regarding structure and function. The first FMO model was developed by Ziegler (Ziegler 1999) based on the crystal structure of E. coli glutathione reductase (Thieme et al. 1981; Mattevi et al. 1991). A representation of this model is shown in Figure 3.1(a). In this model, emphasis is placed on a dimer interface formed between the putative FAD and NADPH domains juxtaposed to the dimer interface of the opposite FMO monomer (Christensen 1999). Peptide residues 321–339 that contain the proposed FMO signature sequence FATGY have been proposed as a linkage between the putative FAD domain and active site and residues 473–494. In a different model developed within the author's laboratory in collaboration with Professor Ellie Adman (University of Washington), we used the structure of NADPH-peroxidase to model human FMO3 (Figure 3.1(b)). The NADPH-peroxidase model is supported by some of the data in the literature regarding the site of mutations. For example, in the NADPH-peroxidase model, the nonsense mutation M66I that is associated with trimethylaminuria maps to a region near the proposed FAD and $NADP^+$ domains. In the glutathione reductase model, the M66I maps to a region quite distal to the proposed cofactor domains. Both models show that the site of N-glycosylation (discussed below) is in a region remote from the putative cofactor binding domains and this is in accord with other evidence suggesting that N-glycosylation is not an essential element of enzyme function.

As described above, most of the structural information about FMO comes from studies of pig FMO1. Purified pig FMO1 contains approximately 15 nmol of FAD/mg of protein and is devoid of haem iron or other metals (Ziegler 1980). Highly purified pig FMO1 generally contains variable amounts of lipid and it is generally very difficult to segregate the enzyme from minute amounts of lipid. It is notable that another FMO (i.e. FMO2) is tightly associated with a chaperone protein although it is not known how widely this phenomenon exists for other FMOs from other species or organisms. Rabbit FMO2 is tightly associated with calreticulin (Guan et al. 1991), however, for mammalian FMOs it is not known what the physiological role of this association is. The visible spectrum of FMO is similar to other flavoproteins (i.e. λ_{max} of 445 nm and 380 nm and shoulder at 480 nm). In the absence of molecular oxygen, NADPH reduces the FAD prosthetic group to provide reduced $FADH_2$ and the UV-vis spectrum is shifted to shorter wavelength (i.e. λ_{max} of 440 nm and 370 nm with no shoulder) (Poulsen and Ziegler 1979; Beaty and Ballou 1981a, b). As described in greater detail below, addition of molecular oxygen to the fully reduced FMO generates a spectrum similar to peroxy flavins found in other flavoproteins or hydroperoxy isoalloxazines that have been chemically synthesised as models of FMO (Kemal and Bruice 1976; Miller 1982). The formation of a relatively stable hydroperoxy flavin species that represents the 'resting state' of the enzyme is remarkable for at least two reasons. First, FMO somehow stabilises the hydroperoxy species under general cellular conditions that are strongly reducing, and second, stabilisation of the hydroperoxy flavin allows the FMO to oxygenate essentially any nucleophilic substrate that has the appropriate

FMO from 1npx.pdb

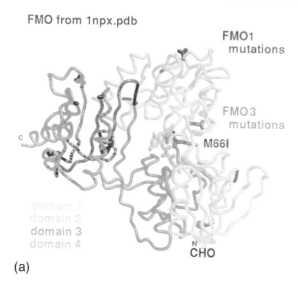

(a)

FMO from 1get.pdb

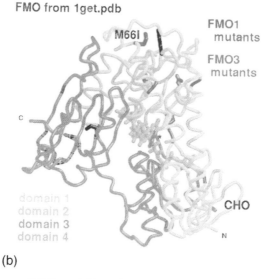

(b)

Figure 3.1 (a) Human FMO3 model based on threading the FMO3 sequence onto the X-ray structure of NADPH-peroxidase (npx). (b) Human FMO3 model based on threading the FMO3 sequence onto the X-ray structure of glutathione reductase (get).

steric dimensions to enter the substrate binding channel. It is this fundamental property of FMOs that distinguish them from other monooxygenases and allow FMO to perform its role as an extremely broad-based mixed function monooxygenase.

The cDNAs of FMO reported in the literature encode for active enzymes of approximately 533–535 amino acids, but examples with 19 or 25 additional C-terminal amino acids have been observed. Several regions of FMO contain relatively highly conserved amino acid residues that are presumably important for structure and function (Kubo *et al.* 1997). For example, the FAD- and NADP$^+$-binding domains (i.e. GXGXXG) near amino acid positions 9–14 and 186–196, respectively, share some similarities to other flavoproteins where the crystal structure is known (Vallon 2000). Site-directed mutagenesis studies of the region 9–14 (i.e. GXGXXV) of rabbit *FMO2* gave a cDNA-expressed enzyme that was devoid of activity and a protein that did not bind FAD (Lawton and Philpot 1993). In contrast to many other monooxygenases from the CYP family, the N-terminal hydrophobic tail does not help to anchor FMO to the membrane. Other regions must be important. Comparison of FMO hydropathy plots showed numerous regions of conserved hydrophobic segments (Cashman 1997). Although the N-terminal region is hydrophobic, the lack of a discernible signal peptide sequence and the obvious importance of the FAD-binding region suggest that this is not a membrane insertion area.

Studies have shown that removal of the hydrophobic amino acid residues of the C-terminus still allow the FMO to associate with a bacterial membrane. Although a direct comparison between bacterial and mammalian membranes for FMO association has not been made, nevertheless it is clear that the hydrophobic portion of the C-terminus is not essential for membrane insertion and enzyme function. The conclusion is that FMO likely has an internal sequence that is essential for membrane association.

Genetic aspects

There are a number of points at which the metabolism of a drug or chemical by FMO could be altered. Some of the points are either poorly understood or have not been investigated. For example, for orally administered drugs, the initial pre-systemic metabolism by FMO in the intestine is poorly understood. It appears that FMO1 is the prominent enzyme in the rabbit intestine (Shehin-Johnson *et al.* 1995) but it is unknown whether FMO1 serves to significantly alter human drug bioavailability such as described for CYP 3A4 or the P-glycoprotein systems. A few drugs (i.e. cimetidine, verapamil and albendazol) that are substrates for FMO may have their bioavailability altered as a consequence of FMO action, but this research area is largely unexplored (Piyapolrungroj *et al.* 2000; Redondo *et al.* 1999). Various studies have shown that expression of FMO is both tissue- and species-dependent (Lemoine *et al.* 1991; Wirth and Thorgeirsson 1978; Duffel *et al.* 1981; Cashman *et al.* 1990; Dannan *et al.* 1986). Although it is common for a tissue to have more than one FMO present, the activity is dominated by the most prominent FMO present. For example, in adult human liver, evidence for FMO3, FMO4 and FMO5 has been obtained but it is FMO3 that is by far the most catalytically important species present (Cashman *et al.* 1995). Of course, in a tissue such as the kidney where multiple forms of FMO are more equally present, the oxygenation of a drug or chemical via FMO is probably determined by the kinetic properties of the particular agent (Ripp *et al.* 1999). Of note is the fact that FMO1 is not functionally present in the adult human liver but is the prominent form of FMO in the foetal liver. This represents an intriguing example of regulation of expression of active FMO protein. The observation that FMO1 is not functionally present in adult human liver underscores another important point: the major form of FMO present in the liver of most commonly used animals is FMO1 and not the form that is the prominent one (i.e. FMO3) found in adult human liver. Thus, some caution should be exercised in comparing preclinical metabolic data obtained in animals with that of adult humans and it is important that an appropriate animal model is used.

In contrast to other monooxygenases of the CYP family, there is little data to suggest that FMOs (other than FMO2) are inducible. FMO2 levels appear to be regulated during development by pregnancy. Gestation increases FMO2 activity in rabbit (Hines *et al.* 1994), mouse placenta (Osimitz and Kulkarni 1982) and pig corpora lutea (Heinze *et al.* 1970). In the rat, some evidence has accumulated that hepatic FMO activity is decreased when animals are placed on a synthetic diet (Kaderlik *et al.* 1991). It is possible that hepatic FMO is maximally present and decreases to a de-induced level in the presence of a synthetic diet or other conditions. Hormones and dietary factors regulate FMO expression but this is done in a species- and tissue-dependent fashion. Traditional receptor-mediated transcriptional regulation that in-volves ligand binding does not appear to play a significant role in FMO expression as it does for CYP, for example. There is one report in the literature that rat FMO1 is induced by treatment of animals with 3-methylcholanthrene (Chung *et al.* 1997) but this has not been independently confirmed. Expression of FMO is likely to be under the auspices of multiple mechanisms. That expression of FMO is tissue-specific comes from the observation that some tissues contain very high levels of FMO mRNA but very low levels of functional protein or FMO activity. For example, human *FMO2* encodes a truncated non-functional protein (Dolphin *et al.* 1998). Interestingly, while Caucasians apparently do not express human FMO2, some individuals of African descent do possess full-length human FMO2. The expression of active human FMO2 will undoubtedly be highly dependent on the ethnicity of the population examined.

Another way that FMO is regulated is through genetic regulation by polymorphisms. Genetic polymorphisms are defined as allelic variations occurring with a pre-valence of at least 1%. Inter-individual variation of enzymic metabolic activity can result in significant population-wide differences in the oxygenation of drugs or xenobiotics. Polymorphisms of monooxygenase genes can exert a dramatic effect on drug metabolism. For example, CYP2D6-mediated debrisoquine 4-hydroxylation is ethnically linked: 'poor metabolisers' make up about 5–10% of the population in the Caucasian population but only about 0.1% of the Asian population (Tucker *et al.* 1977). For those individuals that have the variant gene, the polymorphism causes an exaggerated clinical response to the side effects of debrisoquine. Another example comes from the CYP2C19-mediated 4'-hydroxylation of (*S*)-mephenytoin (Goldstein and de Morais 1994). In Caucasians, the prevalence of the poor (*S*)-mephenytoin metaboliser phenotype is low (approximately 0.2% but quite large in the Asian population (i.e, 15–20%).

In humans, polymorphisms of human *FMO3* were recognised and characterised after observations about the abnormal metabolism of trimethylamine (TMA) (Al-Waiz *et al.* 1987; Hadidi *et al.* 1995; Mitchell *et al.* 1997; Thithapandha 1997; Treacy *et al.* 1998; Dolphin *et al.* 1997; Cashman *et al.* 1997). In normal humans, TMA is metabolised to the polar and non-odorous metabolite trimethylamine N-oxide (TMA N-oxide) that is efficiently excreted in the urine (Al-Waiz *et al.* 1988). For normal individuals, the urinary TMA N-oxide to TMA ratio is 97%:3% (or urinary TMA levels of < 18 μmol/mmol creatinine). For severely affected individuals, the TMA N-oxide:TMA ratio can be as low as 10:90 or almost exactly opposite of the normal condition. Individuals with trimethylaminuria have a diminished capacity to oxidise dietary-derived TMA to its odourless metabolite TMA N-oxide and these people excrete large amounts of TMA in their urine, sweat and breath. Trimethylaminuria patients have been described as suffering from 'fish odour syndrome' because of the fish-like odour (Ayesh *et al.* 1993). Trimethylaminuria is an autosomal recessive inborn error of metabolism that is quite uncommon and non-randomly distributed in the population. It has been relatively well documented in British and Australian populations and it may be more prevalent in North America than currently recognised (Akerman *et al.* 1999). A significant amount of evidence has accumulated that mutations of the human *FMO3* gene are responsible for trimethylaminuria and segregate with the disorder. A genotype–phenotype correlation has emerged. Individuals homozygous or compound heterozygous for the human *FMO3* truncation mutation E305X manifest the most severe phenotype. Another mutation that also causes a severe phenotype is a proline to leucine substitution at codon 153. Homozygotes with this genotype can have a TMA N-oxide:TMA ratio as low as 10%:90% Another causative mutation is a methionine for isoleucine change at codon 66 that causes a more modest change in an individual's ability to metabolise TMA (i.e. TMA levels of 48 μmol/mmol of creatinine) (Treacy *et al.* 1998). There are additional causative mutations and these are listed in Table 3.2.

In addition to the mutations that cause severe trimethylaminuria, evidence for allelic variation of the human *FMO3* gene have been observed that represents pharmacogenetic polymorphisms (Table 3.3). Again, for populations that have a significant number of poor metabolisers it is possible that the individuals may be more susceptible to adverse drug reactions or exaggerated clinical response. Altered substrate activities have been observed for human *FMO3* (Cashman *et al.* 2000) and may be responsible for mild trimethylaminuria (Zschocke *et al.* 1999).

Table 3.2 Nonsense and missense mutations of the human *FMO3* gene associated with trimethylaminuria

Substitution	Location	References
Deletion	Exons 1 and 2	Forrest *et al.* (2000)
A52T	Exon 3	Akerman *et al.* (1999)
N61S	Exon 3	Dolphin *et al.* (1996)
M66I	Exon 3	Treacy *et al.* (1998)
M82T	Exon 3	Dolphin *et al.* (1996)
P153L	Exon 4	Dolphin *et al.* (1997); Treacy *et al.* (1998)
E305X	Exon 7	Treacy *et al.* (1998)
E314X	Exon 7	Akerman *et al.* (1999)
R387L	Exon 7	Akerman *et al.* (1999)
M434I	Exon 9	Dolphin *et al.* (1996)
R492W	Exon 9	Dolphin *et al.* (1996)

Table 3.3 Common polymorphic variation in the Human *FMO3* Gene[a]

Substitution	Location	Prevalence
E158K	Exon 4	K homozygote is present about 17%
V257M	Exon 6	M homozygote is present about 0.5%
E308G	Exon 7	G homozygote is present about 4%

[a] Genotype frequencies determined in a Caucasian population.

The literature is replete with examples of adverse interactions with drugs and/or chemicals mediated by CYP. One such example is the induction of CYP2E1 by ethanol or other related alcohols and ketones that metabolise disulfiram (Antabuse) to toxic species (Guengerich *et al.* 1991). For human FMO evidence for such clear-cut adverse drug interactions has not been reported but some examples have emerged of adverse clinical problems. For example, individuals with trimethylaminuria also suffer from additional metabolic and psychosocial abnormalities including self-esteem, anxiety, clinical depression and addiction to drugs (Todd 1979). Many of these clinical manifestations could arise, at least in part, from abnormal endogenous or xenobiotic metabolism. For example, in an Australian trimethylaminuria cohort several individuals also manifested hypertension and adverse reactions from tyramine, other amines and sulphur-containing medications. Because FMO has been shown to metabolise biogenic amines (Lin and Cashman 1997a,b), deficient FMO metabolism of biogenic amines could contribute to some of the neurochemical effects observed in individuals with trimethylaminuria. One report showed that a trimethylaminuria patient displayed seizures and other behavioural disturbances after subjected to choline loading (McConnell *et al.* 1997). Dietary choline is a major precursor source of TMA. Certain central nervous system drugs that are normally efficiently cleared could produce exaggerated responses for individuals with common polymorphic variants of FMO (Adali *et al.* 1998). For example, the metabolic detoxication of amphetamine and methamphetamine by human FMO3 may be under pharmacogenetic control (Cashman *et al.* 1999b). Anecdotal reports have suggested that tricyclic antidepressants give exaggerated side reactions for individuals suffering from mild or severe trimethylaminuria. Because human FMO3 of the liver is largely responsible for TMA detoxication, hepatic diseases also can exacerbate the trimethylaminuria condition (Fernandez *et al.* 1997; Stransky 1998). In addition, there are some conditions that apparently aggravate the trimethylaminuria condition including menstruation (Zhang *et al.* 1996) and possibly copper deficiency (Blumenthal *et al.* 1980). It is unknown whether a transient trimethylaminuria condition occurs for some children (Mayatepek and Kohlmueller 1998). Associations with such diseases as Prader–Willi syndrome (Chen and Aiello 1993) and Noonon's syndrome (Calvert 1973) have also been linked with trimethylaminuria. Finally, it is possible that small molecules present in *brassica* vegetables can alter the urinary TMA N-oxide to TMA ratio and give a transient trimethylaminuria condition (Fenwick *et al.* 1983). Based on the results of recent studies, it is likely the aggravation of trimethylaminuria by *brassica* vegetables is due to acid condensation products of indole-3-carbinol present in the vegetables. Inhibition of FMO by indole-3-carbinol is discussed in greater detail, below.

Biochemical properties

Below, the biochemical properties of FMO are discussed in some detail to put the monooxygenase system in perspective with other systems. As described above, the mammalian FMOs are a family of gene products that catalyse a remarkable range of oxygenation of nucleophilic nitrogen-, sulphur-, phosphorous- and selenium-containing drugs and xenobiotics to their respective oxides. Although many exceptions are known, generally, formation of polar, oxygenated metabolites can provide a means to terminate the biological activity of a heteroatom-containing compound (Cashman *et al.* 1996). The degree to which a polar, oxygenated metabolite is excreted depends, of course, on further metabolic processes, both oxidative and reductive, and numerous exceptions to the general rule described above have been observed.

There are some biochemical properties unique to the FMO class of monooxygenases. With the possible exception of FMO2, FMOs are unusually sensitive to thermal inactivation and this property often serves as a means to distinguish the contribution to the metabolism of a chemical by FMO from that of other monooxygenases. Thus, procurement of tissue from an animal before the temperature of the animal rises is essential to preserve maximal FMO activity. In the absence of NADPH, about 85% of the activity of most FMOs is lost if the tissue is left standing at 45–55°C for 1–4 minutes. These are conditions sometimes achieved under postmortem conditions (Ziegler 1980). Thermal lability also represents a practical way to distinguish the contribution of FMO from that of CYP to the N- or S-oxidation of a drug or other chemical. Heat inactivation of microsomes at 55°C for one minute in the absence of NADPH largely destroys FMO activity and retains CYP activity. Addition of NADPH to a preparation treated in this fashion allows for CYP to function normally but generally abrogates FMO activity. Of course, heat inactivation of microsomes tends to produce significant quantities of H_2O_2 and it is always a good idea to destroy any H_2O_2 formed by addition of exogenous catalase. The best way to ensure maximal FMO activity is to add NADPH (or an NADPH-generating system) directly to a freshly thawed preparation of enzyme. Even at the customary incubation temperature of 37°C, if enzyme preparations containing FMO activity are allowed to stand for even a few minutes in the absence of NADPH, significant FMO activity can be lost. This problem may have contributed to the fact that many examples of FMO-mediated metabolism were overlooked. The reason for this is that historically, many metabolic reactions were initiated by the addition of NADPH (or an NADPH-generating system) and this procedure inherently destabilised the FMO. If the pre-incubation phase is conducted in the absence of NADPH at 37°C, significant FMO activity can be lost. This has probably led to a general underestimation of the role of FMO in drug and chemical metabolism. With the advent of cDNA-expressed enzymes in drug metabolism, the challenges of the thermal lability of FMO are less of a problem. Today, the investigator is less dependent on the amount of FMO activity lost during post-mortem inactivation because of the availability of the recombinant enzyme. Another advance has come about from the recognition that some recombinant fusion proteins of FMO are considerably more stable than those of the native enzyme (Brunelle *et al.* 1997).

Another fascinating property of FMO is the formation of a relatively stable hydroperoxy flavin intermediate. This is important for at least three inter-related reasons. First, the unusually long-lived hydroperoxy flavin species is remarkably stable and resistant to decomposition and disproportionation. This property allows the FMO to be in an oxygenating mode during essentially the entire lifetime of the catalytic cycle. This may account for the second feature of the hydroperoxy flavin: generally (barring steric limitations), almost any strong nucleophile is oxygenated by FMO and, as it will be described in more detail below, for a class of substrate, generally, all are oxygenated at nearly the same rate. This suggests that the formation of product occurs before the rate-limiting step of the enzyme reaction. Regardless of the mechanistic details, FMO has somehow evolved to stabilise and preserve the integrity of a potentially labile oxygenating agent during the catalytic cycle. Although poorly understood, this ingenious mechanism underscores the potential versatility of the catalyst. In addition, it points to some previously undiscovered molecular property of the FMO active site and substrate-binding region construction that allows the hydroperoxy flavin moiety to exist for long periods of time (on a biological time scale) without decomposition. It is possible that the substrate-binding region is constructed of lipophilic, non-nucleophilic amino acid residues that contribute to stabilising this critical species for FMO catalytic function. Evaluation of this suggestion must await further structural information.

Another important feature of FMO is that, under normal conditions, no detectable production of H_2O_2 or other reactive oxygen species is formed during the catalytic cycle of the enzyme, and therefore minimal FMO-mediated autooxidation of substrate is observed. The conclusion is that formation of hydroperoxy flavin is tightly coupled to formation of oxygenated product unless NADPH is not present. Thus, there is generally an excellent stoichiometry between consumption of one mole of molecular oxygen by FMO and formation of one-half mole oxygenated product and one-half mole of water.

Consequently, monitoring consumption of molecular oxygen with an oxygen electrode or some other oxygen-sensing system can provide a method for determining enzyme kinetics. Of course, for multi-step kinetics or where multiple products are formed it is useful to have the authentic synthetic metabolites and quantify the FMO enzyme reaction products by some separation technique such as HPLC. That FMO does not generate copious amounts of H_2O_2 in the absence of substrate suggests that FMO does not expose the cell to untoward effects of oxidative stress. In addition, and as described below, physiological substrates such as glutathione or other cellular nucleophiles appear to be excluded from the substrate binding channel. This is important because if cellular nucleophiles were continuously oxygenated it would be biologically quite wasteful, lead to cellular stress due to loss of NADPH and possibly contribute to proliferation of cellular reactive oxygen species and oxidative stress. This underscores the role of FMO as a xenobiotic detoxication catalyst. However, under severe cellular stress such as that during conditions of postmortem inactivation, FMO may lose its NADPH cofactor to other apparently more vital cellular function. This may point to the fact that FMO plays an auxiliary role in cellular homeostasis and that under conditions where cellular defence is not essential, FMO participates in cellular survival by providing essential reducing equivalents to important biochemical sites.

Catalytic mechanism

The laboratories of Ziegler and Ballou have characterised the detailed steps of the pig FMO1 catalytic cycle (Poulsen and Ziegler 1979; Beaty and Ballou 1981a,b). It is likely that other FMOs also conform to the same general mechanistic picture although as other FMOs from other species are described this question should be re-examined. The prominent steps of the FMO1 catalytic cycle are shown in Figure 3.2. In the first step of the enzyme reaction (step 1), the fully oxidised flavoprotein (FMO-Fl$_{ox}$) combines with NADPH in a fast step to give the FMO in the reduced form (FMO-FlH$_2$). The K_m for binding of NADPH is in the low micromolar region and the rate constant (i.e. 53 M^{-1}) suggests that it is among the fastest reactions in the cycle. After delivering the reducing equivalents to the flavin, the NADP$^+$ apparently remains proximal to the reduced flavin moiety and possibly serves as a protector or 'gate-keeper' to the complex. This is important as the reduced flavin is not indefinitely stable and reaction of the reduced flavin with molecular oxygen to form the key hydroperoxy flavin (step 2) may require the presence of the NADP$^+$ cofactor in an as yet poorly understood way. Formation of the hydroperoxy flavoenzyme is also rapid (i.e. 45 M^{-1}) and provides the long-lived hydroperoxy flavin oxygenating species that makes up the vast majority of the resting form of the enzyme. The hydroperoxy flavoenzyme is the form of the enzyme that waits in the ground state until an appropriate substrate comes along. For substrates such as dimethylaniline, oxygenation proceeds very rapidly (i.e. bimolecular rate constant of 4700 M^{-1} s^{-1}) with attack of the nucleophilic nitrogen atom on the terminal hydroperoxy flavin oxygen atom (step 3) to produce the product (S–O) and the hydroxy flavoenzyme species (i.e. the pseudobase FMO–FlHOH). The oxygenated product then leaves the product binding region, again, in a very fast step. The next and final step (step 4) is slow and constitutes the overall rate-limiting step of the catalytic cycle. From kinetic measurements, it is not clear whether dehydration of pseudobase FMO–FlOH or desorption of NADP$^+$ is the rate-limiting step but this step is approximately 20-30-fold slower than any of the other steps in the catalytic cycle

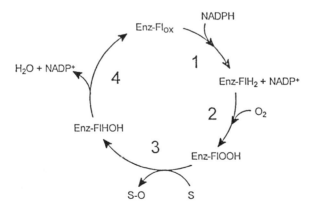

Figure 3.2 Proposed catalytic cycle of FMO. S and S–O represent the substrate and the oxygenated substrate, respectively (adapted from Cashman 1995).

(i.e. 1.9 M^{-1}). It is possible that $NADP^+$ comes away last because $NADP^+$ is a competitive inhibitor of the FMO. As discussed above, under normal conditions, formation of the highly protected hydroperoxy flavin species is tightly coupled to substrate oxygenation. The catalytic mechanism also predicts that because release of product comes before the rate-limiting step all good substrates will have similar V_{max} values. Generally, for very good substrates (low K_m, high V_{max} substrates) this is the case but there are exceptions and it is somewhat dependent on the nature of the substrate. Steric factors may play a role in this general conclusion for a given class of FMO substrate. In summary, detailed kinetic studies are in accord with the mechanism of Figure 3.2 and the proposal does not violate the principles of enzyme saturation and Michaelis–Menten kinetics. Ziegler has proposed that FMO operates by providing a single point of attachment for substrate oxygenation and more complex induced fit interactions are not required (Ziegler 1993). However, the prediction that the hydroperoxy flavin of FMO acts similarly to synthetic isoalloxazine hydroperoxides and that the enzyme simply serves as a reactant in a bimolecular reaction is probably not true for all substrates. As described below, however, additional binding interactions must be at work to produce the various degrees of stereoselectivity observed for the FMOs examined.

Substrate specificity, inhibitors and induction

As the drug discovery pipeline has expanded with the advent of combinatorial chemistry, knowledge of monooxygenase-mediated detoxication should become an ever-increasing component in drug development (Cashman 1996). The substrate specificity of most of the FMOs has been summarised previously (Ziegler 1980, 1990, 1993; Cashman 1995). A model of pig FMO1 was proposed to explain much of the structure function information known at the time (Cashman 1995). This was based on the structure–function relations of a series of 10-[(N,N-dimethylamino)alkyl]-(2-trifluoromethyl)phenothiazine derivatives and other substrates N- and S-oxygenated by FMO1 and FMO3 (Lomri et al. 1993 a–c; Overby et al. 1995; Nagata et al. 1990). Use of the 10-(N,N-dimethylaminopentyl) derivative provides a highly sensitive means of determining FMO activity and N-oxygenation of this substrate is highly correlated with FMO3 immunoreactivity in human liver preparations (Cashman et al. 1993b). Generally, pig FMO1 provides a reasonable starting point for understanding the substrate specificity of animal FMOs because a prominent form of FMO in animal liver is FMO1. However, it appears that human FMO1 is considerably more restricted than pig FMO1 in the size of the nucleophilic heteroatom-containing substrates accepted (Kim and Ziegler 2000). Animal FMO2 differs from pig FMO1 because it is competent to N-oxygenate primary amines. On the other hand, evidence for *stimulation* of pig FMO1 by primary amines has been observed. However, careful examination of FMO1 shows that the enzyme can N-oxygenate primary amines but it does so more than 100-fold less efficiently than FMO3 (Lin and Cashman 1997a,b). Both chlorpromazine and imipramine have been reported to activate human FMO3 toward the oxygenation of good substrates such as methimazole (Overby et al. 1997; Wyatt et al. 1998). As described above, animal FMOs 1, 2, and 3 oxygenate a wide variety of nucleophilic tertiary and secondary amines as well as sulphur-containing compounds. Little is known about FMO4 and FMO6 substrate specificity because of the difficulties associated with cDNA-expression and characterisation. FMO5 is an unusual FMO and evidence for selective N-oxygenation of long-chain aliphatic primary amines has been reported (Overby et al. 1995), but FMO5 does not apparently S-oxygenate the widely used substrate methimazole.

Because FMO3 is probably the most functionally important FMO from the standpoint of human drug and xenobiotic metabolism, most of the discussion in this section will focus on this form of the enzyme. It is notable that the hepatic form of FMO present in rats is FMO1. Thus, rats represent a poor choice to use as FMO models for human hepatic drug development. For mice and dogs where FMO3 levels are more nearly similar to human liver caution should be exercised as there is a gender effect for FMO3 in these species (Ripp et al. 1999). Rabbit liver preparations may overpredict the contribution of FMO3 to drug metabolism because the hepatic levels appear to be present to a greater extent than in human liver. In designing a small animal strategy useful as a small animal model of human hepatic FMO3, the female mouse or dog may be more suitable but gender effects on other pharmacodynamic properties of the drug or xenobiotic to be evaluated may confound the picture.

Traditionally, human FMO3 activity has been described based on studies in human liver microsomes. Today, more studies are emerging that report on human FMO3 activity in other systems including hepatocytes (Fischer and Wiebel 1990; Sherratt and Damani 1989; Rodriguez et al. 1996) and other propagated cells (DiMonte et al. 1991). With the widespread use of recombinant human FMO3 enzyme preparations, a fuller description of substrate specificity is available. When practical, an important approach is to utilise substrates both in vitro and in vivo to characterise human FMO3. Thus, (S)-nicotine (Park et al. 1993), TMA (Treacy et al. 1998), cimetidine (Cashman et al. 1993a), clozapine (Sachse et al. 1999) and ranitidine (Kang et al. 2000) have been used to phenotype various populations for human FMO3 activity. All five chemicals have been shown to be relatively selective probes of human FMO3 activity in vitro. A non-invasive approach is to utilise the tertiary amine TMA that arises from the dietary precursor choline. Each substrate has some advantages and disadvantages. (S)-Nicotine is selectively N-1'-oxygenated by human FMO3 to form exclusively the trans N-1'-oxide both in vitro and in vivo. Animal FMO1 forms a 50:50 mixture of cis and trans nicotine N-1'-oxide. Therefore, the stereochemistry of the product reveals whether FMO1 or FMO3 N-oxygenates nicotine. That only the trans N'1'-oxide diastereomer is formed in humans in vivo suggests that only FMO3 metabolises (S)-nicotine. Despite the usefulness of nicotine as a stereoselective in vitro probe of human FMO3, its use in vivo is somewhat limited due to the relatively high K_m of nicotine. TMA is an excellent substrate for FMO3 in vitro (Cashman et al. 1997; Lang et al. 1998) but because it arises from dietary choline and other sources the TMA levels may vary in vivo and it is therefore important to establish in vivo TMA N-oxide:TMA ratios. Cimetidine S-oxygenation is another selective substrate for human FMO3 presumably because of the nucleophilicity of the sulphur atom and because the imidazole nucleus serves to inhibit CYP-mediated oxidation (Cashman et al. 1993a). Urinary cimetidine S-oxide is mainly present as the (-)-isomer (i.e. (-)−75%: (+)−25% cimetidine S-oxide). The in vivo result was in good agreement with that observed studying the S-oxygenation of cimetidine in vitro. The conclusion is that human FMO3 largely forms (-)-cimetidine S-oxide and human FMO1 forms (+)-cimetidine S-oxide. The bimodal profile of formation of plasma cimetidine S-oxide may be due to absorption differences or due to the action of different FMOs. Another possibility is that presystemic metabolism of cimetidine in the intestine contributes to the variability of cimetidine pharmacokinetics (Lu *et al.* 1998). However, the role of presystemic FMO in the metabolism of drugs and other xenobiotics is an understudied area. Clozapine is a cyclic tertiary amine that is efficiently N-oxygenated by human FMO3 (Tugnait *et al.* 1997). Clozapine and caffeine have been studied *in vivo* as probes of human FMO3 phenotype and genotype (Sachse *et al.* 1999). While clozapine is an effective *in vitro* probe, three common polymorphisms of human FMO3 were not linked to either clozapine or caffeine metabolism. For clozapine, it is possible that the K_m value is too high to serve as a useful *in vivo* marker. The lack of correlation of caffeine with FMO3 genotype and the reported lack of substrate activity for human FMO3 (Rettie and Lang 2000) brings up the issue as to whether caffeine is a useful probe for human FMO3. In view of the lack of a nucleophilic nitrogen atom and the lack of substrate activity, this suggests that caffeine metabolism is not dependent on the human FMO3. Ranitidine N-oxygenation has found use in correlating phenotype with genotype in a Korean population (Park *et al.* 1999). Other nucleophilic heteroatom-containing compounds have been shown to be selectively oxygenated by human FMO3. For example, tamoxifen (Kupfer and Dehal 1996), benzydamine (Ubeaud *et al.* 1999), xanomeline (Ring *et al.* 1999), N-deacetyl ketoconazole (Rodriguez *et al.* 1999) and sulindac sulphide (Hamman *et al.* 2000) are substrates for human FMO3 oxygenation.

To date, few examples of true competitive inhibition of human FMO3 have been reported. Most of the inhibitory effects on FMO have been examples of alternate substrate competitive inhibition. This is the case where a chemical is a better substrate for the compound being studied and addition of the chemical decreases the apparent oxygenation of the compound. Chemicals with low K_m, high V_{max} kinetic parameters for human FMO3 such as methimazole or thiobenzamide generally show alternate substrate competitive inhibition of human FMO3. Recently, a true competitive inhibitor based on dimethylamino stilbene carboxylate was reported (Clement *et al.* 1996). Another compound, indole-3-carbinol and its acid condensation products are potent, competitive inhibitors of human FMO3 (Cashman *et al.* 1999a). Indole-3-carbinol is a dietary constituent of cruciferous vegetables and is degraded to dimers and trimers upon reaching the acidic contents of the stomach. In a study comparing the *in vitro* inhibitory potency with the *in vivo* inhibitory potency, it was shown that the dimer was a potent inhibitor of human FMO3. Rat hepatic and intestinal FMO1 are also inhibited by dietary indole-3-carbinol (Larsen-Su and Williams 1996) and the down-regulation of FMO coupled with the induction of CYP may predispose animals to potential drug–drug interactions (Katchamart *et al.* 2000).

Stereochemical considerations

A number of reports have presented evidence on the stereoselective oxygenation of heteroatom-containing drugs and chemicals by the FMO. Some of this work was recently summarised (Cashman 1998). For FMO, advantage can be taken of the fact that many S- and N-oxides formed by FMO are sufficiently stable to spontaneous racemisation to allow the determination of optical activity after the monooxygenase reaction. For example, sulphoxides are relatively stable to stereomutation and racemise at elevated temperatures (i.e. above 200°C). In general, tertiary amines are more prone to thermal stereomutation than sulphoxides. However, cyclic tertiary amine N-oxides and even some linear amine N-oxides are sufficiently stable to thermal racemisation to assess FMO-mediated stereoselectivity. Because racemisation or decomposition of S- or N-oxides can occur by a number of routes (i.e. photochemical, acid-promoted or elimination reactions), in any studies of quantification of FMO stereoselectivity, it is advisable to chemically synthesise the product and do stability studies on the material that possesses a centre of chirality. Therefore, an important step in the determination of FMO-dependent stereoselectivity is to establish a bioanalytical method to measure the optical purity of the enzyme-catalysed reaction. ^{1}H-NMR spectroscopy in the presence of a chiral auxiliary (i.e. a europium shift reagent, Eu(hfc)$_3$) was used to probe the FMO-mediated stereoselectivity of aryl-1,3-dithiolane S-oxygenation formation (Cashman and Olsen, 1990; Cashman et al. 1990) and the optical purity was established by correlating the circular dichroism absorbance spectra by use of the Cotton effect (Mislow et al. 1965). Another approach is to use gas–liquid chromatography with a chiral stationary phase to separate lipophilic S-oxides. Because of the relatively low volatility and susceptibility to thermal decomposition, chiral phase gas–liquid chromatography is not generally useful in quantifying FMO-mediated stereoselectivity reactions. One possible exception is where an S-oxide or N-oxide can be quantitatively converted into a material that is efficiently chromatographed by gas-liquid chromatography. Two such examples are the thermal rearrangement of FMO metabolites (S)-nicotine N-1'-oxide to an oxazine (Jacob et al. 1986) and the formation of 3,4-dimethoxy styrene from verapamil N-oxide (Cashman, 1989). Of course, the most unambiguous method to determine FMO product stereoselectivity is to determine the X-ray crystal structure of the S- or N-oxide. However, the requirement for significant amounts of high-quality crystals and the amount of time and expense associated with this technique has limited its usefulness.

Commercially available chiral stationary phase HPLC (CSP HPLC) has allowed rapid advances in the characterisation of the stereoselectivity of enzyme-mediated reactions especially when used in conjunction with other methods to determine the absolute configuration. A summary of some of this technology has been presented previously (Cashman 1998). Generally, use of CSP HPLC is the method of choice for the determination of the absolute configuration of FMO products due to the speed, relative lack of expense and accuracy. Only 1–2 µg of FMO metabolite is required in each chromatographic run and this is easily obtained from typical small-scale incubations. In addition, synthetic chemical or enzymic methods are available to stereoselectively form the desired metabolite in enantiomerically enriched form. For example, by use of nonchiral oxidising agents in the presence of a chiral macromolecule, stereoselective S-oxygenation is easily achieved. Treatment of sulphides with sodium meta periodate in the presence of bovine serum albumin often provides multi-milligram quantities of S-oxide with great enantioselectivity (Cashman 1998). With the development of milder and more selective oxidising agents (Boyd et al. 1989), additional reagents should continue to become available to provide increasingly efficient means of obtaining authentic chiral standards. One of the best ways to chemically synthesise small amounts of S- or N-oxides with enantioenriched centres of chirality is with the modified Sharpless chiral oxidation reagent (Pitchen *et al.* 1984). The author's laboratory has had success by using this procedure to synthesise authentic metabolites useful as a standard in developing CSP HPLC for FMO-mediated chemical oxygenations (Cashman 1998). Another useful method for generating FMO-mediated metabolites possessing a centre of chirality is the use of other enzymes. Today, many monooxygenases are available commercially in pure form. Use of different monooxygenases can also provide information whether other enzyme systems give the same product stereochemistry, and this can be valuable in determining the contribution of other monooxygenases in substrate probe stereoselectivity. For example, alkyl-substituted *p*-tolyl sulphides have been shown to be stereoselectively S-oxygenated to S-oxides by FMOs (Rettie *et al.* 1995) (Table 3.4). CYPs or other haemoproteins also S-oxidise these same substrates and in some cases give the same product stereochemistry, in other cases give a distinct one (Pike *et al.* 1999). Thus, FMO1, 2, and 3 give a predominance of the (+)-(*R*) S-oxide but FMO5 and CYP2B and 2C6 mainly produce the (-)-(*S*) S-oxide (Rettie *et al.* 1994) (Table 3.4). This example points out the advantage of using stereochemistry to identify the contribution of a particular enzyme to the formation of a product, but it also suggests that caution should be exercised when multiple enzyme systems can oxidise the same probe substrate, depending on metabolic reaction conditions and substrate concentrations.

Role of FMO in toxicological aspects

In contrast to the CYP field of monooxygenases where some key advances have been made based on the observation that CYP bioactivated a chemical or drug to a toxic metabolite, fewer examples of FMO-mediated bioactivation are available. Rather, as discussed above, FMO has been associated with detoxication processes that convert nucleophilic heteroatom-containing chemicals or drugs into relatively polar, readily excreted metabolites. As described previously by Ziegler, it is possible that FMO evolved to inactivate many chemicals present in plants that would otherwise inhibit

Table 3.4 Stereoselective *S*-oxidation of ethyl *p*-tolyl sulphides and related compounds

Enzyme system	Absolute configuration	Enantiomeric excess	References
Horseradish Peroxidase[a]	(+)-(*R*)	100	Grayson and Rous, (1987).
Hog FMO1[b]	(+)-(*R*)	95	Light *et al.* (1982)
Rabbit FMO1[b]	(+)-(*R*)	99	Rettie *et al.* (1994)
Rabbit FMO2[b]	(+)-(*R*)	91	Rettie *et al.* (1990)
Rabbit FMO3[b]	(+)-(*R*)	v. low	Rettie *et al.* (1994)
Rabbit FMO5[b]	(+)-(*S*)	92	Fisher *et al.* (1995)
Cyclohexanone[b] Monooxygenase	(-)-(*S*)	80	Boyd *et al.* (1989)

[a] The substrate used was methyl *p*-tolyl sulphide.
[b] The substrate used was ethyl *p*-tolyl sulphide.

and destroy CYPs (Ziegler 1990). Thus, conversion of a sulphur-containing chemical to a polar S-oxide that might otherwise be oxidised by and inactivate CYP would constitute a chemoprotective strategy. The evidence for this postulate is that FMO is primarily localised where CYP resides and FMO is recalcitrant to inactivation by many chemicals that inhibit CYP. For example, depending on the structure, thiones (i.e. thioamides, mercaptoimidazoles, thiocarbamides, etc.) are metabolised by both CYP and FMO (Decker et al. 1991, 1992; Decker and Doerge 1991). In the case of 2-mercaptoimidazole (i.e. methimidazole) S-oxidation by CYP or FMO leads to a metabolite (i.e. a sulphenic acid) that covalently modifies CYP (Kedderis and Rickert 1985).

There are numerous examples of reactive metabolites produced by FMO- or CYP-mediated S-oxidative bioactivation that results in CYP inactivation without much effect on FMO. Of course, the presence of thiophiles such as glutathione to form disulphides after reaction with sulphenic acids can attenuate the relative toxicity of sulphenic acids formed by FMO. Once formed, the disulphides can undergo disulphide exchange and produce the parent thione and oxidised glutathione. This is an example of a futile metabolic cycle whereby the substrate is oxidised and after a reductive step is returned to its parent oxidation state. Oxidation of glutathione and consumption of NADPH may make the cell more susceptible to the toxic properties of other reactive metabolites especially if the thione is a low K_m high V_{max} substrate and depletes the cell of glutathione (Mizutani et al. 2000). In summary, judging whether an FMO-dependent oxygenation is a detoxication or bioactivation process is not always a simple exercise. However, on the basis of information available in the literature, the majority of the data suggests that FMO is a detoxication catalyst.

There are a few examples in the literature of how the toxicity of certain chemicals may be different under certain experimental conditions where the expression of FMO is altered. For example, the neurotoxin 1-methyl-4-phenyl-1,2,3,6-tetrahydropyridine (MPTP) is efficiently N-oxygenated by FMO in an apparent detoxication process (Cashman and Ziegler 1986). In a species where low FMO activity is present, the majority of MPTP is metabolised by monoamine oxidase (MAO) to the neurotoxic metabolite 1-methyl-4-phenyl-2,3-dihydropyridinium ion (MPDP$^+$) and N-methyl-4-phenylpyridinium ion (MPP$^+$). In different strains of mice, it is likely that FMO-mediated N-oxygenation of MPTP is a detoxication process leading to a non-toxic, readily excreted product whereas MAO-mediated oxidation results in highly electrophilic metabolites that participate in interruption of cellular function (Chiba et al. 1988). In the 1970s there was the view that the FMO system could N-hydroxylate procarcinogenic arylamines. This is likely to be true for some arylamines but human FMO is likely to be only a minor contributor to the overall N-oxygenation of these types of compounds. However, aliphatic primary amines avoid potentially toxic hydroxylamine formation by efficient N-oxygenation by human FMO3 (Cashman 2000) and this may lead to significant cytoprotection (Clement et al. 2000).

There are a few clinical examples of individuals with deficient FMO activity that result in toxic sequelae. However, to date, aside from the inherited defect of metabolism called trimethylaminuria discussed above, there is no direct link between altered FMO and a human disease state. However, drug–drug interactions may pose a problem for individuals with common variants of FMO.

Acknowledgments

I appreciate the collaborative work of the many co-workers cited in the references from my laboratory. I thank Professor Ellie Adman (University of Washington) for the valuable contribution of Figure 3.1. The author thanks the National Institutes of Health (Grants GM36426 and DK59618) and the TRDRP (Grant 9RT-0196) for their financial support.

References

Adali O, Carver GC and Philpot RM (1998) Modulation of human flavin-containing monooxygenase 3 activity by tricyclic antidepressants and other agents: importance of residue 428. *Arch. Biochem. Biophys.*, **358**, 92–97.

Akerman BR, Lemass H, Chow LM, Lambert DM, Greenberg C, Bibeau C, Mamer OA and Treacy EP (1999) Trimethylaminuria is caused by mutations of the FMO3 gene in a North American population. *Mol. Genet. Metab.*, **68**, 24–31.

Al-Waiz M, Ayesh R, Mitchell SC, Idle JR and Smith RL (1987) A genetic polymorphism of the N-oxidation of trimethylamine in humans. *Clin. Pharmacol. Ther.*, **42**, 588–594.

Al-Waiz M, Ayesh R, Mitchell SC, Idle JR and Smith RL (1988) Trimethylaminuria ('fish-odour syndrome'): a study of an affected family. *Clin. Sci.*, **74**, 231–236.

Ayesh R, Mitchell SC, Zhang A and Smith RL (1993) The fish odour syndrome: biochemical, familial, and clinical aspects. *Br. Med. J.*, **307**, 655–657.

Ball SS and Bruice TC (1983) Oxidation of amines by a 4a-hydroperoxyflavin. *J. Am. Chem. Soc.*, **102**, 6498–6503.

Ballou D, Palmer G and Massey V (1969) Direct demonstration of superoxide anion production during the oxidation of reduced flavin and of its catalytic decomposition by erythrocuprein. *Biochem. Biophys. Res. Commun.*, **36**, 898–904.

Beaty NB and Ballou DP (1981a) The oxidative half-reaction of liver microsomal FAD-containing monooxygenase. *J. Biol. Chem.*, **256**, 4619–4625.

Beaty NB and Ballou DP (1981b) The reductive half-reaction of liver microsomal FAD-containing monooxygenase. *J. Biol. Chem.*, **256**, 4611–4618.

Blumenthal I, Lealman GT and Franklyn PP (1980) Fracture of the femur, fish odour, and copper deficiency in a preterm infant. *Arch. Dis. Child.*, **55**, 229–231.

Boyd DR, Walsh CT and Chen YC (1989) In *Sulfur-Containing Drugs and Related Organic Compounds*, Part A, Volume 2, Damani LA (ed.), Ellis Horwood, Chichester, pp. 67–99.

Brunnelle A, Bi YA, Lin J, Russell B, Luy L, Berkman C and Cashman JR (1997) Characterization of two human flavin-containing monooxygenase (form 3) enzymes expressed in *Escherichia coli* as maltose binding protein fusions. *Drug Metab. Dispos.*, **25**, 1001–1007.

Calvert GD (1973) Trimethylaminuria and inherited Noonan's syndrome. *Lancet*, **1**, 320–321.

Cashman JR (1989) Enantioselective N-oxygenation of verapamil by the flavin-containing monooxygenase. *Mol. Pharmacol.*, **36**, 497–503.

Cashman JR (1995) Structural and catalytic properties of the mammalian flavin-containing monooxygenase. *Chem. Res. Toxicol.*, **8**, 165–181.

Cashman JR (1996) Drug discovery and drug metabolism. *Drug Discovery Today*, **1**, 209–216.

Cashman JR (1997) Monoamine oxidase and flavin-containing monooxygenase. In *Comprehensive Toxicology, Volume 3*, Ch. 6, Guengerich FP (ed.), Elsevier Science, New York, pp. 69–96.

Cashman JR (1998) Stereoselectivity in S- and N-oxygenation by the mammalian flavin-containing monooxygenase and cytochrome P-450 monooxygenases. *Drug Metab. Rev.*, **30**, 675–708.

Cashman JR (2000) Human flavin-containing monooxygenase: substrate specificity and role in drug metabolism. *Curr. Drug Metab.*, **1**, 1037–1045.

Cashman JR and Olsen LD (1990) Stereoselective S-oxygenation of 2-aryl-1,3-dithiolanes by hog liver flavin-containing monooxygenase. *Mol. Pharmacol.*, **38**, 573–585.

Cashman JR and Ziegler DM (1986) Contribution of N-oxygenation to the metabolism of MPTP

1-methyl-4-phenyl-1,2,3,6-tetrahydropyridine by the flavin-containing monooxygenase by various liver preparations. *Mol. Pharmacol.*, **29**, 163–167.

Cashman JR, Olsen LD, Lambright CE and Presas MJ (1990) Enantioselective S-oxygenation of *p*-methoxy phenyl-1,3-dithiolane by various tissue preparations: effect of estradiol. *Mol. Pharmacol.*, **37**, 319–327.

Cashman JR, Park SB, Yang ZC, Washington CB, Gomez DY, Giacomini KM and Brett CM (1993a) Chemical, enzymic, and human enantioselective S-oxygenation of cimetidine. *Drug Metab. Dispos.*, **21**, 587–597.

Cashman JR, Yang Z, Yang L and Wrighton SA (1993b) Stereo- and regioselective N- and S-oxidation of tertiary amines and sulphides in the presence of adult human liver microsomes. *Drug. Metab. Dispos.*, **21**, 492–501.

Cashman JR, Park SB, Berkman CE and Cashman LE (1995) Role of hepatic flavin-containing monooxygenase 3 in drug and chemical metabolism in adult humans. *Chem. Biol. Interact.*, **96**, 33–46.

Cashman JR, Perotti BY and Berkamn CE (1996) Lin J (1996) Pharmacokinetics and molecular detoxication. *Environ. Health Perspect.*, **104**, 23–40.

Cashman JR, Bi YA, Lin J, Youil R, Knight M, Forrest S and Treacy E (1997) Human flavin-containing monooxygenase form 3: cDNA expression of the enzymes containing amino acid substitutions observed in individuals with trimethylaminuria. *Chem. Res. Toxicol.*, **10**, 837–841.

Cashman JR, Xiong Y, Lin J, Verhagen H, van Poppel G, van Bladeren PJ, Larsen-Su S and Williams DE (1999a) *In vitro* and *in vivo* inhibition of human flavin-containing monooxygenase form 3 in the presence of dietary indoles. *Biochem. Pharmacol.*, **58**, 1047–1055.

Cashman JR, Xiong YN, Xu L and Janowsky A (1999b) N-Oxygenation of amphetamine and methamphetamine by the human flavin-containing monooxygenase (form 3): role in bioactivation and detoxication. *J. Pharmacol. Exper. Ther.*, **288**, 1251–1260.

Cashman JR, Akerman BR, Forrest SM and Treacy EP (2000) Population-specific polymorphisms of the human FMO3 gene: signifcance for detoxication. *Drug Metab. Dispos.*, **28**, 169–173.

Chen H and Aiello F (1993) Trimethylaminuria in a girl with Prader-Willi syndrome and del(15)(q11q13). *Am. J. Med. Genet.*, **45**, 335–339.

Chiba K, Kubota E, Miyakawa T, Kato Y and Ishizaki T (1988) Characterization of hepatic microsomal metabolism as an *in vivo* detoxication pathway of 1-methyl-4-phenyl-1,2,3,6-tetrahydropyridine in mice. *J. Pharmacol. Exp. Ther.*, **246**, 1108–1115.

Christensen D (1999) What's that smell? *Science News*, **155**, 316–317.

Chung WG, Park CS, Roh HK and Cha YN (1997) Induction of flavin-containing monooxygenase (FMO1) by a polycyclic aromatic hydrocarbon, 3-methylcholanthrene, in rat liver. *Mol. Cells*, **7**, 738–741.

Clement B, Weide M and Ziegler DM (1996) Inhibition of purified and membrane-bound flavin-containing monooxygenase by (*N*,*N*-dimethylamino)stilbene carboxylates. *Chem. Res. Toxicol.*, **9**, 599–604.

Clement B, Behrens D, Moller W and Cashman JR (2000) Reduction amphetamine hydroxylamine and other aliphastic hydroxylamines by benzamidoxime reductase and human liver microsomes. *Chem Res. Toxicol.*, In press.

Dannan GA, Guengerich FP and Waxman DJ (1986) Hormonal regulation of rat liver microsomal enzymes. Role of gonadal steroids in programming, maintenance and suppression of Δ-4-steroid-5-α-reductase, flavin-containing monooxygenase and sex-specific cytochrome P-450. *J. Biol. Chem.*, **261**, 10728–10735.

Decker CJ and Doerge DR (1991) Rat hepatic microsomal metabolism of ethylenethiourea. Contributions of the flavin-containing monooxgeanse and cytochrome P-450 isozymes. *Chem. Res. Toxicol.*, **4**, 482–489.

Decker CJ, Cashman JR, Sugiyama K, Maltby D and Correia MA (1991) Formation of glutathionyl-spironolactone disulphide by rat liver cytochromes P450 or hog liver flavin-containing monooxygenases: a functional probe of two-electron oxidations of the thiosteroid? *Chem. Res. Toxicol.*, **4**, 669–677.

Decker CJ, Doerge DR and Cashman JR (1992) Metabolism of benzimidazoline-2-thiones by rat hepatic microsomes and hog liver flavin-containing monooxygenase. *Chem. Res. Toxicol.*, **5**, 726–733.

DiMonte D, Wu EY, Irwin I, Delanney LE and Langston JW (1991) Biotransformation of 1-methyl-4-phenyl-1,2,3,6-tetrahydropyridine in primary cultures of mouse astrocytes. *J. Pharmacol. Exper. Therap.*, **258**, 594–600.

Doerge DR and Corbett MD (1984) Primary arylamine oxidation by a flavin hydroperoxide. *Biochem. Pharmacol.*, **33**, 3615–3619.

Dolphin CY, Shephard EA, Povey S, Palmer CN, Ziegler DM, Ayesh R, Smith RL and Phillips IR (1991) Cloning, primary sequence, and chromosomal mapping of a human flavin-containing monooxygenase (FMO1). *J. Bio. Chem.*, **266**, 12379–12385.

Dolphin CT, Shephard EA, Povey S, Smith RL and Phillips IR (1992) Cloning, primary sequence and chromosomal localizator of human FMO2, a new member of the flavin-containing monooxygenate family. *Biochem. J.*, **287**, 261–267.

Dolphin CT, McCombie RR, Smith RL, Shephard EA and Phillips IA (1996) Molecular genetics of human flavin-containing monooxygenases. *ISSX Proceed.*, **10**, 217.

Dolphin CT, Janomohamed A, Smith RL, Shephard EA and Phillips IR (1997a) Missense mutation in flavin-containing monooxygenase 3 gene, FMO3, underlies fish-odour syndrome. *Nat. Genet.*, **17**, 491–494.

Dolphin CT, Riley JH, Smith RL, Shephard EA and Phillips IR (1997b) Structural organisation of the human flavin-containing monooxygenase 3 gene, (FMO3), the favored candidate for fish-odour syndrome, determined directly from genomic DNA. *Genomics*, **46**, 260–267.

Dolphin CT, Beckett DT, Janmohamed A, Cullingford TE, Smith RL, Shephard EA and Phillips IR (1998) The flavin-containing monooxygenase 2 gene (FMO2) of humans, but not of other primates, encodes a truncated, nonfunctional protein. *J. Biol. Chem.*, **273**, 30599–30607.

Duffel MW, Graham JM and Ziegler DM (1981) Changes in dimethylaniline N-oxidase activity of mouse liver and kidney induced by steroid sex hormones. *Mol. Pharmacol.*, **19**, 134–139.

Fenwick GR, Butler EJ and Brewster MA (1983) Are brassica vegetables aggravating factors in trimethylaminura (fish odour syndrome)? [letter]. *Lancet*, **2**, 916.

Fernandez MS, Gutierrez C, Vila JJ, Lopez A, Ibanez V, Sanguesa C, Lluna J and Barrios JE (1997) Congenital intrahepatic portocaval shunt associated with trimethylaminuria. *Pediatr. Surg. Int.*, **12**, 196–197.

Fischer V and Wiebel FJ (1990) Metabolism of fluperlapine by cytochrome P450-dependent and flavin-dependent monooxygenases in continuous cultures of rat and human cells. *Biochem. Pharmacol.*, **39**, 1327–1333.

Fisher MB, Lawton MP, Atta-Asafo-Adjei E, Philpot RM and Rettie AE (1995) Selectivity of flavin-containing monooxygenase 5 for the (S)-sulfoxidethon of short-chain arakyl sulpfides. *Drug Metab. Dispos.*, **23**, 1431–1433.

Forrest SM, Knight M, Akerman BR, Cashman JR and Treacy EP (2000) A novel deletion in the flavin-containing monooxygenase gene (FMO3) in a Greek patient with trimethylaminuria. *Pharmacogenetics*, **11**, 169–174.

Gasser R, Tynes RE, Lawton MP, Korsmeyer KK, Ziegler DM and Philpot RM (1990) The flavin-containing monooxygenase expressed in pig liver: Primary sequence, distribution, and evidence for a single gene. *Biochemistry*, **29**, 119–124.

Gelb BD, Zhang J, Cotter PD, Gershin IF and Desnick RJ (1997) Physical mapping of the human connexin 40 (GJA5), flavin-containing monooxygenase and natriuretic peptide receptor a genes on 1q21. *Genomics*, **39**, 409–411.

Goldstein A and de Morais SM (1994) Biochemistry and molecular biology of the human CYP2C subfamily. *Pharmacogenetics*, **4**, 285–299.

Gonzalez FJ and Nebert DW (1990) Evolution of the P450 gene superfamily: animal–plant 'warfare', molecular drive and human genetic differences in drug oxidation. *Trends Genet.*, **66**, 182–186.

Grayson DH and Rous AJ (1987) Enzymetic preparation of optically-active suphoxides. In *RSC Meeting on Synthesis and Biosynthesis of Natural Products*, Edinburgh.

Guan S, Falick AM and Cashman JR (1990) N-terminus determination: FAD-and NADP+-binding domain mapping of hog liver flavin-containing monooxygenase by tandem mass spectrometry. *Biochem. Biophys. Res. Commun.*, **170**, 937–943.

Guan S, Falick AM, Williams DE and Cashman JR (1991) Evidence for complex formation between rabbit lung flavin-containing monooxygenase and calreticulin. *Biochemistry*, **30**, 9892–9900.

Guengerich FP, Kim DH and Iwasaki M (1991) Role of human cytochrome P450 IIE1 in the oxidation of several low molecular weight cancer suspects. *Chem. Res. Toxicol.*, **4**, 168–179.

Hadidi HF, Cholerton S, Atkinson S, Irshaid YM, Rawashdeh NM and Idle JR (1995) The N-oxidation of trimethylamine in a Jordanian population. *Br. J. Clin. Pharmacol.*, **39**, 179–181.

Hamman MA, Haehner-Daniels BD, Wrighton SA, Rettie AE and Hall SD (2000) Stereoselective sulfoxidation of sulindac sulphide by flavin-containing monooxygenase: Comparison of human liver and kidney microsomes and mammalian enzymes. *Biochem Pharmacol.*, **60**, 7–17.

Heinze E, Hlavica P and Kiese M (1970) N-oxygenation of arylamines in microsomes prepared from corpra lutea of the cycle and other tissues of the pig. *Biochem. Pharmacol.*, **19**, 641–649.

Hines RN, Cashman JR, Philpot RM, Williams DE and Ziegler DM (1994) The mammalian flavin-containing monooxygenases: molecular characterisation and regulation of expression. *Toxicol. Appl. Pharmacol.*, **125**, 1–6.

Hlavica P and Golly L (1991) On the genetic polymorphism of the flavin-containing monooxygenase. In *N-Oxidation of Drugs*, Hlavica P and Damani LA (eds), Chapman and Hall, London, pp. 71–90.

Hodgson E and Levi PE (1992) The role of the flavin-containing monooxygenase (EC 1.14.13.8) in the metabolism and mode of action of agricultural chemicals. *Xenobiotica*, **22**, 1175–1183.

Hodgson E, Cherrington N, Coleman SC, Liu S, Falls JG, Cao Y, Goldstein JE and Rose RL (1998) Flavin-containing monooxygenase and cytochrome P-450-mediated metabolism of pesticides: from mouse to human. *Rev. Toxicol.*, **6**, 231–243.

Jacob P III, Benowitz NL, Yu L and Shulgin AT (1986) Determination of nicotine N-oxide by gas chromatography following thermal conversion of 2-methyl-6-(3-pyridyl)tetrahydro-1,2-oxazine. *Anal. Chem.*, **58**, 2218–2221.

Kaderlik RF, Weser E and Ziegler DM (1991) Selective loss of liver flavin-containing monooxygenases in rats on chemically defined diets. *Prog. Pharmacol. Clin. Pharmacol.*, **3**, 95–103.

Kang JH, Chung WG, Lee KH, Park CS, Shin IC, Roh HK, Dong MS and Baek HM (2000) Phenotypes of flavin-containing monooxygenase activity determined by ranitidine oxidation are positively correlated with genotypes of linked FMO3 gene mutations in a Korean population. *Pharmacogenetics*, **10**, 67–78.

Katchamart S, Stresser DM, Dehal SS, Kupfer DD and Williams DE (2000) Concurrent flavin-containing monooxygenase down-regulation and cytochrome P-450 induction by dietary indoles in rat: implications for drug-drug interaction. *Drug Metab. Dispos.*, **28**, 930–936.

Kedderis GL and Rickert DE (1985) Loss of rat liver microsomal cytochrome P-450 during methimazole metabolism. Role of the flavin-containing monooxygenase. *Drug Metab. Dispos.*, **13**, 58–61.

Kemal C and Bruice TC (1976) Simple synthesis of a 4a-hydroperoxy adduct of a 1,5-dihydroflavine: Preliminary studies of a model for bacterial luciferase. *Proc. Natl. Acad. Sci. USA*, **73**, 995–999.

Kemal C, Chan TW and Bruice TC (1977) Reaction of $^3O_2$ with dihydroflavins. 1. $N^{3,5}$-dimethyl-1,5-dihydrolumiflavin and 1,5-dihydroisoalloxazines. *J. Am. Chem. Soc.*, **99**, 7272–7286.

Kim YM and Ziegler DM (2000) Size limits of thiocarbamides accepted as substrates by human flavin-containing monooxygenase 1. *Drug Metab. Dispos.*, **28**, 1003–1006.

Korsmeyer KK, Guan S, Yang ZC, Falick AM, Ziegler DM and Cashman JR (1998) N-Glycosylation of pig flavin-containing monooxygenase form 1: determination of protein modification by mass spectrometry. *Chem. Res. Toxicol.*, **11**, 1145–1153.

Kubo A, Itoh S, Itoh K and Kamataki T (1997) Determination of FAD-binding domain in flavin-containing monooxygenase. *Arch. Biochem. Biophys.*, **345**, 271–277.

Kupfer D and Sehal SS (1996) Tamoxifen metabolism by microsomal cytochrome P-450 and flavin-containing monooxygenase. *Methods Enzymol.*, **272**, 152–163.

Lang DH, Yeung CK, Peter RM, Ibarra C, Gasser R, Itagaki K, Philpot RM and Rettie AE (1998) Isoform specificity of trimethylamine N-oxygenation by human flavin-containing monooxygenase (FMO) and P450 enzymes: selective catalysis by FMO3. *Biochem. Pharmacol.*, **56**, 1005–1012.

Larsen-Su S and Williams DE (1996) Dietary indole-3-carbinol inhibits FMO activity and the expression of flavin-containing monooxygenase form 1 in rat liver and intestine. *Drug Metab. Dispos.*, **24**, 927–931.

Lawton MP and Philpot RM (1993) Functional characterisation of flavin-containing monooxygenase 1B1 expressed in *Saccharomyces cerevisiae* and *Escherichia coli* and analysis of proposed FAD- and membrane binding domains. *J. Biol. Chem.*, **268**, 5728–5734.

Lawton MP, Cashman JR, Cresteil T, Dolphin C, Elfarra A, Hines RN, Hodgson E, Kimura T, Ozols J, Phillips I, Philpot RM, Poulsen LL, Rettie AE, Williams DE and Ziegler DM (1994) A nomenclature for the mammalian flavin-containing monooxygenase gene family based on amino acid sequence identities. *Arch. Biochem. Biophys.*, **308**, 254–257.

Lemoine A, Williams DE and Cresteil T (1991) Hormonal regulation of microsomal flavin-containing monooxygenase: tissue-dependent expression and substrate specificity. *Mol. Pharmacol.*, **40**, 211–217.

Light DR, Waxman DJ and Walsh C (1982) Studies on the chirality of sulfoxidation catalysed by bacterial flavoenzyme cyclohexanone monooxygenase and hog liver flavin-adenine dinucleotide containing monooxygenase. *Biochemistry*, **21**, 2490–2498.

Lin J and Cashman JR (1997a) Detoxication of tyramine by the flavin-containing monooxygenase: stereoselective formation of the *trans* oxime. *Chem. Res. Toxicol.*, **10**, 842–852.

Lin J and Cashman JR (1997b) N-oxygenation of phenethylamine to the trans-oxime by adult human liver flavin-containing monooxygenase and retroreduction of phenethylamine hydroxylamine by human liver microsomes. *J. Pharmacol. Exp. Ther.*, **282**, 1269–1279.

Lomri N, Thomas J and Cashman JR (1993a) Expression in *Escherichia coli* of the cloned flavin-containing monooxygenase from pig liver. *J. Biol. Chem.*, **268**, 5048–5059.

Lomri N, Yang Z-C and Cashman JR (1993b) Expression in *Escherichia coli* of the flavin-containing monooxygenase from adult human liver. Determination of a distinct tertiary amine substrate specificity. *Chem. Res. Toxicol.*, **6**, 425–429.

Lomri N, Yang Z-C and Cashman JR (1993c) Regio- and stereoselective oxygenations by adult human liver flavin-containing monooxygenase 3. Comparison with forms 1 and 2. *Chem Res. Toxicol.*, **6**, 800–807.

Lu X, Li C and Fleisher D (1998) Cimetidine sulfoxidation in small intestinal microsomes. *Drug Metab. Dispos.*, **26**, 240–242.

Luo Z and Hines RN (1996) Identification of multiple rabbit flavin-containing monooxygenase form 1 (FMO1) gene promoters and observation of tissue-specific DNase I hypersensitive sites. *Arch. Biochem. Biophys.*, **336**, 251–260.

Luo Z and Hines RN (1997) Further characterisation of the major and minor rabbit FMO1 promoters and identification of both positive and negative distal regulatory elements. *Arch. Biochem. Biophys.*, **346**, 96–104.

Mattevi A, Schierbeek A and Hol W (1991) Refined crystal structure of lipoamide dehydrogenase from *Azobacter vinelandii* at 2.2 resolution. A comparison with the structure of glutathione reductase. *J. Mol. Biol.*, **220**, 975–994.

Mayatepek E and Kohlmuller D (1998) Transient trimethylaminuria in childhood. *Acta Paediatr. Int. J. Paediatr.*, **87**, 1205–1207.

McCombie RR, Dolphin CT, Povey S, Phillips IR and Shephard EA (1996) Localization of human flavin-containing monooxygenase genes FMO2 and FMO5 to chromosme 1q. *Genomics*, **15**, 426–429.

McConnell HW, Mitchell SC, Smith RL and Brewster M (1997) Trimethylaminuria associated with seizures and behavioural disturbance: a case report. *Seizure*, **6**, 317–321.

Miller A (1982) A model for FAD-containing monooxygenase: the oxidation of thioanisole derivatives by an isoalloxazine hydroperoxide. *Tetrahedron Lett.*, **23**, 753–756.

Miller A, Bischoff JJ, Bizub C, Luminoso P and Smiley S (1986) Electronic and steric effects in oxidations by isoalloxazine 4a-hydroperoxides. *J. Am. Chem. Soc.*, **108**, 7773–7778.

Mislow K, Green MM, Laur P, Melillo JT, Simmons T and Ternay Jr AL (1965) Absolute configuration and optical rotary power of sulphoxides and sulfinate esters. *J. Am. Chem. Soc.*, **87**, 1958–197.

Mitchell SC, Zhang AQ, Barrett T, Ayesh T and Smith RL (1997) Studies on the discontinuous N-oxidation of trimethylamine among Jordanian, Ecuadorian and New Guinean populations. *Pharmacogenetics*, **7**, 45–50.

Mizutani T, Yoshida K, Murakami M, Shirai M and Kawazoe S (2000) Evidence for the involvement of N-methylthiourea, a ring cleavage metabolite in the hepatotoxicity of methimazole in glutathione-depleted mice: structure-toxicity and metabolic studies. *Chem. Res. Toxicol.*, **13**, 170–176.

Nagata T, Williams DE and Ziegler DM (1990) Substrate specificities of rabbit lung and porcine liver flavin-containing monooxygenases: Differences due to substrate specificity. *Chem. Res. Toxicol.*, **3**, 372–376.

Osimitz TG and Kulkarni AP (1982) Oxidative metabolism of xenobiotics during pregnancy: significance of microsomal flavin-containing monooxygenases. *Biochem. Biophys. Res. Commun.*, **109**, 1164–1171.

Overby LH, Buckpitt AR, Lawton MP, Atta-Asafo-Adjei E, Schulze J and Philpot RM (1995) Characterization of flavin-containing monooxygenase 5 (FMO5) cloned from human and guinea pig: Evidence that the unique catalytic properties of FMO5 are not confined to the rabbit ortholog, *Arch. Biochem. Biophys.*, **317**, 275–284.

Overby LH, Carver GC and Philpot RM (1997) Quantitastion and kinetic properties of hepatic microsomal and flavin-containing monooxygenates 3 and 5 from humans. *Chem. Biol. Interact.*, **106**, 29–45.

Ozols J (1990) Covalent structure of liver microsomal flavin-containing monooxygenase form 1. *J. Biol. Chem.*, **265**, 10289–10299.

Ozols J (1990) Multiple forms of liver microsomal flavin-containing monooxygenases. *Arch. Biochem. Biophys.*, **290**, 103–115.

Ozols J (1994) Isolation and structure of a third form of liver microsomal flavin monooxygenase. *Biochemistry*, **33**, 3751–3757.

Park SB, Jacob P, Benowitz NL and Cashman JR (1993) Stereoselective metabolism of (*S*)-(-)-nicotine in humans: formation of *trans*-(*S*)-(-)-nicotine *N*-1'-oxide. *Chem. Res. Toxicol.*, **6**, 880–888.

Park CS, Chung WG, Kang JH, Roh HK, Lee KH and Cha YN (1999) Phenotyping of flavin-containing monooxygenase using caffeine metabolism of FMO3 gene in a Korean population. *Pharmacogenetics*, **9**, 155–164.

Phillips IR, Dolphin CT, Clair P, Hadley MR, Hutt AJ, McCombie RR, Smith RL and Shephard EA (1995) The molecular biology of the flavin-containing monooxygenases of man. *Chem. Biol. Interact.*, **96**, 17–32.

Philpot R, Overby L and Wyatt M (1996) The flavin-containing monooxygenase gene family: structure and function. *ISSX Proceed.*, **10**, 3.

Pike MG, Martin YN, Mays DC, Benson LM, Naylor S and Lipsky JJ (1999) Roles of FMO and CYP450 in the metabolism in human liver microsomes of S-methyl-*N*,*N*-diethyldithiocarbamate, a disulfiram metabolite. *Alcohol Clin. Exp. Res.*, **23**, 1173–1179.

Pitchen PE, Dunach MN, Deshmukh E and Kagan HB (1984) An efficient asymmetric oxidation of sulphides to sulphoxides. *J. Am. Chem. Soc.*, **106**, 8188–8193.

Piyapolrungroj N, Zhou YS, Li C, Liu G, Zimmermann E and Fleisher D (2000) Cimetidine absorption and elimination in rat small intestine. *Drug Metab. Dispos.*, **28**, 65–72.

Poulsen LL and Ziegler DM (1979) The liver microsomal FAD-containing monooxygenases. Spectral characterisation and kinetic studies. *J. Biol. Chem.*, **254**, 6449–6455.

Redondo PA, Alvarez AI, Garcia JL, Larrode OM, Merino G and Prieto JG (1999) Presystemic metabolism of albendazole: experimental evidence of an efflux of albendazole sulfoxide to intestinal lumen. *Drug Metab. Dispos.*, **27**, 736–740.

Rettie AE and Lang DH (2000) Can caffeine metabolism be used as an in-vivo probe for human flavin-containing monooxygenase activity? *Pharmacogenetics*, **10**, 275–277.

Rettie AE, Bogucki BD, Lim I and Meier GP (1990) Stereoselective sulfoxidation of a series of alkyl *p*-tolyl sulphides by microsomal and purified flavin-containing monooxygenases. *Mol. Pharmacol.*, **37**, 643–651.

Rettie AE, Lawton MP, Sadeque AJ, Meier GP and Philpot RM (1994) Prochiral sulfoxidation as a probe for mulitple forms of the microsomal flavin-containing monooxygeanse: studies with rabbit FMO1, FMO2, FMO3, and FMO5 expressed in *Escherichia coli. Arch. Biochem. Biophys.*, **311**, 369–377.

Rettie AE, Meier GP and Sadeque AJ (1995) Prochiral sulphides as in vitro probes for multiple forms of the flavin-containing monooxygenase. *Chem. Biol. Interact.*, **96**, 3–15.

Ring BJ, Wrighton SA, Aldridge SLK, Hansen K, Haehner B and Shipley LA (1999) Flavin-containing monooxygenase-mediated N-oxidation of the M(1)-muscarinic agonist xanomeline. *Drug Metab. Dispos.*, **27**, 1099–1103.

Ripp SL, Itagaki K, Philpot RM and Elfarra AA (1999) Species and sex differences in expression of flavin-containing monooxygenase and kidney microsomes. *Drug Metab. Dispos.*, **27**, 46–52.

Rodriguez RJ, Davila JC and Acosta D (1996) Establishment of flavin-containing monooxygenases in hepatic microsomes obtained from primary cultures of rat hepatocytes. *Fund. Appl. Toxicol.*, **30**, 215.

Rodriguez RJ, Proteau PJ, Marquez BL, Heatherington CL, Buckholz CJ and O'Connell KL (1999) Flavin-containing monooxygenase-mediated metabolism of N-deacetyl ketoconazole in hepatic microsomes. *Drug Metab. Dispos.*, **27**, 880–886.

Ryerson CC, Ballou DP and Walsh C (1982) Mechanistic studies on cyclohexanone oxygenase. *Biochemistry*, **21**, 2644–2655.

Sachse C, Ruschen S, Dettling M, Schley J, Bauer S, Muller-Oerlinghausen B, Roots I and Brockmoller J (1999) Flavin monooxygenase 3 (FMO3) polymorphism in a white population: all mutation linkage, functional effects on clozapine and caffeine metabolism. *Clin. Pharmacol. Ther.*, **66**, 431–438.

Shehin-Johnson SE, Williams DE, Larsen-Su S, Stresser DM and Hines RN (1995) Tissue-specific expression of flavin-containing monooxygenase (FMO) forms 1 and 2 in the rabbit. *J. Pharmacol. Exper. Therap.*, **272**, 1293–1299.

Shehin-Johnson SE, Palmer KC and Hines RN (1996) Identification of tissue-specific DNase I hypersensitive sites in the flavin-containing monooxygenase form 2 gene. *Drug Metab. Dispos.*, **24**, 891–898.

Sherratt AJ and Damani LA (1989) Activities of cytosolic and microsomal drug oxidases of rat hepatocytes in primary culture. *Drug Metab. Dispos.*, **17**, 20–25.

Stransky L (1998) Fishy odour syndrome after acute hepatitis B. *Dermatosen Beruf Umwelt*, **46**, 79–80.

Suh JK, Poulsen LL, Ziegler DM and Tobertus JD (1996) Molecular cloning and kinetic characterisation of a flavin-containing monooxygenase from *Saccharomyces cerevisiae. Arch. Biochem. Biophys.*, **336**, 268–274.

Thieme R, Pai E, Schrimer R and Schulz G (1981) Three-dimensional structure of glutathione reductase at 2Å resolution. *J. Mol. Biol.*, **152**, 763–782.

Thithapadha A (1997) A pharmacogenetic study of trimethylaminuria in Orientals. *Pharmacogenetics*, **7**, 497–501.

Todd WA (1979) Psychosocial problems as the major complication of an adolescent with trimethylaminuria. *J. Pediatr.*, **94**, 936–937.

Treacy EP, Akerman BR, Chow LML, Youil R, Bibeau C, Lin J, Bruce AG, Knight M, Danks DM, Cashman JR and Forrest SM (1998) Mutations of the flavin-containing monooxygenase gene (FMO3) cause trimethylaminuria, a defect in detoxication. *Hum. Mol. Genet.*, **7**, 839–845.

Tucker GT, Silas JH and Iyun AO (1977) Polymorphic hydroxylation of debrisoquine. *Lancet*, **2**, 718.

Tugnaut M, Haives EM, McKay G, Rettie AE, Harning RL and Midha KK (1997) N-oxygenation of dozapine by flavin-containing monooxygenase. *Drug metab. Dispos.*, **25**, 524–527.

Tynes RE, Sabourin PJ and Hodgson E (1985) Identification of distinct hepatic and pulmonary forms of microsomal flavin-containing monooxygenase in the mouse and rabbit. *Biochem. Biophys. Res. Commun.*, **126**, 1069–1075.

Ubeaud G, Schiller CD, Hurbin F, Jaeck D and Coassolo P (1999) Estimation of flavin-containing monooxygenase activity in intact hepatocytes of rat, hamster, rabbit, dog and human by using N-oxidation of benzydamine. *Eur. J. Pharm. Sci.*, **4**, 255–260.

Vallon O (2000) New sequence motifs in flavoproteins: evidence for common ancestry and structure. *Proteins*, **38**, 95–114.

Williams DE, Ziegler DM, Nordin DJ, Hale SE and Masters BSS (1984) Rabbit lung flavin-containing monooxyenase is immunologically and catalytically distinct from the liver enzyme. *Biochem. Biophys. Res. Commun.*, **125**, 116–122.

Wirth PJ and Thorgeirsson SS (1978) Amine oxidase in mice-sex difference and developmental aspects. *Biochem. Pharmacol.*, **19**, 601–603.

Wyatt MK, Philpot RM, Carver G, Lawton MP and Nikbakht KN (1996) Structural characteristics of flavin-containing monooxygenase genes one and two (FMO1 and FMO2). *Drug Metab. Dispos.*, **24**, 1320–1327.

Wyatt MK, Overby LH, Lawton MP and Philpot RM (1998) Identification of amino acid residues associated with modulation of flavin-containing monooxygenase (FMO) activity of imipramine: structure/function studies in pig and rabbit. *Biochemistry*, **37**, 5930–5938.

Zhang AQ, Mitchell SC and Smith RL (1996) Exacerbation of symptoms of fish-odour syndrome during menstruation [letter]. *Lancet*, **348**, 1740–1741.

Ziegler DM (1980) Microsomal flavin-containing monooxygenase: oxygenation of nucleophilic nitrogen and sulfur compounds. In *Enzymatic Basis of Detoxication, Volume 1*, Jakoby WB (ed.), Academic Press, New York, pp. 201–277.

Ziegler DM (1988) Flavin-containing monooxygenase: catalytic mechanism and substrate specificities. *Drug Metab. Rev.*, **19**, 1–31.

Ziegler DM (1990) Flavin-containing monooxygenases: enzymes adapted for multisubstrate specificity. *Trends Pharmacol. Sci.*, **11**, 321–324.

Ziegler DM (1993) Recent studies on the structure and function of multisubstrate flavin-containing monooxygenases. *Ann. Rev. Pharmacol. Toxicol.*, **33**, 179–199.

Ziegler DM (1999) Flavin-containing monooxygenase family of isozymes. Abstract, Proc. of the First International Workshop on Trimethylaminuria, Bethesda, MD, March 29–30.

Zschocke J, Kohlmueller D, Quak E, Meissner T, Hoffmann GF and Mayatepek E (1999) Mild trimethylaminuria caused by common variants in FMO3 gene. *Lancet*, **354**, 834–835.

4 Amine Oxidases and the Metabolism of Xenobiotics

Keith F. Tipton[1] **and Margherita Strolin Benedetti**[2]

[1] *Trinity College, Dublin, Ireland*
[2] *UCB Pharma, Nanterre Cedex, France*

Introduction

THE AMINE OXIDASE GROUP OF ENZYMES

Mammalian systems contain several different amine oxidases that may be distinguished by their cofactor requirements, substrate specificities and inhibitor sensitivities. The enzymes that are active towards monoamines have been the most extensively studied in terms of their possible involvement in xenobiotic metabolism and these enzymes will be considered first. The monoamine oxidases (amine: oxygen oxidoreductase (deaminating) (flavin-containing); EC 1.4.3.4), which are flavoproteins (FAD), are sensitive to inhibition by acetylenic-amine derivatives such as clorgyline, deprenyl (selegiline) and pargyline but are not inhibited by carbonyl-group reagents such as semicarbazide. The so-called semicarbazide-sensitive amine oxidases (amine: oxygen oxidoreductase (deaminating) (copper-containing); EC 1.4.3.6) contain, in addition to copper, a carbonyl-type group at their active sites. This is now believed to be peptide-bound 3,4,6-trihydroxyphenylethylamine (TOPA) in its quinone form. Theses enzymes are inhibited by semicarbazide but not by the acetylenic monoamine oxidase inhibitors. Both these groups of enzymes catalyse the oxidation of amines according to the overall reaction:

$$RCH_2NR'R'' + O_2 + H_2O \longrightarrow RCHO + NHR'R'' + H_2O_2$$

Although they are involved in the metabolism of some xenobiotics, the products of the reaction, ammonia, hydrogen peroxide and an aldehyde are themselves potentially toxic. The aldehydes formed may be further metabolised by the aldehyde dehydrogenases or aldehyde oxidases to the corresponding carboxylic acids or by the aldehyde reductases to the alcohols, as shown in Figure 4.1.

Enzyme Systems that Metabolise Drugs and Other Xenobiotics. Edited by C. Ioannides.
© 2002 John Wiley & Sons Ltd

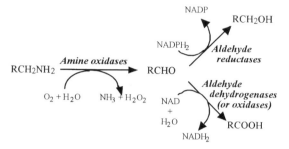

Figure 4.1 General outline scheme of amine metabolism through amine oxidases.

Since aldehydes can be quite reactive and rather difficult to detect, their existence as immediate products of xenobiotic metabolism has, in many cases, been inferred rather than directly demonstrated. Furthermore, some deductions about the involvement of amine oxidases in the metabolism of xenobiotics have been based on little more than the demonstration of the formation of the corresponding carboxylic acid or alcohol. However, since some microsomal monooxgenases can catalyse the conversion of amines to aldehydes, this cannot be regarded as compelling evidence. The fact that some amine oxidase inhibitors also affect cytochrome P-450-dependent monooxygenases (Dupont *et al.* 1987), also means that inhibitor studies should be evaluated to take account of this possibility.

As more is known about the involvements and diverse specificities of the monoamine oxidases, these will be considered first.

The monoamine oxidases

The properties and behaviour of the monoamine oxidases have been extensively reviewed (Abell and Kwan 2000; Shih *et al.* 1999; O'Brien and Tipton 1994; Wouters 1998) and this account will be restricted to factors that may affect, or be affected by, their interaction with xenobiotics.

STRUCTURE OF THE MONOAMINE OXIDASES

The monoamine oxidases oxidise aliphatic and aromatic primary amines and some secondary and tertiary amines. Two isoenzymes of monoamine oxidase, MAO-A and MAO-B, are present in most mammalian tissues. These enzymes were originally distinguished by their sensitivities to inhibition by the acetylenic inhibitors clorgyline and deprenyl and by their substrate specificities. They are now known to be encoded by separate genes, which are both located on chromosome X.

The two enzymes share a relatively high degree of sequence identity. Comparison of the sequences of the different isoenzymes among species shows that there is a greater degree of similarity between the same isoenzyme from different species than between different isoenzymes from the same species. For example, ox and human MAO-B have approximately 90% of amino acid residues identical whereas the overall identity of amino acid sequences for MAO-A and MAO-B from the same species is about 70%. There are discrete and scattered differences in amino acid sequences between ox MAO-A and MAO-B and human MAO-A and MAO-B. The high degree of sequence identity suggests that the genes for MAO-A and MAO-B are derived from a common progenitor gene. Despite the relatively high level of sequence identity between MAO-A and MAO-B from different species, there are species differences between their interactions with some substrates and inhibitors which indicated that it is unwise to attempt to extrapolate from the xenobiotic metabolising activities in experimental animals to the human situation.

SUBSTRATE AND INHIBITOR SPECIFICITIES

Some preferred substrates and inhibitors are shown in Figures 4.2 and 4.3. Typically MAO-A is inhibited by low concentrations of clorgyline and catalyses the oxidative deamination of 5-hydroxytryptamine (5-HT) whereas MAO-B is inhibited by low concentrations of (-)-deprenyl (selegiline) and is active towards benzylamine and 2-phenylethylamine (PEA). These substrate specificities are not absolute. For example, 5-HT is a substrate for MAO-B as well as MAO-A in rat brain and liver. However, in the rat brain the K_m and V_{max} values for MAO-B are six-fold higher and 9-fold lower, respectively, than those for MAO-A. Hence, relatively low concentrations of 5-HT will be oxidised essentially by MAO-A alone, whereas at very high concentrations the MAO-B form could contribute about 10% of the total activity. Similarly, 2-phenylethylamine (PEA) is a substrate for MAO-A but with a considerably higher K_m, and a V_{max} value that is 5-6-fold lower than the corresponding values for MAO-B. Thus, although the A form can contribute to the total activity at very high concentrations of PEA, only the activity of the B form is important at low concentrations of this substrate. In rat heart, which contains very little MAO-B, it has been reported to be possible to detect low levels of benzylamine oxidation by MAO-A. Such factors may be important in understanding the relative contributions of different tissues to the metabolism of xenobiotics.

In most species examined, tyramine is a substrate for both enzymes, as are dopamine and noradrenaline in the human brain. Although tyramine has been frequently regarded as a substrate for both forms of MAO, the kinetic parameters towards this substrate differ. With the enzymes from rat liver mitochondria, for example, the K_m and V_{max} values of MAO-A towards this substrate were 107 μM and 11 nmol.min^{-1}.mg protein^{-1}, respectively, whereas the corresponding values for MAO-B were 579 μM and 20 nmol.min^{-1}.mg protein^{-1}. Therefore, the proportions of the total activity towards tyramine that are contributed by each of the two forms will depend to some extent on the concentration of substrate. Thus about 32% of the total rat liver MAO activity towards 50 μM tyramine will be due to MAO B, but this proportion will rise to 57% at a tyramine concentration of 1 mM. Hence, there is no simple subdivision into specific MAO-A, specific MAO-B and mixed substrates but rather a continuum.

Figure 4.2 indicates the substrate selectivities of the monoamine oxidases for a range of substrates. These refer to the enzymes from human liver and brain. There

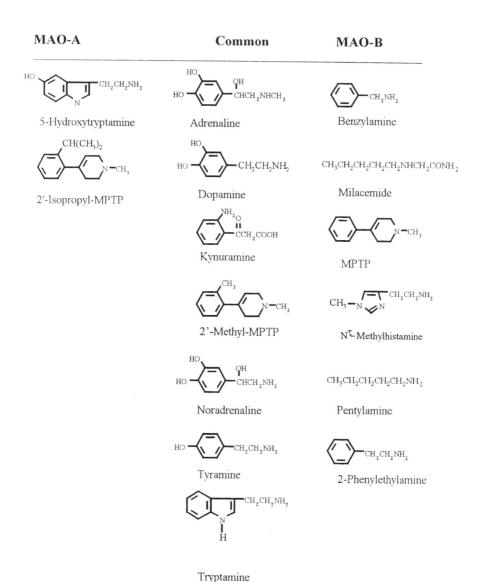

Figure 4.2 Some common monoamine oxidase substrates, arranged according to their preferences for MAO-A or MAO-B or whether they are good substrates for both isoenzymes (common). The classification refers to the monoamine oxidases from human brain and liver; the enzymes from other species may not behave identically.

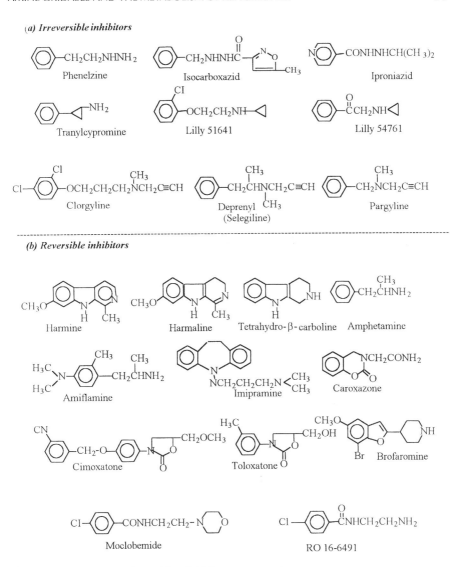

Figure 4.3 Some monoamine oxidase inhibitory compounds.

may, however, be species differences between the specificities of the enzymes and so it cannot be assumed that the behaviour will necessarily be identical in all tissues and species.

The crystal structure of MAO has not yet been determined and, although there have been several attempts at determining structure–activity relationships for substrates and inhibitors, it is still not possible to predict with confidence whether a given compound will be a substrate for either, or both, of the monoamine oxidases.

LOCATION AND DISTRIBUTION

Both MAO-A and MAO-B are tightly associated with the mitochondrial outer membrane. However, significant levels of both MAO-A and MAO-B activities have been found in the microsomal fractions. The properties of the microsomal enzymes appear to be similar to those of the mitochondrial enzyme. MAO has an almost ubiquitous occurrence in the cells of most mammalian species, the most notable exceptions being erythrocytes. Several tissues from different species express essentially only one form of MAO, for example MAO-A predominates in human placenta, rat spleen and heart whereas MAO-B dominates in human blood platelets, ox liver and kidney and pig liver. The distributions of the two isoenzymes in different human tissues are shown in Figure 4.4.

In the rat peripheral nervous system histochemical studies have shown MAO to be localised in the endothelial cells of the endoneurial vessels, the Schwann cells and the neurones in some unmyelinated axons. Microvessels from the blood-brain barrier are rich in MAO-B activity, where it presumably functions as a component of the blood-brain barrier, but a comparison between six mammalian species showed a twenty-five-fold difference in the amount of enzyme present. The two enzymes are rather evenly distributed in the different regions of human brain. Immunohistochemical and

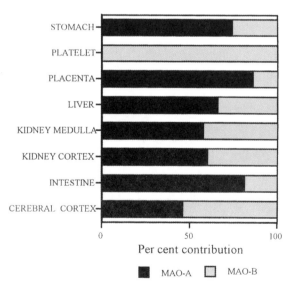

Figure 4.4 The proportions of activities towards tyramine (100 μM) of the two forms of monoamine oxidase (MAO-A and MAO-B) in different human tissues.

pharmacological studies have shown that serotonergic neurones (e.g. cells of the nucleus raphé dorsalis, nucleus centralis superior and glial astrocytes) contain predominantly MAO-B and catecholaminergic neurones (e.g. cells of the substantia nigra, locus coeruleus, nucleus subcoeruleus and the periventricular regions of the hypothalamus) contain predominantly MAO-A. Data obtained with mRNA probes are in general agreement with these results. Positron emission studies in human brain *in vivo*, after intravenous injection of the ^{11}C-labelled inhibitors, clorgyline and (-)-deprenyl showed the anatomical distribution of ^{11}C to parallel the distribution of MAO-A and MAO-B activities in human brain autopsy material.

Relatively high monoamine oxidase activities have also been shown to be present in human liver and kidney (Sullivan *et al.* 1986) and in rat kidney (Squires 1972). These values suggest that MAO may be an important enzyme in first-pass metabolism of amine xenobiotics taken orally. Since only direct administration of amine via an artery will by-pass the lung on the first pass, it is interesting to observe that the activities of MAO-A and MAO-B are significant in that organ. Perfusion studies have shown the lung to be capable of taking up and metabolising amines, particularly those acting as MAO-B substrates such as PEA (Bakhle and Youdim 1979).

REGULATION AND TURNOVER

The levels of MAO-B in blood platelets, and perhaps other tissues, are genetically regulated and appears to reflect personality-type (Garpenstrand *et al.* 2000). MAO-A and MAO-B also differ markedly in their patterns of developmental expression and genetic regulation (Strolin Benedetti *et al.* 1992a), which might affect their xenobiotic-metabolising capacities. In most mammalian species MAO-A appears before MAO-B. Changes in MAO activities during development probably reflects events of cellular differentiation as well as the changing ratios of different cell types within tissues. In the developing rat central nervous system, the increase in MAO-A follows the caudal-rostral pattern of neuronal differentiation. MAO-B activity is at relatively high levels in many tissues at birth, especially the liver. However, the level of this enzyme increases dramatically in the brain after birth probably due to the proliferation of astrocytes.

The process of ageing also appears to involve changes in levels of MAO activity. Increases in human brain MAO-B activity have been observed in several studies, possibly as a result of glial cell proliferation. Results on the effects of age on MAO-A in human brain have been conflicting in that both increases and no change have been reported. In contrast, there is little change in the MAO-A or MAO-B activities in the rat central nervous system from 2 to 28 months of age.

MAO-A, but not MAO-B, activity is elevated in reproductive tissue when levels of progesterone are high. In contrast, oestrogen-treated ovariectomised female rats show increased MAO-B activity and decreased MAO A activity in the cerebellum and in certain areas of the brain stem. Castration leads to a 3-fold increase in MAO-A activity with no change in the MAO-B activity in humans and this increase can be prevented by administration of oestradiol or testosterone. Thyroxine and adrenal corticosteroids both appear to influence MAO-A and MAO-B levels. Studies with cultured human skin fibroblasts revealed that short (6 h) exposures to several hormones, including progesterone, testosterone, triiodothyronine and the glucocorticoid, dexamethasone, reduced the total MAO activity in these cells. In contrast, progesterone, testosterone, corticosterone and dexamethasone were found to increase the synthesis of MAO-A, without affecting MAO-B, in capillary endothelial cells.

When MAO is inhibited by an irreversible inhibitor, the rate of recovery of enzyme activity can be used to estimate the rate of turnover of the enzyme in the tissues. This approach has shown the half-lives of MAO-A and MAO-B in rat liver to be similar, in the range 2.5–3.5 days. The half-life in rat brain was found to be about 10–13 days and in heart from the same species the half-life of MAO-A increased with age. The rates of turnover of the two forms of MAO in rat intestine were found to be different, with the half-life values of 2.2 and 7.5 days being reported for MAO-A and MAO-B, respectively. In a study of the turnover of MAO in baboon brain using positron-emission tomography of the deprenyl-labelled enzyme a half-life of about 30 days was determined. Such a slow rate of turnover in the primate brain would have important implications for the 'wash-out' periods used in studies on the effects of irreversible monoamine oxidase inhibitors in the clinical situation. Since either non-selective or MAO-A selective inhibitors have been widely used to treat depression, and MAO-B inhibitors, such as (-)-deprenyl (selegiline), have proven useful in the treatment of Parkinson's disease and may also have value in Alzheimer's disease, their effects must be taken into account in cases where the monoamine oxidases play a significant role in xenobiotic metabolism.

PHYSIOLOGICAL ROLES

The involvement of any enzyme in xenobiotic metabolism might be expected to impair its normal physiological functions through substrate competition. MAO-A and/ or MAO-B in peripheral tissues such as the intestine, liver, lung and placenta appear to play a protective role in the body by oxidising amines from blood or preventing their entry into the circulation. Intestinal MAO may be responsible for the inactivation of vasoactive amines such as noradrenaline, tyramine, dopamine and 5-HT of dietary origin. Liver MAO-A and MAO-B may be involved in controlling blood levels of pressor amines, including adrenaline, which have escaped deamination by MAO in the platelets or in the gut. MAO-A in the vascular endothelial cells of the lung is believed to be important in eliminating freely circulating 5-HT from the circulation, thereby protecting the heart and vascular system from the effects of the amine. MAO-A in the syncitial trophoblast layer of the placenta and MAO-B in the microvessels of the blood-brain barrier presumably have similar protective functions since these locations constitute metabolic barriers. Consistent with this are the observations that drugs which inhibit MAO induce extreme sensitivity to the dietary amine tyramine (see below) and infusion of relatively high concentrations of pargyline into amniotic fluid results in abortion.

In the central and peripheral nervous system intraneuronal MAO-A and MAO-B have been suggested to protect neurones from exogenous amines and/or to regulate levels of amines synthesised within a neurone. The low levels of MAO-A immunor-eactive material in serotonergic neurones indicates that MAO-A within these neurones might not play a major role in degrading 5-HT (5-hydroxytryptamine; serotonin). The low affinity of MAO-B for the endogenous transmitter indicates that, rather than limiting cytoplasmic levels of 5-HT, the major role of MAO-B in serotonergic neurones may be to eliminate foreign amines and minimise their access to synaptic vesicles. Hence, MAO-B in serotonergic neurones appears capable of enriching the neuroplasmic compartment with 5-HT, relative to many other amines and in conjunction with membrane and vesicle 5-HT uptake systems may contribute to the homogeneity of the neurotransmitter delivered to the synaptic cleft. Noradrenergic neurones contain both MAO-A and MAO-B. Furthermore, noradrenaline is a reasonably good substrate for both forms of the enzyme. However, since the affinity of the vesicular uptake mechanism for noradrenaline is much higher, uptake of noradrenaline into synaptic vesicles should be strongly favoured over degradation by MAO-A or MAO-B.

Xenobiotics that may be metabolised by the monoamine oxidases

Although MAO has been shown to oxidise many xenobiotics, the quantitative importance of its involvement in relation to other metabolic pathways is frequently unknown. High concentrations of xenobiotics that are MAO substrates may affect the metabolism of endogenous amines by competition. However, MAO activity appears to be present in excess in many tissues and a considerable level of inhibition (>80%) is required before substantial changes in brain monoamine concentrations, antidepressant effects and behavioural responses to administered amines, such as PEA, are observed.

XENOBIOTICS ALSO PRODUCED ENDOGENOUSLY: THE TRACE AMINES

These amines are present in relatively low amounts in the tissues and have sometimes been referred to as trace amines or micro-amines. They can be formed by the action of the enzyme aromatic-L-amino-acid decarboxylase (EC 4.1.1.28; sometimes also known as dopa decarboxylase) on the parent amino acids. Only in the case of 2-phenylethylamine, which appears to act by modulating the function of dopamine receptors, has there been much progress in defining physiological roles for these amines. They tend to be produced rather slowly and metabolised very rapidly in the tissues and the quantities arising from extraneous sources can be very much greater than the normal tissue levels.

Since bacterial aromatic-L-amino-acid decarboxylases are able to produce amines from the parent amino acids, it is not surprising that fermented and bacterially enriched foods contain a range of aromatic amines. Thus tyramine, tryptamine, histamine, 2-phenylethylamine, 5-HT and octopamine have all been found, in variable amounts, in foods and beverages such as wines, beers, cheese, salami, fermented cabbage, anchovy paste and soy sauce.

Tyramine is the most notorious of the xenobiotics metabolised by MAO and also the most well studied. This is because of the hypertensive reaction that occurs after ingestion of tyramine-rich foods or beverages by patients under therapy with MAO inhibitors (Blackwell and Marley 1969). Ingested amines such as tyramine are normally metabolised in the peripheral organs with the gastro-intestinal tract appearing to play the dominant role in this first-pass metabolism (Davis *et al.* 1984; Hasan *et al.* 1988). As a result of this, very little ingested amine reaches the circulation. However, if MAO is inhibited, the ingested tyramine will enter the circulation from where it is actively taken up by peripheral adrenergic neurones, displacing stored noradrenaline and giving rise to a hypertensive response that can be fatal (Blackwell 1963). Since some cheeses are particularly rich in tyramine, this effect has become known as the 'cheese reaction'. Because of the widespread occurrence of tyramine in foods and beverages, the diets of patients being treated with such MAO inhibitors has had to be carefully restricted.

Lists of foods and beverages which may be high in tyramine have been complied (Sen 1969; Steward 1976; Da Prada *et al.* 1988). It should be noted, however, that the tyramine content of many foods and beverages can be extremely variable. Thus, Dostert (1984) has shown the tyramine content of beef liver to vary between 5 and 274 mg.g^{-1} depending on the method and period of storage, and Da Prada *et al.* (1988) reported the tyramine content of some cheeses to depend on the state of ripeness. We have found values between 0.19 and 1.31 mg.l^{-1} for different samples of beer from the same producer (Wheatley and Tipton 1987). Such variability may explain wide discrepancies in the reported tyramine content of certain products. For example, earlier reports that Chianti wine had extremely high levels of this amine have not been subsequently confirmed (Da Prada *et al.* 1988).

Hypertensive responses can be elicited by administration of tyramine alone to experimental animals or to human subjects but this effect is greatly potentiated following inhibition of monoamine oxidase. The pressor responses of subjects treated with monoamine oxidase inhibitors can be quite variable, perhaps as a result of individual variations in transport or metabolic efficiency (Bieck *et al.* 1988), and/or to the wide variability of the tyramine content of different products, discussed above. Because of this there is a risk that a patient treated with a monoamine oxidase inhibitor may experiment, by trying different proscribed foods and beverages and find no great adverse effects, only to consume a tyramine-rich sample at a later stage.

MAO-A activity predominates in rat and human (see Figure 4.4) intestine (Strolin Benedetti *et al.* 1983a; Hasan *et al.* 1988). Studies with dog intestinal loops (Davies *et al.* 1984) and with everted intestinal preparations (Hasan *et al.* 1988; Anderson *et al.* 1993) have shown that tissue to be capable of deaminating 70–80% of the tyramine during transport. Selective inhibitors of MAO-A, such as clorgyline, give a cheese reaction that is no less pronounced than that observed in the presence of a non-selective irreversible inhibitor of MAO-A and MAO-B. In contrast, but consistent with the relatively low MAO-B in the gastro-intestinal tract, selective inhibitors of MAO-B, such as *l*-deprenyl, do not potentiate the effects of dietary tyramine. Since MAO-A inhibitors are effective antidepressants whereas MAO-B inhibitors are not, the problem of producing dietary-safe antidepressant MAO inhibitors has been addressed by the development of reversible and competitive MAO-A inhibitors that can be administered in sufficient doses to give an antidepressant effect while being displaced by high concentrations of tyramine. Studies with several reversible MAO-A inhibitors, such as moclobemide and brofaromine, have indeed shown that their ability to cause a cheese reaction can be considerably lower than that observed with an irreversible inhibitor. The combination of a monoamine oxidase inhibitor with an inhibitor of presynaptic amine transport, such as amitriptyline, has been reported to reduce the cheese effect (Pare 1986). Attempts have also been made to combine a monoamine oxidase inhibitor and an amine uptake inhibitor in the same molecule (Tipton *et al.* 1982). However, the pharmacokinetic behaviour of such hybrid molecules might be unsatisfactory if based on an irreversible MAO inhibitor and a reversible uptake inhibitor, such as one of the tricyclic antidepressants.

Long-term administration of MAO-inhibitory antidepressants can result in hypotension. This is believed to result from the gradual accumulation of tyramine and octopamine (see below), derived from it through the action of dopamine-β-hydroxylase (EC 1.14.17.1), which act as false transmitters.

2-Phenylethylamine (PEA) is present in many dietary sources and in particular at high concentrations in some, but not all, types of chocolate. It readily passes through membranes but is efficiently metabolised by both MAO and SSAO (semicarbazide-sensitive amine oxidase). The contribution of lung to the presystemic elimination of PEA was studied by Worland and Ilett (1983). The deamination of 10 μM PEA by whole tissue homogenates from rat suggested that the liver would have the major effect on its presystemic elimination, with a much smaller, approximately equal, contribution from the intestine and lungs. However, a comparison of the areas under the blood concentration–time curve after intra-arterial, intravenous, intraportal and intraduodenal administration of PEA (4 mmol.kg^{-1}) to the rat indicated that first-pass elimination of PEA was largely attributable to the intestine and lung, with a relatively small hepatic contribution.

Tryptamine is a substrate for both monoamine oxidases in human tissues, unlike 5-HT, which is a relatively specific MAO-A substrate. Since the parent amino acid may be administered in relatively high concentrations, as a nutritional supplement, an antidepressant and to assist sleeping, the possibility of increased bacterial and endogenous tryptamine formation cannot be ignored, but there are no quantitative data available on this.

Octopamine acts as a neurotransmitter in some invertebrates and specific octopamine receptors have been cloned from such sources. It has been used as an adrenergic drug but specific receptors have not yet been identified in mammalian systems. It appears to exert its effects by acting as a β_3-adrenoceptor agonist with a very much weaker effect at α_2-receptors. It is metabolised by both monoamine oxidases, but selective inhibition of rat brain MAO-A by clorgyline indicated it to be preferentially oxidised by MAO-A (Lyles 1978; Suzuki *et al.* 1979). As with phenylethylamine and phenylethanolamine, introduction of an hydroxyl group in the β-position in tyramine results in a decreased apparent affinity for MAO-A ($K_m \approx 115$ μM for tyramine and ~ 455 μM for octopamine, with rat brain).

Histamine is not a substrate for MAO. However it may be methylated to N^T-methylhistamine which is a substrate for MAO-B (Figure 4.2).

XENOBIOTIC PRIMARY AMINES (SEE FIGURES 4.2 AND 4.5)

Benzylamine, which is added to some mouthwashes and is also used in organic syntheses, is a highly selective substrate for monoamine oxidase-B and is also oxidised by SSAO. Its metabolite benzaldehyde is a narcotic at high concentrations and the further-oxidation product benzoic acid has antifungal properties. However, there is no

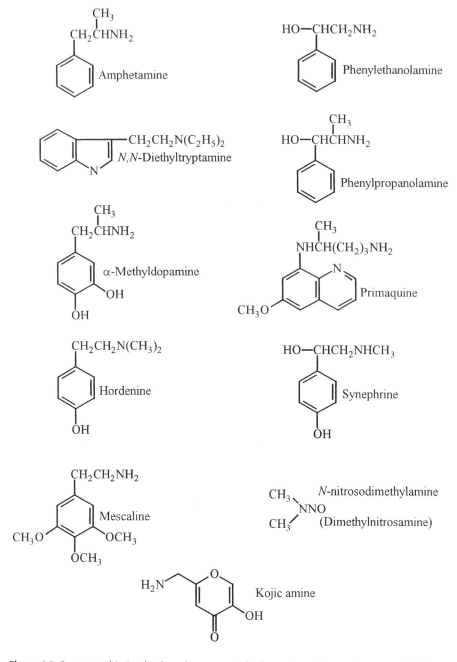

Figure 4.5 Some xenobiotics that have been reported to be amine oxidase substrates or inhibitors.

indication that sufficient quantities of this amine are consumed for these metabolite effects to be significant.

2-Phenylethanolamine is used in commercial processes and also as a topical vasoconstrictor. It is a selective substrate for MAO-B (Suzuki *et al.* 1979). The introduction of an hydroxyl group in the β-position results in it being more selective as a MAO-B substrate than 2-phenylethylamine, as a result of a decreased K_m value. In studies with the *R*- and *S*-enantiomers of 2-phenylethanolamine Williams (1977) reported that MAO-A would act only on the *R*-enantiomer whereas MAO-B would oxidise both enantiomers.

2-Phenylpropanolamine is present in many decongestants and cough medicines. Although tyramine is normally associated with the cheese reaction 2-phenylpropanolamine has also been shown to give rise to a hypertensive reaction in patients treated with monoamine oxidase inhibitors (Dollery *et al.* 1984).

Mescaline: MAO has been shown to play a significant role in the metabolism of mescaline in the mouse, since treatment with the non-selective MAO inhibitors iproniazid and tranylcypromine both resulted in a diminished formation of the acid metabolite, 3,4,5-trimethoxyphenylacetic acid and an increased formation of the alternative metabolite, *N*-acetylmescaline (Shah and Himwich 1971).

Primaquine: This antimalarial drug is metabolised to the corresponding carboxylic acid, carboxyprimaquine. Both MAO and the cytochrome P-450 monoxygenases can catalyse the oxidation of primaquine, but studies with rat liver homogenates indicated that inhibition of MAO with pargyline substantially reduced primaquine metabolism, whereas the cytochrome P-450 monooxygenase inhibitor SKF-525A (proadifen) had a lesser effect (Constantino *et al.* 1999). Thus it appears that, at least *in vitro*, MAO plays the dominant role in the oxidation of this drug.

Amphetamine and other α-methyl substituted amines are not substrates for the monoamine oxidases but are relatively effective reversible inhibitors that are selective towards MAO-A. Inhibitory potency towards MAO-A is stereoselective with the *S*-enantiomer of amphetamine being a more potent inhibitor of rat liver MAO-A than the *R*-enantiomer (Mantle *et al.* 1976), Earlier work, using selectively deuterium-substituted amines, showed that MAO abstracted the pro-*R*-hydrogen (H_{Re}) from the α-carbon of primary amine substrates (Belleau and Moran 1963) and, more recently, Yu *et al.* (1986) and Yu and Davis (1988) (see also Yu 1988) have shown both MAO-A and MAO-B to exhibit the same stereospecificity. The methyl group in *S*-amphetamine occupies the position of the non-abstracted hydon (H_{Si}) in the parent, non-methylated, substrate (2-phenylethylamine). However, the *R*-enantiomer of α-methylbenzylamine, in which the methyl group is in the position of the abstracted hydrogen (H_{Re}) in benzylamine, was reported to be a more potent inhibitor of both MAO-A and MAO-B from rat brain than the *S*-enantiomer (Arai *et al.* 1986). In the case of MAO-B there appears to be little difference in the potencies of the two enantiomers of either amphetamine (Dostert *et al.* 1989) or α-methylbenzylamine (Silverman 1984).

α-Methyldopa was introduced as an antihypertensive drug because of its ability to inhibit aromatic-L-amino-acid decarboxylase and hence decrease catecholamine formation. However, it is a substrate for that enzyme leading to the formation of α-methyldopamine which functions as a false transmitter and, *inter alia*, an inhibitor of MAO.

XENOBIOTICS METABOLISED BY WAY OF PRIMARY AMINES

Propranolol, a β-adrenoceptor blocker used as a cardiac depressant, has not been shown to be a substrate for MAO. However, it is N-dealkylated to the primary amine N-desisopropylpropranolol (DIP) by microsomal enzymes. The aldehyde product of the MAO-catalysed oxidation of DIP by rat liver mitochondria has been identified *in vitro*, as its O-methyloxime after reaction with methoxyamine hydrochloride, and its formation was shown to be inhibited by high, non-selective concentrations of the MAO inhibitor pargyline (Goldszer *et al.* 1981). Figure 4.6 summarises the metabolism of propranolol, by chemical decomposition of the aldehyde or by its further *in vitro* metabolism to the glycol and the acid derivative. Direct formation of the aldehyde from propranolol with loss of isopropylamine has also been shown to occur *in vitro*, but it appears to make a minor contribution to the overall process of oxidative deamination (Bakke *et al.* 1973). As a competitive MAO inhibitor propranolol was found to be selective towards MAO-A with an IC_{50} value towards 5-HT deamination in rat brain homogenates of 260 μM (Milmore and Taylor 1975). These results would be consistent with the compound being a selective substrate for MAO-A, but further studies would be necessary to establish whether that was the case. It appears that the S-enantiomer of propranolol is a better substrate for MAO than the R-enantiomer (Nelson and Bartels 1984). S-propranolol has the same absolute configuration as R-(−)-noradrenaline, which is a substrate predominantly for MAO-A in the rat (Strolin Benedetti *et al.* 1983b). It has not been directly demonstrated whether DIP is a selective substrate for either form of MAO.

A substantial contribution of MAO to propranolol metabolism has also been demonstrated *in vivo*. The carboxylic acid, 2-hydroxy-3-(1′-naphthyloxy)-propionic acid (Figure 4.6) is an important metabolite of propranolol in the human and in several other species, and the glycol metabolite has also been detected (Bargar *et al.* 1983; Walle *et al.* 1983). Side-chain oxidation and glucuronidation have been shown to be the dominant metabolic processes in the dog and human, whereas ring oxidation is dominant in the rat and hamster (Bargar *et al.* 1983; Walle *et al.* 1983). However, it is likely that the relative contributions of the various primary metabolic pathways (ring oxidation, side-chain oxidation and glucuronidation) to the overall propranolol metabolism will also depend on the dose.

Alprenolol and oxprenolol: The *in vivo* metabolism of alprenolol and oxprenolol (Figure 4.6) in the dog has been studied by Walle *et al.* (1981) and that of oxprenolol in the human by Dieterle *et al.* (1986). In the dog, the metabolic fates of alprenolol and oxprenolol were similar to that of propranolol. However, after oral administration of oxprenolol to humans, the direct O-glucuronidation pathway was found to predominate over the ring oxidation and side-chain oxidation pathways. The aldehyde, derived from the primary-amine metabolite, has also been detected following incubation of rat liver mitochondria with oxprenolol *in vitro* (Goldszer *et al.* 1981).

Pronethalol, timolol and other β-blockers: Tocco *et al.* (1980) showed the major metabolic pathway of timolol (Figure 4.6) in the dog to involve oxidation of the basic oxypropanolamine side chain to yield 2-hydroxy-3-([4-(4-morpholinyl)-1,2,5 thia-diazol-3-yloxy)propanoic acid, a pathway that is likely to involve MAO. However, in rat and human the two major metabolites of timolol are more highly oxidised,

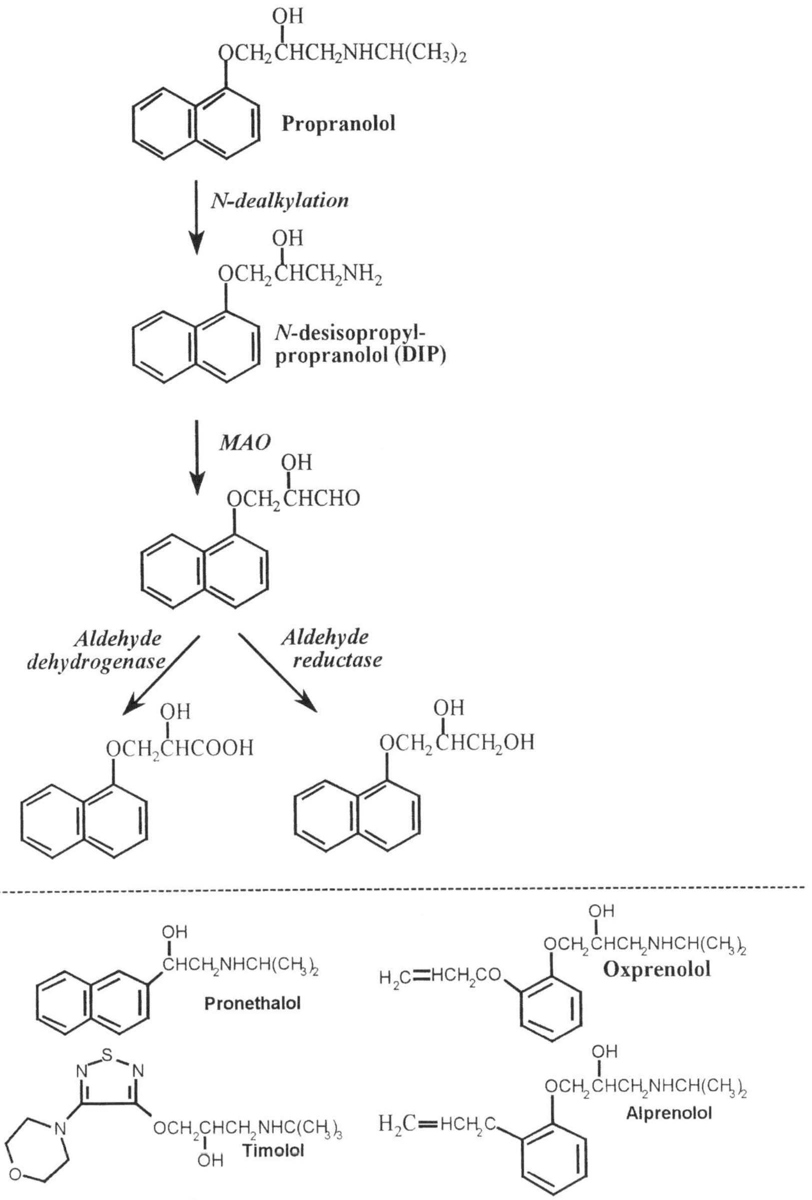

Figure 4.6 Involvement of MAO in the metabolism of propranolol. The structures of some other β-blockers, that are referred to in the text, are also shown.

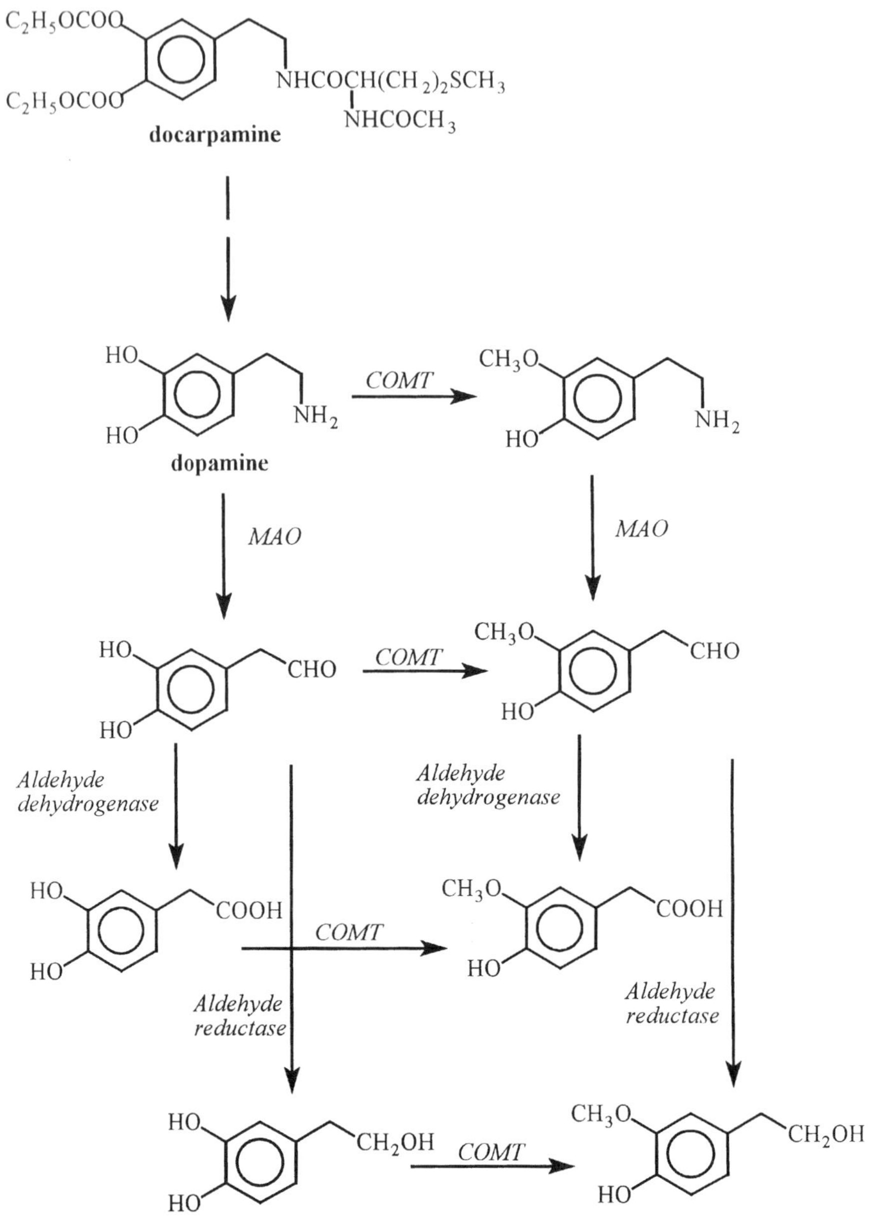

Figure 4.7 Formation of dopamine from docarpamine and its subsequent metabolism. The *O*-methylation reactions catalysed by catechol-*O*-methyltransferase (EC 2.1.1.6; COMT) have *S*-adenosylmethionine as the methyl donor, which is converted to *S*-adenosylhomocysteine.

morpholine ring-opened, species in which the oxypropanolamine side chain is unchanged (Figure 4.6). Bond and Howe (1967) also reported 2-naphthy-1-glycolic acid to be a major metabolite of pronethalol (Figure 4.6) *in vivo*. The probable involvement of MAO in the *in vivo* metabolism of K 5407 and several other β-blocking agents has also been demonstrated (Goldaniga *et al.* 1980).

Docarpamine (*N*-(*N*-acetyl-L-methionyl)-*O,O-bis* (ethoxycarbonyl) dopamine) is used as an orally active prodrug for the delivery of dopamine which can then be oxidatively deaminated by MAO (Figure 4.7), as indicated by the excretion of the dopamine metabolites 3,4-dihydroxyphenylacetic acid (DOPAC) and homovanillic acid (HVA) in human urine (Yoshikawa *et al.* 1990). Studies in dog and rat have shown that the main site of catechol ester hydrolysis is the small intestine, whereas amide hydrolysis occurs in the liver (Yoshikawa *et al.,* 1995). The liver also catalyses extensive sulphoconjugation, although such sulphoconjugated dopamine is thought to be a possible precursor of active free dopamine in plasma (Tano *et al.* 1997).

SECONDARY AND TERTIARY *N*-METHYL AMINES

The MAO-catalysed oxidation of such compounds will yield the corresponding aldehyde plus methylamine. Methylamine is not a substrate for either MAO isoenzyme but, as discussed below, it is oxidatively deaminated by SSAO.

Synephrine is a vasopressor and adrenergic drug which has a somewhat lower K_m value for MAO (~ 250 µM) than its parent primary amine octopamine. Like octopamine it is oxidised by both forms of MAO, with a preference for MAO-B.

Epinine and ibopamine: dopamine has been used in the treatment of congestive heart failure (Goldberg 1989) because of its inotropic and vasodilating properties, but must be administered intravenously because it is poorly absorbed after oral administration (Goldberg 1974). This has led to the development of orally active derivatives of dopamine, such as ibopamine, which is hydrolysed by tissue esterases to epinine (Figure 4.8). Epinine (*N*-methyldopamine) is one of the few modifications of the neurotransmitter molecule dopamine (DA) that still retains agonist activity at the DA1 receptor. It is a substrate for both MAO-A and MAO-B with similar K_m and limiting velocity (V_{max}) values, although the apparent affinities for MAO-A and MAO-B ($K_m \approx 1000$ and 900 µM, respectively, in rat liver) are lower than those for dopamine ($K_m \approx 150$ and 290 µM, respectively). There is no evidence that ibopamine is oxidised by MAO (Strolin Benedetti *et al.* 1998).

Hordenine, the *N,N*-dimethyl derivative of tyramine, has a rather lower apparent affinity for rat liver MAO ($K_m \approx 1449$ µM) than synephrine and the use of selective MAO inhibitors has suggested it to be a selective substrate for MAO-B (Barwell *et al.* 1989).

N,N-Dimethyltryptamine is an hallucinogenic compound that can be isolated from the leaves of some plants, such as *Prstonia Amazonica*. It is a specific substrate for MAO-B whereas the monomethyl derivative is a preferentially oxidised by MAO-A and, as mentioned above, the demethylated product tryptamine is metabolised equally well by both isoenzymes.

Citalopram: This antidepressant drug, a selective serotonin reuptake inhibitor (SSRI), may be demethylated to its *N*-desmethyl- and *N,N*-didesmethyl- (primary amine)

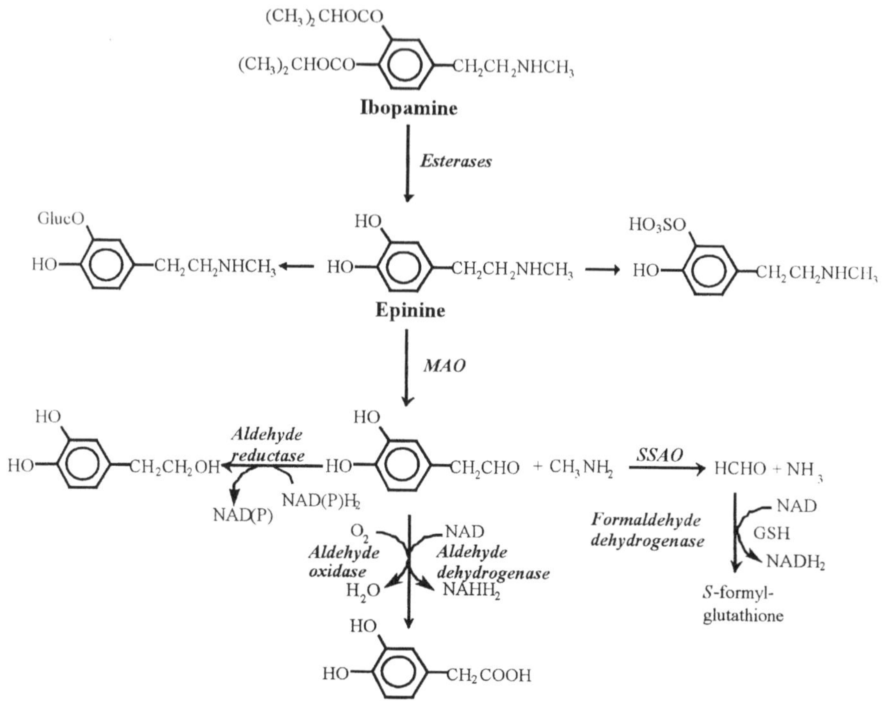

Figure 4.8 Formation of epinine from ibopamine and its subsequent metabolism. The *O*-methylated metabolites (see Figure 4.7) are omitted for clarity.

derivatives by members of the cytochrome P-450 family of monooxygenases. The role of MAO in its metabolism has been studied in detail by Rochat *et al.* (1998). Both the desmethyl-derivatives as well as citalopram itself are substrates for human liver MAO (Figure 4.9) and the predominant role of this enzyme in their metabolism *in vitro* was suggested by the substantial level of inhibition (>90%) by the non-selective MAO inhibitor phenelzine. The urinary excretion of the carboxylic acid metabolites of citalopram, *R*- and *S*-citaloprampropionic acid, was detected after administration of racemic citalopram orally to human subjects, although the extraction procedures used involved the hydrolysis of any glucuroconjugates formed before analysis. The forma-tion of *R*-citaloprampropionic acid from racemic citalopram in human liver microsomal preparations, which also contained mitochondria, was shown to be substantially inhibited (87%) by low concentrations of clorgyline but to be rather insensitive (10.9%) to inhibition by low concentrations of *l*-deprenyl, indicating that MAO-A plays the dominant role in the oxidation of, the *R*-component of this substrate. MAO-A was also dominant in the formation of *R*-citaloprampropionic acid formation from racemic *N*-desmethylcitalopram (65% inhibition by clorgyline) with about 16% inhibition resulting from *l*-deprenyl. In contrast, it appears that MAO B plays the major

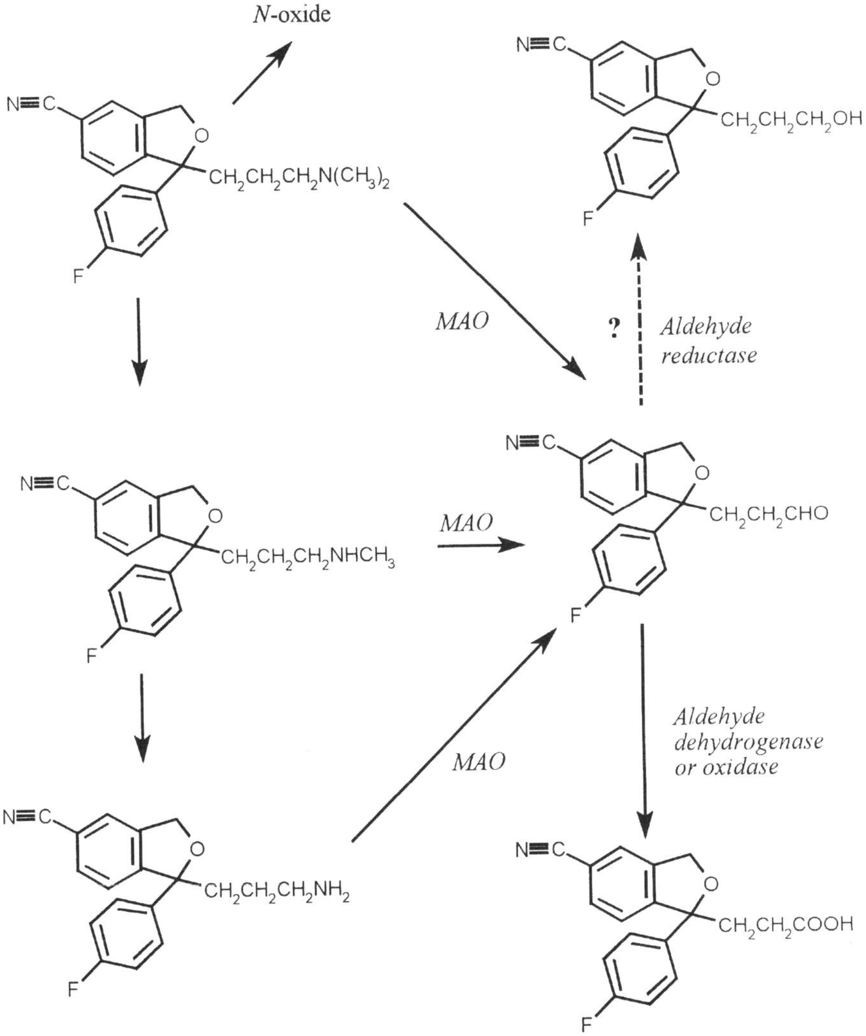

Figure 4.9 Metabolism of citalopram showing the involvement of monoamine oxidase. The primary alcohol product, which would be formed by aldehyde reductase or perhaps alcohol dehydrogenase, has not been detected.

part in metabolism of the primary amine metabolite, *N*,*N*-didesmethylcitalopram, with low concentrations of *l*-deprenyl and clorgyline resulting in 48.8% and 27.4% inhibition of *R*-citaloprampropionic acid formation, respectively. MAO-B also appears to be more important in the metabolism of the *S*-component of racaemic *N*,*N*-didesmethylcitalopram with the formation *S*-citaloprampropionic acid being insensitive to low concentrations of clorgyline but approximately 93% inhibited by low concentrations of *l*-deprenyl. The corresponding figures for the involvements MAO-A and MAO-B in the formation of *S*-citaloprampropionic acid from racemic *N*-desmethylcitalopram were 19.4% and 67.4% respectively, and 55% and 35.5% respectively from racemic citalopram.

A relatively minor role of the cytochrome P-450 monooxygenases in the formation of *S*-citaloprampropionic acid from of *S*-citalopram was indicated by the observation that this process was inhibited by about 20% by the non-selective cytochrome P-450 monooxygenase inhibitor proadifen. This compound also inhibited to some extent the formation of *S*-citaloprampropionic acid from the *S*-enantiomers of *N*-desmethylcitalopram and *N,N*-didesmethycitalopram, by about 25% and 34% respectively. However, these results are not easy to interpret, since proadifen also has a weaker inhibitory effect towards aldehyde oxidase (Robertson and Bland 1993; Watanabe *et al.* 1995). In contrast to these results with the *S*-enantiomers of citalopram and its desmethyl metabolites, the metabolism of the corresponding *R*-enantiomers to *R*-citaloprampropionic acid was increased, up to 2.5 times, by proadifen, suggesting additional possible complexities in the metabolism of this enantiomer.

Triptans: The triptan family of compounds includes 5-HT receptor agonists that have been found to be of value in the relief of migraine. They are believed to act through the stimulation of presynapitc 5-HT$_{1D}$ and related receptors to inhibit the release of calcitonin gene-related peptide and other peptides. In the case of sumatriptan (Figure 4.10) the major excreted metabolite in the human is the corresponding indoleacetic acid derivative and *in vitro* studies with human liver indicated that [^{14}C]-labelled sumatriptan was metabolised by MAO-A; there was no evidence of cytochrome P-450 monooxygenase involvement in its metabolism (Dixon *et al.* 1994). There are, however, species differences in the metabolism of sumatriptan. In humans, the indoleace-

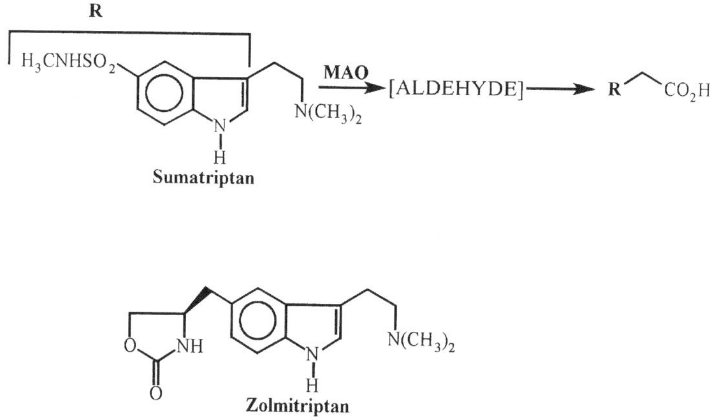

Figure 4.10 The metabolism of sumatriptan (Imitrex) by monoamine oxidase. The structure of zolmitriptan (zomig) is also shown for comparison.

tic acid metabolite is excreted partly as a glucuronide, whereas in animals such conjugation could not be detected. Furthermore, demethylation of the sulphonamide side chain of the drug occurred in rodent and lagomorph species but not in the human (Dixon et al. 1993).

There is also extensive first-pass metabolism of rizatriptan in humans. The major urinary metabolite was detected as being triazolomethyl-indole-3-acetic acid, although small amounts of urinary rizatriptan-$N(10)$-oxide, 6-hydroxy-rizatriptan and 6-hydroxy-rizatriptan sulphate could also be detected after high dosage (Vyas et al. 2000). Studies with the MAO-A selective inhibitor moclobemide indicated MAO-A to be the major metabolising enzyme in humans (Van Haarst et al. 1999). Three major metabolites of zolmitriptan, N-desmethyl-zolmitriptan, zolmitriptan N-oxide and the indoleacetic acid derivative, have been found in vivo. Studies with isolated human hepatocytes and liver microsomes have shown the conversion of zolmitriptan to N-desmethyl-zolmitriptan to be catalysed by CYP1A2 and not by MAO, whereas MAO-A is responsible for further metabolism of N-desmethyl-zolmitriptan (Wild et al. 1999).

Dimethylnitrosamine (N-nitrosodimethylamine) has been reported to act as a substrate for MAO (Lake et al. 1982), but in view of the difference between this substrate and any other known MAO substrate, this requires further investigation.

MAO inhibitors as substrates and MAO substrates as inhibitors

Many specific irreversible inhibitors are intrinsically unreactive compounds that are converted into reactive intermediates by the action of the enzyme itself. The inhibitor first forms a non-covalent complex with the active site of the enzyme and subsequent reaction within that complex leads to the generation of a reactive species that then reacts with the enzyme to form the irreversibly inhibited species. Inhibitors of this type are known as mechanism-based, enzyme-activated, k_{cat}, or suicide inhibitors. They can show a high degree of specificity towards a target enzyme because the generation of the effective inhibitory species from an essentially unreactive compound involves part of the catalytic function of the enzyme itself. Furthermore, the lack of intrinsic reactivity minimises the possibility of unwanted reactions with other tissue components.

Figure 4.11 compares the irreversible inhibition of MAO by the acetylenic inhibitors with the oxidation of a normal amine substrate. In the latter case, the reduction of the enzyme-bound FAD results in the formation of an imine which is rapidly hydrolysed to ammonia plus the corresponding aldehyde. A similar oxidation of the acetylenic amine results in the formation of a doubly unsaturated derivative which acts as a Michael acceptor in reacting with an electron-rich group on the enzyme, in this case position 5 of the isoalloxazine ring of the covalently bound FAD, to form a flavocyanine. Kinetically the inhibition pathway can be represented by the mechanism (see Fowler et al. 1982)

$$E + I \underset{k_{-1}}{\overset{k_{+1}}{\rightleftharpoons}} E.I \xrightarrow{k_{+2}} E.I^* \xrightarrow{k_{+3}} E\text{-}I \qquad (4.1)$$

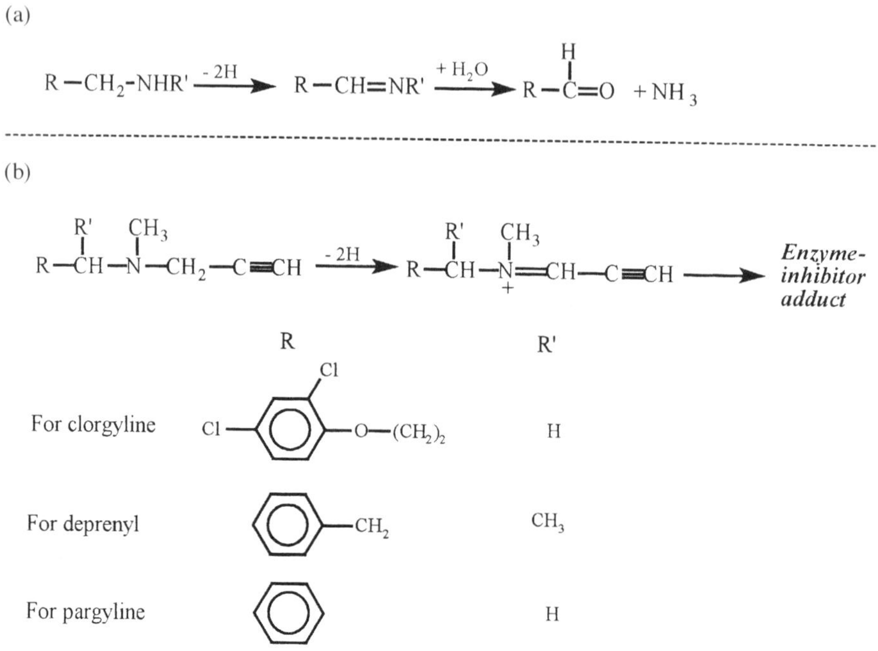

(a)

(b)

Figure 4.11 The oxidation of (a) substrates and (b) acetylenic (mechanism-based) irreversible inhibitors by MAO.

in which the inhibitor first forms a non-covalent complex (E.I) with MAO, and that complex is then transformed, by the action of the enzyme, into an activated species (E.I*) which then reacts with a group on the enzyme to form the irreversibly inhibited species (E-I).

Because mechanism-based inhibitors behave like substrates in binding to the active site of the enzyme and being converted to the reactive species through a process resembling the normal catalytic process of the enzyme, it is not surprising that it is possible for a proportion of the reactive species (EI*) to break down to form product. Compounds that behave in this way have sometimes been referred to as 'suicide-substrates'. In such cases, the formation of product and the mechanism-based inhibition of the enzyme will be competing reactions according to the following extension of the system shown above (Tipton 1989):

$$E + I \underset{k_{-1}}{\overset{k_{+1}}{\rightleftharpoons}} E.I \xrightarrow{k_{+2}} E.I^* \xrightarrow{k_{+3}} E\text{-}I$$

$$\downarrow k_{cat}$$

$$E + \text{Products}$$

(4.2)

The relative effectiveness of a compound to act as a substrate or inhibitor can be defined as the *partition ratio* (*r*), which is the number of mol of product that is produced by 1 mol of enzyme before it is completely inhibited. Thus if a known amount of enzyme, [E], is incubated with excess of the substrate/inhibitor the activity of the enzyme will steadily decrease with time and the amount of product formed when all the enzyme has been inhibited, $[P_\infty]$, can be used to determine the partition ratio, according to the relationship:

$$r - \frac{[P_\infty]}{[E]} = \frac{k_{cat}}{k_{+3}} \tag{4.3}$$

The number of mol of inhibitor necessary to inactivate 1 mol of enzyme will thus be given by $(1 + r)$ and so the initial concentration of inhibitor must be greater than $(1 + r)[E]$ for complete inhibition of the enzyme. If the inhibitor concentration is less than this, the amount of product formed will be given by $r.[E_i]$, where $[E_i]$ represents the concentration of inhibited enzyme.

Clearly compounds that act like this can either be regarded as substrates that also act as inhibitors, if the value of *r* is large, or as inhibitors that also act as substrates, if the value of *r* is small. However, the distinction is essentially arbitrary and there is no agreed criterion as to what constitutes a low, inhibitor, value of *r* and what constitutes a high, substrate, one. Thus the distinction made below is essentially arbitrary.

SUBSTRATES THAT ARE ALSO INHIBITORS

1-Methyl-4-phenyl-1,2,3,6-tetrahydropyridine (MPTP)

MPTP is a selective neurotoxin that causes a Parkinsonism-like syndrome in humans and some other species (for review see Tipton and Singer 1993) as a result of degeneration of the nigrostriatal pathway. MPTP is oxidised to the 1-methyl-4-phenylpyridinium ion (MPP$^+$), which is the active neurotoxin. In the brain the oxidation of MPTP takes place outside the dopaminergic nerve terminals since there appears to be little or no MAO-B within these structures (Westlund *et al.* 1985; O'Carroll *et al.* 1987). The active transport of MPP$^+$ into the dopaminergic nerve endings then results in it being concentrated there.

MPP$^+$ is an inhibitor of mitochondrial energy metabolism, at the level of NADH oxidase (Complex 1). Inhibition of mitochondrial function is not specific to brain mitochondria, but the specific uptake by dopaminergic nerve terminals maintains a sufficiently high concentration for sustained toxicity to result in cell death.

MPTP is oxidised to MPP$^+$ in a two-step process in which the dihydropyridine (1-methyl-4-phenyl-1,2-dihydropyridinium; MPDP$^+$) is formed as an intermediate (Figure 4.12). The oxidation of MPTP to MPDP$^+$ is similar to the initial oxidation step, imine formation, in the oxidation of primary amines by MAO. MAO can also catalyse the conversion of MPDP$^+$ to MPP$^+$ although this reaction can also occur non-enzymically (Singer *et al.*, 1985, 1986). However, MPTP also behaves as a time-dependent irreversible inhibitor as well as a substrate for MAO-B, according to the reaction shown in equation (4.2), which will limit the production of MPP$^+$ (Kreuger *et al.* 1990; Tipton *et al.*, 1986). It also appears that this enzyme may show

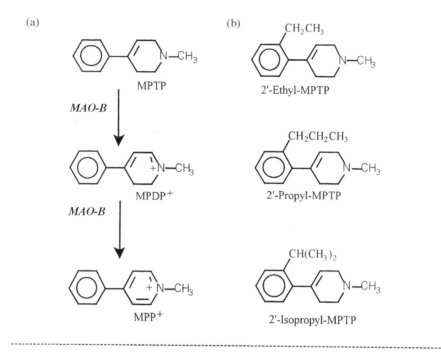

Figure 4.12 (a) The oxidation of 1-methyl-4-phenyl-1,2,3,6-tetrahydropyridine (MPTP) by MAO. (b) The structures of some analogues that are also substrates. (c) Some other metabolites that have been shown to be formed in liver. PTP = (4-phenyl-1,2,3,6-tetrahydropyridine; desmethyl-MPTP), which is also a substrate for MAO, and the carbinolamime metabolite are believed to be formed through the action of cytochrome P-450 monooxygenases on MPTP, whereas a flavin-dependent monooxygenase converts this compound to the N-oxide. The lactam and pyridone metabolites arise from the action of aldehyde oxidase on MPDP$^+$ and MPP$^+$, respectively.

large species-dependent differences in partition ratios with MPTP (Tipton *et al.* 1993; Sullivan and Tipton 1992) which may be one of the contributory factors to the wide species differences in sensitivities towards MPTP toxicity that have been reported (Tipton and Singer 1993).

Inhibitors of MAO-B, such as *l*-deprenyl, protect against the neurotoxicity of MPTP. The discovery that MAO-B was involved in the conversion of MPTP to the active toxin resulted in speculations that an endogenous toxin that was similarly formed through the action of this enzyme was involved in the aetiology of idiopathic Parkinson's disease. However, there is no *a priori* reason to believe that MAO-B is involved in the development of this disease. Studies on the interactions of MPTP analogues with MAO have shown that substitution of alkyl groups at the 2' position of the benzene ring (Figure 4.12) increases the efficiency of oxidation by MAO-A, and lengthening the alkyl group in this position beyond CH_3 decreases the reactivity with MAO-B (Youngster *et al.* 1989). Thus the substitution of an ethyl or an isopropyl group in the 2' position effectively converts a substrate for MAO-B into one that is preferentially oxidised by MAO-A. As might be anticipated from these results, the neurotoxicity of MPTP is prevented by pretreatment with deprenyl but not with clorgyline, that of 2'-methyl-MPTP requires both clorgyline and deprenyl to afford protection and clorgyline alone protects against the neurotoxicity of 2'-ethyl-MPTP (Heikkila *et al.* 1988). Furthermore, although some MAO-B inhibitors, such as *l*-deprenyl, have been shown to protect nerves against, or rescue them from, a number of potentially toxic insults, this occurs at concentrations below those necessary to inhibit the enzyme (for reviews see Tipton 1994; Olanow *et al.* 1998).

Although the conversion of MPTP to MPP^+ appears to be the dominant metabolic process in brain, there are alternative catabolic pathways in liver. Cashman and Ziegler (1986) showed that microsomal oxidation reactions competed with mitochondrial oxidation in rat liver. The microsomal cytochrome P-450 monooxygenases converted MPTP to nor-MPTP (PTP, desmethyl-MPTP), a compound that is not neurotoxic (Sullivan and Tipton 1990, 1992) whereas the primary product of the microsomal flavin-containing monooxygenase activity was MPTP-*N*-oxide. The kinetic parameters for the *N*-oxide formation from MPTP by rat liver microsomes were reported to be: $K_m = 45$ μM and $V_{max} = 4.8$ nmol.min^{-1}.mg of protein^{-1}. For comparison, K_m values for the oxidation of MPTP by monoamine oxidase B from rat and human liver are within the range 65–150 μM and the maximum velocities are similar to those for *N*-oxide formation (Tipton *et al.* 1986; Gessner *et al.* 1986). Cashman and Ziegler (1986) also examined the metabolism of MPTP in whole homogenates of human liver biopsy samples that were supplemented with NADPH. They found MPTP to be oxidised to approximately equal extent to MPTP-*N*-oxide and MPP^+, with nor-MPTP being formed to a lesser extent. These metabolites of MPTP formed *in vitro* by liver tissues are shown in Figure 4.12. The existence of alternative metabolic pathways for MPTP in liver will presumably also contribute to the insensitivity of peripheral tissues to permanent damage from MPP^+. In spite of the observation that the liver microsomal conversion of MPTP to nor-MPTP is some 20 to 25 times slower than *N*-oxide formation (Cashman and Ziegler 1986), the involvement of the cytochrome P-450 system in the metabolic process *in vivo* is illustrated by the significant decrease in the acute neurotoxicity of MPTP that occurs after induction of this system by pretreatment of rats with phenobarbital (100 mg/kg, intraperitoneally for three days; Strolin Benedetti, unpublished results).

Monoamine oxidase has also been reported to catalyse the oxidation of N-methylated 1,2,3,4-tetrahydroisoquinolines and of their 6,7-dihydroxy derivatives to the corresponding quinolinium ions, which have been suggested to be endogenous Parkinsonism-inducing neurotoxins (Naoi *et al.* 1994), although one compound in this series was found to be oxidised by a semicarbazide-sensitive amine oxidase and not by MAO (Naoi *et al.* 1995).

Milacemide

The anticonvulsant drug milacemide (2-*n*-pentylaminoacetamide) (van Dorsser *et al.* 1983) has been shown to be a good substrate for monoamine oxidase-B but to be oxidised only poorly by MAO-A. Furthermore, acute administration of milacemide to rats was found to result in the urinary elimination of glycinamide, which was partly prevented by pretreatment *l*-deprenyl but not by clorgyline (Janssens de Varebeke *et al.* 1988). Oral administration of milacemide (100 mg.kg^{-1}) resulted in increased concentrations of glycine in rat forebrain, cerebellum and medulla (Christophe *et al.* 1983). A significant increase in glycine levels in rat cortex, cerebellum and hippocampus, but not in striatum and substantia nigra, was also reported after intraperitoneal administration of the same dose of milacemide (Chapman and Hart 1988). Thus, milacemide acts as a precursor of glycine in the brain and it has been suggested that this may account for its anticonvulsant actions (Christophe *et al.* 1983). The MAO-B catalysed conversion of milacemide to glycinamide, which subsequently breaks down to glycine, is shown in Figure 4.13. Milacemide also behaves as a time-dependent inhibitor of the enzyme, according to the mechanism shown in equation 4.2 (O'Brien *et al.* 1994a,b). There are, however, pronounced species differences in the partition ratios and kinetic parameters of MAO-B with milacemide and some of its analogues (Sullivan *et al.* 1990; O'Brien *et al.* 1995).

As discussed above, α-methyl-substituted amines are not effective substrates for MAO and the analogue α-methyl-milacemide (2-[(1-methyl)pentyl]aminoacetamide— see Figure 4.13) was found to be a competitive inhibitor with little selectivity towards either form of MAO (O'Brien *et al.* 1991). Since this compound was found to be an effective anticonvulsant in mice but, in contrast to milacemide itself, did not elevate urinary glycine levels, it was concluded that the MAO-catalysed oxidation of milacemide to form glycinamide was not a major factor in its anticonvulsant action but rather served to terminate that activity. This conclusion was supported by studies on the effects of a number of milacemide analogues in which the aminoacetamide portion was retained but the pentyl moiety was replaced with substituted-aromatic residues. Comparison of the abilities of these compounds to act as substrates and inhibitors of MAO revealed no simple correlations with their anticonvulsant activities, as measured by their ability to prevent bicuculline-induced convulsions and death in the mouse (O'Brien *et al.* 1994b). Yu and Davis (1990, 1991a,b) have investigated the potential of the milacemide analogues 2-propyl-1-aminopentane and 2-[(2-propyl)pentyl-amino]acetamide to deliver the anticonvulsant valproate to the brain. Both compounds were shown to be substrates for MAO-B, as shown in Figure 4.13. However,

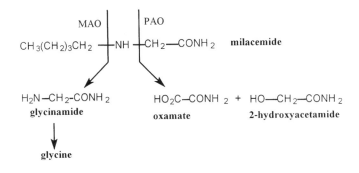

Figure 4.13 Patterns of oxidative cleavage of milacemide and its analogues 2-propyl-1-aminopentane and 2-[(2-propyl)pentylamino]acetamide by amine oxdases. Note: it has been reported that MAO-B from beef liver can form the cleavage products shown for PAO to a very minor extent (Silverman *et al.* 1993). The structure of α-methylmilacemide, which is not a substrate for MAO, is also shown.

although the appearance of valproate in brain could be detected following intraperitoneal administration of either compound, no anticonvulsant activity was manifested, indeed they were found to potentiate the convulsant effects of mercaptopropionic acid.

INHIBITORS THAT ARE ALSO SUBSTRATES

With clorgyline, (−)-deprenyl and pargyline there is no evidence for any product formation, irreversible reaction with the enzyme being apparently stoichiometric (Fowler *et al.* 1982).

Phenelzine (2-phenylethylhydrazine)

This is an irreversible inhibitor of MAO that has been widely used as an antidepressant. However, the conversion of ^{14}C-labelled phenelzine to phenylacetic acid by the rat, both *in vivo* and *in vitro*, in a process that was prevented by the MAO-inhibitors tranylcypromine and pargyline was shown many years ago by Clineschmidt and Horita (1969). Although a mechanism involving the oxidation of the hydrazine group to the corresponding azine derivative was subsequently proposed to explain its action as a mechanism-based inhibitor (Patek and Hellerman 1974), this offered no explanation of the apparent involvement of MAO in the conversion of phenelzine to phenylacetic acid. Studies of the effects of substitution of the hydrogens at carbon-1 of phenelzine by deuterium had shown that the *in vivo* effects of this inhibitor were significantly potentiated (Dyck *et al.* 1983). This might be the result of the isotope hindering the removal of a hydrogen from the carbon-1 position. An oxidation at this position would be expected to result in the formation of the corresponding hydrazone (Tipton and Spires 1971) as an alternative to formation of the inhibitory azine. Subsequent hydrolysis of this compound, perhaps catalysed by the enzyme itself, would result in the formation of the corresponding aldehyde in a manner analogous to that occurring during the oxidation of primary amines by MAO. Studies on the formation of 2-phenylacetaldehyde during the incubation of MAO with phenelzine or 1,1-dideutero-phenelzine, by high-performance liquid chromatography, showed that the deuterium substitution resulted in a decrease in the formation of this product in a way that parallelled the increased potency as a time-dependent inhibitor of the enzyme (Yu and Tipton 1990). These results show that the inhibitory reaction and that leading to product formation are competing reactions according to the scheme shown in equation (4.2), as shown in Figure 4.14. In this mechanism the isotope effect caused by the substitution of deuterium for hydrogen at the side-chain carbon-1 position will result in a decrease in the C−N dehydrogenation without affecting the N−N dehydrogenation, thus favouring the pathway leading to irreversible inhibition over that

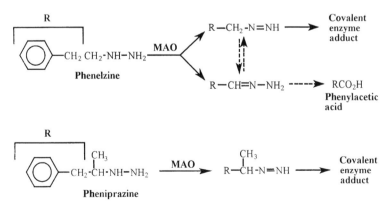

Figure 4.14 Phenelzine ([2phenethyl]hydrazine), as a substrate and inhibitor of MAO and its α-methyl derivative, pheniprazine ([1-methyl-2phenethylhydrazine), as an inhibitor.

leading to product formation. Such deuterium isotope effects may constitute a method for the design of more effective monoamine oxidase inhibitory antidepressants.

A number of different hydrazine derivatives have been used as MAO inhibitory antidepressants (Tipton 1990) but it is not known whether many of these were also substrates for the enzyme. As with other α-methyl substituted amines, the α-methyl derivative of phenelzine, pheniprazine (1-methyl-2-phenylethylhydrazine, see Figure 4.14), is not oxidised by MAO to any detectable extent, although it is a mechanism-based inhibitor with a (relatively weak) selectivity towards MAO-A (Ben Ramadan and Tipton 1999).

MD 780236, MD 240928 and MD 240931

The monoamine oxidase inhibitor 3-(-4((3-chlorophenyl)-methoxy)-phenyl)-5-methy-lamino)methyl-2-oxazolidinone methane sulphonate (MD 780236) follows the mechanism shown in equation (4.2), (Figure 4.15), and a partition ratio of about 530 has been determined with rat liver monoamine oxidase-B (Tipton et al. 1983). The formation of the aldehyde product as well as of the carboxylic acid and alcohol metabolites have been shown both in vivo and in vitro (Strolin Benedetti et al. 1983b; Tipton et al. 1984a). In contrast , MD 780236 acts as a substrate for MAO-A in vitro in the rat (Tipton et al. 1983). Consistent with this, pretreatment of rats with the selective MAO-A inhibitor clorgyline enhances the irreversible inhibition of MAO-B by MD 780236 (Strolin Benedetti and Dow 1983).

MD 780236 is a racaemic compound and the configurations of its R- and S-enantiomers (MD 240928 and MD 240931, respectively) are shown in Figure 4.15. The interaction of the S-enantiomer (MD 240931) with MAO-B behaves according to the 'suicide-substrate' mechanism shown in equation (4.2) and it is this component that will be mainly responsible for the irreversible inhibition of MAO-B by MD 780236 observed in vivo (Dostert et al. 1983). In contrast, with the R-enantiomer (MD 240928), the breakdown of the activated enzyme-inhibitor complex to give products occurs very much more rapidly than the reaction to produce irreversible inhibition, so that it acts essentially as a substrate for MAO-B. It has also been shown that the alcoholic products formed by the metabolism of MD 780236 and its enantiomers are potent MAO-B-selective inhibitors (Dostert et al. 1983; Tipton et al. 1983, 1984b). Thus the reversible inhibition of MAO-B by MD 240928 that has been observed in vivo (Turkish et al. 1988; Dostert et al. 1983) represents it competing as a substrate for the enzyme plus reversible inhibition from any of the alcoholic metabolites that may accumulate.

A comparative study of the metabolic fate of radioactively-labelled MD 240928 showed that urinary elimination accounted for 32% 17% and 57% of the orally administered drug in rat, dog and human, respectively, and that most of the elimination occurred within 24 h. By this time, the unconjugated acid derivative accounted for most of the urinary radioactivity in the three species. The debenzylated derivative was also an important metabolite in rat urine. The alcohol derivative, both free and conjugated, was only as a minor component in the urine of the three species. The debenzylated derivative of the alcohol, accounted for 14–18% of the urinary radio-activity in the dog but was only a minor component in rat and human urine.

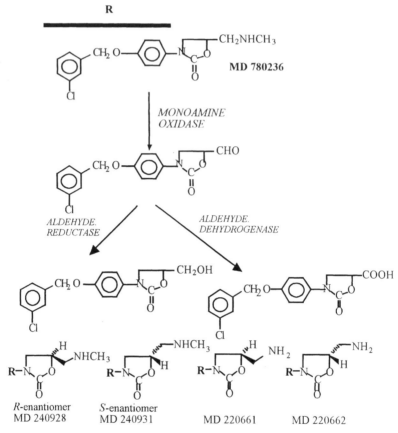

Figure 4.15 Metabolism of MD 780236 {3-[4-((chlorophenyl)-methoxy)-phenyl-5-[(methyl-amino)methyl]-2-oxazolidinone}. The configurations of the R- and S-enantiomers, MD 240928 (almoxatone) and MD 240931, are also shown.

Unchanged drug and its N-demethylated derivative were practically absent from the urine of the three species (see Figure 4.15). The acid derivative was the predominant form found in blood plasma, whereas in brain, which, after 2 to 4 hours, contained 0-2-0.3% of the radioactive dose administered, the alcohol derivative and the primary amine appeared to be the most important metabolites. Lower amounts of the acid derivative and the unchanged drug were, however, also found to be present in brain (Strolin Benedetti *et al.* 1984).

An additional complexity in the interpretation of the actions of MD 780236 and its enantiomers is that they are also rather weak, reversible inhibitors of SSAO (Dostert *et al.* 1984; Kinemuchi *et al.* 1986).

Moclobemide

This compound was developed as a reversible selective MAO-A inhibitor for use as an antidepressant with a minimal cheese effect (Haefely *et al.* 1992). It is a relatively weak MAO inhibitor *in vitro* but is considerably more potent *in vivo*. When preparations of MAO-A are incubated with moclobemide there is an initial low-affinity binding followed by a time-dependent increase in inhibition. This inhibition is, however, reversible by dialysis at 37 °C and, furthermore, if the enzyme-moclobemide mixture is incubated for extended times there is slow recovery of enzyme activity (see Haefely *et al.* 1992) Such behaviour might be explained by moclobemide acting as a very poor substrate for MAO-A according to the mechanism:

$$E + I \underset{k_{-1}}{\overset{k_{+1}}{\rightleftharpoons}} E.I \xrightarrow{k_{+2}} E.I^* \xrightarrow{k_{+3}} E + Products \qquad (4.4)$$

where $k_{+3} \ll k_{+2}$ and the rate governed by k_{+2} is, itself, slow. However, there is extensive metabolism by other enzymes *in vivo* which yields additional MAO-inhibitory products (Cesura *et al.* 1990; Schoerlin and Da Prada, 1990).

RO 16-6491 is also a substrate for MAO-B. After incubation in the presence of the tritiated compound, the radioactivity bound to brain mitochondria and platelet membranes was extracted and analysed by HPLC (Cesura *et al.* 1987). Only a minor peak of radioactivity was eluted as RO 16-6491 (Figure 4.3). The major amount of radioactivity was attributed to the aldehyde derivative (RO 19-7731) whereas a small amount had the same retention time as the acid derivative (RO 11-1903).

The semicarbazide-sensitive amine oxidases

The situation concerning the semicarbazide-sensitive amine oxidases is confusing, as this term covers a rather diverse group of enzymes which also appear to differ significantly between tissues and species. Since they are often distinguished simply in terms of their sensitivity to inhibition by semicarbazide and insensitivity to the acetylenic monoamine oxidase inhibitors, it is, perhaps, not surprising that more than one enzyme is detected and it is not always clear from publications which enzyme is being studied. SSAOs can be found throughout the animal kingdom as well as in plants and micro-organisms. In vertebrates, members of the group include both plasma- and tissue-bound enzymes together with specifically named enzymes such as lysyl oxidase (Smith-Mungo and Kagan 1998), which differs from the other members of this group in having lysine tyrosylquinone rather than TOPAquinone as cofactor (Wang *et al.* 1997), and diamine oxidase (see Callingham *et al.* 1995; Houen 1999; Klinman and Mu 1994; Lewinsohn 1984; Lyles 1995, 1996 for reviews on different aspects of these enzymes).

The plasma- and tissue-bound SSAOs that are active towards primary monoamines have more restricted specificities than the monoamine oxidases; adrenaline, for example, is not a substrate. The best-known substrate for both the plasma- and tissue-bound amine oxidases is the non-physiological amine benzylamine. Indeed, plasma SSAO has sometimes been referred to as benzylamine oxidase. Its specificity for physiologically occurring amines overlaps with that of MAO, and 2-phenylethylamine

(PEA), benzylamine, tyramine and dopamine are all oxidatively deaminated by the plasma and tissue-bound SSAO in most species. Although 5-HT is not a substrate for SSAO from most sources, it is a good substrate for the enzyme from pig and human dental pulp (Nordqvist *et al.* 1982). Octopamine and phenylethanolamine have been reported not to be substrates (Elliott *et al.* 1989b) and, as with the monoamine oxidases, α-methyl substituted amines are not oxidised. It appears that there may be more than one SSAO enzyme in plasma from some species (Boomsma *et al.* 2000; Elliott *et al.* 1992), which further complicates attempts to define specificities.

The fact that blood from many species contains significant amounts of SSAO activity means that caution is necessary in interpreting results obtained with isolated-cell systems that are maintained in media containing foetal calf serum. It has been shown that the behaviour of some compounds that are SSAO substrates can be significantly altered by the enzyme that is present in the culture medium (Conklin *et al.* 1998; Inoue *et al.* 1990).

The tissue-bound enzyme is associated with the plasma membrane and it appears that a proportion of the active sites are exposed to the extracellular milieu (Holt and Callingham 1994). Thus, unlike the situation with MAO, substrates would not have to enter the cell for deamination by either plasma- or tissue-bound SSAO. The rapid disappearance of dopamine and, to a lesser extent, noradrenaline when added to blood plasma from several mammalian species has been attributed to SSAO (Boomsma *et al.* 1993).

Amines that are good substrates for SSAO but are not oxidised by MAO include aminoacetone and methylamine (see Lyles 1995, 1996 for reviews). Its activity, which is high in cardiovascular tissue, is elevated in diabetes, congestive heart failure and following severe burns (Lewinsohn 1977). Elevated levels of SSAO have also been found in the plasma of toxin-induced diabetic rats and sheep (Hayes and Clarke 1990; Elliott *et al.* 1991) and of non-insulin-dependent diabetes in humans (Meszaros *et al.* 1999). SSAO activity is also increased in brown adipose tissue of obese Zucker rats (Barrand and Callingham 1982). A regulatory link between the glucose transporter GLUT 4 and SSAO has recently been reported and this may have important implications for SSAO in non-insulin-dependent diabetes and other disorders involving glucose transport (Enrique-Tarancon *et al.* 1988, 2000). The enzyme is also apparently identical to vascular-adhesion protein 1 (VAP-1), an endothelial glycoprotein that supports adhesion of lymphocytes to hepatic endothelium (Smith *et al.* 1998). It has been postulated that the increased SSAO levels seen in some pathological conditions may represent the need for increased VAP-1 to mediate repair (Kurkijarvi *et al.* 2000; Bono *et al.* 1999).

One of the problems which has limited our understanding of SSAO has been the lack of potent, selective inhibitors of these enzymes. Excluding semicarbazide and related compounds, which inhibit many different enzymes, SSAO tends to share a number of its inhibitors with MAO, making discrimination between amino oxisases *in vivo* difficult and further emphasising the need for new and more specific inhibitors. The compound MDL 72145 ((E)-2-(3′, 4′-dimethoxyphenyl)-3-fluoroallylamine), which has been shown to be a potent irreversible inhibitor of SSAO in rat aorta (Palfreyman *et al.* 1994; Lyles and Fitzpatrick 1985) and brown adipose tissue (Elliott *et al.* 1989a), has been used in a number of studies on the metabolic role of SSAO.

However, it is also a potent inhibitor of MAO-B and affects MAO-A activity to a lesser extent (Zreika *et al.* 1984). The compound MDL 72274A ((*E*)-2-phenyl-3-chloroallyla-mine) appears to be highly selective towards SSAO, whereas MDL 72974A ((*E*)-2-(4-fluorophenethyl)-3-fluoroallylamine) is a potent inhibitor of both MAO and SSAO. However, these compounds have been little used for metabolic studies to date. Another possibility is the anticancer drug procarbazine (*N*-isopropyl-α-(2-methyl hydrazino)-*p*-toluamide hydrochloride), which together with its metabolite mono-methylhydrazine appears to be highly selective for SSAO both *in vivo* and *in vitro* (Holt and Callingham 1995, Holt *et al.* 1992). However, it appears that there may be species and tissue differences in the sensitivities of amine oxidases to inhibition by hydrazine derivatives (Lizcano *et al.* 1996); again indicating the difficulty of extra-polating results obtained in one system to another.

COMPOUNDS THAT ARE SUBSTRATES FOR SSAO AS WELL AS MAO

Xenobiotic and endogenous amines

The involvement of SSAO in the metabolism of amines that are substrates for both enzymes has been less easy to establish because wide species differences in specificity and amount of enzyme present. For example, high activity is found in pig and sheep plasma, the levels are very much lower in human plasma and it is often difficulty to detect any at all in the rat. The levels of the tissue-bound enzyme also vary widely between species but do not parallel those of the plasma enzyme (Boomsma *et al.* 2000). Thus attempts to extrapolate from the situation in experimental animals to that in the human should be interpreted with great caution. Differences in substrate specificity between the SSAO enzymes from different mammalian sources appear to extend to stereospecificity. Alton *et al.* (1995; see also Palcic *et al.* 1995) reported the oxidation of benzylamine by plasma SSAO from bovine, horse, porcine, rabbit, and sheep plasma to involve abstraction of the pro-*S* hydrogen. In contrast, SSAO fom bovine plasma (Yu and Davis 1998) and from human aorta and plasma (Yu *et al.* 1994) has been repored to show no absolute stereospecificity in this respect.

The relatively high levels of SSAO in blood vessels suggest that it may be involved, perhaps in concert with MAO, in regulating the levels of circulating amines. For example, the pressor effects of tyramine in the perfused mesenteric arterial bed from the rat was found to be potentiated if both MAO and SSAO were inhibited. However, the response to the amine was unaffected by inhibition of either enzyme alone (Elliott *et al.* 1989c,d). In contrast, tryptamine-induced contraction of rat aorta was unaffected by SSAO inhibition but enhanced by combined inhibition of MAO and SSAO (Taneja and Lyles 1988). Furthermore, SSAO inhibitors alone can potentiate the contractile effects of sympathomimetic amines on the rat anococcygeus muscle preparation (Callingham *et al.* 1984). However, although SSAO is present in the intestine, it appears from studies with selective inhibitors that it does not play a significant role in limiting the effects of dietary tyramine, even in situations where MAO is inhibited (Hasan *et al.* 1988; Elliott *et al.* 1989d).

The lung from most species, including the rabbit, contains relatively high levels of SSAO activity and since, as discussed above, that organ can play a major role in the metabolism of presystemic elimination of amines such as 2-phenylethylamine, that enzyme might be expected to be involved in the process. However, from studies that used pargyline and semicarbazide, Gewitz and Gillis (1981) concluded that SSAO did not play a significant role in the metabolism of this amine in the perfused rabbit lung.

XENOBIOTIC AMINES

Milacemide analogues

The formation of glycinamide from milacemide appears to be specific to monoamine oxidase since it has been shown not to be a substrate for the semicarbazide-sensitive amine oxidase (Strolin Benedetti et al. 1988). However, the milacemide analogues 2-propyl-1-aminopentane and 2-[(2-propyl)pentylamino]acetamide (Figure 4.13) were both found to be substrates for SSAO from rat aorta (Yu and Davis 1990, 1991a) forming 2-propyl-1-pentaldehyde which could be subsequently oxidised to valproate. The quantitative significance of this, relative to the activity of MAO-B towards these substrates, is unclear. The observation that the acetamide derivative is a substrate was unexpected in view of the commonly-held view that the activity of this enzyme is restricted to primary amines.

Mescaline

Over 40 years ago, Blaschko et al. (1959) showed mescaline to be a substrate for pig plasma SSAO. Indeed, mescaline is more efficiently deaminated than benzylamine in pig plasma (Buffoni and Della Corte 1972). Roth et al. (1977) reported that the mescaline-oxidising activity of rabbit lung homogenates was two to three times greater than that of either liver or kidney and that brain and plasma had comparatively little capacity to metabolise mescaline. They showed mescaline metabolism by the perfused rabbit lung to be sensitive to inhibition by semicarbazide but not by pargyline (1 mM). Since mescaline efflux from the perfused lung was slower than that of its 'metabolite', presumed to be the carboxylic acid, these results were interpreted to indicate that the intact lung efficiently removes perfused mescaline and may be important in the disposition of circulating mescaline in vivo. Unfortunately, the situation in the human is not clear, and mescaline does not appear to be a substrate for the human plasma enzyme. Jacob and Shulgin (1981) reported that a number of thiol analogues of mescaline and isomescaline (2,3,4-trimethoxyphenethylamine) were substrates for beef plasma SSAO but provided no information on the activities of MAO towards them.

Primaquine

The possible contribution of SSAO to the metabolism of this antimalarial drug, which has been shown to be a substrate for the pig plasma enzyme (Blaschko and Hawes 1959), remains uncertain. However, as discussed above, it appears that in rat liver, at least in vitro, the contribution of monoamine oxidase is dominant.

COMPOUNDS THAT ARE SUBSTRATES FOR SSAO BUT NOT FOR MAO

Methylamine

This is produced physiologically from a number of catabolic reactions, such as the breakdown of adrenaline, catalysed by monoamine oxidase, sarcosine and creatine. However, since it is also a common component of some foods and beverages (Lichtenberger et al. 1991; Lin et al. 1984) and is also an atmospheric pollutant and present in cigarette smoke (Yu 1998a), it can be regarded as a xenobiotic that is also produced endogenously. Methylamine is not a substrate for MAO (Yu 1989) but is oxidised by SSAO to form ammonia, hydrogen peroxide and formaldehyde (see Figure 4.16). It has been argued that the high levels of SSAO in lung are to protect against inhaled methylamine and other volatile amines (Lizcano et al. 1990, 1998). However, formaldehyde is an extremely reactive chemical that can produce irreversible adducts with proteins and single-strand DNA, among other harmful cross-linkage reactions (Yu 1998a,b). It is normally metabolised to formate by formaldehyde dehydrogenase in the presence of reduced glutathione. Interestingly, serum does not contain any formaldehyde dehydrogenase and any formaldehyde produced in the blood cannot be metabolised until it is transported into the erythrocytes. This may be significant in terms of formaldehyde-induced toxicity in blood vessels.

The consumption of relatively large amounts of creatine as a nutrition supplement, in attempts to enhance sports performance, would lead to increased methylamine and, hence, formaldehyde and H_2O_2 production from its catabolism. This might underlie some of the deleterious effects of long-term creatine consumption (Yu and Deng 2000).

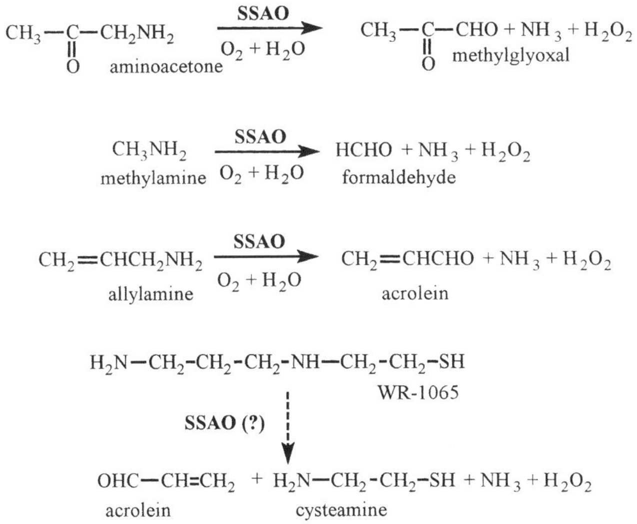

$$CH_3-\underset{\underset{O}{\|}}{C}-CH_2NH_2 \xrightarrow[O_2 + H_2O]{SSAO} CH_3-\underset{\underset{O}{\|}}{C}-CHO + NH_3 + H_2O_2$$

aminoacetone, methylglyoxal

$$CH_3NH_2 \xrightarrow[O_2 + H_2O]{SSAO} HCHO + NH_3 + H_2O_2$$

methylamine, formaldehyde

$$CH_2=CHCH_2NH_2 \xrightarrow[O_2 + H_2O]{SSAO} CH_2=CHCHO + NH_3 + H_2O_2$$

allylamine, acrolein

$$H_2N-CH_2-CH_2-CH_2-NH-CH_2-CH_2-SH$$

WR-1065

$$SSAO\ (?) \downarrow$$

$$OHC-CH=CH_2 + H_2N-CH_2-CH_2-SH + NH_3 + H_2O_2$$

acrolein, cysteamine

Figure 4.16 Oxidation of allylamine, aminoacetone and methylamine by SSAO. The oxidation of WR-1065 (2-(3-aminopropylamino)ethanethiol) is also shown.

Kojic amine

Kojic amine (2-(aminomethyl)-5-hydroxy-4H-pyran-4-one; see Figure 4.5) is a GABA (γ-aminobutyric acid) receptor agonist that has been reported to be a substrate for SSAO (Ferkany *et al.* 1981), but there are no data on the metabolic significance of this. Neither is it known whether MAO also oxidises this compound.

4-S-Cysteaminylphenol (4-S-CAP)

This compound was developed as an inhibitor of melanoma growth. Its cytotoxicity is largely due to the formation of dihydro-1,4-benzothiazine-6,7-dione (dihydro-1,4,benzothiazine-quinone), which is catalysed by tyrosinase and also occurs by autoxidation (Hasegawa *et al.* 1997). However, it appears that SSAO can also catalyse the conversion of this compound to the corresponding aldehyde, which may then be converted to the corresponding acid and alcohol metabolites (Inoue *et al.* 1990), as shown in Figure 4.17. Aldehyde formation in this way appeared to enhance the toxicity of 4-S-CAP. Consistent with the known specificity of SSAO, the α-methyl analogue of 4-S-CAP did not appear to be a substrate.

MD 220661

This oxazolidinone compound is the primary amine derivative of the *R*- enantiomer (MD 240928) of the racemic compound MD 780236 (Figure 4.15). It behaves as a substrate for SSAO, whereas the corresponding *S*-enantiomer (MD 220662) appears to be a simple reversible inhibitor of the enzyme (Dostert *et al.* 1984).

Allylamine

This unsaturated amine is used in the manufacture of pharmaceuticals and vulcanised rubber. Prolonged exposure to allylamine can result in severe necrotic tissue damage with the cardiovascular system being particularly susceptible (Boor and Hysmith 1987). These toxic effects are not due to allylamine (3-aminopropene) itself but result from its metabolism to acrolein by SSAO (Nelson and Boor 1982; Hysmith and Boor 1988; see Figure 4.16). Allylamine is not a substrate for monoamine oxidase and its toxicity in the rat is greatly reduced by pretreatment with semicarbazide (Awasthi and Boor 1993). The hydrogen peroxide generated in the oxidation of allylamine by SSAO also appears to contribute to its toxicity and the addition of catalase to vascular smooth muscle cells *in vitro* partially protects against the toxicity (Ramos *et al.* 1988). Glutathione *S*-transferases catalyse the first step in the detoxification of acrolein (He *et al.* 1999). However, acrolein also activates glutathione *S*-transferase by binding to thiol groups and, although this appears to activate its own detoxification, the resulting glutathione depletion can, in turn, impair the individual's ability to detoxify other xenobiotics and reduce the capacity to remove hydrogen peroxide.

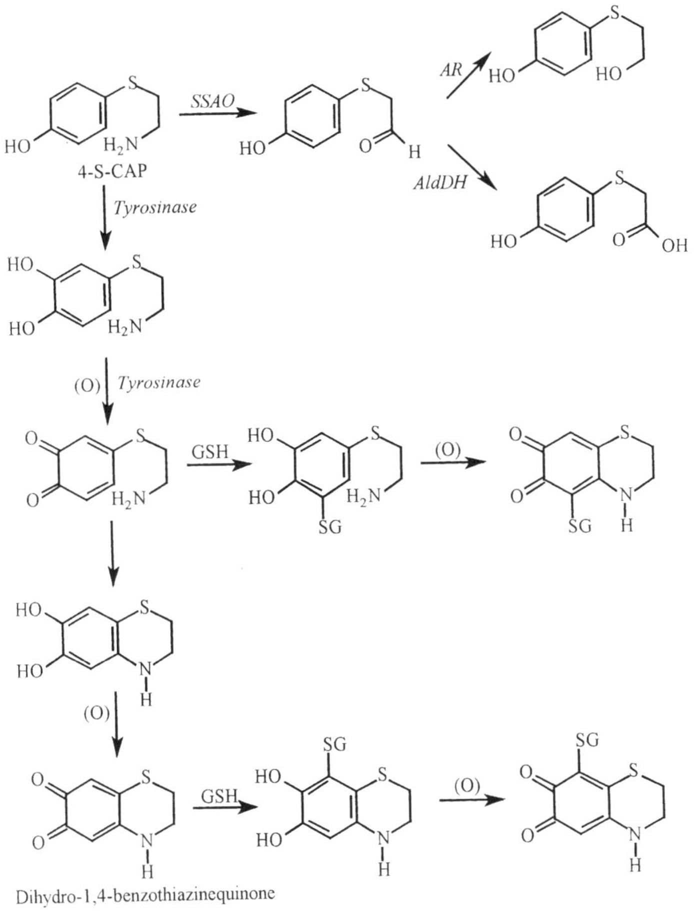

Figure 4.17 Metabolism of the melanocytotoxin 4-S-cysteaminylphenol (4-S-CAP). Generation of reactive oxygen radicals, during the autoxidation reactions and the quinone dihydro-1,4-benzothiazinequinone are believed to contribute most significantly to the cytotoxicity. The aldehyde produced through the action of SSAO is also cytotoxic.

WR-1065

WR-1065 (2-(3-aminopropylamino)ethanethiol) is a cytoprotective drug that is used to protect tissues against the damaging effects of radiation and some anticancer drugs. It is administered as the thiophosphate derivative amifostine (Ethyol; WR-2721; S-2-(3-aminopropylamino)ethyl-phosphorothioate) from which it is formed by the action of alkaline phosphatase (Shaw et al. 1996). It is converted by a 'copper-containing amine oxidase', present in the calf serum used for cell culture, to cysteamine and acrolein (Meier and Issels, 1995 see Figure 4.16). The reaction, which was shown to be accompanied by oxygen consumption and hydrogen peroxide formation, was inhibited by the SSAO inhibitor aminoguanidine. Purified preparations of bovine plasma SSAO were also shown to catalyse the oxidation of WR-1065. The prodrug amifostine was shown not to be a substrate for this reaction. The oxidation of WR-1065 was shown to result in cytotoxicity and depletion of reduced glutathione. The precise details of the reaction are, however, unclear since the oxidative cleavage of WR-1065 would be expected to lead to the formation of cysteamine plus the aminoaldehyde, 3-aminopropanal, and the mechanism of the conversion of this compound to acrolein remains to be clarified. The involvement of one or more additional steps in acrolein formation would be consistent with the very low yield of this compound, relative to that of cysteamine, reported by Meier and Issels (1995). The major pathways of WR-1065 inactivation *in vivo* are the formation of the disulphide between two WR-1065 molecules, and mixed disulphides between WR-1065 and glutathione and cysteine, although cysteamine formation has also been detected (Shaw *et al.* 1996).

Tresperimus

Tresperimus ([4-{(3-aminopropyl)amino]butyl}carbamic acid,2-[[6-[(aminoiminomethyl)-amino]hexyl]amino]-2-oxoethyl ester) is an immunosuppressive drug that is metabolised in blood plasma rather than in the liver (Figure 4.18). The metabolic pathways involved have been studied by Claud *et al.* (2001) and shown to involve the SSAO-catalysed oxidation of the terminal amino group to the corresponding aldehyde, which may then be further oxidised to the carboxylic acid. The aldehyde can also decompose, non-enzymatically, to the desaminopropyl derivative of tresperimus ([4-amino-butyl]carbamic acid,2-[[6-{(aminoiminomethyl)amino}hexyl]amino]-2-oxoethyl ester), which is also a substrate for SSAO giving an aldehyde, which is then further oxidised to the corresponding carboxylic acid. The involvement of SSAO in these processes was confirmed in the rat *in vivo* by its inhibition by hydralazine and in rat and human plasma *in vitro* by its inhibition by semicarbazide and by hydralazine. There was no significant role for polyamine oxidase (see below) in this process, but a minor role for MAO was not excluded. Although the production of the aldehyde derivatives takes place in the plasma, their further oxidation appeared to be a result of intracellular aldehyde dehydrogenase activity. Analysis of the urinary excretion patterns indicated that hydrolytic cleavage of the amide bond in the middle of the molecule to yield guanidinohexylamine (GHA) plus 2-[[[[4-[(3-aminopropyl)-amino]butyl]amino]carbonyl]oxy]-acetic acid (Figure 4.18) also occurred. This appeared to be a minor pathway in the human but was more important in the rat, which has very low plasma SSAO activities. There was no indication that oxidative deamination of GHA occurred.

DIAMINE OXIDASE

Although histamine is a poor substrate for some plasma SSAO enzymes, other diamines are not substrates for it (Buffoni 1966). However, there is a specific tissue-bound SSAO, diamine oxidase (DAO), which is found in high levels in the intestine,

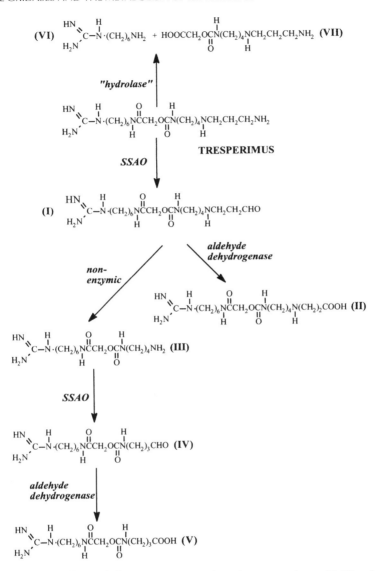

Figure 4.18 Metabolism of the immunosuppressive drug tresperimus ([4-{(3-aminopropyl) amino}butyl}carbamic acid,2-[[6-[(aminoiminomethyl)amino]-hexyl]amino]-2-oxoethyl ester). The products shown are: **(I)** [4-[(3-oxopropyl)-amino]butyl]carbamic acid,2-{[6-[(amino-iminomethyl)amino]hexyl]- amino}-2-oxoethyl ester); **(II)** 1-amino-1-imino 10,13 dioxo 12 oxa 2,9,14,19-tetraazadocosan-22-oic acid; **(III)** [4-aminobutyl]carbamic acid,2-[[6-{(aminoimino-methyl)amino}-hexyl]amino]-2-oxoethyl ester, (desaminopropyl- tresperimus); **(IV)** [4-oxobutyl]-carbamic acid,2-[[6-{(aminoiminomethyl)amino}hexyl]amino]-2-oxoethyl ester; **(V)** 1-amino-1-imino-10,13-dioxo-12-oxa-2,9,14-triazaoctadecan-18-oic acid; **(VI)** N-(6-aminohexyl)-guan-idine (guanidinohexylamine; GHA); **(VII)** 2-[[[[4-[(3-aminopropyl)amino] butyl]amino]-carbonyl]oxy]-acetic acid.

kidney, thymus gland and placenta (Buffoni 1966), that has high activity towards histamine and other diamines, putrescine, cadaverine, spermidine and spermine. A problem in assessing its possible functions in the metabolism of xenobiotics is that it is not always clear whether work that describes an amine-oxidising activity that is inhibited by semicarbazide is referring to SSAO or DAO. However, agmatine (1-amino-4-guanidinobutane, Figure 4.19), has been shown to be a DAO substrate. Agmatine, which is an imidazoline receptor ligand (Holt and Baker 1995; Lortie *et al.* 1996), is produced from arginine by the action of arginine decarboxylase in plants and bacteria. More recently it has also been shown to be produced by rat brain, liver, and kidney (Lortie *et al.* 1996). It is present in some foods and beverages including, for example, beer (Izquierdo-Pulido *et al.* 1996) and, thus, may be regarded as a xenobiotic that is also produced endogenously. Several, but not all, amine oxidases are binding proteins for imidazoline and guanidine compounds (amiloride-binding proteins; Novotny *et al.* 1994; Lizcano *et al.* 1998) but only DAO appears to treat one of these, agmatine, as a substrate.

Polyamine oxidase

This enzyme (N^1-acetylspermidine: oxygen oxidoreductase (deaminating), E.C. 1.5.3.11; PAO) is a flavoprotein that is widely distributed in mammalian (rat and human) tissues, including brain and is also present in blood plasma (see Seiler 1995, 2000 for reviews). It oxidises a secondary amino group in monoacetylspermine and monoacetylspermidine to form, respectively, spermidine and putrescine (Hölttä 1977). The monoamine oxidase inhibitor pargyline is a relatively weak inhibitor of PAO. The drug MDL 72527, an N, N'-bis(2, 3-butadienyl)- derivative of putrescine (Figure 4.20), is a potent and selective inhibitor of PAO (Bey *et al.*1985). Rather surprisingly, PAO has been also reported to be inhibited by semicarbazide (Hölttä, 1977; Kunimoto *et al.* 1989).

The enzyme has been shown to oxidise the antimalarial drug MDL 27695, an N, N'-

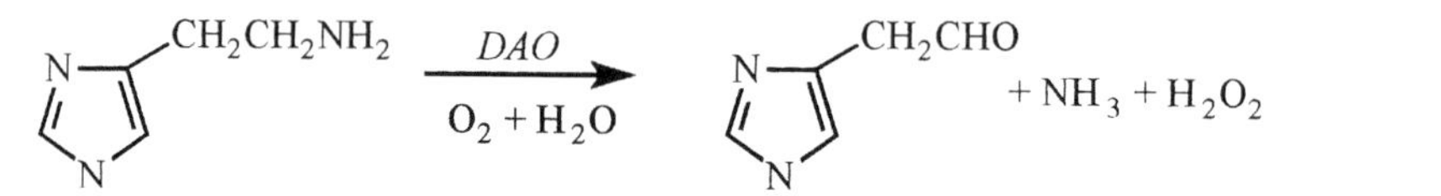

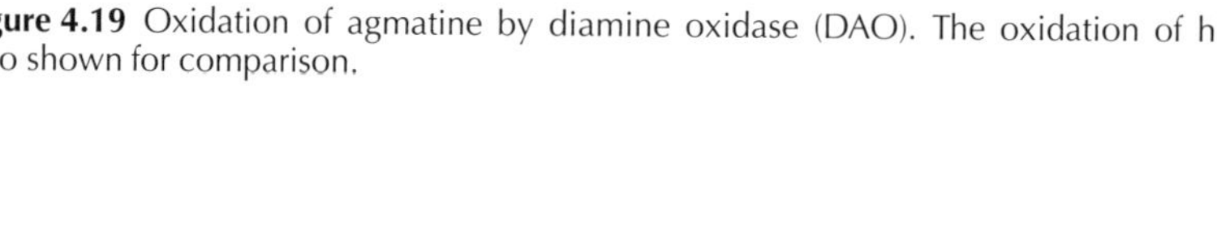

Figure 4.19 Oxidation of agmatine by diamine oxidase (DAO). The oxidation of histamine is also shown for comparison.

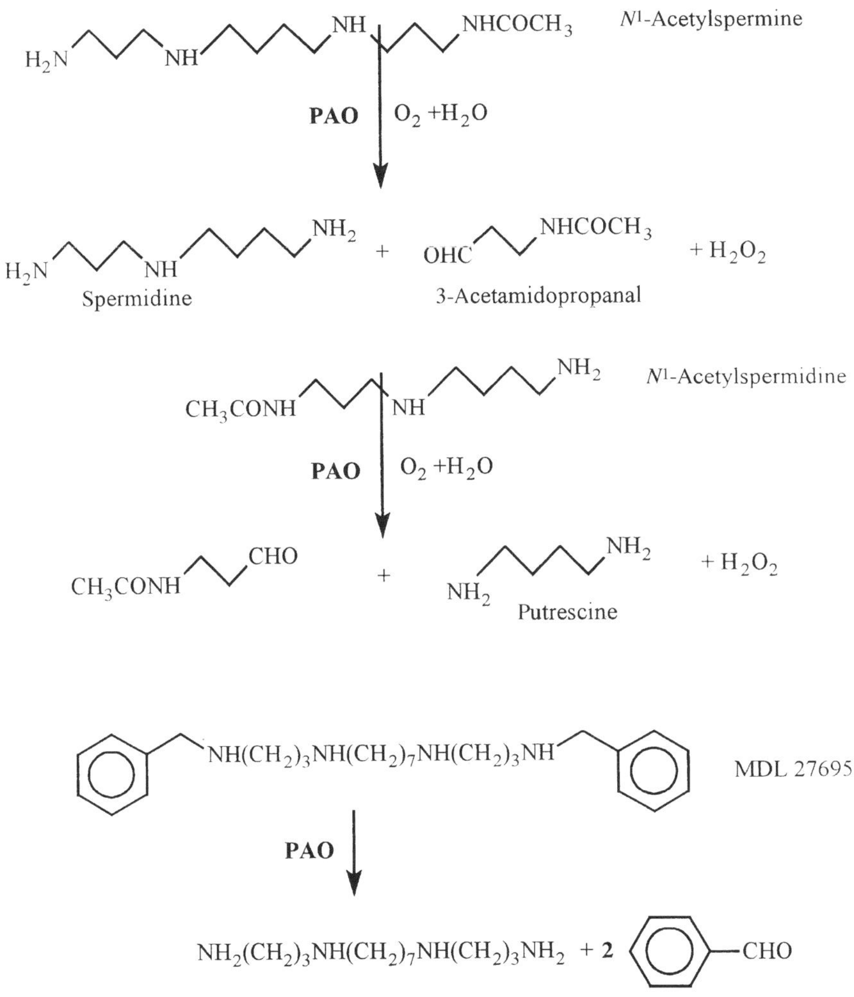

$H_2C{=}C{=}CHCH_2NHCH_2CH_2CH_2CH_2NHCH_2CH{=}C{=}CH_2$ MDL 72527

Figure 4.20 Some reactions catalysed by polyamine oxidase (PAO) and the structure of the PAO inhibitor MDL 72527.

bis(2, 3-benzylpolyamine)- derivative (Figure 4.20), which has also been shown to inhibit the growth of a rat hepatoma cell-line (Bitonti *et al.* 1989, 1990). Another cytotoxic agent *N*(1)-(n-octanesulphonyl)spermine is also oxidised by PAO and studies with the human colon carcinoma-derived CaCo-2 cell line have shown that oxidation serves to diminish its toxicity (Seiler *et al.* 2000).

An unusual reaction catalysed by PAO is the oxidative cleavage of milacemide (Strolin Benedetti *et al.* 1990, 1992b). However, the cleavage pattern is different from that catalysed by MAO in that glycinamide is not formed; the likely reaction products being pentylamine and oxamaldehyde (see Figure 4.13).

Conclusions

This chapter has concentrated on those xenobiotics which are known to be metabolised by the amine oxidases, at least *in vitro*. There are several other cases where too few data are available but it is possible that one or more of the amine oxidases may be involved in the metabolism of specific xenobiotics (for reviews see Dostert *et al.* 1989; Strolin Benedetti and Dostert 1994; Strolin Benedetti and Tipton 1998). Unfortunately we do not yet have sufficient information to be able to predict with certainty whether a compound will be a substrate for a specific amine oxidase in any given species (Wouters 1998). There have been too few metabolic studies to bridge the gap between *in vitro* studies and the behaviour in the, more complex, *in vivo* environment. Furthermore, the consequences that xenobiotic metabolism may have on the normal functions of these enzymes require more detailed study.

References

Abell CW and Kwan SW (2000) Molecular characterization of monoamine oxidases A and B. *Progress in Nucleic Acid Research and Molecular Biology*, **65**, 29–56.

Alton G, Taher TH, Beever RJ and Palcic MM (1995) Stereochemistry of benzylamine oxidation by copper amine oxidases. *Archives of Biochemistry and Biophysics*, 353–361.

Anderson MC, Hasan F, McCrodden JM and Tipton KF (1993) Monoamine oxidase inhibitors and the cheese effect. *Neurochemical Research*, **18**, 1145–1149.

Arai Y, Toyoshima Y and Kinemuchi H (1986) Studies of monoamine oxidase and semicarbazide-sensitive amine oxidase. II. Inhibition by alpha-methylated substrate-analogue monoamines, alpha-methyltryptamine, alpha-methylbenzylamine and two enantiomers of alpha-methylbenzylamine. *Japanese Journal of Pharmacology*, **41**, 191–197.

Awasthi S and Boor PJ (1993) Semicarbazide protection from in vivo oxidant injury of vascular tissue by allylamine. *Toxicology Letters*, **66**, 157–163.

Bakhle YS. and Youdim MBH (1979) The metabolism of 5-hydroxytryptamine and β-phenylethylamine in perfused rat lung and *in vitro*. *British Journal of Pharmacology*, **65**, 147–154.

Bakke OM, Davies DS, Davies L and Dollery CT (1973) Metabolism of propranolol in rat: the fate of the N-isopropyl group. *Life Sciences*, **13**, 1665–1675.

Bargar EM, Walle UK, Bai SA and Walle T (1983) Quantitative metabolic fate of propranolol in the dog, rat, and hamster using radiotracer, high performance liquid chromatography, and gas chromatography-mass spectrometry techniques. *Drug Metabolism and Disposition*, **11**, 266–272.

Barrand MA and Callingham BA (1982) Monoamine oxidase activities in brown adipose tissue of the rat: some properties and subcellular distribution. *Biochemical Pharmacology*, **31**, 2177–2184.

Barwell CJ, Basma AN, Lafi MA and Leake LD (1989) Deamination of hordenine by monoamine oxidase and its action on vasa deferentia of the rat. *Journal of Pharmacy and Pharmacology*, **41**, 421–423.

Belleau B and Moran J (1963) Deuterium isotope effects in relation to the chemical mechanism of monoamine oxidase. *Annals of the New York Academy of Science*, **107**, 822–839.

Ben Ramadan Z and Tipton KF (1999) Some peculiar aspects of monoamine oxidase inhibition. *Neurobiology*, **7**, 159–174.

Bey P, Bolkenius FN, Seiler N and Casara P (1985) *N*-2,3-Butadienyl-1,4-butanediamine derivatives: potent irreversible inactivators of mammalian polyamine oxidase. *Journal of Medicinal Chemistry*, **28**, 1–2.

Bieck PR, Kemmler H, Henriot S and Tipton KF (1988) Evidence for 'local' gastrointestinal effects of MAO inhibition on metabolism and transport of tyramine in human subjects. *Pharmacological Research Communications*, **20**, 129–130.

Bitonti A, Bush TL and McCann PP (1989) Regulation of polyamine biosynthesis in rat hepatoma (HTC) cells by a bisbenzyl polyamine analogue. *Biochemical Journal*, **257**, 769–774.

Bitonti AJ, Dumont JA, Bush TL, Stemerick DM, Edwards ML and McCann PP (1990) Bis(benzyl)-polyamine analogs as novel substrates for polyamine oxidase. *Journal of Bioliological Chemistry*, **265**, 382–388.

Blackwell B (1963) Hypertensive crisis due to monoamine oxidase inhibitors *Lancet*, **ii**, 849–851.

Blackwell B and Marley E (1969) Monoamine oxidase inhibition and intolerance to foodstuffs. *Bibliotheca Nutritio et Dieta*, **11**, 96–110.

Blaschko H and Hawes R (1959). Observations on spermine oxidase of mammalian plasma. *Journal of Physiology*, **145**, 124–131.

Blaschko H, Friedman PJ, Hawes R and Nilsson K (1959) The amine oxidases of mammalian plasma. *Journal of Physiology*, **145**, 384–404.

Bond PA and Howe R (1967) The metabolism of pronethalol. *Biochemical Pharmacology*, **16**, 1261–1280.

Bono P, Jalkanen S and Salmi M (1999) Mouse vascular adhesion protein 1 is a sialoglycoprotein with enzymatic activity and is induced in diabetic insulitis. *American Journal of Pathology*, **155**, 1613–1624.

Boomsma F, Alberts G, Bevers MM, Koning MM, Man in 't Veld AJ and Schalekamp MA (1993) Breakdown of 3,4-dihydroxybenzylamine and dopamine in plasma of various animal species by semicarbazide-sensitive amine oxidase. *Journal of Chromatography*, **621**, 82–88.

Boomsma F, van Dijk J, Bhaggoe UM, Bouhuizen AM and van den Meiracker AH (2000) Variation in semicarbazide-sensitive amine oxidase activity in plasma and tissues of mammals. *Comparative Biochemistry and Physiology*, **C**, *Pharmacology, Toxicology and Endocrinology*, **126**, 69–78.

Boor PJ and Hysmith RM (1987) Allylamine cardiovascular toxicity. *Toxicology*, **44**, 129–145.

Buffoni F (1966) Histaminase and related amine oxidases. *Pharmacological Reviews*, **18**, 1163–1199.

Buffoni F and Della Corte L (1972) Pig plasma benzylamine oxidase. In *Monoamine oxidases— New Vistas. Adv. Biochemical Psychopharmacology*, Costa E and Sandler M (eds), **5**, Raven Press, New York, pp. 133–149.

Callingham BA, McCarry WJ and Barrand MA (1984) Effects of some carbonyl reagents and short-acting MAO inhibitors on semicarbazide-sensitive (clorgyline-resistant) amine oxidase in the rat. In *Monoamine Oxidase and Disease. Prospects for Therapy with Reversible Inhibitors*, Tipton KF, Dostert P and Strolin Benedetti M (eds), Academic Press, London, pp. 595–596.

Callingham BA, Crosbie AE and Rous BA (1995) Some aspects of the pathophysiology of semicarbazide-sensitive amine oxidase enzymes. *Progress in Brain Research*, **106**, 305–321.

Cashman JR and Ziegler DM (1986) Contribution of N-oxygenation to the metabolism of MPTP (1-methyl-4-phenyl-1,2,3,6-tetrahydropyridine) by various liver preparations. *Molecular Pharmacology*, **29**, 163–167.

Cesura AM, Galva MD, Imhof R and Da Prada M (1987) Binding of [3H]Ro 16-6491, a reversible inhibitor of monoamine oxidase type B, to human brain mitochondria and platelet membranes. *Journal of Neurochemistry*, **48**, 170-176.

Cesura AM, Muggli-Maniglio D, Lang G, Imhof R and Da Prada M (1990) Monoamine oxidase inhibition by moclobemide and 2-amino-ethyl carboxamide derivatives: mode of action and kinetic characteristics. *Journal of Neural Transmisssion*, **32**, [Suppl.], 165–70.

Chapman AG and Hart GP (1988) Anticonvulsant drug action and regional neurotransmitter amino acid changes. *Journal of Neural Transmisssion*, **72**, 201–212.

Christophe J, Kutzner R, Hguyen-Bui ND, Damien C, Chatelain P and Gillet L (1983) Conversion of orally administered 2-n-pentylaminoacetamide into glycinamide and glycine in the rat brain. *Life Sciences*, **33**, 533–541.

Claud P, Padovani P, Guichard JP, Artur Y and Lainé R (2001) Involvement of semicarbazide-sensitive amine oxidase (SSAO) in tresperimus metabolism in man and rat. *Drug Metabolism and Disposition*, **29**, 735–741.

Clineschmidt BV and Horita A (1969) The monoamine oxidase catalysed degradation of phenelzine-1-14C – an inhibitor of monoamine oxidase. I. Studies *in vitro*. II. Studies *in vivo*. *Biochemical Pharmacology*, **18**, 1011–1028.

Constantino L, Paixao P, Moreir R, Portela MJ, Do Rosario VE and Iley J (1999) Metabolism of primaquine by liver homogenate fractions. Evidence for monoamine oxidase and cytochrome P450 involvement in the oxidative deamination of primaquine to carboxyprimaquine. *Experimental Toxicology and Pathology*, **51**, 299–303.

Conklin DJ, Langford SD and Boor PJ (1998) Contribution of serum and cellular semicarbazide-sensitive amine oxidase to amine metabolism and cardiovascular toxicity. *Toxicological Sciences*, **46**, 386–392.

Da Prada M, Zürcher G, Wüthrich, I and Haefely WE (1988) On tyramine, food beverages and the reversible MAO inhibitor moclobemide, *Journal of Neural Transmisssion* **26**, [Suppl.], 31–56.

Davis DS, Tasuhara H, Boobis AR and George CF (1984) The effects of reversible and irreversible inhibitors of monoamine oxidase on tyramine deamination in dog intestine. In *Monoamine Oxidase and Disease*. Tipton KF, Dostert P and Strolin Benedetti M (eds), *Prospects for Therapy with Reversible Inhibitors*. Academic Press, London, pp. 443–448.

Dieterle W, Faigle JW, Kung W and Theobald W (1986) The disposition and metabolism of 14C-oxprenolol.HCl in man. *Xenobiotica*, **16**,181–191.

Dixon CM, Saynor DA, Andrew PD, Oxford J, Bradbury A and Tarbit MH (1993). Disposition of sumatriptan in laboratory animals and humans. *Drug Metabolism and Disposition*, **21**, S761–769.

Dixon CM, Park GR and Tarbit MH (1994) Characterization of the enzyme responsible for the metabolism of sumatriptan in human liver. *Biochemical Pharmacology*, **47**, 1253–1257.

Dollery CT, Brown MJ, Davies DS and Strolin Benedetti M (1984) Pressor amines and monoamine oxidase inhibitors. In *Monoamine Oxidase and Disease. Prospects for Therapy with Reversible Inhibitors*, Tipton KF, Dostert P and Strolin Benedetti M (eds), Academic Press, London, pp. 429–441.

Dostert P (1984) Myth and reality of the classical MAO inhibitors, reasons for seeking a new generation. In *Monoamine Oxidase and Disease. Prospects for Therapy with Reversible Inhibitors*. Tipton KF, Dostert P and Strolin Benedetti M (eds). Academic Press, London, pp. 9–24.

Dostert P, Strolin Benedetti M and Guffroy CJ (1983) Different stereoselective inhibition of monoamine oxidase-B by the R- and S-enantiomers of MD 780236. *Journal of Pharmacy and Pharmacology*, **35**, 161–165.

Dostert P, Guffroy C, Strolin Benedetti M and Boucher T (1984) Inhibition of semicarbazide-sensitive amine oxidase by monoamine oxidase B inhibitors from the oxazolidinone series. *Journal of Pharmacy and Pharmacology*, **36**, 782–785.

Dostert P, Strolin Benedetti M and Tipton KF (1989) Interaction of monoamine oxidase with substrates and inhibitors. *Medical Research Reviews*, **9**, 45–89.

Dupont H, Davies DS and Strolin Benedetti M (1987) Inhibition of cytochrome P-450-dependent oxidation reactions by MAO inhibitors in rat liver microsomes. *Biochemical Pharmacology*, **36**, 1651–1657.

Dyck LE, Durden DA, Yu PH, Davis BA and Boulton AA (1983) Potentiation of the biochemical effects of β-phenylethylhydrazine by deuterium substitution. *Biochemical Pharmacology*, **32**, 1519–1522.

Elliott J, Callingham BA and Barrand MA (1989a) In-vivo effects of (E)-2-(3',4'-dimethoxy-phenyl)-3-fluoroallylamine (MDL 72145) on amine oxidase activities in the rat. Selective inhibition of semicarbazide-sensitive amine oxidase in vascular and brown adipose tissues. *Journal of Pharmacy and Pharmacology*, **41**, 37–41.

Elliott J, Callingham BA and Sharman DF (1989b) Semicarbazide-sensitive amine oxidase (SSAO) of the rat aorta. Interactions with some naturally occurring amines and their structural analogues. *Biochemical Pharmacology*, **38**, 1507–1515.

Elliott J, Callingham BA and Sharman DF (1989c) Metabolism of amines in the isolated perfused mesenteric bed of the rat. *British Journal of Pharmacology*, **98**, 507–514.

Elliott J, Callingham BA and Sharman DF (1989d) The influence of amine metabolizing enzymes on the pharmacology of tyramine in the isolated perfused mesenteric arterial bed of the rat. *British Journal of Pharmacology*, **98**, 515–522.

Elliott J, Fowden AL, Callingham BA, Sharman DF and Silver M (1991) Physiological and pathological influences on sheep blood plasma amine oxidase: effect of pregnancy and experimental alloxan-induced diabetes mellitus. *Research in Veterinary Science*, **50**, 334–339.

Elliott J, Callingham BA and Sharman DF (1992) Amine oxidase enzymes of sheep blood vessels and bloodplasma: a comparison of their properties. *Comparative Biochemistry and Physiology, C, Pharmacology, Toxicology and Endocrinology*, **102**, 83–89.

Enrique-Tarancon G, Marti L, Morin N, Lizcano JM, Unzeta M, Sevilla L, Camps M, Palacin M, Testar X, Carpene C and Zorzano A (1998) Role of semicarbazide-sensitive amine oxidase on glucose transport and GLUT4 recruitment to the cell surface in adipose cells. *Journal of Biological Chemistry*, **273**, 8025–8032.

Enrique-Tarancon G, Castan I, Morin N, Marti L, Abella A, Camps M, Casamitjana R, Palacin M, Testar X, Degerman E, Carpene C and Zorzano A (2000) Substrates of semicarbazide-sensitive amine oxidase co-operate with vanadate to stimulate tyrosine phosphorylation of insulin-receptor-substrate proteins, phosphoinositide 3-kinase activity and GLUT4 translocation in adipose cells. *Biochemical Journal*, **350**, 171–180.

Ferkany JW, Andree TH, Clarke DE and Enna J. (1981) Neurochemical effects of kojic amine, a GABAmimetic, and its interaction with benzylamine oxidase. *Neuropharmacology*, **20**, 1177–1182.

Fowler CJ, Mantle TJ and Tipton KF (1982) The nature of the inhibition of rat liver monoamine oxidase types A and B by the acetylenic inhibitors clorgyline, l-deprenyl and pargyline. *Biochemical Pharmacology*, **31**, 3555–3561.

Garpenstrand H, Ekblom J, Forslund K, Rylander G and Oreland L (2000) Platelet monoamine oxidase activity is related to MAOB intron 13 genotype. *Journal of Neural Transmission*, **107**, 523–530.

Gessner W, Brossi A, Bembenek ME, Fritz RR and Abell CW (1986) Studies on the mechanism of MPTP oxidation by human liver monoamine oxidase B. *FEBS Letters*, **199**, 100-102.

Gewitz MH and Gillis CN (1981) Uptake and metabolism of biogenic amines in the developing rabbit lung. *Journal of Applied Physiology*, **50**, 118–122.

Goldaniga G, Montesanti L, Pianezzola E and Valzelli G (1980) Pharmacokinetics and metabolism of a new beta-adrenergic blocking agent, the 1, ter-butyl-amino-3-(1,2,3,4-tetrahydro-1,4-ethano-8-hydroxy-5-naphthoxy)-2-propanol (K 5407). *European Journal of Drug Metabolism and Pharmacokinetics*, **5**, 9–20.

Goldberg LI (1974) Use of sympathomimetic amines in shock. *American Family Physician* **10**, 80–85.

Goldberg LI (1989) Pharmacological bases for the use of dopamine and related drugs in the treatment of congestive heart failure. *Journal of Cardiovascular Pharmacology*, **14**, Suppl 8, S21–28.

Goldszer F, Tindell GL, Walle UK and Walle T (1981) Chemical trapping of labile aldehyde intermediates in the metabolism of propranolol and oxprenolol. *Research Communications in Chemical Pathology and Pharmacology*, **34**, 193–205.

Haefely W, Burkard WP, Cesura AM, Kettler R, Lorez HP, Martin JR, Richards JG, Scherschlicht R and Da Prada M (1992) Biochemistry and pharmacology of moclobemide, a prototype RIMA. *Psychopharmacology* (Berlin), **106**, Suppl: S6–S14.

Haenen RMM, Vermeulen NPE, Tai Tin Tsoi JNL, Ragetli HMN, Timmerman H and Bast A (1988) Activation of the microsomal glutathione-*S*-transferase and reduction of the glutathione dependent protection against lipid peroxidation by acrolein. *Biochemical Pharmacology*, **37**, 1933–1938.

Hasan F, Mc Crodden JM, Kennedy NP and Tipton KF (1988) The involvement of intestinal monoamine oxidase in the transport and metabolism of tyramine. *Journal of Neural Transmission*, **26**, [Suppl.], 1-9.

Hasegawa K, Ito S, Inoue S, Wakamatsu K, Ozeki H, Ishiguro I (1997) Dihydro-1,4-benzothiazine-6,7-dione, the ultimate toxic metabolite of 4-S-cysteaminylphenol and 4-S-cysteaminyl-catechol. *Biochemical Pharmacology*, **53**, 1435-1444.

Hayes BE and Clarke DE (1990) Semicarbazide-sensitive amine oxidase activity in streptozotocin diabetic rats. *Research Communications in Chemical Pathology and Pharmacology*, **69**, 71-83.

He N, Singhal SS, Awasthi S, Zhao T and Boor PJ (1999) Role of glutathione S-transferase 8-8 in allylamine resistance of vascular smooth muscle cells *in vitro*. *Toxicology and Applied Pharmacology*, **158**, 177-185.

Heikkila RE, Kindt MV, Sonsalla PK, Giovanni A, Youngster SK, Mc Keown KA and Singer TP (1988) Importance of monoamine oxidase-A in the bioactivation of neurotoxic analogs of 1-methyl-4-phenyl-1,2,3,6-tetrahydropyridine. *Proceedings of the National Academy of Science (USA)*, **85**, 6172-6176.

Holt A and Baker GB (1995) Metabolism of agmatine (clonidine-displacing substance) by diamine oxidase and the possible implications for studies of imidazoline receptors. *Progress in Brain Research*, **106**, 187-197.

Holt A and Callingham BA (1994) Location of the active site of rat vascular semicarbazide-sensitive amine oxidase. *Journal of Neural Transmisssion*, **41**, 433-437.

Holt A and Callingham BA (1995) Further studies on the ex-vivo effects of procarbazine and monomethylhydrazine on rat semicarbazide-sensitive amine oxidase and monoamine oxidase activities. *Journal of Pharmacy and Pharmacology*, **47**, 837-845.

Holt A, Sharman DF, Callingham BA and Kettler R (1992) Characteristics of procarbazine as an inhibitor in-vitro of rat semicarbazide-sensitive amine oxidase. *Journal of Pharmacy and Pharmacology*, **44**, 487-493.

Höltä E (1977). Oxidation of spermidine and spermine in rat liver: purification and properties of polyamine oxidase. *Biochemistry* (Easton), **16**, 91-100.

Houen G (1999) Mammalian Cu-containing amine oxidases (CAOs): new methods of analysis, structural relationships, and possible functions. *APMIS*, Suppl. **96**, 1-46.

Hysmith RM and Boor PJ (1988). Role of benzylamine oxidase in the cytotoxicity of allylamine toward aortic smooth muscle cells. *Toxicology*, **51**, 133-145.

Inoue S, Ito S, Wakamatsu K, Jimbow K, Fujita K (1990) Mechanism of growth inhibition of melanoma cells by 4-S-cysteaminylphenol and its analogues. *Biochemical Pharmacology*, **39**, 1077-1083.

Izquierdo-Pulido M, Albala-Hurtado S, Marine-Font A and Vidal-Carou MC (1996) Biogenic amines in Spanish beers: differences among breweries. *Zeitschrift für Lebensmittel-Untersuchung und-Forschung*, **203**, 507-511.

Jacob P and Shulgin AT (1981) Sulfur analogues of psychotomimetic agents. Monothio analogues of mescaline and isomescaline. *Journal of Medicinal Chemistry*, **24**, 1348-1353.

Janssens de Varebeke P, Cavalier R, David-Remacle M and Youdim MBH (1988) Formation of the neurotransmitter glycine from the anticonvulsant milacemide is mediated by brain monoamine oxidase B. *Journal of Neurochemistry*, **50**, 1011-1016.

Kinemuchi H, Morikawa F, Ueda T and Arai Y (1986) Studies of monoamine oxidase and semicarbazide-sensitive amine oxidase. I. Inhibition by a selective monoamine oxidase-B inhibitor, MD 780236. *Japanese Journal of Pharmacology*, **41**, 183-189.

Klinman JP and Mu D (1994) Quinoenzymes in biology. *Annual Review of Biochemistry*, **63**, 299-344.

Kunimoto S, Nosaka C, Xu CZ and Takeuchi T (1989) Serum effect on cellular uptake of spermidine, spergualin, 15-deoxyspergualin, and their metabolites by L5178Y cells. *Journal of Antibiotics* (Tokyo), **42**, 116-122.

Kreuger MJ, McKeown K, Ramsay RR, Youngster S and Singer TP (1990) Mechanism-based inactivation of monoamine oxidases A and B by tetrahydropyridines and dihydropyridines. *Biochemical Journal*, **268**, 219-224.

Kurkijarvi R, Yegutkin GG, Gunson BK, Jalkanen S, Salmi M and Adams DH (2000) Soluble vascular adhesion protein 1 accounts for the increased serum monoamine oxidase activity in chronic liver disease. *Gastroenterology*, **119**, 1096-1103.

Lake BG, Harris RA, Collins MA, Cottrell RC, Phillips JC and Gangolli SD (1982) Studies on the metabolism of dimethylnitrosamine in vitro by rat-liver preparations. II. Inhibition by substrates and inhibitors of monoamine oxidase. *Xenobiotica*, **12**, 567–579.

Lewinsohn R (1977) Human serum amine oxidase. Enzyme activity in severely burnt patients and in patients with cancer. *Clinica Chimica Acta*, **81**, 247–256.

Lewinsohn R (1984) Mammalian monoamine-oxidizing enzymes, with special reference to benzylamine oxidase in human tissues. *Brazilian Journal of Medical and Biological Research*, **17**, 223–256.

Lichtenberger LM, Gardner JW, Barreto JC and Morriss FH (1991) Evidence for a role of volatile amines in the development of neonatal hypergastrinemia. *Journal of Pediatric Gastroenterology and Nutrition*, **13**, 342–346.

Lin JK, Chang HW and Lin-Shiau SY (1984) Abundance of dimethylamine in seafoods: possible implications in the incidence of human cancer. *Nutrition and Cancer*, **6**, 148–159.

Lizcano JM, Balsa D, Tipton KF and Unzeta M (1990) Amine oxidase activities in bovine lung. *Journal of Neural Transmission*, **32**, [Suppl.], 341–344.

Lizcano JM, Fernandez de Ariba A, Tipton KF and Unzeta M (1996) Inhibition of bovine lung semicarbazide-sensitive amine oxidase (SSAO) by some hydrazine derivatives. *Biochemical Pharmacology*, **52**, 187–195.

Lizcano JM, Tipton KF and Unzeta M (1998) Purification and characterization of membrane-bound semicarbazide-sensitive amine oxidase (SSAO) from bovine lung. *Biochemical Journal*, **331**, 69–78.

Lortie MJ, Novotny WF, Peterson OW, Vallon V, Malvey K, Mendonca M, Satriano J, Insel P, Thomson SC and Blantz RC (1996) Agmatine, a bioactive metabolite of arginine. Production, degradation, and functional effects in the kidney of the rat. *Journal of Clinical Investigation*, **97**, 413–420.

Lyles GA (1978) Metabolism of octopamine in vitro by monoamine oxidase in some rat tissues. *Life Sciences*, **23**, 223–230.

Lyles GA (1995) Substrate-specificity of mammalian tissue-bound semicarbazide-sensitive amine oxidase. *Progress in Brain Research*, **106**, 293–303.

Lyles GA (1996) Mammalian plasma and tissue-bound semicarbazide-sensitive amine oxidases: biochemical, pharmacological and toxicological aspects. *International Journal of Biochemistry and Cell Biology*, **28**, 259–274.

Lyles GA and Fitzpatrick CMS (1985) An allylamine derivative (MDL 72145) with potent irreversible inhibitory actions on rat aorta semicarbazide-sensitive amine oxidase. *Journal of Pharmacy and Pharmacology*, **37**, 329–335.

Mantle TJ, Tipton KF and Garrett NJ (1976) Inhibition of monoamine oxidase by amphetamine and related compounds. *Biochemical Pharmacology*, **25**, 2073–2077.

Meier T and Issels RD (1995) Degradation of 2-(3-aminopropylamino)-ethanethiol (WR-1065) by Cu-dependent amine oxidases and influence on glutathione status of Chinese hamster ovary cells. *Biochemical Pharmacology*, **50**, 489–496.

Meszaros Z, Karadi I, Csanyi A, Szombathy T, Romics L and Magyar K (1999) Determination of human serum semicarbazide-sensitive amine oxidase activity: a possible clinical marker of atherosclerosis. *European Journal of Drug Metabolism and Pharmacokinetics*, **24**, 299–302.

Milmore JE and Taylor KM (1975) Propranolol inhibits rat brain monamine oxidase. *Life Sciences*, **17**, 1843–1847.

Naoi M, Maruyama W, Niwa T and Nagatsu T (1994) Novel toxins and Parkinson's disease: N-methylation and oxidation as metabolic bioactivation of neurotoxin. *Journal of Neural Transmission*, **41**, [Suppl.], 197–205.

Naoi M, Maruyama W, Zhang JH, Takahashi T, Deng Y and Dostert P (1995) Enzymatic oxidation of the dopaminergic neurotoxin, 1(R), 2(N)-dimethyl-6,7-dihydroxy-1,2,3,4-tetrahydroisoquinoline, into 1,2(N)-dimethyl-6,7-dihydroxyisoquinolinium ion. *Life Sciences*, **57**, 1061–1066.

Nelson TJ and Boor PJ (1982) Allylamine cardiotoxicity—IV. Metabolism to acrolein by cardiovascular tissues. *Biochemical Pharmacology*, **31**, 509–514.

Nelson WL and Bartels MJ (1984) N-dealkylation of propranolol in rat, dog, and man. Chemical and stereochemical aspects. *Drug Metabolism and Disposition*, **12**, 345–352.

Nørdqvist A, Oreland L and Fowler CJ (1982) Some properties of monoamine oxidase and a semicarbazide sensitive amine oxidase capable of the deamination of 5-hydroxytryptamine from porcine dental pulp. *Biochemical Pharmacology*, **31**, 2739–2744.

Novotny WF, Chassande O, Baker M, Lazdunski M and Barbry P (1994) Diamine oxidase is the amiloride-binding protein and is inhibited by amiloride analogues. *Journal of Biological Chemistry*, **269**, 9921–9925.

O'Brien EM and Tipton KF (1994) Biochemistry and mechanism of action of monoamine oxidases A and B. In *Monoamine Oxidase Inhibitors in Neurological Diseases*, Lieberman A, Olanow CW, Youdim MBH and Tipton KF (eds), Marcel Dekker, New York, pp. 31–76.

O'Brien EM, Tipton KF, Strolin Benedetti M, Bonsignori A, Marrari P and Dostert P (1991) Is the oxidation of milacemide by monoamine oxidase a major factor in its anticonvulsant actions? *Biochemical Pharmacology*, **41**, 1731–1737.

O'Brien EM, McCrodden JM, Youdim MBH and Tipton KF (1994a) The interactions of milacemide with monoamine oxidase. *Biochemical Pharmacology*, **47**, 617–623.

O'Brien EM, Dostert P, Pevarello P and Tipton KF (1994b) Interactions of some analogues of the anticonvulsant milacemide with monoamine oxidase. *Biochemical Pharmacology*, **48**, 905–914.

O'Brien EM, Dostert P and Tipton KF (1995) Interactions of some analogues of the anticonvulsant milacemide with monoamine oxidase. *Biochemical Pharmacology*, **48**, 905–914.

O'Carroll A-M, Tipton KF, Sullivan JP, Fowler CJ and Ross SB (1987) Intra and extra synaptosomal deamination of dopamine and noradrenaline by the two forms of human brain monoamine oxidase. Implications for the neurotoxicity of N-methyl-4-phenyl-1,2,3,6-tetrahydropyridine in man. *Biogenic Amines*, **4**, 165–178.

Olanow CW, Mytilineou C and Tatton W (1998) Current status of selegiline as a neuroprotective agent in Parkinson's disease. *Movement Disorders*, **13**, Suppl. 1, 55–58.

Palcic MM, Scaman CH and Alton G (1995) Stereochemistry and cofactor identity status of semicarbazide-sensitive amine oxidases. *Progress in Brain Research*, **106**, 41–47.

Palfreyman MG, McDonald IA, Bey P, Danzin C, Zreika M and Cremer G (1994) Haloallylamine inhibitors of MAO and SSAO and their therapeutic potential. *Journal of Neural Transmission*, **41**, [Suppl.], 407–414.

Pare CM (1986) New pharmacological developments in antidepressants. *Psychopathology*, **19**, Suppl 2, 103–107.

Patek DR and Hellerman L(1974) Mitochondrial monoamine oxidase: mechanism of inhibition by phenylhydrazine and aralkylhydrazine. Role of enzymatic oxidation. *Journal of Biological Chemistry*, **249**, 2373–2380.

Ramos K, Grossman SL and Cox LR (1988). Allylamine-induced vascular toxicity *in vitro*: prevention by semicarbazide-sensitive amine oxidase inhibitors. *Toxicology and Applied Pharmacology*, **95**, 61–71.

Robertson IG and Bland TJ (1993) Inhibition by proadifen of the aldehyde oxidase-mediated metabolism of the antitumour agent acridine carboxamide. *Biochemical Pharmacology*, **45**, 2159–2162.

Rochat B, Kosel M, Boss G, Testa B, Gollet M and Baumann P (1998) Stereoselective biotransformation of the selective serotonin reuptake inhibitor citalopram and its demethylated metabolites by monoamine oxidase in human liver. *Biochemical Pharmacology*, **56**, 15–23.

Roth RA, Roth JA and Gillis CN (1977) Disposition of ^{14}C-mescaline by rabbit lung. *Journal of Pharmacology and Experimental Therapeutics*, **200**, 394-401.

Schoerlin MP and Da Prada M (1990) Species-specific biotransformation of moclobemide: a comparative study in rats and humans. *Acta Psychiatrica Scandinavica*, Suppl, **360**, 108–110.

Seiler N (1995) Polyamine oxidase, properties and functions. *Progress in Brain Research* **106**, 333–344.

Seiler N (2000) Oxidation of polyamines and brain injury. *Neurochemical Research* **25**, 471–490.

Seiler N, Duranton B, Vincent F, Gosse F, Renault J and Raul F (2000) Inhibition of polyamine oxidase enhances the cytotoxicity of polyamine oxidase substrates. A model study with N(1)-(n-octanesulfonyl)spermine and human colon cancer cells. *International Journal of Biochemistry and Cell Biology*, **32**, 703–716.

Sen NP (1969) Analysis and significance of tyramine in foods. *Journal of Food Science*, **34**, 22–26.

Shah NS and Himwich HE (1971) Study with mescaline-8-C14 in mice: effect of amine oxidase inhibitors on metabolism. *Neuropharmacology*, **10**, 547–556.

Shaw LM, Bonner H and Lieberman R (1996) Pharmacokinetic profile of amifostine. *Seminars in Oncology*, **23**, 18–22.

Shih JC, Chen K and Ridd MJ (1999) Monoamine oxidase: from genes to behavior. *Annual Review of Neuroscience*, **22**, 197–217.

Silverman RB (1984) Effect of alpha-methylation on inactivation of monoamine oxidase by N-cyclopropylbenzylamine. *Biochemistry* (Easton), **23**, 5206–5213.

Silverman RB, Nishimura K and Lu X (1993) Mechanism of inactivation of monoamine oxidate-B by the anticonvulsant agent milacemide (2-(n-pentylamino)acetamide). *Journal of the American Chemical Society*, **115**, 4949–4954.

Singer TP, Salach JI and Crabtree D (1985) Reversible inhibition and mechanism-based irreversible inactivation of monoamine oxidases by 1-methyl-4-phenyl-1,2,3,6-tetrahydropyridine (MPTP). *Biochemical and Biophysical Research Communications*, **127**, 707–712.

Singer TP, Salach JI, Castagnoli N and Trevor A (1986) Interactions of the neurotoxic amine 1-methyl-4-phenyl-1,2,3,6-tetrahydropyridine with monoamine oxidases. *Biochemical Journal*, **235**, 785–789.

Smith DJ, Salmi M, Bono P, Hellman J, Leu T and Jalkanen S (1998) Cloning of vascular adhesion protein 1 reveals a novel multifunctional adhesion molecule. *ournal of Experimental Medicine*, **188**, 17–27.

Smith-Mungo LI and Kagan HM (1998) Lysyl oxidase: proprtties, regulation and multiple functions in biology. *Matrix Biology*, **16**, 387–398.

Squires RF (1972) Multiple forms of MAO in intact mitochondria as characterised by selective inhibitors and thermal stability: A comparison of eight mammalian species *Advances in Biochemical Psychopharmacology*, **5**, 355–370.

Steward MM (1976) MAOIs and food: fact and fiction. *Journal of Human Nutrition and Dietetics*, **30**, 415–419.

Strolin Benedetti M and Dostert P (1994) Contribution of amine oxidases to the metabolism of xenobiotics. *Drug Metabolism Reviews*, **26**, 507–535.

Strolin Benedetti M and Dow J (1983) A monoamine oxidase-B inhibitor, MD 780236, metabolized essentially by the A form of the enzyme in the rat. *Journal of Pharmacy and Pharmacology*, **35**, 238–245.

Strolin Benedetti M and Tipton KF (1998) Monoamine oxidases and related amine oxidases as phase I enzymes in the metabolism of xenobiotics. *Journal of Neural Transmisssion*, **52**, [Suppl.], 149–171.

Strolin Benedetti M, Boucher T, Carlsson A and Fowler CJ (1983a) Intestinal metabolism of tyramine by both forms of monoamine oxidase in the rat. *Biochemical Pharmacology*, **32**, 47–52.

Strolin Benedetti M, Dow J, Boucher T and Dostert P (1983b) Metabolism of the monoamine oxidase-B inhibitor, MD 780236 and its enantiomers by the A and B forms of the enzyme in the rat. *Journal of Pharmacy and Pharmacology*, **35**, 837–840.

Strolin Benedetti M, Rovei V, Thiola A and Donath R (1984) Metabolism of MD 240928, a short-acting type B MAOI in rat, dog and man. In *Monoamine Oxidase and Disease. Prospects for Therapy with Reversible Inhibitors*, Tipton KF, Dostert P and Strolin Benedetti M (eds), Academic Press, London, pp. 203–216.

Strolin Benedetti M, Marrari P, Moro E, Dostert P and Roncucci R (1988) Do amine oxidases contribute to the metabolism of milacemide? *Pharmacological Research Communications*, **20**, [Suppl iv], 135–136.

Strolin Benedetti M, Cocchiara G, Colombo M and Dostert P (1990) Does FAD-dependent polyamine oxidase contribute to the metabolism of milacemide? *Journal of Neural Transmisssion*, **32**, [Suppl.], 351–356.

Strolin Benedetti M, Dostert P and Tipton KF (1992a) Developmental aspects of the monoamine-degrading enzyme monoamine oxidase. *Developmental Pharmacology and Therapeutics*, **18**, 191–200.

Strolin Benedetti M, Allievi C, Cocchiara G, Pevarello P and Dostert P (1992b) Involvement of FAD-dependent polyamine oxidase in the metabolism of milacemide in the rat. *Xenobiotica*, **22**, 191–197.

Strolin Benedetti M, Sanson G, Bona L, Gallina M, Persiani S and Tipton KF (1998) The oxidation of dopamine and epinine by the two forms of monoamine oxidase from rat liver. *Journal of Neural Transmisssion*, **52**, [Suppl.], 233–238.

Sullivan JP and Tipton KF (1990) The interactions of monoamine oxidase with some derivatives of 1-methyl-4-phenyl-1,2,3,6-tetrahydropyridine (MPTP). *Journal of Neural Transmisssion*, **29**, [Suppl.], 269–277.

Sullivan JP and Tipton KF (1992) Interactions of the neurotoxin MPTP and its demethylated derivative (PTP) with monoamine oxidase-B. *Neurochemical Research*, **8**, 791–796.

Sullivan JP, McDonnell L, Hardiman OM, Farrell MA, Phillips JP and Tipton KF (1986) The oxidation of tryptamine by the two forms of monoamine oxidase in human tissues. *Biochemical Pharmacology*, **35**, 3255–3260.

Sullivan JP, McCrodden JM and Tipton KF (1990) Enzymological aspects of the action of the parkinsonism-inducing neurotoxin MPTP. *Advances in Behavioural Biology*, **38A**, 227–230.

Suzuki O, Katsumata Y, Oya M and Matsumoto T (1979) Oxidation of phenylethanolamine and octopamine by type A and type B monoamine oxidase. Effect of substrate concentration. *Biochemical Pharmacology*, **28**, 2327–2332.

Taneja DT and Lyles GA (1988). Further studies on the interactions between amine oxidase inhibitors and tryptamine-induced contractions of rat aorta. *British Journal of Pharmacology*, **93**, 253P.

Tano K, Yoshizumi M, Kitagawa T, Hori T, Kitaichi T, Itoh K and Katoh I (1997) Effect of docarpamine, a novel orally active dopamine prodrug, on the formation of free and sulphoconjugated dopamine in patients who underwent cardiac surgery. *Life Sciences*, **61**, 1469–1478.

Tipton, KF (1989) Mechanism-based inhibitors. In *Design of Enzyme Inhibitors as Drugs*, Sandler M and Smith HJ (eds), Oxford University Press, Oxford, pp. 70–93.

Tipton KF (1990) The design and behaviour of selective monoamine oxidase inhibitors. In *Antidepressants: Thirty Years On*, Leonard, B and Spencer (eds) CNS (Clinical Neuroscience) Publishers, London, pp.193–203.

Tipton KF (1994) What is it that *l*-deprenyl (selegiline) might do? *Clinical Pharmacology and Therapeutics*, **56**, 781–796.

Tipton KF and Singer TP (1993) Advances in our understanding of the mechanisms of the neurotoxicity of MPTP and related compounds. *Journal of Neurochemistry*, **61**, 1191–1206.

Tipton KF and Spires IPC (1971) The kinetics of phenethylhydrazine oxidation by monoamine oxidase. *Biochemical Journal*, **125**, 521–524.

Tipton, KF, McCrodden JM, Kalir AS and Youdim MBH (1982) Inhibition of rat liver monoamine oxidase by β-methyl and N-propargyl-amine derivatives. *Biochemical Pharmacology*, **31**, 1251–1255.

Tipton KF, Fowler CJ, McCrodden JM and Strolin Benedetti M (1983) The enzyme-activated irreversible inhibition of type-B monoamine oxidase by 3-(4-[(3-chlorophenyl)methoxy]-phenyl)-5-[(methylamino) methyl]-2-oxazolidinone methanesulphonate (compound MD 780236) and the enzyme-catalysed oxidation of this compound as competing reactions. *Biochemical Journal*, **209**, 235–242.

Tipton KF, McCrodden JM, Henehan GT, Boucher T and Fowler CJ (1984a) The formation of the acidic and alcoholic metabolites of MD 780236. *Biochemical Pharmacology*, **33**, 1377–1378.

Tipton KF, Strolin Benedetti M, McCrodden JM, Boucher T and Fowler CJ (1984b) The inhibition of rat liver monoamine oxidase-B by the enantiomers of MD 780236. In *Monoamine oxidase and Disease. Prospects for Therapy with Reversible Inhibitors*, Tipton KF, Dostert P and Strolin Benedetti M (eds), Academic Press, London, pp. 155–163.

Tipton KF, McCrodden JM and Youdim MBH (1986) Oxidation and enzyme-activated irreversible inhibition of rat liver monoamine oxidase-B by 1-methyl-4-phenyl-1,2,3,6-tetrahydropyridine (MPTP). *Biochemical Journal*, **240**, 379–383.

Tipton KF, McCrodden JM and Sullivan JP (1993) Metabolic aspects of the behaviour of MPTP and some analogues. *Advances in Neurology*, **60**, 186–193.

Tocco DJ, Duncan AE, deLuna FA, Smith JL, Walker RW and Vandenheuvel WJ (1980) Timolol metabolism in man and laboratory animals. *Drug Metabolism and Disposition*, **8**, 236–240.

Turkish S, Yu PH and Greenshaw AJ (1988) Monoamine oxidase-B inhibition: a comparison of *in vivo* and *ex vivo* measures of reversible effects. *Journal of Neural Transmisssion*, **74**, 141–148.

van Dorsser W, Barris D, Cordi A and Roba J (1983) Anticonvulsant activity of milacemide. *Archives Internationales de Pharmacodynamie et de Therapie*, **266**, 239–249.

Van Haarst AD, Van Gerven JM, Cohen AF, De Smet M, Sterrett A, Birk KL, Fisher AL, De Puy ME, Goldberg MR and Musson DG (1999) The effects of moclobemide on the pharmacokinetics of the 5-HT1B/1D agonist rizatriptan in healthy volunteers. *British Journal of Clinical Pharmacology*, **48**, 190–196.

Vyas KP, Halpin RA, Geer LA, Ellis JD, Liu L, Cheng H, Chavez-Eng C, Matuszewski BK, Varga SL, Guiblin AR and Rogers JD (2000) Disposition and pharmacokinetics of the antimigraine drug, rizatriptan, in humans. *Drug Metabolism and Disposition*, **28**, 89–95.

Walle UK, Wilson MJ and Walle T (1981) Propranolol, alprenolol and oxprenolol metabolism in the dog. Identification of N-methylated metabolites. *Biomedical Mass Spectrometry*, **8**, 78–84.

Walle T, Walle UK, Knapp DR, Conradi EC and Bargar EM (1983) Identification of major sulfate conjugates in the metabolism of propranolol in dog and man. *Drug Metabolism and Disposition*, **11**, 344–349.

Wang SX, Nakamura N, Mure M, Klinman JP and Sanders-Loeht J (1997) Characterization of the native lysine tyrosylquinone cofactor of lysyl oxidase by Raman spectroscopy. *Journal of Biological Chemistry*, **272**, 28841–28844.

Watanabe K, Marsunaga T, Yamamoto I and Yashimura H (1995) Oxidation of tolualdehydes to toluic acids by cytochrome P450-dependent aldehyde oxygenase in mouse liver. *Drug Metabolism and Disposition*, **23**, 261–265.

Westlund KN, Denney RM, Kochersperger LM, Rose RM and Abell CW (1985) Distinct monoamine oxidase A and B populations in primate brain *Science*, **230**, 181–183.

Wheatley AM and Tipton KF (1987) Determination of tyramine in alcoholic and non-alcoholic beers by high performance liquid chromatography with electrochemical detection. *Journal of Food. Biochemistry*, **11**, 133–142.

Wild MJ, McKillop D and Butters CJ (1999) Determination of the human cytochrome P450 isoforms involved in the metabolism of zolmitriptan. *Xenobiotica*, **29**, 847–857.

Williams CH (1977) beta-phenylethanolamine as a substrate for monoamine oxidase. *Biochemical Society Transactions*, **5**, 1770–1771.

Worland PJ and Ilett KF (1983) Intestinal contribution to the presystemic elimination of betaphenethylamine in the rat. *Journal of Pharmacy and Pharmacology*, **35**, 636–640.

Wouters J (1998) Structural aspects of monoamine oxidase and its reversible inhibition. *Currents in Medicinal Chemistry*, **5**, 137–162.

Yoshikawa M, Endo H, Komatsu K, Fujihara M, Takaiti O, Kagoshima T, Umehara, M and Ishikawa H (1990) Disposition of a new orally active dopamine prodrug, N-(N-acetyl-L-methionyl)-O,O-bis(ethoxycarbonyl) dopamine (TA-870) in humans. *Drug Metabolism and Disposition*, **18**, 212–217.

Yoshikawa M, Nishiyama S and Takaiti O (1995) Metabolism of dopamine prodrug, docarpamine. *Hypertension Research*, **18**, Suppl 1, S211–S213.

Youngster SK, Sonsalla PK, Sieber B-A and Heikkila RE (1989) Structure-activity study of the mechanism of 1-methyl-4-phenyl-1,2,3,6-tetrahydropyridine (MPTP)-induced neurotoxicity.I. Evaluation of the biological activity of MPTP analogs. *Journal of Pharmacology and Experimental Therapeutics*, **249**, 820–828.

Yu PH (1988) Three types of stereospecificity and the kinetic deuterium isotope effect in the oxidative deamination of dopamine as catalyzed by different amine oxidases. *Biochemistry and Cell Biology*, **66**, 853–861.

Yu PH (1989) Deamination of aliphatic amines of different chain lengths by rat liver monoamine oxidase A and B. *Journal of Pharmacy and Pharmacology*, **41**, 205–208.

Yu PH (1998a) Increase of formation of methylamine and formaldehyde in vivo after administration of nicotine and the potential cytotoxicity. *Neurochemical Research*, **23**, 1205–1210.

Yu PH (1998b) Deamination of methylamine and angiopathy; toxicity of formaldehyde, oxidative stress and relevance to protein glycoxidation in diabetes. *Journal of Neural Transmisssion*, **52**, [Suppl.], 201–216.

Yu PH and Davis BA (1988) Stereospecific deamination of benzylamine catalyzed by different amine oxidases. *International Journal of Biochemistry*, **20**, 1197–1201.

Yu PH and Davis BA (1990) Some pharmacological implications of MAO-mediated deamination of branched aliphatic amines: 2-propyl-1-aminopentane and N-(2-propylpentyl)glycinamide as valproic acid precursors. *Journal of Neural Transmisssion*, **32**, [Suppl.], 89–92.

Yu PH and Davis BA (1991a) Deamination of 2-propyl-1-aminopentane and 2-[(2-propyl)pentyl-amino] acetamide by amine oxidases: formation of valproic acid. *Progress in Neuropsychol-pharmacology and Biological Psychiatry*, **15**, 303–306.

Yu PH and Davis BA (1991b) Simultaneous delivery of valproic acid and glycine to the brain. Deamination of 2-propylpentylglycinamide by monoamine oxidase B. *Molecular and Chememical Neuropathology*, **15**, 37–49.

Yu PH and Deng Y (2000) Potential cytotoxic effect of chronic administration of creatine, a nutrition supplement to augment athletic performance. *Medical Hypotheses*, **54**, 726–728.

Yu PH and Tipton KF (1990) Deuterium isotope effect of phenelzine on the inhibition of rat liver mitochondrial monoamine oxidase activity. *Biochemical Pharmacology*, **38**, 4245–4251.

Yu PH, Zuo DM and Davis BA (1994) Characterization of human serum and umbilical artery semicarbazide-sensitive amine oxidase (SSAO). *Biochemical Pharmacology*, **47**, 1055–1059.

Yu PH, Bailey BA, Durden DA and Boulton AA (1986) Stereospecific deuterium substitution at the alpha-carbon position of dopamine and its effect on oxidative deamination catalyzed by MAO-A and MAO-B from different tissues. *Biochemical Pharmacology*, **15**, 1027–1036.

Zreika M, McDonald IA, Bey P and Palfreyman MG (1984) MDL 72145, an enzyme-activated irreversible inhibitor with selectivity for monoamine oxidase type B. *Journal of Neurochemistry*, **43**, 448–454.

5 Molybdenum Hydroxylases

Christine Beedham

University of Bradford, UK

Introduction

Aldehyde oxidase (AO) and xanthine oxidase (XO) are complex molybdoflavoproteins with similar composition and catalytic properties but which differ in their substrate/inhibitor specificities. Xanthine dehydrogenase (XDH) is thought to be a different functional form of XO, coded by the same gene, which reacts preferentially with NAD^+ in contrast to XO and AO which utilise molecular oxygen (O_2) as an oxidising substrate. In mammals XDH is thought to predominate *in vivo*, but this form undergoes a facile conversion to XO during purification via sulphydryl oxidation or proteolysis (Nishino 1994).

Both proteins show a remarkable degree of sequence identity and Terao *et al.* (1998) have suggested that AO and XO are members of a multigene family, which have originated from a relatively recent duplication event. However, evidence to date indicates that there are fewer members of the 'molybdenum hydroxylase' family than those of the cytochrome P450 superfamily. Nevertheless, the wide range of drugs, xenobiotics and endogenous chemicals that interact with these enzymes, particularly AO, highlight the importance of these enzymes in drug oxidation, detoxication and activation (Beedham 1985, 1987, 1997, 1998).

FUNCTIONALISATION REACTIONS CATALYSED BY MOLYBDENUM HYDROXYLASES

Molybdenum hydroxylases catalyse both oxidation and reduction reactions although the prevalence of the former reactions significantly outweighs the latter *in vivo*. Unlike cytochrome P450, oxidative hydroxylation catalysed by the molybdenum hydroxylases generates reducing equivalents and, although both enzymes utilise molecular oxygen (O_2), the ultimate source of the oxygen atom inserted into substrates is water and not O_2 (Xia *et al.* 1999). In most cases, AO and XO have complementary substrate specificities with microsomal monooxygenases.

With respect to substrates, oxidation involves nucleophilic attack at an electron-

Enzyme Systems that Metabolise Drugs and Other Xenobiotics. Edited by C. Ioannides.
© 2002 John Wiley & Sons Ltd

deficient carbon with the insertion of an oxygen atom to generate either a cyclic lactam or a carboxylic acid from aromatic N-heterocycles and aldehydes respectively (Figure 5.1). N- or S-oxidation has not been reported.

In contrast, N- and S-functional groups are reduced by both AO and XO under anaerobic conditions. Reactions include reduction of N-oxides, sulphoxides, hydro-xamic acids, epoxides and reductive ring cleavage some of which may also be catalysed by cytochrome P450 reductase.

ENZYME STRUCTURE

AO and XO have unusually broad and overlapping substrate specificities indicative of a relatively flexible or accessible binding site. Both enzymes are homodimers, with each subunit containing a molybdopterin cofactor, FAD and two different 2Fe:2S clusters. Insight into the structural organisation of the proteins has been obtained from the crystal stucture of the related molybdo-enzyme, *Desulphovibrio gigas* aldehyde oxidoreductase (Romão *et al.* 1995). The molybdopterin cofactor, initially charac-terised by Rajagopalan and Johnson (1992) and later confirmed by the crystal structure of *gigas* aldehyde oxidoreductase contains mononuclear molybdenum (Mo) coordi-nated to the enzyme via a *cis* dithiolene moiety (Romão *et al.* 1995; Rajagopalan 1997). In addition, Mo is also coordinated to sulphido and oxo ligands in XO and AO with one further ligand, which is either a further oxo ligand, water ligand or probably a metal-coordinated hydroxide (Mo–OH, Figure 5.2) (Xia *et al.* 1999).

REACTION MECHANISM

The overall reaction mechanism can be considered as coupled reductive and oxida-tive half-reactions; hence different electroactive substrates may undergo oxidation or reduction (Kisker *et al.* 1997). Furthermore, oxidation of reducing substrates such as 2-pyrimidinone can provide the reducing equivalents for numerous reduction reactions under anaerobic conditions (see below). Reducing substrates react at the molybdenum centre via a two-electron redox reaction during which the Mo is reduced from Mo (VI) to Mo (IV) (Kisker *et al.* 1997). The most recent mechanism proposed for the reductive half-reaction of XO is shown in Figure 5.2 (Xia *et al.*1999). Base-assisted hydroxylation at an electron deficient carbon via Mo–OH is thought to precede hydride transfer to the sulphido ligand of Mo. Although water is the ultimate source of oxygen atom incorporated into product, it is proposed that the catalytically labile site is Mo–OH although alternative mechanisms invoke either an oxo-ligand or a buried water molecule in the protein as the attacking species (Hille and Sprecher 1987; Howes *et al.* 1996; Bray and Lowe 1997). In all cases, enzyme turnover would generate active enzyme via reaction with solvent (Hille and Sprecher 1987).

$$RH + H_2O \xrightarrow[\text{Oxidative hydroxylation}]{\text{Molybdenum hydroxylases}} ROH + 2H^+ + 2e^-$$

Figure 5.1. Oxidative hydroxylation catalysed by the molybdenum hydroxylases, AO and XO.

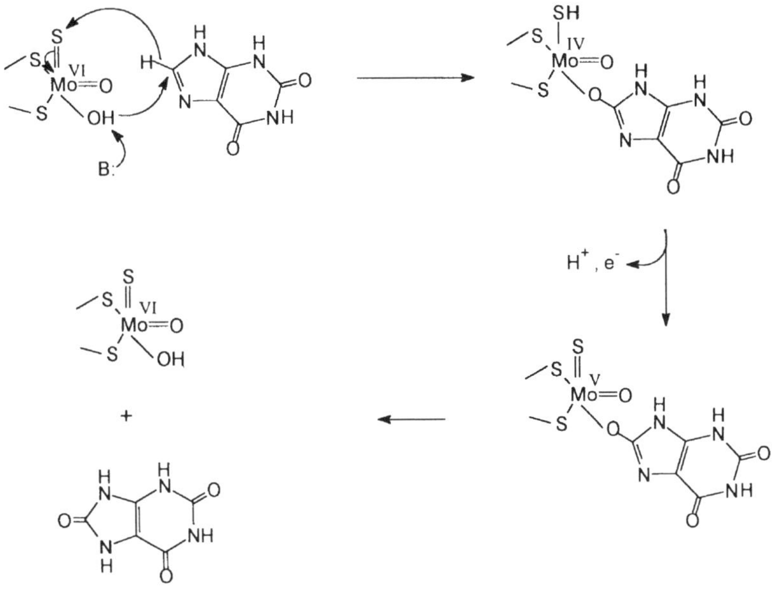

Figure 5.2 Reaction mechanism for the oxidation of xanthine to uric acid catalysed by XO via base-assisted nucleophilic attack (Xia *et al.* 1999).

During the oxidative half-reaction Mo(IV) is re-oxidised via rapid one-electron transfer to an iron-sulphur cluster, generating an intermediate electroparamagnetic Mo(V), and further intramolecular electron transfer to FAD.

RE-OXIDATION OF REDUCED ENZYME

Although oxidising substrates can interact at any of the redox centres, physiological electron acceptors react with the FAD site in XO, XDH and AO (Hille and Nishino 1995). AO and XO generate the reactive oxygen species (ROS), H_2O_2 and O_2^-, during substrate oxidation. The complex kinetics of reduced XO and XDH with their preferred electron acceptors has been reviewed by Hille and Nishino (1995). Reaction of XO with O_2 is biphasic; a fast two-electron transfer which generates hydrogen peroxide (H_2O_2) and a slower one-electron transfer to produce superoxide anion (O_2^-). XO shows negligible activity with NAD^+. Re-oxidation of XDH via NAD^+ is a monophasic two-electron reaction. Surprisingly, the dehydrogenase form is more efficient than XO at generating O_2^- although NAD^+ is the preferred substrate. For rat liver XDH, $K_m^{Oxygen} = 260$ μM whereas $K_m^{NAD} = 8.5$ μM (Saito and Nishino 1989; Nishino 1994). In contrast, AO is without reactivity towards NAD^+ (Li Calzi *et al.* 1995; Turner *et al.* 1995) and lacks an amino sequence characteristic of a NAD^+-binding site, which is conserved in XO proteins (Li Calzi *et al.* 1995; Kurosaki *et al.* 1999; Wright *et al.* 1999a).

Substrate specificity

AO and XO catalyse the oxidation of an extensive range of *N*-heterocycles and aldehydes; of the two enzymes AO has a wider substrate specificity than XO (Beedham 1985, 1987, 1998). This is illustrated in the following section exemplified, where possible, by *in vivo* or clinical data.

OXIDATION

Oxidation of uncharged N-heterocycles

Table 5.1 shows the major groups of uncharged *N*-heterocyclic substrates of AO. A π-deficient ring system containing a *N*-heteroatom is invariably essential for activity

Table 5.1 Uncharged *N*-heterocyclic substrates of AO

Azaheterocyclic nucleus	Substituted drugs/ xenobiotics	AO-generated metabolite(s)
Pyridine	Metyrapone AVS	
Quinoline Isoquinoline	Aminoquinolines Quinine Quinidine	
Phenanthridine		
Acridine	*N*[(2′-Dimethylamino) ethyl]acridine-4-carboxamide (DACA)	
Pyrimidine	2-Pyrimidinones 4-Pyrimidinone IPdR	
Phthalazine	Substitued phthalazines Carbazeran	

Table 5.1 (*continued*)

Azaheterocyclic nucleus	Substituted drugs/ xenobiotics	AO-generated metabolite(s)
Quinazoline	Substituted quinazolines	
Quinoxaline	Substituted quinoxalines Brominidine	
Pteridine	Methotrexate	
Purine	Hypoxanthine, Famciclovir 6-Deoxypenciclovir O^6-Benzylguanine 6-Thioguanine 6-mercaptopurine	
Pyrazolopyrimidine	Zaleplon	

AVS, 2(R,S)-1,2-bis(nicotinamido)propane; IPdR, 5-Iodo-2-pyrimidinone-2'-deoxyribose

towards molybdenum hydroxylases (Beedham 1985, 1987) although unsubstituted pyridine is a very poor substrate for both AO and XO (Krenitsky *et al.* 1972; Stubley and Stell 1980). Of the few pyridine AO substrates, metapyrone is converted to an α–pyridone metabolite by rat liver AO (Usanky and Damani 1983) and AVS, 2(R,S)-1,2-bis(nicotinamido)propane, which is effective in the treatment of vasospasm in dogs following subarachinoid haemorrhage, is metabolised in rabbits to a monopyridone that accounts for ~ 30% total urinary radioactivity (Ishigai *et al.* 1998).

Increased lipophilicity, by fusion of benzene rings to the pyridine nucleus, facilitates binding of substrates to the active site although the position of the hydrophobic groups is critical in governing substrate activity. Quinoline and the isomeric isoquinoline (Table 5.1) give K_m values of 3 mM and 0.2 mM respectively with rabbit liver AO (Stubley *et al.* 1979). 3,4-Benzoquinoline (phenanthridine) has a very high affinity for AO (Stubley and Stell 1980; Rashidi Shagoli 1996), acridine (2,3-benzoquinoline) is a reasonable substrate (McMurtrey and Knight 1984), whereas 5,6- and 7-8-benzoquinolines do not bind to the enzyme (Stubley and Stell 1980). All four cinchona alkaloids, quinine, quinidine, cinchonine and cinchonidine, are oxidised by AO to quinolone metabolites (Beedham et al. 1992). These quinoline-based antimalarials have low K_m values with AO but are coupled with relatively low oxidation rates. Consequently, quinine and quinidine are also substrates for cytochrome P450, principally CYP3A (Spray 1996). In healthy volunteers, 3-hydroxyquinine, quinine glucuronide and 2'-quininone are found as major urinary metabolites of quinine (Figure 5.3) (Wanwimolruk et al. 1995).

In contrast, the acridine anticancer agent N-[(2'-dimethylamino)ethyl]acridine-4-carboxamide, DACA (Figure 5.4), is rapidly converted to an acridone metabolite by human, rat and guinea pig AO (Schofield et al. 2000) but is also hydroxylated by cytochrome P450 in rats and mice (Robertson et al. 1993). However, neither phenolic nor glucuronide metabolites were detected in urine during Phase I clinical trials whereas all the major metabolites contained an acridone nucleus (Schofield et al. 1999). Oxidation of DACA by AO is unusual in that the normal position of oxidation, adjacent to a heterocyclic nitrogen atom, is not available for enzyme attack and oxidation occurs at an alternative electron-deficient site in the molecule (Figure 5.4, Table 5.1).

The diazabenzenes, pyrimidine, pyrazine and pyridazine, show little or no activity with either rabbit liver AO or bovine milk XO (Krenitsky et al. 1972; Stubley and Stell 1980). Oxidation of 2-pyrimidinone to uracil (Figure 5.5) is widely used as an

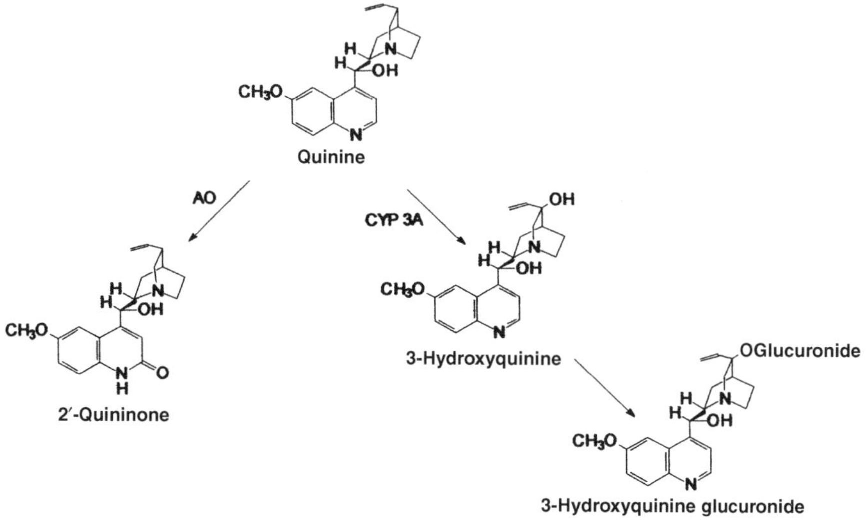

Figure 5.3 Major urinary metabolites of quinine in healthy volunteers (Wanwimolruk et al. 1995).

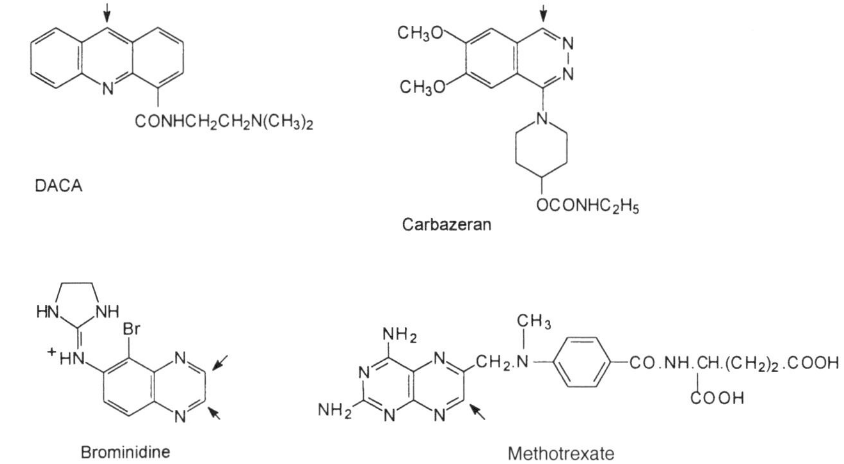

DACA

Carbazeran

Brominidine

Methotrexate

Figure 5.4 *In vivo* substrates of AO.

electron-donating system for AO in *in vitro* reduction studies. Although XO can also catalyse the same reaction, oxidation rates are slow and thus in combined enzyme preparations and liver slices AO is the major 2-pyrimidinone-oxidising enzyme (Oldfield 1998). In fact, 2-pyrimidinone oxidation exhibits biphasic kinetics and it is thought that two AO isozymes are present in guinea pig liver (Yoshihara and Tatsumi 1986; Oldfield 1998). Uracil is also a product of 4-pyrimidinone oxidation which is catalysed by both AO and XO (Oldfield 1998). Results from liver slice incubations show that the latter enzyme predominates in the oxidation reaction (Figure 5.5).

AO-catalysed oxidation of 2-pyrimidinones has been proposed as a bioactivation route for a number of pro-drugs. These include 5-ethynyluracil, a mechanism-based inhibitor of dihydropyrimidine dehydrogenase (Porter *et al.* 1994), 5-fluoro-2-pyrimidinone, a precursor of 5-fluorouracil (Guo *et al.* 1995) and IPdR, which is activated to the radiosensitising nucleoside, 5-iodo-2'-deoxyuridine (Chang *et al.* 1992; Kinsella *et al.* 1998). 5-Fluoro-4-pyrimidinone is also oxidised to 5-fluroruracil but in this case XO

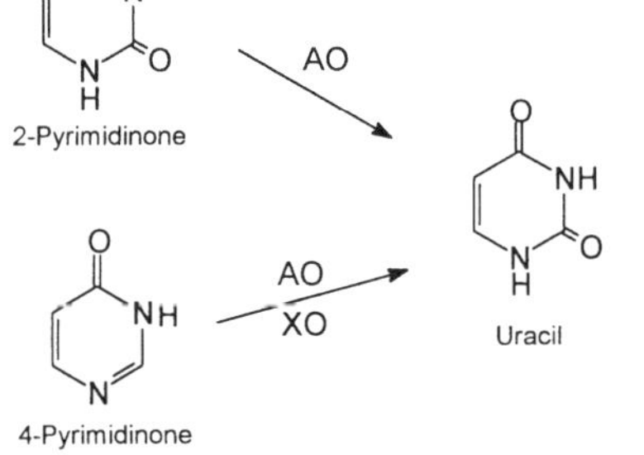

2-Pyrimidinone

4-Pyrimidinone

Uracil

Figure 5.5 Oxidation of 2-pyrimidinone and 4-pyrimidinone to uracil catalysed by AO and XO (Oldfield 1998).

may play the major role (Oldfield 1998). Zaleplon, a pyrazolo[1,5-a]pyrimidine derivative (Figure 5.6) recently introduced as an ultra-short acting hypnotic, undergoes N-dealkylation and oxidation to N-desethylzaleplon and 5-oxo-zaleplon respectively. 5-Oxozaleplon is the major metabolite in human plasma and AO is thought to be responsible for the deactivation step (Kawashima et al. 1999).

Phthalazine, quinazoline, quinoxaline and cinnoline ring systems are all oxidised to lactam metabolites by AO (Table 5.1) (Stubley et al. 1979; Beedham et al. 1990, 1995a). Quinazoline and quinoxaline undergo sequential attack to di-oxo products. Significant interspecies differences in substrate specificity were found although lipophilic substituents facilitated binding of substituted phthalazines and quinazolines to AO. In vivo activity towards this group of compounds is exemplified by carbazeran, a 5,6-dimethoxyphthalazine (Figure 5.4), which was developed as an inotropic agent but was found to undergo complete clearance presystemically via 4-oxidation catalysed by liver AO (Kaye et al. 1984, 1985). However, AO activity is not restricted to liver (see below). The ocular pharmacokinetics and metabolism of the potent ocular hypotensive drug, brominidine (Figure 5.4), has been studied in rabbits. After rapid ocular absorption, brominidine was metabolised to three metabolites, which were characterised as a quinoxaline-2-3-dione derivative and two isomeric quinoxalinones (Acheampong et al. 1995). The same metabolites were also formed in the presence of rabbit and rat liver AO. Diazanaphthalenes show weak or negligible activity towards XO (Krenitsky et al. 1972; Stubley et al. 1979; Beedham et al. 1990).

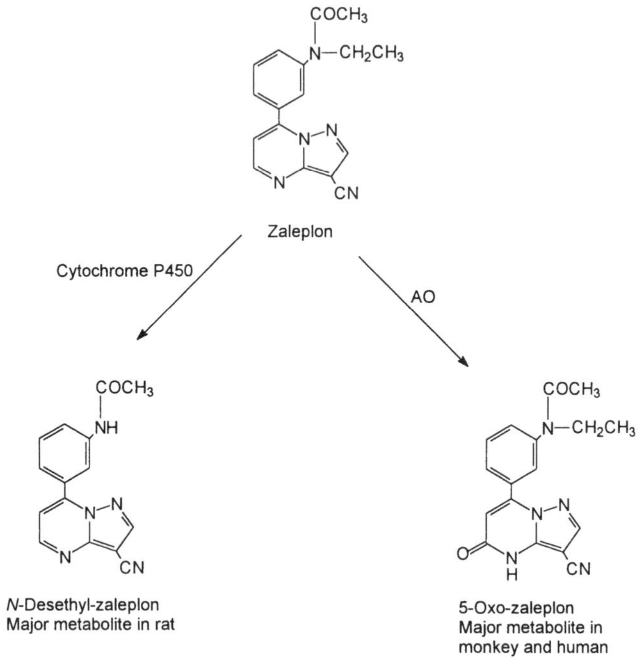

Figure 5.6 Major metabolites of Zaleplon in rat, monkey and human (Chaudhary et al. 1994).

Pteridines are excellent substrates for the molybdenum hydroxylases; AO and XO catalyse oxidation at carbon 2,4, or 7 although not necessarily at the same position or in the same molecule (Krenitsky *et al.* 1972; Hodnett *et al.* 1976). In contrast, oxidation has not been reported at carbon 6. As oxidation can occur at alternative positions and substitution effects are complex, prediction of the site of enzymic attack and nature of attacking enzyme is difficult. *In vitro* studies indicate that XO may predominate in pteridine metabolism, however, the antineoplastic drug, methotrexate (MTX, Figure 5.4), is a substrate for AO (Johns *et al.* 1966; Jordan *et al.* 1999) but a competitive inhibitor of XO (Lewis *et al.* 1984). *In vitro* oxidation rates for the conversion of MTX, a 2,4-diaminopteridine, to 7-hydroxymethotrexate are relatively slow, particularly with human liver enzyme (Jordan *et al.* 1999; Kitamura *et al.* 1999c). However, 7-hydroxymethotrexate is the major MTX metabolite; it is also cytotoxic and may contribute to MTX toxicity (Smeland *et al.* 1996). In fact, the term '7-hydroxy'-MTX is misleading as this metabolite is not a phenol but a cyclic lactam and thus it is not conjugated but excreted directly via the kidney. The cyclic lactam structure is typical of all AO- and XO-generated metabolites (Table 5.1).

Krenitsky's group has reported a number of quantitative investigations into the specificity of rabbit liver AO, bovine milk and human liver XO towards substituted purines (Krenitsky *et al.* 1972, 1986; Hall and Krenitsky 1986). Despite these rigorous and extensive studies, it was concluded that there is no single or simple set of determinants that define substrate specificity of purines towards AO and XO. This is mirrored with other substrate groups (Beedham *et al.* 1990, 1995a). With purines and other ring systems, containing more than two N-atoms, there are multiple oxidation sites and the site of attack varies with different substituents. Each enzyme may catalyse oxidation at the same or a different site. This may indicate that there are multiple productive orientations within the substrate-binding site or that the binding site has considerable flexibility.

Purine is sequentially oxidised, via hypoxanthine and xanthine, to uric acid by XO. Indeed the role of XO in endogenous purine metabolism is well documented (Moriwaki *et al.* 1999). The high efficiency of XO towards hypoxanthine and xanthine *in vitro* and *in vivo* indicates that XO activity towards purine-based drugs would be more significant than that of AO. However, AO has the wider substrate specificity and the results obtained with rabbit liver AO may not reflect the reaction of purines with AO from other species. This is illustrated with the antiviral drug, famciclovir, which is the 6-deoxy-diacetyl ester of the herpes active compound penciclovir (Perry and Wagstaff 1995). After oral administration, famciclovir, a pro-drug, is sequentially deacetylated to 6-deoxypenciclovir followed by AO-catalysed oxidation at carbon 6 to penciclovir. Famciclovir and 6-deoxypenciclovir are efficient substrates for AO whereas they show little or no activity with XO (Clarke *et al.* 1995; Rashidi *et al.* 1997). There are two alternative sites for oxidative attack on 6-deoxypenciclovir and famciclovir, carbon 6 and carbon 8 (Figure 5.7). 6-Deoxypenciclovir is oxidised by rabbit liver AO to 8-oxo-6-deoxypenciclovir and penciclovir in approximately equal amounts (Rashidi *et al.* 1997). Both reactions were completely inhibited by the potent AO inhibitor, menadione, and similar K_m values were obtained for each reaction indicating the participation of a single isozyme. In comparison, with guinea pig, human and rat liver AO, the 6-oxidation pathway is predominant with minimal

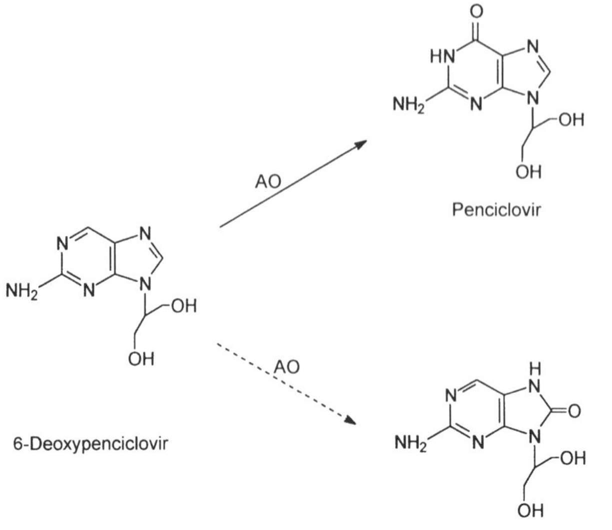

Figure 5.7 AO-catalysed activation of 6-deoxypenciclovir to the antiviral agent, penciclovir, in man and guinea pig (Rashidi *et al.* 1997; Filer *et al.* 1994). (Discontinuous line represents a minor route of metabolism.)

8-hydroxylation. This is consistent with *in vivo* results where penciclovir is the major metabolite accounting for 80-85% dose of famciclovir (Filer *et al.* 1994). AO and XO have also been implicated in the metabolism of other purine-based antivirals such as aciclovir (De Miranda and Good 1992) and 6-deoxycarbovir (Iyer *et al.* 1992).

Nevertheless 8-oxidation of xenobiotic purines is observed *in vivo* and, depending on the substrate, may be catalysed by either AO or XO. The major metabolite of oral 6-mercaptopurine is 6-thiouric acid with an alternative catabolic pathway of *S*-methylation (Figure 5.8). Oxidation is probably catalysed by gut XO and co-administration of allopurinol with 6-mercaptopurine increases bioavailability of the latter drug (Zimm *et al.* 1983). 6-Thioxanthine is thought to be the intermediate in the catabolic pathway, thus oxidation at carbon 2 precedes that at carbon 8 (Zimm *et al.* 1984; Rashidi 1996). However during intravenous infusion, which bypasses intestinal XO, of either 6-mercaptopurine or the more potent 6-thioguanine, the major plasma metabolites are 8-oxo-derivatives; 6-methylmercapto-8-oxopurine and 8-oxothioguanine respectively (KeuzenkampJansen *et al.* 1996; Kitchen *et al.* 1999). Although the formation of the 8-oxopurines could be ascribed to either XO or AO, the difference in metabolite profile between oral and intravenous administration indicates that AO may be the significant enzyme in 8-hydroxylation. Similarly, O^6-benzylguanine, an inactivator of O^6-alkylguanine-DNA alkyltransferase, is also oxidised at carbon 8 by human liver AO (Roy *et al.* 1995).

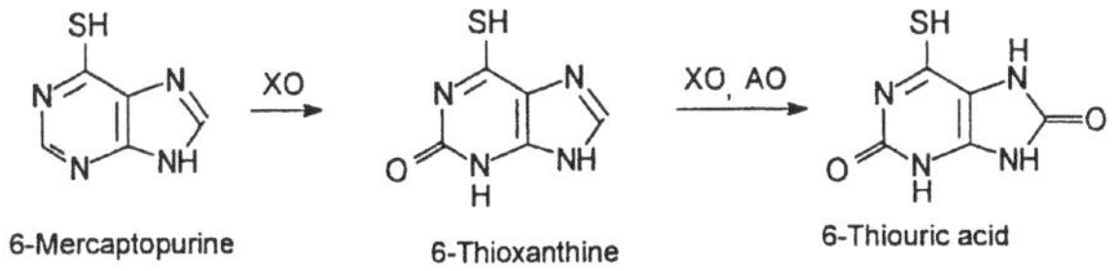

(a) Major metabolic route after oral administration of 6-mercaptopurine

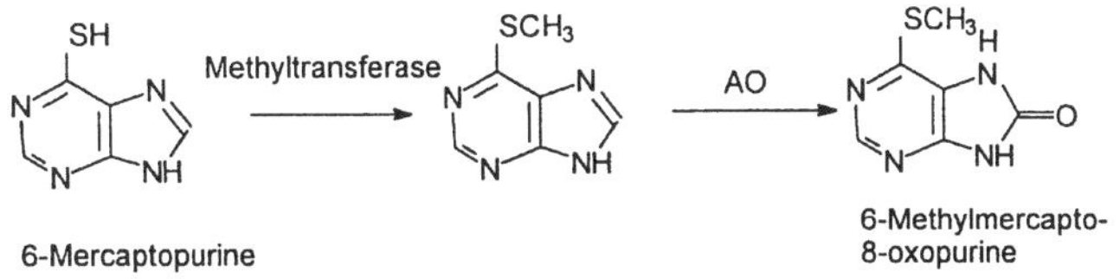

(b) Major metabolic route after intravenous administration of 6-mercaptopurine

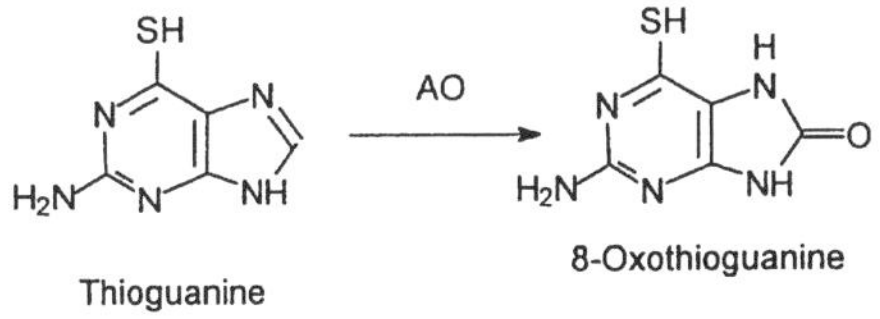

(c) Major metabolic route after intravenous administration of 6-thioguanine

Figure 5.8 Oxidative metabolism of 6-mercaptopurine and 6-thioguanine after oral and intravenous administration (Zimm *et al.* 1984; Rashidi 1996; KeuzenkampJansen *et al.* 1996; Kitchen *et al.* 1999).

Oxidation of *N*-heteroaromatic cations

Quaternisation of a ring nitrogen atom activates a heterocyclic nucleus towards nucleophilic attack. Many heterocyclic cations are, therefore, excellent substrates for AO (Rajagopalan and Handler 1964b; Krenitsky *et al.* 1972; Taylor *et al.* 1984; Beedham *et al.* 1987a) although activity towards XO is only observed at high pH values (Bunting and Gunasekara 1982). Substituent effects and species variation in the oxidation of heterocyclic cations by AO and XO have been reviewed by Beedham (Beedham 1985, 1987).

(a) Oxidation of stable iminium ions

In many substrates, oxidation of N-heteroaromatic iminium ions can occur at two alternative electropositive positions, only one of which is adjacent to the N-hetero-atom (Figure 5.9). N^1-Methylnicotinamide, a niacin catabolite, is simultaneously

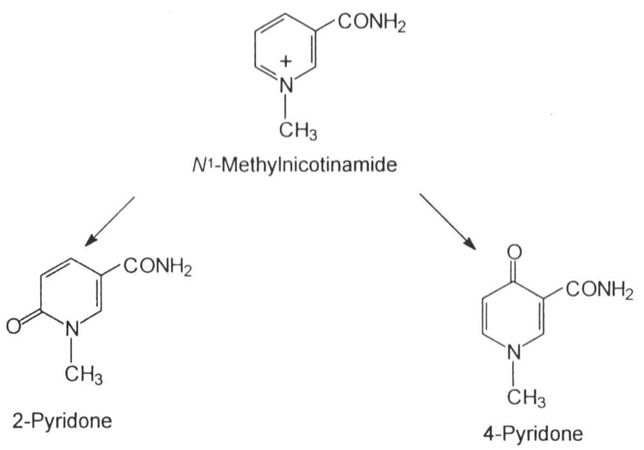

N^1-Methylnicotinamide

2-Pyridone

4-Pyridone

Figure 5.9 Oxidation of N^1-methylnicotinamide to N^1-methyl-2-pyridone-5-carboxamide and N^1-methyl-4-pyridone-5-carboxamide (Shibata 1989; Egashira *et al.*1999).

converted to N^1-methyl-2-pyridone-5-carboxamide and N^1-methyl-4-pyridone-5-car-boxamide (Shibata 1989; Egashira *et al.*1999) and N-alkyl and N-alkylquinolinium salts are oxidised to isomeric 2- and 4-quinolones by AO (Taylor *et al.* 1984; Beedham *et al.* 1987a). In both cases, the position of oxidation varies with species and substituent (Felsted and Chaykin 1967; Ohkubo *et al.* 1983; Shibata, 1989; Taylor *et al.* 1984; Beedham *et al.* 1987a). However, it is not clear whether both reactions are catalysed by one AO isozyme. Ohkubo *et al.* (1983) separated three molybdenum hydroxylase fractions from rat liver, all of which catalysed the oxidation of N^1-methylnicotinamide; XO, which only produced the 2-pyridone, N^1-methylnicoti-namide oxidase I, which preferentially formed the 2-pyridone, and N^1-methylnicoti-namide oxidase II, which only formed the 4-pyridone. The two latter isozymes differed in inhibitor sensitivity, heat stability and pH optimum. In contrast, Wright *et al.* (1999a) have recently reported that rat liver AO is most likely expressed as a single gene but that distinct kinetic forms may arise from variations in redox state.

(b) Oxidation of unstable iminium ions

Iminium ions may also be generated as intermediates during the metabolism of cyclic, aliphatic, secondary and tertiary amines such as indoles, prolidines, piperidines and dihydropyridines via cytochrome P450 or monoamine oxidase (Diaz and Squires 2000; Nguyen *et al.*, 1979; Whittlesea and Gorrod 1993; Rodrigues *et al.* 1994; Lin *et al.* 1996; Wu *et al.* 1988; Yoshihara and Ohta 1998). These unstable iminium ions are highly reactive and can react with glutathione and cellular macromolecules (Whittlesea and Gorrod 1993; Skordos *et al.* 1998) or cause neurotoxicity, e.g. 1-methyl-4-phenyl-2,3-dihydropyridinium, MPDP$^+$ (Wu *et al.* 1988; Yoshihara and Ohta 1998; Yoshihara *et al.* 2000). Further oxidation of iminium ions to cyclic lactams by AO (Figure 5.10) represents an important detoxification step for these compounds and

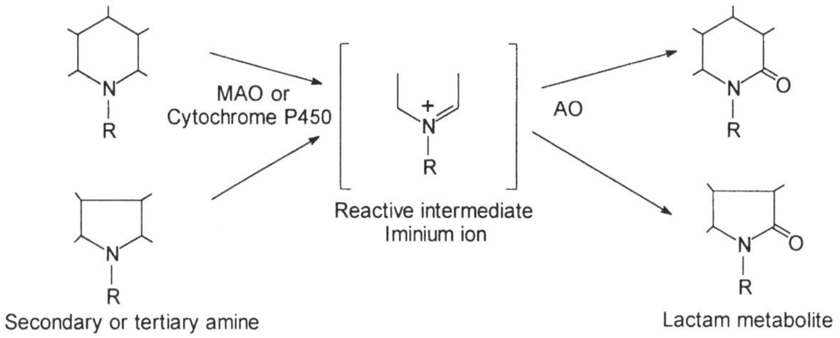

Figure 5.10 Detoxification of reactive iminium ions, generated *in vivo* by cytochrome P450 or MAO, via AO-catalysed oxidation.

differential activity of AO between tissues and species may be significant. Lactam metabolites from aliphatic amines may represent a minor pathway as in the case of thioridazine (Lin *et al.* 1993, 1996) or the major metabolic route as for nicotine and prolintane (Nakajima *et al.* 1996, 2000; Whittlesea and Gorrod 1993). Cotinine formation may also be important in controlling nicotine levels in the brain (Jacob *et al.* 1997).

Oxidation of aldehydes

AO and XO catalyse oxidation of aldehydes to carboxylic acids. However carboxylic acids are also produced via other enzymes, principally NAD-dependent aldehyde dehydrogenases (ALDH). For example, acetaldehyde is produced from ethanol by alcohol dehydrogenase or CYP2E1; subsequent oxidation to acetic acid can either be catalysed by ALDH, AO or XO. Even though the K_m values of acetaldehyde for the molybdenum hydroxylases are relatively high (36 – 130 mM with XO and 1mM for AO) (Fridovich 1966; Morpeth 1983; Rajagopalan and Handler 1964b), it has been proposed that oxidative injury to liver and pancreas during ethanol metabolism is mediated by the ROS generated from the combined activities of AO and XO (Shaw and Jayatilleke 1992; Wright *et al.* 1999b). Increased ROS production in mammary tissue during alcohol metabolism, via alcohol dehydrogenase and XO, is also thought to lead to DNA damage leading to breast cancer (Wright *et al.* 1999b).

Other aliphatic aldehydes are also substrates for AO and XO although K_m values are higher than corresponding values with ALDH. Studies on the substrate specificity of AO and XO towards aliphatic and aromatic aldehydes up to 1987 has been reviewed previously (Beedham 1985, 1987). Hydrophobicity enhances affinity of aldehydes towards AO and many aromatic aldehydes are excellent substrates including benzaldehyde, indole-3-aldehyde, vanillin, retinal and pyridoxal (Johns 1967; Rashidi 1996; Peet 1995; Panoutsopoulos 1994; Huang *et al.* 1999; Krenitsky *et al.* 1972); lower activities are observed with XO. In a systematic study with 11 structurally related benzaldehydes, Panoutsopoulos (1994) has shown that the presence of a 3-hydroxy group is a critical factor influencing binding and oxidation rates. Vanillin (Figure 5.11) was rapidly oxidised by AO with negligible contribution from XO or

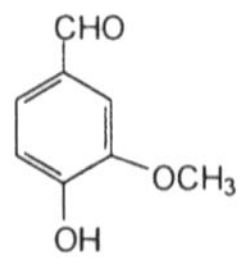

4-Hydroxy-3-methoxy
benzaldehyde (vanillin)

$K_m = 15 - 27 \, \mu M$

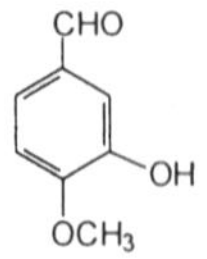

3-Hydroxy-4-methoxy
benzaldehyde (isovanillin)

$K_i = 4 \, \mu M$

Figure 5.11 Differential reactivity of isomeric hydroxy-benzaldehydes towards AO (Panoutsopoulos 1994; Rashidi 1996).

ALDH whereas the isomeric aldehyde, isovanillin, was predominantly transformed by ALDH but was a potent AO inhibitor (Table 5.2). In addition, AO and ALDH both catalyse oxidation of aldehydes derived from phenylethylamines with little contribution from XO. Although it has been proposed that human liver ALDHs are exclusively responsible for the metabolism of biogenic aldehydes and neurotransmitters (Pietruszko *et al.* 1991), we have shown that AO plays a major role in homovanillamine and 5-hydroxylindoleacetaldehyde metabolism in guinea pig liver and brain (Beedham *et al.* 1995b; Peet 1995; Laljee 1998).

With respect to drug metabolism, acid metabolites resulting from oxidation of dimethylamino groups in the nonsteroidal antioestrogen, tamoxifen, and the antidepressant, citalopram (Figure 5.12), are thought to arise from the combined action of either cytochrome P450 or monamine oxidase and AO (Ruenitz and Bai, 1995; Rochat *et al.* 1998).

REDUCTION

In vitro reduction, catalysed by AO or XO, can be demonstrated under hypoxic or anaerobic conditions in the presence of an appropriate electron donor such as 2-pyrimidinone (AO), benzaldehyde (AO) or xanthine (XO). However, extrapolation of *in vitro* results to *in vivo* metabolism is often difficult to assess for the following reasons: (a) many reduction reactions are only observed under strictly anaerobic conditions or are inhibited by high oxygen concentrations, (b) reduction rates are dependent on the availability and concentration of electron donor, (c) there is overlapping substrate specificity between the molybdenum hydroxylases and other reducing enzyme systems such as cytochrome P450 reductase and NADPH-benzoquinone reductase (DT-diaphorase), (d) reduction may also occur in the intestine due to the action of the gut bacteria, and (e) reduction products may not be detected *in vivo* due to re-oxidation back to the parent compound via microsomal cytochrome P450 or flavin monooxygenase.

Nevertheless, AO has a broad substrate specificity and is less sensitive towards oxygen inhibition than XO (Wolpert *et al.* 1973). In addition, AO may also function as the terminal reductase in coupled electron transfer systems with XO or NADPH-

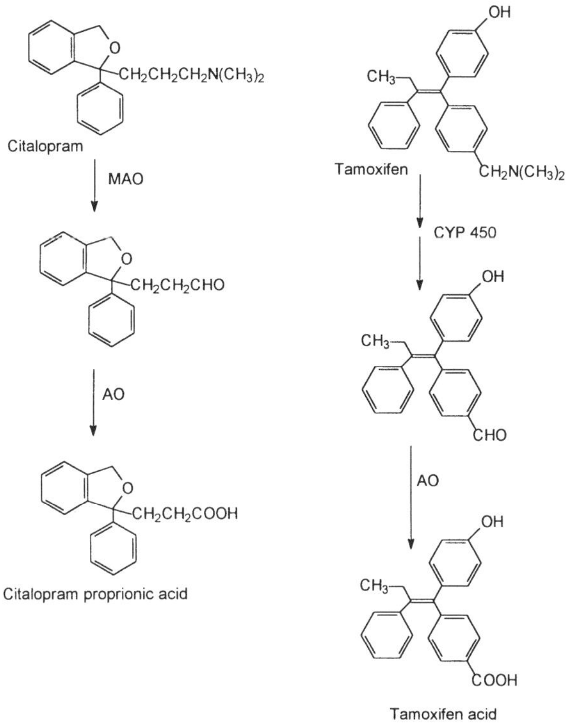

Figure 5.12 The role of AO in the formation of acid metabolites from Citalopram and Tamoxifen.

cytochrome c reductase (Kitamura and Tatsumi 1983a; Tatsumi *et al.* 1982; Kitamura *et al.* 1981). Thus it is likely that, in addition to oxidation reactions, AO-catalysed reduction reactions may be significant *in vivo*. Although there are many reports of AO-catalysed reduction reactions, principally arising from the laboratories of Kitamura, Sugihara and Tatsumi, this area has not been reviewed. Consequently, the major classes of oxidising substrates for AO and XO are described in the following section.

Hepatic AO catalyses the reduction of many *N*-functional groups including nitro groups, *N*-oxides, oximes, azo dyes, *N*-nitroso and *N*-hydroxycarbamoyl substituents. XO shows weak activity as a nitro- and *N*-oxide reductase but may be important in nitrate reduction and subsequent production of nitric oxide (NO).

Nitroreduction

An early investigation by Westerfield *et al.* (1957) described the effect of tungstate administration on the *in vivo* reduction of 4-nitrobenzene sulphonamide in rats. Nitro-reduction to sulphanilamide was decreased in tungsten-treated rats compared to control animals and minimal XO activity was detected in tungsten-treated animals. However, 4-nitrobenzensulphonamide was only a weak substrate for XO and it seems more likely that AO was the molybdenum hydroxylase responsible for the reduction in this case. Indeed, nitrofurazone and the carcinogenic 4-nitroquinoline-N-oxide were reduced to the hydroxylamines by AO with highest nitroreductase activity in rabbit liver and lower activity in rat and guinea pig (Wolpert *et al.* 1973). The relative contribution of the combined molybdenum hydroxylases towards nitrofurazone reduction compared to microsomes was around one third total activity in rat, mouse, hamster and rabbit and around one fifth in guinea pig liver. Nitrated polycyclic hydrocarbons are reduced by bovine milk XO (Bauer and Howard 1990) and rabbit liver AO (Tatsumi *et al.* 1986) to hydroxylamines which may subsequently bind to DNA.

Reduction of nitrates by XO

Reduction of organic nitrates and inorganic nitrate or nitrite to nitric oxide (NO), supported by either xanthine or NADH, is catalysed by XO under anaerobic conditions (Millar *et al.* 1998; Doel *et al.* 2000; Godber *et al.* 2000). It is proposed that XO may be important in the metabolism of organic nitrates such as glyceryl, isoamyl and isobutyl nitrate to NO. Furthermore, the observed inactivation of XO by NO may serve to explain the phenomenon of tolerance with these compounds. Generation of NO from XO is dependent on low O_2 tension and inhibited by oxipurinol. Millar *et al.* (1998) suggested that XO may act as a source of NO derived from endogenous nitrate and nitrite under ischaemic conditions when NO synthase does not function.

N-oxide reduction

XO will also function as an *N*-oxide reductase under anaerobic conditions; thus 3-amino-1,2,4-benzotriazine-1,4-dioxide is reduced by bovine milk XO to the mono-*N*-oxide (Laderoute and Rauth 1986). However, nicotinamide-*N*-oxide, imipramine-*N*-oxide, cyclobenzaprine-*N*-oxide and *S*-(−)-nicotine-1'-*N*-oxide are reduced to their parent amines by rat and rabbit liver AO (Kitamura and Tatsumi 1984a,b; Sugihara *et al.* 1996b). XO also catalyses *S*-(−)-nicotine-1'-*N*-oxide reduction to nicotine *in vitro* although AO is much more efficient in this respect (Sugihara *et al.* 1996b). Interconversion of *S*-(−)-nicotine-1'-*N*-oxide to nicotine *in vivo* may serve to prolong nicotine action via a futile cycle of oxidation-reduction reactions. Nicotine-1'-oxide and cotinine are the major *in vivo* metabolites of *S*-(−)-nicotine-1'-*N*-oxide in humans (Berkman *et al.* 1995).

Oxime reduction

Acetophenone oxime, salicylaldoxime and benzamidoxime are reduced by cytosolic AO under anaerobic conditions whereas no reaction was observed in the presence of electron donors of XO, DT-diaphorase or rabbit liver microsomes (Tatsumi and Ishigai 1987). Acetophenone oxime and salicylaldoxime were converted to the corresponding oxo compounds and benzamidoxime to a ketimine (Figure 5.13). *In vivo* studies in rats indicate that butanal oxime, an anti-skinning agent in varnishes and paints, is reduced to an imine via AO, which undergoes subsequent hydrolysis and conversion

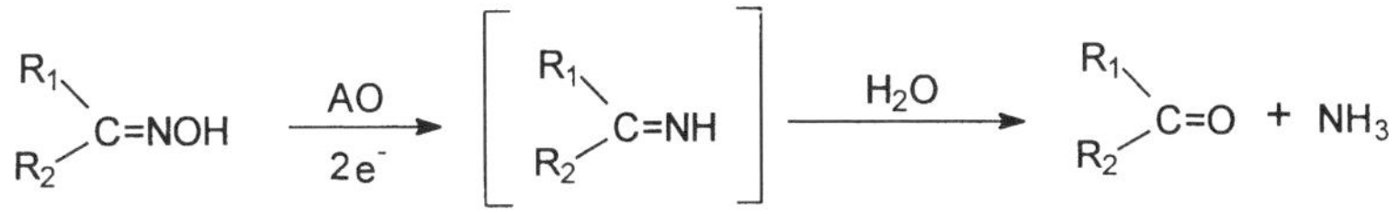

$$R_1 \diagdown C=NOH \xrightarrow[2e^-]{AO} \left[R_1 \diagdown C=NH \right] \xrightarrow{H_2O} R_1 \diagdown C=O + NH_3$$

Figure 5.13 Proposed mechanism for oxime reduction catalysed by AO (Tatsumi and Ishigai 1987).

of butanal to carbon dioxide. The reductive reaction may protect against butanal oxime toxicity via cyanide production in an alternative metabolic pathway catalysed by cytochrome P450 (Mathews *et al*. 1998).

Reduction of azo dyes

Azo dyes are widely used as colourants in drugs, food and beverages (Wolff and Oehme 1974). Detoxification of these potentially carcinogenic compounds is usually associated with reduction of the azo link to a primary amine. However, reductive metabolism of some azo compounds may produce carcinogenic or mutagenic products (Chung 1983). Rabbit liver AO has been shown to possess significant azoreductase activity towards methyl red, methyl orange and other dimethyl aminoazobenzenes (Kitamura and Tatsumi 1983b; Stoddart and Levine 1992). Lipophilic azo dyes, which are readily reduced by microsomal cytochrome P450, were weak substrates whereas water soluble or charged azo dyes were readily reduced by the cytosolic enzyme (Stoddart and Levine 1992).

Reduction of N-functional groups

Rabbit and guinea pig liver AO also catalyse the reduction of aromatic and heterocyclic hydroxamic acids to amides (Sugihara and Tatsumi 1986), nitrosoamines to hydrazines (Tatsumi and Yamada 1982; Tatsumi *et al*. 1983), and *N*-hydroxyurethane to urethane (Sugihara *et al*. 1983) in the presence of an electron donor.

Sulphoxide reduction

AO acts as a sulphoxide reductase under anaerobic conditions e.g. sulindac, sulphinpyrazone, phenothiazine sulphoxide, dibenzyl sulphoxide and diphenylsulphoxide (Lee and Renwick 1995; Yoshihara and Tatsumi 1986, 1990; Tatsumi *et al*. 1982). Reduction of sulindac and sulphinpyrazone produces active sulphide metabolites (Pay *et al*. 1980; Duggan 1981). Lee and Renwick (1995) showed that AO is the main sulindac-reductase in rat and rabbit liver whereas sulindac reduction in kidney cytosol was catalysed either by thioredoxin reductase or a coupled electron transfer system. Diphenyl sulphoxide (DPSO) is reduced to diphenyl sulphide under anaerobic conditions but acts as an inhibitor in the presence of electron acceptors such as molecular O_2 or potassium ferricyanide (Yoshihara and Tatsumi 1986). Although sulphoxides are reduced under hypoxic conditions, re-oxidation of a sulphide to sulphoxide and sulphone metabolites, by microsomal monooxygenases, may occur under normoxia.

Tissue oxygen levels will thus be an important determinant of the overall metabolic fate of a compound. In further studies with DPSO in perfused guinea pig liver, the sulphoxide was exclusively converted to diphenylsulphone under normoxia whereas under hypoxia diphenylsulphone formation decreased in parallel with reducing oxygen concentrations (Yoshihara and Tatsumi 1990). Under hypoxic conditions, the reduction pathway to diphenyl sulphide was only significant in the presence of an AO-reducing substrate such as 2-hydroxypyrimidine or benzaldehyde. Preliminary investigations in rabbit showed that oxidation to diphenyl sulphone and reduction to diphenyl sulphide both occurred after oral administration of diphenylsulphoxide. Chiral inversion of flosequinan (FSO), a peripheral vasodilator, *in vivo* is thought to occur via the formation of flosequinan sulphide (Figure 5.14). Rat liver AO exhibits stereoselective reduction of FSO, reducing R-FSO at a rate ~18 times higher than the S-FSO (Kashiyama *et al.* 1999).

Unlike nitroreduction, XO will not support sulphoxide reduction (Yoshihara and Tatsumi 1986, 1990; Kitamura *et al.* 1999d), although it will transfer electrons from an electron donor such as xanthine to AO in a coupled electron transfer system (Kitamura *et al.* 1981; Kitamura and Tatsumi 1983a).

Reductive dehalogenation

From the above studies it can be seen that XO catalyses a narrower range of reductive reactions than AO, and many of the reactions have much lower reduction rates. A more unusual substrate of XO is 6-bromomethyl-(9H)-purine which undergoes reductive dehalogenation to 6-methylpurine. Not only did XO catalyse the oxidation of 6-bromomethyl-(9H)-purine to its uric acid analogue but also reduced the substrate with concomitant enzyme inhibition by modification of the flavin moiety (Porter 1990).

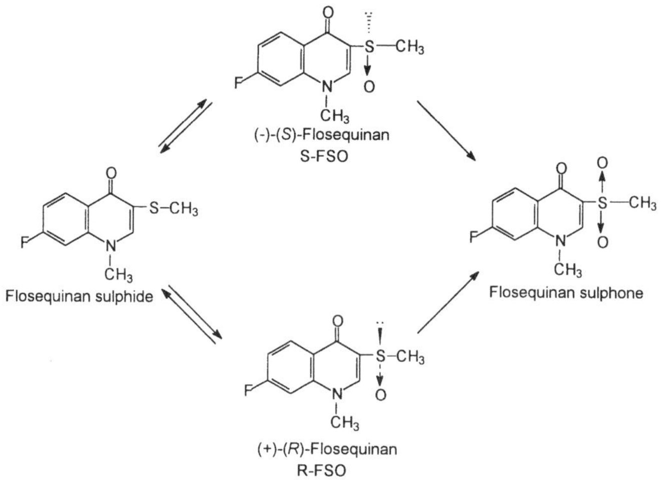

Figure 5.14 In vivo metabolism of flosequinan in the rat (Kashiyama *et al.* 1999).

Reduction *in vivo*

Reduction can be a significant pathway *in vivo*. Zonisamide, a novel anticonvulsant, is primarily metabolised via a reductive pathway during which the benzisoxazole ring is cleaved to 2-sulphamoylectylphenol (Stiff *et al.* 1990). The mechanism, shown in Figure 5.15, is thought to be analogous to oxime reduction involving formation of the intermediate ketamine and subsequent hydrolysis to the final oxo compound with stoichiometric formation of ammonia (Sugihara *et al.* 1996a). Although the reduction reaction can be catalysed by CYP3A (Nakasa *et al.*1993), cytosolic zonisamide reducing activity in rat, rabbit, guinea pig, mice and hamster liver is higher than microsomal activity (Sugihara *et al.* 1996a). It should also be noted that the same group have shown that this reaction can also be catalysed by gut bacteria (Kitamura *et al.* 1996).

Not surpisingly, Yoshihara and Tatsumi (1986) suggest that AO is one of the major sulphoxide reductases in mammals. Based on current evidence this is not an unreasonable conclusion. AO is an effective reductase and does not require a strict anaerobic environment, unlike XO and cytochrome P450 reductase. Localisation within the liver is in the pericentral zone, an area of low oxygen tension. Furthermore, studies with zonisamide indicate that reduction may be significant *in vivo* pathway. However, AO-catalysed reduction may not be restricted to mammals. Kitamura *et al.* (1999a) has shown that Fenthion sulphoxide is reduced to the organophosphorus pesticide, Fenthion, by AO in the hepatopancreas of goldfish, *Carassius auratus*.

Interaction with inhibitors

Inhibitory activity towards the molybdenum hydroxylases appears to mimic that of substrates. There are common inhibitors such as cyanide, arsenite and methanol, which react with the molybdenum cofactor of both enzymes (Rajagopalan and Handler 1964a; Turner *et al.* 1995; Massey and Harris 1997). Many XO inhibitors are substrate analogues based on the structure of allopurinol, a clinically used XO inhibitor. In contrast, a diverse group of inhibitors, including several drugs, are potent *in vitro* inhibitors of AO. However, there are no reported clinical reactions for any of

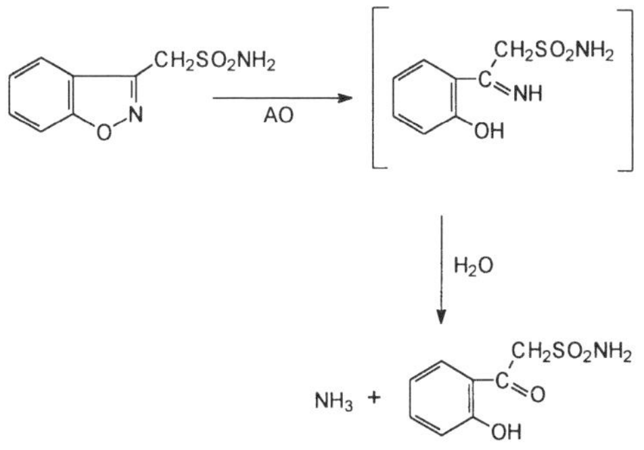

Figure 5.15 Reductive cleavage of Zonisamide catalysed by AO (Sugihara *et al.* 1996).

these compounds. Furthermore, as with AO substrates there is a species variation in the reactivity of AO inhibitors.

INHIBITORS OF AO

Menadione is one of the most potent AO inhibitors (Rajagopalan et al. 1962; Rajagopalan and Handler 1964a); it is universally used in vitro to characterise enzyme activity and is equipotent with oxidation or reduction reactions and with AO from different species (Yoshihara and Tatsumi 1986; Rashidi et al. 1997; Schofield et al. 2000). Interaction is thought to occur at the FAD site (Rajagopalan et al. 1962; Rajagopalan and Handler 1964a; Yoshihara and Tatsumi 1986), which is consistent with the ability of menadione to act as an electron acceptor of XO (Mahler et al. 1955). Consequently, XO-catalysed oxidation rates may be enhanced in the presence of menadione (Rajagopalan et al. 1962; Yoshihara and Tatsumi 1986) and ambiguous results obtained with substrates of both molybdenum hydroxylases in cytosol, hepato-cytes, liver slices and partially purified enzyme fractions. This is illustrated by the oxidation of 4-pyriminidone to uracil (Figure 5.5), catalysed by AO and XO, in guinea pig liver, which is inhibited by 10 μM but not 100 μM menadione (Oldfield 1998). Menadione inhibition has not been studied in vivo but it is unlikely to reduce AO activity because it is rapidly reduced by DT-diaphorase (Thor et al. 1982) or even XO (Mahler et al. 1955).

Table 5.2 shows the most potent AO inhibitors identified in vitro, the majority of which have reported kinetic constants (IC_{50} or K_i values) of the order of 1 μM for either rabbit, rat, guinea pig or human liver AO. In most cases, kinetic constants have been measured against substrate oxidation but, where tested, similar results have been obtained against reduction.

Although the potency of inhibitors such as menadione, β-oestradiol, chorpromazine and amsacrine is similar with animal and human liver, other compounds do not give consistent results. In particular methadone (Robertson and Gamage 1994) and proadifen (SKF 525A) (Robertson and Bland 1993), which were originally identified as potent inhibitors of rat liver enzyme, show little reaction with human or guinea pig hepatic AO (Schofield et al. 2000). In contrast, Rashidi et al. (1995) found a close correlation ($r = 0.96$) between inhibitor reactivity towards guinea pig and human liver AO for 13 drugs with varying inhibitory potency.

INHIBITORS OF XO

Numerous compounds have been tested as XO inhibitors (Hille and Massey 1981); nevertheless there is less structural diversity among XO inhibitors than those of AO as most effective XO inhibitors are purine–based analogues. Structure-activity relation-ships for XO-inhibitors have been reviewed by Beedham (1987). The prototype inhibitor, allopurinol, is a potent mechanism-based inhibitor of XO (Massey et al. 1970) but has weak inhibitory activity towards AO (Hall and Krenitsky 1986; Rashidi et al. 1995; Rashidi 1996). Like menadione, it is extensively used to distinguish between AO and XO activity (Roy et al. 1995; Clarke et al. 1995; Shanmuganathan et al. 1994; Oldfield 1998) although results should be interpreted with caution as

Table 5.2 *In vitro* inhibitors of AO

Inhibitor	Reactivity[a] with			
	Rabbit AO	Rat AO	Guinea Pig AO	Human AO
Menadione	++++	++++	++++	++++
Amsacrine	+++	+++	+++	+++
β-Oestradiol	+++	ND	+++	++++
7-Hydroxy-DACA	ND	++++	++++	++++
Chlorpromazine	ND	ND	++++	++++
Promethazine	ND	ND	+++	+++
Phenothiazine	ND	ND	++++	ND
Hydralazine	+++	ND	++++	+++
Isovanillin	ND	ND	+++	ND
Protocatechuic aldehyde	ND	ND	+++	ND
Dopamine	ND	ND	+++	ND
Quinacrine	+++	ND	ND	ND
Antimycin A	+++	ND	ND	ND
Methadone	ND	++++	+	+
SKF-525A	ND	+++	+	+
Norharman	ND	ND	+++	ND
Benzoquinone	+++	ND	ND	ND
Progesterone	++	ND	ND	ND
Triton X-100	++	ND	ND	ND
Cimetidine	ND	ND	+	++
Disulfiram	ND	ND	++	ND
D-Propoxyphene	ND	++	ND	ND
Amytal	+	ND	ND	ND

[a] Kinetic constants (IC_{50} or K_i value)
+++, < 1 μM; ++, 1 − 10 μM; ++, 10 − 100 μM; +, > 100 μM
ND: Not determined
7-Hydroxy-DACA, 7-Hydroxy-N[(2′-dimethylamino)ethyl]acridin-6-caboxamide
SKF-525A, Proadifen
(Rajagopalan *et al.* 1962; Rajagopalan and Handler 1964a; Johns 1967; Johnson *et al.* 1985; Yoshihara and Tatsumi 1986; Lee and Chan 1988; Panoutsopoulos 1994; Rashidi 1996; Rashidi *et al.* 1995; Laljee 1998; Schofield *et al.* 2000)

allopurinol is oxidised by both enzymes to oxipurinol (Figure 5.16) (Krenitsky *et al.* 1986; Moriwaki *et al.* 1993a; Yamamoto *et al.* 1993). With XO, oxipurinol is strongly bound to the reduced molybdenum cofactor preventing further electron transfer (Massey *et al.* 1970; Spector *et al.* 1986) whereas oxipurinol does not significantly inhibit AO (Hall and Krentisky 1986, Moriwaki *et al.* 1993a).

Inhibition of XO *in vivo* accounts for the hypouricaemic effect of allopurinol in the management of gout (Star and Hochberg 1993). Although the conversion of allopurinol to oxipurinol has been attributed to XO (Murrell and Rapeport 1986), it is more likely that AO is principally responsible for oxipurinol formation (Reiter *et al.* 1990; Moriwaki *et al.* 1993a). In children administered increasing allopurinol doses, Sweet man (1968) found that maximum conversion of allopurinol to oxipurinol coincided with almost complete inhibition of XO activity. In addition to its clinical use in gout, allopurinol is also effective in augmenting the therapeutic effect of drugs metabolised by XO. Co-administration of allopurinol with oral 6-mercaptopurine increases plasma 6-mercaptopurine levels five fold and also enhances drug toxicity (Zimm *et al.* 1983).

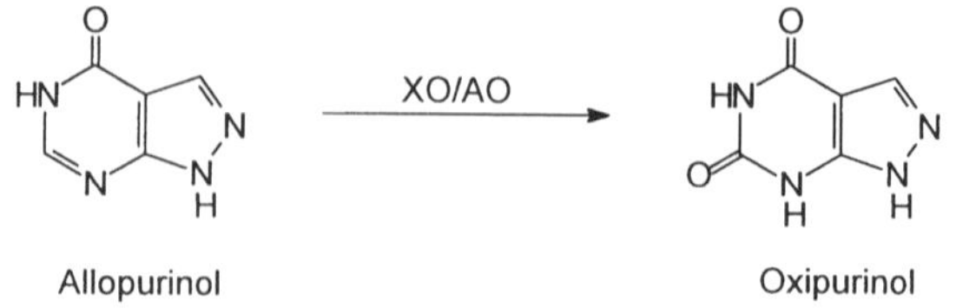

	XO	AO			XO	AO
K_m (μM)	2	380		K_i (μM)	0.08	150

Figure 5.16 Molybdenum hydroxylase-catalysed oxidation of allopurinol to oxipurinol (Hall and Krenitsky 1986; Krenitsky et al 1986).

Allopurinol also has a beneficial effect in immunosuppressive therapy with azathioprine that may be related to XO-inhibition (Chocair et al. 1993). In contrast, there are no reported drug interaction between allopurinol and AO substrates such as famciclovir or thioguanine (Fowles et al. 1994; Hande and Garrow 1996).

BOF-4242, a pyrazolo[1,5-a]triazine (Figure 5.17), is a recently introduced potent XO inhibitor (K_i for (−)-isomer = 1.2 nM) which has a longer duration of action in vivo than allopurinol (Okamoto and Nishino 1995).

Tissue distribution

Tissue distribution of AO and XO has been compared previously in reviews by Beedham (1985, 1987) and Moriwaki et al. (1997,1999). Most of the studies described in the two former reviews are based on substrate specific assays in tissue homogenates or purified enzyme, whereas the later review by Moriwaki et al. (1997) includes histochemical and immunohistochemical techniques to localise enzyme activity. Limited results were available for gene expression in different species and tissues. The present chapter will concentrate on recent studies carried out on the tissue distribution in common laboratory species and humans. Tables 5.3 and 5.4 compare the tissue distribution of XO and AO respectively, using different assay methods.

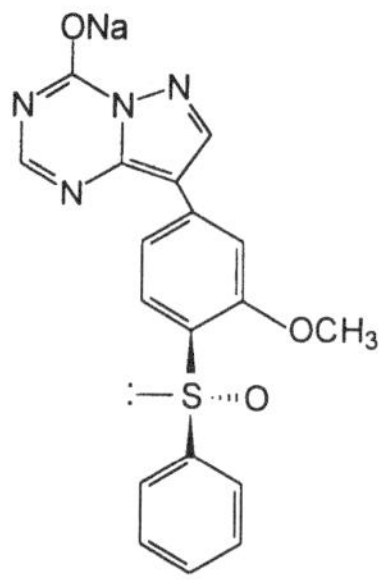

Figure 5.17 (-) Isomer of BOF-4242, pyrazolo[1,5-a]triazine, a tight binding inhibitor of XO (Okamoto and Nishino 1995).

Table 5.3 Tissue distribution of xanthine oxidase

Tissue	Rat	Rabbit	Mouse	Guinea pig	Bovine	Human
Liver	✓✓ (H,I)		✓✓ (E)	✓ (K)	✓ (K,I)	✓ (K,E)
Pericentral	✓✓ (H,I)				✓ (I)	✓ (I,H)
Periportal	✓ (I)					✓✓ (H)
Kupffer Cells					✓ (I)	✓ (I)
Endothelial cells					✓ (I)	
Kidney	✓ (H,I)		✓ (E)	✓ (K)		✓ (K)
Endothelial cells						✓ (I)
Glomerulus Tubules	✓ (H,I)					
Heart	✓ (I)	✓ (K)	✓ (E)		✓ (K)	– (I) ✓ (K)
Adrenals	✓ (H,I)					✓ K
Spleen	✓ (H,I)			✓ (K)		✓ K
Skeletal Muscle	✓ (I)	✓ (K)			✓ (K)	✓ (I,K)
Tongue (epithelial)	✓✓ (H,I)					
Oesophagus	✓✓ (H,I)					
Stomach	✓ (H,I)					
Small Intestine	✓✓✓ (H,I)		✓✓ (E)	✓ (K)		✓✓ (I,K,E)
Large Intestine	✓ (H,I)		✓ (E)	✓ (K)		
Lung	✓✓ (H,I)		✓ (E)	✓ (K)		– (I,K)
Bronchioles	✓✓ (H,I)					
Alveolus	✓✓ (H,I)					
Brain		✓ (K)				– (I) ✓ (K)
Mammary gland			✓ (E)		✓ (I)	✓✓ (I)
Milk					✓✓	✓✓ (I)

✓ XO detected with kinetic measurement (K), Histochemical (H), Immunocytochemistry (I) or mRNA expression (E).
– No activity found.
(Krenitsky et al. 1974; Beedham et al. 1987b; Wagner and Harkness 1989; Kooji et al. 1992; Terao et al. 1992, 1997; Wright et al. 1993; Xu et al. 1994; Moriwaki et al. 1993b, 1998a,b; Linder et al. 1999; Moriwaki et al. 1996).

Enzyme localisation within a tissue may depend on the limitations of a visualisation technique or differences in mRNA expression, transcription or translational events. Thus the comparative tissue distribution can vary with different assay methods. Kinetic measurements (designated K in Tables 5.3 and 5.4) based on specific substrate turnover with a spectroscopic or fluorogenic end-point will give an accurate estimation of functional enzyme within a tissue but may lack sensitivity and give little information on cell-specific localisation. Furthermore, purification of the protein may lead to inactivation of the holoenzyme. While histochemical techniques (H), using enzyme-specific substrates, will also detect functional enzyme, they will differentiate between enzyme activity in different cell sub-types. However, immunohistochemical methods (I) are usually more sensitive to tissues containing lower enzyme activity but there is a risk of cross-reaction with other proteins. Consequently, there is little variation between results obtained from tissues with high molybdenum hydroxylase activity with any of the above techniques (K, H or I). Determination of mRNA expression (E) can give misleading results regarding functional molybdenum hydroxylase activity, as steady-state levels of mRNA do not necessarily correlate with

Table 5.4 Tissue distribution of aldehyde oxidase

Tissue	Rat	Rabbit	Mouse	Guinea Pig	Bovine	Human
Liver	✓✓ (H,I)	✓✓✓ (E,K)	✓ (E,I,K)	✓✓ (K)	✓✓ (I,K,E)	✓ (E)
Pericentral	✓✓ (H,I)			✓✓ (K)		✓ (I)
Periportal	✓ (I)					
Kidney		✓✓ (E,K)		✓ (K)	✓ (I,E)	✓ (E)
Glomerulus Tubules	✓ (H,I)					
Heart	✓ (I)	✓ (E)	✓ (E)			
Spleen			✓ (E,I,K)	✓ (K)	✓ (I,K,E)	
Skeletal Muscle	—					
Tongue (epithelial)	✓ (H,I)					
Oesophagus	✓✓ (I)		✓ (E)		✓ (I)	
Stomach	✓ (H,I)	Trace				
Small Intestine	✓ (I)			✓ (K)		
Large Intestine	✓ (I)			✓ (K)	✓ (I)	
Lung	✓✓ (H,I)	✓✓ (E,K)	✓ (E,I,K)	✓ (K)	✓ (I,K,E)	✓ (E)
Bronchioles	✓✓ (H,I)					
Alveolus	✓ (H,I)					
Brain		✓ (E)	✓✓ (E,I,K)			✓✓ (E)

✓ AO detected with kinetic measurement (K), Histochemical (H), Immunocytochemistry (I) or mRNA expression (E).
– No activity found.
(Krenitsky *et al.* 1974; Wright *et al.* 1995, 1997; Beedham *et al.* 1987b; Li Calzi *et al.* 1995; Bendotti *et al.* 1997; Moriwaki *et al.* 1998a,b; Huang *et al.* 1999; Kurosaki *et al.* 1999)

catalytically active protein (Kurosaki *et al.* 1995, 1999; Li Calzi *et al.* 1995; Linder *et al.* 1999).

Despite the different methods employed, there is a general consensus about relative tissue distribution for each enzyme among all species studied. However, there is some interspecies variation in the expression of XO; thus rabbit tissues show lower XO activity than other species (Wagner and Harkness 1989). In contrast, the specificity of AO varies significantly among species. There are also major differences in tissue distribution between AO and XO (Tables 5.3 and 5.4). Therefore, in humans and mammals, highest XO expression and activities are found in proximal intestine, lactating mammary gland and liver, whereas high AO levels are consistently found in the liver, lung, kidney and brain. Much lower AO activity is present in intestine with no AO mRNA expression in the human mammary gland (Wright *et al.* 1997).

MOLYBDENUM HYDROXYLASE ACTIVITY IN RAT

In rats, XO activity was high in surface epithelium of small intestine, strong to moderate in liver cytoplasm, moderate in surface epithelium of tongue, oesophagus, stomach, bronchioles, alveoli, renal tubules and large intestine but not detected in heart or muscle fibres (Moriwaki *et al.* 1998a). Using enzyme histochemistry, Moriwaki *et al.* (1998a) observed high AO activity in Wistar rat liver with lower activity in surface epithelium of tongue, bronchioles and renal tubules. AO activity in heart, skeletal muscle, stomach, oesophagus small and large intestine were only detected using a more sensitive immunohistochemical technique. Both AO and XO were localised in the pericentral rather than the periportal zone of the liver (Moriwaki *et al.* 1998a,b). Localisation of the enzymes in this area of low oxygen tension may indicate a role for these enzymes in reduction reactions *in vivo*. However, AO distribution in rat liver is strain/animal dependent (see below).

MOLYBDENUM HYDROXYLASE ACTIVITY IN MOUSE

Although moderate XO activity has been found in mouse intestine and liver, expression of XO mRNA is low in all mouse tissues studied (Terao *et al.* 1992). However, after induction with interferon, Terao *et al.* (1992) showed a rapid elevation of mRNA in mouse liver, kidney, small and large intestine, heart, lung but not in spleen and brain. In contrast, the transcipt coding for mouse AO is expressed at highest concentrations in the oesophagus and in liver, lung, heart, testis with lower levels in brain, spinal cord and eye (Kurosaki *et al.* 1999). AO mRNA was not detected in stomach, skin, striated muscle or small and large intestine using Northern blot analysis. Kurosaki *et al.* (1999) did not find a strict correlation between mRNA levels and functional AO protein; thus there was no detectable AO activity in oesophagus whereas brain AO activity was much higher than would be predicted from mRNA accumulation. It is likely that the synthesis of mouse XO and AO is under translational and post-translational control. Within the CNS, Bendotti *et al.* (1997) have shown that there is specific and high expression of the AO gene in mouse choroid plexus. In addition, AO mRNA is localised in the large motor neurones of the nuclei of facial, motor trigemini and hypoglossus nerves and motor neurones of the anterior horns of the spinal cord.

MOLYBDENUM HYDROXYLASE ACTIVITY IN RABBIT

Rabbit liver and intestine exhibits lower oxidative activity towards xanthine than rat, mice or guinea pig (Krenitsky *et al.* 1974) but xanthine dehydrogenase activity has been demonstrated in heart, skeletal muscle and brain (Wagner and Harkness 1989). Using cDNA cloning, Huang *et al.* (1999) have shown that rabbit retinal oxidase is identical to AO and that retinal oxidase mRNA is widely expressed in tissues; mRNA expression was shown to be very high in liver and lung, relatively high in kidney, pancreas, brain stem and spinal cord with lower expression in stomach and muscle.

MOLYBDENUM HYDROXYLASE ACTIVITY IN GUINEA PIG

Quantitative measurement of AO and XO activity in guinea pig tissues has been compared using phenathridine (AO), phthalazine (AO) and xanthine (XO) as specific enzyme substrates (Beedham *et al.* 1987a). Kidney AO showed approximately 40% of liver activity with lower rates obtained for spleen, intestine and lung. XO activity was highest in jejunum, followed by duodenum, ileum, liver, spleen, kidney and lung. Both molybdenum hydroxylases are located in guinea pig liver cytosol and mitochondria (Critchley *et al.* 1992). Substrate/inhibitor studies have identified AO in guinea pig cortex and striatum (Beedham *et al.* 1995b; Laljee 1998).

MOLYBDENUM HYDROXYLASE ACTIVITY IN BOVINE TISSUE

AO distribution is similar in bovine tissues with high expression of mRNA and functional protein in liver, lung, spleen and brain (Li Calzi et al. 1995). Positive immunoreactivity was also observed in kidney, eye, thymus, testis, duodenum, heart and oesophagus although not in striatal muscle and pancreas.

HUMAN MOLYBDENUM HYDROXYLASE ACTIVITY

In contrast to rat liver XO, intense staining of XO protein was found in the cytoplasm of human periportal hepatocytes with little protein expression in pericentral hepatocytes. High XO activity and protein levels were also seen in the small intestine, predominantly in the cytoplasm of enterocytes and goblet cells throughout the villi. XO protein expression is maximal in cells located midway through the length of the villi. In the lactating mammary gland there is high activity in acinar cells with some staining in non-lactating mammary gland (Linder et al. 1999). Immunohistochemical staining showed no activity in human heart, lung and brain although other workers have expressed XO mRNA in these tissues (Xu et al. 1994). Human AO mRNA is expressed in liver, lung, kidney, pancreas, prostate, testis and ovary (Wright et al. 1995). The AO gene is highly expressed in glial cells of the human spinal cord but, unlike mouse mRNA, not in neurones (Berger et al. 1995). This discrepancy could be due to a genuine species variation in brain localisation or differences in sample preparation/hybridisation techniques. In the same study, very little expression of XO mRNA was detected in the spinal cord (Berger et al. 1995). Interestingly XO activity in human brain is increased in tumour tissue (Kökoglu et al. 1990) and in rat olfactory cortex after systemic administration of the excitotoxin, kainic acid (Battelli et al. 1995).

Species variation in molybdenum hydroxylase activity

The molybdenum hydroxylases are found in most organisms throughout the animal kingdom. In 1974 Krenitsky et al., on the basis of their functional and structural similarities, proposed that AO and XO evolved from a common primitive precursor whose gene(s) underwent duplication and subsequent divergent evolution. Recent studies by Terao et al. (1998) indicated a recent duplication event as intron/exon boundaries in human AO and XO were almost identical. Substrate specificity and kinetic properties of XO show less variation between species and tissues than AO (Beedham 1998). This is reflected in the conservation of XO gene sequences whereas there is less recognition between AO cDNAs from various animal species (Terao et al. 1998). Mammalian AO gene structures are similar to reptiles and birds whereas they diverge from that of insects and amphibians (Terao et al. 1998) and although plant and mammalian AOs are structurally related, there is little commonality between amino acid sequences of plants and mammalian AO (Kurosaki et al. 1999).

Despite the similarities between mammalian cDNAs and amino acid sequence (Kurosaki et al. 1999; Wright et al. 1999a) there is still a marked species variation in protein expression and catalytic activity of AO; this has been extensively reviewed by

Krenitsky *et al.* (1974), Moriwaki *et al.* (1998a, 1999) and Beedham (1985, 1987, 1998). Of particular interest in drug metabolism is the variation in enzyme activity in those species most widely used for pre-clinical trials, i.e. dog and rat.

AO ACTIVITY IN DOG AND RAT LIVER

AO activity, measured with a variety of substrates, is very low in dog liver (Rodrigues 1994; Krenitsky *et al.* 1974; Kitamura *et al.* 1999c). Minimal AO activity *in vitro* is mirrored by *in vivo* studies with carbazeran, a potent inotropic agent, in dog. Carbazeran undergoes rapid inactivation in baboon and humans catalysed by AO (Figure 5.4), but this route is absent in dogs (Kaye *et al.* 1984, 1985).

The existence of a marked variation/or lack of AO activity in Sprague–Dawley rats has been noted with many uncharged and charged heterocyclic substrates and aldehydes (Stanlovic and Chaykin 1971; Beedham *et al.* 1987a; Rashidi *et al.* 1997; Sugihara et al. 1995). Striking variations have also been noted in Wistar rats by Gluecksohn–Waelsch *et al.* (1967), who found appreciable AO activity towards N^1-methylnicotinamide in only 36 out of 76 animals. We have recently shown that functional enzyme was absent in 60% of Sprague–Dawley rats (Beedham 1998). Low and variable methotrexate-hydroxylating activity has led to conflicting reports relating to *in vivo* biotransformation in rats (Kitamura *et al.* 1999c).

Recent studies with Zaleplon, a sedative hypnotic, illustrate the marked species differences in AO activity. Zaleplon is metabolised by two competing pathways, 5-oxidation catalysed by AO and N-dealkylation mediated by cytochrome P450 isozymes (Figure 5.6). In mouse, rat and dog, where AO activity is low, the major metabolite is N-desethylzaleplon whereas in monkeys and humans, 5-oxozaleplon is formed (Chaudhary *et al.* 1994; Kawashima *et al.* 1999).

Superimposed on inter-animal variation there are also noticeable strain differences in both rats and mice. Sugihara *et al.* (1995) and Kitamura *et al.* (1999b) demonstrated 63- to 104-fold variation in AO-activity in 12 rat strains using the substrates, benzaldehyde and methotrexate respectively. Highest activity was obtained with SEA:SD rats and lowest with WKA/SEA rats. Expressed protein levels appear to correlate with catalytic activity and K_m values did not vary significantly among the different strains. In a separate study, Kunieda *et al.* (1999) distinguished 11 rat strains with high AO activity and 9 strains with no detectable activity. These authors mapped the AO locus in rats to chromosome 9, which is homologous to mouse chromosome 1 and human chromosome 2q. This is consistent with the loci for AO in the two latter species (Holmes 1979; Berger *et al.*1995). Hybrid crosses between 'high' and 'null' strains apparently showed intermediate activity indicating codominant alleles (Kunieda *et al.* 1999).

Factors affecting molybdenum hydroxylase activity *in vivo*

HORMONAL REGULATION OF AO ACTIVITY

It has long been recognised that hepatic AO activity is two- to four-fold higher in male than female mice and that AO levels may be controlled by androgenic and oestrogenic hormones (Huff and Chaykin 1967, 1968; Gluecksohn-Waelsh *et al.* 1967). Furthermore, in mouse and rat, kinetically distinct AO forms are expressed in male and female liver (Yoshihara and Tatsumi 1997b; Wright *et al.* 1999a) and in both species significantly lower K_m values were obtained for substrate oxidation using the male enzyme compared to the female enzyme.

Treatment of female mice with testosterone proprionate increased hepatic AO activity which is associated with a decrease in the K_m value to the male type (Yoshihara and Tatsumi 1997a,b). In contrast, administration of oestradiol diproprionate significantly decreased enzyme levels in adult male mice (Ventura and Dachtler 1981; Yoshihara and Tatsumi 1997a). Oestradiol is a potent *in vitro* inhibitor of AO (Table 5.2) (Rajagopalan *et al.* 1962; Rashidi *et al.* 1995). Yoshihara and Tatsumi (1997a) have also shown that modulators of growth hormone, such as monosodium glutamate, also influence AO activity and they proposed that, in mice, pituitary growth hormone may be a major regulatory factor of gender differences in hepatic AO activity.

Kurosaki *et al.* (1999) compared AO mRNA, immunoreactive protein and catalytic activity in control male and female mice with those treated with testosterone proprionate. In concordance with previous reports (Yoshihara and Tatsumi 1997a,b), enzyme activity and protein levels were much higher in male than female liver but this was not correlated with gene expression as basal mRNA levels were similar in both sexes. Testosterone treatment increased the AO transcipt around three-fold with significant increases in protein levels/catalytic activity to male levels. Yoshihara and Tasumi (1997b) have suggested that different isozymes are present in male and female mouse liver. It should be noted that gender differences in AO activity appear to be restricted to liver as they are not observed in mouse lung, brain and spinal cord (Kurosaki *et al.* 1999). Furthermore, sex-specific differences may be restricted to mouse liver with little relevance to drug metabolism in humans.

Under saturating substrate conditions, we have found that AO activity does not vary significantly between males and females in guinea pig or rat liver (Beedham *et al.* 1987b; Rashidi 1996). On the other hand, Wright *et al.* (1999a) have shown that rat female hepatic AO exhibits kinetic characteristics distinct from male enzyme although results indicate that both forms are coded by one gene. They posit that male and female rat liver AO are interconverted by redox manipulation of the thiol:disulphide potential of a single protein. However, as only 40% of Sprague–Dawley rats show high AO activity (Rashidi 1996; Beedham 1998) the changes observed by Wright *et al.* (1999a) may reflect intra-strain variation rather that inter-sex differences.

Limited clinical data indicate that the pharmacokinetics of penciclovir, resulting from AO-catalysed oxidation, are similar in male and female volunteers (Pratt *et al.* 1994).

HORMONAL REGULATION OF XO ACTIVITY

Regulation of XO in mice appears to differ from that of AO. Yoshihara and Tatsumi (1997a) found, in the same animals that showed changes in AO activity, XO activity was constant and not affected by testosterone administration. Likewise, guinea pig liver XO did not vary significantly between males and females (Beedham *et al.* 1987b) but results on rat liver XO are contradictory. Levinson (Levinson and Chalker 1980; Decker and Levinson 1982) reported androgen-dependent XO activity in rat liver and we have found that mean XO activity is higher in males (Rashidi 1996). However, in AO-active rats there was no difference in XO activity (see above).

There is also conflicting evidence about gender effects on human XO which may be due to different assay methods or other contributing factors. Guerciolini *et al.* (1991) used a direct, sensitive radiochemical assay to measure XO activity in liver samples from patients undergoing partial hepactectomy or open liver biopsy. XO activity in these samples was found to be significantly higher in male (1.43 ± 0.43 nmol uric acid formed/h per gm tissue) than female liver (1.05 ± 0.38 nmol uric acid formed/h per gm tissue). In contrast Relling *et al.* (1992), using caffeine metabolite ratios, reported significantly higher XO activity in female liver whereas other studies have found no significant gender differences (Kalow and Tang 1991; Vistisen *et al.* 1992; Chung *et al.* 2000). Nevertherless, boys are able to tolerate higher 6-mercaptopurine doses (Hale and Lilleyman 1991) but experience higher treatment failure rates than girls (Chessells *et al.* 1995). This would be consistent with higher oxidation rates to 6-thiouric acid by XO (Figure 5.8) which would switch the metabolic pathway away from activation to the inosine nucleotide.

The validity of using caffeine as an indicator of XO activity has been tested using allopurinol to inhibit caffeine metabolism and by comparing caffeine metabolite ratios with urinary concentration ratios of uric acid to xanthine and hypoxanthine (Grant *et al.* 1986). Caffeine is sequentially metabolised by CYP1A2, N-acetyltransferase and/ or XO with a minor contribution from CYP2A6 (Figure 5.18) (Hamelin *et al.* 1994).

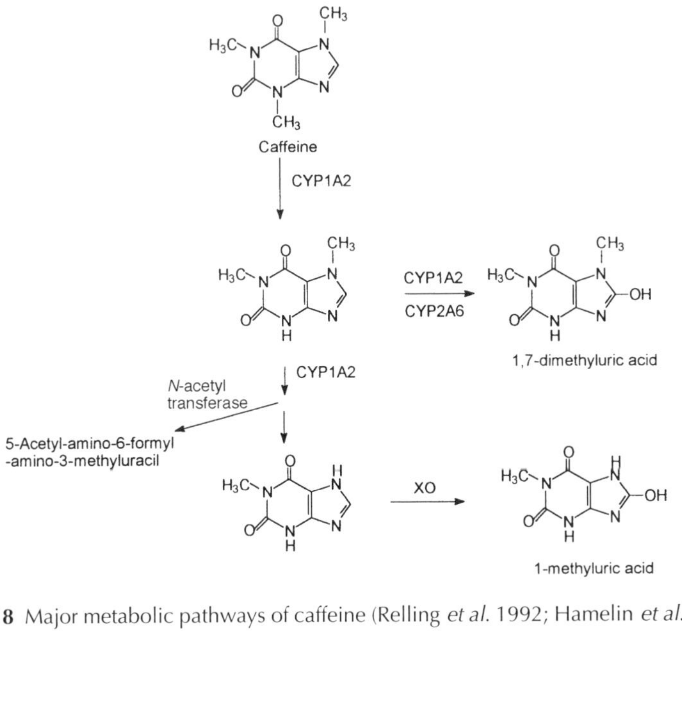

Figure 5.18 Major metabolic pathways of caffeine (Relling *et al.* 1992; Hamelin *et al.* 1994).

Measurement of metabolite ratios in urine arising from the intake of caffeine-containing beverages is a non-invasive method of phenotyping large populations. Nevertheless, results from caffeine studies are not always in agreement as metabolite ratios may be affected by caffeine intake and may not reflect changes in a single enzyme.

ETHNIC DIFFERENCES

Distribution of XO activity appears to be Gaussian (Grant *et al.* 1983; Guerciolini *et al.* 1991; Vistisen *et al.* 1992), although some studies indicate marked interindividual variation in some ethnic groups. Guerciolini *et al.* (1991) reported a 3.5- to 3.9-fold range in hepatic XO activity whereas Chung *et al.* (2000) have recently found up to nine-fold variation in caffeine metabolite ratios in Koreans. In the earlier study, probit analysis suggested that there may be a low activity subgroup in both males and females. In addition, ethnic differences in XO activity are indicated because XO indices were lower in black Americans than white caucasians, with 1% of population having very low XO activity all of whom were black American males (Relling *et al.* 1992). Information on the distribution of AO is not yet available.

EFFECT OF AGE OF MOLYBDENUM HYDROXYLASE ACTIVITY

There is no correlation between age and XO activity using either direct of indirect indicators of enzyme activity (Guerciolini *et al.*1991; Relling *et al.*, 1992; Chung *et al.* 2000). Total clearance of allopurinol does not differ in elderly subjects compared to young controls although clearance of its active metabolite, oxipurinol is significantly reduced (Turnheim *et al.* 1999). Elimination of allopurinol is predominantly via metabolism whereas oxipurinol is cleared by renal excretion. Although Turnheim *et al.* (1999) have suggested that allopurinol metabolism is a function of XO activity, AO is thought to be primarily responsible for the oxidation of allopurinol to oxipurinol (Reiter *et al.* 1990; Moriwaki *et al.* 1993a) and thus it would appear that neither molybdenum hydroxylase varies in elderly volunteers. Similarly, studies with famciclovir in elderly volunteers and patients have shown that similar amounts of penciclovir, the AO-generated active metabolite (Figure 5.7) are excreted in the urine although renal clearance of penciclovir was slightly lower in elderly subjects (Fowles *et al.* 1992; Perry and Wagstaff 1995). As creatinine clearance was also decreased in elderly patients, the changes in penciclovir pharmacokinetics were thought to be due to age-dependent decreases in renal function, which is also observed with oxipurinol excretion.

Hamelin *et al.* (1994), using caffeine as a probe substrate, found that XO activity was higher in children with cystic fibrosis than age-matched volunteers. This may have implication in drug disposition in this population but may be more significant in the pathogenesis of cystic fibrosis due to an increase in the production of ROS.

Genetic factors and molecular biology

AO and XO are located on the same chromosome in humans at a short distance from each other (Terao *et al.* 1998). The candidate gene for AO maps on chromosome 2q33-q35 (Wright *et al.* 1997; Terao *et al.* 1998) whereas the locus for XO is 2p22 or 2p23 (Ichida *et al.* 1993; Xu *et al.* 1994; Berger *et al.* 1995). Lack of XO is a rare autosomal recessive disorder known as hereditary xanthinuria (Shibutani *et al.* 1999). It is characterised by hypouricaemia, hypouricosuria and xanthinuria and has an incidence of about 1 in 70 000 (Harkness *et al.* 1986). Classical xanthinuria was first reported in 1954 by Dent and Philport (1954) but, on the basis of results obtained from xanthinuric patients with allopurinol, has been more recently classified into two or three subtypes. Classical type 1 xanthinuria is characterised by XO deficiency whereas in classical type 2 xanthinuria both XO and AO are lacking (Reiter *et al.* 1990). Consequently, allopurinol is converted to oxipurinol via AO in type 1 patients whereas type 2 patients are unable to form oxipurinol (Reiter *et al.* 1990 and references therein). Ishida and co-workers have identified a nonsense mutation and deletion of the XO gene in patients with classical type 1 xanthinuria confirming that the primary genetic defect in this subgroup is the XO gene (Ichida *et al.* 1997; Levartovsky *et al.* 2000). In contrast, an abnormality in the molybdenum cofactor that is required for both enzymes is thought to be the cause of classical type 2 xanthinuria (Ichida *et al.* 1998). To date, there are no reports of a single deficiency in aldehyde oxidase.

AO and XO activities are also lacking in patients manifesting Molybdenum Cofactor Deficiency (Reiss 2000). This is also a rare autosomal recessive disorder that, unlike xanthinuria which is relatively benign, leads to severe neurological symptoms and an early childhood death. It is caused by defects in the biosynthesis of the molybdopterin cofactor which is required for AO, XO and sulphite oxidase. Although hypouricaemia is observed in Molybdenum Cofactor Deficiency, the pathogenesis of the clinical features resemble those seen in isolated sulphite oxidase deficiency and thus may not be related to AO or XO (Shalata *et al.* 1998, 2000).

References

Acheampong AA, Shackleton M and Tang-Liu DDS (1995) Comparative ocular pharmacokinetics of brominidine after a single dose application to the eyes of albino and pigmented rabbits. *Drug Metabolism and Disposition*, **23**, 703–712.

Battelli MG, Buonamici L, Abbondanza A, Virgili M, Contestabile A and Stirpe F (1995) Excitotoxic increase of xanthine dehydrogenase and xanthine oxidase in the rat olfactory cortex. *Developmental Brain Research*, **86**, 340–344.

Bauer SL and Howard PC (1990) The kinetics of 1-nitropyrene and 3-nitrofluroanthene metabolism using bovine liver xanthine oxidase. *Cancer Letters*, **54**, 37–42.

Beedham C (1985) Molybdenum hydroxylases as drug metabolising enzymes. *Drug Metabolism Reviews*, **16**, 119–156.

Beedham C (1987) Molybdenum hydroxylases: biological distribution and substrate-inhibitor specificity. In *Progress in Medicinal Chemistry, 24*, Ellis GP and West GB (eds), Elsevier Science Publishers (Biomedical Division), Amsterdam, pp. 85–127.

Beedham C (1997) The role of non-P450 enzymes in drug oxidation. *Pharmacy World & Science*, **19**, 1–9.

Beedham C (1998) Molybdenum hydroxylases. In *Metabolism of Xenobiotics*. Gorrod JW, Oeschlager H and Caldwell J (eds), Taylor & Francis, London and New York, pp. 51–58.

Beedham C, Bruce SE, Critchley DJ, Al-Tayib Y and Rance DJ (1987a) Species variation in hepatic aldehyde oxidase activity. *European Journal of Drug Metabolism and Pharmacokinetics*, **12**, 307–310.

Beedham C, Bruce SE and Rance DJ (1987b) Tissue distribution of the molybdenum hydroxy-lases, aldehyde oxidase and xanthine oxidase, in male and female guinea pigs. *European Journal of Drug Metabolism and Pharmacokinetics.* **12**, 303–306.

Beedham C, Bruce SE, Critchley D J and Rance DJ (1990) 1-Substituted phthalazines as probes of the substrate-binding site of mammalian molybdenum hydroxylases. *Biochemical Pharmacology*, **39**, 1213–1221.

Beedham C, Al-Tayib Y and Smith JA (1992) The role of guinea pig and rabbit hepatic aldehyde oxidase in the oxidative *in vitro* metabolism of cinchona alkaloids. *Drug Metabolism and Disposition*, **20**, 889–895.

Beedham C, Critchley DGP and Rance DJ (1995a) Substrate specificity of human liver aldehyde oxidase towards substituted quinazolines and phthalazines: a comparison with hepatic enzyme from guinea pig, rabbit and baboon. *Archives of Biochemistry and Biophysics*, **319**, 481–490.

Beedham C, Peet CF, Carter H, Panoutsopoulos GI and Smith JA (1995b) Role of aldehyde oxidase in biogenic amine metabolism. In *Progress in Brain Research*, Tipton K and Boulton AA (eds), Elsevier, Amsterdam, pp. 345–353.

Bendotti C, Prosperini E, Kurosaki M, Garattini E and Terao M (1997) Selective localisation of mouse aldehyde oxidase mRNA in the choroid plexus and motor neurons. *Neuroreport*, **8**, 2343–2349.

Berger R, Mezey E, Clancy KP, Harta G, Wright RM, Repine JE, Brown RH, Brownstein M and Patterson D (1995) Analysis of aldehyde oxidase and xanthine dehydrogenase/oxidase as possible candidate genes for autosomal recessive familial amyotrophic lateral sclerosis. *Somatic Cell and Molecular Genetics*, **21**, 121–131.

Berkman CE, Park SB, Wrighton SA and Cashman JR (1995) In vitro–in vivo correlations of human (S)-nicotine metabolism. *Biochemical Pharmacology*, **50**, 565–570.

Bray RC and Lowe DJ (1997) Towards the reaction mechanism of xanthine oxidase from EPR studies. *Biochemical Society Transactions*, **25**, 762–767.

Bunting JW and Gunasekara A (1982) An important enzyme-substrate binding interaction for xanthine-oxidase. *Biochimica et Biophysica Acta*, **704**, 444–449.

Chang CN, Doong SL and Cheng YC (1992) Conversion of 5-iodo-2-pyrimidinone-2'-deoxyribose to 5-iodo-deoxyuridine by aldehyde oxidase: implication in hepatotoxic drug design. *Biochemical Pharmacology*, **43**, 2269–2273.

Chaudhary I, DeMaio W and Kantrowitz J (1994) Species comparison of in vitro and in vivo metabolism of CL284,846, a non-benzodiazepine sedative/hypnotic agent. *Pharmaceutical Research* (NY), **11**, 319.

Chessells JM, Richards SM, Bailey CC, Lilleyman JS and Eden OB (1995) Gender and treatment outcome in childhood lymphoblastic leukemia- report from the MRC UKALL trials. *British Journal of Haematology*, **89**, 364–372.

Chocair P, Duley J, Simmonds HA, Cameron JS, Ianhez L, Arap L and Sabbage E (1993) Low-dose allopurinol plus azathioprine cyclosporin prednisolone, a novel immunosuppressive regimen. *Lancet*, **342**, 83–84.

Chung K-T (1983) The significance of aza-reduction in the mutagenesis and carcinogenesis of aza dyes. *Mutation Research*, **114**, 269–281.

Chung WG, Kang JH, Park CS, Cho MH and Cha YN (2000) Effect of age and smoking on in vivo CYP1A2, flavin-containing monooxygenases, and xanthine oxidase activities in Koreans: Determination by caffeine metabolism. *Clinical Pharmacology and Therapeutics*, **67**, 258–266.

Clarke SE, Harrell AW and Chenery RJ (1995) Role of aldehyde oxidase in the *in vitro* conversion of famciclovir to penciclovir in human liver. *Drug Metabolism and Disposition*, **23**, 251–254.

Critchley DJP, Rance DJ and Beedham C (1992) Subcellular localisation of guinea pig hepatic molybdenum hydroxylases. *Biochemical and Biophysical Research Communications*, **185**, 54–59.

De Miranda P and Good SS (1992) Species differences in the metabolism and disposition of antiviral nucleoside analogues: Acyclovir. *Antiviral Chemistry and Chemotherapy*, **3**, 1–8.

Decker DE and Levinson DJ (1982) Quantitation of rat liver xanthine oxidase by radioimmunoassay: a mechanism for sex-specific differences. *Arthritis and Rheumatism*, **25**, 326–332.

Dent CE and Philport GR (1954) Xanthinuria, an inborn error (or deviation) of metabolism. *Lancet*, **i**, 182–185.

Diaz GJ and Squires EJ (2000) Role of aldehyde oxidase in the hepatic *in vitro* metabolism of 3-methylindole in pigs. *Journal of Agriculture and Food Chemistry,* **48**, 833–837.

Doel JJ, Godber BLJ, Goult TA, Eisenthal R and Harrison R (2000) Reduction of organic nitrites to nitric oxide by xanthine oxidase: Possible role in metabolism of nitrovasodilators. *Biochemical and Biophysical Research Communications,* **270**, 880–885.

Duggan DE (1981) Sulindac: therapeutic implications of the prodrug/pharmacophore equilibrium. *Drug Metabolism Reviews,* **12**, 325–337.

Egashira Y, Isagawa A, Komine T, Yamada E, Ohta T, Shibata K and Sanada H (1999) Tryptophan-niacin metabolism in liver cirrhosis rat caused by carbon tetrachloride. *Journal of Nutritional Science and Vitaminology,* **45**, 459–469.

Felsted RL and Chaykin S (1967) N-Methylnicotinamide oxidation in a number of mammals. *Journal of Biological Chemistry,* **242**, 1274–1279.

Filer CW, Allen GD, Brown TA, Fowles SE, Hollis FJ, Mort EE, Prince WT and Ramji JV (1994) Metabolic and pharmacokinetic studies following oral administration of ^{14}C-famciclovir to healthy subjects. *Xenobiotica,* **24**, 357–368.

Fowles SE, Pue MA, Pierce D, Pratt SK, Crome P, Prince WT and Bruce-Jones P (1992) Pharmacokinetics of penciclovir in healthy elderly subjects following a single oral administration of 750 mg famciclovir. *British Journal of Clinical Pharmacology,* **34**, 450P.

Fowles SE, Pratt SK, Laroche J and Prince WT (1994) Lack of a pharmacokinetic interaction between oral famciclovir and allopurinol in healthy volunteers. *European Journal of Clinical Pharmacology,* **46**, 355–359.

Fridovich I (1966) The mechanism of the enzymatic oxidation of aldehydes. *Journal of Biological Chemistry,* **241**, 3126–3128.

Glueksohn-Waelsch S, Greengard P, Quinn GP and Teicher LS (1967) Genetic variations of an oxidase in mammals. *Journal of Biological Chemistry,* **242**, 1271–1273.

Godber BLJ, Doel JJ, Sapkota GP, Stevens CR, Eisenthal R and Harrison R (2000) Reduction of nitrite to nitric oxide catalysed by xanthine oxidoreductase. *Journal of Biological Chemistry,* **275**, 7757–7763.

Grant DM, Tang BK and Kalow W (1983) Variability in caffeine metabolism. *Clinical Pharmacology and Therapeutics,* **33**, 591–602.

Grant DM, Tang BK, Campbell ME and Kalow W (1986) Effect of allopurinol on caffeine disposition in man. *British Journal of Clinical Pharmacology,* **21**, 454–458.

Guerciolini R, Szumlanski C and Weinshilboum RM (1991) Human liver xanthine oxidase: Nature and extent of individual variation. *Clinical Pharmacology and Therapeutics,* **50**, 663–672.

Guo X, Lerner-Tung M, Chen H-X, Chang C-N, Zhu J-L, Chang C-H, Pizzorno G, Lin T-S and Cheng Y-C (1995) 5-Fluoro-2-pyrimidinone, a liver aldehyde oxidase-activated prodrug of 5-fluorouracil. *Biochemical Pharmacology,* **49**, 1111–1116.

Hale JP and Lilleyman JS (1991) Importance of 6-mercaptopurine dose in lymphoblastic-leukemia. *Archives of Disease in Childhood,* **66**, 462–466.

Hall WW and Krenitsky TA (1986) Aldehyde oxidase from rabbit liver: Specificity toward purines and their analogs. *Archives of Biochemistry and Biophysics,* **251**, 36–46.

Hamelin BA, Xu K, Vallé F, Manseau L, Richer M and Lebel M (1994) Caffeine metabolism in cystic fibrosis: Enhanced xanthine oxidase activity. *Clinical Pharmacology and Therapeutics,* **56**, 521–529.

Hande KR, and Garrow GC (1996) Purine antimetabolites. In *Cancer Chemotherapy and Biotherapy: Principles and Practice,* Chabner BA and Longo DL (eds), Lippincott-Raven Publishers, Philadelphia, pp. 235-252.

Harkness RA, McCreanor D, Simpson D and MacFadyen IR (1986) Pregnancy in and incidence of xanthine oxidase deficiency. *Journal of Inherited Metabolic Disorders,* **9**, 407–408.

Hille R and Massey V (1981) Tight binding inhibitors of xanthine oxidase. *Pharmacology and Therapeutics,* **14**, 249–263.

Hille R and Nishino T (1995) Xanthine oxidase and xanthine dehydrogenase. *FASEBJ,* **9**, 995–1003.

Hille R and Sprecher H (1987) On the mechanism of action of xanthine oxidase. Evidence in support of an oxo transfer mechanism in the molybdenum-containing hydroxylases. *Journal of Biological Chemistry,* **262**, 10914–10917.

Hodnett CM, McCormack JS and Sabean JA (1976) Oxidation of selected pteridine derivatives by mammalian xanthine oxidase and aldehyde oxidase. *Journal of Pharmaceutical Sciences*, **68**, 1150–1155.

Holmes RS (1979) Genetics, ontogeny, and testosterone inducibility of aldehyde oxidase isozymes in the mouse: Evidence for two genetic loci (*Aox-1* and *Aox-2*) closely linked on Chromosome 1. *Biochemical Genetics*, **17**, 517–527.

Howes BD, Bray RC, Richards RL, Turner NA, Bennett B and Lowe DJ (1996) Evidence favoring molybdenum–carbon bond formation in xanthine oxidase action: ^{17}O and ^{13}C-ENDOR and kinetic studies. *Biochemistry*, **35**, 1432–1443.

Huang DY, Furukawa A and Ichikawa Y (1999) Molecular cloning of retinal oxidase aldehyde oxidase cDNAs from rabbit and mouse livers and functional expression of recombinant mouse retinal oxidase cDNA in *Escherichia coli*. *Archives Biochemistry and Biophysics*, **364**, 264–272.

Huff SD and Chaykin S (1967) Genetic and androgenic control of N^1-methylnicotinamide oxidase activity in mice. *Journal of Biological Chemistry*, **242**, 1265–1270.

Huff SD and Chaykin S (1968) Kinetics of testosterone action, *in vivo*, on liver N^1-methylnicotinamide oxidase activity in mice. *Endocrinology*, **83**, 1259–1265.

Ichida K, Amaya Y, Noda S, Minoshima T, Hosoya T, Sakai O, Shimizu N and Nishino T (1993) Cloning of the cDNA encoding human xanthine dehydrogenase (oxidase); structural analysis of the protein and chromosomal location of the gene. *Gene*, **133**, 279–284.

Ichida K, Amaya Y, Kamatani N, Nishino T, Hosoya T and Sakai O (1997) Identification of two mutations in human xanthine dehydrogenase gene responsible for classical type I xanthinuria. *Journal of Clinical Investigation*, **99**, 2391–2397.

Ichida K, Yoshida T, Sakuma R and Hosoya T (1998) Two siblings with classical xanthinuria type 1: significance of allopurinol loading test. *International Medicine*, **37**, 77–82.

Ishigai M, Ishitani Y, Orikasa Y, Kamiyama H and Kumaki K (1998) Metabolism of 2(R,S)-1, 2-bis(nicotinamido)propane, a new agent with anti-vasospasm activity, in rats and rabbits. *Arzneimittel-Forschung-Drug Research*, **48**, 429–435.

Iyer KR, Beers SA, Vince R and Remmel RP (1992) Regioselective metabolism of (-)-6-deoxycarbovir by xanthine oxidase and aldehyde oxidases. *ISSX Proceedings*, p. 85.

Jacob P, Ulgen M and Gorrod JW (1997) Metabolism of (-)-(S)-nicotine by guinea pig and rat brain: identification of cotinine. *European Journal of Drug Metabolism and Pharmacokinetics*, **22**, 391–394.

Johns DG (1967) Human liver aldehyde oxidase: Differential inhibition and oxidation of charged and uncharged substrates. *Journal of Clinical Investigation*, **46**, 1492–1505.

Johns DG Iannotti AT, Sartorelli JR, Booth BA and Bertino JR (1966) The relative toxicities of methotrexate and aminopterin. *Biochemical Pharmacology*, **15**, 555–561.

Johnson C, Stubley-Beedham C and Stell JGP (1985) Hydralazine: A potent inhibitor of aldehyde oxidase activity *in vitro* and *in vivo*. *Biochemical Pharmacology*, **34**, 4251–4256.

Jordan CGM, Rashidi MR, Laljee H, Clarke SE, Brown JE and Beedham C (1999) Aldehyde oxidase-catalysed oxidation of methotrexate in the liver of guinea-pig, rabbit and man, *Journal of Pharmacy and Pharmacology*, **51**, 411–418.

Kalow W and Tang BK (1991) Caffeine as a metabolic probe: Exploration of the enzyme-inducing effect of cigarette smoking. *Clinical Pharmacology and Therapeutics*, **49**, 44–48.

Kashiyama E, Yokoi T, Odomi M and Kamataki T (1999) Stereoselective S-oxidation and reduction of flosequinan in rat. *Xenobiotica*, **29**, 815–826.

Kawashima K, Hosoi K, Naruke T, Shiba T, Kitamura M and Watabe T (1999) Aldehyde oxidase-dependent marked species difference in hepatic metabolism of the sedative-hypnotic, zaleplon, between monkeys and rats. *Drug Metabolism and Disposition*, **27**, 422–428.

Kaye B, Offerman JL, Reid JL, Elliot HL and Hillis WS (1984) A species difference in the presystemic metabolism of carbazeran in dog and man. *Xenobiotica*, **14**, 935–945.

Kaye B, Rance DJ and Waring L (1985) Oxidative metabolism of carbazeran *in vitro* by liver cytosol of baboon and man, *Xenobiotica*, **15**, 237–242.

KeuzenkampJansen CW, vanBaal JM, DeAbeu R, deJong JGN, Zuiderent T and Trijbels JMF (1996) Detection and identification of 6-methylmercapto-8-hydroxypurine, a major metabolite of 6-mercaptopurine, in plasma during intravenous administration. *Clinical Chemistry*, **42**, 380–386.

Kinsella TJ, Kunugi KA, Vielhuber KA, Potter DM, Fitzsimmons ME and Collins JM (1998) Preclinical evaluation of 5-iodo-2-pyrimidinone-2'-deoxyribose as a prodrug for 5-iodo-2'-deoxyuridine-mediated radiosensitization in mouse and human tissues. *Clinical Cancer Research*, **4**, 99–109.

Kisker C, Schindelin H and Rees DC (1997) Molybdenum-cofactor-containing enzymes: structure and mechanism. *Annual Reviews in Biochemistry*, **66**, 233–267.

Kitamura S and Tatsumi K (1983a) A sulfoxide reducing enzyme system consisting of aldehyde oxidase and xanthine oxidase. *Chemical and Pharmaceutical Bulletin*, **31**, 760–763

Kitamura S and Tatsumi K (1983b) Azoreductase activity of liver aldehyde oxidase. *Chemical and Pharmaceutical Bulletin*, **31**, 3334–3337.

Kitamura K and Tatsumi S (1984a) Reduction of tertiary amine *N*-oxides by liver preparations: Function of aldehyde oxidase as a major *N*-oxide reductase. *Biochemical and Biophysical Research Communications*, **121**, 749–754.

Kitamura S and Tatsumi K (1984b) Involvement of liver aldehyde oxidase in the reduction of nicotinamide N-oxide. *Biochemical and Biophysical Research Communications*, **120**, 602–606.

Kitamura S, Tatsumi K, Hirata Y and Yoshimura H (1981) Further studies on sulfoxide-reducing enzyme system. *Journal of Pharmaceutical Dynamics*, **4**, 528–533.

Kitamura S, Sugihara K, Kuwasako M and Tatsumi K (1996) The role of mammalian intestinal bacteria in the reductive metabolism of zonisamide. *Journal of Pharmacy and Pharmacology*, **49**, 253–256.

Kitamura S, Kadota T, Yoshida M and Ohta S (1999a) Inter-conversion between fenthion and fenthion sulfoxide in goldfish, *Carassius auratus*. *Journal of Health Science*, **45**, 266–270.

Kitamura S, Nakatani K, Sugihara K and Ohta S (1999b) Strain differences of the ability to hydroxylate methotrexate in rats. *Comparative Biochemistry and Physiology C—Pharmacology Toxicology & Endocrinology*, **122**, 331–336.

Kitamura S, Sugihara K, Nakatani K, Ohta S, O'Hara T, Ninomiya S, Green CE and Tyson CA (1999c) Variation of hepatic methotrexate 7-hydroxylase activity in animals and humans. *UBMB Life*, **48**, 607–611.

Kitchen BJ, Moser A, Lowe E, Balis FM, Widemann B, Anderson L, Strong J, Blaney SM, Berg SL, O'Brien M and Adamson PC (1999) Thioguanine administered as a continuous intravenous infusion to pediatric patients is metabolized to the novel metabolite 8-hydroxy-thioguanine. *Journal of Pharmacology and Experimental Therapeutics*, **291**, 870–874.

Kokoglu E, Belce A, Ozyurt and Tepeler Z (1990) Xanthine oxidase levels in human brain tumours. *Cancer Letters*, **50**, 179–181.

Kooji A, Schijns M, Frederiks WM, Vannooden CJF and James J (1992) Distribution of xanthine oxidoreductase activity in human tissues—a histochemical and biochemical study. *Virchows Archiv B-Cell Pathology including Molecular Pathology*, **63**, 17–23.

Krenitsky TA, Neil SM Elion GB and Hitchings GC (1972). A comparison of the specificities of xanthine oxidase and aldehyde oxidase. *Archives of Biochemistry and Biophysics*, **150**, 585–599.

Krenitsky TA, Tuttle V, Cattau EL and Wang P (1974) A comparison of the distribution and electron acceptor specificities of xanthine oxidase and aldehyde oxidase. *Comparative Biochemistry & Physiology*, **49B**, 687–703.

Krenitsky TA, Spector T and Hall WW (1986) Xanthine oxidase from human liver: Purification and characterization. *Archives of Biochemistry and Biophysics*, **247**, 108–119.

Kunieda T, Kobayashi E, Tachibana M and Ikadai H (1999) A genetic linkage map of rat chromosome 9 with a new locus for variant activity of liver aldehyde oxidase. *Experimental Animals*, **48**, 3–45.

Kurosaki M, Li Calzi M, Scanziani E, Garattini E and Terao M (1995) Tissue and cell specific expression of mouse xanthine oxidoreductase gene in vivo: regulation by bacterial lipopolysaccharide. *Biochemical Journal*, **306**, 225–234.

Kurosaki M, Demontis S, Barzago MM, Garattini E and Terao M (1999) Molecular cloning of the cDNA coding for mouse aldehyde oxidase: tissue distribution and regulation in vivo by testosterone. *Biochemical Journal*, **341**, 71–80.

Laderoute KR and Rauth MA (1986) Identification of two major reduction products of the hypoxic cell toxin 3-amino-1,2,4-benzotriazine-1,4-dioxide. *Biochemical Pharmacology*, **35**, 3417–3420.

Laljee H (1998) In vitro oxidation of N-heterocycles and aldehydes by guinea pig liver and brain. PhD Thesis, University of Bradford.

Lee Y-L and Chan KK (1988) Metabolic interaction between methotrexate and 4'-(9-acrinyl-amino)methansulfon-M-anisidine in the rabbit. *Cancer Research*, **48**, 5706–5111.

Lee SC and Renwick AG (1995) Sulphoxide reduction by rat and rabbit tissues in vitro. *Biochemical Pharmacology*, **49**, 1557–1565.

Levartovsky D, Lagziel A, Sperling O, Liberman U, Yaron M, Hosoya T, Ichida K and Peretz H (2000) *XDH* gene mutation is the underlying cause of classical xanthinuria: A second report. *Kidney International*, **57**, 2215–2220.

Levinson DJ and Chalker D (1980) Rat hepatic xanthine oxidase: age and sex specific differences. *Arthritis and Rheumatism*, **23**, 77–82.

Lewis AS, Murphy L, McCalla Fleary M and Purcell S (1984) Inhibition of mammalian xanthine oxidase by folate compounds and amethopterin. *Journal of Biological Chemistry*, **259**, 12–15.

Li Calzi M, Raviolo C, Ghibaudi E, De Gioia L, Salmona M, Cazzaniga G , Kurosaki, M, Terao M and Garattini E (1995) Purification, cDNA cloning, and tissue distribution of bovine liver aldehyde oxidase. *Journal of Biological Chemistry*, **270**, 31037–31045.

Lin G, Hawes EM, McKay G, Korchinski ED and Midha KK (1993) Metabolism of piperidine-type phenothiazine antipsychotic agents. 4. Thioridazine in dog, man and rat. *Xenobiotica*, **23**, 1059–1074.

Lin G, Chu KW, Damani LA and Hawes EM (1996) Identification of lactams as in vitro metabolites of piperidine-type phenothiazine antipsychotic drugs. *Journal of Pharmaceutical and Biomedical Analysis*, **14**, 727–738.

Linder N, Rapola J, Raivio KO (1999) Cellular expression of xanthine oxidoreductase protein in normal human tissues. *Laboratory Investigation*, **79**, 967–974.

Mahler HR, Fairhurst AS and Mackler B (1955) Studies on metalloflavoproteins IV. The role of the metal. *Journal of American Chemical Society*, **77**, 1514–1521.

Massey V and Harris CM (1997) Milk xanthine oxidoreductase: the first one hundred years. *Biochemical Society Transactions*, **25**, 750–754.

Massey V, Komai H, Palmer G and Elion G B (1970) On the mechanism of inactivation of xanthine oxidase by allopurinol and other pyrazolo[3,4-d] pyrimidines. *Journal of Biological Chemistry*, **245**, 2837–2844.

Mathews JM, Black SR and Burka LT (1998) Disposition of butanal oxime in rat following oral, intravenous and dermal administration. *Xenobiotica*, **28**, 767–777.

McMurtrey KD and Knight TJ (1984) Metabolism of acridine by rat liver enzymes. *Mutation Research*, **140**, 7–11.

Millar TM, Stevens CR, Benjamin N, Eisenthal R, Harrison and Blake DR (1998) Xanthine oxidoreductase catalyses the reduction of nitrates and nitrite to nitric oxide under hypoxic conditions. *FEBS Letters*, **427**, 225–228.

Moriwaki Y, Yamamoto T, Nasako Y, Takahashi S, Suda M, Hiroisha K, Hada T and Higashino K (1993a) *In vitro* oxidation of pyrazinamide and allopurinol by rat liver aldehyde oxidase. *Biochemical Pharmacology*, **46**, 975–981.

Moriwaki Y, Yamamoto T, Suda M, Nasako Y, Takahashi S, Agbedana OE, Hada T, and Higashino K (1993b) Purification and immunohistochemical tissue localisation of human xanthine oxidase. *Biochimica et Biophysica Acta*, **1164**, 327–330.

Moriwaki Y, Yamamoto T, Yamaguchi K, Suda M, Yamakita J, Takahashi S and Higashino K (1996) Immunohistochemical localisation of xanthine oxidase in human tissues. *Acta Histochemica et Cytochemica*, **29**, 153–162.

Moriwaki Y, Yamamoto T and Higashino K (1997) Distribution and pathophysiological role of molybdenum-containing enzymes. *Histology and Histopathology*, **12**, 513–524.

Moriwaki Y, Yamamoto T, Yamakita J, Takahashi S and Higashino K (1998a) Comparative localisation of aldehyde oxidase and xanthine oxidoreductase activity in rat tissues. *Histochemical Journal*, **30**, 69–74.

Moriwaki Y, Yamamoto T, Yamakita J, Takahashi S, Tsutsumi Z and Higashino K (1998b). Zonal distribution of allopurinol-oxidizing enzymes in rat liver. *Advances in Experimental Medicine and Biology*, **431**, 47–50.

Moriwaki Y, Yamamoto T and Higashino K, (1999). Enzymes involved in purine metabolism -A review of histochemical localisation and functional implications. *Histology and Histopathology*, **14**, 1321–1340.

Morpeth FF (1983) Studies on the specificity towards aldehyde substrates and steady-state kinetics of xanthine oxidase. *Biochimica et Biophysica Acta*, **744**, 328–334.

Murrell GAC and Rapeport WG (1986) Clinical pharmacokinetics of allopurinol. *Clinical Pharmacokinetics*, **11**, 343–353.

Nakajima M, Yamamoto T, Nunoya K, Yokoi T, Nagahsima K, Inoue K, Funae Y, Shimada N, Kamataki T and Kuroiwa Y (1996) The role of human cytochrome P4502A6 in C-oxidation of nicotine. *Drug Metabolism and Disposition*, **24**, 1212–1217.

Nakajima M, Yamagishi S, Yamamoto H, Yamamoto T, Kuroiwa Y and Yokoi T (2000) Deficient cotinine formation from nicotine is attributed to the whole deletion of the CYP2A6 gene in humans. *Clinical Pharmacology and Therapeutics*, **67**, 57–69.

Nakasa H, Komiya M, Ohmori S, Rikisha T, Kuichi M and Kitada M (1993) Characterization of human liver aptochrome PU50 involved in the reductive metabolism of zonisamide. *Molecular Pharmacology*, **44**, 216–221.

Nguyen TL, Gruenke LG and Castagnoli N (1979) Metabolic oxidation of nicotine to chemically reactive intermediates. *Journal of Medicinal Chemistry*, **22**, 259–263.

Nishino T (1994) The conversion of xanthine dehydrogenase to xanthine oxidase and the role of the enzyme in reperfusion injury. *Journal of Biochemistry*, **116**, 1–6.

Ohkubo K, Sakiyama S and Fujimura S (1983) Purification and characterization of N^1-methylnicotinamide oxidases I and II separated from rat liver. *Archives of Biochemistry and Biophysics*, **221**, 534–542.

Okamoto K and Nishino T (1995) Mechanism of inhibition of xanthine oxidase with a new tight binding inhibitor. *Journal of Biological Chemistry*, **270**, 7816–7821.

Oldfield S (1998) In vitro metabolism of pyrimidinones by hepatic molybdenum hydroxylases and dihydropyrimidine dehydrogenase. PhD Thesis, University of Bradford.

Panoutsopoulos GI (1994) Hepatic oxidation of aromatic aldehydes. PhD Thesis, University of Bradford.

Pay GF, Wallis RB and Zelaschi D (1980) A metabolite of sulphinpyrazone that is largely responsible for the effect of the drug on the platelet prostaglandin pathway. *Biochemical Society Transactions*, **8**, 727–728.

Peet CF (1995) In vitro oxidation of biogenic aldehydes in guinea pig liver and brain. PhD Thesis, University of Bradford.

Perry CM and Wagstaff AJ (1995) Famciclovir: A review of its pharmacological properties and therapeutic efficacy in Herpes Virus infections. *Drugs*, **50**, 396–415.

Pietruszko R, Kuys G and Ambroziak A (1991) Physiological role of aldehyde dehydrogenase (EC 1.2.1.3). *Alcoholism*, **206**, 101–106.

Porter DJT (1990) Xanthine oxidase catalysed reductive debromination of 6-(bromomethyl)-9H-purine with concomitant covalent modification of the FAD prosthetic group. *Journal of Biological Chemistry*, **265**, 13540–13546.

Porter DJT, Harrington JA, Almond MR, Lowen GT, Zimmerman P and Spector T (1994) 5-Ethynyl-2-(1H)-pyrimidinone: aldehyde oxidase-activation to 5-ethynyluracil, a mechanism-based inhibitor of dihydropyrimidine dehydrogenase. *Biochemical Pharmacology*, **47**, 1165–1171.

Pratt SK, Pue MA, Fairless AJ, Fowles SE, Laroche J, Kumurav R and Prince WT (1994) Lack of an effect of gender on the pharmacokinetics of penciclovir following single oral doses of famciclovir. *British Journal of Clinical Pharmacology*, **37**, 493P.

Rajagopalan KV (1997) The molybdenum cofactors—perspective from crystal structure. *Journal of Biological and Inorganic Chemistry*, **2**, 786–789.

Rajagopalan KV and Handler P (1964a) Hepatic aldehyde oxidase II. Differential Inhibition of electron transfer to various electron acceptors. *Journal of Biological Chemistry*, **239**, 2022–2026.

Rajagopalan KV and Handler P (1964b) Hepatic aldehyde oxidase III. The substrate binding site. *Journal of Biological Chemistry*, **239**, 2027–2035.

Rajagopalan KV and Johnson JL (1992) The pterin molybdenum cofactors. *Journal of Biological Chemistry*, **267**, 10199–10202.

Rajagopalan KV, Fridovich I and Handler P (1962) Hepatic aldehyde oxidase I. Purification and Properties. *Journal of Biological Chemistry*, **237**, 922–928.

Rashidi MR (1996) In vitro oxidation of purines by hepatic molybdenum hydroxylases. PhD Thesis, University of Bradford.

Rashidi MR, Smith JA, Clarke SE and Beedham C (1995) Inhibition of human and guinea-pig liver aldehyde oxidase in-vitro. *Journal of Pharmacy and Pharmacology*, **47**, 1090.

Rashidi MR, Smith JA, Clarke SE and Beedham C (1997) In vitro oxidation of famciclovir and 6-deoxypenciclovir by aldehyde oxidase from human, guinea pig, rabbit and rat liver. *Drug Metabolism and Disposition*, **25**, 805–813.

Reiss AU (2000) Genetics of molybdenum cofactor deficiency. *Journal of Human Genetics*, **106**, 157–163.

Reiter S, Simmonds HA, Zöllner N, Braun SL and Knedel M (1990) Demonstration of a combined deficiency of xanthine oxidase and aldehyde oxidase in xanthinuric patients not forming oxipurinol. *Clinica Chimica Acta*, **187**, 221–234.

Relling MV, Lin J-S, Ayers GD and Evans WE (1992) Racial and gender differences in N-acetyltransferase, xanthine oxidase and CYP1A2* activities. *Clinical Pharmacology and Therapeutics*, **52**, 643–658.

Robertson IGC and Bland TJ (1993) Inhibition by SKF-525A of the aldehyde oxidase-mediated metabolism of the experimental antitumour agent acridine carboxamide. *Biochemical Pharmacology*, **45**, 2159–2162.

Robertson GC and Gamage RSKA (1994) Methadone: a potent inhibitor of rat liver aldehyde oxidase. *Biochemical Pharmacology*, **47**, 584–587.

Robertson IGC, Palmer BD, Paxton JW and Bland TJ (1993) Metabolism of the experimental antitumour agent acridine carboxamide in the mouse. *Drug Metabolism and Disposition*, **21**, 530–536.

Rochat B, Kosel M, Boss G, Testa B, Gillet M and Baumann P (1998) Stereoselective biotransformation of the selective serotonin reuptake inhibitor citalopram and its demethylated metabolites by monoamine oxidases in human liver. *Biochemical Pharmacology*, **56**, 15–23.

Rodrigues AD (1994) Comparison of levels of aldehyde oxidase with cytochrome P450 activities in human liver *in vitro*. *Biochemical Pharmacology*, **48**, 97–200.

Rodrigues AD, Ferrero JL, Amann MT, Rotert GA, Cepa SP, Surber BW, Machinist JM, Tich NR, Sullivan JP, Garvey DS, Fitzgerald MF and Arneric SP (1994) The *in vitro* hepatic metabolism of ABT-418, a cholinergic channel activator, in rats, dogs, cynomolgus monkeys, and humans. *Drug Metabolism and Disposition*, **22**, 788–798.

Romão MJ, Archer M, Moura I, Moura JJG, LeGall J, Engh R, Schneider M, Hof P and Huber R (1995) Crystal-structure of the xanthine oxidase-related aldehyde oxidoreductase from d-gigas. *Science*, **270**, 1170–1176.

Roy SK, Korzekwa KR, Gonzalez FJ, Moschel RC and Dolan ME (1995) Human liver oxidative metabolism of O^6-benzylguanine. *Biochemical Pharmacology*, **50**, 1385–1389.

Ruenitz PC and Bai X (1995) Acidic metabolites of tamoxifen: Aspects of formation and fate in the female rat. *Drug Metabolism and Disposition*, **23**, 993–998.

Saito T and Nishino T (1989) Differences in redox and kinetic properties between NAD-dependent and O_2-dependent types of rat liver xanthine dehydrogenase. *Journal of Biological Chemistry*, **264**, 10015–10022.

Schofield PC, Robertson IGC, Paxton JW, McCrystal MR, Evans BD, Kestell P and Baguley BC (1999) Metabolism of N- [(2'-dimethylamino)ethyl]acridine-4-carboxamide in cancer patients undergoing a phase I clinical trial. *Cancer Chemotherapy and Pharmacology*, **44**, 51–58.

Schofield PC, Robertson IGC and Paxton JW (2000) Inter-species variation in the metabolism and inhibition of N- [(2'-dimethylamino)ethyl]acridine-4-carboxamide (DACA) by aldehyde oxidase. *Biochemical Pharmacology*, **59**, 161–165.

Shalata A, Mandel H, Reiss J, Szargel R, Cohen-Akenine A, Dorche C, Zabot MT, Van Gennip A, Ableing N, Berant M and Cohen N (1998) Localization of a gene for Molybdenum Cofactor Deficiency, on the short arm of chromosome 6, by homozygosity marking. *American Journal of Human Genetics*, **63**, 148–154.

Shalata A, Mandel H, Dorche C, Zabot MT, Shalev S, Hugeirat Y, Arieh D, Ronit Z, Reiss J, Anbinder Y and Cohen N (2000) Prenatal diagnosis and carrier detection for molybdenum cofactor deficiency type A in Northern Israel using polymorphic DNA markers. *Prenatal Diagnosis*, **20**, 7–11.

Shanmuganathan K, Koudriakova T, Nampalli S, Du J, Gallo JM, Schinazi RF and Chu CK (1994) Enhanced brain delivery of an Anti-HIV nucleoside 2'F-ara-ddI by xanthine oxidase mediated biotransformation. *Journal of Medicinal Chemistry*, **37**, 821–827.

Shaw S and Jayatille KCE (1992) The role of cellular oxidases and catalytic iron in the pathogenesis of ethanol-induced liver injury. *Life Science*, **50**, 2045–2052.

Shibata K (1989) Tissue distribution of N^1-methyl-2-pyridone-5-carboxamide- and N^1-methyl-4-pyridone-3-carboxamide-forming N^1-methylnicotinamide oxidases in rats. *Agricultural and Biological Chemistry*, **53**, 3355–3356.

Shibutani Y, Ueo T, Yamamoto T, Takahashi S, Moriwaki Y and Higashino K (1999) A case of classical xanthinuria (type 1) with diabetes mellitus and Hashimoto's thyroiditis. *Clinica et Chimica Acta*, **285**, 183–189.

Skordos KW, Skiles GL, Laycock JD, Lanza DL and Yost GS (1998) Evidence supporting the formation of 2,3-epoxy-3-methylindoline: A reactive intermediate of the pneumotoxin 3-methylindole. *Chemical Research in Toxicology*, **11**, 741–749.

Smeland E, Fuskevåg O, Nymann K, Svendson JS, Olsen R, Lindal S, Bremnes RM and Anarbakke J (1996) High-dose 7-hydroxymethotrexate: Acute toxicity and lethality in a rat model. *Cancer Chemotherapy and Pharmacology*, **37**, 415–422.

Spector T, Hall WW and Krenitsky TA (1986) Human and Bovine Xanthine Oxidases. Inhibition studies with oxipurinol. *Biochemical Pharmacology*, **35**, 3109–3114.

Spray HE (1996) In vitro metabolism of quinine, tolbutamide, codeine and morphine by liver slices. PhD Thesis, University of Bradford.

Stanlovic, M and Chaykin S (1971) Aldehyde oxidase: Catalysis of the oxidation of N-methylnicotinamide and pyridoxal. *Archives of Biochemistry and Biophysics*, **145**, 27–34.

Star AU and Hochberg MC (1993) Prevention and management of gout. *Drugs*, **45**, 212–222.

Stiff DD and Zemaitis MA (1990) Metabolism of the anticonvulsant agent zonisamide in the rat. *Drug Metabolism and Disposition*, **18**, 888–894.

Stoddart AM and Levine WG (1992) Azoreductase activity by purified rabbit liver aldehyde oxidase. *Biochemical Pharmacology*, **43**, 2227–2235.

Stubley C and Stell JG P (1980) Investigation of the binding site of aldehyde oxidase. *Journal of Pharmacy and Pharmacology*, **32**, 51.

Stubley C, Stell JGP and Mathieson DW (1979) The oxidation of azaheterocycles with liver aldehyde oxidase. *Xenobiotica*, **9**, 474–484.

Sugihara K and Tatsumi K (1986) Participation of liver aldehyde oxidase in reductive metabolism of hydroxamic acids to amides. *Archives of Biochemistry and Biophysics*, **247**, 289–293.

Sugihara K, Kitamura S and Tatsumi K (1983) Involvement of liver aldehyde oxidase in conversion of N-hydroxyurethane to urethane. *Journal of Pharmaceutical Dynamics*, **6**, 677–683.

Sugihara K, Kitamura S and Tatsumi K (1995) Strain differences of liver aldehyde oxidase activity in rat. *Biochemistry and Molecular Biology International*, **37**, 861–869.

Sugihara K, Kitamura S and Tatsumi K (1996a) Involvement of mammalian liver cytosols and aldehyde oxidase in reductive metabolism of zonisamide. *Drug Metabolism and Disposition*, **24**, 199–202.

Sugihara K, Kitamura S and Tatsumi K (1996b) S-(-)-Nicotine-1'-N-oxide reductase activity of rat liver aldehyde oxidase. *Biochemistry and Molecular Biology International*, **40**, 535–541.

Sweetman L (1968) Urinary and cerebrospinal fluid oxypurine levels and allopurinol metabolism in the Lesch–Nyhan Syndrome. *Federation Proceedings*, **27**, 1055–1058.

Tatsumi K and Ishigai M (1987). Oxime-metabolising activity of liver aldehyde oxidase. *Archives of Biochemistry and Biophysics*, **253**, 413–418.

Tatsumi K and Yamada H (1982) Reduction of N-nitrosodiphenylamine to the corresponding hydrazine by guinea pig liver preparations. *Chemical and Pharmaceutical Bulletin*, **30**, 3842–3845.

Tatsumi K, Kitamura S and Yamada H (1982) Involvement of liver aldehyde oxidase in sulfoxide reduction. *Chemical and Pharmaceutical Bulletin*, **12**, 4585–4588.

Tatsumi K, Yamada H and Kitamura S (1983) Evidence for the involvement of liver aldehyde oxidase in reduction of nitrosamines to the corresponding hydrazine. *Chemical and Pharmaceutical Bulletin*, **31**, 764–767.

Tatsumi K, Kitamura S and Narai N (1986) Reductive metabolism of aromatic nitro compounds including carcinogens by rabbit liver preparations. *Cancer Research*, **46**, 1089–1093.

Taylor SM, Stubley-Beedham C and Stell JGP (1984) Simultaneous formation of 2- and 4-quinolones from quinolinium cations catalysed by aldehyde oxidase. *Biochemical Journal*, **220**, 67–74.

Terao M, Cazziniga P, Ghezzi P, Bianchi M, Falciani F, Perani P and Garattini E (1992) Molecular cloning of a cDNA for mouse liver xanthine dehydrogenase. *Biochemical Journal*, **283**, 863–870.

Terao M, Kurosaki M, Zanotta S and Garattini E (1997) The xanthine oxidoreductase gene: structure and regulation. *Biochemical Society Transactions*, **25**, 791–796.

Terao M, Kurosaki M, Demontis S, Zanotta S and Garattini E (1998) Isolation and characterization of the human aldehyde oxidase gene: conservation of intron/exon boundaries with the xanthine oxidoreductase gene indicates a common origin. *Biochemical Journal*, **332**, 383–393.

Thor H, Smith MT, Hartzell P, Bellomo G, Jewell SA and Orrenius S (1982) The metabolism of menadione (2-methyl-1,2-naphthoquinone) by isolated hepatocytes. A study of the implications of oxidative stress in intact cells. *Journal of Biological Chemistry*, **257**, 12419–12425.

Turner NA, Doyle WA, Ventom AM and Bray RC (1995) Properties of rabbit liver aldehyde oxidase and the relationship to xanthine oxidase and dehydrogenase. *European Journal of Biochemistry*, **232**, 646–657.

Turnheim K, Kivanek P and Oberbauer R (1999) Pharmacokinetics and pharmocodynamics of allopurinol in elderly and young subjects. *British Journal of Clinical Pharmacology*, **48**, 501–509.

Usanky JI and Damani LA (1983) Oxidation of metyrapone to an α-pyridone metabolite by a mammalian molybdenum hydroxylase. *Journal of Pharmacy and Pharmacology*, **35**, 72.

Ventura SM and Dachtler SL (1981) Effects of sex hormones on hepatic aldehyde oxidase activity in C57BL/6J mice. *Hormone Research*, **14**, 250–259.

Vistisen K, Poulsen HE and Loft S (1992) Foreign compound metabolism capacity in man measured from metabolites of dietary caffeine. *Carcinogenesis*, **13**, 1561–1568.

Wagner M and Harkness RA (1989) Distribution of xanthine dehydrogenase and oxidase activities in human and rabbit tissues. *Biochimica et Biophysica Acta*, **991**, 79–84.

Wanwimolruk S, Wong SM, Zhang H, Coville PF and Walker RJ (1995) Metabolism of quinine in man: Identification of a major metabolite, and effects of smoking and rifampicin treatment. *Journal of Pharmacy and Pharmacology*, **47**, 957–963.

Westerfield WW, Richert DA and Higgins ES (1957) The metabolic reduction of organic nitro groups. *Journal of Chemical Society*, **227**, 379–391.

Whittlesea CMC and Gorrod JW (1993)The enzymology of the in vitro oxidation of prolintane to oxoprolintane, *Journal of Clinical Pharmacology and Therapeutics*, **18**, 357–364.

Wolff AW and Oehme FW (1974) Carcinogenic chemicals in food as an environmental issue. *Journal of American Veterinary and Medical Association*, **164**, 623–629.

Wolpert MK, Althaus JR and Johns DG (1973) Nitroreductase activity of mammalian liver aldehyde oxidase. *Journal of Pharmacology and Experimental Therapeutics*, **185**, 202–213.

Wright RM, Vaitaitis GM, Wilson CM, Repine TB, Terada LS and Repine JE (1993) cDNA cloning, characterisation, and tissue-specific expression of human xanthine dehydrogenase xanthine oxidase. *Proceedings of the National Academy of Sciences of the United States of America*, **90**, 10690–10694.

Wright RM, Vaitaitis GM, Weigel LK, Repine TB, McManamam JL and Repine JE (1995) Identification of the candidate ALS2 gene at chromosome 2q33 as a human aldehyde oxidase gene. *Redox Report*, **1**, 313–321.

Wright RM, Weigel LK, VarellaGarcia M, Vaitaitis G and Repine JE (1997) Molecular cloning, refined chromosomal mapping and structural analysis of the human gene encoding aldehyde oxidase (AOX1), a candidate for the ALS2 gene. *Redox Report*, **3**, 135–144.

Wright RM, Clayton DK, Riley MG, McManaman JL and Repine JE (1999a) cDNA cloning, sequencing and characterization of male and female rat liver aldehyde oxidase (rAOX1)—differences in redox status may distinguish male and female forms of hepatic AOX. *Journal of Biological Chemistry*, **274**, 3878–3886.

Wright RM, McManaman JL and Repine JE (1999b) Alcohol-induced breast cancer: A proposed mechanism. *Free Radical Biology & Medicine*, **26**, 348–354.

Wu E, ShinkaT, Caldera-Munoz P, Yoshizumi H, Trevor A and Castagnoli Jr N (1988) Metabolic studies on the nigrostriatal toxin MPTP and its MAO-B generated dihydropyridinium metabolite MPDP+. *Chemical Research in Toxicology*, **1**, 186–194.

Xia M, Dempski R and Hille R (1999) The reductive half-reaction of xanthine oxidase - Reaction with aldehyde substrates and identification of the catalytically labile oxygen. *Journal of Biological Chemistry*, **274**, 3323–3330.

Xu P, Huecksteadt TP, Harrison R and Hoidal JR (1994) Molecular cloning, tissue expression of human xanthine dehydrogenase. *Biochemical and Biophysical Research Communications*, **199**, 998–1003.

Yamamoto T, Moriwaki Y, Suda M, Nasako Y, Takahashi S, Hiroisha K, Nakano T, Hada T and Higashino K (1993) Effect of BOF-4272 on the oxidation of allopurinol and pyrazinamide invivo—is xanthine dehydrogenase or aldehyde oxidase more important in oxidizing both allopurinol and pyrazinamide? *Biochemical Pharmacology*, **46**, 2277–2284.

Yoshihara S and Ohta S (1998) Involvement of hepatic aldehyde oxidase in conversion of 1-methyl-4-phenyl-2,3-dihydropyridinium (MPDP+) to1-methyl-4-phenyl- 5,6-dihydro-2-pyridone. *Archives of Biochemistry and Biophysics*, **360**, 93–98.

Yoshihara S and Tatsumi K (1986) Kinetic and inhibition studies on reduction of diphenylsulphoxide by guinea pig liver aldehyde oxidase. *Archives of Biochemistry and Biophysics*, **249**, 8–14.

Yoshihara S and Tatsumi K (1990) Metabolism of diphenylsulphoxide in perfused guinea pig liver. Involvement of aldehyde oxidase as a sulphoxide reductase. *Drug Metabolism and Disposition*, **18**, 876–881.

Yoshihara S and Tatsumi K (1997a) Involvement of growth hormone as a regulating factor in sex differences of mouse hepatic aldehyde oxidase. *Biochemical Pharmacology*, **53**, 1099–1105.

Yoshihara S and Tatsumi K (1997b) Purification and characterization of hepatic aldehyde oxidase in male and female mice. *Archives of Biochemistry and Biophysics*, **338**, 29–34.

Yoshihara S, Harada K and Ohta S (2000) Metabolism of 1-methyl-4-phenyl-1,2,3,6-tetrapyridinium (MPTP+) in perfused rat liver. Involvement of hepatic aldehyde oxidase as a detoxification enzyme. *Drug Metabolism and Disposition*, **28**, 538–543.

Zimm S, Collins JM, O'Neill D, Chabner BA and Poplack DG (1983) Inhibition of first-pass metabolism in cancer chemotherapy: Interaction of 6-mercaptopurine and allopurinol. *Clinical Pharmacology and Therapeutics*, **34**, 810–817.

Zimm S, Grygiel JJ, Strong JM, Monks TJ and Poplack DG (1984) Identification of 6-mercaptopurine riboside in patients receiving 6-mercaptopurine as a prolonged intravenous infusion. *Biochemical Pharmacology*, **33**, 4089–4092.

6 Prostaglandin Synthases

Gisela H. Degen, Christoph Vogel and Josef Abel

Institut für Arbeitsphysiologie an der Universität Dortmund, Germany

Introduction

Prostaglandin H synthase (PGHS; EC 1.14.99.1) is a haem-containing protein with two enzymic activities: a cyclooxygenase and a peroxidase. The bifunctional enzyme catalyses key reactions in the biosynthesis of prostanoids (prostaglandins and related compounds). Since these products are important mediators of several physiologic and pathophysiologic processes, there is a great deal of interest in PGHS and its modulation. Moreover, a large and diverse number of chemicals including carcinogens have been found to be oxidised by PGHS-peroxidase or by peroxyl radicals generated during prostanoid biosynthesis. This chapter addresses two major aspects: one is the involvement of the enzyme in the metabolism and bioactivation of xenobiotics, the other relates to compounds affecting the activity and/or expression of PGHS or more precisely its two known isoforms (PGHS-1 and PGHS-2). In addition to compounds which inhibit PGHS-cyclooxygenase activity, there is emerging evidence that various xenobiotics, by inducing predominantly PGHS-2, modulate the synthesis of biologically active prostanoids.

PGHS genes and isozymes

There are two structurally related isoforms of PGHS encoded by separate genes. The two PGHS isozymes called PGHS-1 and PGHS-2 or COX-1 and COX-2 are homodimeric, haem-containing, glycosylated membrane-bound proteins (Smith and DeWitt 1996; Smith *et al.* 1996). PGHS-1, first purified in 1976 (Miyamoto *et al.* 1976) and cloned in 1988 (Merlie *et al.* 1988; DeWitt and Smith 1988, 1990; Yokoyama *et al.* 1988), is considered as the constitutive enzyme and synthesises prostaglandins involved in 'housekeeping' functions. PGHS-2, identified in 1991 by several laboratories, is inducible by various stimuli including proinflammatory cytokines and growth factors implying a role for PGHS-2 in inflammation and control of cell growth (for reviews see DuBois *et al.* 1998; Hershman 1996, 1999).

The PGHS-isozymes differ considerably with respect to their expression and

---

Enzyme Systems that Metabolise Drugs and Other Xenobiotics. Edited by C. Ioannides.
© 2002 John Wiley & Sons Ltd

biology, but have a very similar structure and similar kinetic properties (Garavito and DeWitt 1999; Smith and DeWitt 1996). Both isoforms convert free arachidonic acid to PGH_2; the release of free fatty acid substrate from membrane lipids is predominantly catalysed by phospholipases A_2 (Dennis 1994). PGH_2 is then metabolised by distinct synthases to prostaglandins (e.g. PGE_2, PGD_2, $PGF_{2\alpha}$), prostacyclins (PGI_2) or to thromboxanes (TxA_2). Being key enzymes in the synthesis of these biologically active prostanoids (Figure 6.1) make PGH-synthases a prime target for an important class of pharmaceutical compounds, the non-steroidal antiinflammatory drugs (see below).

PGHS-1 is located on human chromosome 9 (Funk *et al.* 1991). The genomic DNA sequence of human and murine PGHS-1 of approximately 22 kb contain eleven exons and ten introns (Yokoyama and Tanabe 1989; Kraemer *et al.* 1992). The PGHS-1 gene

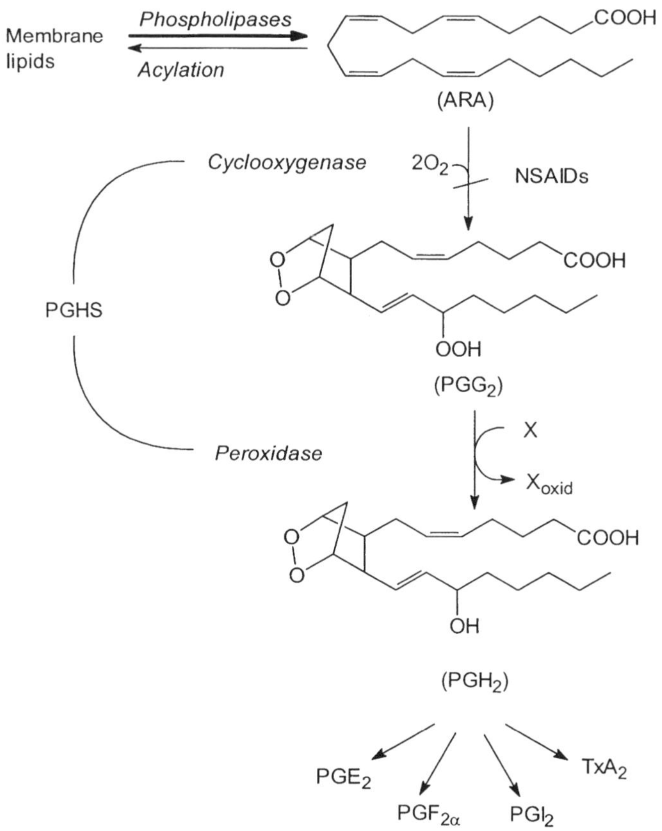

Figure 6.1 Arachidonic acid metabolism by PGHS. PGHS expresses two enzymic activities: its cyclooxygenase catalyses bisoxygenation of free fatty acid substrate, and converts arachidonate to PGG_2. The peroxidase reduces this intermediate to PGH_2, a reaction which requires the presence of reducing cosubstrates (X). Depending upon additional enzymes present in a given cell, the PGH_2 product is converted to prostaglandins, prostacyclins or thromboxanes, biologically active compounds collectively termed prostanoids (see text for more details).

has no TATA box (Kraemer *et al.* 1992), consistent with its constitutive expression as a housekeeping gene in most tissues and cells. Two Sp1 elements have been identified in the PGHS-1 promoter region responsible for the basal PGHS-1 gene transcription (Xu *et al.* 1997). PGHS-1 codes for a protein of approximately 70-72 kDa; the mature processed forms of various species all have 576 amino acids and highly conserved functional domains (Smith and DeWitt 1996). The active enzyme contains haem as a prosthetic group. It exists as a head-to-tail homodimer and is attached to lipid bilayers by means of a unique membrane binding domain which was first recognised by analysis of the X-ray crystal structure of ovine PGHS-1 (Picot *et al.* 1994).

The human PGHS-2 gene was mapped to chromosome 1 (Jones *et al.* 1993). PGHS-2 was first identified on the basis of its structural similarity to PGHS-1 (Kujubu *et al.* 1991; Xie *et al.* 1991). Molecular cloning revealed that the PGHS-2 gene contains ten exons and nine introns, and is considerably smaller (approximately 8 kb) than the PGHS-1 gene (Fletcher *et al.* 1992; Xie *et al.* 1991; Kosaka *et al.* 1994; Tazawa *et al.* 1994; Appleby *et al.* 1994). PGHS-2 transcribes an mRNA of about 4.4 kb in length, which is larger than the 2.8 kb transcript for PGHS-1. The PGHS-2 mRNA contains several RNA instability sequences (AUUUA), and is more rapidly degraded than PGHS-1 mRNA (Evett *et al.* 1993; Ristimaki *et al.* 1994). Cytoplasmic proteins of 90 to 35 kDa have been identified which interact with these instability sequences, and are important in the post-transcriptional control of PGHS-2 (Dixon *et al.* 2000). PGHS-2 obtained from chick, mouse, rat or human code for proteins with 587 amino acids in the mature forms. PGHS-2 often migrates as two bands on SDS-PAGE with apparent molecular masses of approximately 74 and 72 kDA, due to different degrees of glycosylation (Smith and DeWitt 1996). Immunoreactive bands of lower molecular weight (approximately 60 kD) have been observed and seem to correspond to degradation products (Kargman *et al.* 1996).

The subcellular localisation and compartmentation of the isozymes appear to be similar. Both PGHS-1 and PGHS-2 are predominantly associated with the endoplasmic reticulum; they are also found on the nuclear envelope (Spencer *et al.* 1998). Cross-species comparisons of PGHS isozymes show a greater than 85% amino acid identity for human, ovine and rodent homologues, while the PGHS-1 or PGHS-2 sequences from the same organism share about 60% identity (Smith and DeWitt 1996; Feng *et al.* 1993). Yet functional domains and key amino acid residues involved in catalysis are noticeably conserved among the isozymes, and the three-dimensional X-ray crystal structures of PGHS-1 and PGHS-2 are practically superimposable (Picot *et al.* 1994; Kurumbail *et al.* 1996; Luong *et al.* 1996). These and biochemical studies, e.g. on enzyme forms with specific mutations in key amino acid residues, have immensely contributed to our present understanding of the inner workings and catalytic functions of PGHS (reviewed by Marnett and Maddipatti 1991; Smith and DeWitt 1996; Garavito and DeWitt 1999).

Enzyme function in the metabolism and bioactivation of xenobiotics

Despite some differences in inhibitor selectivity and substrate preferences as well as pronounced differences in the expression pattern, PGHS-1 and PGHS-2 are essentially identical in structure and catalytic function. Unless otherwise specified, we refer to them as PGHS in this section, since both isozymes possess the same two catalytic activities. Upon cyclooxygenase catalysed bisoxygenation of arachidonic acid to PGG_2, the endoperoxide-hydroperoxide is reduced to the corresponding alcohol (PGH_2) by the *peroxidase* activity. As depicted in Figure 6.1, this occurs at the expense of an oxidisable cofactor/co-substrate (X). Since the first report on this reaction termed 'co-oxygenation' or 'co-oxidation' (Marnett *et al.* 1975), a large number and variety of chemicals were found to be oxidised by PGHS. In many cases this involves formation of reactive intermediates and/or products with mutagenic activity.

Prior to discussing examples of PGHS-mediated xenobiotic oxidations and bioactivation mechanisms the following points are of interest with respect to the relation of cyclooxygenase to peroxidase:

- Prostanoid biosynthesis is efficiently inhibited at the level of the cyclooxygenase by non-steroidal antiinflammatory drugs (NSAIDs). Well-known examples are aspirin, indomethacin, ibuprofen, diclofenac and many others. These NSAIDs inhibit the cyclooxygenase, but they do not affect the peroxidase activity. Aspirin is the only NSAID that covalently binds to PGHS; it exerts subtly different effects on the isozymes: Acetylation of PGHS-1 blocks the cyclooxygenase (not the peroxidase) activity whereas acetylation of PGHS-2 converts it to a form that still oxygenates arachidonic acid, but at C-15 instead of C-11 (Lecomte *et al.* 1994; O'Neill *et al.* 1994).
- The substrate fatty acid requirements of the cyclooxygenase are very specific, but the peroxidase activity can use a broad range of substrates other than PGG_2 (see below). Thus, PGHS can oxidise chemicals in the presence of organic hydroperoxides or H_2O_2 instead of arachidonate (ARA). Whereas cyclooxygenase inhibition prevents ARA-dependent co-oxidation of xenobiotics *in vitro*, NSAIDs will not necessarily block their PGHS-mediated metabolism in cells which generate lipid peroxides e.g. HPETES (monohydroperoxy polyunsaturated fatty acids) or hydrogen peroxide (Marnett and Maddipatti 1991; Degen 1993a).
- Hydroperoxides have an essential role to play in the oxygenation of the free fatty acid substrate, arachidonic acid, and are considered to be obligatory initiators of the cyclooxygenase reaction (Lu *et al.* 1999). Reducing co-substrates are important for enzyme activity, since they delay considerably the self-inactivation of PGHS. Thus, with purified PGHS, the number of turnovers can be as high as 1300 and as low as 10 depending on the concentration of reductants and other factors (Markey *et al.* 1987; Marshall *et al.* 1987).

MECHANISMS OF XENOBIOTIC OXIDATION BY PGHS

PGHS-dependent bioactivation of xenobiotics can occur by different mechanisms. The peroxidase can directly oxidise chemicals which act as reducing co-substrates for the enzyme. Another mechanism involves secondary oxidant species formed by peroxidase-generated cosubstrate radicals and molecular oxygen. Moreover, peroxyl radicals generated as intermediates during PG-biosynthesis can act as potent oxidant species. Thus, a number of diverse reactions are involved in the oxidation of

Table 6.1 Examples: prostaglandin H synthase and xenobiotic oxidations

Reaction type	Chemical	References
Dehydrogenation	Benzidine	Zenser et al. (1983a)
	Diethylstilboestrol	Degen et al. (1982)
	Phenidone	Marnett et al. (1982)
Demethylation	Aminopyrine	Lasker et al. (1981)
	N-Methylaniline	Sivarajah et al. (1982)
C-Hydroxylation	Phenylbutazone	Siedlik and Marnett (1984); Hughes et al. (1988)
	Oestradiol	Freyberger and Degen (1989a)
N-Oxidation	IQ*	Morrison et al. (1993); Wolz et al. (1995)
	N-Acetylbenzidine	Zenser et al. (1999)
Sulfoxidation	Sulindac sulphide	Egan et al. (1980)
Dioxygenation	Diphenylisobenzofuran	Marnett et al. (1979a)
Epoxidation	B[a]P-7,8-diol	Marnett et al. (1979b); Sivarajah et al. (1979)
	Aflatoxin B$_1$	Battista and Marnett (1985)

* 2-amino-3-methylimidazo[4,5-f]quinoline; benzo[a]pyrene-7,8-dihydrodiol
O- and S-Dealkylations *not* observed so far, neither dealkylation of nitrosamines.

xenobiotics by PGHS (see Table 6.1; and reviews by Eling *et al.* 1990; Marnett 1990; Smith *et al.* 1991; Zenser and Davis 1990).

PGHS are unique among other peroxidase species because they are able to biosynthesise the hydroperoxide substrate for the peroxidase activity. On the other hand, PGHS-peroxidase resembles a classical peroxidase, with its haem moiety being converted to a higher oxidation state analogous to compound I of other peroxidases (Marnett and Maddipati 1991). Subsequent reduction of enzyme to its resting state is coupled to oxidation of electron donors (reducing co-substrates) or direct oxygen

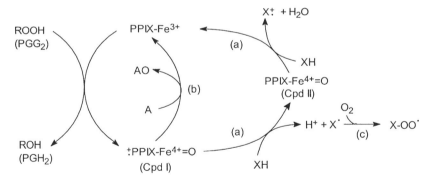

Figure 6.2 Peroxidase catalytic cycle and postulated mechanisms of xenobiotic co-oxidation. The two-electron reduction of hydroperoxide(PGG$_2$ or ROOH) by PGHS yields compound I. It must be reduced to regenerate the resting enzyme, a task accomplished by an as yet unknown endogenous reducing cofactor and/or by xenobiotics which undergo cooxidation. (see text for more details). The two major mechanism which govern cooxidation involve (a) single electron transfers to the oxidised PGHS intermediates, or (b) direct oxygen transfer from the oxygenated haem to a suitable acceptor compound. Certain radical intermediates may also bind oxygen (c) and thus give rise to secondary oxidant species. Moreover, peroxyl radicals are generated during PG-biosynthesis and act as oxidising agents (not depicted here).

transfer to a suitable acceptor compound (see Figure 6.2). Native PGHS with ferric protoporphyrin IX in the Fe(III) state (Fe^{3+} − PPIX) is oxidised to compound I upon reaction with hydroperoxide (PGG$_2$ or ROOH). Compound I contains a hypervalent iron-oxo form (Fe^{4+} = O and porphyrin oxidised to a cation radical), and is transformed by one-electron reduction to another intermediate, compound II (in which iron is Fe^{4+} and the porphyrin fully covalent). Compound II is reduced at the expense of another electron donor molecule (XH) to resting enzyme. For more details on spectral intermediates and the catalytic cycle see also recent papers (Koshkin and Dunford 1999; Wu *et al.* 1999).

Chemicals known to act as **reducing co-substrates** for PGHS-peroxidase include several naturally occurring compounds, but also various drugs and carcinogens, e.g. benzidine, 2-aminofluorene, some nitrofurans, paracetamol (acetaminophen), diethylstilboestrol, and the benzene metabolite hydroquinone. By serving as electron donors, compounds are oxidised to free radicals and/or other reactive intermediates (see Figures 6.3–6.8). Thus, direct oxidation of xenobiotics by PGHS-peroxidase represents a potentially important mechanisms of bioactivation.

The common trait with these chemicals is that during conversion of PGG$_2$ (or reduction of other hydroperoxides) to the corresponding alcohol, the peroxide oxygen is reduced to water rather than being incorporated into the xenobiotic compounds. The overall reaction can be summarised as follows:

$$ROOH + 2\ XH \xrightarrow{\text{Peroxidase}} ROH + 2X' + H_2O \tag{6.1}$$

With regard to the stoichiometry this means that for each molecule of hydroperoxide that is reduced (a two-electron process), two molecules of xenobiotic (X) will be oxidised (by a one-electron process) to electron-deficient derivatives. Yet, with compounds capable of donating another electron (e.g. adrenalin, diethylstilboestrol) the ratio for molecule co-substrate oxidised/hydroperoxide reduced is close to 1 in incubations with purified PGHS enzyme (Freyberger and Degen 1989b; Marnett and Eling 1983). In cells, however, where endogenous cosubstrates compete for cooxidation, the stoichiometry differs (Degen 1993b).

In some cases, PGHS-peroxidase can also catalyse peroxide reduction by direct transfer of the peroxide oxygen to an acceptor molecule (A). This is evident when ^{18}O-labelled peroxide is used in the reaction.

$$R^{18}O^{18}OH + A \xrightarrow{\text{Peroxidase}} RO^{18}H + AO^{18} \tag{6.2}$$

Such a peroxygenase reaction (depicted in Figure 6.2, reaction (b)) has been reported for sulindac and methyl phenyl sulphide which are oxidised to sulphoxides, with a stoichiometry of one molecule xenobiotic being oxidised per molecule of hydroperoxide reduced (Egan *et al.* 1980, 1981). Recently, a peroxygenase-type reaction was also reported by Zenser and collegues with *N*-acetylbenzidine (see below).

By looking in more detail into the PGHS-dependent oxidation of some chemicals, it becomes apparent that quite a range of metabolites and reactive intermediates is generated by various mechanisms. By donating one electron, *reducing cosubstrates* are initially oxidised to either oxygen-, nitrogen-, or carbon-centred radicals.

The first intermediates in the oxidation of **phenols** are phenoxyl radicals. Examples are diethylstilboestrol (DES) and paracetamol (acetaminophen). These cosubstrates can undergo another one-electron oxidation and thus form two-electron oxidation products, i.e. a quinone and quinone-imine, respectively (see Figures 6.3 and 6.4). Metabolism of DES by PGHS (Degen *et al.* 1982) resembles its oxidation by horse-radish peroxidase (HRP), but, this cannot be generalised to all substrates (Table 6.2). DES quinone tautomerises to the stable metabolite Z,Z-dienoestrol; but, reactive DES-intermediates (Figure 3) also bind covalently to proteins (Degen *et al.* 1986; Degen 1990). PGHS-peroxidation results in bioactivation of DES and of steroid oestrogens (Freyberger and Degen 1989a), and this is thought to play a role in their genotoxic effects (see below).

The one-electron oxidation product *N*-acetyl-*p*-benzo-semiquinone imine (Figure 6.4) may be further oxidised to *N*-acetyl-*p*-benzoquinone imine (NAPQI) or dimerise; it can also undergo disproportionation, forming paracetamol and NAPQI. An

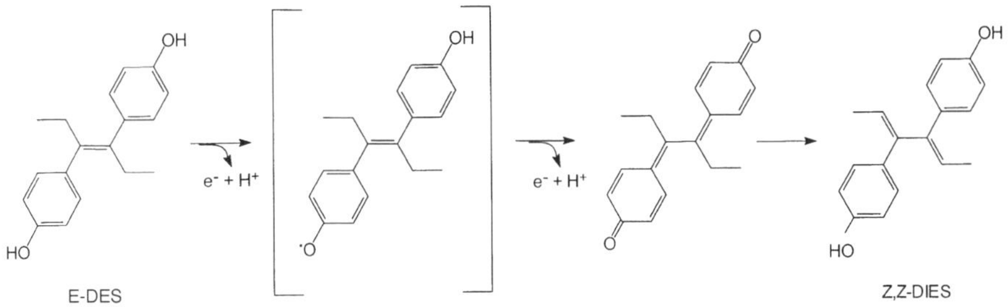

Figure 6.3 Peroxidative metabolism of diethylstilboestrol (E-DES) generates reactive intermediates and the stable product dienestrol (Z,Z-DIES).

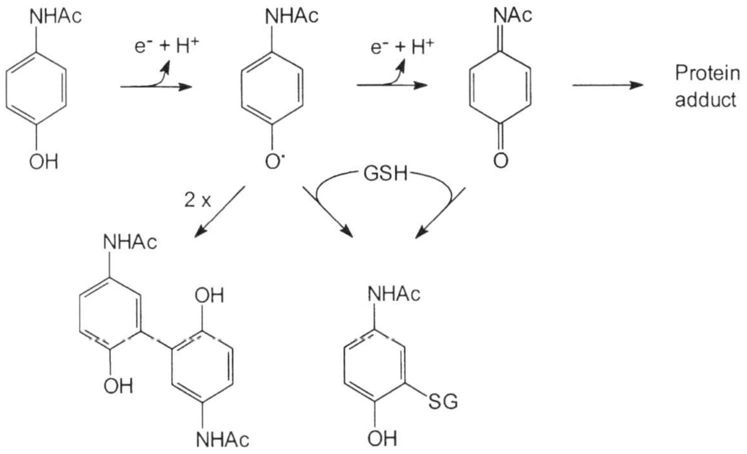

Figure 6.4 PGHS-catalysed oxidation of paracetamol.

interruption after the first step is evidenced by a formation of dimers and other oligomers (Moldeus *et al.* 1985). The latter was decreased by glutathione (GSH) with subsequent formation of the glutathione conjugate in incubations with PGHS or HRP. HRP was proposed to catalyse the one-electron oxidation of this substrate whereas PGHS can catalyse both one- and two-electron oxidations of paracetamol (Potter and Hinson 1987a,b). In contrast, cytochrome P450 metabolism occurs by two-electron oxidation; it did not produce oligomers but NAPQI, and in the presence of GSH resulted only in the formation of the conjugate, 3-(glutathione-S-yl)-paracetamol (Potter and Hinson 1987b). NAPQI is known to bind to proteins and held responsible for the observed toxicity of high doses of analgesics like paracetamol and phenacetin in liver and kidney.

Aromatic amines are another important class of compounds that are oxidised by PGHS. Reported products of benzidine (Figure 6.5) are the free radical cation, the two-electron oxidised benzidinediimine, a charge-transfer complex between imine and parent amine, a dimer (azo-bendizine) and oligomers (Zenser *et al.* 1983a). An *N*-hydroxy product was not found with benzidine, and not detected in similarly detailed studies of peroxidatic metabolism of 2-aminofluorene (Figure 6.5) (reviewed by Eling *et al.* 1990; Zenser and Davis 1990). On the other hand, formation of an *N*-

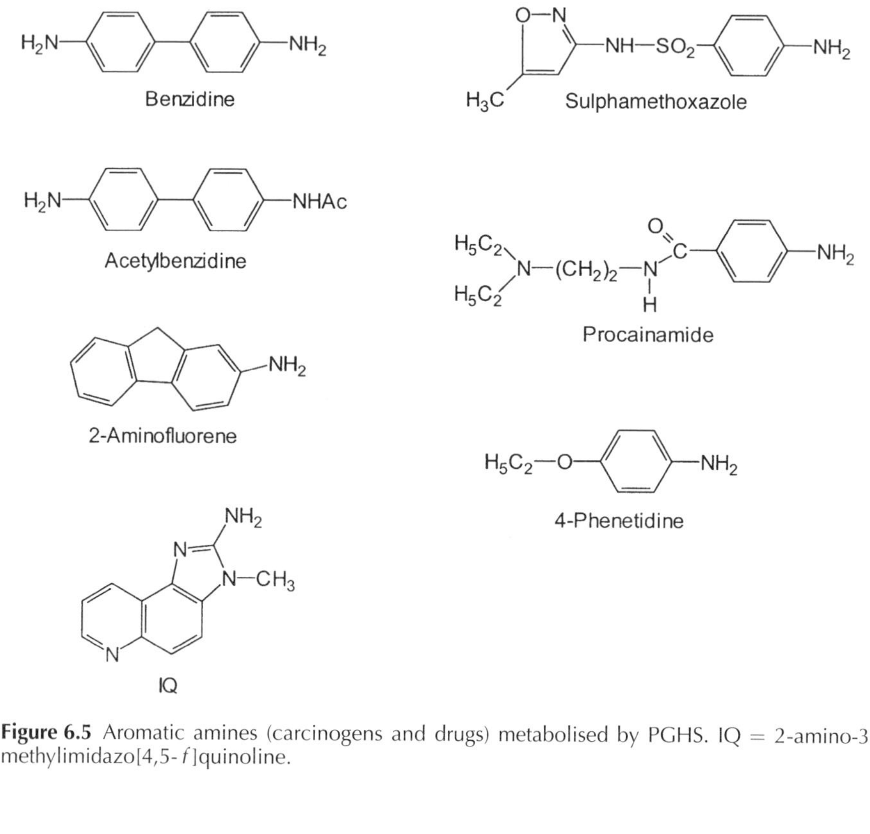

Figure 6.5 Aromatic amines (carcinogens and drugs) metabolised by PGHS. IQ = 2-amino-3-methylimidazo[4,5-*f*]quinoline.

hydroxylamine by PGHS was recently demonstrated with N-acetylbenzidine (ABZ; Zenser et al. 1999). The PGHS-metabolite N'-hydroxy-N-acetylbenzidine (N'-HA) is only observed under reducing conditions; it is converted to 4'-nitro-4-aminobiphenyl by PGHS. The results of mechanistic studies suggest a peroxygenase–type reaction rather than one-electron transfer reactions in the stepwise oxidation of ABZ (Figure 6.5) to N-oxidation products (Zenser et al. 1999). In support of this view, the PGHS metabolite N'-HA was not detected in incubations with HRP (Lakshmi et al. 2000). Oxidation of ABZ by PGHS generates DNA adducts (Lakshmi et al. 1998). The enzyme is thought to catalyse the bioactivation of this benzidine metabolite in the carcinogen's target tissue, the bladder epithelium which contains rather high PGHS-activity (see below).

The carcinogenic arylamine 2-aminofluorene (Figure 6.5) is metabolised by PGHS probably via free radicals to adduct-forming intermediates, some polymeric species, and to azo-bis-fluorene (Boyd and Eling 1984; Krauss and Eling 1985). Formation of nitrofluorene is indicative of its N-oxygenation. In addition, PGHS activates 2-AF and other arylamines to mutagens (for a review see Eling et al. 1990). Another interesting example is the food-borne heterocyclic amine 2-amino-3-methyl-imidazo[4,5-f]quino-line (IQ; Figure 6.5). Metabolism of IQ by PGHS yields azo-IQ and nitro-IQ as stable metabolites (Wolz et al. 1995). PGHS-catalysed oxidation of IQ and related com-pounds results in their bioactivation to mutagenic species (Wild and Degen 1987; Petry et al. 1989). PGHS-mediated metabolism of IQ also yields DNA-adducts (see below).

There are also various arylamines among drugs. Examples of compounds shown to be N-oxygenated by PGHS on their primary amine function are sulphamethoxazole and procainamide (Figure 6.5). The metabolites formed are the hydroxylamine and nitro-derivatives. Since PGHS-mediated formation of drug-modified proteins can oc-cur in phagocytic and antigen-presenting cells, this pathway is considered to be of toxicological importance for adverse reactions to these xenobiotics (Cribb et al, 1990; Goebel et al. 1999).

Many studies provide evidence for a one-electron oxidation of aromatic amines by PGHS peroxidase to N-centred radicals. The fate of such radicals can differ. For instance, PGHS-dependent oxidation of p-phenetidine (Figure 6.5) resulted in the formation of several intensely coloured products due to radical coupling (Moldeus et al. 1985). In incubations with radiolabelled compound, also binding to protein and DNA, was observed which may be due to radicals and/or quinoid metabolites formed upon dimerisation of the intermediates.

The tertiary amine aminopyrine (AP) is comparable in its basicity to aromatic amines. AP is oxidised by PGHS to a rather stable radical cation detectable by ESR and UV/VIS spectroscopy (Eling et al. 1985; Lasker et al. 1981). The radical cation can disproportionate to an iminium cation and its parent compound AP (Figure 6.6). The iminium cation (a two-electron oxidation product) is hydrolysed to the demethy-lated amine and formaldehyde. Also a number of secondary and tertiary N-methylated aromatic amines are N-demethylated by PGHS (Sivarajah et al. 1982). The proposed mechanism of N-demethylation is in accord with the inability of PGHS to catalyse O- and S-dealkylations, reactions which occur by $C(\alpha)$-hydroxylation rather than hetero-atom oxidation.

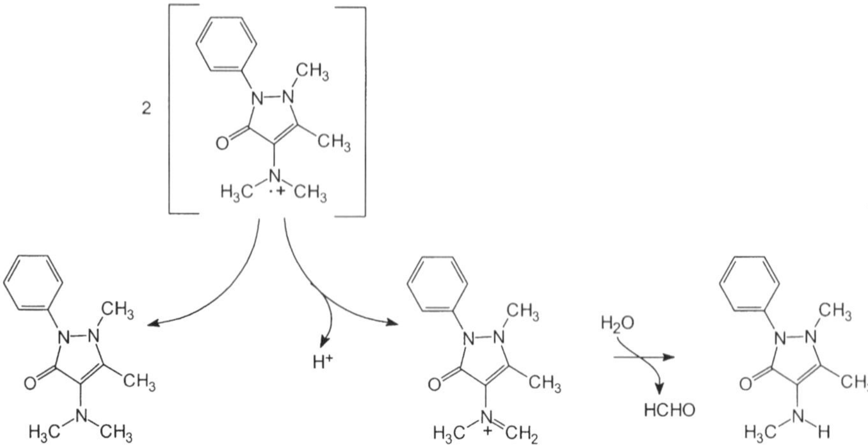

Figure 6.6 PGHS-catalysed oxidation of aminopyrine.

The next examples illustrate reactions of **carbon oxidation** by PGHS. In the case of phenylbutazone (Figure 6.7) and 13-*cis*-retinoic acid, one-electron oxidation by PGHS-peroxidase initially generates the C-centred radicals (Hughes *et al.* 1988; Samokyszyn and Marnett 1987). One molecule of oxygen is added thereby producing peroxyl radicals that undergo secondary reactions. Peroxyl radicals are potent oxidants capable of hydrogen abstraction or epoxidation. Accordingly, 4-hydroperoxy- and 4-hydroxyphenylbutazone as well as 4-hydroxyretinoic acid were isolated as metabolites (Hughes *et al.* 1988; Samokyszyn and Marnett 1987). Moreover, the retinoic acid peroxyl radical adds to the 5,6-double bond of another retinoic acid molecule to form the 5,6-epoxy product (Marnett 1990). Phenylbutazone and retinoic acid exemplify a mechanism of PGHS-mediated drug metabolism depicted in Figure 6.2 as reaction (c). It is based on the PGHS-peroxidase catalysed one-electron oxidation of certain chemicals to C-centred radicals that couple with oxygen to form peroxyl radicals. These secondary oxidants can bioactivate other compounds, e.g. benzo[a]pyrene-7,8-diol (Reed *et al.* 1984). This offers also an explanation for oxidation reactions usually not observed with other peroxidases and a more efficient oxidation of chemicals by PGHS than expected for a stoichiometry of 1:1 or 1:2 for hydroperoxide reduction to xenobiotic oxidation.

Another, although similar, mechanism is probably involved in the PGHS-mediated

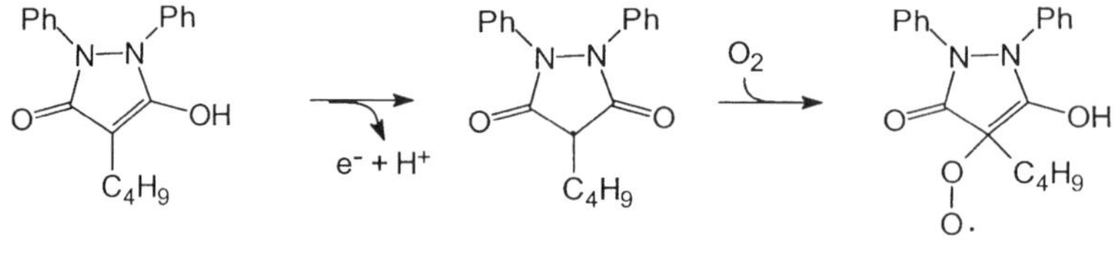

Figure 6.7 PGHS-catalysed oxidation of phenylbutazone.

aryl hydroxylation found with steroid oestrogens (Degen *et al.* 1987). One-electron oxidation of oestradiol (Figure 6.8) produces a phenoxyl radical; its resonance form, a C-centred radical, can trap oxygen thereby producing a peroxyl radical which abstracts hydrogen to form an *ortho*-hydroperoxide. The hydroperoxide is reduced (non-enzymically or by PGHS) to the corresponding alcohol which tautomerises to the catechol product, a process driven by re-establishment of aromaticity. Catechol oestrogens are also cooxidised by PGHS-peroxidase to reactive intermediates, i.e. semiquinone and *ortho*-quinone metabolites that bind to proteins (Freyberger and Degen 1989a).

In essence, chemicals that are reducing co-substrates of PGHS-peroxidase are oxidised by two major mechanisms which involve either single electron transfers to the oxidised enzyme intermediate or direct oxygen transfer from the oxygenated haem to a suitable acceptor compound (Figure 6.2, reactions (a) and (b)). Yet oxygen incorporated into some xenobiotics is not peroxide-derived, but, apparently comes from molecular oxygen (Figure 6.2, reaction (c)). Stepwise one-electron oxidation of chemicals and the formation of xenobiotic peroxyl radicals are illustrated above.

Additional mechanisms exist, since certain compounds are oxidised which are **no** or very poor reducing co-substrates of PGHS-peroxidase. One mechanism of PGHS-dependent bioactivation is indirectly linked to the cyclooxygenase activity of the

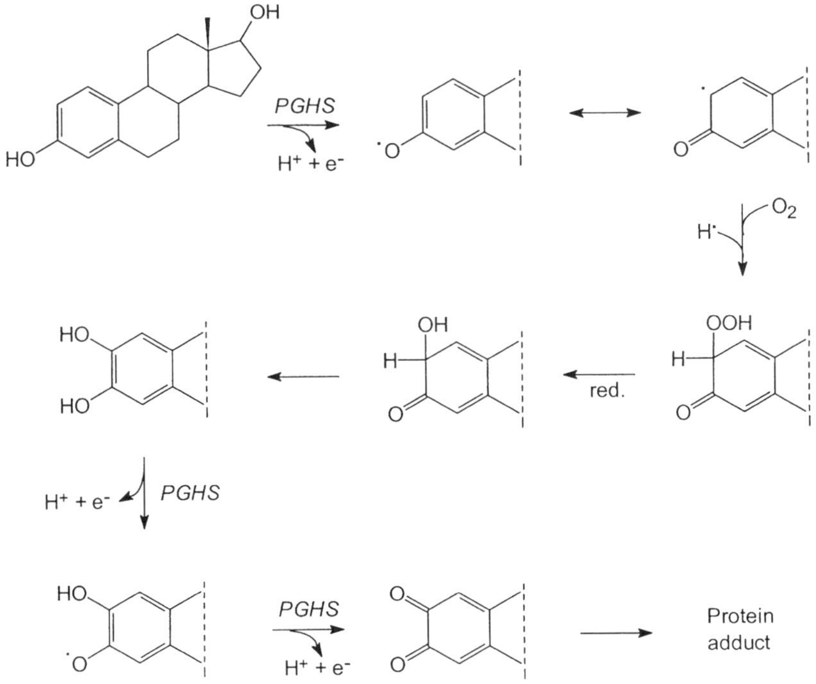

Figure 6.8 PGHS-catalysed oxidation of oestradiol and catecholoestrogens.

enzyme. During conversion of arachidonate to PGG_2, fatty acid peroxyl radicals are formed as intermediates. Similar lipid peroxyl radicals are formed during metabolism of unsaturated fatty acids by lipoxygenases. Peroxyl radicals are potent oxidising agents and rather selective in their reactions (Marnett 1990). They are capable of bioactivating some procarcinogens or promutagens by epoxidation (references in Table 6.1). For example, the carcinogenic mycotoxin aflatoxin B_1, the polycyclic hydrocarbon benzo[a]pyrene (BP), and its metabolite BP-7,8-dihydrodiol (BP-7, 8-diol) are epoxidised by PGHS; but, the stoichiometry is only 0.1:1 for xenobiotic oxidation to hydroperoxide (PGG_2) reduction (Marnett and Eling 1983). Although these reactions are rather sensitive to antioxidant inhibition, peroxyl-radical-dependent epoxidation of BP-7,8-diol has been demonstrated in mouse skin. This was detected by means of a stereochemical approach that can distinguish between peroxyl-radical-dependent and monooxygenase-mediated epoxidation of the (+)-enantiomer of BP-7,8-diol. The findings suggested that peroxyl radicals are important contributors to BP-7,8-diol oxidation in the skin of normal mice, whereas cytochrome P-450 has primacy in β-naphthoflavone-pretreated animals (Marnett 1990; Pruess-Schwartz et al. 1989). These and other studies (see below) indicate that PGHS can serve as a complementary/ alternative metabolic activation enzyme to the cytochrome P-450 enzymes. Although it is unlikely that PGHS plays a major role in systemic drug metabolism, it is thought to be an important determinant of chemical toxicity in extrahepatic tissues (Eling et al. 1990).

As described here and in previous reviews on this topic (Eling et al. 1990; Eling and Curtis 1992; Marnett 1990; Smith et al. 1991; Zenser and Davis 1990) in many cases oxidation of xenobiotics by PGHS results in the formation of reactive intermediates and/or generates products with mutagenic activity. Ram seminal vesicles microsomes (RSVM) lack monooxygenase activity, but contain the highest PGHS activity of all tissues, and are a good source of enzyme for experimental studies. For example, in bacterial mutagenicity assays, arachidonate supplemented RSVM can serve as bioactivating system instead of NADPH-fortified liver microsomes (or 'S-9 mix'). RSVM thus have a function similar to liver preparations used to study cytochrome P450-dependent drug oxidations.

Whereas ram seminal vesicles express only PGHS-1, many tissues contain both PGHS-1 and PGHS-2. Moreover, PGHS-2 expression is inducible by several endogenous factors and certain xenobiotics. The role of PGHS isozymes with regard to xenobiotic oxidation is the same. This is exemplified by studies with purified PGHS-2 and transiently PGHS-2 cDNA transfected COS cells; bioactivation of 2-aminofluorene was demonstrated by formation of DNA adducts and its inhibition by aspirin and indomethacin (Liu et al. 1995) in accord with previous data for 2-AF oxidation by PGHS-1 (RSVM; see above). In studies with the heterocyclic amine IQ (Figure 6.5) and lysates from cells expressing PGHS-1 or PGHS-2 both catalysed IQ-DNA adduct formation (Liu and Levy 1998). As seen with CYP1A2-dependent oxidation, PGHS-isozyme-dependent IQ adduct formation was enhanced by N-acetyltransferases. This observation points towards a basic similarity in IQ-activation by the two oxidising enzymes. Similarities seen in studies on the mutagenic activation of IQ by either PGHS or cytochrome P-450 support this view (Wild and Degen 1987), and apparently the same IQ-DNA adducts are formed (Wolz et al. 2000).

XENOBIOTIC OXIDATION BY PGHS: BIOLOGICAL IMPLICATIONS

PGHS-dependent oxidation of xenobiotics involves multiple mechanisms and can generate a range of reactive intermediates (free radical species, peroxyl radicals, quinones, quinoneimines, epoxides and others). Unless they are detoxified, such metabolites bind to macromolecular nucleophiles and damage critical cellular components (DNA, certain proteins). Thus, and because of its tissue distribution (see below), PGHS is considered as bioactivating enzyme for xenobiotics, in particular in tissues with low cytochrome P-450 activity. The following studies are indicative of such a role for PGHS in various tissue-specific pathologies of certain carcinogens and drugs.

Zenser and colleagues demonstrated PGHS-mediated bioactivation of acetylbenzidine (AZB) and formation of a DNA adduct; N'-(3'-monophospho'-deoxyguanosin-8-yl)-N-acetylbenzidine (Lakshmi et al. 1998). It is thought to play an important role in the initiation of benzidine-induced bladder cancer in humans, since this adduct has been detected in exfoliated bladder cells of benzidine-exposed workers who excrete AZB in their urine (Rothman et al. 1996). It was also detected by postlabelling analysis in N-acetylating and PGHS-competent porcine urothelial cell cultures upon exposure to benzidine or ABZ (Degen et al. 1998b). This observation supports a role for PGHS in the local bioactivation of this carcinogen.

IQ is a multisite carcinogen in animals and along with related food mutagens considered as risk factor for colon cancer in humans. Bioactivation of IQ by PGHS generates products mutagenic in acetyltransferase-proficient S. typhimurium strains, and both IQ and nitro-IQ are genotoxic in ovine seminal vesicle (OSV) cells which express PGHS and N-acetyltransferase (NAT), but lack cytochrome P-450 activity (Degen et al. 1998a). The IQ-derived DNA-adducts in bacterial and mammalian cells were indistinguishable from those observed with P450/NAT-activation (Wolz et al. 2000). Therefore, it is difficult to assess the contribution of PGHS-activation in vivo, e.g. in target sites for IQ carcinogenicity such as colon. For another carcinogenic amine, 2-aminofluorene (2-AF), peroxidase-catalysed formation of two extra adducts was described which differed from N-(deoxyguanosin-8-yl)-2-AF formed by the reaction of N-hydroxy-2-AF with DNA (Krauss and Eling 1985). These uncharacterised adducts were detected in the kidney and bladder epithelia of dogs fed radiolabelled 2-AF (Krauss et al. 1989); their biological importance for amine-induced canine bladder cancer is unknown.

Not only DNA adduct formation but also covalent binding to certain proteins may be a critical reaction for toxicity of xenobiotics. This appears to be the case for paracetamol and related analgesics (see above). Moreover, peroxidase activation of DES yields reactive intermediates which bind preferentially to tubulin and other spindle proteins (Degen 1990), a reaction which could explain its genotoxic (aneuploidogenic) effects. Induction of micronuclei by DES in PGHS-competent cells was inhibited by indomethacin along with a decrease in DES oxidation (Foth et al. 1992). Since PGHS is also present in target cells for neoplastic transformation by synthetic and by steroid oestrogens, cooxidation seems to play a role in their adverse effects (Degen et al. 1983; Schnitzler et al. 1994).

PGHS is also implicated in the toxicitys of sulphonamides and procainamide (see above). PGHS-mediated bioactivation can occur in phagocytic and antigen-presenting cells, and yields drug-modified proteins that are involved in idiosyncratic adverse reactions to these xenobiotics (Cribb *et al.* 1990; Goebel *et al.* 1999). PGHS is further implicated in the teratogenicity of hydantoin and oxazolidenedione anticonvulsants, i.e. phenytoin and dimethadione (Kubow 1992; Wells *et al.* 1989a,b), and teratogenic effects of such structurally diverse chemicals as benzo[a]pyrene (Winn and Wells 1997), isotretinoin (Kubow 1992) and thalidomide (Arlen and Wells 1996). These and additional studies with these compounds (Figure 6.9) revealed PGHS-catalysed formation of reactive intermediates and free radical-mediated oxidative damage to cellular macromolecules as well as evidence for protection against the chemical-induced teratogenicity by aspirin and other PGHS-inhibitors (Kubow and Wells 1989; Parman *et al.* 1998, 1999; Winn and Wells 1997).

As pointed out already, PGHS is primarily thought to catalyse xenobiotic metabolism in extrahepatic tissues and cells with low cytochrome P450 activity. The *in vivo* contribution of PGHS is not readily apparent when peroxidase and peroxyl radical-mediated bioactivation of chemicals finally result in the same products as those generated by cytochrome P450-dependent oxidation. This similarity is seen for DNA adducts of IQ and of acetylbenzidine formed by either activating system (see above). Thus, studies often rely on specific enzyme inhibitors; an example is 3-methylindole (3-MI), a fermentation product of tryptophan, known to cause lung toxicity in several species. 3-MI is oxidised by both cytochrome P450 enzymes or PGHS to reactive intermediates that yield glutathione conjugates and covalent protein binding; the modulation of these was inhibited in Clara cells and alveolar macrophages by the cytochrome P450 inhibitor 1-aminotriazole by 94% and by 24% respectively (Thornton-Manning *et al.* 1993). Bray and colleagues used aspirin and indomethacin to investigate the role of PGHS in the 3-MI-induced pneumotoxicity in goats, and observed a clear protective effect when these compounds were administered prior to 3-MI (Acton *et al.* 1992). The above examples indicate that PGHS participates in the

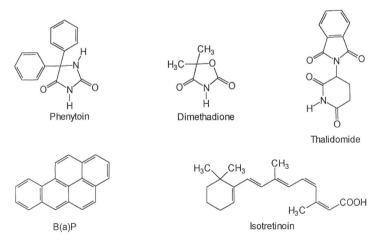

Figure 6.9 Some teratogenic compounds which are oxidised by PGHS. B(a)P, Benzo[a]pyrene.

bioactivation of xenobiotics. PGHS has been also suggested to mediate detoxification reactions, e.g. of styrene (Eling *et al.* 1990; Eling and Curtis 1992).

In summary, PGHS catalyses the bioactivation of a large and diverse number of chemicals, and there is now ample evidence that such reactions contribute to the toxicity of xenobiotics. Another aspect is that of xenobiotics affecting PGHS-activity. Drugs which inhibit cyclooxygenase and agents which induce PGHS, predominantly PGHS-2, by modulating the synthesis of biologically active prostanoids, may contribute to tissue-specific pathologies (see below).

Substrate specificity

PGHS is bifunctional, containing both cyclooxygenase and peroxidase activities. In a strict biochemical sense the term *substrate* refers to fatty acids and hydroperoxides, whereas xenobiotics metabolised by the enzyme are considered as *co-substrates* (or cofactors). Thus, with respect to drug metabolism, the terminology for PGHS-isozymes and, for example, cytochrome P450 enzymes differs.

Whereas PGHS-peroxidases can use various peroxides and a broad range of xenobiotics as co-substrates, the substrate requirements of the cyclooxygenases are very specific. Best substrates for the cyclooxygenase activities of PGHS-1 and PGHS-2 are arachidonate and dihomo-γ-linolenate, i.e. 20-carbon ω-6 polyunsaturated fatty acids. The apparent K_m values with arachidonic acid are about $\sim$5 µM for both isozymes (Meade *et al.* 1993a; Barnett *et al.* 1994; Laneuville *et al.* 1994, 1995). Kinetic studies with purified PGHS-1 show a sigmoidal dependence on arachidonate concentration, whereas PGHS-2 displays simple saturable behaviour (Swinney *et al.* 1997). In addition, PGHS/COX-2 seems to accept a wider range of fatty acids than PGHS/COX-1; in general, 18-carbon polyunsaturated fatty acids are more efficiently oxygenated by PGHS-2 than by PGHS-1. Other ω-3 polyunsaturated fatty acids from fish oil, e.g. docosahexaenoic acid (22:6, w-3) and eicosapentanoic acid (20:5, w-3), are competitive inhibitors of both isozymes without being a substrate (for a review see Smith and DeWitt 1996). Moreover, there are differences in peroxide requirements, with hydroperoxide concentrations needed for activating the cyclooxygenase of PGHS-2 being lower than those needed to activate PGHS-1, which may play a role in their differential cellular control (Capdevila *et al.* 1995; Kulmacz and Wang 1995; Kulmacz 1998). Despite biochemical differences in peroxide requirements and fatty acid substrate specificities, PGHS-isoenzymes are virtually the same in the context of catalytic mechanisms.

Studies with ovine PGHS-1 showed that its peroxidase preferentially reduces fatty acid hydroperoxides, including those generated by lipoxygenases (e.g. 15-HPETE, 12-HPETE). The enzyme is less active with hydrogen peroxide, and almost inactive with tertiary hydroperoxides (Marnett and Maddipatti 1991). The substrate specificity of PGHS-2 peroxidase has not been examined in great detail, but, in various assays, it exhibits an activity that is quantitatively similar to that of PGHS-1 peroxidase (Fletcher et al 1992; Meade et al 1993b; Barnett et al 1994; Laneuville et al 1994).

Since the peroxidase activity of PGHS can function independently of the cyclooxygenase activity, there are convenient assays for measuring the enzyme activity. A typical spectral peroxidase assay employs H_2O_2 and the phenolic compound guaiacol as oxidising and reducing co-substrates, respectively. Marnett's laboratory has evaluated a large number of reducing co-substrates by quantitating their ability to support the reduction of the stable primary hydroperoxide PPHP (5-phenyl-4-pentenyl hydroperoxide) by catalytic amounts of the enzyme (Markey *et al.* 1987). In this HPLC-based assay, the yield of 5-phenyl-4-pentenyl-alcohol indicates PPHP-reduction and reflects the efficiency of a given compound to act as reducing co-substrate for PGHS-peroxidase. The identity of the endogenous reducing co-substrate is not known, and it may vary between cells; candidates for this role are adrenaline, lipoic acid, ascorbate and uric acid (Markey *et al.* 1987). Among the best reductants for PGHS-peroxidase are aromatic diamines (benzidine, tetramethylphenylenediamine) and polyhydroxylated chemicals (diethylstilboestrol, hydroquinone). These compounds are also known to be readily oxidised by other mammalian peroxidases and by horseradish peroxidase (HRP). Although PGHS catalyses a number of oxidations similar to those of HRP, there are also differences between the two peroxidases (Marnett and Maddipati 1991). Examples of chemicals that are good substrates for PGHS but not for HRP and vice versa are shown in Table 6.2.

Although metabolism of DES and some structural analogues by PGHS shows many similarities to HRP-mediated oxidation, there are notable differences in the extent to which some compounds are oxidised, for instance hexestrol (Degen and McLachlan 1985; Degen *et al.* 1986). Moreover, there are differences in the metabolites produced by either HRP or PGHS. The carcinogen 2-naphthylamine (2-NA) is co-oxidised by PGHS to ring-oxygenated products (metabolites atypical for a peroxidase-type reaction) that are clearly different from polymeric non-oxygenated metabolites of 2-NA that are generated with HRP. The formation of the ring-oxygenated metabolites probably occurred by peroxyl radical-mediated attack or direct transfer of peroxide oxygen from PGHS to 2-NA (Curtis *et al.* 1995). The heterocyclic arylamine IQ

Table 6.2 Substrate specificity of peroxidases

Compound	PGHS	HRP	References
Methylphenylsulphide	+	−	Egan *et al.* (1981)
Sulindac sulphide	+	−	Egan *et al.* (1981)
ANFT[a]	+	(−/+)[d]	Lakshmi *et al.* (1992)
HMN[b]	+	−	Zenser *et al.* (1983b)
Lipoic acid	+	−	Markey *et al.* (1987)
Indole acetic acid	+	−	Markey *et al.* (1987)
Amitrol	+	−	Krauss and Eling (1987)
Paracetamol	+	+	Potter and Hinson (1987a,b)
Diethylstilboestrol	+	+	Degen *et al.* (1982)
2-Aminofluorene	+	+	Boyd and Eling (1984)
2-Naphthylamine	+	(+)[d]	Curtis *et al.* (1995)
Hexestrol	+	(−/+)[d]	Degen *et al.* (1986)
IQ[c]	+	(−/+)[d]	Wild and Degen (1987)
Aniline	−	+	Markey *et al.* (1987)
Iodide	−	+	Markey *et al.* (1987)

[a] ANFT: 2-amino-4-(5-nitro-2-furyl)thiazole.
[b] HMN: 3-hydroxymethyl-1-{[3-(5-nitro-2-furyl)allylidene]amino}hydantoin.
[c] IQ: 2-amino-3-methylimidazo[4,5-f]quinoline.
[d] Difference in the type or extent (-/+) of metabolism (see text).

(2-amino-3-methylimidazo[4,5-f]quinoline) is another interesting example: PGHS-mediated oxidation resulted in the formation of mutagenic metabolites whereas such activation was not seen with HRP (Wild and Degen 1987). In conclusion, HRP is not always a suitable model enzyme for PGHS-mediated oxidations of xenobiotics.

Tissue distribution

PGHS is widely distributed in animal and human tissues, but is not present in all cells of a given tissue (Smith 1985). Enzyme activity is particularly abundant in the accessory sex glands, the kidney, the lung, gastrointestinal and urogenitalial tract, and in the vasculature. Early studies on prostanoid biosynthesis in animal tissues (Christ and VanDorp 1972) showed the wide distribution of PGHS in many tissues of various species, but could not discriminate between the expression of isozymes. This is now achievable with suitable techniques (Northern Blotting, RT-PCR (Reverse Transcription - Polymerase Chain Reaction), Western-Blotting, Immunohistochemistry) for PGHS-1 and PGHS-2 mRNAs or proteins.

PGHS-1 is constitutively expressed in almost all tissues. Consistent with its role for the maintenance of physiological functions of the stomach, mediating normal platelet function, and regulating blood flow, high quantities of PGHS-1 are present in epithelial cells of the crypts (Cohn *et al.* 1997), platelets and endothelial cells (Habib *et al.* 1993). High levels of PGHS-1 were found in the gastrointestinal tract (Kargman *et al.* 1996), the kidney medulla (Harris *et al.* 1994) and the interstitium (Smith and DeWitt 1995; Seibert *et al.* 1994). A high constitutive expression of PGHS-1 was found in foetal organs (heart, kidney, lung, brain), and also in the decidual lining of the uterus (Gibb and Sun 1996; Sawdy *et al.* 1997).

PGHS-2 is normally not detectable by immunohistochemistry in most tissues. A basal expression has been observed in the hippocampus, the cingulate cortex of the brain, in trachea and renal cortex (Yamagata *et al.* 1993; Walenga *et al.* 1996; Harris *et al.* 1994). Sheep placenta is a rich source of PGHS-2 (Johnson *et al.* 1995), and PGHS-2 expression has been reported in human amnion epithelial cells and chorion laeve trophoblasts (Gibb and Sun 1996). The expression of PGHS isoenzymes was also examined in maternal uterine tissues of pregnant sheep (Gibb *et al.* 2000). PGHS-1 or PGHS-2 mRNA were not detectable in the myometrium; in endometrial epithelium only PGHS-2 mRNA was expressed, consistent with PGHS-2 protein localisation. The increased expression of PGHS-2 rather than PGHS-1 is thought to be important for the onset of parturition.

The mRNA expression pattern of PGHS-1 and PGHS-2 in human tissue samples was investigated using the sensitive RT-PCR technique. High mRNA levels of PGHS-1 and PGHS-2 were detected in the human prostate and in human lung. The mRNA levels of both isoforms were low in testis, pancreas, kidney, thymus and liver (O'Neill and Ford-Hutchinson 1993). A similar mRNA expression pattern was found in animals; the highest PGHS-1 and particularly PGHS-2 expression was observed in lung, low PGHS-1 and PGHS-2 mRNA levels in thymus, and no PGHS-2 mRNA expression was detected in the liver of rats (Vogel C, Boerboom AM, Degen GH and Abel J, unpublished observations) and mice (Vogel *et al.* 1997). In human liver parenchymal cells, PGHS-2 was detected at higher levels than PGHS-1 (Koga *et al.* 1999). However, since the tissue samples were from patients with hepatocellular carcinoma, this finding is not necessarily indicative of a species difference.

Liver may serve as an example of organ heterogeneity. Hepatocytes contain less than a third of liver cyclooxygenase activity (Mion *et al.* 1995). The majority of hepatic PGHS-1 is found in non-parenchymal cells, i.e. sinusoidal endothelial cells, Kupffer cells, and Ito cells; PGHS-2 is inducible by various stimuli (Gallois et al 1998; Suzuki-Yamamoto *et al.* 1999; Zhang *et al.* 1995).

Species distribution

Numerous reports indicate the ability of many mammalian and non-mammalian species to synthesise prostaglandins in various tissues. Surprisingly, there are few studies on comparative aspects of PG-biosynthesis in animal and human tissues. By far the highest PGHS activity of all species is found in the seminal vesicles of rams, with a ten- to twenty-fold lower activity in those of bulls (Christ and Van Dorp 1972). Much lower prostaglandin secretion by vesicular glands of other species was reported (Silvia et al 1994). This is reflected at the protein level, with up to 15% of the total microsomal protein being PGHS-1 in ovine seminal vesicles, but only 1.5% in the bovine tissue, and much less PGHS (< 0.05% of microsomal protein) in other species including humans (Fischer 1989). With regard to the kidney (where PGHS is highest in inner medulla and papilla) the rabbit tissue showed higher PG-synthesis activity than those of hamster, cow, sheep or pig (Christ and Van Dorp 1972).

Despite these quantitative differences in PGHS-activity between species, on the whole the pattern of PGHS isozyme mRNA expression in human tissues (O'Neill and Ford-Hutchinson 1993) appears to be similar to that reported in several animal species, at least when tissue homogenates are analysed. PGHS-isozyme expression was studied in gastrointestinal (GI-) tract tissues of rat, dog, squirrel monkey, rhesus monkey and human by Western Blotting and immunhistochemical analysis (Kargman *et al.* 1996). It confirmed the presence of PGHS-1 in all GI-tract regions (stomach to caecum) of these species, with higher PGHS-1 protein levels in humans and monkeys than in rats and dogs. In contrast, expression of PGHS-2 protein was absent in most GI-tissues examined; only small amounts were detected, e.g. in rat caecum (Kargman *et al.* 1996).

In studies on the renal localisation of PGHS-isozymes with rats, dogs, monkeys and human kidney specimen Khan *et al.* (1998b) observed a similar regional distribution of PGHS-1 across the species investigated, but, significant interspecies differences in PGHS-2 localisation were noted; a similar distribution between dogs and rats, and between monkeys and humans, respectively, but differences between primate kidneys in comparison to those of dogs and rats. Moreover, PGHS-2 expression was markedly increased in kidneys of volume-depleted rats and dogs, but not in monkeys (Khan *et al.* 1998). In the authors' opinion, interspecies differences in the presence and distribution of PGHS-isozymes may further help to explain the difference in species susceptibility to NSAID-related renal toxicity.

Epidermal expression of PHGS-isozymes was evaluated by immunohistochemistry performed on human and mouse skin biopsy sections, and by Western Blotting of protein extracts from cultures of human keratinocytes (Leong *et al.* 1996). These data and those of others suggest that in human epidermis and in keratinocyte cultures, expression of PGHS-2 occurs as part of the normal keratinocyte differentiation whereas in murine skin, PGHS-2 is not expressed, but inducible (Scholz *et al.* 1995; Müller-Decker *et al.* 1998a,b, 1999). In light of the above observations, further studies on isoenzyme expression and regulation in different species are indicated.

Influence of age and sex

AGE-DEPENDENT EXPRESSION

PGHS-1 expression and activity are considerably higher in large seminal vesicles (>8 g wet weight) of mature uncastrated sheep than in small organs (<3 g wet weight) from young animals. In addition to PGHS-1 levels being related to overall size and functional status of the gland, seasonal fluctuations in enzyme activity were observed (Koburg 1991). Thus, it was suggested that an increase in PGHS-1 tissue levels is due to elevated production of androgens during puberty. The notion of a developmental regulation of PGHS-1 is further supported by studies demonstrating the stimulating effect of the sex steroid testosterone on PGHS-1 expression and the development of seminal vesicles in young rams (Silvia *et al.* 1994).

PGHS-1 expression was also found to increase during the first four weeks after birth in the developing ovine vasculature (Brannon *et al.* 1994). In lungs from foetal lambs to newborn and four-month-old lambs the PGHS-1 and PGHS-2 protein levels increased four- to fivefold (Brannon *et al.* 1998). An intense immunostaining of PGHS-1 was found in endothelial and airway epithelial cells whereas PGHS-2 was not detected. PGHS-2 immunohistochemistry has been performed on human lung tissues from autopsies of foetuses, preterm infants, and term infants; PGHS-2 staining occurred exclusively in the type II pneumocytes, and in ciliated cells (Lassus *et al.* 2000). Consistent with animal studies, the most intense bronchial staining of PGHS-2 was found in foetuses and the least intense in term infants. These findings illustrate the cell-specific expression of PGHS-1 and PGHS-2 in the lung and are indicative of the developmental role of both isoforms in the perinatal lung. A developmental regulation of PGHS-2 has been reported in rat brain (Yamagata *et al.* 1993). Using *in situ* hybridization, PGHS-2 mRNA was detected in regions of the hippocampus on day 1 (postnatal), with increasing levels on day 5, and reaching adult levels on day 16 after birth. Postnatal development of PGHS-2 has been studied also in rat kidney. The isozyme was found in a subset of thick ascending limb of Henle (TAL) cells; positive immunostaining PGHS-2 was observed in a small number of these cells in early postnatal life, increasing from day 5 to day 15, and decreasing thereafter to reach adult levels (Vio *et al.* 1999). During maximal expression, about 20% of TAL cells expressed PGHS-2, but only 2% in the early postnatal period and in adult animals. The transient induction of PGHS-2 appeared to correspond to a de-repression of PGI IS-2 expression secondary to low levels of glucocorticoids (Vio *et al.* 1999).

An age-dependent change in expression has been shown for PGHS-2, although not in the context of early developmental stages as described above. The analysis of PGHS-1 and PGHS-2 revealed a higher PGHS-2 protein and mRNA expression in peritoneal macrophages derived from old mice compared with those from young mice, whereas levels of PGHS-1 protein were not different (Hayek *et al.* 1997). Moreover, older rats exhibit significantly higher PGHS-2 expression (mRNA and protein) in the kidney than young animals (Chung *et al.* 1999). It has been proposed that the increase of PGHS-2 during lifetime is due to increased formation of reactive oxidant species (ROS) which can activate the redox-sensitive transcription factor NF-kB (Nuclear Factor—KappaB) (Kim *et al.* 2000). The age-associated increase in PGHS-2 and PGE_2 production has been sugggested to be involved in the dysregulation of immune functions in aging animals (Hayek *et al.* 1997).

SEX-SPECIFIC DIFFERENCES

A sex-dependent divergence of the PGHS-1- or PGHS-2-mediated cooxidation of xenobiotics is not documented for any species. Aside from findings which imply a regulatory role for sex hormones in PGHS isozyme expression of certain reproductive organs (see below) there are no obvious sex specific differences for PGHS.

Men are known to have a higher risk of developing cardiovascular diseases than premenopausal women. Regarding the vasculoprotective action of oestrogens, it is of interest that oestradiol has been reported to stimulate prostacyclin production in human vascular endothelial cells (Mikkola *et al.* 1995). An *in vivo* study (Stanzcyk *et al.* 1995) found an increase in the urinary prostacyclin metabolite 6-keto-$PGF_{1\alpha}$ and a decrease of urinary thromboxane in postmenopausal women after infusion of oestradiol. On the other hand, oestradiol treatment caused a decrease in PGHS-2 mRNA levels in *vena cava* of ovariectomised rats; oestradiol was also found to inhibit the PGHS-2 increase seen under inflammatory conditions induced by lipopolysac-charide (LPS) (Diel *et al.* 1996). The significance of these observations is not immediately clear, since various other factors are discussed with respect to cardio-vascular diseases.

Regulation of PGHS by endogenous factors (cytokines/hormones)

In general, PGHS-1 failed to respond to cytokines or other stimuli. Under some circumstances, however, PGHS-1 can be induced by certain stimuli which trigger differentiation and/or developmental processes. For instance, experimentally induced differentiation of osteoblasts and monocytes was associated with an increased expression of PGHS-1 (Oshima *et al.* 1991; Smith *et al.* 1993). Treatment of bone marrow-derived mast cells with stem cell factor induced mast cell differentiation and PGHS-1 mRNA accumulation (Samet *et al.* 1995); also stimulation with recombinant c-*kit* ligand (which binds to KIT, a mast/stem cell growth factor receptor) alone or in combination with IL-10 (Interleukin-10) led to increased expression of PGHS-1 transcript (Murakami *et al.* 1995). Treatment of monocytic cells (U937 and THP-1) by TPA resulted in increased levels of PGHS-1 mRNA and protein (Hoff *et al.* 1993; Smith *et al.* 1993). This increase of PGHS-1 expression was associated with enhanced differentiation into a macrophage-like phenotype. A hormonal regulation of PGHS-1 expression has been demonstrated in ovine seminal vesicles (Silvia *et al.* 1994). The castration of young male sheep (wethers) led to decreased levels of PGHS-1 in the involuted accessory sex gland. Administration of testosterone restored the PGHS-1 levels to those seen in age-matched rams and normal seminal vesicle development. The results suggest that PGHS-1 is positively regulated by androgens during the pubertal development of the seminal vesicles (Silvia *et al.* 1994).

The regulation of the PGHS-2 gene by endogenous factors like cytokines and growth factors is documented in numerous studies, and the molecular mechanisms of transcriptional control by various stimuli are an area of active reserach. DNA binding sites for NF-κB, CAAT/enhancer binding protein (C/EBP), ATF/cyclic-AMP response element (CRE), and E-box sequences have been identified in the 5'-flanking region of the PGHS-2 genes of different species (Fletcher *et al.* 1992; Kosaka *et al.* 1994; Inoue *et al.* 1994; Newton *et al.* 1997; Yamamoto *et al.* 1995). They mediate the rapid modulation of PGHS-2 expression upon stimulation of cells with e.g. lipopolysaccharide, hormones, and cytokines (Hwang *et al.* 1997; Inoue *et al.* 1995; Jones *et al.* 1993; Morris and Richards 1996; Reddy *et al.* 2000).

Particularly pro-inflammatory cytokines like interleukin (IL)-1β, IL-1α, tumour necrosis factor (TNF)-α, and interferon (INF)-γ, but also growth factors like epidermal growth factor (EGF), platelet-derived growth factor (PDGF), fibroblast growth factor (FGF), and transforming growth factor (TGF)-α can induce PGHS-2 expression in many cell types (reviewed by Herschman 1996,1999; Smith and De Witt 1996). On the other hand, the anti-inflammatory cytokines IL-10, IL-4, and IL-13 exert an inhibitory effect on PGHS-2 expression (Mertz *et al.* 1994, 1996; Dworski and Sheller 1997; Onoe *et al.* 1996). The effects of the immunosuppressive cytokine TGF-β on PGHS-2 expression are cell-specific; TGF-β suppresses the LPS-induced PGHS-2 expression in macrophages (Reddy *et al.* 1994), but TGF-β stimulates PGHS-2 expression and production of prostaglandins in tetradecanoyl phorbol acetate (TPA-) and IL-1β treated fibroblasts (Gilbert *et al.* 1994a) or epithelial cells (Gilbert *et al.* 1994b). The immunomodulatory suppressive effects of cytokines may be mediated by an altered PGHS-2 expression.

Hormonal regulation of PGHS-2 by gonadotropins was demonstrated by Sirois and Richards (1992) who thereby contributed to the discovery of the inducible isoform. PGHS-2 was induced in granulosa cells of rat preovulatory follicles upon treatment with gonadotropic hormones (Sirois *et al.* 1993; Sirois and Richards 1993). Promoter analysis revealed that the E-box and the C/EBP binding site are involved in the induction of the PGHS-2 gene (Morris and Richards 1996; Liu *et al.* 1999). These and other observations are indicative for an important role of prostaglandins in processes of female reproduction, e.g. ovulation, implantation and parturation (Chakraborty *et al.* 1996; Charpigny *et al.* 1997; Mitchell *et al.* 1995).

Adrenal steroids are important regulators of PGHS-2 expression. The effect of endogenous glucocorticoids is demonstrated by elevated prostanoid levels in adrenalectomised animals; treatment with glucocorticoids completely suppressed the increase of prostanoids (Masferrer *et al.* 1992). Glucocorticoids selectively suppressed PGHS-2, but not PGHS-1, mRNA and protein accumulation (Masferrer *et al.* 1994). Synthetic glucocorticoids such as dexamethasone also block the induction of PGHS-2 mRNA and protein usually seen after stimulation with inducers like serum, IL-1β, PDGF, TPA, and forskolin (O'Banion *et al.* 1992; DeWitt and Meade 1993; Rzymkiewicz *et al.* 1994; Kujubu and Herschman 1992). The strong inhibitory effects of anti-inflammatory glucocorticoids on PGHS-2 expression are mediated via transcriptional and post-transcriptional processes (DeWitt and Meade 1993; Evett *et al.* 1993; Newton *et al.* 1998). The down-regulation of PGHS-2 by glucocorticoids is not mediated via nGRE enhancer elements which are absent in the promoter of the PGHS-2 genes. Interestingly, in contrast to most other cell types, glucocorticoids induce the expression of PGHS-2 in amnion cells (Zakar *et al.* 1995), thus giving another example for a tissue-specific regulation.

There is evidence for a critical role of PGHS-2 in bone resorption and osteoclast formation. Parathyroid hormone (PTH) is a potent activator of osteoclastic bone resorption and increases PGHS-2 transcription in osteoblasts (Tetradis *et al.* 1997). Osteoclast formation stimulated by PTH or by 1,25-dihydroxyvitamin D_3 (1,25-D) is reduced in cell cultures deficient in PGHS-2, and is reversed by exogenous PGE_2. Furthermore, specific inhibitors for PGHS-2 can repress osteoclast formation in wild type cell cultures indicating that PGHS-2 and PGE_2 are crucial factors for the PTH and 1,25-D-stimulated bone resorption (Okada *et al.* 2000).

The action of 17β-oestradiol on the arachidonate metabolism is cell-specific. Treatment of human peripheral monocytes (Miyagi *et al.* 1993) or cultured endometrial stromal cells (Bulun *et al.* 1999) with oestradiol stimulated their PGE_2 production, whereas in cultured bovine coronary endothelial cells the exposure to oestradiol reduced the production of thromboxane B_2 and 6-keto $PGF_{1\alpha}$ (Stewart *et al.* 1999). Since the mRNA expression of PGHS-1 or PGHS-2 was not altered, oestradiol appeared to act at a post-transcriptional level. A similar tissue-specific effect, yet accompanied by changes in PGHS-2, was noted in other studies. Treatment of ovariectomised rats with oestradiol caused a decrease in PGHS-2 mRNA levels in *vena cava*, but an increased expression of PGHS-2 in the uterus of these animals (Diel *et al.* 1996). These and other reports on changes of PGHS isozyme expression in ovine and baboon endometrium during the menstrual cycle point to oestradiol and progesterone as regulators of prostaglandin production (Charpigny *et al.* 1997; Kim *et al.* 1999). The mechanisms responsible for PGHS regulation by these steroid hormones are unknown.

The importance of PGHS-2 in various diseases, like colon carcinoma, neurological disorders, kidney diseases, and inflammation has been extensively reviewed (recent overviews by DuBois, 2000; Halliday *et al.* 2000; Prescott and Fitzpatrick, 2000; Vane *et al.* 1998; Williams *et al.* 1999).

Modulation of PGHS expression by prior exposure to xenobiotics

Similar to endogenous stimuli, xenobiotics which modulate PGHS expression are predominantly affecting the PGHS-2, but not the PGHS-1 isoform. An exception is the tumour promoter tetradecanoyl phorbol acetate (TPA), in some studies TPA treatment was found to induce also, to some extent, PGHS-1 mRNA expression in specific cell types (Hamasaki *et al.* 1993; Kitzler *et al.* 1995). The mechanism of PGHS-1 induction by TPA is unclear. Most studies, however, demonstrate a pronounced induction of PGHS-2 by TPA in the absence of notable effects on PGHS-1 transcription. This differential response to inducers led to the discovery of the PGHS-2 isoform, first described as a so-called 'immediate-early' gene in Swiss 3T3 cells (Lim *et al.* 1987). Numerous studies followed, showing that TPA induces PGHS-2 expression in a wide variety of cell types of different species, including chicken, mouse, rat, and human (Table 6.3).

A range of cells including fibroblasts, epithelial, smooth muscle cells as well as keratinocytes can respond to TPA treatment *in vitro* by increased expression of PGHS-2. Induction of PGHS-2 has been demonstrated also *in vivo* in mouse skin; acute

Table 6.3 Xenobiotics and induction of PGHS-2 expression: Studies with various cell types and in tissues

Cell type and tissue	Xenobiotic	References
Murine fibroblast 3T3 cells	TPA	Kujubu *et al.* (1991)
Murine embryo fibroblasts		Gilbert *et al.* (1994a)
Human monocytes		Hoff *et al.* (1993)
Chick embryo fibroblasts		Xie *et al.* (1991)
Rat aortic smooth muscle cells		Rimarachin *et al.* (1994)
Human vascular endothelial cells		Jones *et al.* (1993)
Rat bronchial epithelial cells		Hamasaki Y *et al.* (1993)
Rat intestinal epithelial cells (RIE-1 and IEC-6)		Du Bois *et al.* (1994); Gilbert *et al.* (1994b)
Human breast cancer cells (MCF-7 and MDA-MB 231)	TPA	Liu and Rose (1996)
Murine keratinocytes (MSCP5) mouse epidermis		Scholz *et al.* (1995)
Murine peritoneal macrophages	Procainamide	Goebel *et al.* (1999)
Immortalised mouse liver cells (ML-457)	WY-14,643, monoethylhexyl phthalate, clofibrate, ciprofibrate ethyl ester	Ledwith *et al.* (1997)
Human mammary epithelial cells (184B5) and human colon carcinoma cells (CaCo-2)	WY-14,643, fatty acids	Meade *et al.* (1999)
Rabbit corneal epithelium	WY-14,643	Bonazzi *et al.* (2000)
Canine kidney cells (MDCK)	TCDD	Kraemer *et al.* (1996)
Human umbilical vascular endothelial cells		Liu *et al.* (1997)
Murine hepatoma cells (Hepa1c1c7)		Puga *et al.* (1997)
Murine peritoneal macrophages		Vogel *et al.* (1997)
C57BL/6 mice lung and spleen	TCDD	Vogel *et al.* (1998)
Primary rat hepatocytes		Vogel *et al.* (2000)
Murine fibroblasts (C3H/M2)		Wölfle *et al.* (2000)
Rat thymocytes and murine thymic lymphoma cells (WEHI 7.1)	Down-regulation by TCDD	Olnes *et al.* (1996)
Oral epithelial cells	Benzo[a]pyrene	Kelley *et al.* (1997)
Human airway epithelial cells	Residual oil fly ash	Samet *et al.* (1996)
Rat alveolar macrophages	Silica	Chen *et al.* (1997)

Notes: TPA: Tetradecanoyl phorbolacetate; TCDD, 2,3,7,8-tetrachlorodibenzo-p-dioxin; WY-14,643, [4-chloro-6-(2,3-xylidino)-2-pyrimidyl-thio] acetic acid.

inflammation and epidermal hyperplasia evoked by wounding or by TPA treatment resulted in a transient induction of PGHS-2 expression while PGHS-1 remained unchanged (Scholz *et al.* 1995). Futher studies by the same group imply an important role of PGHS-2 in both regenerative hyperplasia and during multistage carcinogenesis in the mouse model (Müller-Decker *et al.* 1995, 1998a,b).

The molecular mechanisms involved in the TPA-mediated induction of the PGHS-2 gene have been studied by several groups (Xie and Herschman 1995, 1996; Inoue *et al.* 1995; Pilbeam *et al.* 1997; Wadleigh and Herschman 1999). These papers and reviews (Herschman 1999; Vogel 2000) also cover studies on factors and elements important in the transcriptional control of this gene. It may suffice to point out that the exogenous chemical TPA—by mimicking endogenous factors and using established signal transduction pathways—can interfere with the regulation of PGHS-2 expression.

More recently, also non-TPA-type compounds, including several rodent liver and skin tumour promoters were shown to modulate PGHS-2 expression (Table 6.3). Specifically, the peroxisome proliferators, WY-14,643, monoethylhexyl phthalate, clofibrate, ciprofibrate ethyl ester, and eicosatetraynoic acid elevated PGHS-2 mRNA and protein levels *in vitro* (Ledwith *et al.* 1997). In contrast to TPA, these peroxisome proliferators caused little or no increase in PGE$_2$ levels, and inhibited the serum-induced synthesis of PGE$_2$ in immortalised mouse liver cells (ML-457). These compounds act cell specifically on PGHS-2 and prostaglandin metabolism via both positive and inhibitory mechanisms. Ledwith *et al.* (1997) also reported induction of PGHS-2 by thapsigargin, okadaic acid and calcium ionophore, but not by phenobarbital or dehydroepiandrosterone sulphate.

An induction of PGHS-2 by WY-14,643 with the ability to activate the peroxisome proliferator-activated receptor (PPAR) was also shown in rabbit corneal epithelial cells (Bonazzi *et al.* 2000). The PGHS-2 induction was independent of prostaglandin synthesis inhibition (by NSAIDs) and not associated with a parallel increase in PGE$_2$ accumulation. Meade *et al.* (1999) identified a peroxisome proliferator response element in the human PGHS-2 promoter which seems to be responsible for the enhanced PGHS-2 expression seen after treatment of colon and mammary epithelial cells with various peroxisome proliferators including WY-14,643, and several fatty acids.

Evidence for PGHS-2 modulation has been reported for 2,3,7,8-tetrachlorodibenzo-*p*-dioxin (TCDD) and for other chemicals. Treatment of murine macrophage cultures with procainamide or the potent inducer LPS/INF-γ resulted in clearly increased expression of PGHS-2 and enhanced production of PGE$_2$; with both stimuli the effect was more pronouced in macrophages from C57BL/6 mice than in those from A/J mice (Goebel *et al.* 1999). The treatment of peritoneal macrophages from two mice strains (C57BL/6 and DBA/2) with TCDD resulted in a strain-dependent induction of PGHS-2 mRNA whereas PGHS-1 levels remained unchanged (Vogel *et al.* 1997). Kraemer *et al.* (1996) found that TCDD led to a marked increase of PGHS-2 in MDCK cells. An induction of PGHS-2 by TCDD was also shown in primary human epithelial cells (Liu *et al.* 1997) and in a murine hepatoma cell line (Puga *et al.* 1997). Similar TCDD effects on PGHS-2 as shown in these *in vitro* studies are also documented *in vivo*. A dose-dependent induction of PGHS-2 was observed in lung and spleen of TCDD-sensitive C57BL/6 mice, but no or weak induction in the organs of insensitive DBA/2 mice (Vogel *et al.* 1997, 1998). The results indicated an involvement of the Ah-receptor (AhR) in the TCDD-mediated induction of PGHS-2. However, the molecular mechanism by which this modulation occurs was still unclear. Transfection studies with reporter plasmids containing XRE (Xenobiotic Responsive Elements) elements of the PGHS-2 promoter were inconsistent regarding the functional activity of these XRE elements. In transfected Hepa 1c1c7 cells, the AhR failed to activate the transcription of the PGHS-2 gene (Kraemer *et al.* 1996), whereas in thymocytes the XRE element appears to be necessary for down-regulation of PGI IS-2 (Olnes *et al.* 1996).

Recently, the mechanism of the PGHS-2 activation by TCDD has been studied in more detail. Promoter analysis revealed that the *cis*-acting element C/EBP is involved, rather than the AhR-binding XRE motif identified in the 5′-flanking region of the human, rat and mouse PGHS-2 gene (Fletcher *et al.* 1992; Sirois *et al.* 1993; Hla and Neilson 1992). There is evidence from *in vitro* and *in vivo* studies that the AhR-associated c-Src tyrosine kinase (Enan and Matsumura 1996) takes part in the activation of PGHS-2. The upregulation of PGHS-2 by TCDD was blocked by c-Src inhibitors herbimycin and geldanamycin, and in c-Src deficient mice, PGHS-2 failed to respond to TCDD treatment (Vogel *et al.* 1998, 2000).

Thus, the regulation of PGHS-2 by TCDD differs from the classical AhR/Arnt-pathways described for the induction of several xenobiotic-metabolising enzymes, e.g. cytochromes P450s (Nebert *et al.* 2000; Chapter 2). The exposure of oral epithelial cells to the carcinogen benzo[*a*]pyrene (B[a]P) was found to induce PGHS-2 (Table 6.3); the non-carcinogen benzo[e]pyrene (B[e]P) had no effect on PGHS-2 expression (Kelley *et al.* 1997). Since B[a]P, not B[e]P, is known as an AhR agonist, a similar mechanism as described above for TCDD can be invoked for the induction of PGHS-2 by B[a]P.

Another class of xenobiotic agents such as mineral dust and silica have been shown to interfere with the arachidonic acid cascade (Table 6.3). Treatment of rat alveolar macrophages with silica led to transcriptional activation of the PGHS-2 gene; the effect has been attributed to silica-induced NF-κB activation (Chen *et al.* 1997). Exposure of human airway epithelial cells to the particulate air pollutant residual oil fly ash (ROFA) resulted in an increased PGHS-2 expression and $PGE_2/PGF_{2\alpha}$ production (Samet *et al.* 1996, 1999). ROFA contains significant quantities of transition metals which can generate reactive oxygen species, and thus induce PGHS-2 via the NF-κB signal transduction pathway. Ozone can cause neutrophilic airways inflammation. After inhalation of ozone, increased levels of PGE_2 and thromboxane B_2 were found in bronchoalveolar lavage fluid (Hazucha *et al.* 1996). Although a direct effect on the PGHS-2 expression was not described, it is likely to occur since in other studies the inducible effect of reactive oxygen species and oxidative stress on PGHS-2 expression has been demonstrated and considered to contribute to inflammatory reactions (Feng *et al.* 1995). These findings suggest that induction of PGHS 2 may be a crucial factor in mediating adverse health effects (e.g. lung inflammation and pulmonary injury) of particles and other air pollutants.

Collectively, there is evidence that, in addition to endogenous factors, several exogenous agents are capable of modulating PGHS-2 expression. Table 6.3 lists xenobiotics found to increase PGHS-2 expression. Related to chemical toxicity, this is important, since induction of PGHS-2 could result: (a) in an inappropriate synthesis of prostanoids, and (b) an increased peroxidative metabolism of xenobiotics.

NSAIDs and other PGHS inhibitors

PGHS as a key enzyme in the synthesis of biologically active prostanoids (Figure 6.1) is a prime target for an important class of pharmaceutical compounds, the non-steroidal antiinflammatory drugs (NSAIDs). Well-known examples are aspirin, indo-methacin, ibuprofen, diclofenac, piroxicam, sulindac, and many others. Whereas many NSAIDs inhibit PG-synthesis by both isozymes to a similar extent, new drugs have been developed which are much more potent inhibitors of PGHS-2 than of PGHS-1. When the existence of PGHS-2 was discovered, pharmaceutical research focused on designing such drugs, because the PGHS-2-dependent PG-synthesis is thought to be predominantly responsible for inflammatory processes. Although NSAIDs-specific for PGHS-2 with good efficacy in a low dose-range offer advantages with respect to unwanted side effects, like ulcerogenicity and nephrotoxicity, other undesirable effects cannot be eliminated so far. A detailed discussion on the selectivity of the compounds and the mechanisms by which NSAIDs inhibit the cyclooxygenase isozymes are beyond the scope of this chapter. Interested readers are referred to recent reviews that also cover the role of prostanoids in physiology and in certain diseases (DuBois *et al.* 1998; Hawkey 1999; Vane *et al.* 1998; Taketo 1998a; Wolfe *et al.* 1999).

Of particular interest is the role of PGHS and its inhibitors in cancer. PGHS-2 expression is markedly increased in several tumours, e.g. human colorectal cancer and human invasive transitional cell carcinoma of the bladder, and PGHS-inhibitors show antitumour activity in animal models of these forms of cancer (Mohammed *et al.* 1999; Williams *et al.* 1999; Oshima *et al.* 1996). Moreover, epidemiological studies have shown that chronic administration of PGHS-inhibiting drugs can effectively suppress the development of several tumours (Giovannucci *et al.* 1995; Thun 1996; Taketo 1998b; Williams *et al.* 1999). The underlying mechanisms are not fully under-stood yet.

The potency of NSAIDs to suppress the synthesis of prostanoids is based on their ability to bind to the active domain of the PGHS protein and thereby inhibiting PGHS cyclooxygenase activity (Bhattacharyya *et al.* 1996; Garavito and DeWitt 1999). More recently, it was described that NSAIDs could act through an additional mechanism to inhibit PG-biosynthesis; aspirin and sulindac markedly suppressed the mRNA expres-sion of cytosolic phospholipase A_2 ($cPLA_2$) in NIH3T3 cells (Yuan *et al.* 2000), an acyl esterase which catalyses the release of arachidonic acid from cellular phospholipids (Figure 6.1). The authors proposed that this pathway contributes, at least in part, to the cancer chemopreventive effects of NSAIDs.

Synthetic glucocorticoids such as dexamethasone inhibit prostanoid synthesis by down-regulating the expression of PGHS-2, an effect found in most tissues except amnion. The modulation of PGHS-2 expression by glucocorticoids has already been discussed in more detail (see above).

In addition to drugs, natural plant compounds such as certain stilbenes and flavonoids can also affect arachidonate metabolism by PGHS. Resveratrol (*trans-*

3,4',5-trihydroxystilbene), a phytoalexin found in grapes and other food products, exerted chemopreventive activity in assays of multi-stage carcinogenesis (Jang et al. 1997). Resveratrol was found to inhibit the cyclooxygenase and hydroperoxidase activity of PGHS (Johnson and Maddipati 1998). In addition, resveratrol affects arachidonic acid release, and can suppress the PGHS-2 induction mediated by LPS and TPA (Martinez and Moreno 2000; Subbaramaiah et al. 1999). Similar to resveratrol, various flavonoid derivatives such as oroxylin A, wogonin, skullcapflavone II, tectorigenin and iristectorigenin A show inhibitory activities on PGHS and on lipoxygenases (You et al. 1999). Silymarin is used clinically for the treatment of inflammatory liver diseases (Flora et al. 1998). Like resveratrol, silymarin possesses cancer-preventive effects in different in vivo and in vitro carcinogenesis models (Anderson et al. 1998; Lahiri-Chatterjee et al. 1999). Silymarin acts specifically on PGHS-2 with no change in constitutive PGHS-1 protein levels (Zhao et al. 1999). One proposed mechanism of the PGHS-2-specific inhibition by flavonoids is inhibition of activation nuclear factor-kappa B (Chen et al. 2000).

The above examples illustrate that a variety of naturally occurring xenobiotics can affect PGHS activity and expression. In contrast to agents which induce PGHS-2 expression, natural plant ingredients appear to inhibit PG-biosynthesis by various mechanisms. On the other hand, also man-made chemicals, e.g. o-phenylphenol and its metabolite, have been found to inhibit PGHS activity (Freyberger and Degen 1998).

Genetics and pharmacogenetics

The reason for the existence the two PGHS-isozymes is as yet unknown. PGHS-1 and PGHS-2 knockout animals were constructed by gene targeting techniques to study physiological function and pathophysiological states. Langenbach et al. (1995) have shown that mice genetically deficient in PGHS-1 exhibit normal life spans and had no impairment of health. These findings were surprising in view of the constitutive expression and physiological importance of PGHS-1, since severe gastric bleeding and renal dysfunction are seen upon chronic treatment of normal animals with NSAIDs, an experimental model equivalent to gene-deletion studies. The PGHS-1 null mice produced only 1% of the normal PG levels. They failed to produce viable offspring attributable to a disturbed parturition in female animals (Langenbach et al. 1995). On the other hand, only about 60% of the pups lacking PGHS-2 expression survived to weaning. The PGHS-2 knockout mice had serious renal developmental deficiencies, peritonitis, and 25% of the animals began to die after 3 weeks of age. About 75% of surviving pups lived to one year of age (Morham et al. 1995). Another PGHS-2-deficient mouse showing similar characteristics has been described by Dinchuk et al. (1995). In accordance with the results from the PGHS-2 gene-deletion studies are recent findings that a selective inhibitor for PGHS-2 (SC58236, administered during pregnancy until weaning) impairs glomerulogenesis and development of the renal cortex in both mice and rats (Komhoff et al. 2000).

The above findings reveal that prostaglandins derived via the individual PGHS isoforms have distinct as well as common functions. It appears that deficiency of PGHS-2 has more profound effects than deficiency of PGHS-1. Recent data from PGHS-deficient mice indicate that both PGHS-1 as well as PGHS-2 are mediating inflammatory responses, and that both isoforms have significant roles in carcinogenesis (Langenbach *et al.* 1999a,b). Thus, the terms for PGHS isoforms as 'housekeeping' and/or 'response' genes may not be entirely precise.

There are no known polymorphisms for PGHS-1. Only two reports are documenting a genetic polymorphism for the PGHS-2 gene. Molecular cloning of the human PGHS-2 gene and digestion of genomic DNA of 78 individuals with Hind III revealed that a polymorphism is present in about 5% of the population (Jones *et al.* 1993). By sequence analysis of genomic DNA, a silent mutation in exon 3 of the human PGHS-2 was detected (Spirio *et al.* 1998). Phenotypic changes like altered enzyme activity or gene inducibility of PGHS-2 as a consequence of the identified polymorphisms have not been reported so far.

Acknowledgements

The authors gratefully acknowledge the efforts of a dedicated group of graduate students and postdoctoral fellows who studied several aspects of prostaglandin synthases, and the Deutsche Forschungsgemeinschaft for past and present financial support of research carried out in the authors' laboratories. Special thanks are due to Dr Petra Janning for help in preparing the illustrations and Dr Volker Mostert for a critical reading of the manuscript.

References

Acton KS, Boermans HJ and Bray TM (1992) The role of prostaglandin H synthase in 3-methylindole-induced pneumotoxicity in goat. *Comp. Biochem. Physiol.*, **101C**, 101–108.

Anderson D, Dobrzynska MM, Basaran N, Basaran A and Yu TW (1998) Flavonoids modulate comet assay responses to food mutagens in human lymphocytes and sperm. *Mutat. Res.*, **402**, 269–277.

Appleby SB, Ristimaki A, Neilson K, Narko K and Hla T (1994) Structure of the human cyclooxygenase-2 gene. *Biochem. J.*, **302**, 723–727.

Arlen RR and Wells PG (1996) Inhibition of thalidomide teratogenicity by acetylsalicylic acid: evidence for prostaglandin H synthase-catalysed bioactivation of thalidomide to a teratogenic reactive intermediate. *J Pharmacol. Exp. Ther.*, **277**, 1649–1658.

Barnett J, Chow J, Ives D, Chiou M, Mackenzie R, Osen E, Nguyen B, Tsing S, Bach C, Freire J, Chan H, Sigar E and Ramesha C (1994) Purification, characterization and selective inhibition of human prostaglandin G/H synthase 1 and 2 expressed in the baculovirus system. *Biochim Biophys. Acta.*, **1209**, 130–139.

Battista JR and Marnett LJ (1985) Prostaglandin H synthase-dependent epoxidation of aflatoxin B$_1$. *Carcinogenesis*, **5**, 719–723.

Bhattacharyya DK, Lecomte M, Rieke CJ, Garavito M and Smith WL (1996) Involvement of arginine 120, glutamate 524, and tyrosine 355 in the binding of arachidonate and 2-phenylpropionic acid inhibitors to the cyclooxygenase active site of ovine prostaglandin endoperoxide H synthase-1. *J. Biol. Chem.*, **271**, 2179–2184.

Bonazzi A, Mastyugin V, Mieyal PA, Dunn MW and Laniado-Schwartzman M (2000) Regulation of cyclooxygenase-2 by hypoxia and peroxisome proliferators in the corneal epithelium. *J. Biol. Chem.*, **275**, 2837–2844.

Boyd JA and Eling TE (1984) Evidence for a one-electron mechanism of 2-amino-flourine oxidation by prostaglandin H synthase and horseradish peroxidase. *J. Biol. Chem.*, **259**, 13885–13996.

Brannon TS, North AJ, Wells LB and Shaul PW (1994) Prostacyclin synthesis in ovine pulmonary artery is developmentally regulated by changes in cyclooxygenase-1 gene expression. *J. Clin. Invest.*, **93**, 2230–2235.

Brannon TS, MacRitchie AN, Jaramillo MA, Sherman TS, Yuhanna IS, Margraf LR and Shaul PW

(1998) Ontogeny of cyclooxygenase-1 and cyclooxygenase-2 gene expression in ovine lung. *Am. J. Physiol.*, **274**, 66–71.

Bulun SE, Zeitoun K, Takayama K, Noble L, Michael D, Simpson E, Johns A, Putman M and Sasano H (1999) Estrogen production in endometriosis and use of aromatase inhibitors to treat endometriosis. *Endocr. Relat. Cancer*, **6**, 293–301.

Capdevila JH, Morrow JD, Belosludtsev YY, Beauchamp DR, DuBois RN, and Falck JR (1995) The catalytic outcomes of the constitutive and the mitogen inducible isoforms of prostaglandin H_2.synthase are markedly affected by glutathione and glutathione peroxidase. *Biochemistry*, **34**, 3325–3337.

Chakraborty I, Das SK, Wang J, and Dey SK (1996) Developmental expression of the cyclooxygenase-1 and cyclooxygenase-2 genes in the peri-implantation mouse uterus and their differential regulation by the blastocyst and ovarian steroids. *J. Mol. Endocrinol.*, **16**, 107–122.

Charpigny G, Reinaud P, Tamby JP, Creminon C, Martal J, Maclouf J, and Guillomot M (1997) Expression of cyclooxygenase-1 and -2 in ovine endometrium during the estrous cycle and early pregnancy. *Endocrinology*, **138**, 2163–2167.

Chen F, Sun S, Kuhn DC, Gaydos LJ, Shi X, Lu Y and Demers LM (1997) Involvement of NF-kappaB in silica-induced cyclooxygenase II gene expression in rat alveolar macrophages. *Am. J. Physiol.*, **272**, 779–786.

Chen Y, Yang L and Lee TJ (2000) Oroxylin A inhibition of lipopolysaccharide-induced iNOS and COX-2 gene expression via suppression of nuclear factor-kappaB activation. *Biochem. Pharmacol.*, **59**, 1445–1457.

Christ EJ and Van Dorp DA (1972) Comparative aspects of prostaglandin biosynthesis in animal tissue. *Biochim. Biophys. Acta*, **270**, 537–541.

Chung HY, Kim HJ, Shim KH and Kim KW. (1999) Dietary modulation of prostanoid synthesis in the aging process: role of cyclooxygenase-2. *Mech. Ageing Dev.*, **111**, 97–106.

Cohn SM, Schloemann S, Tessner T, Seibert K and Stenson WF (1997) Crypt stem cell survival in the mouse intestinal epithelium is regulated by prostaglandins synthesised through cyclooxygenase–1. *J. Clin. Invest.*, **99**, 1367–1379.

Cribb AE, Miller M, Tesoro A, Spielberg SP (1990) Peroxidase-dependent oxidation of sulphonamides by monocytes and neutrophils from humans and dogs. *Mol. Pharmacol.*, **38**, 744–751.

Curtis JF, Tomer K, McGown S, and Eling TE (1995) Prostaglandin H synthase-catalysed ring oxygenation of 2-naphthylamine: evidence for two distinct oxidation pathways.*Chem. Res. Toxicol.*, **6**, 875–883.

Degen GH (1990) Role of prostaglandin synthase in mediating genotoxic and carcinogenic effects of estrogens. *Environm. Health Perspect.*, **88**, 217–223.

Degen GH (1993a) Prostaglandin-H-synthase containing cell lines as tools for studying the metabolism and toxicity of xenobiotics. *Toxicology*, **82**, 243–256.

Degen GH (1993b) Ovine seminal vesicle cell cultures, a tool for studies of carcinogen activation by prostaglandin H synthase. In *Eicosanoids and other Bioactive Lipids in Cancer, Inflammation and Radiation*, (Nigam S, Honn K, Marnett LJ and Walden TL (eds)), Kluwer, Norwell, USA, pp. 419–421.

Degen GH and McLachlan JA (1985) Peroxidase-mediated in vitro metabolism of diethylstilboestrol and structural analogs with different biological activities. *Chem. Biol. Interact.*, **54**, 363–375.

Degen GH, Eling TE, and McLachlan JA (1982) Oxidative metabolism of diethylstilboestrol by prostaglandin synthetase. *Cancer Res.*, **42**, 919–923.

Degen GH, Wong A, Eling TE, Barrett JC, and McLachlan JA (1983) Involvement of prostaglandin synthetase in the peroxidative metabolism of diethylstilboestrol in Syrian hamster embryo fibroblast cell cultures. *Cancer Res.*, **43**, 922–996.

Degen GH, Metzler M, and Sivarajah K (1986) Co-oxidation of diethylstilboestrol and structural analogs by prostaglandin synthase. *Carcinogenesis*, **7**, 137–142.

Degen GH, Jellinck PH and Hershcopf RJ (1987) Prostaglandin H synthase catalyses regiospecific release of tritium from estradiol. *Steroids*, **49**, 561–580.

Degen GH, Wolz E, Gerber M and Pfau W (1998a) Bioactivation of 2-amino-3-methyl-imidazo-[4,5- *f*]quinoline (IQ) by prostaglandin-H synthase. *Arch. Toxicol.*, **72**, 183–186.

Degen GH, Schlattjan JH, Mähler S, Föllmann W and Bolt HM (1998b) *Metabolismus* und Genotoxizität von Benzidin in Urothel- und Samenblasen-Zellkulturen. In *Gesundheitsgefahren durch biologische Arbeitsstoffe. Neuro-, Psycho-und Verhaltenstoxizität. Dokumentationsband zur 38. Jahrestagung der Deutschen Gesellschaft für Arbeitsmedizin und Umweltmedizin in Wiesbaden vom 11.-14.05*, Hallier E and Bünger J (eds). DGAUM, Lübeck, pp. 603–604.

Dennis EA (1994) Diversity of group types, regulation and function of phospholipase A_2. *J. Biol. Chem.*, **269**, 1057–1060.

DeWitt DL and Meade EA (1993) Serum and glucocorticoid regulation of gene transcription and expression of the prostaglandin H synthase-1 and prostaglandin H synthase-2 isozymes. *Arch. Biochem. Biophys.*, **306**, 94–102.

DeWitt DL and Smith WL (1988) Primary structure of prostaglandin G/H synthase from sheep vesicular gland determined from the complementary DNA sequence. *Proc. Natl. Acad. Sci. USA*, **85**, 1412–1416.

DeWitt DL and Smith WL (1990) Cloning of sheep and mouse prostaglandin endoperoxide synthases. *Methods Enzymol.*, **187**, 469–479.

Diel P, Hegele-Hartung C, Burton G, Kauser K, and Fritzemeier KH (1996) Tissue specific effects of oestrogens on expression of COX II, iNOS and IL-6 mRNA in V. cava and uterus of ovex rats. *Exp. Clin Endocrinol. Diabet.*, **104**, Suppl. 1, 171.

Dinchuk JE, Car BD, Focht RJ, Johnston JJ, Jaffee BD, Covington MB, Contel NR, Eng VM, Collins RJ, Czerniak PM, Gorry SA and Trzaskos JM (1995) Renal abnormalities and an altered inflammatory response in mice lacking cyclooxygenase II. *Nature*, **378**, 406–409.

Dixon DA, Kaplan CD, McIntyre TM, Zimmerman GA and Prescott SM (2000) Post-transcriptional control of cyclooxygenase-2 gene expression. The role of the 3′-untranslated region. *J. Biol. Chem.*, **275**, 11750–11757.

DuBois RN (2000) Review article: cyclooxygenase—a target for colon cancer prevention. *Aliment. Pharmacol. Ther.*, **14**, Suppl. 1, 64–67.

DuBois RN, Awad J, Morrow J, Roberts LJ 2nd, and Bishop PR (1994) Regulation of eicosanoid production and mitogenesis in rat intestinal epithelial cells by transforming growth factor-alpha and phorbol ester. *J. Clin. Invest.*, **93**, 493–498.

DuBois RN, Abrahamson SB, Crofford L, Gupta RA, Simon LS, Van De Putte LB, and Lipsky PE (1998) Cyclooxygenase in biology and disease. *FASEB J.*, **12**, 1063–1073.

Dworski R and Sheller JR (1997) Differential sensitivities of human blood monocytes and alveolar macrophages to the inhibition of prostaglandin endoperoxide synthase-2 by interleukin-4. *Prostaglandins*, **53**, 237–251.

Egan RW, Gale PH, VandenHeuvel WJ, Baptista EM, and Kuehl FA Jr (1980) Mechanism of oxygen transfer by prostaglandin hydroperoxidase. *J. Biol. Chem.*, **255**, 323–326.

Egan RW, Gale PH, Baptista EM, Kennicott KL, VandenHeuvel WJ, Walker RW, Fagerness PE, and Kuehl FA Jr (1981) Oxidation reactions by prostaglandin cyclooxygenase-hydroperoxidase. *J. Biol. Chem.*, **256**, 7352–7361.

Eling TE and Curtis JF (1992) Xenobiotic metabolism by prostaglandin H synthase. *Pharmacol. Ther.*, **53**, 261–273.

Eling TE, Mason RP, and Sivarajah K (1985) The formation of aminopyrine cation radical by the peroxidase activity of prostaglandin H synthase and subsequent reactions of the radical. *J. Biol. Chem.*, **260**, 1601–1607.

Eling TE, Thompson DC, Foureman GL, Curtis JF and Hughes MF (1990) Prostaglandin H synthase and xenobiotic oxidation. *Annu. Rev. Pharmacol. Toxicol.*, **30**, 1–45.

Enan E and Matsumura F (1996) Identification of c-Src as the integral component of the cytosolic Ah receptor complex, transducing the signal of 2,3,7,8-tetrachlorodibenzo-p-dioxin (TCDD) through the protein phosphorylation pathway. *Biochem. Pharmacol.*, **52**, 1599–1612.

Evett GE, Xie W, Chipman JG, Robertson DL and Simmons DL (1993) Prostaglandin G/H synthase isoenzyme 2 expression in fibroblasts: regulation by dexamethasone, mitogens, and oncogenes. *Arch. Biochem. Biophys.*, **306**, 169–177.

Feng L, Sun W, Xia Y, Tang WW, Chanmugam P, Soyoola E, Wilson CB and Hwang D (1993) Cloning two isoforms of rat cyclooxygenase: differential regulation of their expression. *Arch. Biochem. Biophys.*, **307**, 361–368.

Feng L, Xia Y, Garcia GE, Hwang D and Wilson CB (1995) Involvement of reactive oxygen intermediates in cyclooxygenase-2 expression induced by interleukin-1, tumour necrosis factor-alpha, and lipopolysaccharide. *J. Clin. Invest.*, **95**, 1669–1675.

Fischer B (1989) Gewinnung eines gegen Prostaglandin-H-Synthase (PHS) gerichteten polyklonalen Antiserums und PHS-Nachweis in unterschiedlichen Geweben verschiedener Spezies mit Hilfe immunologischer Methoden. Diploma Thesis, University of Würzburg.

Fletcher BS, Kujubu DA, Perrin DM and Herschman HR (1992) Structure of the mitogen-inducible TIS10 gene and demonstration that the TIS10-encoded protein is a functional prostaglandin G/H synthase. *J. Biol. Chem.*, **267**, 4338–4344.

Flora K, Hahn M, Rosen H and Bennet K (1998) Milk thistle (*Sibylum marianum*) for the therapy of liver disease. *American Journal of Gastroenterology*, **93**, 139–143.

Foth J, Schnitzler R, Jager M, Koob M, Metzler M and Degen GH (1992) Characterization of sheep seminal vesicle cells—a new tool for studying genotoxic effects in vitro. *Toxicology in Vitro*, **6**, 219–225.

Freyberger A and Degen GH (1989a) Covalent binding to proteins of reactive intermediates resulting from prostaglandin H synthase catalysed cooxidation of stilbene and steroid estrogens. *J. Biochem. Toxicol.*, **4**, 95–103.

Freyberger A and Degen GH (1989b) Studies on the stoichiometry of estrogen oxidation catalysed by purified prostaglandin H synthase holoenzyme. *J. Steroid Biochem.*, **33**, 473–481.

Freyberger A and Degen GH (1998) Inhibition of prostaglandin H synthase by *o*-phenylphenol. *Arch. Toxicol.*, **72**, 637–644.

Funk CD, Funk LB, Kennedy ME, Pong AS and Fitzgerald GA (1991) Human platelet/erythroleukemia cell prostaglandin G/H synthase: cDNA cloning, expression, and gene chromosomal assignment. *FASEB J.*, **5**, 2304–2312.

Gallois C, Habib A, Tao J, Moulin S, Maclouf J, Mallat A and Lotersztajn S (1998) Role of NF-kappaB in the antiproliferative effect of endothelin-1 and tumour necrosis factor-alpha in human hepatic stellate cells. Involvement of cyclooxygenase-2. *J. Biol. Chem.*, **273**, 23183–23190.

Garavito RM and DeWitt DL (1999) The cyclooxygenase isoforms: structural insights into the conversion of arachidonic acid to prostaglandins. *Biochim Biophys. Acta*, **1441**, 278–287.

Gibb W and Sun M (1996) Localization of prostaglandin H synthase type 2 protein and mRNA in term human fetal membranes and decidua. *J. Endocrinol.*, **150**, 497–503.

Gibb W, Sun M, Gyomorey S, Lye SJ and Challis JR (2000) Localization of prostaglandin synthase type-1 (PGHS-1) mRNA and prostaglandin synthase type-2 (PGHS-2) mRNA in ovine myometrium and endometrium throughout gestation. *J. Endocrinol.*, **165**, 51–58.

Gilbert RS, Reddy ST, Kujubu DA, Xie W, Luner S and Herschman HR (1994a) Transforming growth factor beta 1 augments mitogen-induced prostaglandin synthesis and expression of the TIS10/prostaglandin synthase 2 gene both in Swiss 3T3 cells and in murine embryo fibroblasts. *J. Cell. Physiol.*, **159**, 67–75.

Gilbert RS, Reddy ST, Targan S and Herschman HR (1994b) TGF-beta 1 augments expression of the TIS10/prostaglandin synthase-2 gene in intestinal epithelial cells. *Cell. Mol. Biol. Res.*, **40**, 653–660.

Giovannucci E, Egan KM, Hunter DJ, Stampfer MJ, Colditz GA, Willett WC and Speizer FE (1995) Aspirin and the risk of colorectal cancer in women. *N. Engl. J. Med.*, **333**, 609–614.

Goebel C, Vogel C, Sachs B, Schraa S, Abel J, Uetrecht J, Degen GH and Gleichmann E (1999) Procainamide, a drug causing lupus, induces prostaglandin H synthase-2 and formation of T-sensitizing drug metabolites in mouse macrophages. *Chem. Res. Toxicol.*, **12**, 488–500.

Habib A, Creminon C, Frobert Y, Grassi J, Pradelles P and Maclouf J (1993) Demonstration of an inducible cyclooxygenase in human endothelial cells using antibodies raised against the carboxyl-terminal region of the cyclooxygenase-2. *J. Biol. Chem.*, **268**, 23448–23454.

Halliday G, Robinson SR, Shepherd C and Kril J (2000) Alzheimer's disease and inflammation: a review of cellular and therapeutic mechanisms. *Clin. Exp. Pharmacol. Physiol.*, **27**, 1–8.

Hamasaki Y, Kitzler J, Hardman R, Nettesheim P and Eling TE (1993) Phorbol ester and epidermal growth factor enhance the expression of two inducible prostaglandin H synthase genes in rat tracheal epithelial cells. *Arch. Biochem. Biophys.*, **304**, 226–234.

Harris RC, McKanna JA, Akai Y, Jacobson HR, Dubois RN and Breyer MD (1994) Cyclooxygen-
ase-2 is associated with the macula densa of rat kidney and increases with salt restriction.
J. Clin. Invest., **94**, 2504–2510.
Hawkey CJ (1999) COX-2 inhibitors. *Lancet*, **353**, 307–314.
Hayek MG, Mura C, Wu D, Beharka AA, Han SN, Paulson KE, Hwang D and Meydani SN
(1997) Enhanced expression of inducible cyclooxygenase with age in murine macrophages.
J. Immunol., **159**, 2445–2451.
Hazucha MJ, Madden M, Pape G, Becker S, Devlin R, Koren HS, Kehrl H and Bromberg PA
(1996) Effects of cyclo-oxygenase inhibition on ozone-induced respiratory inflammation and
lung function changes. *Eur. J. Appl. Physiol.*, **73**, 17–27.
Hemler M and Lands WE (1976) Purification of the cyclooxygenase that forms prostaglandins.
Demonstration of two forms of iron in the holoenzyme. *J. Biol. Chem.*, **251**, 5575–5579.
Herschman HR (1996) Prostaglandin synthase 2. *Biochim. Biophys. Acta.*, **1299**, 125–140.
Herschman HR (1999) Function and regulation of prostaglandin synthase 2. *Adv. Exp. Med.
Biol.*, **469**, 3–8.
Hla T and Neilson K (1992) Human cyclooxygenase-2 cDNA. *Proc. Natl. Acad. Sci. USA*, **89**,
7384–7388.
Hoff T, DeWitt D, Kaever V, Resch K and Goppelt-Struebe M (1993) Differentiation-associated
expression of prostaglandin G/H synthase in monocytic cells. *FEBS Lett.*, **320**, 38–42.
Hughes MF, Mason R and Eling TE (1988) Prostaglandin hydroperoxide-dependent oxidation of
phenylbutazone: relationship to inhibition of prostaglandin cyclooxygenase. *Molec. Pharmacol.*, **34**, 186–193.
Hwang D, Jang BC, Yu G and Boudreau M (1997) Expression of mitogen-inducible cyclooxygenase induced by lipopolysaccharide: mediation through both mitogen-activated protein kinase
and NF-kappaB signaling pathways in macrophages. *Biochem. Pharmacol.*, **54**, 87–96.
Inoue H, Nanayama T, Hara S, Yokoyama C and Tanabe T (1994) The cyclic AMP response
element plays an essential role in the expression of the human prostaglandin-endoperoxide
synthase 2 gene in differentiated U937 monocytic cells. *FEBS Lett.*, **350**, 51–54.
Inoue H, Yokoyama C, Hara S, Tone Y and Tanabe T (1995) Transcriptional regulation of human
prostaglandin-endoperoxide synthase-2 gene by lipopolysaccharide and phorbol ester in
vascular endothelial cells. *J. Biol. Chem.*, **270**, 24965–24971.
Jang M, Cai L, Udeani GO, Slowing KV, Thomas CF, Beecher CW, Fong HH, Farnsworth NR,
Kinghorn AD, Mehta RG, Moon RC and Pezzuto JM (1997) Cancer chemopreventive activity
of resveratrol, a natural product derived from grapes. *Science.*, **275**, 218–220.
Johnson JL and Maddipati KR (1998) Paradoxical effects of resveratrol on the two prostaglandin
H synthases. *Prostaglandins Other Lipid Mediat.*, **56**, 131–143.
Johnson JL, Wimsatt J, Buckel SD, Dyer RD and Maddipati KR (1995) Purification and
characterization of prostaglandin H synthase-2 from sheep placental cotyledons. *Arch.
Biochem. Biophys.*, **324**, 26–34.
Jones DA, Carlton DP, McIntyre TM, Zimmerman GA and Prescott SM (1993) Molecular cloning
of human prostaglandin endoperoxide synthase type II and demonstration of expression in
response to cytokines. *J. Biol. Chem.*, **268**, 9049–9054.
Kargman S, Charleson S, Cartwright M, Frank J, Riendeau D, Mancini J, Evans J and O'Neill G
(1996) Characterization of prostaglandin G/H synthase 1 and 2 in rat, dog, monkey and
human gastrointestinal tracts. *Gastroenterology*, **111**, 445–454.
Khan KN, Venturini CM, Bunch RT, Brassard JA, Koki AT, Morris DL, Trump BF, Maziasz TJ and
Alden CL (1998) Interspecies differences in renal localisation of cyclooxygenase isoforms:
implications in nonsteroidal antiinflammatory drug-related nephrotoxicity. *Toxicol Pathol.*, **26**,
612–620.
Kelley DJ, Mestre JR, Subbaramaiah K, Sacks PG, Schantz SP, Tanabe T, Inoue H, Ramonetti JT
and Dannenberg AJ (1997) Benzo[a]pyrene up-regulates cyclooxygenase-2 gene expression
in oral epithelial cells. *Carcinogenesis*, **18**, 795–799.
Kim JJ, Wang J, Bambra C, Das SK, Dey SK and Fazleabas AT (1999) Expression of cyclooxygen-
ase-1 and -2 in the baboon endometrium during menstrual cycle and pregnancy. *Endocrinology*, **140**, 2672–2678.
Kim H, Kim K, Yu B and Chung H (2000) The effect of age on cyclooxygenase-2 gene expres-

sion. NF-kappaB activation and Ikappa/Balpha degradation. *Free. Radic. Biol. Med.*, **28**, 683–692.

Kitzler J, Hill E, Hardman R, Reddy N, Philpot R and Eling TE (1995) Analysis and quantitation of splicing variants of the TPA-inducible PGHS-1 mRNA in rat tracheal epithelial cells. *Arch. Biochem. Biophys.*, **316**, 856–863.

Koburg J (1991) Analyse der Prostaglandin-H-Synthase in Schafsamenblasen mit biochemischen und immunologischen Methoden. Doctoral Thesis, University of Würzburg.

Koga H, Sakisaka S, Ohishi M, Kawaguchi T, Taniguchi E, Sasatomi K, Harada M, Kusaba T *et al.* (1999) Expression of cyclooxygenase-2 in human hepatocellular carcinoma: relevance to tumour differentiation. *Hepatology*, **29**, 688–696.

Komhoff M, Wang JL, Cheng HF, Langenbach R, McKanna JA, Harris RC and Breyer MD (2000) Cyclooxygenase-2-selective inhibitors impair glomerulogenesis and renal cortical development. *Kidney Int.*, **57**, 414–422.

Koshkin V and Dunford HB (1999) Coupling of the peroxidase and cyclooxygenase reactions of prostaglandin H synthase. *Biochim Biophys Acta*, **1430**, 341–348.

Kosaka T, Miyata A, Ihara H, Hara S, Sugimoto T and Takeda O (1994) Characterization of the human gene PTGS2 encoding prostaglandin endoperoxide synthase-2. *Eur. J. Biochem.*, **221**, 889–897.

Kraemer SA, Meade EA and DeWitt DL (1992) Prostaglandin endoperoxide synthase gene structure: identification of the transcriptional start site and 5′-flanking regulatory sequences. *Arch. Biochem. Biophys.*, **293**, 391–400.

Kraemer SA, Arthur KA, Denison MS, Smith WL and DeWitt DL (1996) Regulation of prostaglandin endoperoxide H synthase-2 expression by 2,3,7,8,-tetrachlorodibenzo-p-dioxin. *Arch. Biochem. Biophys.*, **330**, 319–328.

Krauss RS and Eling TE (1985) Formation of unique arylamine: DNA adducts from 2-aminofluorene activated by prostaglandin H synthase. *Cancer Res.*, **45**, 1680–1686.

Krauss RS and Eling TE (1987) Macromolecular binding of the thyroid carcinogen 3-amino-1,2,4-triazole (amitrole) catalysed by prostaglandin H synthase, lactoperoxidase and thyroid peroxide H. *Carcinogenesis*, **8**, 659–664.

Krauss RS, Angerman-Stewart J, Eling TE, Dooley KL and Kadlubar FF (1989) The formation of 2-aminofluorene-DNA adducts *in vivo*: evidence for peroxidase-mediated activation. *J. Biochem. Toxicol.*, **4**, 111–117.

Kubow S (1992) Inhibition of phenytoin bioactivation and teratogenicity by dietary n-3 fatty acids in mice. *Lipids*, **27**, 721–728.

Kubow S and Wells PG (1989) In vitro bioactivation of phenytoin to a reactive intermediate by prostaglandin synthetase, horseradish peroxidase and thyroid peroxidase. *Mol. Pharmacol.*, **35**, 504–511.

Kulmacz RJ (1998) Cellular regulation of prostaglandin H synthase catalysis. *FEBS Letters*, **430**, 154–157.

Kulmacz RJ and Wang (1995) Comparison of hydroperoxide initiator requirements for the cyclooxygenase activities of prostaglandin H synthase-1 and -2. *J. Biol. Chem.*, **270**, 24019–24023.

Kujubu DA and Herschman HR (1992) Dexamethasone inhibits mitogen induction of the TIS10 prostaglandin synthase/cyclooxygenase gene. *J. Biol. Chem.*, **267**, 7991-7994.

Kujubu DA, Fletcher BS, Varnum BC, Lim RW and Herschman HR (1991) TIS10, a phorbol ester tumour promoter-inducible mRNA from Swiss 3T3 cells, encodes a novel prostaglandin synthase/cyclooxygenase homologue. *J. Biol. Chem.*, **266**, 12866–12872.

Kurumbail RG, Stevens AM, Gierse JK, McDonald JJ, Stegeman RA, Part JY, Gildehaus D, Miyashiro JM, Penning TD, Jerbert K, Isakson PC and Stallings WC (1996) Structural basis for selective inhibition of cyclooxygenase-2 by anti-inflammatoty agents. *Nature*, **384**, 644–648.

Lahiri-Chatterjee M, Katlyar SK, Mohan RR and Agarwal R (1999) A flavonoid antioxidant, silymarin, affords exceptionally high protection against tumour promotion in the SENCAR mouse skin tumourigenesis model. *Cancer Res.*, **59**, 622–632.

Lakshmi VM, Zenser TV and Davis BB (1998) N′-(3′-monophospho-deoxyguanosin-8-yl)-N-acetylbenzidine formation by peroxidative metabolism. *Carcinogenesis*, **19**, 911–917.

Lakshmi VM, Hsu FF, Davis BB and Zenser TV (2000) Sulfinamide formation following peroxidatic metabolism of N-acetylbenzidine. *Chem. Res. Toxicol.*, **13**, 96–102.

Laneuville O, Breuer DK, DeWitt DL, Hla T, Funk CD and Smith WL (1994) Differential inhibition of human prostaglandin endoperoxide H synthases-1 and -2 by nonsteroidal anti-inflammatory drugs. *J. Pharmacol. Exp. Ther.*, **271**, 927–934.

Laneuville O, Breuer DK, Xu N, Huang ZH, Gage DA, Watson JT, Lagarde M, DeWitt DL and Smith WL (1995) Fatty acid substrate specificities of human prostaglandin endoperoxide H synthases-1 and -2. Formation of 12-hydroxy-(9Z, 13E/Z, 15Z)-octadecatrienoic acids from alpha-linolenic acid. *J. Biol. Chem.*, **270**, 19330–19336.

Langenbach R, Morham SG, Tiano HF, Loftin CD, Ghanayem BI, Chulada PC, Mahler JF, Lee CA, Goulding EH, Kluckman KD, Kim HS and Smithies O (1995) Prostaglandin synthase 1 gene disruption in mice reduces arachidonic acid-induced inflammation and indomethacin-induced gastric ulceration. *Cell*, **83**, 483–492.

Langenbach R, Loftin CD, Lee C and Tiano H (1999a) Cyclooxygenase-deficient mice. A summary of their characteristics and susceptibilities to inflammation and carcinogenesis. *Ann. N.Y. Acad. Sci.*, **889**, 52–61.

Langenbach R, Loftin C, Lee C and Tiano H (1999b) Cyclooxygenase knockout mice: models for elucidating isoform-specific functions. *Biochem. Pharmacol.*, **58**, 1237–1246.

Lasker JM, Sivarajah K, Mason R, Kalyanaraman B, Abou-Donia MH and Eling TE (1981) A free radical mechanism of prostaglandin synthase-dependent aminopyrine demethylation. *J. Biol Chem.*, **256**, 7764–7767.

Lassus P, Wolff H and Andersson S (2000) Cyclooxygenase-2 in human perinatal lung. *Pediatr. Res.*, **47**, 602–605.

Lecomte M, Laneuville O, Ji C, DeWitt DL and Smith WL (1994) Acetylation of human prostaglandin H synthases-2 (cyclooxygenase-2) by aspirin. *J. Biol. Chem.*, **269**, 13207–13215.

Ledwith BJ, Pauley CJ, Wagner LK, Rokos CL, Alberts DW and Manam S (1997) Induction of cyclooxygenase-2 expression by peroxisome proliferators and non-tetradecanoylphorbol 12,13-myristate-type tumour promoters in immortalised mouse liver cells. *J. Biol. Chem.*, **272**, 3707–3714.

Leong J, Hughes-Fulford M, Rakhlin N, Habib A, Maclouf J and Goldyne ME (1996) Cyclooxygenases in human and mouse skin and cultured human keratinocytes: association of COX-2 expression with human keratinocyte differentiation. *Exp. Cell. Res.*, **224**, 79–87.

Lim RW, Varnum BC and Herschman HR (1987) Cloning of tetradecanoyl phorbol ester-induced 'primary response' sequences and their expression in density-arrested Swiss 3T3 cells and a TPA non-proliferative variant. *Oncogene*, **1**, 263–270.

Liu Y and Levy GN (1998) Activation of heterocyclic amines by combinations of prostaglandin H synthase-1 and -2 with N-acetyltransferase 1 and 2. *Cancer Letters*, **133**, 115–123.

Liu XH and Rose DP (1996) Differential expression and regulation of cyclooxygenase-1 and -2 in two human breast cancer cell lines. *Cancer Res.*, **56**, 5125–5127.

Liu Y, Levy GN and Weber WW (1995) Activation of 2-aminofluorene by prostaglandin endoperoxide H synthase-2. *Biochem. Biophys. Res. Commun.*, **215**, 346–354.

Liu Y, Levy GN and Weber WW (1997) Induction of human prostaglandin endoperoxide H synthase-2 (PHS-2) mRNA by TCDD. *Prostaglandins*, **53**, 1–10.

Liu J, Antaya M, Boerboom D, Lussier JG, Silversides DW and Sirois J (1999) The delayed activation of the prostaglandin G/H synthase-2 promoter in bovine granulosa cells is associated with down-regulation of truncated upstream stimulatory factor-2. *J. Biol. Chem.*, **274**, 35037–35045.

Lu GQ, Tsai AL, Van Wart HE and Kulmacz RJ (1999) Comparison of the peroxidase reaction kinetics of prostaglandin H synthase-1 and -2. *J. Biol. Chem.*, **274**, 16162–16167.

Luong C, Miller A, Barnett J, Chow J, Ramesha C and Browner MF (1996) Flexibility of the NSAID binding site in the structure of human cyclooxygenase-2. *Nat. Struct. Biol.*, **3**, 927–933.

Markey CM, Alward A, Weller PE and Marnett LJ (1987) Quantitative studies of hydroperoxide reduction by prostaglandin H synthase. *J. Biol. Chem.*, **262**, 6266–6279.

Marnett LJ (1990) Prostaglandin synthase-mediated metabolism of carcinogens and a potential role for peroxyl radicals as reactive intermediates. *Environm. Health Perspect.*, **88**, 5–12.

Marnett LJ and Eling TE (1983) Cooxidation during prostaglandin biosynthesis: a pathway for the metabolic oxidation of xenobiotics. *Rev. Biochem. Toxicol.*, **5**, 135–172.

Marnett LJ and Maddipatti KR (1991) Prostaglandin H Synthase. In *Peroxidases, Chemistry and Biology*, Everse J, Everse K, Grisham M, (eds), CRC Press, Boca Raton, FL, pp. 293–334.

Marnett LJ, Wlodawer P and Samuelsson B (1975) Cooxygenation of organic substrates by the prostaglandin synthetase of sheep vesicular glands. *J. Biol. Chem.*, **250**, 8510–8517.

Marnett LJ, Bienkowski MJ and Pagels WR (1979a) Oxygen 18 investigation of the prostaglandin synthetase-dependent co-oxidation of diphenylisobenzofuran. *J. Biol. Chem.*, **254**, 5077–5082.

Marnett LJ, Johnson JT and Bienkowski MJ (1979b) Arachidonic acid-dependent metabolism of 7,8-dihydroxy-7,8-dihydrobenzo[a]pyrene by ram seminal vesicles. *FEBS Letters*, **106**, 13–16.

Marnett LJ, Siedlik PH and Fung LW (1982) Oxidation of phenidone and BW755C by prostaglandin endoperoxide synthetase. *J. Biol. Chem.*, **257**, 6957–6964.

Marshall PJ, Kulmacz RJ and Lands WEM (1987) Constraints on prostaglandin biosynthesis in tissues. *J. Biol. Chem.*, **262**, 3510–3517.

Martinez J and Moreno JJ (2000) Effect of resveratrol, a natural polyphenolic compound, on reactive oxygen species and prostaglandin production. *Biochem. Pharmacol.*, **59**, 865–870.

Masferrer JL, Seibert K, Zweifel B and Needleman P (1992) Endogenous glucocorticoids regulate an inducible cyclooxygenase enzyme. *Proc. Natl. Acad. Sci. USA*, **89**, 3917–3921.

Masferrer JL, Reddy ST, Zweifel BS, Seibert K, Needleman P, Gilbert RS and Herschman HR (1994) In vivo glucocorticoids regulate cyclooxygenase-2 but not cyclooxygenase-1 in peritoneal macrophages. *J. Pharmacol. Exp. Ther.*, **270**, 1340–1344.

Meade EA, Smith WL and DeWitt DL (1993a) Expression of the murine prostaglandin (PGH) synthase-1 and PGH synthase-2 isozymes in cos-1- cells. *J. Lipid Mediators*, **6**, 119–129.

Meade EA, Smith WL and DeWitt DL (1993b) Differential inhibition of prostaglandin endoperoxide synthase (cyclooxygenase) isozymes by aspirin and other non-steroidal antiinflammatory drugs. *J. Biol. Chem.*, **268**, 6610–6614.

Meade EA, McIntyre TM, Zimmerman GA and Prescott SM (1999) Peroxisome proliferators enhance cyclooxygenase-2 expression in epithelial cells. *J. Biol. Chem.*, **274**, 8328–8334.

Merlie JP, Fagan D, Mudd J and Needleman P (1988) Isolation and characterization of the complementary DNA for sheep seminal vesicle prostaglandin endoperoxide synthase (cyclooxygenase). *J. Biol. Chem.*, **263**, 3550–3553.

Mertz PM, DeWitt DL, Stetler-Stevenson WG and Wahl LM (1994) Interleukin 10 suppression of monocyte prostaglandin H synthase-2. Mechanism of inhibition of prostaglandin-dependent matrix metalloproteinase production. *J. Biol. Chem.*, **269**, 21322–21329.

Mertz PM, Corcoran ML, McCluskey KM, Zhang Y, Wong HL, Lotze MT, DeWitt DL, Wahl SM and Wahl LM (1996) Suppression of prostaglandin H synthase-2 induction in human monocytes by in vitro or in vivo administration of interleukin 4. *Cell. Immunol.*, **173**, 252–260.

Mikkola T, Turunen P, Avala K, Orpana A, Viinika L and Ylikorkola O (1995) 17β-Estradiol stimulates prostacyclin, but not endothelin-1, production in human vascular endothelial cells. *J. Clin. Endocrinol. Metab.*, **80**, 1832–1836.

Mion F, Jasuja R and Johnston DE (1995) The contribution of hepatocytes to prostaglandin synthesis in rat liver. *Prostaglandins Leukot. Essent. Fatty Acids*, **53**, 109–115.

Mitchell MD, Romero RJ, Edwin SS and Trautman MS (1995) Prostaglandins and parturition. *Reprod. Fertil. Dev.*, **7**, 623–632.

Miyagi M, Morishita M and Iwamoto Y (1993) Effects of sex hormones on production of prostaglandin E2 by human peripheral monocytes. *J. Periodontol.*, **64**, 1075–1078.

Miyamoto T, Ogino N, Yamamoto S and Hayaishi O (1976) Purification of prostaglandin endoperoxide synthetase from bovine vesicular gland microsomes. *J. Biol. Chem.*, **251**, 2629–2636.

Mohammed SI, Knapp DW, Bostwick DG, Foster RS, Khan KN, Masferrer JL, Woerner BM, Snyder PW and Koki AT (1999) Expression of cyclooxygenase-2 (COX-2) in human invasive transitional cell carcinoma (TCC) of the urinary bladder. *Cancer Res.*, **59**, 5647–5650.

Moldeus P, Larsson R and Ross D (1985) Involvement of prostaglandin synthase in the metabolic activation of acetaminophen and phenacetin. In *Arachidonic Acid Metabolism and Tumor Initiation*, Marnett LJ (ed) Martinus Nijhoff, Boston, pp. 171–198.

Morham SG, Langenbach R, Loftin CD, Tiano HF, Vouloumanos N, Jennette JC, Mahler JF, Kluckman KD, Ledford A, Lee CA and Smithies O (1995) Prostaglandin synthase 2 gene disruption causes severe renal pathology in the mouse. *Cell*, **83**, 473–482.

Morris JK and Richards JS (1996) An E-box region within the prostaglandin endoperoxide synthase-2 (PGS-2) promoter is required for transcription in rat ovarian granulosa cells. *J. Biol. Chem.*, **271**, 16633–16643.

Morrison LD, Eling TE and Josephy PD (1993) Prostaglandin H synthase-dependent formation of the direct-acting mutagen 2-nitro-3-methylimidazo[4,5-*f*]-quinoline (nitro-IQ) from IQ. *Mutation Research*, **302**, 45–52.

Müller-Decker K, Scholz K, Marks F and Fürstenberger G (1995) Differential expression of prostaglandin-H synthase isozymes during multistage carcinogenesis in mouse epidermis. *Mol. Carcinog.*, **12**, 31–41.

Müller-Decker K, Kopp-Schneider A, Marks F, Seibert K and Fürstenberger G (1998a) Localization of prostaglandin H synthase isoenzymes in murine epidermal tumours: suppression of skin tumour promotion by inhibition of prostaglandin H synthase-2. *Mol. Carcinog.*, **23**, 36–44.

Müller-Decker K, Scholz K, Neufang G, Marks F and Fürstenberger G (1998b) Localization of prostaglandin-H synthase-1 and -2 in mouse skin: implications for cutaneous function. *Exp. Cell Res.*, **242**, 84–91.

Müller-Decker K, Reinerth G, Krieg P, Zimmermann R, Heise H, Bayerl C, Marks F and Fürstenberger G (1999) Prostaglandin-H-synthase isozyme expression in normal and neoplastic human skin. *Int. J. Cancer*, **82**, 648–656.

Murakami M, Matsumoto R, Urade Y, Austen KF and Arm JP (1995) c-kit ligand mediates increased expression of cytosolic phospholipase A2, prostaglandin endoperoxide synthase-1, and hematopoietic prostaglandin D2 synthase and increased IgE-dependent prostaglandin D2 generation in immature mouse mast cells. *J. Biol. Chem.*, **270**, 3239–3246.

Nebert DW, Roe AL, Dieter MZ, Solis WA, Yang Y and Dalton TP (2000) Role of the aromatic hydrocarbon receptor and [Ah] gene battery in the oxidative stress response, cell cycle control, and apoptosis. *Biochem. Pharmacol.*, **59**, 65–85.

Newton R, Kuitert LM, Bergmann M, Adcock IM and Barnes PJ (1997) Evidence for involvement of NF-kappaB in the transcriptional control of COX-2 gene expression by IL-1beta. *Biochem. Biophys. Res. Commun.*, **237**, 28–32.

Newton R, Seybold J, Kuitert LM, Bergmann M and Barnes PJ (1998) Repression of cyclooxygenase-2 and prostaglandin E2 release by dexamethasone occurs by transcriptional and post-transcriptional mechanisms involving loss of polyadenylated mRNA. *J. Biol. Chem.*, **273**, 32312–32321.

O'Neill GP and Ford-Hutchinson AW (1993) Expression of mRNA for cyclooxygenase-1 and cyclooxygenase-2 in human tissues. *FEBS Letter*, **330**, 156–160.

O'Neill GP, Mancini JA, Kargman S, Yergey J, Kwan MY, Falguyeret JP, Oullet M, Cromlish W, Culp S, Evans JF, Fordhutchinson AW and Vickers PJ (1994) Overexpression of human prostaglandin G/H synthase-1 and -2 by recombinant vaccinia virus: inhibition by nonsteroidal anti-inflammatory drugs and biosynthesis of 15-hydroxyeicosatetraenoic acid. *Mol. Pharmacol.*, **45**, 245–254.

Okada Y, Lorenzo JA, Freeman AM, Tomita M, Morham SG, Raisz LG and Pilbeam CC (2000) Prostaglandin G/H synthase-2 is required for maximal formation of osteoclast-like cells in culture. *J. Clin. Invest.*, **105**, 823–832.

Olnes MJ, Verma M and Kurl RN (1996) 2,3,7,8-tetrachlorodibenzo-p-dioxin modulates expression of the prostaglandin G/H synthase-2 gene in rat thymocytes. *J. Pharmacol. Exp. Ther.*, **279**, 1566–1573.

Onoe Y, Miyaura C, Kaminakayashiki T, Nagai Y, Noguchi K, Chen QR, Seo H, Ohta H, Nozawa S, Kudo I and Suda T (1996) IL-13 and IL-4 inhibit bone resorption by suppressing cyclooxygenase-2-dependent prostaglandin synthesis in osteoblasts. *J. Immunol.*, **156**, 758–764.

Oshima T, Yoshimoto T, Yamamoto S, Kumegawa M, Yokoyama C and Tanabe T (1991) cAMP-dependent induction of fatty acid cyclooxygenase mRNA in mouse osteoblastic cells (MC3T3-E1). *J. Biol. Chem.*, **266**, 13621–13626.

Oshima M, Dinchuk JE, Kargman SL, Oshima H, Hancock B, Kwong E, Trzaskos JM, Evans JF and Taketo MM (1996) Suppression of intestinal polyposis in Apc delta716 knockout mice by inhibition of cyclooxygenase 2 (COX-2). *Cell*, **87**, 803–809.

Parman T, Chen G and Wells PG (1998) Free radical intermediates of phenytoin and related teratogens. Prostaglandin H synthase-catalysed bioactivation, electron paramagnetic resonance spectrometry, and photochemical product analysis. *J Biol. Chem.*, **273**, 25079–25088.

Parman T, Wiley MJ and Wells PG (1999) Free radical-mediated oxidative DNA damage in the mechanism of thalidomide teratogenicity. *Nat. Med.*, **5**, 582–585.

Petry TW, Josephy PD, Pagano DA, Zeiger E, Knecht KT and Eling TE (1989) Prostaglandin hydroperoxidase-dependent activation of heterocyclic aromatic amines. *Carcinogenesis*, **10**, 2201–2207.

Picot D, Loll PJ and Garavito RM (1994) The X-ray crystal structure of the membrane protein prostaglandin H_2 synthase-1. *Nature*, **367**, 243–249.

Pilbeam C, Rao Y, Voznesensky O, Kawaguchi H, Alander C, Raisz L and Herschman H (1997) Transforming growth factor-beta1 regulation of prostaglandin G/H synthase-2 expression in osteoblastic MC3T3-E1 cells. *Endocrinology*, **138**, 4672–4682.

Potter DW and Hinson JA (1987a) The 1- and 2-electron oxidation of acetaminophen catalysed by prostaglandin H synthase. *J. Biol. Chem.*, **262**, 974–980.

Potter DW and Hinson JA (1987b) Mechanism of acetaminophen oxidation to N-acetyl-p-benzoquinone imine by horseradish peroxidase and cytochrome P-450. *J. Biol. Chem.*, **262**, 966–973.

Prescott SM and Fitzpatrick FA (2000) Cyclooxygenase-2 and carcinogenesis. *Biochim. Biophys. Acta*, **1470**, M69–M78.

Pruess-Schwartz D, Nimesheim A and Marnett LJ (1989) Peroxyl radical- and cytochrome P-450-dependent metabolic activation of (+)-7,8-dihydroxy-7,8-dihydrobenzo(a)pyrene in mouse skin in vitro and in vivo. *Cancer Res.*, **49**, 1732–1737.

Puga A, Hoffer A, Zhou S, Bohm JM, Leikauf GD and Shertzer HG (1997) Sustained increase in intracellular free calcium and activation of cyclooxygenase-2 expression in mouse hepatoma cells treated with dioxin. *Biochem. Pharmacol.*, **54**, 1287–1296.

Reddy ST, Gilbert RS, Xie W, Luner S and Herschman HR (1994) TGF-beta 1 inhibits both endotoxin-induced prostaglandin synthesis and expression of the TIS10/prostaglandin synthase 2 gene in murine macrophages. *J. Leukoc. Biol.*, **55**, 192–200.

Reddy ST, Wadleigh DJ and Herschman HR (2000) Transcriptional regulation of the cyclooxygenase-2 gene in activated mast cells. *J. Biol. Chem.*, **275**, 3107–3113.

Reed GA, Brooks EA and Eling TE (1984) Phenylbutazone-dependent epoxidation of 7,8-dihydroxy-7,8-dihydrobenzo[a]pyrene. A new mechanism for prostaglandin H synthase-catalysed oxidations. *J. Biol. Chem.*, **259**, 5591–5595.

Rimarachin JA, Jacobson JA, Szabo P, Maclouf J, Creminon C and Weksler BB (1994) Regulation of cyclooxygenase-2 expression in aortic smooth muscle cells. *Arterioscler. Thromb.*, **14**, 1021–1031.

Ristimaki A, Garfinkel S, Wessendorf J, Maciag T and Hla T (1994) Induction of cyclooxygenase-2 by interleukin-1 alpha. Evidence for post-transcriptional regulation. *J. Biol. Chem.*, **269**, 11769–11775.

Rothman N, Bhatnagar VK, Hayes RB, Zenser TV, Kashyap SK, Butler MA, Bell DA, Lakshmi V, Jaeger M, Kashyap R, Butler MA, Bell DA, Lakshmi V, Jaeger M, Kashyap R, Hirronen A, Schulte PA, Doremeci M, Hsu F, Parich DJ, Davis BB and Talaska G (1996) The impact of interindividual variation in NAT2 activity on benzidine urinary metabolites and urothelial DNA adducts in exposed workers. *Proc. Natl. Acad. Sci. USA*, **93**, 5084–5089.

Rzymkiewicz D, Leingang K, Baird N and Morrison AR (1994) Regulation of prostaglandin endoperoxide synthase gene expression in rat mesangial cells by interleukin-1 beta. *Am. J. Physiol.*, **266**, 39–45.

Samet JM, Fasano MB, Fonteh AN and Chilton FH (1995) Selective induction of prostaglandin G/H synthase I by stem cell factor and dexamethasone in mast cells. *J. Biol. Chem.*, **270**, 8044–8049.

Samet JM, Reed W, Ghio AJ, Devlin RB, Carter JD, Dailey LA, Bromberg PA and Madden MC (1996) Induction of prostaglandin H synthase 2 in human airway epithelial cells exposed to residual oil fly ash. *Toxicol. Appl. Pharmacol.*, **141**, 159–168.

Samet JM, Ghio AJ, Costa DL and Madden MC (1999) Increased expression of cyclooxygenase 2 mediates oil fly ash-induced lung injury. *Exp. Lung Res.*, **26**, 57–69.

Samokyszyn VM and Marnett LJ (1987) Hydroperoxide-dependent cooxidation of 13-*cis*-retinoic acid by prostaglandin H synthase. *J. Biol. Chem.*, **262**, 14119–14133.

Sawdy R, Slater D, Fisk N, Edmonds DK and Bennett P (1997) Use of a cyclo-oxygenase type-2-selective non-steroidal anti-inflammatory agent to prevent preterm delivery. *Lancet*, **350**, 265–266.

Scholz K, Fürstenberger G, Müller-Decker K and Marks F (1995) Differential expression of prostaglandin-H synthase isoenzymes in normal and activated keratinocytes in vivo and in vitro. *Biochem J.*, **309**, 263–269.

Schnitzler R, Foth J, Degen GH and Metzler M (1994) Induction of micronucleiy stilbene-type and steroidal estrogens in Syrian hamster embryo and ovine seminal vesicle cells in vitro. *Mutation Research*, **311**, 85–93.

Seibert K, Zhang Y, Leahy K, Hauser S, Masferrer J, Perkins W, Lee L and Isakson P (1994) Pharmacological and biochemical demonstration of the role of cyclooxygenase 2 in inflammation and pain. *Proc. Natl. Acad. Sci. USA.*, **91**, 12013–12017.

Siedlik PH and Marnett LJ (1984) Oxidizing radical generation by prostaglandin H synthase. *Methods in Enzymol.*, **105**, 412–416.

Silvia WJ, Brockman JA, Kaminski MA, DeWitt DL and Smith WL (1994). Prostaglandin endoperoxide synthase in seminal vesicles. *Mol. Androl.*, **6**, 197–207.

Sirois J and Richards JS (1992) Purification and characterization of a novel, distinct isoform of prostaglandin endoperoxide synthase induced by human chorionic gonadotropin in granulosa cells of rat preovulatory follicles. *J. Biol. Chem.*, **267**, 6382–6388.

Sirois J and Richards JS (1993) Transcriptional regulation of the rat prostaglandin endoperoxide synthase 2 gene in granulosa cells. *J. Biol. Chem.*, **268**, 21931–21938.

Sirois J, Levy LO, Simmons DL and Richards JS (1993) Characterization and hormonal regulation of the promoter of the rat prostaglandin endoperoxide synthase 2 gene in granulosa cells. *J. Biol. Chem.*, **268**, 12199–12206.

Sivarajah K, Mukhtar H and Eling TE (1979) Arachidonic acid-dependent metabolism of (±)-trans-7,8-dihydroxy-7,8-dihydrobenzo(a)pyrene (BP-7,8-diol) to 7,10/8,9-diols. *FEBS Letters*, **106**, 17–20.

Sivarajah K, Lasker JM, Eling TE and Abou-Donia MH (1982) Metabolism of N-alkyl compounds during the biosynthesis of prostaglandins. *Molec. Pharmacol.*, **21**, 133–141.

Smith WL (1985) Cellular and subcellular compartmentation of prostaglandin and thromboxane synthesis. In: *Biochemistry of Arachidonic Acid Metabolism* Lands WEM (ed). Martinus Nijhoff Publ., Boston, USA, pp. 77–95.

Smith WL and DeWitt DL (1995) Biochemistry of prostaglandin endoperoxide H synthase-1 and synthase-2 and their differential susceptibility to nonsteroidal anti-inflammatory drugs. *Semin. Nephrol.*, **15**, 179–194.

Smith WL and DeWitt DL (1996) Prostaglandin endoperoxide H synthases -1 and -2. *Adv Immunol.*, **62**, 167–215.

Smith BJ, Curtis JF and Eling TE (1991) Bioactivation of xenobiotics by prostaglandin H synthase. *Chem. Biol. Interactions*, **79**, 245–264.

Smith CJ, Morrow JD, Roberts LJ 2d and Marnett LJ (1993) Differentiation of monocytoid THP-1 cells with phorbol ester induces expression of prostaglandin endoperoxide synthase-1 (COX-1). *Biochem. Biophys. Res. Commun.*, **192**, 787–793.

Smith WL, Garavito RM and DeWitt DL (1996) Prostaglandin endoperoxide H synthases (cyclooxygenases)-1 and -2. *J. Biol. Chem.*, **271**, 33157–33160.

Spencer AG, Woods JW, Arakawa T, Singer II and Smith WL (1998) Subcellular localisation of prostaglandin endoperoxide H synthases-1 and -2 by immunoelectron microscopy. *J. Biol. Chem.*, **273**, 9886–9893.

Spirio LN, Dixon DA, Robertson J, Robertson M, Barrows J, Traer E, Burt RW, Leppert MF, White R and Prescott SM (1998) The inducible prostaglandin biosynthetic enzyme, cyclooxygenase 2, is not mutated in patients with attenuated adenomatous polyposis coli. *Cancer Res.*, **58**, 4909–4912.

Stanczyk FZ, Rosen GF, Ditkoff EC, Vijod AG, Bernstein L and Lobo RA (1995) Influence of estrogen on prostacyclin and thromboxane balance in postmenopausal women. *J. N. Am. Menopause Soc.*, **2**, 137–143.

Stewart KG, Zhang Y and Davidge ST (1999) Estrogen decreases prostaglandin H synthase products from endothelial cells. *J. Soc. Gynecol. Investig.*, **6**, 322–327.

Subbaramaiah K, Michaluart P, Chung WJ, Tanabe T, Telang N and Dannenberg AJ (1999) Resveratrol inhibits cyclooxygenase-2 transcription in human mammary epithelial cells. *Ann. N.Y. Acad. Sci.*, **889**, 214–223.

Suzuki-Yamamoto T, Yokoi H, Tsuruo Y, Watanabe K and Ishimura K (1999) Identification of prostaglandin F-producing cells in the liver. *Histochem. Cell Biol.*, **112**, 451–456.

Swinney DC, Mak AY, Barnett J and Ramesha CS (1997) Differential allosteric regulation of Prostaglandin H synthase 1 and 2 by arachidonic acid. *J. Biol. Chem.*, **272**, 12393–123948.

Taketo MM (1998a) Cyclooxygenase-2 inhibitors in tumourigenesis (part I). *J. Natl. Cancer Inst.*, **90**, 1529–1536.

Taketo MM (1998b) Cyclooxygenase-2 inhibitors in tumourigenesis (part II). *J. Natl. Cancer Inst.*, **90**, 1609–1620.

Tazawa R, Xu XM, Wu KK and Wang LH (1994) Characterization of the genomic structure, chromosomal location and promoter of human prostaglandin H synthase-2 gene. *Biochem. Biophys. Res. Commun.*, **203**, 190–199.

Tetradis S, Pilbeam CC, Liu Y, Herschman HR and Kream BE (1997) Parathyroid hormone increases prostaglandin G/H synthase-2 transcription by a cyclic adenosine 3′,5′-monophosphate-mediated pathway in murine osteoblastic MC3T3-E1 cells. *Endocrinology*, **138**, 3594–3600.

Thornton-Manning JR, Nichols WK, Manning BW, Skiles GL and Yost GS (1993) Metabolism and bioactivation of 3-methylindole by Clara cells, alveolar macrophages, and subcellular fractions from rabbit lungs. *Toxicol. Appl. Pharmacol.*, **122**, 182–190.

Thun MJ (1996) NSAID use and decreased risk of gastrointestinal cancers. *Gastroenterol. Clin. North Am.*, **25**, 333–348.

Vane JR, Bakhle YS and Botting RM (1998) Cyclooxygenases 1 and 2. *Annu. Rev. Pharmacol Toxicol.*, **38**, 97–120.

Vio CP, Balestrini C, Recabarren M and Cespedes C (1999) Postnatal dvelopment of cyclooxygenase-2 in the rat kidney. *Immunopharmacology*, **44**, 205–210

Vogel C (2000) Prostaglandin H synthases and their importance in chemical toxicity. *Current Drug Metabolism*, **1**, 391–404.

Vogel C, Schuhmacher US, Degen GH, Goebel C and Abel J (1997) Differential effects of 2,3,7,8-tetrachlorodibenzo-p-dioxin on the expression of prostaglandin-H synthase isoenzymes in mouse tissues. *Adv. Exp. Med. Biol.*, **433**, 139–143.

Vogel C, Schuhmacher US, Degen GH, Bolt HM, Pineau T and Abel J (1998) Modulation of prostaglandin H synthase-2 mRNA expression by 2,3,7,8-tetrachlorodibenzo-p-dioxin in mice. *Arch. Biochem. Biophys.*, **351**, 265–271.

Vogel C, Boerboom AMFJ, Baechle C, El-Bahay C, Kahl R, Degen GH and Abel J (2000) Regulation of prostaglandin endoperoxide H synthase-2 induction by dioxin in rat hepatocytes: clue for a c-Src mediated pathway. *Carcinogenesis*, **21**, 2267–2274.

Wadleigh DJ and Herschman HR (1999) Transcriptional regulation of the cyclooxygenase-2 gene by diverse ligands in murine osteoblasts. *Biochem. Biophys. Res. Commun.*, **264**, 865–870.

Walenga RW, Kester M, Coroneos E, Butcher S, Dwivedi R and Statt C (1996) Constitutive expression of prostaglandin endoperoxide G/H synthetase (PGHS)-2 but not PGHS-1 in human tracheal epithelial cells in vitro. *Prostaglandins*, **52**, 341–359.

Wells PG, Zubovits JT, Wong ST, Molinari LM and Ali S (1989a) Modulation of phenytoin teratogenicity and embryonic covalent binding by acetylsalicylic acid, caffeic acid and α-phenyl-N-t-butylnitrone: Implications for bioactivation by prostaglandin synthetase. *Toxicol. Appl. Pharmacol.*, **97**, 192–202.

Wells PG, Nagai MK and Greco GS (1989b) Inhibition of trimethadione and dimethadione teratogenicity by the cyclooxygenase inhibitor acetylsalicylic acid: a unifying hypothesis for the teratogenic effects of hydantoin anticonvulsants and structurally related compounds. *Toxicol. Appl. Pharmacol.*, **97**, 406–414.

Wild D and Degen GH (1987) Prostaglandin H synthase-dependent mutagenic activation of heterocyclic aromatic amines of the IQ-type. *Carcinogenesis*, **8**, 541–545.

Williams C, Shattuck-Brandt RL and DuBois RN (1999) The role of COX-2 in intestinal cancer. *Ann N Y Acad. Sci.*, **889**, 72–83.

Winn LM and Wells PG (1997) Evidence for embryonic prostaglandin H synthase-catalysed bioactivation and reactive oxygen species-mediated oxidation of cellular macromolecules in phenytoin and benzo[a]pyrene teratogenesis. *Free Radic. Biol. Med.*, **22**, 607–621.

Wölfle D, Marotzki S, Dartsch D, Schaefer W and Marquardt H (2000) Induction of cyclooxygenase expression and enhancement of malignant cell transformation by 2,3,7,8-tetrachlorodibenzo-p-dioxin. *Carcinogenesis*, **21**, 15–21.

Wolfe MM, Lichtenstein DR and Singh G (1999) Gastrointestinal toxicity of nonsteroidal antiinflammatory drugs. *N. Engl. J. Med.*, **340**, 1889–1899.

Wolz E, Wild D and Degen GH (1995) Prostaglandin-H synthase mediated metabolism and mutagenic activation of 2-amino-3-methylimidazo[4,5- f]quinoline (IQ). *Arch. Toxicol.*, **69**, 171–179.

Wolz E, Pfau W and Degen GH (2000) Bioactivation of the food mutagen 2-amino-3-methylimidazo-[4,5-f]quinoline (IQ) by prostaglandin-H synthase and by monooxygenases: DNA adduct analysis. *Food Chem. Toxicol.*, **38**, 513–522.

Wu G, Wei C, Kulmacz RJ, Osawa Y and Tsai A (1999) A mechanistic study of self-inactivation of the peroxidase activity in prostaglandin H synthase. *J. Biol. Chem.*, **274**, 9231–9237.

Xie W and Herschman H R (1995) v-Src induces prostaglandin synthase 2 gene expression by activation of the c-Jun N-terminal kinase and the c-Jun transcription factor. *J. Biol. Chem.*, **270**, 27622–27628.

Xie W and Herschman H R (1996) Transcriptional regulation of prostaglandin synthase 2 gene expression by platelet-derived growth factor and serum. *J. Biol. Chem.*, **271**, 31742–31748.

Xie WL, Chipman JG, Robertson DL, Erikson RL and Simmons DL (1991) Expression of a mitogen-responsive gene encoding prostaglandin synthase is regulated by mRNA splicing. *Proc. Natl. Acad. Sci. USA.*, **88**, 2692–2696.

Xu XM, Tang JL, Chen X, Wang LH and Wu KK (1997) Involvement of two Sp1 elements in basal endothelial prostaglandin H synthase-1 promoter activity. *J. Biol. Chem.*, **272**, 6943–6950.

Yamagata K, Andreasson KI, Kaufmann WE, Barnes CA and Worley PF (1993) Expression of a mitogen-inducible cyclooxygenase in brain neurons: regulation by synaptic activity and glucocorticoids. *Neuron.*, **11**, 371–386.

Yamamoto K, Arakawa T, Ueda N and Yamamoto S (1995) Transcriptional roles of nuclear factor kappa B and nuclear factor-interleukin-6 in the tumour necrosis factor alpha-dependent induction of cyclooxygenase-2 in MC3T3-E1 cells. *J. Biol. Chem.*, **270**, 31315–31320.

Yokoyama C and Tanabe T (1989) Cloning of human gene encoding prostaglandin endoperoxide synthase and primary structure of the enzyme. *Biochem. Biophys. Res. Commun.*, **165**, 888–894.

Yokoyama C, Takai T and Tanabe T (1988) Primary structure of sheep prostaglandin endoperoxide synthase deduced from cDNA sequence. *FEBS Letters*, **231**, 347–351.

You KM, Jong HG and Kim HP (1999) Inhibition of cyclooxygenase/lipoxygenase from human platelets by polyhydroxylated/methoxylated flavonoids isolated from medicinal plants. *Arch. Pharm. Res.*, **22**, 18–24.

Yuan CJ, Mandal AK, Zhang Z and Mukherjee AB (2000) Transcriptional regulation of cyclooxygenase-2 gene expression: novel effects of nonsteroidal anti-inflammatory drugs. *Cancer Res.*, **60**, 1084–1091.

Zakar T, Hirst JJ, Milovic JE and Olson DM (1995) Glucocorticoids stimulate the expression of prostaglandin endoperoxide H synthase-2 in amnion cells. *Endocrinology*, **136**, 1610–1619.

Zenser TV and Davis BB (1990) Oxidation of xenobiotics by prostaglandin H synthase. In *Toxic Interactions* (Goldstein RS, Hewitt WR, Hook JB, (eds)), Academic Press, San Diego, pp. 61–85.

Zenser TV, Mattammal MB, Wise RW, Rice JR and Davis BB (1983a) Prostaglandin H synthase-catalysed activation of benzidine: a model to assess pharmacologic intervention of chemical carcinogenesis. *J. Pharmacol. Exp. Ther.*, **227**, 545–550.

Zenser TV, Cohen SM, Mattammal MB, Wise RW, Rapp NS and Davis BB (1983b) Prostaglandin hydroperoxidase-catalysed activation of certain N-substituted aryl renal and bladder carcinogens. *Environm. Health Perspect.*, **49**, 33–41.

Zenser TV, Lakshmi VM, Hsu FF and Davis BB (1999) Peroxygenase metabolism of N-acetyl-benzidine by prostaglandin H synthase. *J. Biol. Chem.*, **274**, 14850–14856.

Zhang F, Warskulat U, Wettstein M, Schreiber R, Henninger HP, Decker K and Häussinger D (1995) Hyperosmolarity stimulates prostaglandin synthesis and cyclooxygenase-2 expression in activated rat liver macrophages. *Biochem. J.*, **312**, 135–143.

Zhao J, Sharma Y and Agarwal R (1999) Significant inhibition by the flavonoid antioxidant silymarin against 12-O-tetradecanoylphorbol 13-acetate-caused modulation of antioxidant and inflammatory enzymes, and cyclooxygenase 2 and interleukin-1alpha expression in SENCAR mouse epidermis: implications in the prevention of stage I tumour promotion. *Mol. Carcinog.*, **26**, 321–333.

7 Lipoxygenases

Arun P. Kulkarni

Florida Toxicology Research Center, USA

Introduction

Biotransformation of xenobiotics represents the key element in the understanding of the pharmacological efficacy of drugs and toxicity of pesticides, industrial chemicals, environmental contaminants/pollutants and other chemicals. Most, if not all, organic chemicals undergo oxidation in the body. Depending upon the chemical, oxidation may result in an increase or decrease in the biological response of the parent compound. The contribution of enzymes such as microsomal cytochrome P450 (P450), flavin-containing monooxygenase (MFMO), prostaglandin H synthase (PGS) and few others in the oxidation of xenobiotics is well documented. About twelve years ago, the potential importance of the lipoxygenase (LO) pathway in xenobiotic oxidation was recognised (Kulkarni and Cook 1988a,b). Several publications from different laboratories have now established LO as one of the major enzymes of xenobiotic oxidation. This chapter primarily focuses on the available literature on the subject. The discussion is intended to be provocative, to promote future research rather than be definitive as this area of investigation is just beginning to evolve.

Biochemistry and properties of lipoxygenases

Arachidonic acid, an essential polyunsaturated fatty acid, is oxidised in the body by three different enzymes, i.e. P450, LO and PGS (Figure 7.1). Functionally, these enzymes of arachidonic acid cascade differ from each other in the oxygenation process. P450 is a monooxygenase and inserts one of the two atoms of O_2 into arachidonic acid while LO, being a dioxygenase, incorporates both atoms of O_2 into arachidonic acid. PGS, on the other hand, is a bis-dioxygenase and adds four atoms of oxygen to arachidonic acid. The second difference resides in the catalytic centre of these enzymes. It is noteworthy that both P450 and PGS are microsomal haemoproteins while LOs are non-haemo iron proteins found mainly in the cytosol.

LOs are ubiquitous in aerobic organisms. They occur in many species of algae (Gerwick 1994), plants (Siedow 1991; Grechkin 1998), aquatic invertebrates (De

Enzyme Systems that Metabolise Drugs and Other Xenobiotics. Edited by C. Ioannides.
© 2002 John Wiley & Sons Ltd

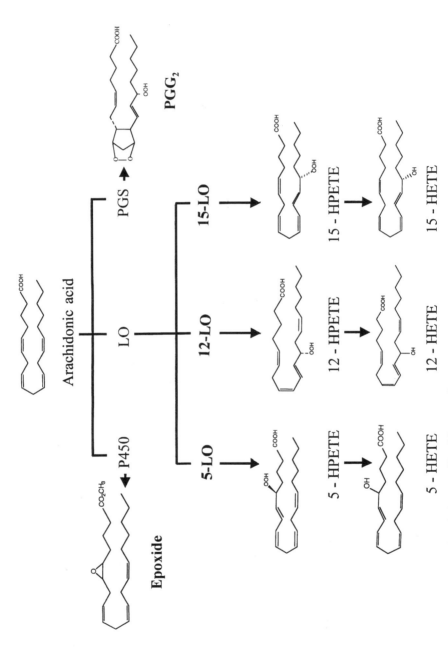

Figure 7.1 Metabolism of arachidonic acid by different pathways.

Petrocellis and Di Marzo 1994) and vertebrates (Yamamoto 1992). In mammals, a significant amount of LO activity occurs in blood cells and in many tissues (Table 7.1). Several excellent reviews on LO are available. For a more complete understanding of the LO system, the reader is encouraged to consult review(s) covering general information (Decker 1985; Schewe *et al.* 1986; Malle *et al.* 1987; Holtzman 1991, 1992; Gardner 1996) or specific aspects such as molecular biology and catalytic properties (Yamamoto 1992; Funk 1996; Kuhn and Thiele 1999), structure and function (Nelson and Seitz 1994; Gaffney 1996), regio- and stereo-chemistry (Lehmann 1994), 5-LO (Ford-Hutchinson *et al.* 1994; Bell and Harris 1999; Silverman and Drazen 1999; Steinhiber 1999), 12-LO (Yamamoto *et al.* 1997; Dailey and Imming 1999; Tang and Honn 1999) and the role in health and disease (Samuelsson *et al.* 1987; Konig *et al.* 1990).

DIOXYGENASE ACTIVITY

LOs catalyse regio- and stereo-specific dioxygenation of polyunsaturated fatty acids containing the 1(Z), 4(Z)-pentadiene system. The dioxygenation process involves hydrogen abstraction to produce a fatty acid radical (L•), radical rearrangement to form a conjugated diene and oxygen insertion to produce a fatty acid peroxyl radical intermediate (LOO•) before lipid hydroperoxide (LOOH) is released in the medium (Figure 7.2). Although the arachidonate molecule has several sites for oxygenation, LO selects one of them for the initial hydrogen abstraction. According to the current nomenclature, the classification of LOs is based on the positional specificity of arachidonate oxygenation. Mammalian cells contain at least three major types of LOs, i.e. 5-, 12- and 15-(S)LO. The dioxygenation of arachidonate by 5-LO yields 5-(S)hydroperoxy-6,8,11,14-(E,Z,Z,Z)-eicosatetraenoic acid (5-HPETE) while 12-LO and 15-LO convert arachidonate into cytotoxic 12-HPETE and 15-HPETE, respectively (Figure 7.1). Inside the cell, hydroperoxides are either reduced to the corresponding hydroxy derivatives, i.e. 5-HETE, 12-HETE and 15-HETE (Figure 7.1), or serve as intermediates in the genesis of a wide array of bioactive molecules such as leukotrienes (LTs), lipoxins, hepoxilins and other products by LO or other enzymes (Figure 7.1). Among plants, the most studied LOs are from soybean (SLO). Of the multiple known forms of SLO, L-1 is routinely used in xenobiotic oxidation studies. It is referred to hereafter simply as SLO.

The quantitative methods commonly used to assay the dioxygenase activity of LO include an estimation of oxygen uptake by means of a Clark electrode, spectrophotometric recording of conjugated diene formation absorbing at 234 nm and radiometry of ^{14}C or ^{3}H-arachidonic acid metabolites after their separation by TLC or HPLC. Haemoglobin (Hb), a common contaminant in the preparations of mammalian tissue LOs, poses a serious problem since it exhibits both pseudo-dioxygenase (Kuhn *et al.* 1981) and pseudo-peroxidase activities. A recently developed method using $ZnSO_4$ (Hover and Kulkarni 2000a) essentially eliminates this obstacle without affecting the dioxygenase and co-oxidase activities in the preparations of human term placental LO (HTPLO).

The positional specificity displayed by different LOs is not absolute as was once perceived. For example, SLO is a 15-LO with arachidonate but it generates

Table 7.1 Occurrence of lipoxygenase activity in mammalian tissues

Tissue	Species	LO	References
Heart	Rabbit	15-LO	Bailey et al. (1995)
Lymph node	Pig	12-LO	Shinjo et al. (1986)
Lung	Man	5-,12-,15-LO	Roy and Kulkarni, (1999)
Nasal polyp	Man	15-LO	Bioque et al. (1992)
Kidney	Man	5-LO	Stewart et al. (1997)
	Rat	5-,12-LO	Oyekan et al. (1997)
	Rabbit, Pig	LO	Stewart et al. (1993)
Urinary bladder	Pig	12-LO	Shinjo et al. (1986)
Liver	Man	5-,12-,15-LO	Roy and Kulkarni (1996b)
Gall bladder	Pig	12-LO	Shinjo et al. (1986)
Muscle (biceps femoris)	Pig	12-LO	Gata et al. (1996)
Dental pulp	Rat	12-, 15-LO	Doli et al. (1991)
Buccal cavity epithelial cells	Man	12-LO	Green (1989a)
Stomach (gastric mucosa)	Rat	5-,12-LO	Stein et al. (1991)
Small intestine	Pig	12-LO	Shinjo et al. (1986)
Intestinal epithelial cells	Rat	12-LO	Kamitani et al. (1999)
Caecum	Pig	12-LO	Shinjo et al. (1986)
Spleen	Pig	5-, 12-LO	Shinjo et al. (1986)
Colon, rectum	Man	15-LO	Ikawa et al. (1999)
Pancreatic cell lines	Man	5-, 12-LO	Ding et al. (1999)
Pacreatic islets	Rat	5-,12-,15-LO	Yamamoto et al. (1983)
Skin (keratinocytes)	Man	15-LO	Green (1989b)
Skin (epidermis)	Rat	12-,15-LO	Lomnitski et al. (1993)
Brain (different regions)	Many	5-,12-LO	Simmet and Peskar (1990)
	Pig	12-LO	Shinjo et al. (1986)
Ocular tissues	Man, Rabbit, Cynomolgus monkey, Rhesus monkey	5-, 12-LO	Kulkarni and Srinivasan (1989)
Thyroid gland	Pig	12-LO	Shinjo et al. (1986)
Thymus	Pig	12-LO	Shinjo et al. (1986)
Testis	Rat	12-LO	Shahin et al. (1978)
	Rat	15-LO	Lomnitski et al. (1993)
Prostate cancer cells	Man	5-LO	Myers and Ghosh (1999)
Vesicular gland	Sheep	LO	Portoghese et al. (1975)
	Bovine	5-,12-,15-LO	Halevy and Sklan (1987)
Leydig cells	Rat	LO	Mele et al. (1997)
Breast	Man	12-LO	Natarajan and Nadler (1998)
Preovulatory follicles (granulosa cells)	Man	5-,12-LO	Pridham et al. (1990)
Uterus	Man	12-LO	Flatman et al. (1986)
Placenta	Man	5-,12-,15-LO	Joseph et al. (1993)
Embryonic tissues	Man	5-,12-,15-LO	Joseph et al. (1994)
Foetal tissues	Man	5-,12-,15-LO	Datta and Kulkarni (1994a)
Endometrium	Man	12-LO	Ihno et al. (1993)
Decidua	Man	12-LO	Ihno et al. (1993)

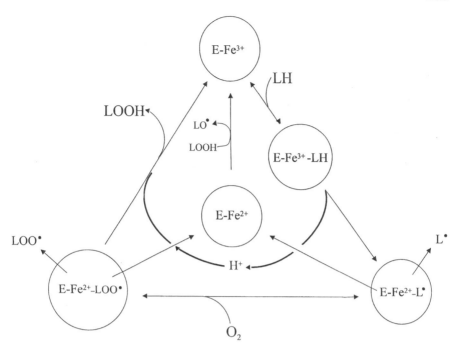

Figure 7.2 Catalytic cycle for the fatty acid peroxidation by lipoxygenase.

13-hydroperoxy octadecadienoic acid (13-HPOD) from linoleic acid. Some mammalian LOs also display similar dual positional specificity and can be classified as 12/15-LOs. The exact number of LOs in mammals is not known. Significant advances in genetic and molecular biology techniques have unveiled the existence of multiple forms of 12-LO and 15-LO (Kuhn and Thiele 1999; Yamamoto *et al.* 1999). It is certain that many more forms and subspecies will be discovered in the future. Human LOs have been purified and characterised in detail from different blood cells (Kuhn and Borngraber 1999; Yamamoto *et al.* 1999). Limited data also exist on the partial purification and properties of LOs from human lung (Roy and Kulkarni 1999), liver (Roy and Kulkarni 1996b), term placenta (Joseph *et al.* 1993), intrauterine conceptual tissues (Joseph *et al.* 1994; Datta *et al.* 1995), and other tissues.

In contrast to a strict requirement for a free fatty acid exhibited by the cyclooxygenase activity of PGS, the dioxygenase activity LOs displays a rather broad substrate specificity. Fatty acids containing two or more double bonds serve as substrates for LO. Certain LOs can utilise unusual substrates. For example, 5-LO from rat basophilic leukaemia cells oxygenates 5- and 6- fluoro- arachidonate (Nave *et al.* 1991) while LOs from different sources can accept 5-, 8-, 12-, 15-keto derivatives of arachidonate (Wiseman and Nichols 1988). Cholesterol esters of polyenoic fatty acids are oxygenated by purified rabbit reticulocyte 15-LO (Belkner *et al.* 1991). Similarly, 12-LO from porcine leukocytes can oxidise 1-palmitoyl-2-arachidonyl-sn-glycero-3-phosphocholine (Takahashi *et al.* 1993). More complex substrates such as biomembranes are also attacked. For example, porcine 12-LO can oxygenate mitochondrial and endoplasmic membranes (Takahashi *et al.* 1993) while rabbit 15-LO oxygenates mitochondria and erythrocyte ghosts (Schnurr *et al.* 1995). Mounting evidence has established that the initiation of atherosclerosis is a free radical process that involves oxidative modification of low density lipoprotein (LDL) by 15-LO (Berliner and Heinecke 1996; Steinberg 1999; Yamashita *et al.* 1999). SLO can also mediate this reaction (Cathcart *et al.* 1991). Membrane-enclosed organelles, a defining characteristic of eukaryotic cells, are lost during maturation as well as differentiation of cells such as reticulocytes, central fibre cells of the eye lens and keratinocytes. Lipid peroxidation catalysed by 15-LO provides a precise mechanism by which this tightly regulated membrane degradation occurs in a timely manner with required specificity (Van Leyen *et al.* 1998).

Although most LOs are cytosolic proteins, debate continues regarding subcellular localisation of some LOs. With a rise in the cellular cytoplasmic calcium concentration, 5-LO migrates to the nuclear envelope. Earlier it appeared that a docking protein called 'Five Lipoxygenase Activating Protein (FLAP)' provides the assistance in the translocation and anchoring of 5-LO into nuclear membranes. Further studies have revealed that FLAP is a shuttling protein which transfers arachidonate to 5-LO and its expression is essential for LTS biosynthesis. This translocation phenomenon also occurs with 12-LO (Hagmann *et al.* 1993) and 15-LO (Brinckmann *et al.* 1998).

The addition of either Ca^{2+} or ATP results in a strong stimulation of dioxygenase activity of HTPLO (Joseph *et al.* 1993) and human liver LO (Roy and Kulkarni 1996b). A similar response is displayed by other LOs. Since iron in the catalytic centre of LOs exists mainly in the inactive Fe^{2+} state, an initial lag period of several seconds is commonly noted while assaying the dioxygenase activity. LOOH ($\sim$10 µM) shortens this lag period by converting the inactive Fe^{2+} into the active Fe^{3+} form. Similar spontaneous activation of both dioxygenase and co-oxidase activities of SLO (Kulkarni *et al.* 1989, 1990) and human liver LO (Roy and Kulkarni 1996b) occurs by nanomolar H_2O_2 in the presence of linoleic acid. High LOOH concentration usually causes autoinactivation of LOs.

CO-OXIDASE ACTIVITY

Besides dioxygenation of fatty acids, the same LO protein also oxidises xenobiotics (Kulkarni and Cook 1988a). The ability of LO to couple xenobiotic oxidation with lipid peroxyl radical or lipid hydroperoxide generation during peroxidation of polyunsaturated fatty acids is termed as the co-oxidase activity (Roy and Kulkarni 1996b). This aerobic reaction is called 'co-oxidation' and the xenobiotic oxidised is referred to as co-substrate. LOs are unique in that they synthesise themselves the oxidants (lipid peroxyl radical intermediates and/or hydroperoxides) needed for xenobiotic co-oxidation. Thus benzo[a]pyrene-7,8-dihydrodiol (BP-diol) epoxidation by SLO can be observed in the presence of either linoleic acid (Byczkowski and Kulkarni 1989), arachidonic acid (Hughes *et al.* 1989) or 15-HPETE (Hughes *et al.* 1989). Similarly, benzidine and other hydrogen donors are co-oxidised by SLO in the presence of either linoleic acid (Kulkarni and Cook 1988a) or via the peroxidase-like activity supported by H_2O_2 (Kulkarni and Cook 1988b). On the other hand, high rates of glutathione

(GSH) oxidation can only be observed in the presence of polyunsaturated fatty acids (Roy *et al.* 1995; Kulkarni and Sajan 1997, 1999) while the reaction is essentially absent in the presence of either H_2O_2 (Yang and Kulkarni 2000) or 13-HPOD (Roy *et al.* 1995). Thus, the term 'co-oxidase activity' implies LO-mediated xenobiotic oxidation supported by the fatty acid peroxyl radicals and/or lipid hydroperoxide while the terms 'pseudo-peroxidase, peroxidase-like activity or hydroperoxidase activity of LO' signify the reactions noted in the presence of lipid hydroperoxides, H_2O_2, or synthetic organic hydroperoxides.

Different approaches have been applied to detect free radical metabolites generated by LO during xenobiotic co-oxidation. The formation of a relatively stable free radical metabolite of the xenobiotic can be observed directly using spectrophotometry as in the case of aminopyrine (Perez-Gilabert *et al.* 1997; Yang and Kulkarni 1998) and phenothiazines (Perez-Gilabert *et al.* 1994a,b; Rajadhyaksha *et al.* 1999). Since most xenobiotic free radicals are unstable and survive for a very short period, indirect methods are used for their detection. For example, ESR studies are essential to detect the spin trap-adducts of relatively unstable phenoxyl radicals from various phenols (Van der Zee *et al.* 1989). In a few cases, free radicals undergo rapid dimerisation to stable metabolites which are amiable to spectrophotometry, as in the case of benzidine diimine formation from benzidine (Kulkarni and Cook 1988a,b) or HPLC to estimate azobis(biphenyl) formed from two 4-aminobiphenyl radicals (Datta *et al.* 1997). The quantitation of formaldehyde, resulting from the decomposition of nitrogen cation free radical, the initial metabolite, during N-demethylation of xenobiotics (Yang and Kulkarni 1998; Hu and Kulkarni 1998; Rajadhyaksha *et al.* 1999; Hover and Kulkarni 2000b,c) offers yet another indirect tool to evaluate the reactions involving free radicals generated by LOs.

Proposed mechanisms of xenobiotic co-oxidation

It is envisioned that at least four distinct mechanisms are involved in the oxidation of xenobiotics catalysed by LOs in the presence of either polyunsaturated fatty acids or hydroperoxides.

PEROXYL RADICAL-MEDIATED REACTIONS

The fatty acid peroxyl radicals ($LOO^\bullet$) serve as potent, direct acting oxidising agents. As shown in Figure 7.3, the hydrogen abstraction from a donor molecule represents one of the prominent reactions displayed by lipid peroxyl radicals. This process yields a free radical as the product of one electron oxidation of xenobiotic and lipid hydroperoxide. LO-dependent oxygenation of the acceptor molecules is another important attribute noted with peroxyl radicals. Peroxyl radicals are efficient epoxidising agents and they donate their terminal oxygen atoms to an aliphatic double bond to give rise to an epoxide (Figure 7.3) as noted in the LO-catalysed epoxidation of aldrin (Naidu *et al.* 1991a) and BP-diol (Byczkowski and Kulkarni 1989). In the case of sulphur-containing compounds, the LO-dependent oxygenation of a cosubstrate results in sulphoxidation as reported for thiobenzamide (Naidu and Kulkarni 1991) or

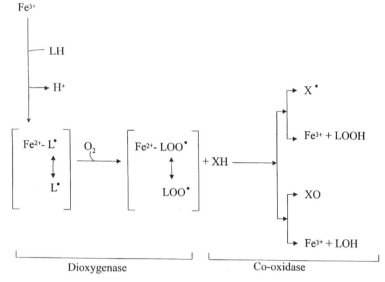

Figure 7.3 Peroxyl radical-dependent xenobiotic co-oxidation by lipoxygenase.

desulphuration (displacement of sulphur atom by oxygen atom), as in the case of parathion (Naidu *et al.* 1991b).

Multiple sources exist which supply LOO$^\bullet$ to the LO system (Figure 7.2).

(1) LOO$^\bullet$ are generated as intermediates during dioxygenation of fatty acids by LO before LOOH is released into the medium. This represents the primary source of LOO$^\bullet$ for xenobiotic oxidation.
(2) LO has been described as a leaky system. The hydrogen abstraction from a polyunsaturated fatty acid during dioxygenation by LO results in the formation of a fatty acid radical (L$^\bullet$). A portion of this L$^\bullet$ pool escapes into the medium where its reaction with O_2 yields LOO$^\bullet$.
(3) The third pathway of peroxyl radical generation originates from the decomposition of LOOH by LO under aerobic conditions (Hughes *et al.* 1989).

HYDROPEROXIDE-DEPENDENT REACTIONS

LOs oxidise various reducing agents, which are capable of reducing the Fe(III) enzyme to the Fe(II) form (Figure 7.4), in the presence of LOOH. The hydroperoxide-dependent xenobiotic oxidation is an indirect process that requires participation of LO in the hydrogen abstraction. As shown in Figure 7.4, hydroperoxide regenerates the active (Fe^{3+}) enzyme in the cycle and the xenobiotic undergoes a one electron oxidation by LO to a free radical. It is noteworthy that the reduction of hydroperoxide during LO-mediated xenobiotic co-oxidation does not yield the corresponding hydroxy derivative of fatty acid, as noted with PGS. Instead, the hydroperoxide undergoes a homolytic cleavage by LO to generate an alkoxy radical (LO$^\bullet$) and hydroxide ion (Van

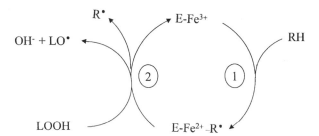

Figure 7.4 Lipid hydroperoxide-dependent xenobiotic co-oxidation by lipoxygenase.

der Zee *et al*. 1989; Cucurou *et al*. 1991). The studies conducted with SLO indicate that qualitatively, the enzyme lacks a strict specificity toward peroxide. Thus far, investigators have employed 5-HPETE (Cucurou *et al*. 1991), 15-HPETE (Hughes *et al*. 1989), and 13-HPOD (Mansuy *et al*. 1988; Riendeau *et al*. 1991; Chamulitrat *et al*. 1992) in studies on the LO-mediated xenobiotic oxidation. Kulkarni and Cook (1988b) were the first to demonstrate that H_2O_2 can be substituted for LOOH to support the co-oxidase activity of SLO. Subsequent reports have confirmed these observations for SLO (Perez-Gilabert *et al*. 1994a,b; Rosei *et al*. 1994; Datta *et al*. 1997) and extended to human tissue LO (Datta *et al*. 1997). A recent study from our laboratory (Hover and Kulkarni 2000d) has revealed that synthetic organic peroxides such as tert-butyl hydroperoxide or cumene hydroperoxide can also support hydroperoxidase activity of SLO. The experimental data on phenothiazine N-demethylation by SLO suggest that these hydroperoxides are as efficient as H_2O_2.

ELECTRON TRANSFER-DEPENDENT REACTIONS

In some cases, xenobiotic co-oxidation occurs indirectly. This involves generation of what is called a 'Shuttle Oxidant' from a good substrate by LO in the primary reaction which non-enzymatically oxidises another chemical. This is exemplified by the secondary hyperoxidation of benzidine (Hu and Kulkarni 2000) and L-dopa (Persad *et al*. 2000) by the primary phenothiazine cation radicals generated by SLO in the presence of H_2O_2 (Figure 7.5).

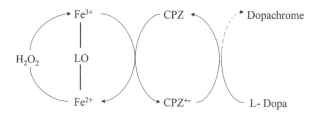

Figure 7.5 Lipoxygenase-mediated hyperoxidation of dopa in the presence of chlorpromazine.

OTHER MECHANISMS

In a few cases, xenobiotics are directly oxidised by purified LO in the absence of exogenous unsaturated fatty acid or hydroperoxide. For a long time, it has been known that acetylenic fatty acids such as 5,8,11-eicosatriynoic acid (ETI) and 5,8,11,14-eicosatetraynoic acid (ETYA) serve as suicide substrates and are directly oxidised by LO. The other reported examples of xenobiotic co-oxidation in the absence of fatty acid or lipid peroxide include the SLO-mediated oxidative conversion of hexanal phenylhydrazone into its a-azo hydroperoxide (Galey *et al*. 1988), 2-[(4'-hydroxy-3'-methoxy)-phenoxy]-4-(4"-hydroxy-3"-methoxy-phenyl)-8-hydroxy-6-oxo-3-oxabicyclo [3.3.0]-7-octene formation from curcumin (Schneider *et al*. 1998), the hydroperoxide generation from alkenes (Novak 1999) and several others. Additionally, it has been shown that the SLO-generated peroxyl radicals of LDL can co-oxidise α-tocopherol and probucol to their respective phenoxyl radicals, in the absence of exogenous fatty acid (Kalyanaraman *et al*. 1992). SLO can utilise tissue microsomes as a source of oxidisable lipid substrate during xenobiotic co-oxidation. Thus, the addition of increasing amount of SLO to incubates containing either human (Smith *et al*. 1995) or mouse (Rioux and Castonguay 1998) lung microsomes can oxidise 4-(methylnitrosamino)-1-(3-pyridyl)-1-butanone (NNK) to keto aldehyde and keto alcohol and the metabolite production is suppressed by nordihydroguaiaretic acid (NDGA), an inhibitor of LO (Smith *et al*. 1995).

FATE OF XENOBIOTIC FREE RADICALS

The fate of the primary free radicals generated during one-electron oxidation of reducing cosubstrate by LO depends on the nature of the radical and the surrounding environment. They may (1) undergo coupling reactions to yield dimers, or oligomers; (2) undergo further oxidation to yield two electron oxidation products (e.g. aminopyrine or benzidine); (3) serve as a shuttle oxidant and react with another compound to generate secondary metabolite(s) as shown for the phenothiazine-benzidine combination, (4) react with endogenous or exogenous thiol and trigger thiol pumping, i.e. reduction of a free radical back to the parent chemical and the generation of GS$^\bullet$ or (5) react with macromolecules to produce protein, DNA, or RNA adducts.

Chemicals oxidized

ENDOGENOUS OR PHYSIOLOGICALLY RELEVANT CHEMICALS

A number of endogenous chemicals are oxidised by LO (Table 7.2). In the presence of linoleic acid, SLO oxidises both NADH and NADPH at the rate of $\sim$350 nmoles/min/mg protein (Roy *et al*. 1994). The blockade of the reduction of ferricytochrome-C or nitroblue tetrazolium during LO-mediated NDA(P)H oxidation by superoxide dismutase suggests the generation of superoxide anions (Figure 7.6). The noted apparent K_m values of 13 µM and 117 µM for NADH and NADPH, respectively, are within physiological range. O' Donnell and Kuhn (1997) have recently described that at near physiological concentration, both NADH and NADPH are co-oxidised by 15-LO

Table 7.2 Oxidation of naturally occurring chemicals via lipoxygenase pathway

Chemical	Enzyme	Cofactor	References
Glutathione	SLO	PUFAs	Roy et al. (1995)
	SLO	LA	Kulkarni and Sajan, (1997)
	HTPLO	LA	Kulkarni and Sajan, (1999)
Proline	SLO	LA	Byczkowski et al. (1991)
NADH	SLO	PUFAs	Roy et al. (1994)
NADPH	SLO	PUFAs	Roy et al. (1994)
Ascorbic acid	SLO	LA,AA	Roy and Kulkarni (1996a)
Dopa	SLO	H_2O_2	Rosei et al. (1994)
α-Methyl dopa	SLO	H_2O_2	Rosei et al. (1994)
Dopamine	SLO	H_2O_2	Rosei et al. (1994)
N-Acetyl dopamine	SLO	H_2O_2	Rosei et al. (1994)
Noradrenaline	SLO	H_2O_2	Rosei et al. (1994)
Adrenaline (Epinephrine)	SLO	H_2O_2	Rosei et al. (1994)
DL-α-Tocopherol	SLO	13-HPOD	Mansuy et al. (1988)
Vit.E derivatives (10)	SLO	13-HPOD	Cucurou et al. (1991)
5,6-Dihydroxyindole	SLO	H_2O_2	Blarzino et al. (1999)
5,6-Dihydroxyindole-2-carboxylic acid	SLO	H_2O_2	Blarzino et al. (1999)
β-Carotene	SLO	LA	Wu et al. (1999)
	Potato 5-LO	LA	Aziz et al. (1999)
Retinol	SLO	LA	Waldmann and Schreier (1995)
β-Ionone	SLO	LA	Waldmann and Schreier (1995)
4-Hydroxy-β-ionone	SLO	LA	Waldmann and Schreier (1995)

SLO, Soybean lipoxygenase-1; HTPLO, Human term placental lipoxygenase; PUFAs, Polyunsaturated fatty acids; LA, Linoleic acid; AA, Arachidonic acid.

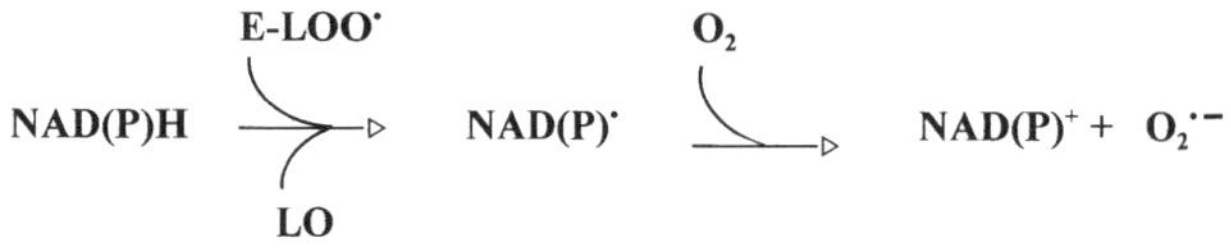

Figure 7.6 NAD(P)H oxidation and superoxide anion generation by lipoxygenase.

purified from rabbit reticulocytes in the presence of linoleic acid. The reaction was not observed with 13-HPOD.

Reduced glutathione (GSH) is the most abundant non-protein intracellular thiol. In mammals, GSH regulates 5-and 15-LO activities in human polymorphonuclear leukocytes and lymphocytes (Hatzelmann and Ullrich 1987; Claesson et al. 1992; Jakobsson et al. 1992). It is noteworthy that GSH is an extremely poor substrate for horseradish peroxidase (Harman et al. 1986) and is not oxidised by purified PGS in the presence of arachidonate (Eling et al. 1986). In fact, the authors reported that GSH itself inhibits the cyclooxygenase activity of the enzyme. In contrast to these reports, it is interesting to note that GSH serves as an excellent substrate for SLO in the presence of polyunsaturated fatty acids (Roy et al. 1995; Kulkarni and Sajan 1997). It is accompanied by a low rate of superoxide anion formation (Roy et al. 1995). The reaction is directly linked with the fatty acid peroxyl radical generation by LO. Negligible GSH oxidation by SLO occurs in the presence of 13-HPOD or H_2O_2 (Roy *et al.* 1995; Yang and Kulkarni 2000). Human tissue LO (e.g. HTPLO) also displays this ability (Kulkarni and Sajan 1999). The reaction occurs at pH 7.4 at a significant rate in the presence of physiologically relevant concentrations of fatty acid and GSH (Kulkarni and Sajan, 1999). The spectrophotometric assay of dioxygenase activity of SLO revealed that the addition of increasing concentration of GSH to the incubation media results in a proportional decline in 13-HPOD accumulation (Roy *et al.* 1995). However, oxygen consumption under identical conditions proceeds unaltered. This suggests that the dioxygenase activity of SLO is not inhibited by GSH, and the fatty acid peroxyl radicals are diverted away from hydroperoxide formation. The final fatty acid metabolites resulting from the reaction were not identified in this study. Considering the data on NAD(P)H and GSH oxidation and concurrent reactive oxygen species production, it is tempting to speculate that LO plays a significant contributory role in the genesis of cellular oxidative stress.

Ascorbic acid, an essential dietary vitamin, serves as a readily available reducing agent in mammalian cells. Ascorbate is easily oxidised by SLO in the presence of linoleic or arachidonic acid but not in the reaction media supplemented with either 13-HPOD or H_2O_2 (Roy and Kulkarni 1996a). This suggests that ascorbate is not a substrate for the hydroperoxidase activity of LO. The lack of 13-HPOD-dependent ascorbate oxidation has also been reported for potato tuber 5-LO (Cucurou *et al.* 1991). The absence of superoxide generation suggests that fatty acid peroxyl radicals co-oxidise ascorbate to free radicals which disproportionate to yield dehydro-L-ascorbate (Roy and Kulkarni 1996a). The ascorbate addition markedly decreases the apparent rate of LO-mediated xenobiotic oxidation due either to a competitive inhibition and/or reduction of xenobiotic free radical back to the parent compound. Interestingly, at low concentrations, ascorbate itself apparently does not reduce the iron in the LO to inactivate dioxygenase activity. On the other hand, several phenols and arylamines enhance the basal rate of ascorbate oxidation in the linoleic acid-coupled reactions mediated by SLO (Roy and Kulkarni 1996a). It is interesting to note that the palmitate ester of ascorbate serves as a substrate for the hydroperoxidase activity of SLO and potato tuber 5-LO (Cucurou *et al.* 1991).

Other physiologically relevant compounds oxidised via the LO pathway (Table 7.2) include noradrenaline, *N*-acetyldopamine, adrenaline, and dopa to the corresponding melanin pigments (Rosei *et al.* 1994). Both 5-*S*-cysteinyl-dopa and 5-S-cysteinyl dopamine ultimately give rise to pheomelanin (Mosca *et al.* 1996). Catecholic tetrahydroisoquinolines such as salsolinol, tetrahydropapaveroline, laudanosoline and apomorphine are easily oxidised by SLO in the presence of H_2O_2 to their respective melanins (Mosca *et al.* 1998). Linoleic acid supports oxidation of dopamine, serotonin and norepinephrine by rat brain cytosolic LO (Byczkowski *et al.* 1992). Additionally, hydroxylation of proline by SLO in the presence of linoleic acid is also known (Byczkowski *et al.* 1991).

ENVIRONMENTAL CHEMICALS

Polycyclic aromatic hydrocarbons (PAHs) are a class of ubiquitous environmental pollutants, many of which are mutagenic, carcinogenic and teratogenic. The prototype of this class of compounds is benzo[a]pyrene (BP) that requires double epoxidation to produce the ultimate toxicant, the diolepoxide (Figure 7.7). Although the role played by P450 in the initial epoxidation of the aromatic system is well accepted, it now appears that the LO pathway may be more important in the final epoxidation step. Sevanian and Peterson (1989) reported arachidonate- or 13-HPOD- dependent BP-caused cytotoxicity and mutagenesis in Chinese hamster lung fibroblasts (V79 cells). Although V79 cells possess both PGS and LO activities, nordihydroguaiaretic acid (NDGA) but not indomethacin, the inhibitor of PGS, exhibited the inhibitory response. This suggests an involvement of LO pathway in the induction of BP toxicity. Both P450 and PGS may not be involved in the BP activation by human or rat colonic mucosal microsomes since an addition of arachidonate, linoleate or their hydroperoxides enhanced the process 4-5 fold whereas NADPH had no effect (Craven and DeRubertis 1980; Craven et al. 1983). Nemoto and Takayama (1984) studied the covalent binding of BP to proteins following activation by microsomal and cytosolic enzymes in rat liver and lung. With microsomes, linoleic acid was more effective than arachidonate, whereas, with cytosol linoleic acid, linolenic acid and arachidonate but not oleic acid supported the reaction; arachidonate being the best cofactor. Indomethacin did not inhibit BP activation but NDGA and quercetin significantly reduced the binding. The linoleic acid-dependent binding with liver and lung microsomes was >3 and >16-fold greater respectively than that after incubation with NADPH. The authors opined that LO plays a dominant role in BP activation in liver and lung. In line with this postulate, Adriaenssens et al. (1983) did not observe a decrease in covalent binding of BP to DNA by indomethacin in mice. Aspirin, also an inhibitor of PGS, did not alter the number of pulmonary adenomas in mice treated with BP suggesting that PGS does not activate this carcinogen in vivo. Byczkowski and Kulkarni (1992) studied the LO-mediated co-oxidation of [^{14}C]BP in rat lung cytosol using linoleic acid. The oxidation yielded 1,6-dione, 3,6-dione and 6,12-dione of BP, with the 6,12-dione production being predominant. The quinones derived from BP may either bind covalently to tissue macromolecules or further undergo redox reactions.

Scharping et al. (1992) studied epoxidation of BP-7,8-dihydrodiol (BP-diol), the proximal carcinogen, in human liver microsomes and by SLO in the presence and absence of arachidonate. The reaction was inhibited by NDGA. The SLO-mediated reaction was augmented by the addition of phenylbutazone. Apparently, phenylbuta-zone is oxidised by LO to a carbon-centred free radical which traps molecular oxygen to form a peroxyl radical that effects epoxidation of BP-diol. The crucial direct evidence that purified LO can mediate the final BP bioactivation step is available. Byczkowski and Kulkarni (1989) were the first to demonstrate that BP-diol can be activated by SLO in the presence of linoleic acid. BP-trans-anti-7,8,9,10-tetrahydrote-trol, the product of hydrolytic breakdown of ultimate mutagenic BP-anti-7,8-dihydro-diol-9,10-epoxide was detected as the major metabolite by radiometry combined with HPLC. The peroxy radical derived from linoleic acid during dioxygenation by SLO was reported to be responsible for the epoxidation of BP-diol. Subsequently, Hughes et al. (1989) confirmed these findings and noted that the SLO-catalysed BP-diol epoxidation can also occur in the presence of arachidonate, γ-linolenic acid and 15-HPETE. Based on oxygen consumption studies, it was postulated that 15-HPETE-is reduced to an alkoxyl radical and hydroxyl anion. The radical rearranges to an allylic epoxyl radical which reacts with molecular oxygen to form a peroxyl radical that

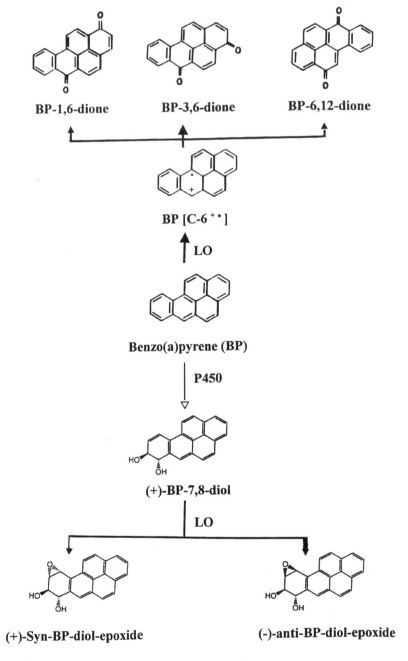

Figure 7.7 Lipoxygenase-mediated oxidation of benzo[a]pyrene.

serves as the oxidant for the epoxidation reaction. The attempts to detect 15-HPETE-derived peroxyl radical by ESR were unsuccessful, but the formation of two carbon-centred radicals either at C11 or C13, the precursors of this species, was noted. Although the fatty acid metabolites were not analysed, a simultaneous formation of epoxy-hydroxy, dihydroxy and trihydroxy fatty acids was proposed (Hughes *et al.* 1989). These observations with SLO raise a question whether LO-mediated BP-diol bioactivation occurs in human lung. The possibility is strongly indicated since LO capable of xenobiotic oxidation occurs in human (Roy and Kulkarni 1999) and animal lungs (Nemoto and Takayama 1984, Kulkarni *et al.* 1992). Indomethacin does not inhibit arachidonate-dependent BP-diol epoxidation by either hamster trachea or human bronchus explant in culture (Reed *et al.* 1984). This provides compelling evidence that the activation proceeds via the LO pathway and not by PGS. Joseph *et al.* (1994) have documented that purified LO from human intrauterine conceptual tissues and HTPLO can easily epoxidise radioactive BP-diol to several metabolites. The production of BP-*trans*-anti-tetrol by LO was about eight-times above the control value noted with boiled enzyme preparations. These results explain the report (Manchester *et al.* 1988) on the detection of DNA adducts of BP in human placentas of non-smokers which lack both P450 and PGS activities (Kulkarni 1996, 2000).

A potent tobacco-specific carcinogen, NNK, induces lung tumours in animals and is a suspected human carcinogen. Smith *et al.* (1995) opined that P450 is enzymes are only partially responsible for the activation of NNK in human lung microsomes. They noted an increased formation of keto aldehyde and keto alcohol by the addition of exogenous arachidonate and the inhibition of their formation by NDGA, thus suggesting an involvement of the LO pathway in the NNK oxidation by human lung microsomes. The observed NDGA-caused modest decline in the rate of keto alcohol and keto aldehyde production by patas monkey lung microsomes are in harmony with this postulate (Smith *et al.* 1997). More recently, Rioux and Castonguay (1998) and Castonguay *et al.* (1998) have provided the necessary direct evidence that purified SLO mediates NNK oxidation by a-carbon hydroxylation and *N*-oxidation.

Occupational exposure to arylamines such as 4-aminobiphenyl (4-ABP) is known to be associated with an elevated risk of bladder cancer. 4-ABP also occurs in environmental tobacco smoke. Interestingly, the concentration of 4-ABP is 30 times greater in the sidestream smoke than in the mainstream smoke. Thus, under certain conditions, non-smokers can be exposed to relatively high levels of 4-ABP. Several reports have documented an increased risk of haematopoietic malignancies and lung cancers in adults to be related to smoking by the mother during pregnancy. It is widely accepted that oxidative metabolism of 4-ABP is essential to exert its carcinogenic effect. Datta *et al.* (1997) have demonstrated that both SLO and HTPLO can oxidise 4-ABP. HPLC and mass spectroscopy were used to isolate and identify the metabolite, respectively. The major metabolite 4,4-azobis(biphenyl), was reported to arise via a mechanism involving an initial one-electron oxidation of 4-ABP to a free radical species. When the specific activity data were normalised on the basis of units of dioxygenase activity, HTPLO was found to be twice as efficient as SLO in the 4-ABP oxidation. These results explain, at least partly, the reported 4-ABP-adducts in the non-smokers' placentas, a tissue which is essentially deficient in both P450 and PGS capable of xenobiotic oxidation (Kulkarni 1996, 2001).

Although the use of benzidine is banned in the USA, benzidine-based dyes are still being used in many countries in textile and other industries. Therefore, cancer incidence and mortality due to exposure to aromatic amines are still important public health issues on a global scale. Besides uroepithelial cancers, few isolated reports have implicated human exposure to benzidine and other arylamines in the cancers of other organs such as pancreas (Anderson *et al.* 1997), kidneys and liver (Morikawa *et al.* 1997). In rats, mice and hamsters, benzidine causes hepatocarcinogenicity (Whysner *et al.* 1996). Benzidine undergoes bioconversion to electrophiles that covalently bind to DNA, induce mutations and initiate carcinogenesis. Benzidine is a poor substrate for the NADPH-dependent oxidation by rat hepatic P450 and MFMO and it does not generate protein-bound and water-soluble metabolites (Zenser *et al.* 1979). Prior acetylation seems to be involved. Although PGS-catalysed bioactivation of benzidine explains bladder carcinogenicity, it can not explain hepatocarcinogenicity because PGS occurs in negligible amounts in the liver (Smith and Marnett 1991). Benzidine serves as an excellent substrate for co-oxidation by SLO in the presence of either linoleic acid or H_2O_2 (Kulkarni and Cook 1988a,b; Kulkarni *et al.* 1989; Hu and Kulkarni 2000). Initial one-electron oxidation of benzidine yields a nitrogen-centred cation radical which may either undergo a second one-electron oxidation to produce benzidine di-imine or disproportionate to produce parent diamine and benzidine di-imine. Both the cation radical and the di-imine are electrophilic derivatives of benzidine capable of covalent binding to macromolecules. Similar to SLO, extensive benzidine oxidation to reactive benzidine di-imine can be observed with animal LO isolated from lung (Kulkarni *et al.* 1992), brain (Naidu *et al.* 1992), liver (Roy and Kulkarni 1994) and embryo (Roy *et al.* 1993). The purified LO from human term placenta (Joseph *et al.* 1993), conceptual tissues (Datta *et al.* 1995), adult lung (Roy and Kulkarni 1999) and liver (Roy and Kulkarni 1996b) also catalyse this reaction at a high rate.

2-Aminofluorene (2-AF) is a potent carcinogen and teratogen to which humans may be exposed from environmental, industrial and dietary sources. Metabolic activation is obligatory for 2-AF toxicity. 2-AF activation by P450 and MFMO plays some role in hepatocarcinogenesis. However, extrahepatic tissues such as mammary gland and Zymbal gland, which contain very low levels of these enzymes, are also targets for 2-AF-induced cancer. Since 2-AF is a poor substrate for PGS, its bioactivation via the LO pathway was examined (Roy and Kulkarni 1991). In the presence of linoleic acid, SLO catalysed oxidation of 2-AF at the rate of 521 nmol/min/nmol of SLO. Arachidonate, linolenic acid, and *cis*-11,14-eicosadienoic acid were less than half as effective as linoleate. The experiments conducted with radioactive substrate revealed the generation of electrophilic 2-AF intermediates by SLO which bind covalently in significant amounts to either bovine serum albumin or calf thymus DNA. SLO-mediated 2-AF bioactivation can be blocked by ETI, ETYA, NDGA, gossypol, and esculetin, the classical LO inhibitors, and by butylated hydroxyanisole (BHA) and butylated hydroxytolyne (BHT) clearly suggesting that LO may serve as an alternate pathway to P450 in 2-AF bioactivation in mammalian extrahepatic tissues.

Aflatoxin B_1 (AFB), a substituted coumarin produced as a secondary metabolite by *Aspergillus flavus* and *A. parasiticus*, is a known contaminant in various food products. AFB is a potent hepatocarcinogen and teratogen in humans. It is believed that a requisite event in AFB toxicity is prior oxidation at the 8,9-vinyl ether bond to form the AFB-8,9-epoxide, an electrophilic species which is the ultimate carcinogenic metabolite. Linoleic acid, linolenic acid and arachidonic acid strongly inhibit NADPH-dependent hepatic microsomal AFB activation (Firozi *et al.* 1986; Ho *et al.* 1992). This leads to a logical question as to whether P450 serves as the sole catalyst. PGS can epoxidise AFB but cannot explain hepatocarcinogenicity since liver contains biologically non-significant levels of this enzyme (Smith and Marnett 1991). Earlier, Amstad and Cerruti (1983) reported that the binding of AFB metabolites to DNA in mouse embryo fibroblasts is inhibited by inhibitors of LO. Purified SLO can mediate AFB activation (Liu and Massey 1992). A recent study by Roy and Kulkarni (1997) has provided the necessary evidence that partially purified adult human liver LO can epoxidise AFB in the presence of polyunsaturated fatty acids. Thus, LO clearly represents an additional or alternate pathway of AFB bioactivation in the human liver. AFB contamination occurs in grain dust and handling of grains in large quantity can result in a significant pulmonary exposure to AFB. Available data indicate that AFB epoxidation can be effected by human lung cytosolic LO (Donnelly *et al.* 1996) as well as by the cytosolic LO from guinea pig lungs and kidney (Liu and Massey 1992). Similarly, the experimental data on AFB bioactivation collected for the purified LO from human intrauterine conceptual tissues and HTPLO (Datta and Kulkarni 1994a) explain, at least partly, the developmental toxicity associated with the exposure to this fungal alkaloid during pregnancy.

The colon carcinogen 1,2-dimethylhydrazine is activated by P450 to the proximate carcinogen methylazoxymethanol. Either NAD-dependent oxidation or fatty acid supported co-oxidation releases the final methylating agent and formaldehyde. Craven *et al.* (1985) reported that rat colonic mucosal LO, but not P450, mediates this reaction in the presence of arachidonate, linoleate and 15-HPETE.

INDUSTRIAL CHEMICALS

Styrene is a high-volume industrial chemical used in the manufacture of plastics, resins, synthetic rubber and insulators. As in the past, a high potential of worker exposure to low levels of styrene monomer can be anticipated in the future. Many metabolic studies have established that prior styrene oxidation to styrene-7,8-oxide is essential to observe toxicity, since it binds covalently to cellular macromolecules. P450, haemoglobin, myoglobin, and horseradish peroxidase have been shown to mediate this epoxidation reaction. For reasons unclear at present, PGS does not catalyse styrene epoxidation (Stock *et al.* 1986). On the other hand, Belvedere *et al.* (1983) observed that the LO pathway serves as an additional route for epoxidation of styrene in the presence of arachidonic acid. The rate of the SLO-mediated reaction was about 4-fold greater than that noted with the NADPH-dependent P450-mediated hepatic microsomal reaction. The peroxyl radicals of arachidonate or those derived from hydroperoxide degradation are expected to mediate this reaction.

Acrylonitrile (ACN) is another important industrial chemical widely used in the manufacture of plastics, rubber, and acrylic fibres. Residues of ACN are found in water, food, and cigarette smoke. ACN is carcinogenic in the rat but the potential carcinogenic risk to humans is uncertain. The mechanism by which ACN initiates tumour formation is unknown; however, the available evidence suggests that metabolic activation is required. ACN undergoes epoxidation to produce reactive 2-cyanoethylene oxide (CEO) and ultimately cytotoxic cyanide. The proposal that P450 is responsible for bioactivation of ACN seems inconsistent in view of the following:

(1) The covalent binding of ACN to brain, lung and liver microsomes from the untreated rats is not significantly increased by the addition of NADPH (Roberts *et al.* 1989).
(2) In the physiologically based dosimetry study, Gargas *et al.* (1995) concluded that an extrapolation of *in vitro* data to the whole animal does not agree since the V_{max} for ACN epoxidation by rat liver P450 accounts for <8% of the *in vivo* estimate for the CEO formation.
(3) In animals, ACN carcinogenicity is also observed in Zymbal' s gland, stomach, brain and uterus that contain very low or undetectable levels of P450. The detailed study conducted by Roy and Kulkarni (1999) revealed that LO pathway may be involved. SLO and partially purified human lung LO preparations, which predominantly contain 15-LO, and smaller amounts of 5-LO and 12-LO, were found to be capable of extensive ACN metabolism to CEO and cyanide in the presence of polyunsaturated fatty acids (Figure 7.8). Among the fatty acids tested, linoleic acid was most effective in supporting epoxidation of ACN to CEO by the human lung LO. Interestingly, the human lung enzyme was an approximately sixfold better catalyst than SLO in converting ACN to cyanide. Significant covalent binding of the radioactivity derived from [^{14}C]-ACN to bovine serum albumin and calf thymus DNA occurred when the reaction media contained either active SLO or human lung LO. These reactions were strongly inhibited by NDGA and the antioxidant butylated hydroxytoluene (BHT). The experimental evidence clearly suggests that the LO pathway may be of toxicological relevance in the ACN bioactivation in humans.

PESTICIDES

Pesticides represent a group of potentially dangerous chemicals of diverse structures deliberately distributed in our ecosystem. Few *in vitro* studies have documented the LO-mediated oxidative metabolism of pesticides (Table 7.3). Low levels of residues of now abandoned organochlorine insecticides still occur in human tissues, milk and the food we consume (Kulkarni and Mitra 1990). Scientific interest is further amplified since some of these pesticides are suspected of being endocrine disruptors. Epoxidation of aldrin to dieldrin is a model reaction noted with P450 and PGS. The demonstration that SLO (Naidu *et al.* 1991a) can easily oxidise aldrin to dieldrin clearly implicates LO as an additional pathway for this epoxidation. The rate of SLO-mediated reaction was 8-20-fold greater than different isoforms of P450. Parathion, an organophosphorous insecticide, is oxidised by SLO (Figure 7.9) (Naidu *et al.* 1991b). The fatty acid peroxyl radicals and to some extent those derived from the13-HPOD decomposition were proposed to donate oxygen to parathion to produce initially a sulphur-oxygen intermediate. The breakdown of this unstable intermediate releases the desulphuration product, paraoxon, and dearylation metabolites, *p*-nitrophenol

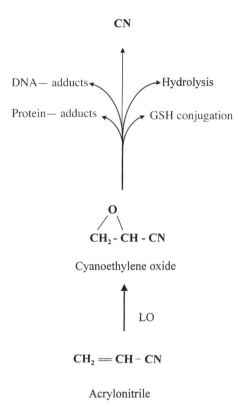

Figure 7.8 Lipoxygenase-mediated oxidative metabolism of acrylonitrile.

and diethyl phosphate or diethyl phosphorothioic acid. The displacement of a sulphur atom by oxygen (desulphuration) results in the bioactivation of parathion since paraoxon is a very potent anticholinesterase. Again, the rate of paraoxon formation by SLO was up to 20 times greater than the P450-dependent reaction. It is noteworthy that cytosolic rat brain LO also yields similar results (Naidu *et al.* 1991b). Recently Hu and Kulkarni (1998) have documented that several pesticides undergo SLO-catalysed N-demethylation to release formaldehyde in the presence of H_2O_2 (Table 7.3). The highest specific activity was noted with aminocarb. A significant suppression of the formaldehyde production by GSH, DTT (dithiothreitol), BHT, and BHA suggests a free radical nature of the aminocarb N-demethylation. HTPLO can also efficiently catalyse this reaction in the presence of linoleic acid (Hover and Kulkarni 2000c).

DRUGS

Aminopyrine is extensively used by many investigators as a prototype xenobiotic in the mechanistic studies on N-dealkylation catalysed by P450, PGS and horseradish peroxidase. According to Agundez *et al.* (1995), *in vivo* metabolism of aminopyrine in

Table 7.3 Oxidation of pesticides by lipoxygenase

Pesticide	Enzyme	Cofactor	Reaction	Reference
Aldrin	SLO	Linoleic acid	Epoxidation	Naidu *et al.* (1991a)
Parathion	SLO	Linoleic acid	Desulphuration, Dearylation	Naidu *et al.* (1991b)
Aminocarb	SLO	H_2O_2	N-demethylation	Hu and Kulkarni (1998)
	SLO, HTPLO	Linoleic acid	N-Demethylation	Hover and Kulkarni (2000c)
Zectran	SLO	H_2O_2	N-Demethylation	Hu and Kulkarni (1998)
	SLO, HTPLO	Linoleic acid	N-Demethylation	Hover and Kulkarni (2000c)
Dicrotophos	SLO	H_2O_2	N-Demethylation	Hu and Kulkarni (1998)
	SLO, HTPLO	Linoleic acid	N-Demethylation	Hover and Kulkarni (2000c)
Chlordimeform	SLO	H_2O_2	N-Demethylation	Hu and Kulkarni (1998)
	SLO, HTPLO	Linoleic acid	N-Demethylation	Hover and Kulkarni (2000c)
Famphur	SLO	H_2O_2	N-Demethylation	Hu and Kulkarni (1998)
Formetanate	SLO	H_2O_2	N-Demethylation	Hu and Kulkarni (1998)
Pirimicarb	SLO	H_2O_2	N-Demethylation	Hu and Kulkarni (1998)
Tetramethiuram	SLO	H_2O_2	N-Demethylation	Hu and Kulkarni (1998)

SLO, soybean lipoxygenase; HTPLO, Human term placental lipoxygenase

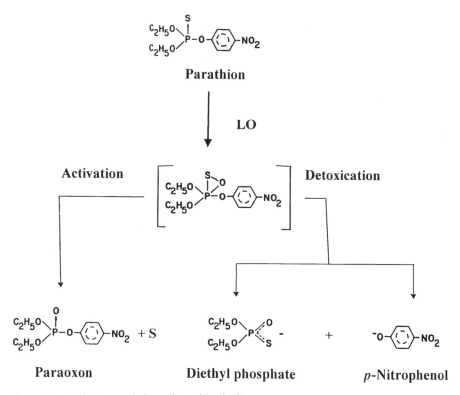

Figure 7.9 Oxidative metabolism of parathion by lipoxygenase.

humans exhibits a ~200-fold variation and this is not related to gender, intake of caffeine or alcohol, or P450 polymorphism. Since LO activity occurs in the liver and many other tissues, a hypothesis that this pathway may be involved was tested using SLO in the presence of H_2O_2 (Perez-Gilabert *et al*. 1997; Yang and Kulkarni 1998). As compared to P450 and PGS, aminopyrine *N*-demethylation by SLO occurs at a much higher rate (823 nmol/min/mg protein) (Yang and Kulkarni 1998). The formaldehyde production is strongly inhibited by NDGA and gossypol, clearly implicating LO as the catalyst. A nitrogen-centred free radical cation, the expected initial one-electron oxidation product, can be observed by spectrophotometry (Perez-Gilabert *et al*. 1997; Yang and Kulkarni 1998). The rate of accumulation of this radical species and formaldehyde depends on pH, the amount of the enzyme, and the concentration of aminopyrine and H_2O_2. Ascorbate, GSH, and DTT markedly suppress the radical formation supporting the contention that a free radical mechanism is involved in the *N*-dealkylation of aminopyrine via the LO pathway (Figure 7.10). It was postulated that the cation radical is further converted to an iminium cation either by deprotonation or hydrogen atom abstraction. Subsequent hydrolysis of the iminium cation yields monomethylamine and formaldehyde (Perez-Gilabert *et al*. 1997; Yang and Kulkarni 1998).

$$- e^- \qquad\qquad - (H \text{ or } H^+) \qquad\qquad + H_2O$$

$$R_2N\text{-}CH_3 \longrightarrow R_2N\text{-}CH_3^{+\bullet} \longrightarrow R_2N^+\text{=}CH_2 \longrightarrow R_2NH_2 + HCHO$$

Figure 7.10 Mechanism for N-demethylation of xenobiotics by lipoxygenase.

Tricyclic antidepressants, such as imipramine and closely related compounds, represent a group of drugs widely used today in the USA and other countries for the treatment of major depression. Their remarkable efficacy in alleviating depression is well established. However, few side-effects such as cardiovascular toxicity and species-specific teratogenicity in animals have been noted. Imipramine, a prototype of this class, undergoes extensive metabolism in the body. However, it is not clear which enzyme(s) is responsible for its oxidation. The published data on the role played by P450 are not convincing in view of the facts that (1) the expected increase in the rate of imipramine oxidation by brain or liver microsomes is not observed in rats pre-treated with P450 inducers such as phenobarbital and β-naphthoflavone; (2) imipra-mine behaves as a P450 inhibitor and inactivates P450 since reactive intermediates generated during its oxidation quickly form stable complexes with P450 and accumu-late in microsomes with time; (3) a large inter-individual variation occurs in the blood levels of imipramine in patients. The role of MFMO in imipramine oxidation is also debatable since the purified enzyme does not catalyse this reaction unless detergent is added. Since LO activity occurs in the liver of rat (Roy and Kulkarni 1994) and human (Roy and Kulkarni 1996b, 1997) and in the rat brain (Byczkowski et al. 1992; Naidu et al. 1992), its involvement in imipramine oxidation can be suspected. An exploration of this hypothesis revealed that imipramine is an excellent substrate for dealkylation by SLO in the presence of H_2O_2 (Hu et al. 1999). Desipramine, the expected product of imipramine mono N-demethylation, was identified by HPLC. When incubated separately, desipramine also yielded formaldehyde, suggesting that imipramine under-goes double dealkylation by SLO. Under identical conditions, not all the antidepres-sants tested are equally oxidised. The most rapid oxidation ($\sim$210 nmol/min/mg SLO) was observed with trimipramine while desipramine, clomipramine, diltiazem, amitrip-tyline and doxepin exhibited a relatively low oxidation rate (10 nmol/min/mg SLO) (Table 7.4).

Currently, phenothiazines enjoy wide acceptance as relatively safe and efficacious drugs for the treatment of psychotic illnesses. Other therapeutic applications include their use as tranquilisers, sedatives, antiemetics, antimicrobials, etc. Some reports point out alterations in the endocrine function and untoward effects on the cardiovascular and reproductive systems associated with acute exposure. Phenothiazines undergo extensive oxidative metabolism in the body. It is generally accepted that oxidation of phenothiazines to relatively stable nitrogen-centred cation radicals is the first essential step responsible for their biological activity. In each case, the cation radical serves as an intermediate that yields the phenothiazine sulphoxide and other metabolites. The ability of P450, haemoglobin, myeloperoxidase and horseradish peroxidase to metabo-lise phenothiazines is well established. Recently, Perez-Gilabert et al. (1994a,b) identified an involvement of a non-haem iron protein-dependent pathway. The authors reported the formation of cation-free radical from phenothiazines by SLO in the presence of H_2O_2. Besides SLO, purified HTPLO also generates cation radicals from

Table 7.4 Oxidation of drugs by lipoxygenase

Drug	Reaction	Enzyme	Cofactor	Reference
Aminopyrine	N-Demethylation	SLO	H_2O_2	Perez-Gilabert et al. (1997)
		SLO	H_2O_2	Yang and Kulkarni (1998)
Chlorpromazine	N-Demethylation	HTPLO, SLO	Linoleic acid	Rajadhyaksha et al. (1999)
		SLO	H_2O_2	Hover and Kulkarni (2000c)
		HTPLO, SLO	CHP, TBHP	Hover and Kulkarni (2000d)
Promazine	N-Demethylation	HTPLO, SLO	Linoleic acid	Rajadhyaksha et al. (1999)
		SLO	H_2O_2	Hover and Kulkarni (2000c)
		HTPLO, SLO	CHP, TBHP	Hover and Kulkarni (2000d)
Promethazine	N-Demethylation	HTPLO, SLO	Linoleic acid	Rajadhyaksha et al. (1999)
		SLO	H_2O_2	Hover and Kulkarni (2000c)
		HTPLO, SLO	CHP, TBHP	Hover and Kulkarni (2000d)
Triflupromazine	N-Demethylation	SLO	H_2O_2	Rajadhyaksha et al. (1999)
Trimeprazine	N-Demethylation	HTPLO, SLO	Linoleic acid	Hover and Kulkarni (2000c)
		SLO	H_2O_2	Rajadhyaksha et al. (1999)
		HTPLO, SLO	CHP, TBHP	Hover and Kulkarni (2000d)
Trifluoperazine	N-Demethylation	SLO	H_2O_2	Rajadhyaksha et al. (1999)
Imipramine	N-Demethylation	SLO	H_2O_2	Hu et al. (1999)
Desipramine	N-Demethylation	SLO	H_2O_2	Hu et al. (1999)
Trimipramine	N-Demethylation	SLO	H_2O_2	Hu et al. (1999)
Clomipramine	N-Demethylation	SLO	H_2O_2	Hu et al. (1999)
Amitriptyline	N-Demethylation	SLO	H_2O_2	Hu et al. (1999)
Diltiazem	N-Demethylation	SLO	H_2O_2	Hu et al. (1999)
Doxepin	N-Demethylation	SLO	H_2O_2	Hu et al. (1999)
Phenytoin	Oxidation	SLO	Linoleic acid	Yu and Wells (1995)
Diethylstilboestrol	Oxidation	SLO	H_2O_2	Nunez-Delicado et al. (1997b)
Isoproterenol	Oxidation	SLO	H_2O_2	Rosie et al. (1994)
Oxyphenbutazone	Hydroxylation	SLO	Linoleic acid	Portoghese et al. (1975)
Phenidone	Oxidation	SLO, 5-LO	13-HPOD	Cucurou et al. (1991)

SLO, soybean lipoxygenase-1; HTPLO, human term placental lipoxygenase; CHP, cumene hydroperoxide; TBHP, tert-butyl hydroperoxide; 13-HPOD, 13-hydroperoxy octadecadienoic acid.

phenothiazines (Rajadhyaksha *et al.* 1999). Phenothiazines are easily demethylated releasing formaldehyde in the H_2O_2 supplemented incubation media containing either SLO or HTPLO (Table 7.4) (Rajadhyaksha *et al.* 1999). Among the phenothiazines tested, promazine is oxidised by SLO at the highest rate (828 nmol/min/mg protein) while the slowest oxidation rate is exhibited by triflupromazine (12 nmol/ min/mg protein). A similar structure–activity response was displayed by HTPLO. Cumene hydroperoxide and tert-butyl hydroperoxide also support the reaction (Table 7.4) with equal or better efficiency (Hover and Kulkarni 2000d). In this case, promethazine appears to be a better substrate than chlorpromazine (CPZ). Additional experimental evidence (Hover and Kulkarni 2000c) revealed that although H_2O_2 and synthetic hydroperoxides are better cofactors, polyunsaturated fatty acids can also support *N*-dealkylation of phenothiazines by SLO and HTPLO. Linoleic acid is up to eight times more efficient in supporting *N*-dealkylation of CPZ than either γ-linolenic acid or arachidonic acid. Although CPZ, promazine, promethazine and trimeprazine are oxidised in the linoleic acid-supported HTPLO-catalysed reaction (Table 7.4), promethazine seems to be the best substrate while promazine, the most resistant.

Phenytoin is an efficacious anticonvulsant and a known human teratogen. The SLO-mediated arachidonate- or linoleate-dependent oxidation to free radical species is accompanied by covalent binding to proteins which can be suppressed by LO inhibitors such as NDGA, quercetin, BW755C, and ETYA (Kubow and Wells 1988; Yu and Wells 1995). Cyclophosphamide is a prodrug whose toxicity and therapeutic efficacy depends on its metabolic activation. Kanekal and Kehrer (1994) observed co-oxidation of cyclophosphamide with SLO and 15-LO from rabbit reticulocytes in the presence of linoleic acid. This results in the generation of unstable tautomer of 4-hydroxy intermediate, an active metabolite, which breaks down to release acrolein phosphoramide mustard in the incubation medium.

Diethystilboestrol (DES), a human transplacental carcinogen, is widely used in livestock and in few other therapeutic applications. It is oxidised by P450, and serves as an excellent reducing substrate for various peroxidases including PGS. DES-quinone, one of the metabolites of DES, binds to DNA and is presumed to be the ultimate toxicant. Although DES-quinone formation by human tissue LO has not yet been reported, Nunez-Delicado *et al.* (1997b) noted that SLO is capable of the initial one-electron oxidation of DES to DES semiquinone in the presence of H_2O_2. Subsequently, dismutation of two molecules of DES semiquinone yields one molecule each of DES-quionone and DES. Isoproterenol, a β-adrenoreceptor agonist, is widely used as a bronchodilator in the treatment for asthma. This two-electron donor compound serves as a co-substrate for the hydroperoxidase activity of SLO in the presence of H_2O_2 and is easily oxidised to a final stable product, aminochrome (Rosei *et al.* 1994; Nunez-Delicado *et al.* 1996). A sequential generation of *o*-quinone and leukaminochrome as intermediates was postulated in the aminochrome production (Nunez-Delicado *et al.* 1996).

OTHER CHEMICALS

Several model chemicals, drugs and other chemicals of diverse structures are also known to be oxidised via the LO pathway (Table 7.5).

Table 7.5 Oxidation of other compounds via lipoxygenase pathway

Chemical	Enzyme (source)	Cofactor	References
N-Methyl aniline	SLO and HTPLO	H_2O_2	Hover and Kulkarni (2000b)
N,N-Dimethyl aniline	SLO and HTPLO	H_2O_2	Hover and Kulkarni (2000b)
N,N,N',N'-Tetramethyl benzidine	SLO and HTPLO	H_2O_2	Hover and Kulkarni (2000b)
N,N-Dimethyl-p-phelylenediamine	SLO and HTPLO	H_2O_2	Hover and Kulkarni (2000b)
N,N-Dimethyl-3-nitroaniline	SLO and HTPLO	H_2O_2	Hover and Kulkarni (2000b)
N,N-Dimethyl-p-toluidine	SLO and HTPLO	H_2O_2	Hover and Kulkarni (2000b)
Phenol	SLO	13-HPOD	Mansuy et al. (1988)
p-Aminophenol	SLO, Potato 5-LO	13-HPOD	Cucurou et al. (1991)
1,3-Dihydroxybenzene (Resorcinol)	SLO, Potato 5-LO	13-HPOD	Cucurou et al. (1991)
1,2-Dihydroxybenzene(pyrocatechol)	SLO, Potato 5-LO	13-HPOD	Cucurou et al. (1991)
Guaiacol (O-methoxyphenol)	SLO	13-HPOD	Streckert and Stan (1975)
	SLO, Potato 5-LO	13-HPOD	Cucurou et al. (1991)
	SLO	H_2O_2	Kulkarni and Cook (1988a,b)
	SLO	H_2O_2	Fontana et al. (1997)
1,2,3-Trihydroxybenzene (Pyrogallol)	SLO	H_2O_2	Kulkarni and Cook (1988a,b)
	SLO	H_2O_2	Fontana et al. (1997)
4-Methyl catechol	SLO	H_2O_2	Nunez-Delicado et al. (1999)
4-tert-Butyl catechol	SLO	H_2O_2	Nunez-Delicado et al. (1999)
4-tert-Octyl catechol	SLO	H_2O_2	Nunez-Delicado et al. (1999)
Trolox C	SLO	H_2O_2	Nunez Delicado et al. (1997a)
2,2,5,7,8-Pentamethylchroman-6-ol	SLO	H_2O_2	Nunez-Delicado et al.(1997c)
Hydrocaffeic acid	SLO	H_2O_2	Mosca et al. (1996)
Salsolinol	SLO	H_2O_2	Mosca et al. (1996)

continued overleaf

Table 7.5 (continued)

Chemical	Enzyme (source)	Cofactor	References
Tetrahydropapaveroline	SLO	H_2O_2	Mosca et al. (1996)
Dopa methyl ester	SLO	H_2O_2	Mosca et al. (1996)
Thiobenzamide	SLO	Linoleic acid	Naidu and Kulkarni (1991)
ABTS	SLO	H_2O_2	Kulkarni and Cook (1988a,b)
	SLO	H_2O_2	Fontana et al. (1997)
Scopoletin	SLO	H_2O_2	Fontana et al. (1997)
Homovanillic acid	SLO	H_2O_2	Fontana et al. (1997)
Pentadienols (8 chemicals)	SLO	–	Zhang and Kyler (1989)
N-Phenyllinoleamide	Mouse macrophage		Bioque et al. (1995)
N,N,N',N'-Tetramethyl-1,4-phenylenediamine	SLO	Linoleic acid	Kulkarni and Cook (1988a)
	SLO	H_2O_2	Kulkarni and Cook (1988b)
3,3′,5,5′-Tetramethylbenzidine	SLO	Linoleic acid	Kulkarni and Cook (1988a)
	SLO	H_2O_2	Kulkarni and Cook (1988b)
3,3′-Dimethoxybenzidine	SLO	Linoleic acid	Kulkarni and Cook (1988a)
	SLO	H_2O_2	Kulkarni and Cook (1988b)
p-Phenylenediamine	SLO	Linoleic acid	Kulkarni and Cook (1988a)
	SLO	H_2O_2	Kulkarni and Cook (1988b)
4-Hydroxy-β-ionone	SLO	Linoleic acid	Kulkarni and Cook (1988b)
trans-4-Hydroxy-2-nonenal	SLO	PUFA, H_2O_2	Waldmann and Schrieber (1995)
		9-,13-HPOD	
3Z-Nonenal	SLO	13-HPOD	Chen and Chung (1996)
Pinolenic acid	15-LO	–	Gardner and Grove (1998)
Crocin	SLO	Linoleic acid	Kuklev et al. (1993)
2-Bromorthanolamine	Rat medullary interstitial cells	–	Spaapen et al. (1980)
			Grieve et al. (1990)

SLO, soybean lipoxygenase-1; HTPLO, human term placental lipoxygenase; 13-HPOD, 13-hydroperoxy octadecadienoic acid; ABTS, 2,2-azinobis(3-ethylbenzo-thiazoline-6-sulphonic acid).

Drug–chemical interactions

Unintentional human exposure to multiple chemicals from the working or living environment is unavoidable and common. Also, under certain medical situations, some patients simultaneously consume multiple drugs. The major concern over such concurrent exposure to chemicals is an exaggerated or diminished biological response, which is always difficult to predict or understand. To provide a logical explanation, an assessment of the metabolic fate of each component of the mixture has been a focal point of many investigations. At least three examples of drug–chemical interactions involving LO can be cited. Hu and Kulkarni (2000) tested a hypothesis that under certain conditions, the primary free radical metabolites of efficient substrates for LO may stimulate the generation of reactive species from other chemicals in the secondary reaction. The evidence presented indicates that SLO-generated CPZ cation radical ($CPZ^{+\bullet}$) serves as a shuttle oxidant which is capable of simultaneous oxidation of several other chemicals. Such a metabolic interaction results in a 42-fold stimulation of benzidine diimine formation from benzidine by SLO in the H_2O_2 supplemented incubation media. Seven other phenothiazines also display similar phenomenon and stimulate benzidine oxidation, albeit to a lesser extent. $CPZ^{+\bullet}$ also stimulates the oxidation of tetramethyl benzidine, o-dianisidine, guaiacol, pyrogallol, phenylenediamine, and tetramethyl phenylenediamine (Table 7.6). The highest degree of $CPZ^{+\bullet}$-caused stimulation (94-fold) was noted in the Wursters Blue radical formation from tetramethyl phenylenediamine while the least degree of enhancement was noted with guaiacol oxidation by SLO.

L-Dopa deficiency in brain neurones is a hallmark of Parkinsonism. Although levodopa is the therapeutic agent commonly used to treat the disease, a combination of CPZ and L-dopa is contraindicated. Many investigators have invoked a receptor mechanism to explain this undesirable drug–drug interaction. However, the results of the metabolic study (Persad *et al.* 2000) conducted with SLO may provide an alternate explanation since it was observed that the oxidation of dopa to dopachrome is stimulated ~25-fold by $CPZ^{+\bullet}$ generated by SLO. *In vivo*, dopa undergoes decarboxylation to produce dopamine. However, *in vitro*, $CPZ^{+\bullet}$ also accelerates dopamine

Table 7.6 Stimulation of xenobiotic oxidation by soybean lipoxygenase in the presence of chlorpromazine

Xenobiotic	Final concentration (μM)	Fold stimulation	References
Benzidine	10	42	Hu and Kulkarni (2000)
Tetramethyl benzidine	25	8	Hu and Kulkarni (2000)
o-Dianisidine	25	2	Hu and Kulkarni (2000)
Guaiacol	100	4	Hu and Kulkarni (2000)
Pyrogallol	10	24	Hu and Kulkarni (2000)
Phenylenediamine	25	25	Hu and Kulkarni (2000)
Tetramethyl phenylenediamine	50	94	Hu and Kulkarni(2000)
L-Dopa	1000	25	Persad *et al.* (2000)

The reaction media contained 200 nM SLO, 1.0 mM chlorpromazine and 1.0 mM H_2O_2 and the indicated concentration of the test xenobiotic.

oxidation significantly. Thus, it appears that the hyperoxidation of both dopa and dopamine via the LO pathway may contribute, at least in part, to the exhaustion of cellular reserves and thereby may impart dopamine deficiency as observed in the brains of the Parkinsonism patients. An enhancement of H_2O_2-dependent SLO-catalysed oxidation of dopa, 5-S-cysteinyl-dopa and 5-S-cysteinyl dopamine by catechols such as NDGA, salsolinol, tetrahydropapaveroline, dopa methyl ester, hydrocaffeic acid, and caffeic acid (Mosca *et al.* 1996) represents another example of metabolic modulation by LO. Dopa was ineffective while NDGA-caused enhancement in the initial rate of oxidation was five-fold with dopa, 14-fold with 5-*S*-cysteinyl-dopa and 2-fold with 5-S-cysteinyl dopamine. The authors (Mosca *et al.* 1996) proposed that the mechanism of enhancement involves redox cycling of catechols. Thus, the quinones produced by SLO rapidly interact with the test substrate causing its non-enzymatic oxidation.

Modulation of lipoxygenase activity

The anecdotal reports support the notion that LO is an inducible enzyme. For example, Coffey *et al.* (1996) reported that effect of 2-h human exposures to 0.4 ppm ozone results in a 8-fold increase in the bronchoalveolar lavage fluid LTC_4 content. Exposure of rabbits to phosgene results in a 10-fold increase in the synthesis of LO-generated arachidonate metabolites in lungs (Guo *et al.* 1990). The lung vascular injury caused by intraperitoneal injection of endotoxin precipitates into high levels of lung 5-HETE and LTC_4 in rats (Chang *et al.* 1989). Bailey *et al.* (1993) reported that in rabbits fed an atherogenic diet containing 1% cholesterol for 14 weeks, 15-LO levels in heart, aortic adventitia, and lung, but not liver, are increased up to 100-fold. The subcutaneous treatment of rabbits for 5-6 days with the haemolytic agent, phenylhydrazine, results in >1000-fold increase in 15-LO levels in heart, lung and aorta (Bailey *et al.* 1993, 1995). Similar results were obtained when severe anaemia was induced in rats by frequent bleeding (Bailey *et al.* 1996). Infusion of the oxidant tert-butyl hydroperoxide in isolated rabbit lung elevates LTB_4, C_4, D_4, and E_4 production by 2 to 3-fold (Farrukh *et al.* 1988). A significant rise in 15-HETE occurs when human bronchial epithelial cells are exposed to toluene diisocyanate (Mattoli *et al.* 1990). Acrolein exposure to cultured bovine tracheal epithelial cells leads to elevation of 12-HETE and 15-HETE production (Doupnik and Leikauf 1990).

An exposure of A/J mice to NNK for 7 weeks in the drinking water causes a >2-fold increase in the plasma concentration of LTB_4 (Castonguay *et al.* 1998; Rioux and Castonguay 1998) and this effect can be blocked by the co-treatment with the LO inhibitor, A-79175. The authors proposed that NNK-derived intermediates induce the expression of 5-LO (Castonguay *et al.* 1998). Feeding oxidised palm oil to rats for 3 days results in more than 2-fold increase in the liver LO activity (Pereira and Das 1991). A significant increase in the production of 15-HETE and LTB_4 was reported in the livers of rats treated with the diabetogen, streptozotocin by Rosello-Catafau *et al.* (1994). A recent report indicates that the streptozotocin-induced experimental diabetes is also associated with an increase in the renal LO activity (Stewart *et al.* 1998).

Perchellet and Perchellet (1989) reported that daily application of 12-*O*-tetradecanoylphorbol-13-acetate (TPA) results in up to 4-fold increase in the lipid hydroperoxide levels in mouse skin and the effect is blocked by the LO inhibitors but not by the cyclooxygenase inhibitors. According to Jiang *et al.* (1994), a maximal induction of mouse epidermal 8-LO results 24 h after TPA application. The oral treatment of mice with TMK688, a LO inhibitor, not only inhibits LO induction but also protects mice from skin carcinogenesis induced by 7,12-dimethylbenz[*a*]anthracene and TPA plus BP (Jiang *et al.* 1994). In human erythroleukaemia cells, the membrane fraction contains about 90% of the total cellular 12-LO activity and the pretreatment of the cells for 3 days with TPA causes a marked, time-dependent increase in membrane-bound 12-LO activity and protein (Hagmann *et al.* 1993). Phorbol 12-myristate 13-acetate also causes a significant increase in the expression of microsomal 12-LO activity and mRNA in human epidermoid carcinoma A431 cells (Liaw *et al.* 1998). Dexamethasone treatment has been found to result in a selective upregulation of 5-LO in human mast cells (Colamorea *et al.* 1999) and in DMSO-stimulated HL-60 cells (Zaitsu *et al.* 1998). The increased expression of 5-LO and FLAP in human monocytes and THP-1 cells can be noted with the dexamethasone treatment (Riddick *et al.* 1997). Reports describing an increase in the LT production following an exposure to TPA and Ca^{2+} ionophore A23187 in isolated perfused rat liver (Hagmann *et al.* 1989) or to ethanol by rat hepatocytes (Peres *et al.* 1984) in primary culture are also available. Bioque *et al.* (1992) reported that a 2-hour exposure of mouse peritoneal macrophages to *N*-phenyllinoleamide increases the LO-mediated metabolism of exogenous arachidonate to 12-HETE. According to Brungs *et al.* (1994), vitamin D_3 addition to culture medium causes a 4-fold induction of 5-LO mRNA, a 14-fold increase in 5-LO protein and a 38-fold upregulation of 5-LO activity of intact HL-60 cells after differentiation in the presence of DMSO and serum protein for 4 days. Hamasaki and Miyazaki (1991) observed that overnight incubation of the calcium-ionophore-stimulated RBL-1 cells with retinoic acid enhances LTC_4 production by >28-fold and LTD_4 by >7-fold. However, the authors concluded that the induction of LTC_4 synthase, but not 5-LO was responsible for these results.

Lipoxygenase inhibition

Several hundred chemicals have been synthesised and screened during the past few years as selective inhibitors for 5-, 12-, or 15-LO. Many reviews covering this subject are available (Batt 1992; Negro *et al.* 1997; Cornicelli and Trivedi 1999). Unfortunately, there is no universally accepted unified approach to evaluate the relative potency of different chemicals to cause inhibition of LOs. The absence of a standardised protocol has resulted in a wide variation in reports in terms of the enzyme source, assay procedure and end points measured to establish the potency of candidate compounds.

To establish the role of a specific enzyme in xenobiotic oxidation *in vivo*, most investigators, use so-called 'selective inhibitors'. However, a serious problem of non-specificity, often ignored by many, exists. Thus, for example, SKF-525 and metyrapone, which are considered as selective inhibitors of P450, also inhibit SLO with respective IC_{50} values of 40 and 150 µM (Pretus *et al.* 1985). Indomethacin and several non-steroidal anti-inflammatory drugs (NSAID), are widely believed as the specific PGS inhibitors, but they cause a significant denaturation of rat hepatic microsomal P450 (Falzon *et al.* 1986). Siegel *et al.* (1980b) have reported that aspirin, indomethacin, sodium salicylate, phenylbutazone, ibuprofen, naproxen and sulindac, the PGS inhibitors, can also inhibit the LO in rat neutrophils from carrageenan pleural exuate. The reports are not consistent with indomethacin. Thus, indomethacin causes a three-fold increase in LTB_4 formation (Docherty and Wilson 1987) in the calcium ionophore stimulated human neutrophils along with an increase in the 5-, 12-, and 15-HETE formation. Indomethacin blocks the LO pathway in hamster isolated lungs (Uotila *et al.* 1981). Apparently, indomethacin, aspirin and salicylate inhibit the conversion of 12-HPETE to 12-HETE in the LO preparations of human blood platelets (Siegel *et al.* 1979, 1980a). According to Sircar *et al.* (1983), indomethacin and other anti-inflammatory drugs inhibit SLO while Marcinkiewicz *et al.* (1985) observed no decrease in the rate of oxygen consumption during arachidonate peroxidation by SLO. Indomethacin caused $\sim$ 50% inhibition of covalent binding of activated phenytoin by SLO (Yu and Wells 1995) while it enhanced significantly the binding of arachidonate-dependent activated BP in the cytosol of rat liver and lung (Nemoto and Takayama 1984). Acetylenic fatty acids, e.g. ETI and ETYA, are widely accepted as dual inhibitors of PGS and LO pathways. However, they can also effectively inhibit different glutathione *S*-transferases (Datta and Kulkarni 1994b). NDGA, a commonly used LO inhibitor also inhibits PGS (Smith and Marnett 1991). Considering these reports, it is imperative that investigators exercise extreme caution while interpreting the results and drawing conclusions. Using this approach, it is nearly impossible to establish the role played by LO in xenobiotic oxidation *in vivo*.

Despite numerous difficulties (see above), continued efforts in this area have led to the discovery of some potent LO inhibitors which exhibit high potency and isoform selectivity. According to Suzuki *et al.* (1997), compound YT-18 (2,3-dihydro-2,4,6,7-tetramethyl-2-[(4-phenyl-1-piperazinyl)methyl]-5-benzofuranamine) selectively inhibits 5-LO from human and porcine leukocytes and RBL cells but has almost no effect on 12-, 15-LO and cyclooxygenase-1 and -2. Gorins *et al.* (1996) tested a series of (carboxyalkyl)benzyl propargyl ethers as inhibitors of 12-LO from porcine leukocyte cytosol. The most potent acetylenic (carboxyalkyl)benzyl ethers did not inhibit human platelet 12-LO, human neutrophil 5-LO, rabbit reticulocyte 15-LO or soybean 15-LO. Some examples of so called selective LO inhibitors are shown in Figure 7.11.

Since a large number of chemicals inhibit LO activity, several mechanisms have been proposed to explain their mode of action. Thus a chemical may inhibit LO activity by serving as an antioxidant, iron chelator, substrate analogue, FLAP inhibitor, blocker of LO induction etc. However, the discussion here is limited to those inhibitors which are co-oxidised in the process (Table 7.7). The phenolic antioxidants BHT and BHA, which break free radical chain reactions, also block xenobiotic metabolism by different LOs. The ESR data reported by Kagan *et al.* (1990) demonstrate that seven BHT homologues undergo one-electron oxidation by SLO in the presence of linoleic acid to generate free radical species. It was proposed that phenoxyl radicals are formed from an interaction of phenols with the SLO-derived lipid peroxyl radicals. According to Schilderman *et al.* (1993), linoleate-supported metabolism of 2-tert-butyl(1,4)hydroquinone, a demethylated metabolite of antioxidant BHA, by SLO involves a two-electron oxidation process that directly yields 2-tert-butyl(1,4)paraquinone without semiquinone radical or oxygen radical formation.

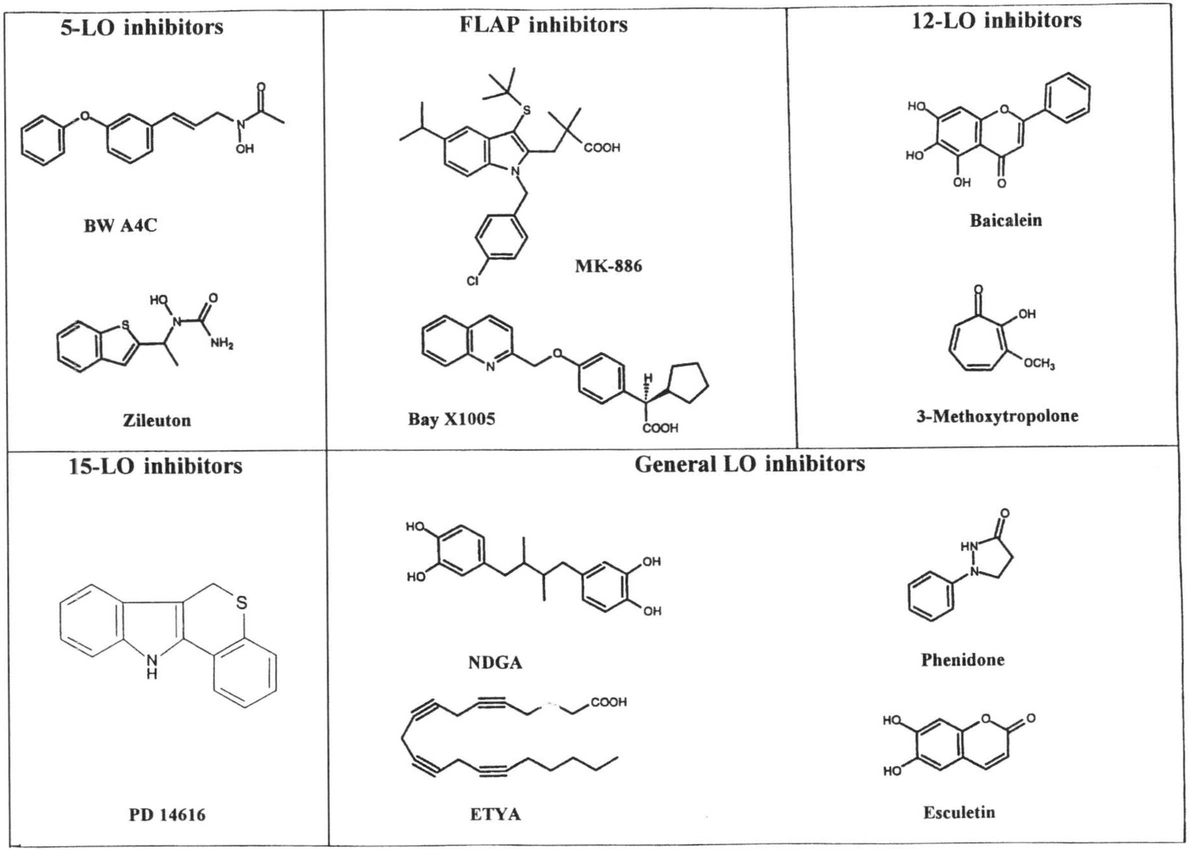

Figure 7.11 Some inhibitors of lipoxygenase.

Table 7.7 Oxidation of inhibitors and related compounds by lipoxygenase

Chemical	Enzyme	Co-factor	References
N-(4-Chlorophenyl)-N-hydroxy-N'-(3-chlorophenyl)urea (CPHU)	SLO	LA, 13-HPOD	Chamulitrat et al. (1992)
	SLO	13-HPOD	Falgueyret et al. (1992)
	SLO	13-HPOD	Desmarais et al. (1994)
N-[(E)-3-(3-Phenoxyphenyl)prop-2-enyl]acetohydroxamic acid (BWA4C)	5-LO	13-HPOD	Riendeau et al. (1991)
N(1-Benzo[b]thien-2-ylethyl)-N-hydroxyurea (Zileuton)	SLO	LA, 13-HPOD	Chamulitrat et al. (1992)
	SLO, 5-,12-LO	13-HPOD	Falgueyret et al. (1992)
	SLO	LA, 13-HPOD	Chamulitrat et al. (1992)
	SLO, 5-,12-LO	13-HPOD	Falgueyret et al. (1992)
Dimehtylhydroxylamine	SLO	LA, 13-HPOD	Chamulitrat et al. (1992)
Isopropylhydroxylamine	SLO	LA, 13-HPOD	Chamulitrat et al. (1992)
Cyclohexylhydroxylamine	SLO	LA, 13-HPOD	Chamulitrat et al. (1992)
Desferal	SLO	LA, 13-HPOD	Chamulitrat et al. (1992)
N-Hydroxyurea	SLO	LA, 13-HPOD	Chamulitrat et al. (1992)
5-Hydroxy-2-phenethyl-2,3-dihydrobenzofuran	5-LO	13-HPOD	Riendeau et al. (1991)
5-Hydroxy-2-phenethyl-6-(3-phenoxypropyl)-2,3-dihydrobenzofuran	5-LO	13-HPOD	Riendeau et al. (1991)
7-Chloro-4-hydroxy-2-(4-methoxyphenyl)-methyl-3-methylbenzofuran	5-LO	13-HPOD	Riendeau et al. (1991)
7-Chloro-4-hydroxy-2-[(4-methoxyphenyl)-methyl]-3-methyl-5-propylbenzofuran	5-LO	13-HPOD	Riendeau et al. (1991)
N-(4-Chlorophenyl)-N-hydroxy-N'-(3-chlorophenyl)urea	5-LO	13-HPOD	Riendeau et al. (1991)
N-(3,4-Dichlorophenyl)-N-hydroxy-N'-(3-chlorophenyl)urea	5-LO	13-HPOD	Riendeau et al. (1991)
N-Methyl-2-[4-(2,4,6-trimethylphenyl)phenyl]propenehydroxamic acid	5-LO	13-HPOD	Riendeau et al. (1991)

N-(1-benzo[b]thien-2-ylethyl)- N-hydroxyurea (A-64077)	5-LO	13-HPOD	Riendeau et al. (1991)
Phenidone	SLO	13-HPOD	Mansuy et al. (1988)
BW 755C	SLO	13-HPOD	Mansuy et al. (1988)
BW755C	SLO	15-HPETE	Reynolds (1988)
BW540C	SLO	15-HPETE	Reynolds (1988)
BW A4C	SLO	15-HPETE	Reynolds (1988)
BW A137C	SLO	15-HPETE	Reynolds (1988)
1-Phenyl-3-amino-2-pyrazoline	SLO	13-HPOD	Mansuy et al. (1988)
Nordihydroguaiaretic acid (NDGA)	SLO	13-HPOD	Mansuy et al. (1988)
	SLO	H_2O_2	Mosca et al. (1996)
Phenylhydrazine	SLO	13-HPOD	Mansuy et al. (1988)
2-Hydrazinopyridine	SLO	13-HPOD	Mansuy et al. (1988)
N-Alkylhydroxylamines (7 compounds)	SLO	LA, 13-HPOD	Clapp et al. (1985)
(E)Hexanal phenylhydrazone	SLO	LA	Galey et al. (1988)
2,2,5,7,8-Pentamethylchroman-6-ol	SLO	H_2O_2	Nunez-Delicado et al. (1997c)
N-Hydroxyamphetamine	SLO	13-HPOD	Mansuy et al. (1988)
N-Methylbenzaldehyde-4'-bromophenylhydrazone	SLO	13-HPOD	Mansuy et al. (1988)
N'-Phenylbenzoylhydrazine	SLO	13-HPOD	Mansuy et al. (1988)
Caffeic acid	SLO	H_2O_2	Mosca et al. (1996)

SLO, soybean lipoxygenase; LA, Linoleic acid, 13-HPOD, 13-hydroperoxy-9,11-octadecadienoic acid.

Vitamin A derivatives are free radical scavenger antioxidants and serve as LO inhibitors. Hypervitaminosis-A has been linked with birth defects in humans and animals. The parent retinoid *per se* may not be teratogenic, and bioactivation is required. Lomnitski *et al.* (1993) found a significant linear correlation between SLO-2 inhibition and fatty acid peroxyl radical-dependent bleaching of β-carotene. Wu *et al.* (1999) tentatively identified apocarotenal, epoxycarotenal, apocarotenone and epoxycarotenone among several metabolites of β-carotene generated by SLO and pea LO. Retinol displays a high affinity towards LO and behaves as a competitive inhibitor of the enzyme (Lomnitski *et al.* 1993). According to Datta and Kulkarni (1996), all-*trans*-retinol acetate is an excellent substrate for co-oxidation for SLO and HTPLO in the presence of linoleic acid. It was proposed that the peroxyl radical of linoleic acid generated by LO attacks the π-electrons of the C=C bond and produces 5,6-epoxide of all-*trans*-retinol acetate. The formation of the 5,6-epoxy metabolite has been noted during the SLO-catalysed co-oxidation of all-*trans*-retinoic acid (Matsui *et al.* 1994), retinol, and β-ionone (Waldmann and Schreier 1995). All-trans-retinol acetate is also co-oxidised by the purified LO from chickpeas (Sanz *et al.* 1992). Vitamin E (α-tocopherol) and its seven analogues serve as substrates for the hydroperoxidase activity of SLO and potato tuber 5-LO in the presence of 13-HPOD (Cucurou *et al.* 1991). However, the oxidation rates exhibit some LO-specific differences. Trolox C, a phenolic antioxidant, was found to undergo an H_2O_2-supported multistep oxidation process catalysed by SLO (Nunez-Delicado *et al.* 1997a). The authors proposed that the one-electron oxidation generates the phenoxyl radicals from Trolox-C which, following dismutation, results in the formation cross-conjugated ketodiene. Hydrolysis of the ketodiene finally yields Trolox-C quinone. However, Trolox-C oxidation is not supported by 13-HPOD with either SLO or potato tuber 5-LO (Cucurou *et al.* 1991).

Reynolds (1988) demonstrated that SLO is rapidly inactivated when incubated with arachidonate and either NDGA, the aminopyrazolines BW 755C or BW 540C or the acetohydroxamic acid derivatives BW A4C and BW A137C. 15-HPETE was as effective as arachidonate in promoting inactivation, but linoleic acid and 13-HPOD were much less effective. SLO inhibition was linked to the pseudoperoxidase activity of the enzyme (Reynolds 1988). Using a spectrophotometric assay, Riendeau *et al.* (1991) examined the ability of purified 5-LO from porcine leukocytes to degrade 13-HPOD in the presence of derivatives of diphenyl-*N*-hydroxybenzofurans, 4-hydroxybenzofurans and 5-hydroxydihydrobenzofurans (Table 7.7). A strong stimulation of pseudoperoxidase reaction could be detected only with very effective inhibitors of LTB$_4$ biosynthesis by human leukocytes. The results indicated that *N*-hydroxyurea and benzofuranol derivatives can function as reducing agents for the enzyme. *N*-(4-Chlorophenyl)-*N*-hydroxy-*N'*-(3-chlorophenyl)urea (CPHU) serves as a reducing agent and stimulates 13-HPOD-supported hydroperoxidase activity of the recombinant human 5-LO, porcine leukocyte 12-LO and SLO (Falgueyret *et al.* 1992; Desmarais *et al.* 1994). RP-HPLC data support the proposal put forth by the authors that nitroxide radical is formed during the SLO-mediated metabolism of CPHU in the presence of 13-HPOD. ESR studies conducted by Chamulitrat *et al.* (1992) provided the direct evidence that NOH of the hydroxamate group of CPHU, *N*–[(E)-3-(3-phenoxyphenyl)prop-2-enyl]acetohydroxamic acid (BW A4C) and *N*-(1-benzo(b)thien-2-ylethyl)-*N*-hydroxyurea (Zileuton) is oxidised by SLO to form their corresponding nitroxides when incubated in the presence of linoleic acid. In addition, the authors provided ESR evidence and documented the formation of expected nitroxide metabolites as the one-electron oxidation products of dimethylhydroxylamine, isopropylhydroxylamine, desferal, N-hydroxyurea and cyclohexylhydroxylamine in the incubation media containing SLO and 13-HPOD. N-hydroxyurea is an animal teratogen while zileuton, a 5-LO inhibitor is a drug for the treatment of asthma. As an 5-LO inhibitor, diphenyl disulphide was a thousand-fold more potent than diethyldisulphide and it is noteworthy that glutathione, a typical reducing thiol, is almost inactive in inhibiting this enzyme even at 80 mM (Egan and Gale 1985). Other chemical classes reported as LO inhibitors include n-alcohols and n-alkylthiols (Kuninori et al. 1992), flavonoids (Yoshimoto et al. 1983; Kim et al. 1998; You et al. 1999) and many others. Some of these chemicals are expected to be oxidised by LO.

Lipoxygenase-mediated glutathione conjugation of xenobiotics

The formation of GSH conjugate, a result of an interaction between an electrophile of either endo- or exo-biotic origin and the sulphydryl group in the GSH molecule, is one of the most common reactions encountered in the human body. Currently, many believe that glutathione transferase (GST) is the only pathway responsible for an enzymatic generation of thioethers from xenobiotics in different mammalian tissues. Kulkarni and Sajan (1997) were first to report GSH conjugation of ethacrynic acid (EA), a diuretic drug, by SLO in the presence of arachidonic acid, linoleic acid and γ-linolenic acid. Spectrophotometric, TLC, HPLC, radiometry and MS analyses of the reaction media indicated that both SLO and GST produce an identical adduct, i.e. EA-SG. The rate of EA-SG formation was up to 1650-fold greater than that observed with different purified isozymes of mammalian GSTs. A recent study (Kulkarni and Sajan 1999) has further established that human tissue LO (HTPLO) is also capable of extensive EA-SG formation. A marked blockade of EA-SG formation by NDGA, ETI, esculetin and gossypol clearly implicates LO involvement in the reaction. The observations that the reaction is significantly suppressed by BHT, BHA and spin traps suggest the free radical nature of the reaction. Two possible mechanisms of EA-SG formation were proposed. As shown in Figure 7.12, it is envisioned that GSH is first oxidised by LO to GS$^\bullet$. In the second step, the GS$^\bullet$ directly attacks the C$=$C bond in the EA molecule to generate a carbon-centred radical of EA which reacts with another molecule of GSH to finally yield EA-SG. According to the second mechanism, EA is presumed to be first oxidised by LO to EA$^\bullet$. A spontaneous interaction of this cation radical with GSH is expected to result in EA-SG. Noteworthy is the fact that the reaction occurs at a significant rate under the physiologically relevant concentrations of GSH and fatty acid, and pH. These results strongly suggest that in vivo thioether formation from certain chemicals may occur via the LO pathway.

Another example of GSH conjugation of xenobiotics via LO pathway includes p-aminophenol (PAP). Although the mechanism(s) responsible for PAP nephrotoxicity is not yet established, oxidative metabolism and subsequent conjugation with GSH are believed to be the key steps involved. For several reasons, many investigators have dismissed the involvement of microsomal P450 and PGS in the PAP bioactivation process. On the other hand, the ESR study conducted with SLO has shown the

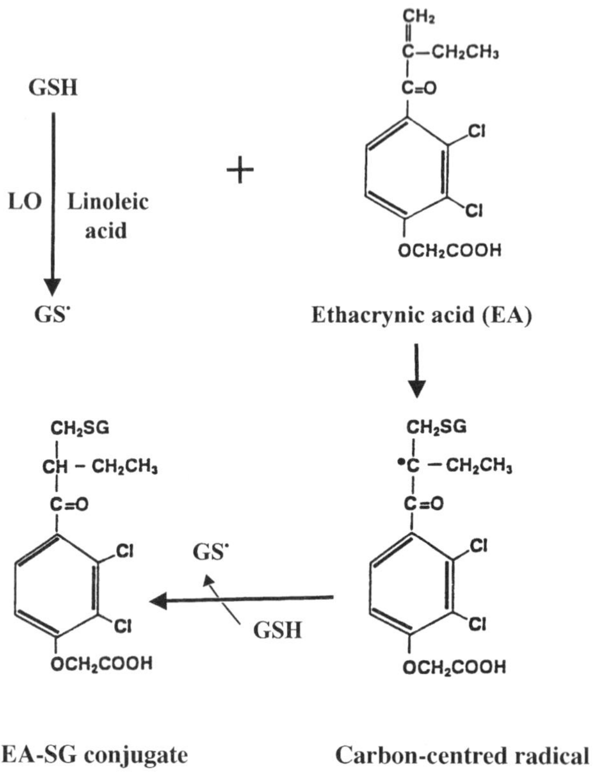

Figure 7.12 Conjugation of ethacrynic acid with glutathione by lipoxygenase.

formation of a short-lived 4-aminophenoxyl radical as the initial one-electron oxidation product of PAP in the reaction media supplemented with linoleic acid (Van der Zee *et al.* 1989). Furthermore, one-electron oxidation of PAP by SLO also occurs in the presence of 13-HPOD (Mansuy *et al.* 1988; Cucurou *et al.* 1991). Recently, Yang and Kulkarni (2000) noted SLO-mediated formation of GSH conjugates from PAP in the presence of H_2O_2. The LO inhibitors and free radical scavengers markedly decreased the rate of SLO-mediated GS-PAP formation. Since LO activity occurs both in the livers of rats (Roy and Kulkarni 1994) and humans (Roy and Kulkarni 1996b, 1997), and in human kidney (Oyekan *et al.* 1997; Stewart *et al.* 1997), a role for this pathway in nephrotoxicity of PAP is expected.

In vivo evidence

Gathering *in vivo* evidence to document LO involvement in xenobiotic oxidation has been troublesome. Currently, the use of so-called 'selective inhibitors' is popular, considering the non-selectivity of enzyme inhibitors (discussed above), the conclusions reached may be debatable. Despite this, the results of many studies suggest that

LO pathway plays a contributory role in xenobiotic oxidation *in vivo*. Castonguay *et al.* (1998) and Rioux and Castonguay (1998) investigated the preventive efficacy of PGS and 5-LO inhibitors against NNK carcinogenesis in female A/J mice. A-79175, the 5-LO inhibitor, was found to be a stronger inhibitor of lung tumourigenesis than the PGS inhibitor aspirin. Both lung tumour multiplicity and incidence were inhibited by A-79175. MK-886, which binds to FLAP and inhibits 5-LO also decreased the mean tumour volume (Rioux and Castonguay 1998). The authors proposed that besides P450, LO activates NNK (Castonguay *et al.* 1998; Rioux and Castonguay 1998). The incubation of 82-132 and LM2 murine lung tumour cells with MK-886 and A-79715 decreased NNK-caused cell proliferation in a concentration-dependent manner (Rioux and Castonguay 1998). The authors opined that an inhibition of NNK activation by LO may be the mechanism responsible for the observed effects. NDGA pretreatment diminishes the bladder toxicity of cyclophosphamide in male ICR mice (Frasier and Kehrer 1993). A significant decrease in the excretion of acrolein equivalents into urine was observed during the first 6 h following cyclophosphamide dosing. Although these results suggest that the LO pathway may be involved, the authors concluded that the protective action of NDGA and indomethacin is not due to interference with the metabolism of cyclophosphamide. An inhibition of the LO pathway also prevents DMBA-caused cancer of skin (Katiyar *et al.* 1992; Jiang *et al.* 1994) and mammary gland (Noguchi *et al.* 1993; Kitagawa and Noguchi 1994) suggesting *in vivo* carcinogen bioactivation by LO.

It has been documented that the preparations of rodent embryos (Vanderhoek and Klein 1988; Roy *et al.* 1993) and human intrauterine conceptual tissues during early gestation period (Joseph *et al.* 1994; Datta and Kulkarni 1994a; Datta *et al.* 1995) possess LO activity capable of xenobiotic oxidation. Yu and Wells (1995) observed that pretreatment of CD-1 mice with ETYA, a dual PGS and LO inhibitor, results in a dose-related decrease in the incidence of phenytoin-induced foetal cleft palates and resorptions. This reduction in phenytoin teratogenicity was considerably greater than that previously reported for acetylsalicylic acid, which inhibits PGS. Thus, these results provide strong evidence that LO pathway may be more important in the bioactivation of phenytoin *in vivo*.

Conclusions

Although the available information reviewed here reflects a promising start in the understanding of the role played by the LO pathway in xenobiotic oxidation, a lot needs to be accomplished. Few examples document oxidation, epoxidation, desulphuration, dearylation, sulphoxidation and dealkylation of xenobiotics; however, much remains to be investigated as regards the spectrum of reactions catalysed by different LOs. At present, the data are too sparse to establish clearly *in vivo* xenobiotic oxidation via the LO pathway. More data are needed to establish the relative significance of LO-mediated co-oxidation of chemicals in the light of other competing pathways. Although it appears that the use of inhibitors to selectively block the oxidative pathways of xenobiotic oxidation may finally turn out to be futile, an exploration of stereochemical differences in the metabolite generation may be beneficial. Exploitation of cell culture techniques and *ex-vivo* models for organs culture represents another fruitful approach that would provide a good start in this direction. Inducibility of LO and LO-mediated GSH conjugation of xenobiotics represent important areas which deserve serious attention as they may explain some of the toxicological puzzles. Finally, one should bear in mind that although laboratory animal data are useful for the advancement of science, one should not forget the critical need for human data since the science of toxicology is meant to serve human interests and not those of the rats.

References

Adriaenssens PI, Sivarajah K, Boorman GA, Eling TE and Anderson MW (1983) Effect of aspirin and indomethacin on the formation of benzo(a)pyrene-induced pulmonary adenomas and DNA adducts in A/HeJ mice. *Cancer Research*, **43**, 4762–4767.

Agundez JA, Martinez C and Benitez J (1995) Metabolism of aminopyrine and derivatives in man: in vivo study of monomorphic and polymorphic metabolic pathways. *Xenobiotica*, **25**, 417–427.

Amstad P and Cerruti P (1983) DNA binding of aflatoxin B_1-2,3-oxide: evidence for its formation in rat liver in vivo and by human liver microsomes. *Biochemical and Biophysical Research Communications*, **60**, 1036–1043.

Anderson KE, Hammons GJ, Kadlubar FF, Potter JD, Kederlik KR, Ilett KF, Minchin RF, Teitel CH, Chou HC, Martin MV, Guengerich FP, Barine GW, Lang NP and Peterson LA (1997) Metabolic activation of aromatic amines by human pancreas. *Carcinogenesis*, **18**, 1085–1092.

Aziz S, Wu Z and Robinson DS (1999) Potato lipoxygenase catalysed co-oxidation of β-carotene. *Food Chemistry*, **64**, 227–230.

Bailey JM, Makheja AM and Simon T (1993) 15-Lipoxygenase induction as an index of oxidative stress and atherogenesis. *Biochemical Society Transactions*, **21**, 406S.

Bailey JM, Makheja AM, Lee R and Simon T (1995) Systemic activation of 15-lipoxygenase in heart, lung, and vascular tissues by hypercholesterolemia: relationship to lipoprotein oxidation and atherogenesis. *Atherosclerosis*, **113**, 247–258.

Bailey JM, Lee R and Simon T (1996) Activation of 12- and 15-lipoxygenase in the cardio-pulmonary system by hemorrhagic events. *Biochemical Society Transactions*, **24**, 435S.

Batt DG (1992) 5-Lipoxygenase inhibitors and their anti-inflammatory activities. *Progress in Medicinal Chemistry*, **29**, 1–63.

Bell RL and Harris RR (1999) The enzymology and pharmacology of 5-lipoxygenase and 5-lipoxygenase activating protein. *Clinical Reviews in Allergy and Immunology*, **17**, 91–109.

Belkner J, Wiesner R, Kuhn H and Lankin VZ (1991) The oxygenation of cholesterol esters by the reticulocyte lipoxygenase. *FEBS Letters*, **279**, 110–114.

Belvedere G, Tursi F, Elovaara E and Vainio H (1983) Styrene oxidation to styrene oxide coupled with arachidonic acid oxidation by soybean lipoxygenase. *Toxicology Letters*, **18**, 39–44.

Berliner JA and Heinecke JW (1996) The role of oxidised lipoproteins in atherogenesis. *Free Radicals in Biology and Medicine*, **20**, 707–727.

Bioque G, Bulbena O, Gomez G, Rosello-Catafau J and Gelpi E. (1992) Influence of N-phenyllinoleamide from toxic oil samples on the lipoxygenase metabolism of exogenous arachidonic acid in mouse peritoneal macrophages. *Prostaglandins Leukotrienes and Essential Fatty Acids*, **47**, 187–191.

Bioque G, Abian J, Bulbena O, Rosello-Catafau J and Gelpi E (1995) Mass spectrometric identification of N-phenyllinoleamine metabolites in mouse peritoneal macrophages. *Rapid Communications in Mass Spectrometry*, **9**, 753–760.

Blarzino C, Mosca L, Foppoli C, Coccia R, DeMarco C and Rosei MA (1999) Lipoxygenase/H_2O_2-catalysed oxidation of dihydroxyindoles: synthesis of melanin pigments and study of their antioxidant properties. *Free Radical Biology & Medicine*, **26**, 446–453.

Brinckmann R, Schnurr K, Heydeck D, Rosenbach T, Kolde G and Kuhn H (1998) Membrane translocation of 15-lipoxygenase in hematopoietic cells is calcium dependent and activates the oxygenase activity of the enzyme. *Blood,* **91**, 64–74.

Brungs M, Radmark O, Samuelsson B and Steinhilber D (1994) On the induction of 5-lipoxygenase expression and activity in HL-60 cells: effects of vitamin-D3, retinoic acid, DMSO and TGF-beta. *Biochemical and Biophysical Research Communications,* **205**, 1572–1580.

Byczkowski JZ and Kulkarni AP (1989) Lipoxygenase-catalysed epoxidation of benzo(a)pyrene-7,8-dihydrodiol. *Biochemical and Biophysical Research Communications,* **159**, 1199–1205.

Byczkowski JZ and Kulkarni AP (1992) Linoleate-dependent co-oxygenation of benzo(a)pyrene-7,8-dihydrodiol by rat cytosolic lipoxygenase. *Xenobiotica,* **22**, 609–618.

Byczkowski JZ, Ramgoolie PJ and Kulkarni AP (1991) Proline hydroxylation by soybean lipoxygenase. *Biochemistry International,* **25**, 639–646.

Byczkowski JZ, Ramgoolie J and Kulkarni AP (1992) Some aspects of activation and inhibition of rat brain lipoxygenase. *International Journal of Biochemistry,* **24**, 1691–1695.

Castonguay A, Rioux N, Duperron C and Jalbert G (1998) Inhibition of lung tumourigenesis by NSAIDS : A working hypothesis. *Experimental Lung Research,* **24**, 605–615.

Cathcart MK, McNally AK and Chisolm GM (1991) Lipoxygenase-mediated transformation of human low density lipoprotein to an oxidised and cytotoxic complex. *Journal of Lipid Research,* **32**, 63–70.

Chamulitrat W, Mason, RP and Riendeau D (1992) Nitroxide metabolites from alkylhydroxylamines and N-hydroxyurea derivatives resulting from reductive inhibition of soybean lipoxygenase. *The Journal of Biological Chemistry,* **267**, 9574–9579.

Chang SW, Westcott JY, Pickett WC, Murphy RC and Voelkel NF (1989) Endotoxin-induced lung injury in rats: role of eicosanoids. *Journal of Applied Physiology,* **66**, 2407–2418.

Chen HJC and Chung FL (1996) Epoxidation of trans-4-hydroxy-2-nonenal by fatty acid hydroperoxides and hydrogen peroxide. *Chemical Research in Toxicology,* **9**, 306–312.

Claesson HE, Odlander B and Jakobsson PJ (1992) Leukotriene B_4 in the immune system. *International Journal of Immunopharmacology,* **14**, 441–449.

Clapp CH, Banerjee A and Rotenberg SA (1985) Inhibition of soybean lipoxygenase 1 by N-alkylhydroxylamines. *Biochemistry,* **24**, 1826–1830.

Coffey MJ, Wheeler CS, Gross KB, Eschenbacher WL, Sporn PHS and Peters-Golden M (1996) Increased 5-lipoxygenase metabolism in the lungs of human subjects exposed to ozone. *Toxicology,* **114**, 187–197.

Colamorea T, Di Paola R, Macchia F, Guerrese MC, Tursi A, Butterfield JH, Caiaffa MF, Haeggstrom JZ and Macchia L (1999) 5-Lipoxygenase upregulation by dexamethasone in human mast cells. *Biochemical and Biophysical Research Communications,* **265**, 617–624.

Cornicelli JA and Trivedi BK (1999) 15-Lipoxygenase and its inhibition: A novel therapeutic target for vascular disease. *Current Pharmaceutical Design,* **5**, 11–20.

Craven PA and DeRubertis FR (1980) Fatty acid induced drug and carcinogen metabolism in rat and human colonic mucosa: A possible link to the association of high dietary fat intake and colonic carcinogenesis. *Biochemical and Biophysical Research Communications,* **94**, 1044–1051.

Craven PA, DeRubertis FR and Fox JW (1983) Fatty acid-dependent benzo(a)pyrene oxidation in colonic mucosal microsomes: Evidence for a distinct metabolic pathway. *Cancer Research,* **43**, 35–40.

Craven PA, Neidig M and DeRubertis FR (1985) Fatty acid-stimulated oxidation of methylazoxymethanol by rat colonic mucosa. *Cancer Research,* **45**, 1115–1121.

Cucurou C, Battioni JP, Daniel R and Mansuy D (1991) Peroxidase-like activity of lipoxygenase: different substrate specificity of potato 5-lipoxygenase and soybean 15-lipoxygenase and particular affinity of vitamin E derivatives for the 5-lipoxygenase. *Biochimica et Biophysica Acta,* **1081**, 99–105.

Dailey LA and Imming P (1999) 12-Lipoxygenase: Classification, possible therapeutic benefits from inhibition and inhibitors. *Current Medicinal Chemistry,* **6**, 389–398.

Datta K and Kulkarni AP (1994a) Oxidative metabolism of Aflatoxin B_1 by lipoxygenase purified from human term placenta and intrauterine conceptal tissues. *Teratology,* **50**, 311-317.

Datta K and Kulkarni AP (1994b) Inhibition of mammalian hepatic glutathione S-transferases by acetylenic fatty acids. *Toxicology Letters*, **73**, 157–165.

Datta K and Kulkarni AP (1996) Co-oxidation of t-retinol acetate by human term placental lipoxygenase and soybean lipoxygenase. *Reproductive Toxicology*, **10**, 105–112.

Datta K, Joseph P, Roy S, Srinivasan SN and Kulkarni AP (1995) Peroxidative xenobiotic oxidation by partially purified peroxidase and lipoxygenase in human foetal tissues at 10 weeks of gestation. *General Pharmacology*, **26**, 107–112.

Datta K, Sherblom PM and Kulkarni AP (1997) Co-oxidative metabolism of 4-aminobiphenyl by lipoxygenase from soybean and human term placenta. *Drug Metabolism and Disposition*, **25**, 196–205.

Decker K (1985) Eicosanoids, signal molecules of liver cells. *Seminars in Liver Diseases*, **5**, 175–190.

De Petrocellis L and Di Marzo V (1994) Aquatic invertebrates open up new perspectives in eicosanoid research: biosynthesis and bioactivity. *Prostaglandins, Leukotrienes and Essential Fatty Acids*, **51**, 215–229.

Desmarais SR, Riendeau D and Gresser MJ (1994) Inhibition of soybean lipoxygenase-1 by a diaryl-N-hydroxyurea by reduction of the ferric enzyme. *Biochemistry*, **33**, 13391–13400.

Ding XZ, Iversen P, Cluck MW, Knezetic JA and Adrian TE (1999) Lipoxygenase inhibitors abolish proliferation of human pancreatic cancer cells. *Biochemical and Biophysical Research Communications*, **261**, 218–223.

Docherty JC and Wilson TW (1987) Indomethacin increases the formation of lipoxygenase products in calcium ionophore stimulated human neutrophils. *Biochemical and Biophysical Research Communications*, **148**, 534–538.

Doli T, Anamura S, Shirakawa M, Okamoto H and Tsujimoto A (1991) Inhibition of lipoxygenase by phenolic compounds. *Japanese Journal of Pharmacology*, **55**, 547–550.

Donnelly PJ, Stewart RK, Ali SL, Conlan AA, Reid KR, Petsikas D and Massey TE (1996) Biotransformation of aflatoxin B$_1$ in human lung. *Carcinogenesis*, **17**, 2487–2494.

Doupnik CA and Leikauf GD (1990) Acrolein stimulates eicosanoid release from bovine airway epithelial cells. *American Journal of Physiology*, **259**, L222–L229.

Egan RW and Gale PH (1985) Inhibition of mammalian 5-lipoxygenase by aromatic disulfides. *The Journal of Biological Chemistry*, **260**, 11554–11559.

Eling TE, Curtis JF, Harmann LS and Mason RP (1986) Oxidation of glutathione to its thiyl free radical metabolite by prostaglandin H synthase. A potential endogenous substrate for the hydroperoxidase. *Journal of Biological Chemistry*, **261**, 5023–5028.

Falgueyret JP, Desmarais S, Roy PJ and Riendeau D (1992) N-(4-chlorophenyl)-N-hydroxy-N'-(3-chlorophenyl)urea, a general reducing agent for 5-, 12-, and 15-lipoxygenases and a substrate for their pseudoperoxidase activities. *Biochemistry and Cell Biology*, **70**, 228–236.

Falzon M, Neilsch A and Burke MD (1986) Denaturation of cytochrome P-450 by indomethacin and other non-steroidal anti-inflammatory drugs: Evidence for a surfactant mechanism and a selective effect of a *p*-chlorophenyl moiety. *Biochemical Pharmacology*, **35**, 4019–4024.

Farrukh IS, Michael JR, Peters SP, Sciuto AM, Adkinson EW, Summer WR and Gurtner (1988) The role of cyclooxygenase and lipoxygenase mediators in oxidant-induced lung injury. *American Review of Respiratory Diseases*, **137**, 1343–1349.

Firozi P, Aboobaker VS and Bhattacharya RK (1986) Modulation by certain factors of metabolic activation of aflatoxin B$_1$ as detected in vitro in a simple fluorimetric assay. *Chemico-Biological Interactions*, **59**, 173–184.

Flatman S, Hurst JS, McDonald-Gibson RG, Jonas GEG and Slater TF (1986) Biochemical studies on a 12-lipoxygenase in human uterine cervix. *Biochimica et Biophysica Acta*, **883**, 7–14.

Fontana M, Costa M, Mosca L and Rosei MA (1997) A specific assay for discriminating between peroxidase and lipoxygenase activities. *Biochemistry and Molecular Biology International*, **42**, 163–168.

Ford-Hutchinson AW, Gresser M and Young RN (1994) 5-Lipoxygenase. *Annual Review of Biochemistry*, **63**, 383–417.

Frasier L and Kehrer JP (1993) Effect of indomethacin, aspirin, nordihydroguairetic acid, and piperonyl butoxide on cyclophosphamide-induced bladder damage. *Drug and Chemical Toxicology*, **16**, 117–133.

Funk CD (1996) The molecular biology of mammalian lipoxygenases and the quest for eicosanoid functions using lipoxygenase-deficient mice. *Biochimica et Biophysica Acta*, **1304**, 65–84.

Gaffney BJ (1996) Lipoxygenases: Structural principles and spectroscopy. *Annual Review of Biophysical and Biomolecular Structures*, **25**, 431–459.

Gata JL, Pinto MC and Marcias P (1996) Lipoxygenase activity in pig muscle: purification and partial characterization. *Journal of Agriculture and Food Chemistry*, **44**, 2573–2577.

Galey JB, Bombard S, Chopard C, Girerd JJ, Lederer F, Thang DC, Nam NH, Mansuy D and Chottard JC (1988) Hexanal phenyl hydrazone is a mechanism-based inactivator of soybean lipoxygenase-1. *Biochemistry*, **27**, 1058–1066.

Gardner HW (1996) Lipoxygenase as a versatile biocatalyst. *Journal of American Official Chemists Society*, **73**, 1347–1357.

Gardner HW and Grove MJ (1998) Soybean lipoxygenase-1 oxidises 3Z-nonenal. *Plant Physiology*, **116**, 1359–1366.

Gargas ML, Anderson ME, Teo SK, Batra R, Fennell TR and Kedderis GL (1995) A physiologically based dosimetry description of acrylonitrile and cyanoethylene oxide in the rat. *Toxicology and Applied Pharmacology*, **134**, 185–194.

Gerwick WH (1994) Structure and biosynthesis of marine algal oxylipins. *Biochimica et Biophysica Acta*, **1211**, 243–255.

Gorins G, Kuhnert L, Johnson CR and Marnett LJ (1996) (Carboxyalkyl) benzyl propargyl ethers as selective inhibitors of leukocyte-type 12-lipoxygenases. *Journal of Medicinal Chemistry*, **39**, 4871–4878.

Grechkin A (1998) Recent developments in biochemistry of the plant lipoxygenase pathway. *Progress in Lipid Research*, **37**, 317–352.

Green FA (1989a) Lipoxygenase activities of the epithelial cells of the human buccal cavity. *Biochemical and Biophysical Research Communications*, **160**, 545–551.

Green FA (1989b) Generation and metabolism of lipoxygenase products in normal and membrane-damaged cultured human keratinocytes. *Journal of Investigative Dermatology*, **93**, 486–491.

Grieve EM, Whiting PH and Hawksworth GM (1990) Arachidonic acid-dependent metabolism of 2-bromoethanolamine to a toxic metabolite in rat medullary interstitial cell cultures. *Toxicology Letters*, **53**, 225–226.

Guo YL, Kennedy TP, Michael JR, Sciuto AM, Ghio AJ, Adkinson NF and Gurtner GH (1990) Mechanism of phosgene-induced lung toxicity: role of arachidonate mediators. *Journal of Applied Physiology*, **69**, 1615–1622.

Hagmann W, Parthe S and Kaiser I (1989) Uptake, production and metabolism of cysteinyl leukotrienes in the isolated perfused rat liver. *Biochemical Journal*, **261**, 611–616.

Hagmann W, Kagawa D, Renaud C and Honn KV (1993) Activity and protein distribution of 12-lipoxygenase in HEL cells. Induction of membrane-association by phorbol ester TPA, modulation of activity by glutathione and 13-HPOD, and Ca2+-dependent translocation to membranes. *Prostaglandins*, **46**, 471–477.

Halevy O and Sklan D (1987) Inhibition of arachidonic acid oxidation by β-carotene, retinol and α-tocopherol. *Biochimica et Biophysica Acta*, **918**, 304–307.

Hamasaki Y and Miyazaki S (1991) Retinoic acid stimulates peptide leukotriene-syntheses in rat basophilic leukemia-1 (RBL-1) cells. *Biochimica et Biophysica Acta*, **1082**, 126–129.

Harman LS, Carver DK, Schrieber J and Mason RP (1986) One and two electron oxidation of reduced glutathione by peroxidases. *The Journal of Biological Chemistry*, **261**, 1642–1648.

Hatzelmann A and Ullrich V (1987) Regulation of 5-lipoxygenase activity by the glutathione status in human polymorphonuclear leukocytes. *European Journal of Biochemistry*, **169**, 175–184.

Ho TA, Coutts TM, Rowland IR and Alldrick AJ (1992) Inhibition of the metabolism of the mutagens occurring in food by arachidonic acid. *Mutation Research*, **269**, 279–284.

Holtzman MJ (1991) Arachidonic acid metabolism. Implications of biological chemistry for lung function and disease. *American Review of Respiratory Diseases*, **143**, 188–203.

Holtzman MJ (1992) Arachidonic acid metabolism in airway epithelial cells. *Annual Review of Physiology*, **54**, 303–329.

Hover C and Kulkarni AP (2000a) A simple and efficient method for hemoglobin removal from mammalian tissue cytosol by zinc sulfate and its application to the study of lipoxygenase. *Prostaglandins Leukotrienes and Essential Fatty Acids*, **62**, 97–105.

Hover C and Kulkarni AP (2000b) Lipoxygenase-mediated hydrogen peroxide-dependent N-demethylation of N,N'-dimethylaniline and related compounds. *Chemico-Biological Interactions*, **124**, 191–203.

Hover CG and Kulkarni AP (2000c) Human term placental lipoxygenase-mediated N-demethylation of phenothiazines and pesticides in the presence of polyunsaturated fatty acids. *Placenta*, **21**, 646–653.

Hover CG and Kulkarni AP (2000d) Hydroperoxide specificity of plant and human tissue lipoxygenase: An in vitro evaluation using N-demethylation of phenothiazines. *Biochimica et Biophysica Acta*, **1475**, 256–264.

Hu J and Kulkarni AP (1998) Soybean lipoxygenase-catalysed demethylation of pesticides. *Pesticide Biochemistry and Physiology*, **61**, 145–153.

Hu J and Kulkarni AP (2000) Metabolic fate of chemical mixtures. I. Shuttle oxidant effect of lipoxygenase-generated radical of chlorpromazine and related phenothiazines on the oxidation of benzidine and other xenobiotics. *Teratogenesis, Carcinogenesis and Mutagenesis*, **20**, 195–208.

Hu J, Sajan M and Kulkarni AP (1999) Lipoxygenase-mediated N-demethylation of imipramine and related tricyclic antidipressants in the presence of hydrogen peroxide. *International Journal of Toxicology*, **18**, 251–257.

Hughes MF, Chamulitrat W, Mason RP and Eling TE (1989) Epoxidation of 7,8-dihydroxy-7,8-dihydrobenzo(a)pyrene via a hydroperoxide-dependent mechanism catalysed by lipoxygenase. *Carcinogenesis*, **10**, 2075–2080.

Ihno Y, Ishihara O and Kinoshita K (1993) Synthesis of 12-hydroxyeicosatetraenoic acid by human enometrium and decidua. *Prostaglandins, Leukotrienes and Essential Fatty Acids*, **49**, 609–613.

Ikawa H, Kamitani H, Calvo BF, Foley JF and Eling TE (1999) Expression of 15-lipoxygenase-1 in human colorectal cancer. *Cancer Research*, **59**, 360–366.

Jakobsson PJ, Steinhilber D, Odlander B, Radmark O, Claesson HE and Samuelsson B (1992) On the expression and regulation of 5-lipoxygenase in human lymphocytes. *Proceedings of National Academy of Sciences USA*, **89**, 3521–3525.

Jiang H, Yamamoto S and Kato R (1994) Inhibition of two-stage skin carcinogenesis as well as complete skin carcinogenesis by oral administration of TMK688, a potent lipoxygenase inhibitor. *Carcinogenesis*, **15**, 807–812.

Joseph P, Srinivasan SN and Kulkarni AP (1993) Purification and partial characterization of lipoxygenase with dual catalytic activities from human term placenta. *Biochemical Journal*, **293**, 83–91.

Joseph P, Srinivasan SN, Byczkowski JZ and Kulkarni AP (1994) Bioactivation of benzo(a)pyrene-7,8-dihydrodiol catalysed by lipoxygenase purified from human term placenta and conceptal tissues. *Reproductive Toxicology*, **8**, 307–313.

Kagan VE, Serbinova EA and Packer L (1990) Generation and recycling of radicals from phenolic antioxidants. *Archives of Biochemistry and Biophysics*, **280**, 33–39.

Kalyanaraman B, Darley-Usmar VM, Wood J, Joseph J and Parthasarathy S (1992) Synergistic interaction between the probucol phenoxyl radical and ascorbic acid in inhibiting the oxidation of low density lipoprotein. *The Journal of Biological Chemistry*, **267**, 6789–6795.

Kamitani H, Ikawa H, Hsi LC, Watanabe T, DuBois RN and Eling TE (1999) Regulation of 12-lipoxygenase in rat intestinal epithelial cells during differentiation and apoptosis induced by sodium butyrate. *Archives of Biochemistry and Biophysics*, **368**, 45–55.

Kanekal S and Kehrer JP (1994) Metabolism of cyclophosphamide by lipoxygenases. *Drug Metabolism and Disposition*, **22**, 74–78.

Katiyar SK, Agarwal R, Wood GS and Mukhtar H (1992) Inhibition of 12-O-tetradecanoyl-phorbol-13-acetate-caused tumour promotion in 7,12-dimethylbenz(a)anthrene-initiated Sencar mouse skin by a polyphenolic fraction isolated from green tea. *Cancer Research*, **52**, 6890–6897.

Kim HP, Mani I, Iversen L and Ziboh VA (1998) Effect of naturally occurring flavonoids and biflavonoids on epidermal cyclooxygenase and lipoxygenase from guinea pigs. *Prostaglandins, Leukotrienes and Essential Fatty Acids*, **58**, 17–24.

Kitagawa H and Noguchi M (1994) Comparative effects of piroxicam and esculetin on incidence, proliferation, and cell kinetics of mammary carcinomas induced by 7,12-dimethylbenz(a)-anthracene in rats on high and low fat diet. *Oncology*, **51**, 401–410.

Konig W, Schonfeld W, Raulf M, Koller M, Knoller J, Scheffer J and Brom J (1990) The neutrophil and leukotrienes—role in health and disease. *Eicosanoids*, **3**, 1–22.

Kubow S and Wells PJ (1988) In vitro evidence for lipoxygenase-catalysed bioactivation of phenytoin. *Pharmacologist*, **30**, A74.

Kuhn H and Borngraber S (1999) Mammalian 15-lipoxygenases. Enzymatic properties and biological implications. *Advances in Experimental Medicine and Biology*, **447**, 5–28.

Kuhn H and Thiele BJ (1999) The diversity of the lipoxygenase family. Many sequence data but little information on biological significance. *FEBS Letters*, **449**, 7–11.

Kuhn H, Gotze R, Schewe T and Rapoport SM (1981) Quasi-lipoxygenase activity of haemo-globin. A model for lipoxygenases. *European Journal of Biochemistry*, **120**, 161–168.

Kuklev DV, Imbs AB, Long FK and Bezuglov VV (1993) Products of the lipoxygenase oxidation of pinolenic acid. *Russian Journal of Bioorganic Chemistry*, **19**, 745–747.

Kulkarni AP (1996) Role of xenobiotic metabolism in developmental toxicity. In *Handbook of Developmental Toxicology*, Hood R. (ed.), CRC Press, Boca Raton, FL, pp. 383–421.

Kulkarni AP (2001) Role of biotransformation in conceptual toxicity of drugs and other chemicals. *Current Pharmaceutical Design*, **7**, 833–857.

Kulkarni AP and Cook DC (1988a) Hydroperoxidase activity of lipoxygenase: A potential pathway for xenobiotic metabolism in the presence of linoleic acid. *Research Communications in Chemical Pathology and Pharmacology*, **61**, 305–314.

Kulkarni AP and Cook DC (1988b) Hydroperoxidase activity of lipoxygenase: Hydrogen peroxide-dependent oxidation of xenobiotics. *Biochemical and Biophysical Research Communications*, **155**, 1075–1081.

Kulkarni AP and Mitra A (1990) Pesticide contamination of food in the United States. Chapter 9 In *Food Contamination from Environmental Sources*, Nriagu JO and Simons MS (eds), Vol. 23, Advances in Environmental Science and Technology, Wiley, New York, pp. 257–293.

Kulkarni AP and Sajan M (1997) A novel mechanism of glutathione conjugation formation by lipoxygenase: A study with ethacrynic acid. *Toxicology and Applied Pharmacology*, **143**, 179–188.

Kulkarni AP and Sajan M (1999) Glutathione conjugation of ethacrynic acid by human term placental lipoxygenase. *Archives of Biochemistry and Biophysics*, **371**, 220–227.

Kulkarni PS and Srinivasan BD (1989) Cyclooxygenase and lipoxygenase pathways in anterior uvea and conjunctiva. In *The Ocular Effects of Prostaglandins and Other Eicosanoids*, Alan R. Liss, New York, pp. 39–52.

Kulkarni AP, Chaudhuri J, Mitra A and Richards IS (1989) Dioxygenase and peroxidase activities of soybean lipoxygenase: Synergistic interaction between linoleic acid and hydrogen peroxide. *Research Communications in Chemical Pathology and Pharmacology*, **66**, 287–296.

Kulkarni AP, Mitra A, Chaudhuri J, Byczkowski JZ and Richards IS (1990) Hydrogen peroxide: A potent activator of dioxygenase activity of soybean *lipoxygenase. Biochemical and Biophysical Research Communications*, **166**, 417–423.

Kulkarni AP, Cai Y and Richards IS (1992) Rat pulmonary lipoxygenase: Dioxygenase activity and role in xenobiotic oxidation. *International Journal of Biochemistry*, **24**, 255–261.

Kuninori T, Nishiyama J, Shirakawa M and Shimoyama A (1992) Inhibition of soybean lipoxygenase-1 by n-alcohols and n-alkylthiols. *Biochimica et Biophysica Acta*, **1125**, 49–55.

Lehmann WD (1994) Regio- and stereochemistry of the dioxygenation reaction catalysed by (S)-type lipoxygenases or by the cyclooxygenase activity of prostaglandin H synthases. *Free Radical Biology & Medicine*, **16**, 241–253.

Liaw YW, Liu YW, Chen BK and Chang WC (1998) Induction of 12-lipoxygenase expression by phorbol 12-myristate 13-acetate in human epidermoid carcinoma A431 cells. *Biochimica et Biophysica Acta*, **1389**, 23–33.

Liu L and Massey TE (1992) Bioactivation of aflatoxin B1 by lipoxygenases, prostaglandin H synthase and cytochrome P450 monooxygenase in guinea pig tissues. *Carcinogenesis*, **13**, 533–539.

Lomnitski L, Bar-Natan R, Sklan D and Grossman S (1993) The interaction between β-carotene and lipoxygenase in plant and animal systems. *Biochimica et Biophysica Acta*, **1167**, 331–338.

Malle E, Leis HJ, Karadi I and Kostner GM (1987) Lipoxygenases and hydroperoxy/hydroxy-eicosatetraenoic acid formation. *International Journal of Biochemistry*, **19**, 1013–1022.

Manchester DK, Weston A, Choi J, Trivers GE, Fennessey PV, Quintanna E, Farmer PB, Mann DL and Harris CC (1988) Detection of benzo(a)pyrene diol epoxide-DNA adducts in human placenta. *Proceedings of National Academy of Sciences USA*, **85**, 9243–9247.

Mansuy D, Cucurou C, Biatry B and Battioni JP (1988) Soybean lipoxygenase-catalysed oxidations by linoleic acid hydroperoxide: Different reducing substrates and dehydrogenation of phenidone and BW755C. *Biochemical and Biophysical Research Communications*, **151**, 339–346.

Marcinkiewicz E, Duniec Z and Robak J (1985) Salazosulfapyridine and non-steroidal anti-inflammatory drugs do not inhibit soybean lipoxygenase. *Biochemical Pharmacology*, **34**, 148–149.

Matsui K, Kajiwara T, Hatanaka A, Waldamann D and Schreiber P (1994) 5,6-Epoxidation of all-trans-retinoic acid with soybean lipoxygenase-2 and lipoxygenase-3. *Bioscience, Biotechnology and Biochemistry*, **58**, 140–145.

Mattoli S, Mezzetti M, Fasoli A, Patalano F and Allegra L (1990) Nedocromil sodium prevents the release of 15-hydroxyeicosatetraenoic acid from human epithelial cells exposed to toluene diisocyanate in vitro. *International Archives of Allergy and Applied Immunology*, **92**, 16–22.

Mele PG, Dada LA, Paz C, Neuman I, Cymeryng CB, Mendez CF, Finkielstein CV, Maciel CF and Podesta EJ (1997) Involvement of arachidonic acid and the lipoxygenase pathway in mediating luteinizing hormone-induced testosterone synthesis in rat Leydig cells. *Endocrine Research*, **23**, 15–26.

Morikawa Y, Shiomi K, Ishihara Y and Matsura M (1997) Triple primary cancers involving kidney, urinary bladder and liver in a dye worker. *American Journal of Industrial Medicine*, **31**, 44–49.

Mosca L, Foppoli C, Coccia R and Rosei MA (1996) Pheomelanin production by the lipoxygenase-catalysed oxidation of 5-S-cysteinyldopa and 5-S-cysteinyldopamine. *Pigment Cell Research*, **9**, 117–125.

Mosca L, Blarzino C, Coccia R, Foppoli C and Rosei MA (1998) Melanins from tetrahydroisoquinolines: spectroscopic characteristics, scavenging activity and redox transfer properties. *Free Radicals in Biology and Medicine*, **24**, 161–167.

Myers CE and Ghosh J (1999) Lipoxygenase inhibition in prostate cancer. *European Urology*, **35**, 395–398.

Naidu AK and Kulkarni AP (1991) Role of lipoxygenase in xenobiotic metabolism: Sulfoxidation of thiobenzamide by purified soybean lipoxygenase. *Research Communications in Chemical Pathology and Pharmacology*, **71**, 175–188.

Naidu AK, Naidu AK and Kulkarni AP (1991a) Aldrin epoxidation: Catalytic potential of lipoxygenase coupled with linoleic acid oxidation. *Drug Metabolism and Disposition*, **19**, 758–763.

Naidu AK, Naidu AK and Kulkarni AP (1991b) Role of lipoxygenase in xenobiotic oxidation: Parathion metabolism catalysed by highly purified soybean lipoxygenase. *Pesticide Biochemistry and Physiology*, **41**, 150–158.

Naidu AK, Naidu AK and Kulkarni AP (1992) Dioxygenase and hydroperoxidase activities of rat brain cytosolic lipoxygenase. *Research Communications in Chemical Pathology and Pharmacology*, **75**, 347–356.

Natarajan R and Nadler J (1998) Role of lipoxygenase in breast cancer. *Frontiers in Bioscience*, **3**, E81–E88.

Nave JF, Jacobi D, Gaget C, Dulery B and Ducep JB (1991) Evaluation of 5- and 6-fluoro derivatives of arachidonic acid as substrates and inhibitors of 5-lipoxygenase. *Biochemical Journal*, **278**, 549–555.

Negro JM, Miralles JC, Ortiz JL, Funes E and Garcia A (1997) Biosynthesis inhibitors for leukotrienes in bronchial asthma. *Allergology and Immunopathology*, **25**, 209–216.

Nelson MJ and Seitz SP (1994) The structure and function of lipoxygenase. *Current Opinions in Structural Biology*, **4**, 878–884.

Nemoto N and Takayama S (1984) Arachidonic acid-dependent activation of benzo[a]pyrene to bind to proteins with cytosolic and microsomal fractions from rat liver and lung. *Carcinogenesis*, **5**, 961–964.

Noguchi M, Kitagawa H, Miyazaki I and Mizukami Y (1993) Influence of esculetin on incidence, proliferation, and cell kinetics of mammary carcinomas induced by 7,12-dimethylbenz(a)-anthracene in rats on high and low-fat diets. *Japanese Journal of Cancer Research*, **84**, 1010–1014.

Novak MJ (1999) New minimal substrate structural requirements in the enzymatic peroxidation of alkenes with soybean lipoxygenase. *Bioorganic and Medicinal Chemistry Letters*, **9**, 31–34.

Nunez-Delicado E, Perez-Gilabert M, Sanchez-Ferrer A and Garcia-Carmona F (1996) Hydroperoxidase activity of lipoxygenase: a kinetic study of isoproterenol oxidation. *Biochimica et Biophysica Acta*, **1293**, 17–22.

Nunez-Delicado E, Sanchez-Ferrer A and Garcia-Carmona F (1997a) A kinetic study of the one-electron oxidation of Tolox C by the hydroperoxidase activity of lipoxygenase. *Biochimica et Biophysica Acta*, **1335**, 127–134.

Nunez-Delicado E, Sanchez-Ferrer A and Garcia-Carmona F (1997b) Hydroperoxidative oxidation of diethylstilbestrol by lipoxygenase. *Archives of Biochemistry and Biophysics*, **348**, 411–414.

Nunez-Delicado E, Sanchez-Ferrer A and Garcia-Carmona F (1997c) Cyclodextrins as secondary antioxidants: synergism with ascorbic acid. *Journal of Agricultural Food Chemistry*, **45**, 2830–2835.

Nunez-Delicado E, Sojo MM, Sanchez-Ferrer A and Garcia-Carmona F (1999) Hydroperoxidase activity of lipoxygenase in the presence of cyclodextrins. *Archives of Biochemistry and Biophysics*, **367**, 274–280.

O'Donnell VB and Kuhn H (1997) Co-oxidation of NADH and NADPH by a mammalian 15-lipoxygenase: inhibition of lipoxygenase activity at near-physiological NADH concentrations. *Biochemical Journal*, **327**, 203–208.

Oyekan A, Balazy M and McGiff JC (1997) Renal oxygenases: differential contribution to vasoconstriction induced by ET-1 and ANG II. *American Journal of Physiology*, **273**, R293–300.

Perchellet EM and Perchellet JP (1989) Chartacterization of the hydroperoxide response observed in mouse skin treated with tumour promoters in vivo. *Cancer Research*, **49**, 6193–6201.

Pereira TA and Das NP (1991) Assay of liver cytosol lipoxygenase by differential pulse polarography. *Analytical Biochemistry*, **197**, 96–100.

Peres HD, Roll FJ, Bissell DM and Goldstein IM (1984) Production of chemotactic activity for polymorphonuclear leukocytes by cultured rat hepatocytes exposed to ethanol. *Journal of Clinical Investigation*, **74**, 1350–1357.

Perez-Gilabert M, Sanchez-Ferrer A and Garcia-Carmona F (1994a) Lipoxygenase-catalysed oxidation of chlorpromazine by hydrogen peroxide at acidic pH. *Biochimica et Biophysica Acta*, **1214**, 203–208.

Perez-Gilabert M, Sanchez-Ferrer A and Garcia-Carmona F (1994b) Enzymatic oxidation of phenothiazines by lipoxygenase/H_2O_2 system. *Biochemical Pharmacology*, **47**, 2227–2232.

Perez-Gilabert M, Sanchez-Ferrer A and Garcia-Carmona F (1997) Oxidation of aminopyrine by the hydroperoxidase activity of lipoxygenase: A new proposed mechanism of N-demethylation. *Free Radical Biology & Medicine*, **23**, 548–555.

Persad AS, Stedeford TJ and Kulkarni AP (2000) The hyperoxidation of L-dopa by chlorpromazine in a lipoxygenase catalysed reaction. *Toxicologist*, **54**, 1746A.

Portoghese PS, Svanborg K and Samuelsson B (1975) Oxidation of oxyphenbutazone by sheep vesicular gland microsomes and lipoxygenase. *Biochemical and Biophysical Research Communications*, **63**, 748–755.

Pretus HA, Ignarro LJ, Ensley HE and Feigen LP (1985) Inhibition of soybean lipoxygenase by SKF 525-A and metyrapone. *Prostaglandins*, **30**, 591–598.

Pridham D, Lei ZM, Chegini N, Rao CV, Yussman MA and Cook CL (1990) Light and electron microscope immunocytochemical localization of 5- and 12-lipoxygenases and cyclooxygenase enzymes in human granulosa cells from preovulatory follicles. *Prostaglandins, Leukotrienes and Essential Fatty Acids*, **39**, 231–238.

Rajadhyaksha A, Reddy V, Hover C and Kulkarni AP (1999) N-demethylation of phenothiazines by lipoxygenase from soybean and human term placenta in the presence of hydrogen peroxide. *Teratogenesis, Carcinogenesis and Mutagenesis*, **19**, 211–222.

Reed GA, Grafstrom RC, Krauss RS, Autrup H and Eling TE (1984) Prostaglandin H synthase-dependent co-oxidation of (±)-7,8-dihydroxy-7,8-dihydrobenzo[a]pyrene in hamster trachea and human bronchus explants. *Carcinogenesis*, **5**, 955–960.

Reynolds CH (1988) Inactivation of lipoxygenase by inhibitors in the presence of 15- hydroperoxyeicosatetraenoic acid. *Biochemical Pharmacology*, **37**, 4531–4537.

Riddick CA, Ring WL, Baker JR, Hodulik CR and Bigby TD (1997) Dexamethasone increases expression of 5-lipoxygenase and its activating protein in human monocytes and THP-1 cells. *European Journal of Biochemistry*, **246**, 112–118.

Riendeau D, Falgueyret JP, Guay J, Ueda N and Yamamoto S (1991) Pseudoperoxidase activity of 5-lipoxygenase stimulated by potent benzofuranol and N-hydroxyurea inhibitors of the lipoxygenase reaction. *Biochemical Journal*, **274**, 287–292.

Rioux N and Castonguay A (1998) Inhibitors of lipoxygenase: a new class of cancer chemopreventive agents. *Carcinogenesis*, **19**, 1393–1400.

Roberts AE, Lacy SA, Pilon D, Turner MJ and Rickert DE (1989) Metabolism of acrylonitrile to 2-cyanoethylene oxide in F-344 rat liver microsomes, lung microsomes and lung cells. *Drug Metabolism and Disposition*, **17**, 481–486.

Rosei MA, Blarzino C, Foppoli L, Mosca L and Coccia R (1994) Lipoxygenase-catalysed oxidation of catecholamines. *Biochemical and Biophysical Research Communications*, **200**, 344–350.

Rosello-Catafau J, Hotter G, Closa D, Ortiz MA, Pou-Torello JM, Gimeno M, Bioque G and Gelpi E (1994) Liver lipoxygenase arachidonic acid metabolites in streptozotocin-induced diabetes in rats. *Prostaglandins, Leukotrienes and Essential Fatty Acids*, **51**, 411–413.

Roy P and Kulkarni AP (1996a) Oxidation of ascorbic acid by lipoxygenase: effect of selected chemicals. *Food and Chemical Toxicology*, **34**, 563–570.

Roy P and Kulkarni AP (1999) Cooxidation of acrylonitrile by soybean lipoxygenase and partially purified human lung lipoxygenase. *Xenobiotica*, **29**, 511–531.

Roy P, Roy S, Mitra A and Kulkarni AP (1994) Superoxide generation by lipoxygenase in the presence of NADH and NADPH. *Biochimica et Biophysica Acta*, **1214**, 171–179.

Roy P, Sajan M and Kulkarni AP (1995) Lipoxygenase mediated glutathione oxidation and superoxide generation. *Journal of Biochemical Toxicology*, **10**, 111–120.

Roy S and Kulkarni AP (1991) Lipoxygenase: A new pathway for 2-aminofluorene bioactivation. *Cancer Letters*, **60**, 33–39.

Roy S. and Kulkarni AP (1994) Dioxygenase and co-oxidase activities of rat hepatic cytosolic lipoxygenase. *Journal of Biochemical Toxicology*, **9**, 171–179.

Roy S and Kulkarni AP (1996b) Isolation and some properties of dioxygenase and co-oxidase activities of adult human liver cytosolic lipoxygenase. *Journal of Biochemical Toxicology*, **11**, 161–174.

Roy S and Kulkarni AP (1997) Aflatoxin B$_1$ epoxidation catalysed by partially purified adult human liver lipoxygenase. *Xenobiotica*, **27**, 231–241.

Roy S, Mitra AK, Hilbelink DR, Dwornik JJ and Kulkarni AP (1993) Lipoxygenase activity in rat embryos and its potential significance in xenobiotic oxidation. *Biology of Neonate*, **63**, 297–302.

Samuelsson B, Dahlen SE, Lindgren JA, Rouzer CA, Serhan CN (1987) Leukotrienes and lipoxins: Structures, biosynthesis, and biological effects. *Science*, **237**, 1171–1176.

Sanz LC, Perez AG and Olias JM (1992) Purification and catalytic properties of chickpea lipoxygenases. *Phytochemistry*, **31**, 2967–2972.

Scharping CE, McManus ME and Holder GM (1992) NADPH-supported and arachidonic acid-supported metabolism of the enantiomers of trans-7,8-dihydrobenzo(a)pyrene-7,8-diol by human liver microsomal samples. *Carcinogenesis*, **13**, 1199–1207.

Schewe T, Rapoport SM and Kuhn H (1986) Enzymology and physiology of reticulocyte lipoxygenase: Comparison with other lipoxygenases. *Advances in Enzymology and Related Areas of Molecular Biology*, **58**, 191–272.

Schilderman PAEL, van Maanen JMS, ten Vaarwerk FJ, Lafleur MVM, Westmijze EJ, ten Hoor F and Kleinjans JCS (1993) The role of prostaglandin H synthase-mediated metabolism in the induction of oxidative DNA damage by BHA metabolites. *Carcinogenesis*, **14**, 1297–1302.

Schneider C, Amberg A, Feurle J, Roß, Roth M, Toth G and Schreier P (1998) 2-{(4″-Hydroxy-3′-methoxy)-phenoxy]-4-(4″-hydroxy-3″-methoxy-phenyl)-8-hydroxy-6-oxo-3-oxabiclo[3.3.0]-7-octene: unusual product of the soybean lipoxygenase-catalysed oxygenation of curcumin. *Journal of Molecular Catalysis B: Enzymatic*, **4**, 219–227.

Schnurr K, Kuhn H, Rapoport SM and Schewe T (1995) 3,5-Di-t-butyl-4-hydroxytoluene (BHT) and probucol stimulate selectively the reaction of mammalian 15-lipoxygenase with biomembranes. *Biochimica et Biophysica Acta*, **1254**, 66–72.

Sevanian A and Peterson H (1989) Induction of cytotoxicity and mutagenesis is facilitated by fatty acid hydroperoxidase activity in Chinese hamster lung fibroblast (V79 cells). *Mutation Research*, **224**, 185–196.

Shahin I, Grossman S and Sredni B (1978) Lipoxygenase-like enzyme in rat testis microsomes. *Biochimica et Biophysica Acta*, **529**, 300–308.

Shinjo F, Yoshimoto T, Yokoyama C, Yamamoto S, Izumi SI, Komatsu N and Watanabe K (1986) Studies on porcine arachidonate 12-lipoxygenase using its monoclonal antibodies. *The Journal of Biological Chemistry*, **261**, 3377–3381.

Siedow JN (1991) Plant lipoxygenase: structure and function. *Annual Review of Plant Physiology and Plant Molecular Biology*, **42**, 145–188.

Siegel MI, McConnell RT and Cuatrecasas P (1979) Aspirin-like drugs interfere with arachidonate metabolism by inhibition of the 12-hydroperoxy-5,8,10,14-eicosatetraenoic acid peroxidase activity of the lipoxygenase pathway. *Proceedings of the National Academy of Sciences, USA*, **76**, 3774–3778.

Siegel MI, McConnell RT, Porter NA and Cuatrecasas P (1980a) Arachidonate metabolism via lipoxygenase and 12L-hydroperoxy-5,8,10,14-eicosatetraenoic acid peroxidase sensitive to anti-inflammatory drugs. *Proceedings of the National Academy of Sciences, USA*, **77**, 308–312.

Siegel MI, McConnell RT, Porter NA, Selph JL, Truax JF, Vanegar B and Cuatrecasas P (1980b) Aspirin-like drugs inhibit arachidonic acid metabolism via lipoxygenase and cyclo- oxygenase in rat neutrophils from carrageenan exudates. *Biochemical and Biophysical Research Communications*, **92**, 688–695.

Silverman ES and Drazen JM (1999) The biology of lipoxygenase: Function, structure, and regulatory mechanisms. *Proceedings of the Association of American Physicians*, **111**, 525–536.

Simmet T and Peskar BA (1990) Lipoxygenase products of polyunsaturated fatty acid metabolism in the central nervous system: Biosynthesis and putative functions. *Pharmacological Research*, **22**, 667–682.

Sircar JC, Schewnder CF and Johnson EA (1983) Soybean lipoxygenase inhibition by nonsteroidal antiinflammatory drugs. *Prostaglandins*, **25**, 393–397.

Smith WL and Marnett LJ (1991) Prostaglandin endoperoxide synthase: structure and catalysis. *Biochimica et Biophysica Acta*, **1083**, 1–17.

Smith TJ, Stoner GD and Yang CS (1995) Activation of 4-(methylnitrosamino)-1-(3-pyridyl)-1-butanone (NNK) in human lung microsomes by cytochrome P450, lipoxygenase and hydroperoxides. *Cancer Research*, **55**, 5566–5573.

Smith TJ, Liao A, Liu Y, Jones AB, Anderson LM and Yang CS (1997) Enzymes involved in the bioactivation of 4-(methylnitrosamino)-1-(3-pyridyl)-1-butanone in patas monkey lung and liver microsomes. *Carcinogenesis*, **18**, 1577–1584.

Spaapen LJM, Verhagen J, Veldink GA and Vliegenthart (1980) The effect of modification of sulfhydryl groups in soybean lipoxygenase-1. *Biochimica et Biophysica Acta*, **618**, 153–162.

Stein T, Bailey B, Auguste LJ and Wise L (1991) Measurement of prostaglandin G/H synthase and lipoxygenase activity in the stomach wall by HPLC. *Biochromatography*, **10**, 222–225.

Steinberg D (1999) At last, direct evidence that lipoxygenases play a role in atherogenesis. *Journal of Clinical Investigation*, **103**, 1487–1488.

Steinhiber D (1999) 5-Lipoxygenase: A target for antiinflammatory drugs revisited. *Current Medicinal Chemistry*, **6**, 71–85.

Stewart VC, Tisocki K, Bell JA, Whiting PH and Hawksworth GM (1993) Species differences in prostaglandin synthetase/lipoxygenase-mediated cooxidation in the kidney. *British Journal of Clinical Pharmacology*, **36**, 164P–165P.

Stewart VC, Whiting PH, Bell JA and Hawksworth GM (1997) Human renal lipoxygenase. *Biochemical Society Transactions*, **25**, S623.

Stewart VC, Whiting PH, Bell JA and Hawksworth GM (1998) Streptozotocin-induced diabetes is associated with increased renal lipoxygenase activity. *Biochemical Society Transactions*, **26**, S134.

Stock BH, Bend JR and Eling TE (1986) The formation of styrene glutathione adduct catalysed by prostaglandin H synthase. A possible new mechanism for the formation of glutathione conjugates. *The Journal of Biological Chemistry*, **261**, 5959–5964.

Streckert G and Stan HJ (1975) Conversion of linoleic acid hydroperoxide by soybean lipoxygenase in the presence of guaiacol: Identification of the reaction products. *Lipids*, **10**, 847–854.

Suzuki H, Miyauchi D and Yamamoto S (1997) A selective inhibitor of arachidonate 5-lipoxygenase scavenging peroxide activator. *Biochemical Pharmacology*, **54**, 529–532.

Takahashi Y, Glasgow WC, Suzuki H, Taketani Y, Yamamoto S, Anton M, Kuhn H and Brash AR (1993) Investigation of the oxygenation of phospholipids by the porcine leukocyte and human platelet arachidonate 12-lipoxygenases. *European Journal of Biochemistry*, **218**, 165–171.

Tang K and Honn KV (1999) 12-(S)-HETE in cancer metastasis. *Advances in Experimental Medicine and Biology*, **447**, 181–191.

Uotila P, Mannisto J, Simberg N and Hartiala K (1981) Indomethacin inhibits arachidonic acid metabolism via lipoxygenase and cyclo-oxygenase in hamster isolated lungs. *Prostaglandins and Medicine*, **7**, 591–599.

Vanderhoek JY and Klein KL (1988) Lipoxygenases in rat embryo tissues. *Proceedings of the Society for Experimental Biology and Medicine*, **188**, 370–374.

Van der Zee J, Eling TE and Mason RP (1989) Formation of free radical metabolites in the reaction between soybean lipoxygenase and its inhibitors. An ESR study. *Biochemistry*, **28**, 8363–8367.

Van Leyen K, Duvoisin RM, Engelhardt H and Wiedmann M (1998) A function for lipoxygenase in programmed organelle degradation. *Nature*, **395**, 392–395.

Waldmann D and Schreier P (1995) Stereochemical studies of epoxides formed by lipoxygenase-catalysed co-oxidation of retinol, β-ionone and 4-hydroxy-β-ionone. *Journal of Agricultural and Food Chemistry*, **43**, 626–630.

Whysner J, Verna L and Williams GM (1996) Benzidine mechanistic data and risk assessment: species- and organ-specific metabolic activation. *Pharmacology and Therapeutics*, **71**, 107–126.

Wiseman JS and Nichols JS (1988) Ketones as electrophilic substrates of lipoxygenase. *Biochemical and Biophysical Research Communications*, **154**, 544–549.

Wu Z, Robinson DS, Hughes RK, Casey R, Hardy D and West SI (1999) Co-oxidation of β-carotene catalysed by soybean and recombinant pea lipoxygenases. *Journal of Agriculture and Food Chemistry*, **47**, 4899–4906.

Yamamoto S (1992) Mammalian lipoxygenases: molecular structure and functions. *Biochimica et Biophysica Acta*, **1128**, 117–131.

Yamamoto S, Ishii M, Nakadate T, Nakaki T and Kato R (1983) Modulation of insulin secretion by lipoxygenase products of arachidonic acid. *The Journal of Biological Chemistry*, **258**, 12149–12152.

Yamamoto S, Suzuki H, Ueda N (1997) Arachidonate 12-lipoxygenases. *Progress in Lipid Research*, **36**, 23–41.

Yamamoto S, Suzuki H, Nakamura M and Ishimura K (1999) Arachidonate 12-lipoxygenase isozymes. *Advances in Experimental Medicine and Biology*, **447**, 37–44.

Yamashita H, Nakamura A, Noguchi N, Niki E and Kuhn H (1999) Oxidation of low density lipoprotein and plasma by 15-lipoxygenase and free radicals. *FEBS Letters*, **445**, 287–290.

Yang X and Kulkarni AP (1998) N-dealkylation of aminopyrine catalysed by soybean lipoxygenase in the presence of hydrogen peroxide. *Journal of Biochemical Toxicology*, **12**, 175–183.

Yang X and Kulkarni AP (2000) Lipoxygenase-mediated biotranformation of p-aminophenol in the presence of glutathione. Possible conjugate formation. *Toxicology Letters*, **111**, 253–261.

Yoshimoto T, Furukawa M, Yamamoto S, Horie T and Watanabe-Kohno S (1983) Flavonoids: Potent inhibitors of arachidonate 5-lipoxygenase. *Biochemical and Biophysical Research Communications*, **116**, 612–618.

You KM, Jong HG and Kim HP (1999) Inhibition of cyclooxygenase/lipoxygenase from human platelets by polyhydroxylated/methoxylated flavonoids isolated from medicinal plants. *Archives of Pharmacological Research*, **22**, 18–24.

Yu WK and Wells PG (1995) Evidence for lipoxygenase-catalysed bioactivation of phenytoin to a teratogenic reactive intermediate:in vitro studies using linoleic acid-dependent soybean lipoxygenase, and in vivo studies using pregnant CD-1 mice. *Toxicology and Applied Pharmacology*, **131**, 1–12.

Zaitsu M, Hamasaki Y, Yamamoto S, Kita M, Hayasaki R, Muro E, Kobayashi I, Matsuo M, Ichimaru T and Miyazaki S (1998) Effect of dexamethasone on leukotriene synthesis in DMSO-stimulated HL-60 cells. *Prostaglandins, Leukotrienes and Essential Fatty Acids*, **59**, 385–393.

Zenser VT, Mattammal MB and Davis BB (1979) Co-oxidation of benzidine by renal medullary prostaglndin cyclooxygenase.*Journal of Pharmacology and Experimental Therapeutics*, **211**, 460–464.

Zhang P and Kyler KS (1989) Enzymatic asymmetric hydroxylation of pentadienols using soybean lipoxygenase. *Journal of the American Chemical Society*, **111**, 9241–9242.

8 UDP-Glucuronosyltransferases

K.W. Bock

University of Tübingen, Germany

Introduction

Glucuronidation represents one of the most important phase-II biotransformation reactions converting thousands of lipophilic endobiotics and xenobiotics (drugs, dietary plant constituents, etc.) and their phase-I metabolites into hydrophilic and excretable conjugates (Dutton 1980; see Chapter 1). For many compounds such as plant constituents which already contain functional groups ($-OH$, $-COOH$, $-SH$, $-NH_2$) glucuronidation represents the primary biotransformation reaction. Glucuronidation is catalysed by a supergene family of UDP-glucuronosyltransferases (UGTs) which are integral proteins of the membranes of the endoplasmic reticulum and the nuclear envelope. UGTs are present in many tissues of vertebrates (mammals, fish and—although no sequences have been published—in amphibia, reptiles and birds; Dutton 1980).

In general, glucuronidation occurs in concert with other biotransformation reactions which have been termed phase-I (functionalisation reactions), phase-II (conjugation) and phase-III (export of conjugates from cells). As shown in Figure 8.1, lipophilic compounds (X) (entering cells by passive diffusion and by uptake carriers such as OATPs (organic anion transporting proteins), a process which has been termed phase 0) are usually converted in phase I, mainly by cytochromes P450 (CYPs), into a number of electrophilic and nucleophilic metabolites. Reactive electrophiles are often conjugated by glutathione *S*-transferases. When reactive metabolites accumulate in cells they may interact with cellular macromolecules such as DNA (a reaction which may initiate carcinogenesis) or with proteins (in some cases initiating autoimmune diseases; see Chapter 1). Nucleophilic metabolites are mainly conjugated by UGTs, sulphotransferases, etc. Some phenolic metabolites such as benzo[a]pyrene diphenols are readily autoxidised to electrophilic quinones which, in turn, may undergo redox cycling leading to oxidative stress. It is noteworthy that some acyl glucuronides are electrophilic and known to react with cellular proteins. Conjugates with glucuronic acid, sulphate or glutathione need to be excreted from cells by ATP-dependent export pumps such as multidrug resistance proteins (MRPs). It is noteworthy that transport by

Enzyme Systems that Metabolise Drugs and Other Xenobiotics. Edited by C. Ioannides.
© 2002 John Wiley & Sons Ltd

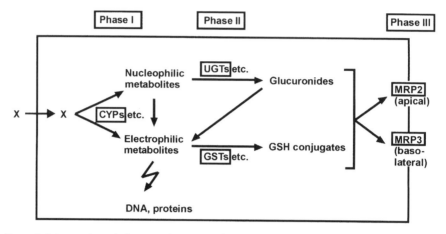

Figure 8.1 Integration of glucuronidation into the biotransformation system of lipophilic endo- and xenobiotics. CYPs, cytochromes P450; UGTs, UDP-glucuronosyltransferases; GSTs, glutathione S-transferases; MRPs, multidrug resistance proteins.

MRPs often determine the disposition of glucuronides: MRP2 has been characterised as an apical export pump which in hepatocytes is exclusively incorporated into canalicular membranes, and therefore secretes conjugates via the bile into the intestine. In contrast, MRP3 secretes glucuronides via the basolateral membrane into the blood. Concerted actions of phase I, II and III enzymes/proteins (the biotransformation system) are supported by their occasional coinduction by xenobiotics; for example, by Ah receptor agonists and by phenobarbital-type inducers. Biotransformation of lipophilic compounds probably represents a detoxification process which is essential for life.

UGTs catalyse the transfer of glucuronic acid from UDP-glucuronic acid to phenols, hydroxylamines, carboxylic acids etc . In this way glucuronides with differing pH stability are formed (Figure 8.2(A)). Whereas ether glucuronides are quite stable (a), N-glucuronides of some arylamines such as 2-naphthylamine and 4-aminobiphenyl including N-glucuronides of their hydroxylamines are acid-labile at pH < 7 (b) (Beland and Kadlubar 1990). Ester (acyl) glucuronides are unstable at neutral and alkaline pH > 7 (c) (Benet et al. 1993).

Based on kinetic data and chemical modification of the enzyme protein, a general acid-base Sn2 mechanism has been proposed for transfer of phenolic aglycones to glucuronic acid (Figure 8.2(B)). A general base (B) of the active site may protonate phenolic compounds facilitating their transfer to glucuronic acid. The base may involve a charge relay system between the catalytic aspartate–glutamate residue and the histidyl residue. The leaving group (UDP) would then be protonated by an acidic amino acid residue of the active site. The catalytic cycle would be completed by a proton exchange by the basic and acidic catalytic residues (Radominska-Pandya et al. 1999).

Application of molecular biology techniques to UGTs led to an explosion of our

A

(a) R - OH   + UDPGA ⟶ R - OGA   + UDP

(b) R · N(OH)(H)   + UDPGA ⟶ R · N(OH)(GA)   + UDP

(c) R · C(=O)(OH)   + UDPGA ⟶ R · C(=O)(OGA)   + UDP

B

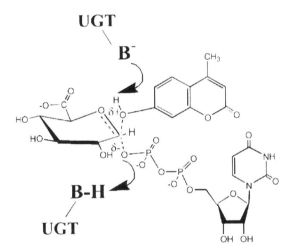

Figure 8.2 UGT reactions. (a) Reactions with differing pH stabilities of resulting glucuronides. UDPGA, UDP-glucuronic acid. (b) Mechanism proposed for the glucuronidation of phenolic compounds (Reproduced from Radominska-Pandya *et al.* 1999 *Drug Metabolism Reviews*, **31**, 817–899. Published by Marcel Dekker, Inc., New York).

knowledge about UGT isoforms and about their functions in the metabolism of endobiotics, drugs, dietary plant constituents and in detoxification of carcinogens. A remarkable tissue-specific expression of UGTs was found. It is becoming increasingly clear that the biotransformation system not only deals with drugs or other xenobiotics but also controls the levels of key endogenous compounds such as hormone receptor ligands (Nebert 1991). Conjugation of endogenous substrates needs to be explored more extensively.

The present review focuses on human UGT isoforms, their functions and their regulation by xenobiotics. Human UGTs are compared to those characterised in experimental animals. Recently excellent reviews have appeared (Burchell *et al.* 1997; Radominska-Pandya *et al.* 1999; Tukey and Strassburg 2000; King *et al.* 2000). A hypothetical model for the integration of UGTs into endoplasmic reticulum membranes is discussed which may have important implications in UGT activities (Radominska-Pandya *et al.* 1999). For recent additions of UGT sequences, the reader should consult the UDP glucuronosyltransferase homepage (http:»www.unisa.edu.au/ pharm_medsci/gluc_trans/default.htm; e-mail address of the developers of the homepage: Peter.Mackenzie@Flinders.edu.au or Ross.McKinnon@unisa.edu.au. It is hoped that the UGT homepage will increasingly be used as a medium to communicate established UGT isoforms.)

Historical perspectives

Soon after the discovery of UGTs (Dutton and Storey 1953; Dutton 1956, 1997) evidence for their multiplicity was obtained from clinical observations on non-haemolytic familial hyperbilirubinaemias such as Crigler–Najjar syndromes I and II (Axelrod *et al.* 1957; Schmid *et al.* 1957; Arias 1962; Arias *et al.* 1969). In liver microsomes from patients, bilirubin glucuronidation was undetectable whereas glucuronidation of other substrates was unimpaired. Multiplicity was also supported by differential induction of rat UGT activities by treatment of rats with 3-methylcholanthrene or phenobarbital (Bock *et al.* 1973; Wishart 1978a). These studies were substantiated by differential appearance of similar groups of UGT activities in late-foetal and neonatal development (Wishart 1978b). Purification of membrane-bound UGTs proved to be very difficult due to their strong interaction with phospholipids and their existence as strongly interacting oligomers in microsomal membranes. Nevertheless, purification helped to isolate UGT sequences from rats (Jackson *et al.* 1984; Mackenzie *et al.* 1984; Iyanagi *et al.* 1986) and humans (Harding *et al.* 1988). The growing number of sequences allowed a nomenclature system based on their evolutionary divergence (Burchell *et al.* 1991). This system allowed not only the naming of glucuronosyl transferases in vertebrates (fish and mammals) but also finding related glycosyltransferases in invertebrates (*Caenorhabditis elegans, Drosophila melanogaster*), bacteria, yeasts and plants (Mackenzie *et al.* 1997).

Nomenclature, UGT domain structure and polymorphisms

UDP-GLUCURONOSYLTRANSFERASE FAMILIES 1 AND 2

Based on evolutionary divergence, mammalian microsomal UDP-glucuronosyltransferases (EC 2.4.1.17) have been grouped into two distinct families: family 1 includes bilirubin and phenol UGTs, and family 2 includes steroid UGTs (Burchell *et al.* 1991). However, it is obvious from Tables 8.1 and 8.2 that both family 1 and 2 members are involved in steroid glucuronidation. For naming each gene, it is recommended that the root symbol UGT for human (Ugt for mouse), denoting '<u>U</u>DP-glucuronosyltransfer-ase,' be followed by an Arabic number representing the family, a letter designating the subfamiliy, and an Arabic numeral denoting the individual gene within the family or subfamily, e.g. 'human UGT1A6'.

Interestingly, human family 1 UGT isoforms are formed from a large UGT1 gene locus, spanning over 200 kb, containing more than a dozen promoters/first exons which are joined by exon sharing with their common exons 2 through 5 (Figure 8.3). Hence the different family 1 members have identical C-terminal halves of the UGT protein but different N-terminal halves. The UGT1 gene locus may have evolved by exon 1 duplication. Each first exon is regarded as a distinct gene and numbered according to the distance from the common exons (e.g. UGT1A1, UGT1A2). The human gene complex is present at chromosome 2 (2q37). The gene locus appears to be conserved between humans and experimental animals. Hence, orthologous genes are found at similar distances from the common exons. For example, the major bilirubin UGT of humans, rats and other species is encoded by exon 1 next to the common exons. The phenol UGT conjugating planar phenols is encoded by the sixth exon.

Family 2 consists of two subfamilies. UGT2 enzymes are encoded by six exons. Despite this difference from UGT1 proteins, both families share a high degree of similarity in the C-terminal end. The genes of subfamily 2B may have evolved by gene duplication. They are clustered on human chromosome 4q13. Three isoforms are clustered within a 192 kb region in a provisional order of UGT2B7 - 2B4 - 2B15 (Monaghan *et al.* 1994; Riedy *et al.* 2000). UGT2A1 has been identified as a major protein in bovine and rat olfactory epithelium which conjugates a broad substrate spectrum including odorants (Lazard *et al.* 1991). Recently, the human orthologue has been cloned and also mapped to chromosome 4 (4q13; Jedlitschky *et al.* 1999). It shows an identity of 87% with the rat UGT2A1 and of 43–62% with other human

Human

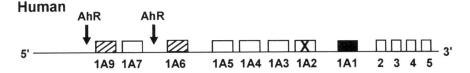

Rat

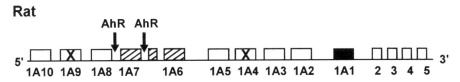

Figure 8.3 Schematic diagram of the human and rat UGT1 gene locus (Ritter *et al.* 1992; Iyanagi 1991, respectively). This gene spanning more that 200 kb consists of at least a dozen promoters/first exons which are linked by exon-sharing with their common exons 2-5. Exons indicated by crosses are pseudogenes. AhR indicates genes controlled by the Ah receptor. The rat UGT1A6 exon 1 is preceeded by a non-coding exon 1a (Emi *et al.* 1996). Recently it was found that UGT1A9 is the next isoform following UGT1A7 in the human UGT1 gene locus which—as illustrated in the figure—is incomplete (Ida S. Owens, National Institutes of Health, Bethesda, USA, personal communication).

UGT isoforms. In addition to odorants, it conjugates some steroids, especially androgens and some drugs. UGT2B isoforms are the most abundant. UGT1B is a minor subfamily that currently contains only one representative, cloned from fish (UGT1B; Mackenzie *et al*. 1997).

UGT DOMAIN STRUCTURE

All human UGTs have a common domain structure. In the variable *N*-terminal domain, a signal peptide has been identified (Blobel and Dobberstein 1975) which determines the transfer of UGTs to the endoplasmic reticulum (ER) followed by a variable *N*-terminal domain. At the conserved *C*-terminal domain, a single transmembrane fragment is found which is followed by a stop transfer signal (Blobel 1980). Evidence has been obtained that the cofactor UDP-glucuronic acid interacts with both the *N*- and *C*-terminal domains (Figure 8.4; Radominska-Pandya *et al*. 1999).

GLUCURONOSYL AND GLYCOSYLTRANSFERASES

It has become evident that UDP-glucuronosyltransferases may be part of a larger family of proteins (both membrane-bound and cytosolic) that preferentially use other

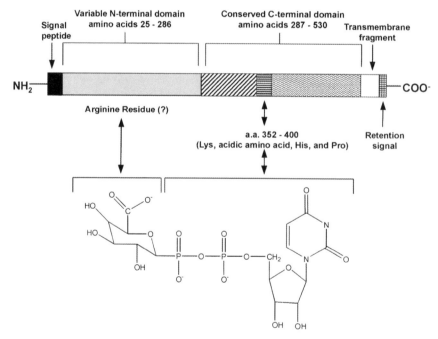

Figure 8.4 Domain structure of UGTs and postulated interactions between specific amino acids of the UGT active site and UDPGA (Reproduced from Radominska-Pandya *et al*. 1999 *Drug Metabolism Reviews*, **31**, 817–899. Published by Marcel Dekker, Inc., New York).

nucleotide sugars, including UDP-glucose and UDP-galactose (Mackenzie *et al.* 1997). Some of these glycosyltransferases reveal a striking homology in their C-terminal halves in which the following bona fide 'signature sequence' of amino acids has been found which determines their membership in the supergene family (h, hydrophobic residue; x, non-specific residue):

<pre>
        FhTHGGxxShxExhxxGVPhhxxPhxxDQ
          S   A   G   T
               C
</pre>

This conserved signature sequence probably represents the binding site for UDP. (It is noteworthy that there is a number of UDP-*N*-acetylglucosamine transferases which are not members of this glycosyltransferase supergene family.) Once the preferred nucleotide has been identified, the enzymes should be termed UDP-glucuronosyl-transferases (UGTs), UDP-glucosyltransferases, etc. Inclusion of glucuronosyltrans-ferases in the larger superfamily of glycosyl-transferases (Mackenzie *et al.* 1997) is preliminary because of the limited identification of many invertebrate sequences (Tukey and Strassburg 2000). Nevertheless, the similarity between the mammalian olfactory UGT2A1 and the *Drosophila* olfactory glucosyltransferase, between steroid glucuronosyltransferases and the baculovirus ecdysone glucosyltransferase, etc. ap-pears striking enough to suggest a broader superfamily and to stimulate further studies on the evolutionary relationship between UDP-glucuronosyltransferases and the superfamily of glycosyltransferases.

POLYMORPHISMS AND ALLELIC VARIANTS

An increasing number of allelic variants and polymorphisms of UGT isoforms is being discovered and a nomenclature system for the variants has been proposed (Mackenzie *et al.* 1997).

UGT1A1

Several important inherited hyperbilirubinaemias have been described (Schmid *et al.* 1957; Arias *et al.* 1969). Note that the clinical disorder does not always match the genotype.

Crigler–Najjar syndromes, type I and II

The inheritable and severe defects were analysed with respect to the gene structure of the UGT1 locus. Human UGT1A1 is the primary isoform responsible for bilirubin metabolism and maintenance of normal levels of serum bilirubin. A spectrum of mutations in the common exons 2–5 or of exon 1 is responsible for the phenotypic profiles (UGT1A1*2 - 27, Mackenzie *et al.* 1997). The development of 'chimera-plasty', a form of gene transplantation, is being evaluated as potential therapy (Gura 1999).

Gilbert's syndrome

This is characterised by a mild unconjugated hyperbilirubinaemia. It appears to be present in *ca* 5% of the Caucasian population. The basis for this syndrome is an atypical mutation of the TATA box region of the UGT1A1 promoter. The variant allele consists of seven TA repeats in the $A(TA)_n TAA$ motif whereas six TA repeats characterise the common allele (UGT1A1*1). The presence of the $A(TA)_7 TAA$ allele (UGT1A1*28) was found to decrease UGT1A1 gene expression *in vivo*. Homozygote individuals carrying the $A(TA)_7 TAA$ allele show significantly higher plasma levels of unconjugated bilirubin caused by a 30% reduction of UGT1A1 gene transcription (Bosma *et al.* 1995; Monaghan *et al.* 1996). Additional UGT1A1 alleles have been reported exclusively in an African–American population who have five or eight repeats (termed UGT1A1*33 and UGT1A1*34, respectively; Beutler *et al.* 1998). In general, there is a correlation between the number of repeats and the bilirubin levels. Evidence has been obtained recently that the $(TA)_8$ allele may be associated with lower oestradiol glucuronidation, higher oestrogen levels and a higher risk in breast cancer (Guillemette *et al.* 2000). Lower glucuronidation in Gilbert patients may also be the reason for more frequent side-effects after treatment with the topoisomerase inhibitor irinotecan due to lower glucuronidation rates of its major metabolite SN-38 (Gupta *et al.* 1994; Iyer *et al.* 1998).

UGT1A6

Two missense mutations were uncovered leading to E181A and R184S mutations of the UGT1A6 gene (UGT1A6*2, Ciotti *et al.* 1997). This genotype is present in about one third of the population and leads to lower glucuronidation of simple phenols than in carriers of the wild-type allele. The functional consequences are not yet clear. Co-occurence of UGT1A6*2 with the mutation of the UGT1A1 promoter (UGT1A1*28) has also been observed (Lampe *et al.* 1999).

UGT2B4

Several polymorphic UGT2B4 isoforms have been described (2B4-D^{458}, 2B4-E^{458}, etc.) which differ in the conjugation of the bile acid hyodeoxycholic acid. They are also involved in the conjugation of androgen metabolites, catechol oestrogens and eugenol (Lévesque *et al.* 1999).

UGT2B7

Variants have been described which poorly glucuronidate *S*-oxazepam (Patel *et al.* 1995). Recently, stable expression of both UGT2B7 (H^{268}) and UGT2B7 (Y^{268}) revealed differences in UGT activities towards several substrates with similar K_m values. However, no difference in oxazepam glucuronidation was found (Coffman *et al.* 1998).

UGT2B15

Variants of UGT2B15 (D^{85}) and UGT2B15 (Y^{85}) have been described (Lévesque *et al.* 1997). No difference in substrate specificity but slightly higher V_{max}-values were found when Y^{85} was compared with D^{85}.

Tissue distribution

LIVER, OESOPHAGUS, GASTROINTESTINAL TRACT AND KIDNEY

The liver is the major site of glucuronidation in the living organism. In this organ, UGTs are differentially expressed in the centrilobular and periportal zones (Ullrich *et al.* 1984; Saarikoski *et al.* 1998). It is becoming increasingly clear that the small intestine plays also a significant role in the first-pass metabolism of orally administered drugs such as morphine (Lin *et al.* 1999). Some UGT isoforms are expressed in a remarkably tissue-specific fashion in the gastrointestinal tract. In fact, hUGT1A7 has been found to be expressed only in gastric tissue and in the oesophagus, the latter also expressing UGT1A8, 1A9 and 1A10 (Strassburg *et al.* 1997a, 1999). UGT1A8 and UGT1A10 are expressed in the colon but not in the liver (Mojarrabi and Mackenzie 1998a; Strassburg *et al.* 1998). Evidence has been obtained that some isoforms (UGT1A1, UGT1A6, UGT2B4 and UGT2B7) are intriguingly expressed in a polymorphic manner in the small intestine (Strassburg *et al.* 2000). UGTs are expressed at other external surfaces such as skin and lung, as well as in kidney. The level of UGT1A9 in the kidney appears to be higher than in human liver (McGurk *et al.* 1998). Furthermore, remarkable species differences have been observed. For example, rat kidney is capable of glucuronidating bilirubin in contrast to the human kidney (McGurk *et al.* 1998).

OLFACTORY EPITHELIUM

An isoform, UGT2A1, has been found selectively in rat and human olfactory epithelium (Lazard *et al.* 1991; Jedlitschky *et al.* 1999). This isoform is involved in the metabolism of a variety of endo- and xenobiotics including steroid hormones and odorants. Inactivation of the latter compounds may facilitate removal of odorant signals thereby terminating their action. The comparative evolutionary approach made possible the identification of a related UDP-glucosyltransferase in the olfactory organ of *Drosophila melanogaster* (Wang *et al.* 1999).

STEROIDOGENIC TISSUES

It is becoming increasingly clear that UGTs play a role not only in overall pharmaco-kinetics but also in local protection of cells against toxicants. For example, the broadly expressed UGT1A6 has been detected in steroidogenic tissues such as testis and ovary (Münzel *et al.* 1994; Becedas *et al.* 1998). In the testis it appears to be expressed in Sertoli cells and in spermatogonia (Brands *et al.* 2000). The blood–testis barrier generated by Sertoli cells does not sufficiently protect spermatogonia against lipophilic compounds. This may explain why UGT1A6 is expressed in the spermatogonia themselves.

In addition to a protective role in steroidogenic tissues, UGTs may control endogenous receptor ligands such as testosterone or dihydrotestosterone, as suggested by Nebert (1991). This has been shown in a prostate carcinoma cell line LNCaP (Bélanger et al. 1998). These steroid target cells have been shown to express both UGT2B15 and UGT2B17, the latter being expressed at a lower level but being inducible by antioxidants. For example, treatment of the cells with the soy bean isoflavone biochanin A increased testosterone glucuronide formation and reduced formation of the androgen-dependent prostate-specific antigen, suggesting that UGTs are responsible for control of this endogenous receptor ligand (Sun et al. 1998).

BRAIN

Recently, expression of UGTs (UGT1A6 and UGT2B7) has been detected in rat and human brain (Martinasevic et al. 1998; King et al. 1999), particularly in Purkinje cells of the cerebellum and hippocampal pyramidal cells (King et al. 1999; Brands et al. 2000). UGT1A6 has been shown to glucuronidate serotonin. However, more work is needed to establish whether UGTs control the levels of catecholamines. UGT activities have also been found in brain microvessel fractions which are derived from the blood-brain barrier (Suleman et al. 1998). Expression of UGT2B7 in brain may be interesting since it is the only UGT isoform which forms morphine 6-glucuronide. This glucuronide is biologically active and a strong analgesic (Paul et al. 1989; Osborne et al. 1992). It has been found to be more potent on μ opioid receptors than morphine itself. Formation of morphine 6-glucuronide in brain tissue from morphine would circumvent the blood-brain barrier which is poorly penetrable by morphine 6-glucuronide from the blood side (Wu et al. 1997).

Substrate specificity

UGTs conjugate a remarkable diversity of endobiotics and xenobiotics containing a number of different functional groups (e.g. $-OH$, $-COOH$, $-NH_2$ and $-SH$). Most glucuronides are biologically inactive. However, some are bioactive (morphine 6-glucuronide) or even toxic (oestrogen D-ring glucuronides and acyl glucuronides). Substrate hydrophobicity is essential for glucuronidation by UGTs. For lipophilic compounds the membrane may be involved in the transport of the substrate to the active site of the enzyme (see below).

Heterologous expression of UGT isoforms has been a powerful tool in determining their substrate specificity (Guengerich et al. 1997; Townsend et al. 1999). The results of these studies are summarised in Tables 8.1 and 8.3 for family 1 members and Tables 8.2 and 8.4 for family 2 members. Quantitative comparison of glucuronidation rates of UGT isoforms is not yet possible since the protein levels of recombinant UGT isoforms are seldom known.

For UGT1A5, 2B10 and 2B11 no substrates have been found so far (July) 2000. The substrates have been grouped here into endobiotics (Tables 8.1 and 8.2) and xenobiotics (Tables 8.3 and 8.4). Endobiotic substrates have been grouped as steroids [C18 steroids (oestradiol), C19 steroids (dihydrotestosterone) and C21 steroids (pregnandiol)] and as a miscellaneous group of other endobiotics such as bilirubin, bile acids

Table 8.1 Endobiotic substrates of human UGT1 isoforms

Isoform	Substrate	
	Steroids	Others
UGT1A subfamily		
UGT1A1	Oestriol	Bilirubin
	β-Oestradiol	atRA[a]*
	2-Hydroxyoestriol	5,6-Epoxy-atRA[a]*
	2-Hydroxyoestrone	
	2-Hydroxyoestradiol	
UGT1A3	2-Hydroxyoestrone	5,6-Epoxy-atRA*
	2-Hydroxyoestradiol	4-OH-atRA
	Oestrone	Lithocholic acid
UGT1A4	Androsterone	
	Epiandrosterone	
	5α-Androstane-3α, 17β-diol	
	5β-Androstane-3α, 11α, 17β-diol	
	5α-Pregnan-3α, 20α-diol	
	5α-Pregnan-3β, 20β-diol	
	5-Pregnene-3β-ol-20-one	
UGT1A6		Serotonin[b]
UGT1A7	not known	
UGT1A8	Oestrone	
	2-Hydroxyoestrone	
	4-Hydroxyoestrone	
	2-Hydroxyoestradiol	
	4-Hydroxyoestradiol	
	Dihydrotestosterone	
	5α-Androstane-3α, 17β-diol	
UGT1A9		Thyroxine
		Reverse triiodothyronine
UGT1A10	2-Hydroxyoestrone	
	4-Hydroxyoestrone	
	Dihydrotestosterone	

For references see Radominska-Pandya *et al.* (1999) unless indicated.
*atRA, all-*trans*-retinoic acid
[a] Radominska-Pandya *et al.* (1997) [b] King *et al.* (1999)

(hyodeoxycholic acid), etc. Xenobiotic substrates have been grouped as marketed drugs, dietary plant constituents (coumarin derivatives, flavonoids etc.) and carcinogens including benzo[a]pyrene phenols and arylamines. A list of 350 substrates tested by utilising recombinant UGTs is found in the addendum to the review of Tukey and Strassburg (2000).

UGT FAMILY 1 MEMBERS (TABLES 8.1 AND 8.3)

UGT1A1

This major isoform is the only human UGT involved in the conjugation of bilirubin. Normally it has to glucuronidate 200–400 mg of bilirubin formed daily from the

Table 8.2 Endobiotic substrates of human UGT2 isoforms

Isoform	Substrate	
	Steroids	Others
UGT2B subfamily		
UGT2B4 alleles	Androsterone	Hyodeoxycholic acid
	5α-Androstane-3α, 17β-diol	
	5β-Pregnan-11α, 17β-diol-20-one	
	5β-Pregnan-11α, 17β-triol-20-one	
UGT2B7 (Y)	4-Hydroxyoestrone	Hyodeoxycholic acid
	4-Hydroxyoestradiol	
	Androsterone	
	Epitestosterone	
	5β-Androstane-3α, 17β-diol	
	5β-Pregnan-3α-ol-11,20-dione	
	5β-Pregnen-11α-l-3,20-dione	
UGT2B7 (H)	Androsterone	Lithocholic acid
	Epitestosterone	Hyodeoxycholic acid
	4-Hydroxyoestrone	atRA[a]*
	2-Hydroxyoestrone	5,6-epoxy-atRA[a]*
	4-Hydroxyoestradiol	4-OH-atRA[a]*
	Oestriol	4-OH-atRA[a]
		Linoleic acid[b]
		Linoleic acid 9,10-diol
		Linoleic acid 12,13-diol
		Arachidonic acid
		Phytanic acid
UGT2B15	Androsterone	
	Dihydrotestosterone	
UGT2B17	Androsterone	
	Dihydrotestosterone	
	Testosterone	

For references see Radominska-Pandya *et al.* (1999) unless indicated.
*atRA, all-*trans*-retinoic acid.
[a]Samokyszyn *et al.* (2000) [b]Jude *et al.* (2000)

catabolism of haem. As already discussed, UGT activity of UGT1A1 is a major factor determining the blood level of bilirubin.

This enzyme is also involved in the conjugation of retinoic acid, oestrogens and ethinyloestradiol at the 3-OH position (Ebner and Burchell 1993). Senafi *et al.* (1994) have shown that UGT1A1 conjugates dietary phenolic plant constituents such as the flavonoid quercetin and anthraquinones. The enzyme also conjugates opioids such as buprenorphine and nalorphine (King *et al.* 1996) and SN-38, a major metabolite of the chemotherapeutic topoisomerase inhibitor irinotecan (Iyer *et al.* 1998).

UGT1A3 and UGT1A4

These isoforms are known for the conjugation of tertiary amines to quarternary ammonium glucuronides, including important drugs such as imipramine, cyprohepta-dine, ketotifene, etc. (Green and Tephly 1998). Rat UGT1A4 is a pseudogene. This

Table 8.3 Xenobiotic substrates of human UGT1A isoforms

Isoform	Drugs	Plant constituents	Carcinogens, others
UGT1A1	Ethinyloestradiol Buprenorphin SN-38	Quercetin Naringenin 4-Methylumbelliferone	1-Naphthol
UGT1A3	Cyproheptadine	Scopoletin Naringenin 4-Methylumbelliferone	
UGT1A4	Amitriptyline Imipramine Clozapine	Diosgenin Tigogenin	Benzidine
UGT1A6	Paracetamol[a]	Methylsalicylate 4-Methylumbelliferone[a]	3-OH-BaP BaP-3,6-diphenol[c] 1-Naphthylamine[d] 2-Naphtlylamine[d] N-OH-2-naphthylamine[d]
UGT1A7	SN-38[b]		7-OH-BaP[e] N-OH-PhIP[e]
UGT1A8		Scopoletin Naringenin Eugenol	N-OH-PhIP[e]
UGT1A9	Propofol Paracetamol[a]	Quercetin Alizarin 4-Methylumbelliferone[a]	3-OH-BaP BaP-3,6-diphenol[c] BaP-7,8-dihydrodiol N-OH-PhIP[e]
UGT1A10	Mycophenolic acid		

For references see Radominska-Pandya *et al.* (1999) unless indicated.
BaP, benzo[a]pyrene; PhIP, 2-amino-1-methyl-6-phenylimidazo[4,5-b]pyridine
[a]Bock *et al.* (1993) [b]Ciotti *et al.* (1999b) [c]Gschaidmeier *et al.* (1995) [d]Orzechowski *et al.* (1994) [e]Strassburg *et al.* (1999).
SN-38: 7-Ethyl-10-hydroxycamptothecin.

Table 8.4 Xenobiotic substrates of human UGT2B isoforms

Isoform	Drugs	Plant constituents	Carcinogens, others
UGT2B4 alleles		Eugenol	
UGT2B7	Morphine Naloxone Codeine Buprenorphine Carboxylic-acid-containing drugs		BaP-7,8-dihydrodiol[a]
UGT2B15		Eugenol Scopoletin Naringenin 4-Methylumbelliferone	
UGT2B17		Eugenol	

For references see Radominska-Pandya *et al.* (1999)
BaP, Benzo[a]pyrene
[a]Jin *et al.* (1993b)

may be the reason why rats cannot form quarternary ammonium glucuronides. Human UGT1A4 is more active than UGT1A3. UGT1A3 is also involved in the conjugation of catechol oestrogens and UGT1A4 in the conjugation of androsterone and pregnanediol (Table 8.1).

UGT1A6

This broadly expressed isoform is mostly involved in the conjugation of simple planar phenols. Serotonin can be considered as an endogenous substrate of UGT1A6 (King et al. 1999). It has been shown to be a major enzyme in the conjugation of paracetamol (acetaminophen; Bock et al. 1993). Paracetamol glucuronidation is enhanced in cigarette smokers (Bock et al. 1987, 1994). UGT1A6 is also involved in the conjugation of a variety of planar primary amines such as 1- and 2-naphthylamine and the corresponding hydroxylamines (Orzechowski et al. 1994). Both the human and rat orthologues are able to form conjugates of benzo[a]pyrene (BaP) phenols and diphenols such as BaP-3,6-diphenol to monoglucuronides. Rat UGT1A6, but not human UGT1A6, forms both BaP-3,6-diphenol mono- and diglucuronides (Gschaidmeier et al. 1995).

UGT1A7

In humans this isoform is expressed in a remarkably tissue-specific manner. It was found in the oesophagus and gastric tissue but not in liver (Strassburg et al. 1999). It conjugates 7-hydroxy-BaP and 4-hydroxy-PhIP (2-amino-1-methyl-6-phenylimidazo[4,5-b]pyridine). In the rat it is expressed in liver and intestine where it very efficiently conjugates both planar and bulky phenols (Burchell et al. 1997). In particular, it is involved in the detoxification of benzo[a]pyrene (BaP) metabolites such as 3-OH-BaP- and BaP-7,8-dihydrodiol (Grove et al. 1997).

UGT1A9

The human isoform is important in the hepatic conjugation of both planar and bulky phenols (Ebner and Burchell 1993). It conjugates endogenous substrates such as thyroxin and drugs such as the anaesthetic propofol. Paracetamol is also conjugated but with much lower affinity than by UGT1A6 (Bock et al. 1993).

UGT1A8 and UGT1A10

Both isoforms are expressed in the intestine but not in the liver. Endogenous substrates of UGT1A8 are catechol oestrogens and androgens, coumarins, flavonoids and anthraquinones (Mojarrabi and Mackenzie 1998a; Cheng et al. 1999). Both isoforms have been shown to conjugate metabolites of the potential colon carcinogen PhIP (Nowell et al. 1999). UGT1A10 has been demonstrated to conjugate the immunosuppressive agent mycophenolic acid (Mojarrabi and Mackenzie 1998b).

UGT FAMILY 2 MEMBERS (TABLES 8.2 AND 8.4)

UGT2B4

Several UGT2B4 alleles have been described which are expressed in the liver and extrahepatic tissues. They are involved in the conjugation of hyodeoxycholic acid, androgen metabolites, catechol oestrogens and eugenol (Lévesque *et al.* 1999). More work is needed to characterise the substrate specificity of the different polymorphic variants since alleles have been described which appeared to be selective for hyodeoxycholic acid (Pillot *et al.* 1993).

UGT2B7

This isoform represents a major UGT isoform of family 2 which is expressed in the liver and many other organs. Two allelic variants have been described: UGT2B7 (Y) and UGT2B7 (H). They conjugate a wide variety of drugs such as morphine, (non-steroid antiinflammatory drugs) and zidovudine (Barbier *et al.* 2000), and endogenous compounds such as catechol oestrogens at 4-OH, androsterone, bile acids such as hyodeoxycholic acid, retinoids, fatty acids such as linoleic acid and in particular hydroxy fatty acids (Jin *et al.* 1993a; Coffman *et al.* 1998; Radominska-Pandya *et al.* 1999).

UGT2B15 and UGT2B17

UGT2B15 and 2B17 are expressed in the liver and the prostate (Green *et al.* 1994; Beaulieu *et al.* 1996). They conjugate dihydrotestosterone, other androgens and various plant constituents. UGT2B17 has been found to be inducible by antioxidants (Sun *et al.* 1998) and by cytokines (Lévesque *et al.* 1998).

UGTs INVOLVED IN METABOLISM AND DISPOSITION OF CARCINOGENS

Glucuronidation plays a major role in detoxification of carcinogenic compounds, for example of the tobacco-specific nitrosamine NNAL [4-(methylnitrosamino)-1-(3-pyridyl)-1-butanol]. Several UGT isoforms are involved in its detoxification such as UGT1A6, 1A7, 1A9 and 1A10 (Nguyen *et al.* 2000). In the rat, NNAL has been shown to be conjugated by UGT2B1 (Ren *et al.* 1999). A number of UGTs are also involved in detoxification of polycyclic aromatic hydrocarbons (PAHs; Bock 1991; Bock and Lilienblum 1994) and of simple and heterocyclic arylamines (Lee Chiu and Huskey 1998).

Benzo(a)pyrene (BaP)

The role of glucuronidation in the metabolism of PAHs is discussed using BaP as example. Selective pathways are listed in Figure 8.5. Extensive studies of the carcinogenicity of BaP metabolites revealed that BaP-7,8-dihydrodiol-8,9-epoxide is one of the ultimate carcinogens leading to DNA adducts (Conney 1982). As an intermediate in this pathway BaP-7,8-dihydrodiol is formed. This compound was found to be a

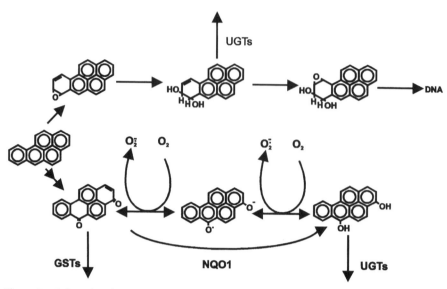

Figure 8.5 Selected pathways of benzo[a]pyrene (BaP) metabolism. Upper pathway, formation of a major ultimate carcinogen (Conney 1982), lower pathway, prevention of BaP quinone toxicity by the actions of glutathione S-transferases (GSTs), NADPH quinone oxidoreductase-1 (NQO1) and UGTs, BAP-3,6-quinone as example.

substrate of rat UGT1A7 (Grove *et al.* 1997) and of human UGT1A9 and UGT2B7 (Jin *et al.* 1993b). The mutant Gunn rat strain does not form any functional UGT1 proteins (Hu and Wells 1992). It shows higher BaP DNA adducts than congenic Wistar rats, indicating that glucuronidation by family 1 members is able to reduce the formation of DNA adducts.

A major fraction of BaP metabolites is excreted in rat bile. For example, when BaP was injected intravenously, 68% of the metabolites were excreted during 6 hours, and of this amount 34% and 9% represented glucuronide and sulphate conjugates, respectively. The remaining 25% were probably glutathione conjugates (Boroujerdi *et al.* 1981). BaP-3,6-diphenol diglucuronide was identified as a major glucuronide secreted into rat bile (Bevan and Sadler 1992; Yang *et al.* 1999).

Phenolic metabolites were found to be conjugated more efficiently than BaP-7,8-dihydrodiol. Conjugation of diphenols may efficiently reduce or prevent toxic quinone/quinol redox cycles (Lilienblum *et al.* 1985). Glucuronidation also reduced the mutagenicity of BaP-3,6-quinone in the Ames test (Bock *et al.* 1990a).

Rat UGT1A7 has been shown to prevent the cytotoxicity of BaP-3,6-quinone (Grove *et al.* 2000). BaP-3,6-diphenol is conjugated to mono- and diglucuronides. Several UGTs including UGT1A6 are involved in the formation of monoglucuronides. Diglucuronides are formed by rat UGT1A7 (Bock *et al.* 1999) and human UGT1A9 (Gschaidmeier *et al.* 1995). They are not formed by human UGT1A6. As an exception, rat UGT1A6 forms both mono- and diglucuronides of BaP-3,6-diphenol. This example may be interesting since the monoglucuronide has to change its position in the UGT binding site for diglucuronide formation (see below).

Arylamines

In studies on the metabolism of 2-naphthylamine and 4-aminobiphenyl, it has been shown that glucuronides determine the target of their carcinogenicity, the urinary bladder. For example, 2-naphthylamine is oxidised to N-hydroxy-2-naphthylamine which is efficiently conjugated. The resulting N-hydroxy-N-glucuronide represents a stable transport form and can be considered as a proximal carcinogen. It decomposes at the slightly acidic pH of urine to the hydroxylamine and is protonated to a nitrenium ion, which readily reacts with DNA, and may potentially initiate bladder cancer (Beland and Kadlubar 1990; see chapter 1). The N-hydroxy-N-glucuronide of 2-naphthylamine was shown to be conjugated by human UGT1A6 and UGT1A9 facilitating detoxification of this reactive compound within the cell (Orzechowski et al. 1994). Benzidine and metabolites have been shown to be efficiently conjugated by UGT1A4 and UGT1A9 (Ciotti et al. 1999a).

Recently, mutagenic heterocyclic arylamines such as PhIP have been discovered which are formed at trace levels in food such as meat and fish in typical household cooking practices (Sugimura and Sato 1983). They also have been found to be oxidised to hydroxylamines and conjugated with glucuronic acid to N–OH–N–glucuronides (Alexander et al. 1991). The N–OH–N–glucuronide of PhIP has been found to be acid-stable and not hydrolysed by E. coli β-glucuronidase (Kaderlik et al. 1994a). It is a major metabolite in humans (Malfatti et al. 1999), and hence this pathway is a major detoxifying pathway. N-OH-PhIP has been found to be O-acetylated by N-acetyltransferases. Evidence was obtained that reactive N-acetoxy-PhIP is transported via the bloodstream and may lead to DNA adduct formation in the colon via this route (Kaderlik et al. 1994b). Several family 1 members (UGT1A3, 1A8, 1A9, 1A10) have been shown to conjugate metabolites of PhIP, and may effectively prevent bioactivation by acetylation or sulphation (Nowell et al. 1999).

Reactive acyl glucuronides as chemically reactive intermediates

Glucuronides are generally considered to be biologically inactive. Therefore, it is important to stress that acyl glucuronides are chemically reactive at pH > 7. Acyl glucuronides are formed from a number of carboxyl-containing endobiotics (bilirubin, lithocholic acid, retinoic acid) and drugs such as NSAIDs, clofibrate and the anti-epileptic drug valproic acid. They are known to undergo 'acyl migration' along the hydroxyl groups at C2–C4 of glucuronic acid. These glucuronides are β-glucuronidase-resistant and therefore escape the usual urinary analysis of glucuronides. It has been suggested that these β-glucuronidase-resistant glucuronides of NSAIDs may be transport forms through the small intestine, explaining their action in the colon (Dickinson 1998).

Acyl glucuronides have attracted considerable interest because they form protein adducts and are possibly responsible for immune reactions. It is striking that many drugs withdrawn from the market (benoxaprofen, zomepirac, etc.) belong to this class (Benet *et al.* 1993). Adducts may be generated by two mechanisms (Figure 8.6):

(1) Nucleophilic displacement. In this mechanism the drug covalently binds to the protein and the glucuronic acid moiety is liberated.
(2) Imine mechanism: After acyl migration, the aldehyde group of the ring-open tautomer condenses with a lysine group on the protein to form an imine. At the completion of the reaction, the adduct contains the drug together with the glucuronic acid moiety.

Covalent binding of drugs to proteins may generate epitopes which may initiate autoimmune disease. One of the targets of covalent binding has been shown to be the UGT protein itself (Terrier *et al.* 1999). This covalent binding of the substrate leads to inhibition of UGT activity. Reactivity depends on the stability of the acyl glucuronide which varies considerably. For example, telmisartan acyl glucuronide appears to be much more stable than other acyl glucuronides (Ebner *et al.* 1999).

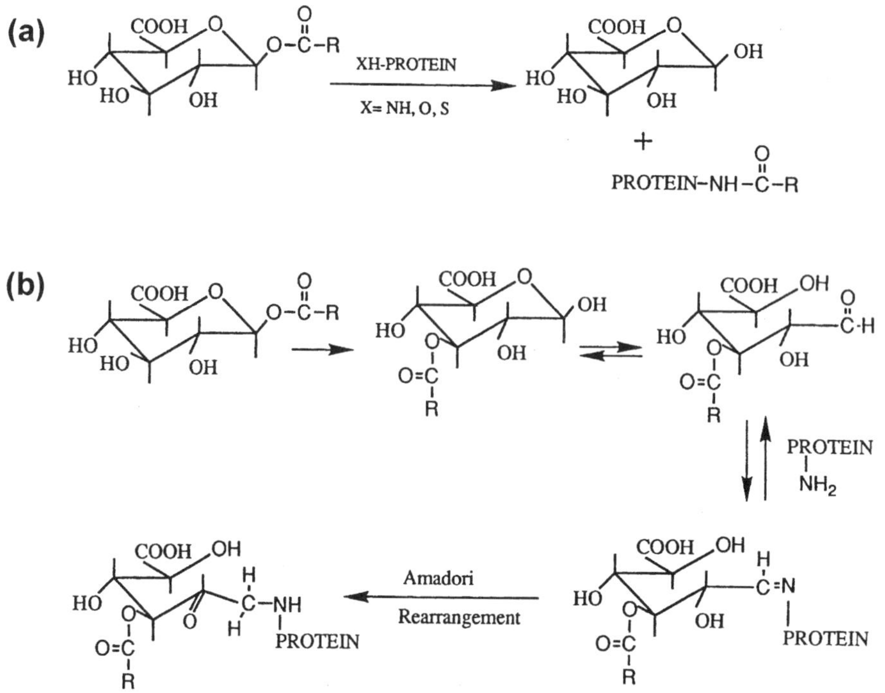

Figure 8.6 Postulated mechanisms for the irreversible binding of carboxylic acids to proteins via their glucuronides. (a) Nucleophilic displacement mechanism, leading to an acylated protein and liberation of D-glucuronic acid. (b) Imine mechanism, by which the glucuronic acid is part of the adduct (Reproduced from Benet *et al.* 1993. *Life Sciences*, **53**, 141–146. Published by Elsevier Science, Oxford).

Factors regulating formation and export of glucuronide from cells

MEMBRANE TOPOLOGY AND COFACTOR SUPPLY

After many years of discussion of the conformation (Berry and Hallinan 1974) and the compartmentation hypothesis (Zakim *et al.* 1988), the advent of UGT sequences provided compelling evidence for the localisation of UGTs behind a lipid barrier in the lumenal part of the endoplasmic reticulum (ER). As already discussed, UGT sequences are preceded by a signal peptide (Blobel and Dobberstein 1975) mediating the integration of the UGT polypeptide with the ER membrane (Figure 8.4). This signal peptide is cleaved and the protein *N*-glycosylated on the lumenal site of the ER. The polypeptide is retained in the ER by a transmembrane domain and stop transfer signal (Blobel 1980) at the *C*-terminus of the protein. However, experiments with truncated proteins revealed that the variable *N*-terminal domain (amino acids, 25–286) containing the aglycone binding site may also be involved in strong interactions with the lipid bilayer. In fact, deletion of the signal peptide and the stop transfer signal did not prevent membrane targeting and insertion of UGT1A6, suggesting the presence of internal topogenic elements able to translocate and retain the isoform in the ER membrane (Ouzzine *et al.* 1999).

Current information about the structure of the active site of UGTs has been discussed recently (Radominska-Pandya *et al.* 1999). It provided the basis for the proposed acid-base catalytic mechanism discussed in Figure 8.2(B) and the UDPGA binding illustrated in Figure 8.4. The proposed model is supported by three-dimensional structure data at 1.8 Å resolution of *E. coli* galactose-1-phosphate uridylyltransferase (Wedekind *et al.* 1995) which belongs to the glycosyltransferase supergene family including UGTs. More work is needed to elucidate the structure of the active site and in particular the hydrophobic pockets of the multiple aglycone-binding sites of UGTs, for example those encoded by the exons 1 of family 1 members.

Localisation of UGTs on the lumenal part of ER membranes raises the question as to how the cofactor UDP-glucuronic acid (UDPGA), synthesized in the cytosol, is transported through the membrane. This transport is the reason for the known 'latency' of UGT activity. The activity is low in native microsomes and can be activated by various procedures: for example, addition of detergents, of pore-forming agents such as alamethicin, repeated freezing and thawing, spontaneous activation by endogenous phospolipases generating detergents such as lysophosphatidylcholines, etc. (Dutton 1980). Transporters may exist both for transport of the cofactor UDPGA and for glucuronides produced intralumenally. These transporters have not been characterised. Recent findings suggest the participation of two asymmetric antiporters: UDPGA influx is coupled to UDP-*N*-acetylglucosamine efflux, and UDP-*N*-acetylglucosamine influx is coupled to UMP efflux (Bossuyt and Blanckaert 1995). The latter nucleotide is a product of UDP. Hence, these antiport pathways may link UDPGA influx with UGT activity. The reason for UDP-*N*-acetylglucosamine being a physiologic activator of UGT activity has not been elucidated. Recently, evidence was provided for an UDPGA/glucuronide antiport (Bánhegyi *et al.* 1996).

Interestingly, accumulating evidence suggests that the active enzyme consists of dimers of 2 UGT polypeptide chains. Homo- and heterodimers have been found by crosslinking and co-immunoprecipitation (Meech and Mackenzie 1997; Ikushiro *et al.*

1995). Oligomer formation between UGT monomers is also supported by radiation-inactivation analysis of liver microsomal UGTs (Peters *et al.* 1984; Vessey and Kempner 1989; Gschaidmeier and Bock 1994). Based on these findings, a dynamic topological model has been proposed in which dimerisation of UGT monomers at the *C* terminus may lead to the formation of a channel that allows access of UDPGA to the enzyme active site, possibly in conjunction with highly specialised transport proteins (Bossuyt and Blanckaert 1995; Hirschberg *et al.* 1998). The produced glucuronides may be expelled into the cytosol via the same channel used for the entry of UDPGA (Figure 8.7). The lipophilic aglycone may diffuse through the bilayer and bind to the active site. This may explain the correlation between UGT activities and substrate lipophilicity (Illing and Benford 1976; Bock and Lilienblum 1994).

Radiation-inactivation analysis revealed the existence of tetramers for the formation of the diglucuronide of BaP-3,6-diphenol (Gschaidmeier and Bock 1994). Cooperation of several functional UGTs may be necessary, in this case since the position of the monoglucuronide at the active site has to be changed for diglucuronide formation. Monoglucuronides may be expelled from one UGT unit, and neighbouring UGTs may then accept the monoglucuronide as substrate for diglucuronide formation.

The cofactor UDPGA is formed by UDP-glucose dehydrogenase from UDP-glucose. The latter is readily formed either from glycogen or from glucose. The level of UDPGA in liver cells was found to be 0.2–0.3 μmoles/g liver tissue, in the range of the K_m values for UGTs (Bock and White 1974). Therefore, any change in the cellular level of UDPGA may affect glucuronidation. For example, addition of ethanol inhibits UDP-glucose dehydrogenase and lowers the level of UDPGA (Moldeus *et al.* 1978). However, in the liver of fed rats high glucuronidation rates did not decrease the intracellular UDPGA concentration, suggesting efficient regeneration of the cofactor (Bock and White 1974).

GLUCURONIDE TRANSPORTERS (MRPs), BILIARY TRANSPORT OF GLUCURONIDES AND ENTEROHEPATIC CYCLING

Multispecific transporters for glucuronides (as well as for sulphate and glutathione conjugates) have been characterised (Jedlitschky *et al.* 1996). Interestingly, they are

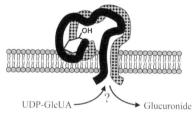

UDP-GlcUA ——— ? ——→ Glucuronide

Figure 8.7 Hypothetical model of UGT topology. The catalytic unit is formed by dimerisation of UGT monomers. Lipophilic aglycones diffuse through the bilayer and bind to the active site formed by interactions of two monomers. UDP-GlcUA = UDPGA may gain access to the active site via a proteinaceous channel formed when the monomers dimerise. After conjugation, the glucuronide is expelled into the cytosol via the same channel used for entry of UDPGA (Reproduced from Radominska-Pandya *et al.* 1999 *Drug Metabolism Reviews*, **31**, 817–899. Published by Marcel Dekker, Inc., New York).

expressed in different parts of epithelial cells such as hepatocytes and thus determine the disposition of glucuronides. For example, MRP1 and MRP3 are present in the basolateral membrane whereas MRP2 is present exclusively in the apical membranes of cells (Keppler and König 1997; Kool *et al.* 1997). Their functions have been characterised for etoposide glucuronide, oestradiol-17β glucuronide (Jedlitschky *et al.* 1996), and for bilirubin mono- and diglucuronides (Jedlitschky *et al.* 1997). Absence of MRP2 is the cause of the Dubin–Johnson syndrome, which is characterised by heritable conjugated hyperbilirubinaemia. MRPs are also differentially expressed in other polarised epithelial cells; for example, in intestinal epithelia and Caco 2 cell monolayers (Walle *et al.* 199; Bock *et al.* 2000).

MRPs may determine glucuronide formation in cells since accumulation of glucuronides inhibits the reversible UGT reaction (Bock and White 1974). As discussed in the Introduction section, MRPs also determine the fate of glucuronides. For example, SN-38, a major metabolite of the camptothecin topoisomerase inhibitor irinotecan, has been shown to be rapidly converted to glucuronides which are excreted using MRP2 via the bile into the intestine (Sugiyama *et al.* 1998). After hydrolysis by bacterial β-glucuronidases, the aglycone may be re-absorbed and undergo extensive enterohepatic cycling. This leads to prolongation of the drug's biological half-life and accumulation of the compounds in the organism. However, in intestinal cells glucuronides are released by MRP2 into the intestinal lumen. Subsequent hydrolysis by β-glucuronidases leads to reabsorption of the aglycones and to futile cycles. These futile cycles may explain in part gastrointestinal toxicity (diarrhoea) of the potent chemotherapeutic irinotecan (Gupta *et al.* 1994).

Transcriptional regulation by xenobiotics

In general, differential transcription of different UGT genes largely determines the protein levels and activities of different UGT isoforms. Marked species- and tissue-specific differences exist in constitutive or basal expression, in developmental and xenobiotic-induced expression of UGT isoforms. Factors responsible for tissue-specific expression (such as HNF1 [hepatic nuclear factor 1], C/EBP[CCAAT enhancer binding protein]α and β etc.) are beginning to be unravelled (Hansen *et al.* 1998; Ishii *et al.* 2000; Lee *et al.* 1997). UGTs may also be regulated by interleukins in disease states, as shown for UGT2B17 (Lévesque *et al.* 1998). This chapter deals with induction of UGT isoforms by xenobiotics such as phenobarbital-type inducers, Ah receptor ligands and antioxidants/electrophiles.

PHENOBARBITAL-TYPE AND OTHER INDUCERS

Rat UGT1A1 appears to be induced by phenobarbital (Burchell *et al.* 1997); contradictory findings have been reported and need to be clarified (Ikushiro *et al.* 1995). It has been clearly shown that UGT2B1 is induced by phenobarbital (Mackenzie 1986). This isoform has been shown to conjugate chloramphenicol and morphine and is responsible for the enhanced UGT activities in liver microsomes of phenobarbital-treated rats (Bock *et al.* 1973). The human UGT1A1 has also been shown to be phenobarbital-inducible (Sutherland *et al.* 1993). Preliminary experiments provide evidence that this isoform is probably controlled by the CAR/RXR [constitutively active receptor/retinoid X receptor] element first identified in the promotor region of phenobarbital-inducible CYP2B genes (Kawamoto *et al.* 1999).

Clofibrate and dexamethasone are known to induce UGTs (Emi *et al.* 1995; Ikushiro *et al.* 1995). However, the induction mechanisms have not been elucidated.

Ah RECEPTOR AGONISTS

Experimental animals

Early genetic studies with 3-methylcholanthrene-responsive and -non-responsive mouse strains suggested induction of liver UGTs via the Ah receptor (Owens 1977). These findings were substantiated by the absence of UGT1A6 expression in Ah receptor knockout mice (Fernandez-Salguero *et al.* 1995). Extensive studies in rat liver showed that basal expression of UGT1A6 in rat liver is low, but it is > tenfold induced by 3-methylcholanthrene treatment (Iyanagi *et al.* 1986; Emi *et al.* 1996). Marked induction by TCDD (2,3,7,8-tetrachlorodibenzo-*p*-dioxin) was restricted to the liver whereas high constitutive expression was preponderant in extrahepatic tissues (Münzel *et al.* 1994). Hence, responsiveness to Ah receptor-type inducing agents appears to correlate inversely with the level of constitutive expression. UGT1A7 was also shown to be inducible by 3-methylcholanthrene (Emi *et al.* 1995; Metz and Ritter 1998).

Rabbit UGT1A6 was found to be constitutively expressed in the liver and not further induced by TCDD (Lamb *et al.* 1994), indicating species-dependent differences in UGT1A6 expression. Similarly, microsomal UGT activities towards 1-naphthol and 4-methylumbelliferone, major substrates of UGT1A6, were high in dog liver and only moderately induced by β-naphthoflavone; but induction of UGT activities was observed in dog intestine (Richter von *et al.* 2000). Lack of induction by Ah receptor agonists is probably due to high constitutive expression of liver UGTs in rabbits and dogs.

Humans

UGT1A6 is constitutively expressed in human livers. Induction by TCDD studied in human hepatocyte cultures appears to be moderate with high interindividual variability (Münzel *et al.* 1996). Induction of UGT1A6 by PAH-type inducers is supported by *in vivo* findings that paracetamol glucuronidation (paracetamol being a substrate of UGT1A6, Bock *et al.* 1993) is increased in cigarette smokers (Mucklow *et al.* 1980; Bock *et al.* 1987, 1994). Glucuronidation of propranolol is also increased in smokers (Walle *et al.* 1987). In the Caco-2 cell model, UGT1A6 induction by TCDD is clearly detectable and appears to be Ah receptor-dependent; i.e. the Ah receptor/Arnt (Aryl hydrocarbon nuclear transferase) complex binds to a consensus GTGCG DNA core sequence, the xenobiotic response element (XRE, Münzel *et al.* 1998). Interestingly, a second UGT1 family member, hUGT1A9, also appears to be inducible by TCDD (Münzel *et al.* 1999), similar to at least two rat UGT isoforms (rUGT1A6 and rUGT1A7) which are inducible by polycyclic aromatic hydrocarbons.

ANTIOXIDANTS/ELECTROPHILES

UGTs, for example rat UGT1A6, are known to be inducible by antioxidant-type or phase II enzyme inducers (Buetler et al. 1995). Recently, evidence has been obtained in the Caco-2 cell model that three human UGTs (UGT1A6, UGT1A9 and UGT2B7) are induced by the prototype antioxidant t-butylhydroquinone (TBHQ, Münzel et al. 1999). Interestingly, antioxidant-inducible proteins also include the apical conjugate export pump MRP2 (multidrug resistance protein 2, Bock et al. 2000). Since paracetamol is a substrate of human UGT1A6 (Bock et al. 1993), it is conceivable that increased paracetamol glucuronide excretion found in subjects on a Brussels sprouts and cabbage diet (Pantuck et al. 1984) may be due to antioxidant-type induction of UGT1A6 in the intestine and liver.

Antioxidant-type induction may be considered as an adaptive response to oxidative or electrophilic stress which is triggered by a sublethal dose of a variety of antioxidants/prooxidants. Increased glucuronidation has been observed after treatment with TBHQ (a metabolite of the widely used antioxidant BHA [2(3)-tert-butyl-4-hydroxyanisole] and with ethoxyquine (Bock et al. 1980). Antioxidant-type inducers also include the phase II enzyme inducer oltipraz (4-methyl-5-(2-pyrazinyl)-1,2-dithiole-3-thione, Kessler and Ritter 1997; Grove et al. 1997), a variety of plant flavonoids such as quercetin and polyphenols of green tea (Sohn et al. 1994).

For glutathione S-transferase Ya (Rushmore and Pickett 1990) and NAD(P)H-quinone oxidoreductase (NQO1, Jaiswal 1994; Venugopal and Jaiswal 1996) it has been shown that antioxidant-type inducers activate a novel redox-sensitive signal transduction pathway. This pathway includes transcription factors such as the zipper proteins Nrf1, Nrf2 and c-Jun, which bind to antioxidant response elements (AREs) or electrophile response elements (EpREs) in the regulatory region of the induced enzymes. Mutational analysis identified TGACNNNGC as the core of the ARE sequences. Activation of this novel antioxidant-responsive pathway is of interest in connection with the efforts of chemoprevention of cancer (Wattenberg 1983; Talalay et al. 1995). Prochaska and Talalay (1988) termed the induction of phase II enzymes by antioxidants 'monofunctional induction' in contrast to Ah receptor agonists which induce both phase I (e.g. CYP1A1) and phase II enzymes, the latter termed 'bifunctional induction'. In the case of NQO1, evidence has been presented that its induction by β-naphthoflavone (BNF) is mediated by the 'ARE mechanism' (Venugopal and Jaiswal 1999). As an agonist of the Ah receptor, BNF leads to the induction of CYP1A1. Efficient metabolism of BNF by CYP1A1 may generate antioxidant/electrophile stress which triggers NQO1 induction. Findings with Nrf2- deficient mice suggest that Nrf2 regulates the induction of phase II enzymes, including some UGTs such as UGT1A6 (Itoh et al. 1997; Masayuki Yamamoto, Institute for Basic Medical Sciences, University of Tsukuba, Japan, personal communication). It is therefore conceivable that BNF and PAH induction of some UGT isoforms is mediated via an ARE-like mechanism (Figure 8.8). This is supported by recent transient transfection experiments using UGT1A6 reporter gene plasmids containing 3kb of 5'-flanking region (Münzel et al. 1996) in which ARE-like domains were found. Treatment of these transfectants with t-butylhydroquinone showed consistent 2-fold induction (Schmohl S, Münzel PA, Bock-Hennig BS and Bock KW, unpublished results). Furthermore, evidence for a linkage between

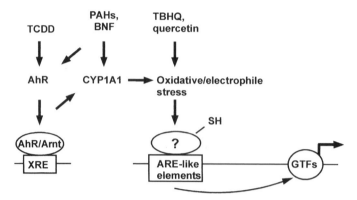

Figure 8.8 Hypothetical mechanism for transcriptional regulation of human UGT1A6 by xenobiotics. XRE, xenobiotic responsive element; ARE antioxidant responsive element; TBHQ, t-butylhydroquinone; PAHs, polycyclic aromatic hydrocarbons; BNF, β-naphthoflavone; GTFs, general transcription factors.

XRE- and ARE-mediated UGT induction has been obtained (Bock *et al.* 1998). Experiments were conducted with rat 5L hepatoma cells in which induction of UGT activity was detected after treatment with the selective Ah receptor agonist TCDD or the prototypical antioxidant t-butylhydroquinone. In the 5L hepatoma cell mutant lacking the Ah receptor (Weiß *et al.* 1996), UGT induction by both types of inducers were abolished, whereas after stable integration of the Ah receptor both types of induction reappeared.

Alteration of UGTs in disease

It has been known for quite some time that glucuronidation of some drugs (e.g. oxazepam) is little affected in liver cirrhosis whereas the glucuronidation of others (e.g. zidovudine) is decreased, similar to cytochrome P450-dependent drug metabolism (Hoyumpa and Schenker 1991; Kroemer and Klotz 1992; Furlan *et al.* 1999). Only a few studies on the effect of disease states on particular UGT isoforms have been carried out. Since cytokines have been shown to affect gene expression of UGTs, it is likely that inflammation and infections affect glucuronidation (Lévesque *et al.* 1998).

UGT GENE EXPRESSION IN CARCINOGENESIS AND ITS ROLE IN TOXIN RESISTANCE

Early studies showed increased UGT activities in rat liver foci and nodules (Bock *et al.* 1982; Yin *et al.* 1982) associated with a toxin-resistance phenotype consisting of decreased phase I and increased phase II activities. Northern Blot analysis using a selective UGT1A6 probe indicated that UGT1A6 expression was persistently enhanced (Bock *et al.* 1990b). In mouse liver, both UGT-positive and UGT-negative foci were detected (Bock *et al.* 1989). In humans mostly down-regulation of UGTs has been detected in liver adenomas and carcinomas but not in focal nodular hyperplasia (Strassburg *et al.* 1997b). UGTs were differentially affected. For example, the UGT1A6 isoform was not affected whereas a recent study showed increased UGT2B4 in liver tumours (Kondoh *et al.* 1999). The mechanisms responsible for altered enzyme expression in hepatocarcinogenesis are still unknown.

AUTOANTIBODIES IN VIRAL DISEASE

Patients with chronic hepatitis D often show serum liver–kidney microsomal antibodies type 3 (LKM-3). Surprisingly, UGTs were found to be candidate antigens. Anti-UGT1 antibodies were detected in all LKM-3-positive sera from patients with hepatitis D. Sera from patients with hepatitis B did not react with UGT protein (Philipp *et al.* 1994).

UGTs in experimental animals

This section highlights differences in experimental animals compared to humans. For information on UGT sequences in experimental animals the reader should consult the UGT homepage (see the Introduction).

RAT

Rat UGTs have been extensively studied. The first UGT sequences were characterised from this species. Substrate specificity of rat liver UGTs has been listed in a recent review (Burchell *et al.* 1997). Basal expression of rUGT1A6 is low in rat liver but is increased > 10-fold by Ah receptor agonists (Iyanagi *et al.* 1986). However, there is high constitutive expression in extrahepatic tissues (Münzel *et al.* 1994). Differences in tissue- and species-specific expression of UGT1A6 have already been discussed. Major UGT isoforms in rat liver appear to be the bilirubin-conjugating UGT1A1 and UGT2B1 (Emi *et al.* 1995). UGT2B1 has a broad substrate specificity including morphine, which is conjugated to the 3-glucuronide but not to the 6-glucuronide (Coffman *et al.* 1998), chloramphenicol and the pulmonary carcinogen 4-(methylnitrosamino)-1-(3-pyridyl)-1-butanol = NNAL (Ren *et al.* 1999).

MOUSE

Mouse UGTs of family 1 (UGT1A1 and UGT1A6) and of family 2 (UGT2B5) have been cloned. In Ah receptor knock-out mice both basal expression of UGT1A6 and induction by Ah receptor agonists were absent (Fernandez-Salguero *et al.* 1995).

CAT

The domestic cat is exquisitely sensitive to adverse effects of many drugs such as paracetamol (acetaminophen) and phenolic toxins that are normally glucuronidated before elimination (Gregus *et al.* 1983; Savides *et al.* 1984). Deficient glucuronidation of planar phenolic compounds has also been identified in other carnivora such as the lion, African civet, forest genet and spotted hyaena (Capel *et al.* 1972; Caldwell *et al.* 1975). To investigate the evolutionary basis for deficient glucuronidation, UGTs of family 1 have been cloned. Whereas cat UGT1A1 was found to be expressed in the liver as a major isoform, a number of stop codons and deletions have been found in the cat UGT1A6 structural gene which is unlikely to produce functional protein (Court and Greenblatt 2000). A similar defect was found in the margay (*Leopardus wiedii*). This gene defect may have arisen via disruption of the UGT1A6 gene in a common ancestor of these species. From an evolutionary standpoint, these species differences may reflect the highly carnivorous nature of the feline diet, and resultant low degree of selection pressure from phytoalexins.

DOG

No dog UGT sequence has been so far published (July 2000). However, one pheno-barbital-inducible morphine UGT has been purified (Oguri *et al.* 1996). Dog liver UGT activities towards morphine, 4-methylumbelliferone and 1-naphthol were found to be 10-fold higher than in human liver, suggesting extensive first-pass metabolism of phenolic compounds in the dog (Richter von *et al.* 2000). Metabolism of retigabin was dominated by *N*-glucuronidation in the dog but not in rats (Hiller *et al.* 1999).

GUINEA PIG

Two guinea pig UGT isoforms (UGT2B21 and UGT2B22) have recently been cloned; they are involved in the conjugation of morphine to both morphine 3- and 6-glucuronides (Miyoshi *et al.* 1998). Interestingly, while UGT2B21 formed the 3-glucuronide but not the 6-glucuronide, UGT2B22 did not have catalytic activity. However, coexpression of both UGT2B21 and 2B22 led to the formation of both morphine 3- and 6-glucuronides, suggesting that hetero-oligomers of the two UGTs were necessary to form morphine 6-glucuronide.

MONKEY

UGT2B9 has been cloned and characterised; it shares a 89% similarity in amino acid sequence to human UGT2B7. UGT2B9 has been shown to conjugate endogenous substrates such as 3-hydroxyandrogens and many drugs such as profene-NSAIDs, fibrates, straight chain fatty acids at the carboxylic acid moiety, and morphine to the morphine 3- and 6-glucuronides (Green *et al.* 1997; Beaulieu *et al.* 1998). Recently, another isoform (UGT2B23) has been characterised which is active on androgens and oestrogens (Barbier *et al.* 1999).

PIG

A liver UGT has been purified and its phospholipid-dependence was studied (Magda-lou *et al.* 1982).

RABBIT

UGTs of families 1 and 2 have been cloned (Tukey *et al.* 1993; Lamb *et al.* 1994). In contrast to rodents, the rabbit UGT1A4 and UGT1A7 are able to form quarternary *N*-glucuronides of imipramine (Bruck *et al.* 1997). Imipramine quarternary *N*-glucuronidation was found to occur at twice the rate of humans (Coughtrie and Sharp 1991).

Conclusions

Molecular biology technology provides a solid foundation to establish individual isoforms of a supergene family of UDP-glucuronosyltransferases expressed in most tissues of vertebrates such as mammals and fish. It also enables the inclusion of glucuronosyltransferases in a much larger supergene family of glycosyltransferases expressed in invertebrates, yeast, bacteria and plants. Interestingly, it was found that all human UGT family 1 members are derived by exon-sharing from one large gene locus located at chromosome 2. Individual isoforms are expressed in a remarkably tissue-specific manner. More work is needed to identify expressed UGTs in particular cell types, for example in the brain or steroidogenic tissues. Furthermore, the evolutionary relationship between glucuronosyl- and related glycosyltransferases has to be substantiated.

Recombinant technology enabled the characterisation of the function of individual UGT isoforms. Most of them show a broad specificity for a variety of xenobiotics (drugs, dietary plant constituents and toxicants). In addition, endobiotics are increasingly recognised as substrates for individual UGT isoforms. UGTs not only inactivate these xeno- and endobiotics but also play a critical role in generating bioactive or even toxic compounds.

Recent studies provide evidence for a dynamic topological model in which UGTs are acting as oligomers in membranes. The model may have implications for the transport of lipophilic aglycones and the cofactor UDP-glucuronic acid to the active site and the subsequent release of glucuronides. More work is needed to substantiate this model, to characterise glucuronide transporters present in the apical and basolateral plasma membrane of epithelial cells, and to investigate the disposition of glucuronides in the organism.

The factors responsible for transcriptional regulation of UGTs are beginning to be characterised, including xenobiotic-activated transcription factors such as the aryl hydrocarbon receptor as well as general enhancers such as HNF1, C/EBPα and β.

Further work along these lines will improve our knowledge of similarities and differences between glucuronidation in experimental animals and humans. Further studies of this major metabolic reaction may facilitate the exrapolation of endo- and xenobiotic biotransformation from experimental models to individual human beings in health and disease.

(1) Studies on the human UGT1 locus (Fig. 8.3) have been completed (Gong QH *et al.*, Pharmacogenetics 11, 357–368, 2001).
(2) A novel hUGT2A2 has been described which is expressed in the gastrointestinal tract (Tukey RH and Strassburg CP, Mol. Pharmacol. 59, 405–414, 2001).

(3) Evidence for transporters of UDP-glucuronic acid to the active site of UGTs in the lumen of endoplasmic reticulum membranes (Fig. 8.7) has been obtained (Kawakita M *et al.*, Abstracts, 10. Int. Workshop on Glucuronidation and the UDP-glucuronosyltransferases, Hemeji, Institute of Technology, Hyogo, Japan, 2001).

(4) Evidence for contribution of Nrf-2 to antioxidant-type induction of mouse UGT1A6 (Fig. 8.8) has been reported by Chan K and Kan YU (Proc. Natl. Acad. Sci. 96, 1231–1236, 1999) and by Enomoto A *et al.* (Tox. Sci. 59, 169–177, 2001).

Acknowledgments

The author thanks the following colleagues for helpful discussions and providing reprints and preprints: Drs Michael H. Court (Department of Pharmacology, Tufts University, School of Medicine, Boston, Massachussetts, USA); Sylvie Fournel-Gigleux (UMR 7561, Faculty of Medicine, Vandoeuvre-les-Nancy, France); Peter Mackenzie (Department of Clinical Pharmacology, Flinders University of South Australia, Adelaide, Australia); Ida S. Owens (Heritable Disorders Branch, National Institutes of Child Health and Human Development, National Institutes of Health, Bethesda, MD 20892, USA); Anna Radominska-Pandya (Department of Biochemistry and Molecular Biology, University of Arkansas for Medical Sciences, Little Rock, Arkansas, USA); Christian Strassburg (Department of Gastroenterology and Hepatology, Hannover Medical School, Hannover, Germany); and Thomas R. Tephly (Department of Pharmacology, University of Iowa, Iowa City, Iowa, USA). I also thank Mrs A von Bank and Mrs E. Schenk for typing the manuscript and the Deutsche Forschungsgemeinschaft and the Bundesministerium für Forschung und Technologie for financial support of the work carried out in the author's laboratory.

References

Alexander J, Wallin H, Rossland OJ, Solberg KE, Holme JA, Becher G, Andersson R and Grivas S (1991) Formation of a glutathione conjugate and a semistable transportable glucuronide conjugate of N2-oxidised species of 2-amino-1-methyl-6-phenylimidazo[4,5-b]pyridine (PhIP) in rat liver. *Carcinogenesis*, **12**, 2239–2245.

Arias IM (1962) Chronic unconjugated hyperbilirubinaemia without overt signs of hemolysis in adolescents and adults. *Journal of Clinical Investigation*, **41**, 2233–2245.

Arias IM, Gartner LM, Cohen M, Ben Ezzer J and Levi AJ (1969) Chronic nonhemolytic unconjugated hyperbilirubinaemia with glucuronyl transferase deficiency. *American Journal of Medicine*, **47**, 395–409.

Axelrod J, Schmid R and Hammaker L (1957) A biochemical lesion in congenital non-obstructive, non-hæmolytic jaundice. *Nature*, **180**, 1426–1427.

Bánhegyi G, Braun L, Marcolongo P, Csala M, Fulceri R, Mandl J and Benedetti A (1996) Evidence for an UDP-glucuronic acid/phenol glucuronide antiport in rat liver microsomal vesicles. *Biochemical Journal*, **315**, 171–176.

Barbier O, Levesque E, Bélanger A and Hum DW (1999) UGT2B23, a novel uridine diphosphate-glucuronosyltransferase enzyme expressed in steroid target tissues that conjugates androgen and estrogen metabolites. *Endocrinology*, **140**, 5538–5548.

Barbier O, Turgeon D, Girard C, Green MD, Tephly TR, Hum DW and Bélanger A (2000) 3'-Azido-3'-deoxythimidine (AZT) is glucuronidated by human UDP-glucuronosyltransferase 2B7 (UGT2B7). *Drug Metabolism and Disposition*, **28**, 497–502.

Beaulieu M, Lévesque E, Hum DW and Bélanger A (1996) Isolation and characterization of a novel cDNA encoding a human UDP-glucuronosyltransferase active on C_{19} steroids. *Journal of Biological Chemistry*, **271**, 22855–22862.

Beaulieu M, Lévesque E, Barbier O, Turgeon D, Bélanger G, Human DW and Bélanger A (1998) Isolation and characterization of a Simian UDP-glucuronosyltransferase UGT2B18 active on 3-hydroxyandrogens. *Journal of Molecular Biology*, **275**, 785–794.

Becedas L, Lundgren B and De Pierre JW (1998) Characterization of the UDP-glucuronosyltransferase isoenzyme expressed in rat ovary and its regulation by gonadotropins. *Biochemical Journal*, **332**, 51–55.

Beland FA and Kadlubar FF (1990) Metabolic activation and DNA adducts of aromatic amines and nitroaromatic hydrocarbons. In *Chemical Carcinogenesis and Mutagenesis I*, Cooper CS and Grover PL (eds) Handbook of Experimental Pharmacology 94/I, Springer-Verlag, Berlin, pp. 267–325.

Bélanger A, Hum DW, Beaulieu M, Lévesque E, Guillemette C, Tchernof A, Bélanger G, Turgeon D and Dubois S (1998) Characterization and regulation of UDP-glucuronosyltransferases in steroid target tissues. *Journal of Steroid Biochemistry and Molecular Biology*, **65**, 301–310.

Benet LZ, Spahn-Langguth H, Iwakawa S, Volland C, Mizuma T, Mayer S, Mutschler E and Lin ET (1993) Predictability of the covalent binding of acidic drugs in man. *Life Sciences*, **53**, 141–146.

Berry C and Hallinan T (1974) 'Coupled transglucuronidation': A new tool for studying the latency of UDP-glucuronyl transferase. *FEBS Letters*, **42**, 73–76.

Beutler E, Gelbart T and Demina A (1998) Racial variability in the UDP-glucuronosyltransferase 1 (UGT1A1) promoter: A balanced polymorphism for regulation of bilirubin metabolism. *Proceedings of the National Academy of Sciences USA*, **95**, 8170–8174.

Bevan DR and Sadler VM (1992) Quinol diglucuronides are predominant conjugated metabolites found in bile of rats following intratracheal instillation of benzo(a)pyrene. *Carcinogenesis*, **13**, 403–407.

Blobel G (1980) Intracellular protein topogenesis. *Proceedings of the National Academy of Sciences USA*, **77**, 1496–1500.

Blobel G and Dobberstein B (1975) Transfer of proteins across membranes. I. Presence of proteolytically processed and unprocessed nascent immunoglobulin light chains on membrane-bound ribosomes of murine myeloma. *Journal of Cell Biology*, **67**, 835–851.

Bock KW (1991) Roles of UDP-glucuronosyltransferases in chemical carcinogenesis. *CRC Critical Reviews in Biochemistry and Molecular Biology*, **26**, 129–150.

Bock KW and Lilienblum W (1994) Roles of uridine diphosphate glucuronosyl-transferases in chemical carcinogenesis. In *Conjugation/Deconjugation Reactions in Drug Metabolism and Toxicity*, Kauffman FC (ed.), Handbook of Experimental Pharmacology, Vol. 112, Springer-Verlag, Berlin, pp. 391–428.

Bock KW and White INH (1974) UDP-glucuronosyltransferase in perfused rat liver and in microsomes: Influence of phenobarbital and 3-methylcholanthrene. *European Journal of Biochemistry*, **46**, 451–459.

Bock KW, Fröhling W, Remmer H and Rexer B (1973) Effects of phenobarbital and 3-methylcholanthrene on substrate specificity of rat liver microsomal UDP-glucuronyltransferase. *Biochimica et Biophysica Acta*, **327**, 46–56

Bock KW, Kahl R and Lilienblum W (1980) Induction of rat hepatic UDP-glucuronosyltransferases by dietary ethoxyquin. *Naunyn–Schmiedeberg's Archives of Pharmacology*, **310**, 249–252.

Bock KW, Lilienblum W, Pfeil H and Eriksson LC (1982) Increased uridine diphosphate-glucuronosyltransferase activity in preneoplastic liver nodules and Morris hepatomas. *Cancer Research*, **42**, 3747–3752.

Bock KW, Wiltfang J, Blume R, Ullrich D and Bircher J (1987) Paracetamol as a test drug to determine glucuronide formation in man. Effects of inducers and of smoking. *European Journal of Clincal Pharmacology*, **31**, 677–683

Bock KW, Kobusch A-B and Fischer G (1989) Heterogeneous alterations of UDP-glucuronosyltransferases in mouse hepatic foci. *Journal of Cancer Research and Clinical Oncology*, **115**, 285–289.

Bock KW, Lipp H-P and Bock-Hennig BS (1990a) Induction of drug-metabolizing enzymes by xenobiotics. *Xenobiotica*, **20**, 1101–1111.

Bock KW, Münzel PA, Röhrdanz E, Schrenk D and Eriksson LC (1990b) Persistently increased expression of a 3-methylcholanthrene-inducible phenol uridine diphosphate-glucuronosyltransferase in rat hepatocyte nodules and hepatocellular carcinomas. *Cancer Research*, **50**, 3569–3573.

Bock KW, Forster A, Gschaidmeier H, Brück M, Münzel P, Schareck W, Fournel-Gigleux S and Burchell B (1993) Paracetamol glucuronidation by recombinant rat and human phenol UDP-glucuronosyltransferases. *Biochemical Pharmacology*, **45**, 1809–1814.

Bock KW, Schrenk D, Forster A, Griese E-U, Mörike K, Brockmeier D and Eichelbaum M (1994) The influence of environmental and genetic factors on CYP2D6, CYP1A2 and UDP-glucuronosyltransferases in man using sparteine, caffeine, and paracetamol as probes. *Pharmacogenetics*, **4**, 209–218.

Bock KW, Gschaidmeier H, Heel H, Lehmköster T, Münzel PA, Raschko F and Bock-Hennig B (1998) Ah receptor-controlled transcriptional regulation and function of rat and human UDP-glucuronosyltransferase isoforms. *Advances in Enzyme Regulation*, **38**, 207–222.

Bock KW, Raschko FT, Gschaidmeier H, Seidel A, Oesch F, Grove AD and Ritter JK (1999) Mono- and diglucuronide formation from benzo(a)pyrene and chrysene diphenols by AHH-1 cell-expressed UDP-glucuronosyltransferase UGT1A7. *Biochemical Pharmacology*, **57**, 653–656.

Bock KW, Eckle T, Ouzzine M, Fournel-Gigleux S (2000) Coordinate induction by antioxidants of UDP-glucuronosyltransferase UGT1A6 and the apical conjugate export pump MRP2 (multidrug resistance protein 2) in Caco-2- cells. *Biochemical Pharmacology*, **59,** 467–470

Boroujerdi M, Kung H-C, Wilson AGE and Anderson MW (1981) Metabolism and DNA binding of benzo(a)pyrene in vivo in the rat. *Cancer Research*, **41**, 951–957.

Bosma PJ, Chowdhury JR, Bakker C, Gantla S, de Boer A, Oostra BA, Lindhout D, Tytgat GN, Jansen PL, Oude-Elferink RP and Chowdhury NR (1995) The genetic basis of the reduced expression of bilirubin UDP-glucuronosyltransferase 1 in Gilbert's syndrome. *New England Journal of Medicine*, **333**, 1171–1175.

Bossuyt X and Blanckaert N (1995) Mechanism of stimulation of microsomal UDP-glucuronosyltransferase by UDP-N-acetylglucosamine. *Biochemical Journal*, **305**, 321–328.

Brands A, Münzel PA and Bock KW (2000) In situ hybridization studies of UDP-glucuronosyltransferase UGT1A6 expression in rat testis and brain. *Biochemical Pharmacology*, in press.

Bruck M, Li Q, Lamb JG and Tukey RH (1997) Characterization of rabbit UDP-glucuronosyltransferase UGT1A7: Tertiary amine glucuronidation is catalysed by UGT1A7 and UGT1A4. *Archives of Biochemistry and Biophysics*, **344**, 357–364.

Burchell B, Nebert DW, Nelson DR, Bock KW, Iyanagi T, Jansen PLM, Lancet D, Mulder GJ, Chowdhury RJ, Siest G, Tephly TR and Mackenzie PI (1991) The UDP-glucuronosyltransferase gene superfamily: Suggested nomenclature based on evolutionary divergence. *DNA and Cell Biology*, **7**, 487–494.

Burchell B, McGurk K, Brierley CH and Clarke DJ (1997) UDP-glucuronosyl-transferases. *Comprehensive Toxicology*, **3**, 401–435.

Buetler TM, Gallagher EP, Wang C, Stahl DL, Hayes JD and Eaton DL (1995) Induction of phase I and phase II drug-metabolizing enzyme mRNA, protein, and activity by BHA, ethoxyquin, and oltipraz. *Toxicology and Applied Pharmacology*, **135**, 45–57.

Caldwell J, French MR, Idle JR, Renwick AG, Bassir O, Williams RT (1975). Conjugation of foreign compounds in the elephant and hyaena. *FEBS Letters*, **60**, 391-395.

Capel ID, French MR, Millburn P, Smith RL, Williams RT (1972) The fate of [14C]phenol in various species. *Xenobiotica*, **2**, 25–34

Cheng Z, Radominska-Pandya A and Tephly TR (1999) Studies on the substrate specificity of human intestinal UDP-glucuronosyltransferases 1A8 and 1A10. *Drug Metabolism and Disposition*, **27**, 1165–1170.

Ciotti M, Marrone A, Potter C and Owens IS (1997) Genetic polymorphism in the human UGT1A6 (planar phenol) UDP-glucuronosyltransferase: pharmacological implications. *Pharmacogenetics*, **7**, 485–495.

Ciotti M, Lakshmi VM, Basu N, Davis BB, Owens IS and Zenser TV (1999a) Glucuronidation of benzidine and its metabolites by cDNA-expressed human UDP-glucuronosyltransferases and pH stability of glucuronides. *Carcinogenesis*, **20**, 1963–1969.

Ciotti M, Basu N, Brangi M, Owens IS (1999b). Glucuronidation of 7-ethyl-10-hydroxycamptothecin (SN-39) by the human UDP-glucuronosyltransferases encoded at the UGT1 locus. *Biochemical and Biophysical Research Communications*, **260**, 199–202.

Coffman BL, King CD, Rios GR and Tephly TR (1998) The glucuronidation of opioids, other xenobiotics, and androgens by human UGT2B7Y(268) and UGT2B7H(268). *Drug Metabolism and Disposition*, **26**, 73–77.

Conney AH (1982) Induction of microsomal enzymes by foreign chemicals and carcinogenesis by polycyclic aromatic hydrocarbons: GH Clowes memorial lecture. *Cancer Research*, **42**, 4875–4917.

Coughtrie MWH and Sharp S (1991) Glucuronidation of imipramine in rabbit and human liver microsomes: assay conditions and interaction with other tertiary amine drugs. *Biochemical Pharmacology*, **42**, 1497–1501.

Court MH and Greenblatt DJ (2000) Molecular genetic basis for deficient acetaminophen glucuronidation by cats: UGT1A6 is a pseudogene, and evidence for reduced diversity of expressed hepatic UGT1A isoforms. *Pharmacogenetics*, **10**, 355–369.

Dickinson RG (1998) Do acyl glucuronides play a role in the antiproliferative effects of NSAIDs on colon cancer cells? 9th International Workshop on Glucuronidation and UDP-Glucuronosyltransferases, Brisbane, Australia, Oral 20.

Dutton GJ (1956) Uridine diphosphate glucuronic acid as glucuronyl donor in the synthesis of 'ester', aliphatic and steroid glucuronides. *Biochemical Journal*, **64**, 693–701.

Dutton GJ (1980) *Glucuronidation of Drugs and Other Compounds*. CRC Press, Boca Raton, FL.

Dutton GJ (1997) Raising the colors: Personal reflections on the glucuronidation revolution 1950-1970. *Drug Metabolism Reviews*, **29**, 997–1024.

Dutton GJ and Storey IDE (1953) The isolation of a compound of uridine diphosphate and glucuronic acid from liver. *Biochemical Journal*, **53**, 37–38.

Ebner T and Burchell B (1993) Substrate specificities of two stably expressed human liver UDP-glucuronosyltransferases of the UGT1 gene family. *Drug Metabolism and Disposition*, **21**, 50–55.

Ebner T, Heinzel G, Prox A, Beschke K and Wachsmuth H (1999) Disposition and chemical stability of telmisartan 1-O-acylglucuronide. *Drug Metabolism and Disposition*, **27**, 1143–1149.

Emi Y, Ikushiro S, Iyanagi T (1995) Drug-responsive and tissue-specific alternative expression of multiple first exons in rat UDP-glucuronosyltransferase family 1 (UGT1) gene complex. *Journal of Biochemistry*, **117**, 392–399.

Emi Y, Ikushiro S, Iyanagi T (1996) Xenobiotic responsive element-mediated transcriptional activation in the UDP–glucuronosyltransferase family 1 gene complex. *Journal of Biological Chemistry*, **271**, 3952–3958.

Fernandez-Salguero P, Pineau T, Hilbert DM, McPhail T, Lee SST, Kimura S, Nebert DW, Rudikoff S, Ward JM and Gonzalez FJ (1995) Immune system impairment and hepatic fibrosis in mice lacking the dioxin-binding Ah receptor. *Science*, **268**, 722–726.

Furlan V, Demirdjian S, Bourdon O, Magdalou J and Taburet A-M (1999) Glucuronidation of drugs by hepatic microsomes derived from healthy and cirrhotic human livers. The *Journal of Pharmacology and Experimental Therapeutics*, **289**, 1169–1175.

Green MD and Tephly TR (1998) Glucuronidation of amine substrates by purified and expressed UDP-glucuronosyltransferase proteins. *Drug Metabolism and Disposition*, **26,** 860–867.

Green MD, Oturu EM and Tephly TR (1994) Stable expression of a human liver UDP-glucuronosyltransferase (UGT2B15) with activity toward steroid and xenobiotic substrates. *Drug Metabolism and Disposition*, **22**, 799–805.

Green MD, Bélanger G, Hum DW, Bélanger A and Tephly TR (1997) Glucuronidation of opioids, carboxylic acid-containing drugs, and hydroxylated xenobiotics catalysed by expressed monkey UDP-glucuronosyltransferase 2B9 protein. *Drug Metabolism and Disposition*, **25**, 1389–1394.

Gregus Z, Watkins JB, Thompson TN, Harvey MJ, Rozman K and Klaassen CD (1983) Hepatic phase I and phase II biotransformation in quail and trout: Comparison to other species commonly used in toxicity testing. *Toxicology and Applied Pharmacology*, **67**, 430-441.

Grove AD, Kessler FK, Metz RP and Ritter JK (1997) Identification of a rat oltipraz-inducible UDP-glucuronosyltransferase (UGT1A7) with activity towards benzo[a]pyrene-7,8-dihydrodiol. *Journal of Biological Chemistry*, **272**, 1621–1627.

Grove AD, Llewellyn GC, Kessler FK, White KL Jr, Crespi CL, Ritter JK (2000) Differential protection by rat UDP-glucuronosyltransferase 1A7 against benzo(a)pyrene-3,6-quinone versus benzo(a)pyrene-induced cytotoxic effects in human lymphoblastoid cells. *Toxicology and Applied Pharmacology*, **162**, 34–43.

Gschaidmeier H and Bock KW (1994) Radiation inactivation analysis of microsomal UDP-glucuronosyltransferases catalysing mono- and diglucuronide formation of 3,6-dihydroxybenzo(a)pyrene and 3,6-dihydroxychrysene. *Biochemical Pharmacology*, **48**, 1545–1549.

Gschaidmeier H, Seidel A, Oesch F, Burchell B and Bock KW (1995) Formation of mono- and diglucuronides and of other glycosides of benzo(a)pyrene-3,6-quinol by V79 cell-expressed human phenol UDP-glucuronosyltransferases of the UGT1 gene complex. *Biochemical Pharmacology*, **49**, 1601–1606.

Guengerich FP, Parikh A, Johnson EF, Richardson TH, von Wachenfeldt C, Cosme J, Jung F, Strassburg CP, Manns MP, Tukey RH, Pritchard M, Fournel-Gigleux S and Burchell B (1997) Heterologous expression of human drug-metabolizing enzymes. *Drug Metabolism and Disposition*, **25**, 1234–1241.

Guillemette C, Millikan RC, Newman B and Housman DE (2000) Genetic polymorphisms in uridine diphospho-glucuronosyltransferase 1A1 and association with breast cancer among African Americans. *Cancer Research*, **60**, 950–956.

Gupta E, Lestingi TM, Mick R, Ramirez J, Vokes EE and Ratain MJ (1994) Metabolic fate of irinotecan in humans: Correlation of glucuronidation with diarrhea. *Cancer Research*, **54**, 3723–3725.

Gura T (1999) Repairing the genome's spelling mistakes. *Science*, **285**, 316–318.

Hansen AJ, Lee Y-H, Sterneck E, Gonzalez FJ and Mackenzie PI (1998) C/EBPα is a regulator of the UDP glucuronosyltransferase UGT2B1 gene. *Molecular Pharmacology*, **53**, 1027–1033.

Harding D, Fournel-Gigleux S, Jackson MR and Burchell B (1988) Cloning and substrate specificity of a human phenol UDP-glucuronosyltransferase expressed in COS-7 cells. *Proceedings of the National Academy of Sciences USA*, **85**, 8381–8385.

Hiller A, Nguyen N, Strassburg CP, Li Q, Jainta H, Pechstein B, Ruus P, Engel J, Tukey RH and Kronbach T (1999) Retigabine N-glucuronidation and its potential role in enterohepatic circulation. *Drug Metabolism and Disposition*, **27**, 605–612.

Hirschberg CB, Robbins PW and Abeijon C (1998) Transporters of nucleotide sugars, ATP, and nucleotide sulphate in the endoplasmic reticulum and Golgi apparatus. *Annual Reviews in Biochemistry*, **67**, 49–69.

Hoyumpa AM and Schenker S (1991) Is glucuronidation truly preserved in patients with liver disease? *Hepatology*, **13**, 786–795.

Hu Z and Wells PG (1992) In vitro and in vivo biotransformation and covalent binding of benzo[a]pyrene in Gunn and RHA rats with a genetic deficiency in bilirubin uridine diphosphate-glucuronosyltransferase. *Journal of Pharmacology and Experimental Therapeutics*, **263**, 334–342.

Ikushiro S-I, Emi Y and Iyanagi T (1995) Identification and analysis of drug-responsive expression of UDP-glucuronosyltransferase family 1 (UGT1) isozyme in rat hepatic microsomes using anti-peptide antibodies. *Archives of Biochemistry and Biophysics*, **324**, 267–272.

Illing JPA and Benford D (1976) Observations on the accessibility of acceptor substrates to the active centre of UDP-glucuronosyltransferase in vitro. *Biochimica et Biophysica Acta*, **429**, 768–779.

Ishii Y, Hansen AJ and Mackenzie PI (2000) Octamer transcription factor-1 enhances hepatic nuclear factor-1a mediated activation of the human UDP glucuronosyltransferase 2B7. *Molecular Pharmacology*, **57**, 940–947.

Itoh K, Chiba T, Takahashi S, Ishii T, Igarashi K, Katoh Y, Oyake T, Hayashi N, Satoh K, Hatayama I, Yamamoto M and Nabeshima Y (1997) An Nrf2/small Maf heterodimer mediates the induction of phase II detoxifying enzyme genes through antioxidant response elements. *Biochemical and Biophysical Research Communications*, **236**, 313–322.

Iyanagi T (1991) Molecular basis of multiple UDP-glucuronosyltransferase isoenzyme deficiencies in the hyperbilirubinemic rat (Gunn rat). *Journal of Biological Chemistry*, **266**, 24048–24052.

Iyanagi T, Haniu M, Sogawa K, Fujii-Kuriyama Y, Watanabe S, Shively JE and Anan KF (1986)

Cloning and characterization of cDNA encoding 3-methylcholanthrene inducible rat mRNA for UDP-glucuronosyltransferase. *Journal of Biological Chemistry*, **261**, 15607–15614.

Iyer L, King CD, Whitington PF, Green MD, Roy SK, Tephly TR, Coffman BL and Ratain MJ (1998) Genetic predisposition to the metabolism of irinotecan (CPT-11): Role of uridine diphosphate glucuronosyltransferase isoform 1A1 in the glucuronidation of its active metabolite (SN-38) in human liver microsomes. *Journal of Clinical Investigations*, **101**, 847–854.

Jackson MR, McCarthy LR, Corser RB, Barr GC and Burchell B (1984) Cloning of cDNAs coding for rat hepatic microsomal UDP-glucuronyltransferases. *Gene*, **34**, 147–153.

Jaiswal AK (1994) Antioxidant response element. *Biochemical Pharmacology*, **48**, 439–444.

Jedlitschky G, Leier I, Buchholz U, Barnouin K, Kurz G and Keppler D (1996) Transport of glutathione, glucuronate, and sulphate conjugates by the MRP gene-encoded conjugate export pump. *Cancer Research*, **56**, 988–994.

Jedlitschky G, Leier I, Buchholz U, Hummel-Eisenbeiss J, Burchell B and Keppler D (1997) ATP-dependent transport of bilirubin glucuronides by the multidrug resistance protein MRP1 and its hepatocyte canalicular isoform MRP2. *Biochemical Journal*, **327**, 305–310.

Jedlitschky G, Cassidy AJ, Sales M, Pratt N and Burchell B (1999) Cloning and characterization of a novel human olfactory UDP-glucuronosyltransferase. *Biochemical Journal*, **340**, 837–843.

Jin C, Miners JO, Lillywhite KJ and Mackenzie PI (1993a) Complementary deoxyribonucleic acid cloning and expression of a human liver uridine diphosphate-glucuronosyltransferase glucuronidating carboxylic acid-containing drugs. *The Journal of Pharmacology and Experimental Therapeutics*, **264**, 475–479.

Jin C-J, Miners JO, Burchell B and Mackenzie PI (1993b) The glucuronidation of hydroxylated metabolites of benzo[a]pyrene and 2-acetylaminofluorene by cDNA-expressed human UDP-glucuronosyltransferases. *Carcinogenesis*, **14**, 2637–2639.

Jude AR, Little JM, Freeman JP, Evans JE, Radominska-Pandya A and Grant DF (2000) Linoleic acid diols are novel substrates for human UDP-glucuronosyltransferases. *Archives of Biochemistry and Biophysics*, in press.

Kaderlik KR, Mulder GJ, Turesky RJ, Lang NP, Teitel CH, Chiarelli MP and Kadlubar FF (1994a) Glucuronidation of N-hydroxy heterocyclic amines by human and rat liver microsomes. *Carcinogenesis*, **15**, 1695–1701.

Kaderlik KR, Minchin RF, Mulder GJ, Ilett KF, Daugaard-Jenson M, Teitel CH and Kadlubar FF (1994b) Metabolic activation pathway for the formation of DNA adducts of the carcinogen 2-amino-1-methyl-6-phenylimidazo[4,5-b]pyridine (PhIP) in rat extrahepatic tissues. *Carcinogenesis*, **15**, 1703–1709.

Kawamoto T, Sueyoshi T, Zelko I, Moore R, Washburn K and Negishi M (1999) Phenobarbital-responsive nuclear translocation of the receptor CAR in induction of the CYP2B gene. *Molecular and Cellular Biology*, **19**, 6318–6322.

Keppler D and König J (1997) Expression and localisation of the conjugate export pump encoded by the MRP2 (cMRP/cMOAT) gene in liver. *FASEB Journal*, **11**, 509–516.

Kessler FK and Ritter JK (1997) Induction of rat liver benzo(a)pyrene-trans- 7,8-dihydrodiol glucuronidating activity by oltipraz and β-naphthoflavone. *Carcinogenesis*, **18**, 107–114.

King CD, Green MD, Rios GR, Coffman BL, Owens IS, Bishop WP and Tephly TR (1996) The glucuronidation of exogenous and endogenous compounds by stably expressed rat and human UDP-glucuronosyltransferase 1.1. *Archives of Biochemistry and Biophysics*, **332**, 92–100.

King CD, Rios GR, Assouline JA and Tephly TR (1999) Expression of UDP-glucuronosyltransferases (UGTs) 2B7 and 1A6 in the human brain and identification of 5-hydroxytryptamine as a substrate. *Archives of Biochemistry and Biophysics*, **365**, 156–162.

King CD, Rios GR, Green MD and Tephly TR (2000) UDP-Glucuronosyltransferases, *Current Drug Metabolism*, in press.

Kondoh N, Wakatsuki T, Ryo A, Hada A, Aihara T, Horiuchi S, Goseki N, Matsubara O, Takenaka K, Shichita M, Tanaka K, Shuda M and Yamamoto M (1999) Identification and characterization of genes associated with human hepatocellular carcinogenesis. *Cancer Research*, **59**, 4990–4996.

Kool M, de Haas M, Scheffer GL, Scheper RJ, van Eijk MJT, Juijn JA, Baas F and Borst P (1997) Analysis of expression of cMOAT (MRP2), MRP3, MRP4, and MRP5, homologues of the multidrug resistance-associated protein gene (MRP1), in human cancer cell lines. *Cancer Research*, **57**, 3537–3547.

Kroemer HK and Klotz U (1992) Glucuronidation of drugs. A re-evaluation of the pharmacological significance of the conjugates and modulating factors. *Clinical Pharmacokinetics*, **23**, 292–310.

Lamb JG, Straub P and Tukey RH (1994) Cloning and characterization of cDNAs encoding mouse Ugt1.6 and rabbit UGT1.6: Differential induction by 2,3,7,8-tetrachlorodibenzo-p-dioxin. *Biochemistry*, **33**, 10513–10520.

Lampe JW, Bigler J, Horner NK and Potter JD (1999) UDP-glucuronosyltransferase (UGT1A1*28 and UGT1A6*2) polymorphisms in Caucasians and Asians: relationships to serum bilirubin concentrations. *Pharmacogenetics*, **9**, 341–349.

Lazard D, Zupko K, Poria Y, Nef P, Lazarovits J, Horn S, Khen M and Lancet D (1991) Odorant signal termination by olfactory UDP glucuronosyl transferase. *Nature*, **349**, 790–793.

Lee Chiu S-H and Huskey SW (1998) Species differences in N-glucuronidation. *Drug Metabolism and Disposition*, **26**, 838–847.

Lee Y-H, Sauer B, Johnson PF and Gonzalez FJ (1997) Disruption of the c/ebpa gene in adult mouse liver. *Molecular and Cellular Biology*, **17**, 6014–6022.

Lévesque E, Beaulieu M, Green MD, Tephly TR, Bélanger A and Hum DW (1997) Isolation and characterization of UGT2B15(Y85): a UDP-glucuronosyltransferase encoded by a polymorphic gene. *Pharmacogenetics*, **7**, 317–325.

Lévesque E, Beaulieu M, Guillemette C, Hum DW and Belanger A (1998) Effect of interleukins in UGT2B15 and UGT2B17 steroid uridine diphosphate-glucuronosyl-transferase expression and activity in the LNCaP cell line. *Endocrinology*, **139**, 2375–2381.

Lévesque E, Beaulieu M, Hum DW and Bélanger A (1999) Characterization and substrate specificity of UGT2B4(E458): a UDP-glucuronosyltransferase encoded by a polymorphic gene. *Pharmacogenetics*, **9**, 207–216.

Lilienblum W, Bock-Hennig BS and Bock KW (1985) Protection against toxic redox cycles between benzo(a)pyrene-3,6-quinone and its quinol by 3-methylcholanthrene-inducible formation of the quinol mono- and diglucuronide. *Molecular Pharmacology*, **27**, 451–458.

Lin JH, Chiba M and Baillie TA (1999) Is the role of the small intestine in first-pass metabolism overemphasized? *Pharmacological Reviews*, **51**, 135–157.

Mackenzie PI (1986) Rat liver UDP-glucuronosyltransferase. *Journal of Biological Chemistry*, **261**, 6119–6125.

Mackenzie P, Gonzalez FJ and Owens IS (1984) Cloning and characterization of DNA complementary to rat liver UDP-glucuronosyltransferase mRNA. *Journal of Biological Chemistry*, **259**, 12153–12160.

Mackenzie PI, Owens IS, Burchell B, Bock KW, Bairoch A, Bélanger S, Fournel-Gigleux S, Green M, Hum DW, Iyanagi T, Lancet D, Louisot P, Magdalou J, Chowdhury JR, Ritter JK, Schachter H, Tephly TR, Tipton KF and Nebert DW (1997) The UDP-glucuronosyltransferase gene superfamily: Recommended nomenclature update based on evolutionary divergence. *Pharmacogenetics*, **7**, 255–269.

Magdalou J, Hochman Y, Zakim D (1982) Factors modulating the catalytic specificity of a pure form of UDP-glucuronosyltransferase. *Journal of Biological Chemistry*, **257**, 13624–13629.

Malfatti MA, Kulp KS, Knize MG, Davis C, Massengill JP, Williams S, Nowell S, MacLeod S, Dingley KH, Turteltaub KW, Lang NP and Felton JS (1999) The identification of [2-14C]2-amino-1-methyl-6-phenylimidazo[4,5-b]pyridine metabolites in humans. *Carcinogenesis*, **20**, 705–713.

Martinasevic MK, King CD, Rios GR and Tephly TR (1998) Immunohistochemical localisation of UDP-glucuronosyltransferases in rat brain during early development. *Drug Metabolism and Disposition*, **26**, 1039–1041.

McGurk KA, Brierley CH and Burchell B (1998) Drug glucuronidation by human renal UDP-glucuronosyltransferases. *Biochemical Pharmacology*, **55**, 1005–1012.

Meech R and Mackenzie PI (1997) UDP-glucuronosyltransferase, the role of the amino terminus in dimerisation. *Journal of Biological Chemistry*, **43**, 26913–26917.

Metz RP and Ritter JK (1998) Transcriptional Activation of the UDP-glucuronosyltransferase 1A7 gene in rat liver by aryl hydrocarbon receptor ligands and oltipraz. *Journal of Biological Chemistry*, **273**, 5607–5614.

Miyoshi A, Tsuruda K, Tsuda M, Nagamatsu Y, Matsuoka S, Yoshisue K, Shimizudani-Tanaka M, Ishii Y, Ohgiya S and Oguri K (1998) 9th International Workshop on Glucuronidation and UDP-Glucuronosyltransferases, Brisbane, Australia.

Mojarrabi B and Mackenzie PI (1998a) Characterization of two UDP glucuronosyltransferases that are predominantly expressed in human colon. *Biochemical and Biophysical Research Communications*, **247**, 704–709.

Mojarrabi B and Mackenzie PI (1998b) The human UDP glucuronosyltransferase, UGT1A10, glucuronidates mycophenolic acid. *Biochemical and Biophysical Research Communications*, **238**, 775–778.

Moldéus P, Andersson B and Norling A (1978) Interaction of ethanol oxidation with glucuronidation in isolated hepatocytes. *Biochemical Pharmacology*, **27**, 2583–2588.

Monaghan G, Clarke DJ, Povey S, See CG, Boxer M and Burchell B (1994) Isolation of human YAC contig encompassing a cluster of UGT2 genes and its regional localisation to chromosome 4Q13. *Genomics*, **23**, 496–499.

Monaghan G, Ryan M, Seddon R, Hume R and Burchell B (1996) Genetic variation in bilirubin UDP-glucuronosyltransferase gene promoter and Gilbert's syndrome. *The Lancet*, **347**, 578–581.

Mucklow JC, Fraser HS, Bulpitt, CJ, Kahn C, Mould G and Dollery CT (1980) Environmental factors affecting paracetamol metabolism in London factory and office workers. *British Journal of Clinical Pharmacology* **10**, 67–74.

Münzel PA, Brück M and Bock KW (1994) Tissue-specific constitutive and inducible expression of rat phenol UDP-glucuronosyltransferase. *Biochemical Pharmacology*, **47**, 1445–1448.

Münzel PA, Bookjans G, Mehner G, Lehmköster T and Bock KW (1996) Tissue-specific TCDD-inducible expression of human UDP-glucuronosyltransferase UGT1A6. *Archives of Biochemistry and Biophysics*, **335**, 205–210.

Münzel PA, Lehmköster T, Brück M, Ritter JK and Bock KW (1998) Aryl hydrocarbon receptor-inducible or constitutive expression of human UDP glucuronosyltransferase UGT1A6. *Archives of Biochemistry and Biophysics*, **350**, 72–78.

Münzel PA, Schmohl S, Heel H, Kälberer K, Bock-Hennig B and Bock KW (1999) Induction of human UDP glucuronosyltransferases (UGT1A6, UGT1A9, and UGT2B7) by t-butylhydroquinone and 2,3,7,8-tetrachlorodibenzo-p-dioxin in Caco-2 cells. *Drug Metabolism and Disposition*, **27**, 569–573.

Nebert DW (1991) Proposed role of drug-metabolizing enzymes. Regulation of steady state levels of the ligands that effect growth, homeostasis, differentiation and neuroendocrine functions. *Molecular Endocrinology*, **5**, 1203–1214.

Nguyen N, Stassburg CP, Kuehl G, Murphey S and Tukey RH (2000) Metabolism of the nicotine by-product NNAL by human UDP-glucuronoyltransferases (UGTs). *Toxicological Sciences*, **54**, 215–216.

Nowell SA, Massengill JS, Williams S, Radominska-Pandya A, Tephly TR, Cheng Z, Strassburg CP, Tukey RH, MacLeod SL, Lang NP and Kadlubar FF (1999) Glucuronidation of 2-hydroxyamino-1-methyl-6-phenylimidazo[4,5-b]pyridine by human microsomal UDP-glucuronosyltransferases: identification of specific UGT1A family isoforms involved. *Carcinogenesis*, **20**, 1107–1114.

Oguri K, Kurogi A, Yamabe K-I, Tanaka M, Yoshisue K, Ishii Y and Yoshimura H (1996) Purification of a phenobarbital-inducible UDP-glucuronosyltransferase isoform from dog liver which catalyses morphine and testosterone glucuronidation. *Archives of Biochemistry and Biophysics*, **325**, 159–166.

Orzechowski A, Schrenk D, Bock-Hennig BS and Bock KW (1994) Glucuronidation of carcinogenic arylamines and their N-hydroxy derivatives by rat and human phenol UDP-glucuronosyltransferases of the UGT1 gene complex. *Carcinogenesis*, **15**, 1549–1553.

Osborne R, Thompson P, Joel S, Trew D, Patel N and Slevin M (1992) The analgesic activity of morphine-6-glucuronide. *British Journal of Clinical Pharmacology*, **34**, 130–138.

Ouzzine M, Magdalou J, Burchell B and Fournel-Gigleux S (1999) An internal signal sequence mediates the targeting and retention of the human UDP-glucuronosyltransferase 1A6 to the endoplasmic reticulum. *Journal of Biological Chemistry*, **274**, 31401–31409.

Owens IS (1977) Genetic regulation of UDP-glucuronosyltransferase induction by polycyclic aromatic compounds in mice. *Journal of Biological Chemistry* **252**, 2827–2833.

Pantuck EJ, Pantuck CB, Anderson KE, Wattenberg LW, Conney AH and Kappas A (1984) Effect of Brussels sprouts and cabbage on drug conjugation. *Clinical Pharmacology and Therapeutics*, **35**, 161–169.

Patel M, Tang BK, Grant DM and Kalow W (1995) Interindividual variability in the glucuronidation of (S) oxazepam contrasted with that of (R) oxazepam. *Pharmacogenetics*, **5**, 287–297.

Paul D, Standifer KM, Inturrisi CE and Pasternak GW (1989) Pharmacological characterization of morphine-6β-glucuronide, a very potent morphine metabolite. *Journal of Pharmacology and Experimental Therapeutics*, **251**, 477–483.

Peters WHM, Jansen PLM and Nauta H (1984) The molecular weights of UDP-glucuronyltransferase determined with radiation-inactivation analysis. A molecular model of bilirubin UDP-glucuronyltransferase. *Journal of Biological Chemistry*, **259**, 11701–11705.

Philipp T, Durazzo M, Trautwein C, Alex B, Straub P, Lamb JG, Johnson EF, Tukey RH and Manns MP (1994) Recognition of uridine diphosphate glucuronosyl transferases by LKM-3 antibodies in chronic hepatitis D. *The Lancet*, **344**, 578–581.

Pillot T, Ouzzine M, Fournel-Gigleux S, Lafaurie C, Radominska A, Burchell B, Siest G and Magdalou J (1993) Glucuronidation of hyodeoxycholic acid in human liver. Evidence for a selective role of UDP-glucuronosyltransferase 2B4. *Journal of Biological Chemistry*, **268**, 25636–25642.

Prochaska HJ and Talalay P (1988) Regulatory mechanisms of monofunctional and bifunctional anticarcinogenic enzyme inducers in murine liver. *Cancer Research*, **48**, 4776–4782.

Radominska A, Little JM, Lehman PA, Samokyszyn V, Rios GR, King CD, Green MD and Tephly TR (1997) Glucuronidation of retinoids by rat recombinant UDP-glucuronosyltransferase 1.1 (Bilirubin UGT). *Drug Metabolism and Disposition*, **25**, 889–892.

Radominska-Pandya A, Czernik PJ, Little JM, Battaglia E and Mackenzie PI (1999) Structural and functional studies of UDP-glucuronosyltransferases. *Drug Metabolism Reviews*, **31**, 817–899.

Ren Q, Murphy SE, Dannenberg AJ, Park JY, Tephly TR and Lazarus P (1999) Glucuronidation of the lung carcinogen 4-(methylnitrosamino)-1-(3-pyridyl)-1-butanol (NNAL) by rat UDP-glucuronosyltransferase 2B1. *Drug Metabolism and Disposition*, **27**, 1010–1016.

Richter von O, Orzechowski A, Münzel P, Bock-Hennig B and Bock KW (2000) Induction of UDP-glucuronosyltransferase activity towards morphine in the intestine, but not in liver, of β-naphthoflavone-treated *dogs*. *Naunyn–Schmiedeberg's Archives of Pharmacology*, **361**, 4, R 137.

Riedy M, Wang J-Y, Miller AP, Buckler A, Hall J and Guida M (2000) Genomic organization of the UGT2b gene cluster on human chromosome 4q13. *Pharmacogenetics* **10**, 251–260.

Ritter JK, Chen F, Sheen YY, Tran HM, Kimura S, Yeatman MT and Owens IS (1992) A novel complex locus UGT1 encodes human bilirubin, phenol, and other UDP-glucuronosyltransferase isozymes with identical carboxyl termini. *Journal of Biological Chemistry*, **267**, 3257–3261.

Rushmore TH and Pickett CB (1990) Transcriptional regulation of the rat glutathione S-transferase Ya subunit gene. *Journal of Biological Chemistry*, **265**, 14648–14653.

Saarikoski ST, Ikonen TS, Oinonen T, Lindros KO, Ulmanen I and Husgafvel-Pursiainen (1998) Induction of UDP-glucuronosyltransferase family 1 genes in rat liver: Different patterns of mRNA expression with two inducers, 3-methylcholanthrene and β-naphthoflavone. *Biochemical Pharmacology*, **56**, 569–575.

Samokyszyn VM, Gall WE, Zawada G, Freyaldenhoven MA, Chen G, Mackenzie PI, Tephly TR and Radominska-Pandya A (2000) 4-Hydroxyretinoic acid, a novel substrate for human liver microsomal UDP-glucuronosyltransferase(s) and recombinant UGT2B7. *Journal of Biological Chemistry*, **275**, 6908–6914.

Savides MC, Oehme FW, Nash SL, Leipold HW (1984). The toxicity and biotransformation of single doses of acetaminophen in dogs and cats. *Toxicology and Applied Pharmacology*, **74**, 26–34.

Schmid R, Axelrod J and Hammaker J (1957) Congenital defects in bilirubin metabolism. *Journal of Clinical Investigations*, **36**, 927.

Senafi SB, Clarke DJ and Burchell B (1994) Investigation of the substrate specificity of a cloned expressed human bilirubin UDP–glucuronosyltransferase: UDP-sugar specificity and involvement in steroid and xenobiotic glucuronidation. *Biochemical Journal*, **303**, 233–240.

Sohn OS, Surace A, Fiala ES, Richie Jr. JP, Colosimo S, Zang E and Weisburger JH (1994) Effects of green and black tea on hepatic xenobiotic metabolizing systems in the male F344 rat. *Xenobiotica*, **24**, 119–127.

Strassburg CP, Oldhafer K, Manns MP and Tukey RH (1997a) Differential expression of the UGT1A locus in human liver, biliary, and gastric tissue: Identification of UGT1A7 and UGT1A10 transcripts in extrahepatic tissue. *Molecular Pharmacology*, **52**, 212–220.

Strassburg CP, Manns MP and Tukey RH (1997b) Differential down-regulation of the UDP-glucuronosyltransferase 1A locus is an early event in human liver and biliary cancer. *Cancer Research*, **57**, 2979–2985.

Strassburg CP, Manns MP and Tukey RH (1998) Expression of the UDP-glucuronosyltransferase 1A locus in human colon. *Journal of Biological Chemistry*, **273**, 8719–8726.

Strassburg CP, Strassburg A, Nguyen N, Li Q, Manns QL and Tukey RH (1999) Regulation and function of family 1 and family 2 UDP-glucuronosyltransferase genes (UGT1A6, UGT2B2) in human oesophagus. *Biochemical Journal*, **338**, 489–498.

Strassburg CP, Kneip S, Topp J, Obermayer-Straub P, Barut A, Tukey RH and Manns MP (2000) Polymorphic gene expression and interindividual variation of UDP-glucuronosyltransferase activity in human small intestine. *Journal of Biological Chemistry*, in press.

Sugimura T and Sato S (1983) Mutagens-carcinogens in foods. *Cancer Research*, (Suppl.) **43**, 2415s–2421s.

Sugiyama Y, Kato Y and Chu X (1998) Multiplicity of biliary excretion mechanisms for the camptothecin derivative irinotecan (CPT-11), its metabolite SN-38, and its glucuronide: Role of canalicular multispecific organic anion transporter and P-glyco-protein. *Cancer Chemotherapy and Pharmacology*, **42**, S44–S49.

Suleman FG, Abid A, Gradinaru D, Daval J-L, Magdalou J and Minn A (1998) Identification of the uridine diphosphate glucuronosyltransferase isoform UGT1A6 in rat brain and in primary cultures of neurons and astrocytes. *Archives of Biochemistry and Biophysics*, **358**, 63–67.

Sun X-Y, Plouzek CA, Henry JP, Wang TTY and Phang JM (1998) Increased UDP-glucuronosyltransferase activity and decreased prostate specific antigen production by biochanin A in prostate cancer cells. *Cancer Research*, **58**, 2379–2384.

Sutherland L, Ebner T and Burchell B (1993) The expression of UDP-glucuronosyltransferases of the UGT1 family in human liver and kidney and in response to drugs. *Biochemical Pharmacology*, **45**, 295–301.

Talalay P, Fahey JW, Holtzclaw WD, Prestera T and Zhang Y (1995) Chemoprotection against cancer by phase 2 enzyme induction. *Toxicology Letters*, **82/83**, 173–179.

Terrier N, Benoit E, Senay C, Lapicque F, Radominska-Pandya A, Magdalou J and Fournel-Gigleux S (1999) Human and rat liver UDP-glucuronosyltransferases are targets of ketoprofen acylglucuronide. *Molecular Pharmacology*, **56**, 226–234.

Townsend AJ, Kiningham KK, St. Clair D, Tephly TR, Morrow CS and Guengerich FP (1999) Symposium overview: Characterization of xenobiotic metabolizing enzyme function using heterologous expression systems. *Toxicological Sciences*, **48**, 143–150.

Tukey RH and Strassburg CP (2000) Human UDP-glucuronosyltransferases: Metabolism, expression, and disease. *Annual Review of Pharmacology and Toxicology 2000*, **40**, 581–616.

Tukey RH, Pendurthi UR, Nguyen NT, Green MD and Tephly TR (1993) Cloning and characterization of rabbit liver UDP-glucuronosyltransferase cDNAs. *Journal of Biological Chemistry*, **268**, 15260–15266.

Ullrich D, Fischer G, Katz N and Bock KW (1984) Intralobular distribution of UDP-glucuronosyltransferase in livers from untreated, 3-methylcholanthrene- and phenobarbital-treated rats. *Chemico-Biological Interactions*, **48**, 181–190.

Venugopal R and Jaiswal AK (1996) Nrf1 and Nrf2 positively and c-Fos and Fra1 negatively regulate the human antioxidant response element-mediated expression of NAD(P)H: quinone oxidoreductase1 gene. *Proceedings of the National Academy of Sciences, USA*, **93**, 14960–14965.

Venugopal R and Jaiswal AK (1999) Antioxidant response element-mediated 2,3,7,8-tetrachlorodibenzo-p-dioxin (TCDD) induction of human NAD(P)H: Quinone oxidoreductase 1 gene expression. *Biochemical Pharmacology*, **58**, 1649–1655.

Vessey DA and Kempner ES (1989) In situ structural analysis of microsomal UDP-glucuronyltransferases by radiation inactivation. *Journal of Biological Chemistry*, **264**, 6334–6338.

Walle T, Walle UK, Cowart TD, Conradi EC and Gaffney TE (1987) Selective induction of propranolol metabolism by smoking: Additional effects on renal clearance of metabolites. *Journal of Pharmacology and Experimental Therapeutics*, **241**, 928–933.

Walle UK, Galijatovic A and Walle T (1999) Transport of the flavonoid chrysin and its conjugated metabolites by the human intestinal cell line Caco-2. *Biochemical Pharmacology*, **58**, 431–438.

Wang Q, Hasan G and Pikielny CW (1999) Preferential expression of biotransformation enzymes in the olfactory organs of *Drosophila melanogaster*, the antennae. *Journal of Biological Chemistry*, **274**, 10309–10315.

Wattenberg LW (1983) Inhibition of neoplasia by minor dietary constituents. *Cancer Research (Suppl)*, **43**, 2448s–2453s.

Wedekind JE, Frey PA and Rayment I (1995) Three-dimensional strucuture of galactose-1-phosphate uridylyltransferase from *Escherichia coli* at 1.8 Å resolution. *Biochemistry*, **34**, 11049–11061.

Weiβ C, Kolluri SK, Kiefer F and Göttlicher M (1996) Complementastion and Ah receptor deficiency in hepatoma cells: Negative feedback regulation and cell cycle control by the Ah receptor. *Experimental Cell Research*, **226**, 154–163.

Wishart GJ (1978a) Demonstration of functional heterogeneity of hepatic uridine diphosphate glucuronosyltransferase activities after administration of 3-methylcholanthrene and phenobarbital to rats. *Biochemical Journal*, **174**, 671–672.

Wishart GJ (1978b) Functional heterogeneity of UDP-glucuronosyltransferase as indicated by its differential development and inducibility by glucocorticoids. *Biochemical Journal*, **174**, 485–489.

Wu D, Kang Y-S, Bickel U and Pardridge WM (1997) Blood-brain barrier permeability to morphine-6-glucuronide is markedly reduced compared with morphine. *Drug Metabolism and Disposition*, **25**, 768–771.

Yang Y, Griffiths WJ, Midtvedt T, Sjövall J, Rafter J and Gustafsson J-A (1999) Characterization of conjugated metabolites of benzo[a]pyrene in germ-free rat urine by liquid chromatography/electrospray tandem mass spectrometry. *Chemical Research in Toxicology*, **12**, 1182–1189.

Yin Z, Sato K, Tsuda H and Ito N (1982) Changes in activities of uridine diphosphate-glucuronyltransferases during chemical hepatocarcinogenesis. *Gann*, **73**, 239–248.

Zakim D and Dannenberg AJ (1992) How does the microsomal membrane regulate UDP-glucuronosyltransferases? *Biochemical Pharmacology*, **43**, 1385–1393.

Zakim D, Cantor M and Eibl H (1988) Phospholipids and UDP-glucuronosyltransferase. Structure/function relationships. *Journal of Biological Chemistry*, **263**, 5164–5169.

9 Glutathione S-transferases

Philip J. Sherratt and John D. Hayes

University of Dundee, UK

Introduction

Glutathione S-transferase (GST; EC 2.5.1.18) isoenzymes are ubiquitously distributed in nature, being found in organisms as diverse as microbes, insects, plants, fish, birds and mammals (Hayes and Pulford 1995). The transferases possess various activities and participate in several different types of reaction. Most of these enzymes can catalyse the conjugation of reduced glutathione (GSH) with compounds that contain an electrophilic centre through the formation of a thioether bond between the sulphur atom of GSH and the substrate (Chasseaud 1979; Mannervik 1985). In addition to conjugation reactions, a number of GST isoenzymes exhibit other GSH-dependent catalytic activities including the reduction of organic hydroperoxides (Ketterer *et al.* 1990) and isomerisation of various unsaturated compounds (Benson *et al.* 1977; Jakoby and Habig 1980). These enzymes also have several non-catalytic functions that relate to the sequestering of carcinogens, intracellular transport of a wide spectrum of hydrophobic ligands, and modulation of signal transduction pathways (Listowsky 1993; Adler *et al.* 1999; Cho *et al.* 2001).

Glutathione S-transferases represent a complex grouping of proteins. Two entirely distinct superfamilies of enzyme have evolved that possess transferase activity (Hayes and Strange 2000). The first enzymes to be characterised were the cytosolic, or soluble, GSTs (Boyland and Chasseaud 1969; Mannervik 1985). To date at least 16 members of this superfamily have been identified in humans (Board *et al.* 1997, 2000; Hayes and Strange 2000). On the basis of their degree of sequence identity, the soluble mammalian enzymes have been assigned to eight families, or classes, designated Alpha (α), Mu (μ), Pi (π), Sigma (σ), Theta (θ), Zeta (ζ), Omega (ω) and Kappa (κ) (Mannervik *et al.* 1985; Meyer *et al.* 1991; Meyer and Thomas 1995; Pemble *et al.* 1996; Board *et al.* 1997, 2000). Four additional classes of this superfamily, called Beta (β), Delta (δ), Phi (ϕ) and Tau (τ) are represented in bacteria, insects and plants (Hayes and McLellan 1999), but discussion of these non-mammalian GSTs is beyond the scope of this chapter. The second, more recently defined superfamily is composed of microsomal transferases, and has been designated *membrane-associated proteins* in

Enzyme Systems that Metabolise Drugs and Other Xenobiotics. Edited by C. Ioannides.
© 2002 John Wiley & Sons Ltd

eicosanoid and glutathione metabolism, or MAPEG for short (Jakobsson *et al.* 1999a). In humans, the MAPEG superfamily has at least six members (Jakobsson *et al.* 2000).

Evolution of a large number of soluble GST and MAPEG members has allowed diversification of function, regulation and subcellular localisation in the two super-families. The soluble GSTs appear to be involved primarily in the metabolism of foreign chemicals, such as carcinogens, environmental pollutants and cancer che-motherapeutic drugs, as well as the detoxication of potentially harmful endogenously derived reactive compounds (Hayes and Pulford 1995). Many endogenous GST substrates are formed as a consequence of modification of macromolecules by reactive oxygen species, and the transferases are therefore considered to serve an antioxidant function (Mannervik 1986; Hayes and McLellan 1999). A few soluble GSTs are also involved in the synthesis and inactivation of prostaglandins. By contrast, MAPEG members are not principally involved in detoxication reactions, but are rather involved in the biosynthesis of leukotrienes and prostanoids, endogenous lipid signal-ling molecules (Jakobsson *et al.* 1999a). Thus, collectively, the catalytic actions of GST isoenzymes contribute to cellular detoxication processes and to autocrine and paracrine regulatory mechanisms.

Historical perspective of the research field

The transferases were first studied because of their involvement in the metabolism of xenobiotics rather than because of their contribution to the biosynthesis of leuko-trienes or prostaglandins. Specifically, in 1961, extracts from rat liver were reported to catalyse the conjugation of GSH with either 1,2-dichloro-4-nitrobenzene (DCNB) (Booth *et al.* 1961) or with bromosulphophthalein (Coombes and Stakelum 1961). Once protein purification schemes were devised, it became apparent that one subunit belonging to class Mu of the soluble GST superfamily was responsible for both of these two activities (Habig *et al.* 1974; Mannervik and Jensson 1982). A separate line of investigation into cytosolic proteins in rat liver that bind carcinogens, steroids and bilirubin, led to the first description of class Alpha GST, though at the time the enzyme(s) was called 'ligandin' (Ketterer *et al.* 1967; Litwack *et al.* 1971; Hayes *et al.* 1979). Recognition that 1-chloro-2,4-dinitrobenzene (CDNB) is a more general transferase substrate facilitated identification of Alpha-, Pi-, Sigma- and Kappa-class GST, as well as other members of the Mu-class family (Habig *et al.* 1976; Kitahara *et al.* 1984; Urade *et al.* 1987; Harris *et al.* 1991). Class Theta transferases were first purified using 1-menaphthyl sulphate and 1,2-epoxy-3-(*p*-nitrophenoxy)propane as substrates (Fjellstedt *et al.* 1973; Gillham 1973; Hiratsuka *et al.* 1990; Meyer *et al.* 1991). More recently, the class Zeta and Omega GST were characterised using a bioinformatics approach (Board *et al.* 1997, 2000).

The first evidence for the existence of distinct membrane associated transferases was provided some twenty years after the original demonstration of soluble GST activity in rat liver. The membrane-associated enzyme was initially designated microsomal GST, and it was recognised to be functionally unique because its ability to conjugate CDNB with GSH is increased by covalent modification of the protein with the thiol agent *N*-ethylmaleimide (Morgenstern *et al.* 1979, 1980). The activation of the CDNB-GSH-conjugating activity of the rat microsomal GST occurs through alkylation of Cys-49

(DeJong *et al.* 1988), though it can also be achieved by limited proteolysis at either Lys-4 or Lys-41 (Morgenstern *et al.* 1989). Importantly, molecular cloning of the enzyme demonstrated that it shares no sequence identity with soluble GST (DeJong *et al.* 1988). McLellan *et al.* (1988) first reported the purification of human microsomal GST, now called MGST-I. A further human MAPEG member, called FLAP (5-lipoxygenase activating protein), was discovered through isolation of a membrane protein required to allow 5-lipoxygenase to convert arachidonic acid to 5-hydroperoxy-8,11,14-*cis*-6-*trans*-eicosatetraenoic acid and leukotriene A$_4$ (Dixon *et al.* 1990; Miller *et al.* 1990). Others were discovered through the characterisation of microsomal transferases responsible for leukotriene C$_4$ synthesis (Lam *et al.* 1994; Welsch *et al.* 1994; Jakobsson *et al.* 1996, 1997; Scoggan *et al.* 1997). The final member, identified because of its regulation by the p53 tumour suppressor protein, has been variously called *p53* inducible gene 12, or PIG12 (Polyak *et al.* 1997), MGST-I-like I (Jakobsson *et al.* 1999a), or prostaglandin E synthase (Jakobsson *et al.* 1999b).

Functions of glutathione S-transferases

CATALYTIC PROPERTIES OF GST

Classically, GST enzymes have been considered to play a major part in phase II of drug-metabolism where they contribute to cell survival by detoxication of foreign compounds. In this role, GST action follows phase I of drug-metabolism which is often catalysed by members of the cytochrome P450 (CYP) supergene family (Klaassen 1996). The CYP enzymes catalyse the introduction of a functional group, such as an epoxide, into an otherwise chemically inactive xenobiotic. This functional group offers an electrophilic centre that is attacked by reduced glutathione (GSH), the incoming nucleophile, in a reaction catalysed by GST (Figure 9.1). The addition of GSH to the molecule gives it a molecular 'flag' which allows the xenobiotic-conjugate to be removed from the cell during phase III of drug-metabolism, a process which requires the participation of drug transporters such as multi-drug resistance associated protein (MRP) (Hayes and McLellan 1999). Once transported out of the cell, the peptide portion of the GSH-conjugate is subjected to peptidase attack, by γ-glutamyltransferase and either aminopeptidase M or cysteinylglycine dipeptidase, to yield a cysteinyl conjugate which is in turn N-acetylated to form a mercapturic acid (Boyland and Chasseaud 1969; Jakoby and Habig 1980). It is the mercapturic acid that is typically the final metabolite that is eliminated from the body in urine.

Compounds that undergo GST-catalysed conjugation with GSH include epoxide-containing compounds, alkyl- and aryl-halides, isothiocyanates, α,β-unsaturated carbonyls and quinones (Hayes and Pulford 1995). Not all xenobiotics that are substrates for GST require to be activated by CYP. It is likely that some are activated by cyclooxygenases (Marnett 1994) or by interaction with free radicals during oxidative stress (Trush and Kensler 1991). Others, such as isothiocyanates, may either be formed during digestion of vegetables (Verhoeven *et al.* 1997) or are generated as products of combustion (Klaassen 1996).

In addition to synthesising glutathione S-conjugates, GSTs catalyse the reduction of peroxide-containing compounds that may otherwise be toxic to the cell (Mannervik

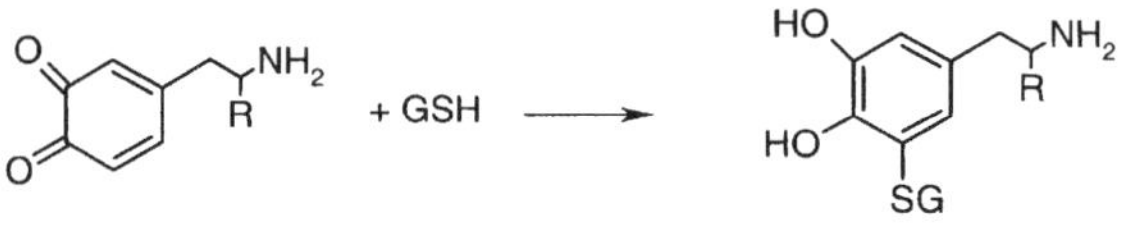

Figure 9.1 Examples of GST catalysed reactions. The GST substrates shown are as follows: 1, aflatoxin B_1-8,9-epoxide; 2, benzylisothiocyanate; 3, dibromoethane; 4, maleylacetoacetate; 5, a model o-quinone.

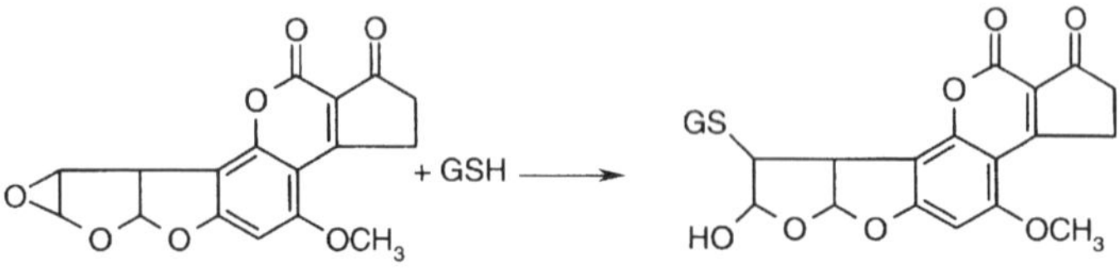

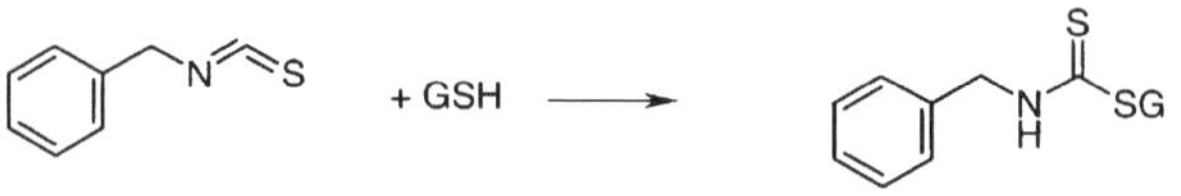

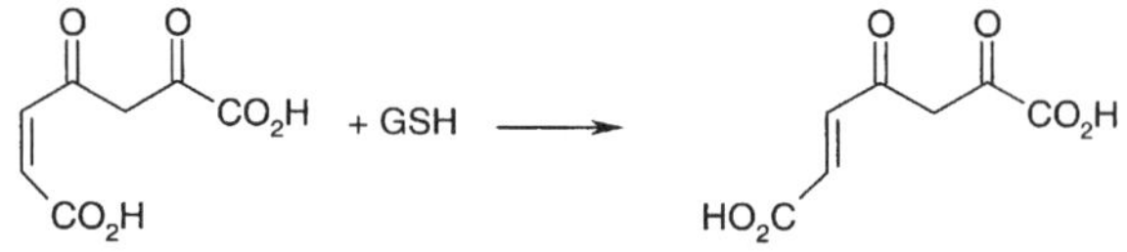

1986). The peroxidase activity of GST is not dependent on selenium but it does require GSH. It is a two-step reaction. The first step is an enzymic reduction of the peroxide to an alcohol, with the concomitant production of hydroxylated glutathione (GSOH). The second step entails the spontaneous reaction of GSOH with a molecule of GSH to yield water and oxidised glutathione (GSSG). Examples of this type of substrate include hydroperoxides of fatty acids and phospholipids. A related reaction is the GSH-dependent reduction of organic nitrate esters to alcohols and inorganic nitrite (Jakoby and Habig 1980).

GSTs also serve an important role in the isomerisation of many biologically important molecules. The transferases can catalyse cis–trans isomerisation reactions or the movement of a double bond within a polycyclic molecule (Benson et al. 1977;

Keen and Jakoby 1978). Physiological examples include the conversion of 13-*cis*-retinoic acid to all-*trans*-retinoic acid, a reaction that results in an increase in affinity of the retinoid for its receptor (Chen and Juchau 1997). The GSH-dependent conversion of prostaglandin (PG) H_2 to either PGD_2 or PGE_2 are other examples of isomerisation reactions in which GSTs are involved (Urade *et al.* 1995; Jakobsson *et al.* 1999b). Also, the isomerisation of maleylacetoacetate to furmaylacetoacetate, a step in the degradation of tyrosine, represents a GST-catalysed reaction (Fernández-Cañón and Peñalva 1998) (Figure 9.1).

BINDING OF NON-SUBSTRATE LIGANDS BY GST

In their non-enzymic ligand-binding capacity, GST isoenzymes serve a variety of functions involved in carcinogen-detoxication and intracellular transport of a wide spectrum of substances. It has been recognised for many years that a number of class Alpha GSTs can bind covalently reactive metabolites formed from 3-methylcholanthrene and azo-dye carcinogens (Litwack *et al.* 1971; Coles and Ketterer 1990). In these instances, GST is thought to sequester genotoxic compounds in a suicide type of reaction that prevents xenobiotics from interacting with DNA.

The soluble transferases bind many compounds in a non-covalent fashion. These ligands, which are typically lipophilic in nature and are not substrates for a conjugation reaction, include steroid hormones, thyroid hormones, bile acids, bilirubin, free-fatty acids and numerous drugs (Listowsky 1993). The biological consequence of these interactions between GST and non-substrate ligands is for the most part uncertain but the transferases may assist other enzymes in catalytic reactions, they may influence nuclear hormone receptor activity, or they may act to transport small molecules around the cell for excretion.

The finding that class Alpha, Mu and Pi GSTs have a high affinity for many glutathione S-conjugates is consistent with the theory that these enzymes do not release xenobiotics to diffuse across the cell once catalysis has occurred (Meyer 1993). However, it remains to be demonstrated whether either non-substrate ligand or glutathione S-conjugates are released by GST following an interaction with a drug-transporter protein such as MRP at the plasma membrane. GSTs are known to occupy a number of compartments within the cell including the nucleus, cytoplasm, mitochondrion and endoplasmic reticulum. Furthermore, drug treatments can cause GST to move between the cytoplasm and the nuclear compartments (Sherratt *et al.* 1998). It therefore appears possible that these enzymes could control the movement of xenobiotics within the cell, or the passage of xenobiotics across cells.

Among the MAPEG superfamily, FLAP appears to act as a carrier protein for arachidonic acid (Mancini *et al.* 1993). Although FLAP does not appear to exhibit catalytic activity, it acts to facilitate the conversion by 5-lipoxygenase of arachidonic acid to leukotriene A_4. FLAP serves as a useful reminder that the binding of non-substrate ligands by GST, and subsequent interaction with other proteins, may be essential for the efficient function of enzyme systems that are unrelated to the transferases.

INVOLVEMENT OF GST IN PROTEIN-PROTEIN INTERACTIONS

Recent studies have shown that soluble GST can bind to other intracellular proteins and modulate their function. In a model presented by the research groups of Ronai and Tew, it has been proposed that class Pi GST subunits form a complex with c-Jun N-terminal Kinase (JNK), and in so doing maintain it in an inactive form (Adler *et al.* 1999). It is believed that under non-stressed conditions, the binding of GSTP1 to JNK helps to block signalling along the stress kinase pathway. According to this hypothesis, when cells are subjected to oxidative stress, class Pi GST subunits dissociate from the kinase and allow JNK to become activated.

Using a tet-off-inducible system to control levels of class Pi GST in NIH 3T3 cells, Yin *et al.* (2000) confirmed that expression of the transferase could reduce the extent of JNK phosphorylation. In these experiments, expression of class Pi GST also caused a reduction in the phosphorylation of mitogen-activated protein (MAP) kinase kinase 7, but an increase in the phosphorylation of MAP kinase kinase 4, p38, extracellular receptor kinase and inhibitor of κ-kinase. The expression of the transferase in 3T3 fibroblasts made the cells more resistant to death caused by exposure to H_2O_2, and this is suggested to be attributable to modulation of stress kinase responses (Yin *et al.* 2000).

Class Pi GST is not unique in modulating signal transduction. A class Mu transferase can bind and inhibit apoptosis signal-regulating kinase 1 (Cho *et al.* 2001)

The human Omega class GSTO1-1 has been shown by Board and his colleagues to interact with the calcium channel ryanodine receptors that are found in the endoplasmic reticulum of various cells (Dulhunty *et al.* 2000). These workers have suggested that this transferase may protect cells against apoptosis in cardiac muscle that contains the ryanodine receptor 2 channel.

This represents an emerging and potentially exciting area of research that adds to the diverse roles played by GST within the cell. Importantly, it suggests that certain transferases can act as sensors of oxidative stress within cells and help coordinate the regulation of stress kinases. Future research is likely to reveal the existence of additional interactions between GST and cellular proteins, as well as details about the putative role of GST as a sensor of oxidative and chemical stress.

Metabolism of xenobiotics and endobiotics by GST

GST AND DETOXICATION

The ability of GSTs to inactivate potentially cytotoxic and genotoxic compounds has been the most thoroughly studied aspect of their function within the cell. The transferases play an important role in the detoxication of a broad spectrum of noxious chemicals that may lead to mutagenic events or cytotoxicity (Coles and Ketterer 1990; Hayes and Pulford 1995). Examples of such compounds include the ultimate carcinogens produced from aflatoxin B_1 and benzo(a)pyrene that are formed as a consequence of phase I drug metabolism (Figure 9.1). These are examples where conjugation of the compound with GSH leads to the production of a harmless metabolite that is readily eliminated from the cell. GSTs can also detoxify a number of man-made compounds that are common environmental pollutants. Examples are pesticides widely used in agricultural farming. The transferases are capable of metabolising dichlorodiphenyltrichloroethane (DDT), atrazine, lindane and methyl parathion either by catalysing formation of GSH-conjugates or by dehalogenation activity.

The detoxication reactions listed above are of benefit to the host, and remove potentially toxic chemicals from the cell. In cancer therapy, however, the defence provided by GST within a tumour cell is a potential problem to the host rather than a benefit. Certain anticancer drugs such as 1,3-bis(2-chloroethyl)-1-nitrosourea (BCNU), chlorambucil, cyclophosphamide, melphalan and thiotepa are detoxified by GST (Tew 1994; Hayes and Pulford 1995). A large body of literature suggests that GST overexpression in tumours is an important mechanism of acquired resistance to cancer chemotherapeutic agents, particularly if it is associated with overexpression of MRP (Morrow et al. 1998).

Apart from detoxication of foreign compounds, there are a number of harmful endogenous compounds formed as by-products of normal metabolism that are GST substrates. The process of aerobic respiration can lead to the production of reactive oxygen species (the superoxide anion O_2^-, hydrogen peroxide H_2O_2, and the hydroxyl radical •OH) (Finkel and Holbrook 2000). Attack of membrane lipids by free radicals leads to the formation of lipid peroxidation products which can propagate a chain reaction of lipid peroxidation in an aerobic environment, that will ultimately end in membrane destruction (Slater 1984). GST from both the soluble and MAPEG supergene families have the capacity to reduce these compounds rendering them harmless. They also act to detoxify downstream products of oxidative damage such as the reactive aldehydes 4-hydroxynonenal and acrolein (Hayes and McLellan 1999). Furthermore, GST can reduce DNA hydroperoxides such as thymine peroxide (Bao et al. 1997), an event that may be important in vivo as GSTs appear to be present within the nucleus of a cell. Another important endogenous function of GST is in the detoxication of o-quinones derived from catecholamines (Baez et al. 1997) (Figure 9.1). A member of the class Mu GST gene family has been shown to provide a neuro-protective metabolic pathway for dopamine and dopa o-quinones. This metabolism prevents the formation of aminochrome and dopachrome respectively that can lead to neurodegeneration.

GST AND BIOACTIVATION

Not all reactions catalysed by GST lead to the formation of a less toxic product. In a number of instances, the action of GST has been shown to lead to a more reactive and potentially carcinogenic intermediate. Compounds that undergo bioactivation are synthetic and do not occur naturally within the environment. This is best exemplified by the case of short-chain alkyl-halides. During the industrial age, a number of such chemicals have been developed that have various useful properties, in particular that of being excellent solvents. The American NTP report in 1986 that dichloromethane (DCM), a solvent used extensively by both the consumer and industry, is carcinogenic in mice but not in other rodents led to one of the first toxicology investigations to entail detailed mechanistic studies (Green 1997). The studies revealed that in the mouse a member of the class Theta GST gene family was capable of metabolising

DCM to form a highly reactive S-chloromethylglutathione conjugate. This intermediate still contains an electrophilic centre with the remaining chloride and the addition of GSH is thought to facilitate its ability to move around the cell. Methanes that contain only a single halide substituent undergo the same conjugation reaction catalysed by class Theta GST but are not toxic to the cell (Chamberlain *et al.* 1998). This is not an isolated incidence, as another commonly used dihaloalkane, dibromoethane, is also activated through substitution of thiolate for halide (van Bladeren 1980) (Figure 9.1). The intermediate formed from this compound, and similar compounds, spontaneously cyclises to form an episulphonium anion that is an even more potent mutagen than the dihalomethane metabolites. These dihaloalkanes are carcinogenic in all rodent model systems. The man-made bifunctional electrophiles that are activated by GST are listed in Table 9.1. Certain alkenes undergo more classical metabolism requiring activation by CYP to introduce the electrophilic centre before their mutagenic potential is increased further by GST. Examples of such compounds include butadiene and isoprene, both of which are used in the rubber industry (Guengerich *et al.* 1995).

There are also examples of compounds that may initially be detoxified by GST-catalysed conjugation but then undergo spontaneous reversal of the initial conjugate to regenerate the toxic xenobiotic (Baillie and Slatter 1991; Baillie and Kassahun 1994). Isothiocyanates are a group of xenobiotics that undergo a GST-catalysed reversible conjugation with GSH to form thiocarbamates (Kolm *et al.* 1995). Following conjugation, the thiocarbamate is secreted from the cell of synthesis, but as it is labile and the reaction is reversible, the parent isothiocyanate can be regenerated. If this occurs at a site that lacks GST, toxicity may result (Bruggeman *et al.* 1986).

The conjugation of haloalkenes with GSH can also lead to toxicity (Anders *et al.* 1988; Dekant *et al.* 1994). In this case the toxicity is not due to the glutathione S-conjugate itself, but results from the formation of reactive metabolites produced from the cysteine S-conjugate. Specifically, unstable thiols can be formed by the action of cysteine conjugate β-lyase (see Chapter 1). Since this enzyme is found in high amounts in the kidney, renal damage is usually observed when unstable thiols are produced.

The ability of GST to activate compounds to a cytotoxic intermediate may be beneficial in certain cases. This property of GST can be turned to therapeutic advantage for the treatment of malignant disease. A number of soluble GSTs, in particular class-Pi transferase, are overexpressed in many human tumours and may contribute to drug resistance. One approach to circumvent this problem is the development of modulators of GST activity, such as specific enzyme inhibitors, or strategies to deplete GSH as methods of reducing GST activity in cancer cells. Another approach has been to use the increase in GST activity in cancer cells to achieve enhanced activation of cytotoxic prodrugs within the tumour. TER 286 [γ-glutamyl-α-amino-β((2-ethyl-N,N,N',N'-tetrakis(2-chloroethyl)phosphorodiamidate)-sulphonyl)-propionyl-(R)-(-)phenylglycine] is a latent drug that is activated to a nitrogen mustard alkylating agent by a reaction catalysed by GST (Morgan *et al.* 1998). The nitrogen mustard then spontaneously yields an aziridium ring compound that can alkylate DNA (Figure 9.2). This compound has been found to be effective against drug-resistant breast and colon cancer cells and has entered clinical trials.

In their activities with respect to toxicity, GSTs can be considered to be both an

Table 9.1 Compounds activated by glutathione S-transferases

Type of activation	Class of compound	Examples
Direct metabolism to reactive intermediate	Halogenated alkanes Oxidised alkenes	Dichloromethane, dibromoethane, trichloropropane, butadiene diepoxide, isoprene diepoxide
Reversible reactions	Isothiocyanates Isocyanates	Benzyl isothiocyanate, phenethyl isothiocyanate methylisocyanate, N-(1-methyl-3,3-diphenylpropyl)isocyanate
β-Elimination of oxidised sulfhydryl of GSH-conjugate	Latent pro-drugs	TER 286
β-Lyase-dependent activation of cysteinyl-conjugates derived from GSH conjugates	Halogenated alkenes	Hexachlorobutadiene, tetrachloroethene

TER 286: γ-glutamyl-α-amino-β((2-ethyl-N,N,N',N'-tetrakis(2-chloroethyl)phosphorodiamidate)-sulphonyl)propionyl-(R)-(−)phenylglycine

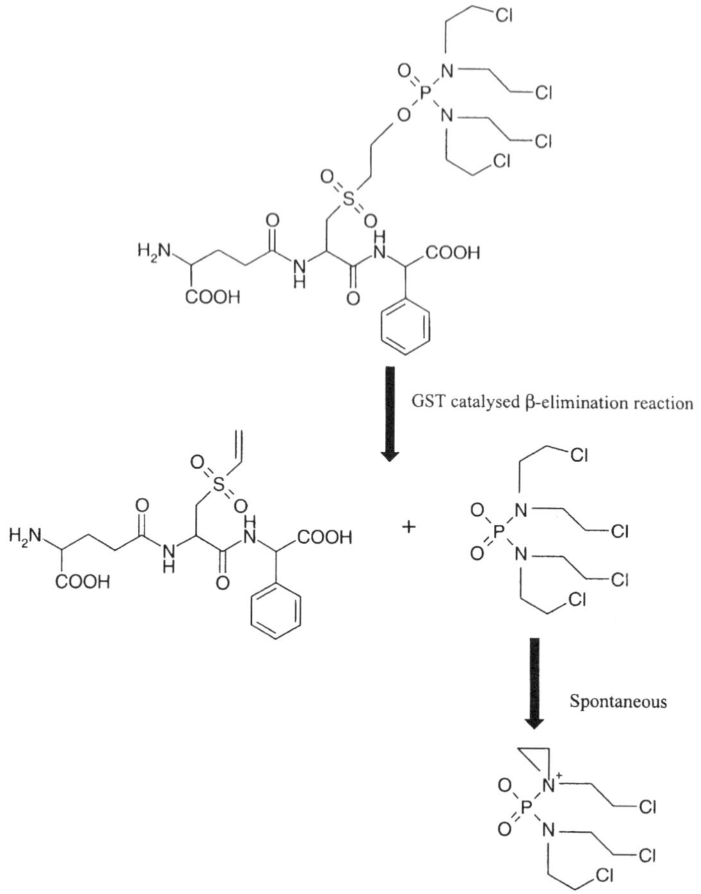

Figure 9.2 Bioactivation of TER 286. TER 286: γ-glutamyl-α-amino-β-((2-ethyl-*N*,*N*,*N'*,*N'*-tetrakis (2-chloroethyl)phosphorodiamidate)-sulphonyl)propionyl-(*R*)-(-)phenylglycine.

advantage and a disadvantage. Thus, an in-depth knowledge of the activity of these enzymes would be helpful during early development of modern drug-treatment strategies. An example of this detailed foreknowledge would have been of value before the use of troglitazone, a member of the thiazolidinedione family of peroxisome proliferator activated receptor-γ (PPARγ) agonist insulin-sensitisers that is used to treat non-insulin-dependent diabetes mellitus. Extended clinical use of this compound can lead to rare instances of hepatic damage where occasional fatalities occur due to liver failure (Kohlroser *et al.* 2000). Early suggestions are that a reaction involving CYP isoenzymes can lead to the formation of functionalised intermediates that can be conjugated with GSH (Baillie and Kassahun 2000; Kassahun *et al.* 2001). Which of the intermediates formed during the metabolism of troglitazone is responsible for the toxic effects is uncertain. As our understanding of GST and other drug-metabolising enzymes increases at the mechanistic level, so our ability to develop safe and effective therapeutics will be enhanced.

GST AND BIOSYNTHETIC METABOLISM

Apart from detoxifying chemicals in the defence of the cell, GSTs play a crucial part in the synthesis of biologically important endogenous molecules. Transferases in the MAPEG superfamily act as prostaglandin E_2 and leukotriene C_4 synthases, and therefore play a central role in the production of mediators of inflammation and hypersensitivity. Members of this superfamily possess some substrate overlap that gives rise to a degree of degeneracy (Table 9.2). In instances where a particular leukotriene C_4 synthase is expressed at low levels in a given tissue, another MAPEG enzyme may compensate for its absence. Certain members of the soluble GST supergene family are also involved in prostanoid biosynthesis (Table 9.2). Class Alpha GST catalyse production of prostaglandin $F_{2\alpha}$. The class Sigma transferase is also known as a GSH-dependent prostaglandin D_2 synthase and initiates the pathway that leads to the eventual formation of the J_2-series of prostaglandins that are important signalling molecules as they are ligands for PPARγ.

In addition to synthesising prostaglandins, class Alpha, Mu and Pi GST catalyse the conjugation of PGA_2 and PGJ_2 with GSH (Bogaards *et al.* 1997). The prostaglandin conjugates are then eliminated from the cell by the MRP transporter. It is therefore envisaged that GST isoenzymes regulate the half-life of certain prostanoids in cells thereby attenuating their biological actions.

Apart from their involvement in prostanoid biology, GSTs are known to contribute to the metabolism of other endogenous compounds. The class Zeta transferase plays an important role in the degradation of tyrosine. It has been identified as the maleylacetoacetate isomerase that is responsible for the conversion of maleylaceto-acetate to furnarylacetoacetate (Fernández-Cañón and Peñalva 1998). In humans, malfunction of this pathway can lead to serious pathological changes in the liver, kidney and peripheral nerves.

Genetic and biochemical properties of the glutathione S-transferase system

GST SUPERGENE FAMILIES

Glutathione S-transferase enzymes have evolved on at least two separate occasions. Two distinct multi-gene families exist, the soluble GST superfamily and the MAPEG superfamily. Transferases from the soluble GST and MAPEG families have no similarity at the level of primary structure. There are significant differences in the sizes of the protein subunits in the two superfamilies. The soluble transferases are all dimeric proteins, and those GST subunits that are most closely related can form heterodimers (Hayes and Pulford 1995). By contrast, MAPEG enzymes are trimeric (Hebert *et al.* 1997; Schmidt-Krey *et al.* 2000) and there is no evidence that they can form heterotrimers.

As can be seen from Table 9.3, GST genes are located on a large number of different chromosomes. Genes encoding human soluble GSTs that are members of the same class (i.e. they share greater than 50% sequence identity) all appear to be clustered on

Table 9.2 Involvement of human GST isoenzymes in leukotriene and prostaglandin metabolism

Superfamily	Isoenzyme	Biosynthesis involving isomerisation, reduction, or non-catalytic binding	Conjugation reactions with GSH
Soluble	GSTA1-1	$PGH_2 \rightarrow PGE_2$; $PGH_2 \rightarrow PGF_{2\alpha}$	Conjugation of PGA_2 and PGJ_2
	GSTA2-2	$PGH_2 \rightarrow PGD_2$; $PGH_2 \rightarrow PGF_{2\alpha}$	–
	GSTM1-1	–	Conjugation of PGA_2 and PGJ_2
	GSTP1-1	–	Conjugation of PGA_2 and PGJ_2
	GSTS1-1	$PGH_2 \rightarrow PGD_2$	–
MAPEG	MGST-I-like I	$PGH_2 \rightarrow PGE_2$	–
	MGST-II	Reduction of 5-HPETE	$LTA_4 \rightarrow LTC_4$
	MGST-III	Reduction of 5-HPETE	$LTA_4 \rightarrow LTC_4$
	LTC_4S	–	$LTA_4 \rightarrow LTC_4$
	FLAP	Binding of arachidonic acid	–

5-HPETE, (*S*)-5-hydroperoxy-8,11,14-*cis*-6-*trans*-eicosatetraenoic acid; LT, leukotriene; PG, prostaglandin

Table 9.3 Genetic and biochemical properties of human glutathione S-transferases

Superfamily	Class	Chromosomes	Enzyme	Substrates
Soluble	Alpha	6p12	GSTA1-1	CDNB; 7-chloro-4-nitrobenzo-2-oxa-1,3-diazole; Δ^5-androstene-3,17-dione
			GSTA2-2	CDNB; 7-chloro-4-nitrobenzo-2-oxa-1,3-diazole; cumene hydroperoxide
			GSTA3-3	Not determined
			GSTA4-4	Ethacrynic acid; 4-hydroxynonenal; 4-hydroxydecenal
Soluble	Mu	1p13.3	GSTM1-1	CDNB; trans-4-phenyl-3-buten-2-one; aflatoxin B_1-epoxide; trans-stilbene oxide
			GSTM2-2	High with CDNB; 1,2-dichloro-4-nitrobenzene; aminochrome
			GSTM3-3	Low towards CDNB; H_2O_2
			GSTM4-4	Not determined
			GSTM5-5	Low towards CDNB
Soluble	Pi	11q13	GSTP1-1	CDNB; acrolein; adenine propenal; BPDE; benzyl isothiocyanate; ethacrynic acid
Soluble	Sigma	4q21-22	GSTS1-1	PGD_2 synthase
Soluble	Theta	22q11	GSTT1-1	1,2-epoxy-3-(p-nitrophenoxy)propane; dichloromethane; dibromoethane
			GSTT2-2	1-menaphthyl sulphate; cumene hydroperoxide

continued overleaf

Table 9.3 (continued)

Superfamily	Class	Chromosomes	Enzyme	Substrates
Soluble	Zeta	14q24.3	GSTZ1-1	Dichloroacetate; fluoroacetate; maleIylacetoacetate
Soluble	Omega	10q23-25	GSTO1-1	Thioltransferase (very low with CDNB and 7-chloro-4-nitrobenzo-2-oxa-1,3-diazole)
Soluble	Kappa	not determined	GSTK1-1	Not determined
MAPEG	(Microsomal)	12p13.1-13.2	MGST-I	CDNB; 7-chloro-4-nitrobenzo-2-oxa-1,3-diazole; 4-nitrobenzyl chloride; cumene hydroperoxide
		9q34.3	MGST-I-like I	PGE_2 synthase
		4q28-31	MGST-II	CDNB; leukotriene C_4 synthase; 5-HPETE
		1q23	MGST-III	Leukotriene C_4 synthase; 5-HPETE
		5q35	LTC_4S	Leukotriene C_4 synthase
		13q12	FLAP	5-lipoxygenase-activating protein (binds arachidonic acid)

BPDE, benzo(a)pyrene diol epoxide: CDNB, 1-chloro-2,4-dinitrobenzene; 5-HPETE, (S)-5-hydroperoxy-8,11,14-cis-6-trans-eicosatetraenoic acid; PGD_2, prostaglandin D_2; PGE_2, prostaglandin E_2

the same chromosome (Pearson *et al.* 1993; Xu *et al.* 1998; Coggan *et al.* 1998). The fact that the Alpha, Mu and Theta GST classes each contain multiple isoenzymes indicates that these families have undergone several relatively recent gene duplication events (Figure 9.3). By contrast, the lack of extensive homology amongst MAPEG members suggests that they have not undergone similar recent gene duplication (Figure 9.4). Possibly the physiological functions of the MAPEG family were largely established, in an evolutionary sense, many years before those of the soluble GST.

At present, the precise relationship between the mitochondrial class Kappa enzyme and other transferases is unclear. Certainly, the dendrogram analysis shown in Figure 9.3 indicates that GSTK1 shares closest homology with the class Omega subunit. As pointed out by Pemble *et al.* (1996), the GSTK1 subunit does not contain a SNAIL/TRAIL motif that is usually present in soluble GST superfamily members, and resides between residues 60–80. Determination of the crystal structure of GSTK1-1, along with identification of the residue involved in forming the glutathione thiolate anion and its catalytic mechanism, will help clarify the evolutionary history of this enzyme.

FUNCTION OF GST ISOENZYMES

Characterisation of the biochemical activities of GST revealed that individual isoenzymes can metabolise a spectrum of electrophilic compounds. In general, individual transferases display overlapping substrate specificities. However, a significant number of GST subunits possess unique catalytic features which supports the notion that each gene evolved to allow detoxication and/or transport of distinct xenobiotics

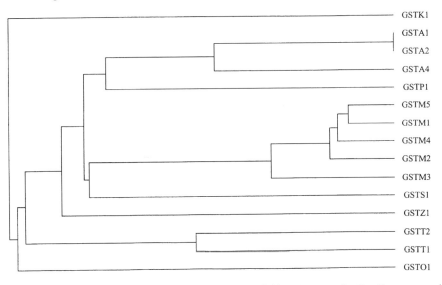

Figure 9.3 An average distance tree of the human soluble GST superfamily. The structural relationship between the different classes of GST was determined by multisequence alignment using Clustalw. The alignment was used to calculate an average distance tree using Jalview. Both applications were accessed through the European Bioinformatics Institute internet server.

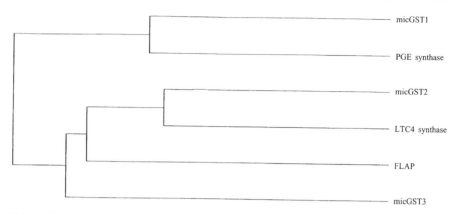

Figure 9.4 An average distance tree of the human MAPEG superfamily. This was constructed as describe in the legend to Figure 9.3. In this dendrogram the following GST abbreviations have been used: micGST1, MGST-I; PGE Synthase, MGST-I-like 1 (or PIG12); LTC4 Synthase, LTC$_4$S; FLAP, 5-lipoxygenase activating protein; micGST3, MGST-III.

or endobiotics. For example, characteristic activities for a few of the human trans-ferases are as follows: GSTA1-1, isomerisation of Δ^5-androstene-3,17-dione; GSTA2-2, reduction of cumene hydroperoxide; GSTA4-4, conjugation of 4-hydroxynonenal with GSH; GSTM1-1, conjugation of *trans*-stilbene oxide; GSTM2-2, conjugation of DCNB with GSH; GSTP1-1, conjugation of benzo[a]pyrene diol epoxide with GSH; GSTS1-1, isomerisation of PGH$_2$ to PGD$_2$; GSTT1-1, conjugation of 1,2-epoxy-3-(p-nitrophenoxy)propane with GSH; GSTT2-2, conjugation of 1-menaphthyl sulphate with GSH; GSTZ1-1, isomerisation of malelyacetoacetate; MGST-I-like I (or PIG12), synthesis of PGE$_2$.

It is interesting that even though the two GST superfamilies occupy different subcellular compartments, both contain members that are active towards aryl-halides (CDNB), organic hydroperoxides (cumene hydroperoxide), and prostaglandin H$_2$ (PGD$_2$, PGE$_2$ and PGF$_{2\alpha}$ synthases). This suggests that during evolution both super-families have been subject to similar selection pressures.

POLYMORPHIC EXPRESSION OF GST

In the human, a significant number of genetic polymorphisms among the soluble GST have been described (for a review, see Hayes and Strange 2000). Importantly, variation in GST alleles is very common in the population and will presumably make a significant contribution to inter-individual differences in drug metabolism. Gene deletions have been reported for *GSTM1* and *GSTT1*, and alterations in amino acid coding sequences have been demonstrated for *GSTA2*, *GSTM1*, *GSTP1*, *GSTT2* and *GSTZ1* (Table 9.4). One of the allelic forms of *GSTT2* encodes a truncated protein (Coggan *et al.* 1998). An allelic variation occurs in intron 6 of *GSTM3* with one form of the gene lacking a YYI transcription factor binding site (Inskip *et al.* 1995). Allelic variations have also been found among MAPEG members, though these occur in the non-coding regions of *MGST-I*, *LTC$_4$S* and *FLAP* (Table 9.4).

For the most part, polymorphisms in individual *GST* genes do not obviously confer a markedly increased risk of cancer. Typically, odds ratios associated with any single variant *GST* allele and the development of particular neoplastic diseases are found to be less than 3.0 (Hayes and Strange 2000). However, combinations of variant GST alleles, either with other polymorphic GST or with alleles of other detoxication or antioxidant genes, are likely to have an additive effect in conferring predisposition to

Table 9.4 Polymorphic human glutathione *S*-transferases

Class or superfamily	Gene	Allele	Alterations in gene or in nucleotides	Protein or amino acids affected
Alpha	GSTA2	GSTA2*A	C335, A629	Thr^{112}, Glu^{210}
		GSTA2*B	G335, C629	Ser^{112}, Ala^{210}
Mu	GSTM1	GSTM1*A	G519	Lys^{173}
		GSTM1*B	C519	Asn^{173}
		GSTM1*0	Gene deletion	No protein
		GSTM1*1×2	Gene duplication	Overexpression
	GSTM3	GSTM3*A	Wildtype	Wildtype protein
		GSTM3*B	3 bp deletion in intron 6	Primary structure unaltered
	GSTM4	GSTM4*A	Wildtype	Wildtype
		GSTM4*B	Changes in introns	Unchanged
Pi	GSTP1	GSTP1*A	A313, C341, C555	Ile^{105}, Ala^{114}, Ser^{185}
		GSTP1*B	G313, C341, T555	Val^{105}, Ala^{114}, Ser^{185}
		GSTP1*C	G313, T341, T555	Val^{105}, Val^{114}, Ser^{185}
		GSTP1*D	A313, T341	Ile^{105}, Val^{114}
Theta	GSTT1	GSTT1*A	Unique gene	Unique protein
		GSTT1*0	Gene deletion	No protein
	GSTT2	GSTT2*A	A415	Met^{139}
		GSTT2*B	G415	Ile^{139}
Zeta	GSTZ1	GSTZ1*A	A94; A124; C245	Lys^{32}; Arg^{42}; Thr^{82}
		GSTZ1*B	A94; G124; C245	Lys^{32}; Gly^{42}; Thr^{82}
		GSTZ1*C	G94; G124; C245	Glu^{32}; Gly^{42}; Thr^{82}
		GSTZ1*D	G94; G124; T245	Glu^{32}; Gly^{42}; Met^{82}
MAPEG	MGST1	MGST1*A	T598 (non-coding 3′)	Wildtype
		MGST1*B	G598 (non-coding 3′)	Unchanged
	LTC_4S	LTC_4S*A	A-444 (promoter)	Wildtype
		LTC_4S*B	C-444 (promoter)	Increase in protein levels
	FLAP	FLAP*A	No *Hind*III site in intron II	Wildtype
		FLAP*B	T → C forming *Hind*III site	Unchanged

The nucleotide number quoted is that found in the cDNA. The amino acid number includes the initiator methionine.

degenerative disease. The most dramatic example of this reported to date occurs in breast cancer. Hirvonen and colleagues have found that in premenopausal women, combinations of the GSTM3*B allele with the GSTT1*00 and GSTP1*AA genotypes appear to have a twenty-six-fold increased risk of developing advanced breast cancer when compared with other GST genotypes (Hirvonen et al. 2001). Substantially increased risk of advanced breast cancer was also seen in GSTM1*00 individuals with GSTP1*AA and either the GSTM3*B allele or the GSTT1*00 genotype (Hirvonen et al. 2001).

GST polymorphisms not only influence susceptibility to disease, but they also appear to influence responsiveness to cancer chemotherapeutic agents. In breast cancer, patients that have two copies of GSTP1*B and/or GSTP1*C have better survival than those with two copies of GSTP1*A and/or GSTP1*D (Sweeney et al. 2000). These authors postulated that the GSTP1*B and GSTP1*C alleles encode enzymes with less activity towards anti-cancer drugs than the enzymes encoded by GSTP1*A and GSTP1*D. However, different GSTP1 subunits may show differences in their interaction with stress kinases.

Biological control of GST

TISSUE-SPECIFIC REGULATION

In most species examined to date, the transferases are expressed in an organ-specific fashion. In the rat, class Alpha GSTs are found in largest amounts in liver, kidney and small intestine, class Mu GSTs are found in liver, lung, heart, spleen, thymus, brain and testis, and class Pi GST is present in most extrahepatic tissues (Hayes and Mantle 1986; Li et al. 1986; Abramovitz and Listowsky 1987). The rat class Sigma GST is expressed in spleen, class Theta GSTs are found in liver, testis, adrenal gland, kidney and lung, and class Kappa has been identified in liver (Urade et al. 1987; Harris et al. 1991; Watabe et al. 1996). In the mouse, expression of class Alpha, Mu and Pi GSTs has also been shown to differ markedly in liver, lung, kidney, spleen, small intestine, heart, brain, testis and ovary (Pearson et al. 1988; McLellan et al. 1992; Mitchell et al. 1997). Class Theta GSTs are present in mouse liver and lung (Mainwaring et al. 1996).

Table 9.5 summarises information about the tissue-specific expression of the human soluble GSTs. In this species, class Alpha are found in substantial amounts in liver, kidney and testis, with some in intestine, pancreas and lung (Mannervik and Widersten 1995; Coles et al. 2000). Human class Mu are found primarily in liver, skeletal muscle, heart, brain and testis, but each subunit shows its own distinct tissue-specific pattern of expression (Takahashi et al. 1993; Rowe et al. 1997). Class Pi and class Theta GSTs are widely distributed in human tissues (Sherratt et al. 1997) (Figure 9.5). Interestingly, in the human, GST T1-1 is present in red blood cells where it is postulated to act as a 'sink' for dihaloalkanes that can be bioactivated, thereby possibly preventing genotoxic damage in other cell types.

Less is known about the distribution of the MAPEG enzymes than the soluble GSTs and most of the available data about their expression relates to the human. Among MAPEG members, MGST-I is thought to serve a detoxication role, and it is therefore appropriate that it is present in high amounts in the liver and kidney. It is, however,

Table 9.5 Tissue distribution of human glutathione S-transferases

Superfamily	Class	Protein	Organ
Soluble	Alpha	GSTA1	Testis ≈ liver ≫ kidney ≈ adrenal > pancreas
		GSTA2	Liver ≈ testis ≈ pancreas > kidney > adrenal > brain
		GSTA3	Placenta
		GSTA4	Small intestine ≈ spleen > liver ≈ kidney > brain
Soluble	Mu	GSTM1	Liver > testis > brain > adrenal ≈ kidney > lung
		GSTM2	Brain ≈ skeletal muscle ≈ testis > heart > kidney
		GSTM3	Testis ≫ brain ≈ small intestine > skeletal muscle
		GSTM4	Testis
		GSTM5	Brain, heart, lung, testis
Soluble	Pi	GSTP1	Brain > heart ≈ lung ≈ testis > kidney ≈ pancreas
Soluble	Sigma	GSTS1	Foetal liver, bone marrow
Soluble	Theta	GSTT1	Kidney ≈ liver > small intestine > brain ≈ prostate
		GSTT2	liver
Soluble	Zeta	GSTZ1	Foetal liver, skeletal muscle
Soluble	Omega	GSTO1	Liver ≈ heart ≈ skeletal muscle > pancreas > kidney
Soluble	Kappa	GSTK1	Liver (mitochondria)
MAPEG	(Microsomal)	MGST-I	Liver ≈ pancreas > prostate > colon ≈ kidney > brain
		MGST-I-like I	Testis > prostate > small intestine ≈ colon
		MGST-II	Liver ≈ skeletal muscle ≈ small intestine > testis
		MGST-III	Heart > skeletal muscle ≈ adrenal gland, thyroid
		LTC$_4$S	Platelets ≈ lung > liver
		FLAP	Lung ≈ spleen ≈ thymus ≈ PBL ≫ small intestine

Data based on references cited in the text.
PBL: Peripheral blood leukocytes

also found in pancreas, prostate and brain (Estonius *et al.* 1999; Lee and DeJong 1999). The MGST-I-like I (PIG12) is a prostaglandin E$_2$ synthase and is found in highest amounts in the testis and prostate, but it is also expressed in the gastrointestinal tract (Jakobsson *et al.* 1999b). The MGST-II, MGST-III, LTC$_4$S and FLAP isoenzymes are all involved in the synthesis of leukotriene C$_4$ but display different patterns of expression, being found in varying amounts in platelets, lung, skeletal muscle, adrenal gland, spleen, small intestine and liver (Jakobsson *et al.* 1996, 1997; Scoggan *et al.* 1997).

The information about the two GST superfamilies in rat, mouse and human tissues has been obtained from combinations of enzyme purification, Western Blotting and multiple-tissue Northern Blots. With the exception of class Alpha, Mu and Pi transferases, relatively little is known about the cell types that express the different soluble GST and MAPEG isoenzymes as few immunohistochemical or *in situ* hybridisation studies have been reported. This type of information will be important in efforts to unravel the functions of the two superfamilies.

HORMONAL REGULATION OF GST

In rodents, significant differences in the hepatic expression of class Alpha, Mu and Pi GST have been seen in male and female animals. The livers of male rats contain more

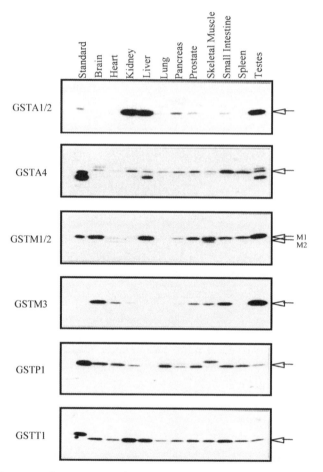

Figure 9.5 Tissue-specific expression of GST isoenzymes in human organs. Extracts from human organs were subjected to immunoblotting as described by Sherratt *et al.* (1997). The identity of the sample loaded in each lane is indicated at the top of the figure. In the left-hand margin, the identity of the anti-serum used to probe the blot is indicated. The horizontal arrows shown in the right-hand margin indicate the mobility of the authentic cross-reacting GST polypeptides.

rGSTA2, rGSTM1 and rGSTM2 than do livers from female rats, a difference that is dependent on pituitary function (Staffas *et al.* 1992). More recently, it has been found that female rats express a higher level of the rGSTA5 subunit in the liver than do male rats, and that male rats down-regulate the enzyme between week 5 and week 10 of life (Hayes *et al.* 1994). Interestingly, the higher level of expression of rGSTA5 in female rats appears to account for the sex-specific difference in sensitivity to aflatoxin B_1 (for a review on aflatoxin B_1 see Hayes *et al.* 1991). The hepatic differences in expression of class Alpha and class Mu transferases in the rat have been attributed to the sexually dimorphic secretion of growth hormone (Staffas *et al.* 1998).

Unlike rats and humans, mice express class Pi GST in hepatocytes. This class of transferase is found in male mouse liver at substantially greater levels than in female mouse liver (McLellan and Hayes 1987; Mitchell *et al.* 1997). Again, the sex differences in content of class Pi GST in mouse liver appear to be due to growth hormone. Importantly, male Lit/Lit mice, that have a specific defect in the production of growth hormone while still being able to synthesise testosterone, express mGSTP1/2 in the liver at levels that are comparable to that found in the livers of female mice (Dolan 1990). This suggests that it is not testosterone but growth hormone that regulates class Pi GST in mouse hepatocytes.

In rodents, sex-specific expression of GST is not restricted to the liver. Marked differences in the transferases can be observed in the rat adrenal gland. Removal of the pituitary causes a 14-fold increase in rGSTM2 in the adrenal gland of the female rat, whereas the same procedure in male rats produces only a 2.7-fold increase in this subunit. Treatment of the hypophysectomised rats with ACTH largely prevents overexpression of rGSTM2 (Mankowitz *et al.* 1990; Staffas *et al.* 1992). In the mouse, major gender-related differences in the GST content of extrahepatic tissues such as kidney and heart have been reported (Mitchell *et al.* 1997). In the case of cardiac tissue, the mGSTA3 subunit is expressed in the heart of female mice but is absent from that of male mice (Mitchell *et al.* 1997).

Evidence exists for hormonal regulation of GST in humans, with transferase levels generally being higher in the female than the male. The level of hGSTA1-1 in liver from females is higher than in males (Mulder *et al.* 1999). Gender differences have also been reported in human colon (Singhal *et al.* 1992) as well as human renal cortex and lung (Temellini *et al.* 1995). However, such sex differences have not been observed by all investigators (Loguercio *et al.* 1996a, 1996b). There is also a literature showing that class Alpha, Mu and Pi transferases are developmentally regulated in human liver, kidney, lung, spleen and adrenal gland (Fryer *et al.* 1986; Faulder *et al.* 1987).

MODULATION OF GST EXPRESSION BY SELENIUM DEFICIENCY

Selenium is an essential trace element and is incorporated covalently into a number of proteins as selenocysteine. In total there are about 20 selenoproteins including glutathione peroxidase, phospholipid hydroperoxide glutathione peroxidase, type I iodothyronine 5′-deiodinase and thioredoxin reductase. Since these enzymes are dependent on selenium for catalysis, Se deficiency results in a dramatic loss of peroxidase and reductase activities. In contrast, examination of transferase levels in the livers of rats showed that 6 weeks of being placed on a Se-deficient diet is sufficient to approximately double the amount of rGSTA1/2, rGSTA3 and rGSTM1 (Arthur *et al.* 1987). A more recent investigation has demonstrated that Se deficiency preferentially induces rGSTA2 in the liver rather than rGSTA1, and that rGSTA5 and rGSTT1 are induced more markedly than other transferase (McLeod *et al.* 1997). Furthermore, it was found that the extent of GST induction caused by Se-deficiency could be diminished by treatment with the antioxidant N-acetylcysteine, suggesting that oxidative stress is at least partially responsible for induction of GST subunits (McLeod *et al.* 1997).

Hepatic GST activity has also been observed to be elevated in mice placed on a

Se-deficient diet (Reiter and Wendel 1985). However, it is currently unknown which murine genes are induced in this situation. It is not known whether the expression of human GST is influenced by absence of Se in the diet.

REGULATION OF RODENT GST BY FOREIGN COMPOUNDS

Transferase activity in various tissues of rodents can be increased by treatment of the animal by drugs. Figure 9.6 shows induction of GST subunits in rat liver by various xenobiotics including phenolic antioxidants, phenobarbital, β-naphthoflavone and indole-3-carbinol. The organs where GST isoenzymes are inducible typically include liver, small intestine, stomach, oesophagus, kidney and lung. In a review of the literature, more than one hundred xenobiotics have been listed as being capable of inducing GST in rats or mice (Hayes and Pulford 1995). This bewildering number of inducing agents can be simplified by recognising that at the genetic level they influence expression of transferases through a limited number of *cis*-acting elements, usually found in the 5'-flanking region of *GST* genes. At least five enhancers that respond to foreign compounds have been identified that are relevant to induction of GST. Specifically, these are the ARE, GPEI, XRE, PBREM and GRE enhancers, and are described in more detail below.

Metabolisable antioxidants, metabolisable polycyclic aromatic hydrocarbons, di-phenols, quinones, isothiocyanates, dithiolethiones and Michael reaction acceptors induce rodent class Alpha, Mu, Pi, Sigma and Theta transferases. Inducible expression of rat *GSTA2* and mouse *Gsta1* occurs through the NF-E2-related factor 2 (Nrf2) transcription factor and the antioxidant responsive element (ARE, $5'-^A/_GTGAC/_T$ $NNNGC^A/_G-3'$) (Rushmore *et al.* 1991; Prestera *et al.* 1993; Itoh *et al.* 1997; Wasserman and Fahl 1997; Hayes *et al.* 2000). This element is also probably responsible for the induction of rGSTA5 by cancer chemopreventive agents (Pulford and Hayes 1996).

The glutathione transferase P enhancer I (GPEI), a palindromic element in which each half comprises an AP-1 site with a single base pair mis-match (Sakai *et al.* 1988; Okuda *et al.* 1989, 1990), has been shown to respond to *tert*-butylhydroquinone *in vitro* (Favreau and Pickett 1995). Despite its similarity to an AP-1 binding site, GPEI-driven gene expression does not appear to be mediated by c-Jun or c-Fos (Morimura *et al.* 1992). It has been speculated that GPEI is involved in induction of the rGSTP1 subunit by coumarin, phenolic antioxidants and *trans*-stilbene oxide in rat liver (Sherratt *et al.* 1998; Kelly *et al.* 2000). Using transgenic rats, Suzuki *et al.* (1996) have shown that GPEI is responsible for the induction of rGSTP1 in the liver by lead. It is also responsible for overexpression of rGSTP1 in hepatic preneoplastic nodules produced by chemical carcinogenesis (Suzuki *et al.* 1995).

Planar aromatic compounds and indoles transcriptionally activate *rGSTA2* through the aryl hydrocarbon (Ah) receptor and the single xenobiotic responsive element (XRE) in the gene promoter (Rushmore and Pickett 1990). Figure 9.6 shows that indole-3-carbinol does not induce rGSTP1 in rat liver.

Glucocorticoids, such as dexamethasome, can either induce or repress GST expression in rodent liver (Dolan 1990; Waxman *et al.* 1992; Prough *et al.* 1996). The variable effect of dexamethasone on GST expression depends on the age of the experimental animal, with hepatic GST in younger animals being inducible by

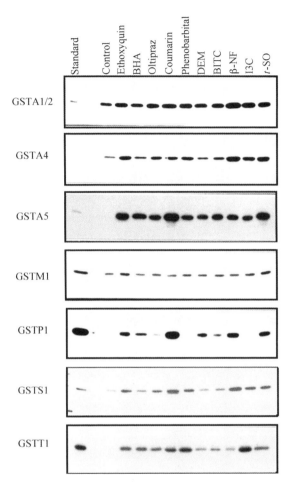

Figure 9.6 Regulation of GST isoenzymes in rat liver. Twelve-week-old male Fischer rats were placed on either a control diet, or a diet containing a xenobiotic. The treatments have been described by Sherratt *et al.* (1998). Briefly, they were as follows: 0.5% ethoxyquin for 14 days; 0.75% butylated hydroxyanisole (BHA) in diet for 14 days; 0.075% oltipraz in diet for 14 days; 0.5% coumarin in diet for 14 days; 0.1% phenobarbital in drinking water for 7 days; 0.5% diethylmaleate (DEM) in diet for 14 days; 0.5% benzylisothiocyanate (BITC) in diet for 14 days; 200 mg/kg β-naphthoflavone (β-NF) intraperitoneal injection daily for 7 days; 0.5% indole-3-carbinol (I3C) in diet for 14 days; 400 mg/kg *trans*-stilbene oxide (*t*-SO) intraperitoneal injection daily for 3 days. Samples of liver cytosol were prepared from these animals and portions (4 μg protein) were subjected to Western Blotting. The xenobiotic treatments are shown at the top of the figure. The antibodies against rat class Alpha, Mu, Pi, Sigma and Theta that were used to probe the blots are shown in the left-hand side of the figure.

glucocorticoids. In the adult rat, dexamethasone is able to inhibit induction of GSTA2 by the planar aromatic hydrocarbon benzanthracene (Falkner *et al.* 1998). Presumably regulation of GST by dexamethasone is mediated by the glucocorticoid and pregnane X receptors, and occurs through the glucocorticoid responsive element (GRE). Examination of the 5′-flanking region of the rat *GSTA2* gene has revealed the presence of a palindromic GRE and several GRE half sites (Falkner *et al.* 1998). Furthermore, the negative effect of dexamethasone on the expression of *GSTA2* in the rat involves the palindromic GRE.

Phenobarbital and 1,4-bis[2-(3,5-dichloropyridyloxy)]benzene induce a number of GST isoenzymes in rat and mouse liver (Hayes *et al.* 1979; Di Simplico *et al.* 1989). Phenobarbital induces rGSTA1/2, rGSTA4, rGSTA5, rGSTS1 and rGSTT1 but not rGSTP1 in rat liver (Figure 9.6). In the case of induction of cytochrome P450 genes, induction by these compounds occurs through the nuclear orphan receptors CAR (constitutive active receptor) and RXR (retinoid X receptor) and the phenobarbital responsive enhancer module (PBREM) (Honkakoski *et al.* 1998). It is probable that phenobarbital induction of GST is mediated by a similar mechanism.

REGULATION OF HUMAN GST BY FOREIGN COMPOUNDS

In the human, the regulation of GST by xenobiotics has been studied in colon and by cell culture techniques. Quantification of the transferases in the rectum of human volunteers who were placed on a diet containing 300 g daily of Brussels sprouts revealed a modest 15% increase in class Alpha GST and 30% increase in class Pi GST (Nijhoff *et al.* 1995). Using primary human hepatocytes treated with various xenobiotics, Northern Blotting has shown that dithiolethiones, 3-methylcholanthrene and phenobarbital can increase steady-state mRNA levels for hGSTA1/2 (Morel *et al.* 1993). Treatment of human primary hepatocytes with the isothiocyanate sulphoraphane has also been shown to increase the level of mRNA for hGSTA1/2 (Mahéo *et al.* 1997). The notion that GSTs are inducible in the liver is supported by the work of Dierickx (1994) who demonstrated increased levels of hGSTA1 in HepG2 cells treated with picolines. Similarly, in HT29 human colon carcinoma cells, GST activity towards CDNB can be increased by treatment with allyl sulphide and benzyl isothiocyanate (Kirlin *et al.* 1999) and in MCF7 breast cancer cells following treatment with catechol (Sreerama *et al.* 1995) or 2,5-bis(2-hydroxybenzylidene)cyclopentanone (Dinkova-Kostova *et al.* 2001). Also, GST activity is elevated in HT29 cells following treatment with aspirin (Patten and DeLong 1999). The above data indicate that human GST genes may be regulated by foreign compounds in the liver and gastrointestinal tract. However, it remains to be demonstrated whether enhancers such as the ARE, GPEI, XRE, PBREM or GRE are involved.

REGULATION OF GST BY ENDOBIOTICS

In rat liver RL34 cells, the rGSTP1 subunit is inducible by the cyclopentenone prostaglandin 15-deoxy-$\Delta^{12,14}$-PGJ$_2$ (Kawamoto *et al.* 2000). This prostanoid is an α,β-unsaturated carbonyl and its induction of rGSTP1 appears to occur through GPEI and may involve c-Jun. The cytotoxic product of lipid peroxidation, 4-hydroxynonenal, is also an α,β-unsaturated carbonyl and it too is an excellent inducer of rGSTP1 in RL34 cells (Fukuda *et al.* 1997).

The fact that GPEI and the ARE share sequence identity (see Hayes *et al.* 1999) raises the important question of whether either 15-deoxy-$\Delta^{12,14}$-PGJ$_2$ or 4-hydroxynonenal are endogenous compounds that stimulate ARE-driven transcription as well as GPEI-driven transcription. Certainly, Prestera *et al.* (1993) have shown that 1-cyclopenten-2-one is a reasonably good inducer of ARE-driven reporter gene expression in HepG2 cells.

Future research directions

Over the past ten years, improvements in protein purification, molecular cloning and bioinformatics have lead to our current understanding of the diversity of the soluble GST and MAPEG superfamilies. However, much remains to be learnt about the *in vivo* functions of the transferases. To date, only one mouse with targeted disruption of a GST gene(s) has been reported. This animal lacks *Gstp1* and *Gstp2*, and though phenotypically normal, it is more sensitive to skin carcinogenesis when treated with 7,12-dimethylbenzanthracene and 12-*O*-tetradecanoylphorbol-13-acetate (Henderson *et al.* 1998). Interestingly, the mutant mouse has proved to be more resistant than the wild-type mouse to acetaminophen (paracetamol) poisoning, presumably because of failure to deplete hepatic GSH pools when it is treated with high doses of the drug (Henderson *et al.* 2000). It is anticipated that the generation of additional gene knockout murine lines will help clarify the biological functions of both GST superfamilies. Clearly, substantially more work is required in this area in order to elucidate the contribution that individual transferases make to the metabolism and disposition of drugs.

A significant number of soluble GST have been crystallised and their three-dimensional structures determined (Reinemer *et al.* 1991; Mannervik and Widersten 1995; Armstrong 1997). With a better knowledge of the mechanisms responsible for catalysis and the active site residues involved in substrate binding, it has become possible to engineer these proteins to exhibit novel properties and unique specificities. For example, Gulick and Fahl (1995) employed forced evolution to select a rat class Alpha GST with increased activity for the nitrogen mustard mechlorethamine. A phage display approach has been used to obtain a mutant of a human class Alpha GST with increased activity towards ethacrynic acid (Hansson *et al.* 1997). More recently, Mannervik and his colleagues have been able to confer the ability to metabolise 4-hydroxynonenal on an enzyme with little activity for this α,β-unsaturated carbonyl by mutating residues associated with the β-$\alpha1$ loop, the $\alpha4$ helix and the $\alpha9$ helix (Nilsson *et al.* 2000). Clearly, this approach has numerous applications in medicine and biotechnology.

Concluding comments

This chapter provides an overview of the functions, genetics and regulation of the two GST superfamilies. Recent studies have begun to address the endogenous functions of GST within the cell, such as their contribution to leukotriene and prostaglandin biosynthesis. Emerging evidence suggests that at least two soluble GST subunits may modulate signal transduction pathways and another influences ryanodine receptor function. This represents a departure from the classical view that GSTs are solely enzymes involved in phase II of drug-metabolism, and indicates they are intimately involved in cellular stress responses. The polymorphic expression of GST represents a significant risk factor in the development of malignant disease, in particular breast cancer, and influences responsiveness to chemotherapy. Over the past 10 years enormous advances have been made in understanding how xenobiotics regulate gene expression. Data are accumulating that induction of GST genes is mediated by Nrf2, c-Jun, the Ah receptor, CAR and possibly the glucocorticoid receptor. Although GSTs have been the subject of scientific research for about forty years, and great advances have been made in our knowledge of their structure, it is clear that much remains to be learnt about the biological functions of these enzymes. Establishing the role of GST in the response of the host to drugs remains a challenge for pharmacologists and toxicologists alike.

Acknowledgements

We are indebted to Professor Philip G. Board and Dr Jack Flanagan for their helpful advice. We also thank the Biotechnology and Biological Sciences Research Council (94/F08200) for financial support.

References

Abramovitz M and Listowsky I (1987) Selective expression of a unique glutathione S-transferase Yb$_3$ gene in rat brain. *The Journal of Biological Chemistry*, **262**, 7770–7773.

Adler V, Yin Z, Fuchs SY, Benezra M, Rosario L, Tew KD, Pincus MR, Sardana M, Henderson CJ, Wolf CR, Davis RJ and Ronai Z (1999) Regulation of JNK signaling by GSTp. *The EMBO Journal*, **18**, 1321–1334.

Anders MW, Lash L, Dekant W, Elfarra AA and Dohn DR (1988) Biosynthesis and biotransformation of glutathione S-conjugates to toxic metabolites. *CRC Critical Reviews in Toxicology*, **18**, 311–341.

Armstrong RN (1997) Structure, catalytic mechanism, and evolution of the glutathione transferases. *Chemical Research in Toxicology*, **10**, 2–18.

Arthur JR, Morrice PC, Nicol F, Beddows SE, Boyd R, Hayes JD and Beckett GJ (1987) The effects of selenium and copper deficiencies on glutathione S-transferase and glutathione peroxidase in rat liver. *Biochemical Journal*, **248**, 539–544.

Baez S, Segura-Aguilar J, Widersten M, Johansson A-S and Mannervik B (1997) Glutathione transferases catalyse the detoxication of oxidized metabolites (*o*-quinones) of catecholamines and may serve as an antioxidant system preventing degenerative cellular processes. *Biochemical Journal*, **324**, 25–28.

Baillie TA and Kassahun K (1994) Reversibility in glutathione-conjugate formation. *Advances in Pharmacology*, **27**, 163–181.

Baillie TA and Kassahun K (2000) Biological reactive intermediates in drug discovery and development. A perspective from the Pharmaceutical Industry. Biological Reactive Intermediates, Paris.

Baillie TA and Slatter JG (1991) Glutathione: a vehicle for the transport of chemically reactive metabolites in vivo. *Accounts of Chemical Research*, **24**, 264–270.

Bao Y, Jemth P, Mannervik B and Williamson G (1997) Reduction of thymine hydroperoxide by phospholipid hydroperoxide, glutathione peroxidase and glutathione transferases. *FEBS Letters*, **410**, 210–212.

Benson AM, Talalay P, Keen JH and Jakoby WB (1977) Relationship between the soluble glutathione-dependent Δ^5-3-ketosteroid isomerase and the glutathione S-transferases of the liver. *Proceedings of the National Academy of Sciences, USA*, **74**, 158–162.

Board PG, Baker RT, Chelvanayagam G and Jermiin LS (1997) Zeta, a novel class of glutathione transferases in a range of species from plants to humans. *Biochemical Journal*, **328**, 929–935.

Board PG, Coggan M, Chelvanayagam G, Easteal S, Jermiin LS, Schulte GK, Danley DE, Hoth LR, Griffor MC, Kamath AV, Rosner MH, Chrunyk BA, Perregaux DE, Gabel CA, Geoghegan KF and Pandit J (2000) Identification, characterization and crystal structure of the Omega class glutathione transferases. *The Journal of Biological Chemistry*, **275**, 24798–24806.

Bogaards JJP, Venekamp JC and van Bladeren PJ (1997) Stereoselective conjugation of prostaglandin A2 and prostaglandin J2 with glutathione, catalysed by the human glutathione S-transferases A1-1, A2-2, M1a-1a, and P1-1. *Chemical Research in Toxicology*, **10**, 310–317.

Booth J, Boyland E and Sims P (1961) An enzyme from rat liver catalysing conjugations with glutathione. *Biochemical Journal*, **79**, 516–524.

Boyland E and Chasseaud LF (1969) The role of glutathione and glutathione S-transferases in mercapturic acid biosynthesis. *Advances in Enzymology*, **32**, 173–219.

Bruggeman IM, Temmink JHM and van Bladeren PJ (1986) Glutathione- and cysteine-mediated cytotoxicity of allyl and benzyl isothiocyanate. *Toxicology and Applied Pharmacology*, **83**, 349–359.

Chamberlain MP, Lock EA, Gaskell BA and Reed CJ (1998) The role of glutathione *S*-transferase- and cytochrome P450-dependent metabolism in the olfactory toxicity of methyl iodide in the rat. *Archives of Toxicology*, **72**, 420–428.

Chasseaud LF (1979) The role of glutathione and glutathione S-transferases in the metabolism of chemical carcinogens and other electrophilic agents. *Advances in Cancer Research*, **29**, 175–274.

Chen H and Juchau MR (1997) Glutathione *S*-transferases act as isomerases in isomerization of 13-*cis*-retinoic acid to all-*trans*-retinoic acid *in vitro*. *Biochemical Journal*, **327**, 721–726.

Cho S-G, Lee YH, Park H-S, Ryoo K, Kang KW, Park J, Eom S-J, Kim MJ, Chang T-S, Choi S-Y, Shim J, Kim Y, Dong M-S, Lee M-J, Kim SG, Ichijo H and Choi E-J (2001) Glutathione S-transferase Mu modulates the stress-activated signals by suppressing apoptosis signal-regulating kinase 1. *The Journal of Biological Chemistry*, **276**, 12749–12755.

Coggan M, Whitbread L, Whittington A and Board P (1998) Structure and organization of the human Theta-class glutathione S-transferase and D-dopachrome tautomerase gene complex. *Biochemical Journal*, **334**, 617–623.

Coles B and Ketterer B (1990) The role of glutathione and glutathione transferases in chemical carcinogenesis. *Critical Reviews in Biochemistry and Molecular Biology*, **25**, 47–70.

Coles BF, Anderson KE, Doerge DR, Churchwell MI, Lang NP and Kadlubar FF (2000) Quantitative analysis of interindividual variation of glutathione S-transferase expression in human pancreas and the ambiguity of correlating genotype with phenotype. *Cancer Research*, **60**, 573–579.

Coombes B and Stakelum GS (1961) A liver enzyme that conjugates sulfobromophthalein with glutathione. *Journal of Clinical Investigation*, **40**, 981–988.

DeJong JL, Morgenstern R, Jörnvall H, De Pierre JW and Tu C-PD (1988) Gene expression of rat and human microsomal glutathione S-transferases. *The Journal of Biological Chemistry*, **263**, 8430–8436.

Dekant W, Vamvakas S and Anders MW (1994) Formation and fate of nephrotoxic and cytotoxic glutathione S-conjugates: cysteine conjugate β-lyase pathway. *Advances in Pharmacology*, **27**, 115–162.

Dierickx PJ (1994) The influence of picolines on glutathione transferase activity and subunit composition in human liver derived Hep G2 cells. *Biochemical Pharmacology*, **48**, 1976–1978.

Dinkova-Kostova AI, Massiah MA, Bozak RE, Hicks RJ and Talalay P (2001) Potency of Michael reaction acceptors as inducers of enzymes that protect against carcinogenesis depends on their reactivity with sulfhydryl groups. Proceedings of the National Academy of Sciences, USA, **98**, 3404–3409.

Dixon RFF, Diehl RE, Opas E, Rands E, Vickers PJ, Evans JF, Gillard JW and Miller DK (1990) Requirement of a 5-lipoxygenase-activating protein for leukotriene synthesis. *Nature (London)*, **343**, 282–284.

Di Simplicio P, Jensson H and Mannervik B (1989) Effects of inducers of drug metabolism on basic hepatic forms of mouse glutathione transferase. *Biochemical Journal*, **263**, 679–685.

Dolan C (1990) Regulation of mouse hepatic glutathione S-transferases. PhD thesis, University of Edinburgh.

Dulhunty A, Gage P, Curtis S, Chelvanayagam G and Board P (2000) The glutathione transferase structural family includes a nuclear chloride channel and a ryanodine receptor calcium release channel modulator. *The Journal of Biological Chemistry*, **275**.

Estonius M, Forsberg L, Danielsson O, Weinander R, Kelner MJ and Morgenstern R (1999) Distribution of microsomal glutathione transferase 1 in mammalian tissue. A predominant alternate first exon in human tissues. *European Journal of Biochemistry*, **260**, 409–413.

Falkner KC, Rushmore TH, Linder MW and Prough RA (1998) Negative regulation of the rat glutathione S-transferase A2 gene by glucocorticoids involves a canonical glucocorticoid consensus sequence. *Molecular Pharmacology*, **53**, 1016–1026.

Faulder CG, Hirrell PA, Hume R and Strange RC (1987) Studies on the development of basic, neutral and acidic isoenzymes of glutathione S-transferase in human liver, adrenal, kidney and spleen. *Biochemical Journal*, **241**, 221–228.

Favreau LV and Pickett CB (1995) The rat quinone reductase antioxidant response element. Identification of the nucleotide sequence required for basal and inducible activity and detection of antioxidant response element-binding proteins in hepatoma and non-hepatoma cell lines. *The Journal of Biological Chemistry*, **270**, 24468–24474.

Fernández-Cañón JM and Peñalva MA (1998) Characterization of a fungal maleylacetoacetate isomerase gene and identification of its human homologue. *The Journal of Biological Chemistry*, **273**, 329–337.

Finkel T and Holbrook NJ (2000) Oxidants, oxidative stress and the biology of aging. *Nature (London)*, **408**, 239–247.

Fjellstedt TA, Allen RH, Duncan BK and Jakoby WB (1973) Enzymatic conjugation of epoxides with glutathione. *The Journal of Biological Chemistry*, **248**, 3702–3707.

Fryer AA, Hume R and Strange RC (1986) The development of glutathione S-transferase and glutathione peroxidase activities in human lung. *Biochimica et Biophysica Acta*, **883**, 448–453.

Fukuda A, Nakamura Y, Ohigashi H, Osawa T and Uchida K (1997) Cellular response to the redox active lipid peroxidastion products: induction of glutathione S-transferase P by 4-hydroxy-2-nonenal. *Biochemical and Biophysical Research Communications*, **236**, 505–509.

Gillham B (1973) The mechanism of the reaction between glutathione and 1-menaphthyl sulphate catalysed by a glutathione S-transferase from rat liver. *Biochemical Journal*, **135**, 797–804.

Green T (1997) Methylene chloride induced mouse liver and lung tumours: an overview of the role of mechanistic studies in human safety assessment. *Human and Experimental Toxicology*, **16**, 3–13.

Guengerich FP, Thier R, Persmark M, Taylor JB, Pemble SE and Ketterer B (1995) Conjugation of carcinogens by θ class glutathione S-transferases: mechanisms and relevance to variations in human risk. *Pharmacogenetics*, **5**, S103–S107.

Gulick AM and Fahl WE (1995) Forced evolution of glutathione S-transferase to create a more efficient drug detoxication enzyme. *Proceedings of the National Academy of Sciences, USA*, **92**, 8140–8144.

Habig WB, Pabst MJ and Jakoby WB (1974) Glutathione S-transferases. The first enzymatic step in mercapturic acid formation. *The Journal of Biological Chemistry*, **249**, 7130–7139.

Habig WH, Pabst MJ and Jakoby WB (1976) Glutathione S-transferase AA from rat liver. *Archives of Biochemistry and Biophysics*, **175**, 710–716.

Hansson LO, Widersten and Mannervik B (1997) Mechanism-based phage display selection of active-site mutants of human glutathione transferase A1-1 catalysing S_NAr reactions. *Biochemistry*, **36**, 11252–11260.

Harris JM, Meyer DJ, Coles B and Ketterer B (1991) A novel glutathione transferase (13-13) isolated from the matrix of rat liver mitochondria having structural similarity to class Theta enzymes. *Biochemical Journal*, **278**, 137–141.

Hayes JD and Mantle TJ (1986) Use of immuno-blot techniques to discriminate between the glutathione S-transferase Yf, Yk, Ya, Yn/Yb and Yc subunits and to study their distribution in extra-hepatic tissues. Evidence for three immunochemically-distinct groups of mammalian glutathione S-transferase isoenzymes. *Biochemical Journal*, **233**, 779–788.

Hayes JD and McLellan LI (1999) Glutathione and glutathione-dependent enzymes represent a co-ordinately regulated defence against oxidative stress. *Free Radical Research*, **31**, 273–300.

Hayes JD and Pulford DJ (1995) The glutathione S-transferase supergene family: regulation of GST and the contribution of the isoenzymes to cancer chemoprotection and drug resistance. *Critical Reviews in Biochemistry and Molecular Biology*, **30**, 445–600.

Hayes JD and Strange RC (2000) Glutathione S-transferase polymorphisms and their biological consequences. *Pharmacology*, **61**, 154–166.

Hayes JD, Strange RC and Percy-Robb IW (1979) Identification of two lithocholic acid-binding proteins; separation of ligandin from glutathione S-transferase B. *Biochemical Journal*, **181**, 699–708.

Hayes JD, Judah DJ, McLellan LI and Neal GE (1991) Contributions of the glutathione S-transferases to the mechanisms of resistance to aflatoxin B_1. *Pharmacology and Therapeutics*, **50**, 443–472.

Hayes JD, Nguyen T, Judah DJ, Petersson DG and Neal GE (1994) Cloning of cDNAs from fetal rat liver encoding glutathione S-transferase Yc polypeptides: the Yc_2 subunit is expressed in adult rat liver resistant to the hepatocarcinogen aflatoxic B_1. *The Journal of Biological Chemistry*, **269**, 20707–20717.

Hayes JD, Ellis EM, Neal GE, Harrison DJ and Manson MM (1999) Cellular response to cancer chemopreventive agents: contribution of the antioxidant responsive element to the adaptive response to oxidative and chemical stress. In *Cellular Responses to Stress*, Downes CP, Wolf CR and Lane DP (eds), Biochemical Society Symposium, Volume 64, Portland Press, London, pp. 141–168.

Hayes JD, Chanas SA, Henderson CJ, McMahon M, Sun C, Moffat GJ, Wolf CR and Yamamoto M (2000) The Nrf2 transcription factor contributes both to the basal expression of glutathione S-transferases in mouse liver and to their induction by the chemopreventive synthetic antioxidants, butylated hydroxyanisole and ethoxyquin. *Biochemical Society Transactions*, **28**, 33–41.

Hebert H, Schmidt-Krey I, Morgenstern R, Murata K, Hirai T, Mitsuoka K and Fujiyoshi Y (1997) The 3.0 Å projection structure of microsomal glutathione transferase as determined by electron crystallography of $p2_12_12$ two-dimensional crystals. *Journal of Molecular Biology*, **271**, 751–758.

Henderson CJ, Smith AG, Ure J, Brown K, Bacon EJ and Wolf CR (1998) Increased skin tumorigenesis in mice lacking pi class glutathione S-transferases. *Proceedings of the National Academy of Sciences, USA*, **95**, 5275–5280.

Henderson CJ, Wolf CR, Kitteringham N, Powell H, Otto D and Park BK (2000) Increased resistance to acetaminophen hepatotoxicity in mice lacking glutathione S-transferase Pi. *Proceedings of the National Academy of Sciences, USA*, **97**, 12741–12745.

Hirvonen A, Bouchardy C, Mitrunen K, Kataja V, Eskelinen, Kosma V-M, Saarikoski ST, Jourenkova N, Anttila S, Dayer P, Uusitupa M and Benhamou S (2001) Polymorphic GSTs and cancer predisposition. *Chemico-Biological Interactions*, **133**, 75–80.

Hiratsuka A, Sebata N, Kawashima K, Okuda H, Ogura K, Watabe T, Satoh K, Hatayama I, Tsuchida S, Ishikawa T and Sato K (1990) A new class of rat glutathione S-transferase Yrs-Yrs inactivating reactive sulfate esters as metabolites of carcinogenic arylmethanols. *The Journal of Biological Chemistry*, **265**, 11973–11981.

Honkakoski P, Moore R, Washburn KA and Negishi M (1998) Activation by diverse xenochemicals of the 51-base pair phenobarbital-responsive enhancer module in the *CYP2B10* gene. *Molecular Pharmacology*, **53**, 597–601.

Inskip A, Elexperu-Camiruaga J, Buxton N, Dias PS, MacIntosh J, Campbell D, Jones PW, Yengi L, Talbot JA, Strange RC and Fryer AA (1995) Identification of polymorphism at the glutathione S-transferase GSTM3 locus: evidence for linkage with GSTM1*A. *Biochemical Journal*, **312**, 713–716.

Ito K, Chiba T, Takahashi S, Ishii T, Igarashi K, Katoh Y, Oyake T, Hayashi N, Satoh K, Hatayama I, Yamamoto M and Nebeshima Y-I (1997) An Nrf2/small Maf heterodimer mediates the induction of phase II detoxifying genes through antioxidant response elements. *Biochemical Biophysical Research Communications*, **236**, 313–322.

Jakobsson P-J, Mancini JA and Ford-Hutchinson AW (1996) Identification and characterization of a novel human microsomal glutathione S-transferase with leukotriene C_4 synthase activity and significant sequence identity to 5-lipoxygenase-activating protein and leukotriene C_4 synthase. *The Journal of Biological Chemistry*, **271**, 22203–22210.

Jakobsson P-J, Mancini JA, Riendeau D and Ford-Hutchinson AW (1997) Identification and characterization of a novel microsomal enzyme with glutathione-dependent transferase and peroxidase activities. *The Journal of Biological Chemistry*, **272**, 22934–22939.

Jakobsson P-J, Morgenstern R, Mancini J, Ford-Hutchinson A and Persson B (1999a) Common structural features of MAPEG- a widespread superfamily of membrane associated proteins with highly divergent functions in eicosanoid and glutathione metabolism. *Protein Science*, **8**, 689–692.

Jakobsson P-J, Thorén S, Morgenstern R and Samuelsson B (1999b) Identification of human prostaglandin E synthase: A microsomal, glutathione-dependent, inducible enzyme, constituting a potential novel drug target. *Proceedings of the National Academy of Sciences, USA*, **96**, 7220–7225.

Jakobsson P-J, Morgenstern R, Mancini J, Ford-Hutchinson A and Persson B (2000) Membrane-associated proteins in eicosanoid and glutathione metabolism (MAPEG). A widespread superfamily. *American Journal of Respiratory and Critical Care Medicine*, **161**, S20–S24.

Jakoby WB and Habig WH (1980) Glutathione Transferases. In *Enzymatic Basis of Detoxication*, Vol II, Jakoby WB (ed.), Academic Press, New York, pp. 63–94.

Kassahun K, Pearson PG, Tang W, McIntosh L, Leung K, Elmore C, Dean D, Wang R, Doss G and Baillie TA (2001) Studies on the metabolism of triglitazone to reactive intermediates *in vitro* and *in vivo*. Evidence for novel biotransformation pathways involving quinone methide formation and thiazolidinedione ring scission. *Chemical Research in Toxicology*, **14**, 62–70.

Kawamoto Y, Nakamura Y, Naito Y, Torii Y, Kumagai T, Osawa T, Ohigashi H, Satoh K, Imagawa M and Uchida K (2000) Cyclopentenone prostaglandins as potential inducers of phase II detoxification enzymes. 15-Deoxy-$\Delta^{12,14}$-prostaglandin J2-induced expression of glutathione S-transferases. *The Journal of Biological Chemistry*, **275**, 11291–11299.

Keen JH and Jakoby WB (1978) Glutathione transferases. Catalysis of nucleophilic reactions of glutathione. *The Journal of Biological Chemistry*, **253**, 5654–5657.

Kelly VP, Ellis EM, Manson MM, Chanas SA, Moffat GJ, McLeod R, Judah DJ, Neal GE and Hayes JD (2000) Chemoprevention of aflatoxin B_1 hepatocarcinogenesis by coumarin, a natural benzopyrone that is a potent inducer of AFB_1-aldehyde reductase, the glutathione S-transferase A5 and P1 subunits, and NAD(P)H:quinone oxidoreductase in rat liver. *Cancer Research*, **60**, 957–969.

Ketterer B, Ross-Mansell P and Whitehead JK (1967) The isolation of carcinogen-binding protein from livers of rats given 4-dimethylaminoazobenzene. *Biochemical Journal*, **103**, 316–324.

Ketterer B, Meyer DJ, Taylor JB, Pemble S, Coles B and Fraser G (1990) GSTs and protection against oxidative stress. In *Glutathione S-Transferases and Drug Resistance*, Hayes JD, Mantle TJ and Pickett CB (eds), Taylor and Francis, Bristol, pp. 97–109.

Kirlin WG, Cai J, DeLong MJ, Patten EJ and Jones DP (1999) Dietary compounds that induce cancer preventive phase 2 enzymes activate apoptosis at comparable doses in HT29 colon carcinoma cells. *Journal of Nutrition*, **129**, 1827–1835.

Kitahara A, Satoh K, Nishimura K, Ishikawa T, Ruike K, Sato K, Tsuda H and Ito N (1984) Changes in molecular forms of rat hepatic glutathione S-transferase during chemical hepatocarcinogenesis. *Cancer Research*, **44**, 2698–2703.

Klaassen CD (1996) *Casarett and Doull's Toxicology. The Basic Science of Poisons*, 5th edition, McGraw-Hill, New York.

Kohlroser J, Mathai J, Reichheld J, Banner BF and Bonkovsky HL (2000) Hepatotoxicity due to troglitazone: report of two cases and review of adverse events reported to the United States Food and Drug Administration. *American Journal of Gastroenterology*, **95**, 272–276.

Kolm RH, Danielson UH, Zhang Y, Talalay P and Mannervik B (1995) Isothiocyanates as substrates for human glutathione transferases: structure-activity studies. *Biochemical Journal*, **311**, 453–459.

Lam BK, Penrose JF, Freeman GJ and Austen KF (1994) Expression cloning of a cDNA for human leukotriene C_4 synthase, an integral membrane protein conjugating reduced glutathione to leukotriene A_4, *Proceedings of the National Academy of Sciences USA*, **91**, 7663–7667.

Lee SH and DeJong J (1999) Microsomal GST-I: genomic organization, expression, and alternative splicing of the human gene. *Biochimica et Biophysica Acta*, **1446**, 389–396.

Li N, Reddanna P, Thyagaraju K, Reddy C and Tu C-PD (1986) Expression of glutathione S-transferases in rat brain. *The Journal of Biological Chemistry*, **261**, 7596–7599.

Listowsky I (1993) High capacity binding by glutathione S-transferases and glucocorticoid resistance. In *Structure and Function of Glutathione Transferases*, Tew KD, Pickett CB, Mantle TJ, Mannervik B and Hayes JD (eds), CRC Press, Boca Raton, FL, pp. 199–209.

Litwack G, Ketterer B and Arias IM (1971) Ligandin: a hepatic protein which binds steroids, bilirubin, carcinogens, and a number of exogenous anions. *Nature (London)*, **234**, 466–467.

Loguercio C, Taranto D, Vitale LM, Beneduce F and Blanco CD (1996a) Effect of liver cirrhosis and age on the glutathione concentration in the plasma, erythrocytes and gastric mucosa of man. *Free Radical Biology and Medicine*, **20**, 483–488.

Loguercio C, Taranto D, Beneduce F, Vitale LM and DelleCave M (1996b) Age affects glutathione content and glutathione transferase activity in human gastric mucosa. *Italian Journal of Gastroenterology and Hepatology*, **28**, 477–481.

Mahéo K, Morel F, Langouët S, Kramer H, Le Ferrec E, Ketterer B and Guillouzo A (1997) Inhibition of cytochromes P-450 and induction of glutathione S-transferases by sulforaphane in primary human and rat hepatocytes. *Cancer Research*, **57**, 3649–3652.

Mainwaring GW, Williams SM, Foster JR, Tugwood J and Green T (1996) The distribution of Theta-class glutathione S-transferases in the liver and lung of mouse, rat and human. *Biochemical Journal*, **318**, 297–303.

Mancini JA, Abramovitz M, Cox ME, Wong E, Charleson S, Perrier H, Wang Z, Prasit P and Vickers PJ (1993) 5-lipoxygenase-activating protein is an arachidonate binding protein. *FEBS Letters*, **318**, 277–281.

Mankowitz L, Castro VM, Mannervik B, Rydström J and DePierre JW (1990) Increase in the amount of glutathione transferase 4-4 in the rat adrenal gland after hypophysectomy and down-regulation by subsequent treatment with adrenocorticotrophic hormone. *Biochemical Journal*, **265**, 147–154.

Mannervik B (1985) The isoenzymes of glutathione transferase. *Advances in Enzymology and Related Areas of Molecular Biology*, **57**, 357–417.

Mannervik B (1986) Glutathione and the evolution of enzymes for detoxication of products of oxygen metabolism. *Chemica. Scripta*, **263**, 281–284.

Mannervik B and Jensson H (1982) Binary combinations of four protein subunits with different catalytic specificities explain the relationship between six basic glutathione S-transferases in rat liver cytosol. *The Journal of Biological Chemistry*, **257**, 9909–9912.

Mannervik B and Wildersten M (1995) Human glutathione transferases: classification, tissue distribution, and functional properties. In *Advances in Drug Metabolism in Man*, Pacifici GM and Fracchia GN (eds), European Commission, pp. 407–459.

Mannervik B, Ålin P, Guthenberg C, Jensson H, Tahir MK, Warholm M and Jörnvall H (1985) Identification of three classes of cytosolic glutathione transferase common to several mammalian species: correlation between structural data and enzymatic properties. *Proceedings of the National Academy of Sciences USA*, **82**, 7202–7206.

Marnett LJ (1994) Generation of mutagens during arachidonic acid metabolism. *Cancer and Metastasis Reviews*, **13**, 303–308.

McLellan LI and Hayes JD (1987) Sex-specific constitutive expression of the pre-neoplastic marker glutathione S-transferase, YfYf, in mouse liver. *Biochemical Journal*, **245**, 399–406.

McLellan LI, Wolf CR and Hayes JD (1988) Human microsomal glutathione S-transferase: its involvement in the conjugation of hexachloro-1,3-butadiene. *Biochemical Journal*, **258**, 87–93.

McLellan LI, Harrison DJ and Hayes JD (1992) Modulation of glutathione S-transferase and glutathione peroxidase by the anticarcinogen butylated hydroxyanisole in murine extrahepatic organs. *Carcinogenesis*, **13**, 2255–2261.

McLeod R, Ellis EM, Arthur JR, Neal GE, Judah DJ, Manson MM and Hayes JD (1997) Protection conferred by selenium deficiency against aflatoxin B$_1$ in the rat is associated with the hepatic expression of an aldo-keto reductase and a glutathione S-transferase subunit that metabolize the mycotoxin. *Cancer Research*, **57**, 4257–4266.

Meyer DJ (1993) Significance of an unusually low K$_m$ for glutathione in glutathione transferases of the α, μ and π classes. *Xenobiotica*, **23**, 823–834.

Meyer DJ and Thomas M (1995) Characterization of rat spleen prostaglandin-H D-isomerase as a sigma class GSH transferase. *Biochemical Journal*, **311**, 739–742.

Meyer DJ, Coles B, Pemble SE, Gilmore KS, Fraser GM and Ketterer B (1991) Theta, a new class of glutathione transferases purified from rat and man. *Biochemical Journal*, **274**, 409–414.

Miller DK, Gillard JW, Vickers PJ, Sadowski S, Léveillé C, Mancini JA, Charleson P, Dixon RAF,

Ford-Hutchinson AW, Fortin R, Gauthier JY, Rodkey J, Rosen R, Rouzer C, Sigal IS, Strader CD and Evans JF (1990) Identification and isolation of a membrane protein necessary for leukotriene production. *Nature (London)*, **343**, 278–281.

Mitchell AE, Morin D, Lakritz J and Jones AD (1997) Quantitative profiling of tissue- and gender-related expression of glutathione S-transferase isoenzymes in the mouse. *Biochemical Journal*, **325**, 207–216.

Mosel F, Fardel O, Meyer DJ, Langouet S, Gilmore KS, Meurier B, Tu C-PD, Kensler TW, Ketterer B and Guillouzo A (1993) Preferential increase of glutathione S-transferase class α transcripts in cultured human hepatocytes by phenobarbital, 3-methylcholanthrene and dithiolethiones. *Cancer Research*, **53**, 231–234.

Morgan AS, Sanderson PE, Borch RF, Tew KD, Niitsu Y, Takayama T, Von Hoff DD, Izbicka E, Mangold G, Paul C, Broberg U, Mannervik B, Henner WD and Kauvar LM (1998) Tumor efficacy and bone marrow-sparing properties of TER286, a cytotoxin activated by glutathione S-transferase. *Cancer Research*, **58**, 2568–2575.

Morgenstern R, DePierre JW and Ernster L (1979) Activation of microsomal glutathione transferase activity by sulfhydryl reagents. *Biochemical and Biophysical Research Communications*, **87**, 657–663.

Morgenstern R, Meijer J, DePierre JW and Ernster L (1980) Characterization of rat liver microsomal glutathione transferase activity. *European Journal of Biochemistry*, **104**, 167–174.

Morgenstern R, Lundquist G, Jörnvall H and DePierre JW (1989) Activation of rat liver microsomal glutathione transferase by limited proteolysis. *Biochemical Journal*, **260**, 577–582.

Morimura S, Okuda A, Sakai M, Imagawa M and Muramatsu M (1992) Analysis of glutathione transferase P gene regulation with liver cells in primary culture. *Cell Growth and Differentiation*, **3**, 685–691.

Morrow CS, Smitherman PK, Diah SK, Schreider E and Townsend AJ (1998) Coordinated action of glutathione S-transferases (GSTs) and Multidrug Resistance Protein I (MPRI) in antineoplastic drug detoxification. *The Journal of Biological Chemistry*, **273**, 20114–20120.

Mulder TPJ, Court DA and Peters WHM (1999) Variability of glutathione S-transferase α in human liver and plasma. *Clinical Chemistry*, **45**, 355–359.

Nishoff WA, Grabben MJAL, Nagengast FM, Janses JBMJ, Verhagen H, van Poppel G and Peters WHM (1995) Effects of consumption of Brussels sprouts on intestinal and lymphocytic glutathione S-transferases in humans. *Carcinogenesis*, **16**, 2125–2128.

Nilsson LO, Gustafsson A and Mannervik B (2000) Redesign of substrate-selectivity determining modules of glutathione transferase A1-1 installs high catalytic efficiency with toxic alkenal products of lipid peroxidation. *Proceedings of the National Academy of Sciences, USA*, **97**, 9408–9412.

Okuda A, Imagawa M, Maeda Y, Sakai M and Muramatsu M (1989) Structural and functional analysis of an enhancer GPEI having a phorbol 12-O-tetradecanoate 13-acetate responsive element-like sequence found in the rat glutathione transferase P gene. *The Journal of Biological Chemistry*, **264**, 16919–16926.

Okuda A, Imagawa M, Sakai M and Muramatsu M (1990) Functional cooperativity between two TPA responsive elements in undifferentiated F9 embryonic stem cells. *The EMBO Journal*, **9**, 1131–1135.

Patten EJ and DeLong MJ (1999) Effects of sulindac, sulindac metabolites, and aspirin on the activity of detoxification enzymes in HT-29 human colon adenocarcinoma cells. *Cancer Letters*, **147**, 95–100.

Pearson WR, Reinhart J, Sisk SC, Anderson KS and Adler PN (1988) Tissue-specific induction of murine glutathione transferase mRNAs by butylated hydroxyanisole. *The Journal of Biological Chemistry*, **263**, 13324–13332.

Pearson WR, Vorachek WR, Xu S-j, Berger R, Hart I, Vannis D and Patterson D (1993) Identification of class-mu glutathione transferase genes *GSTM1-GSTM5* on human chromosome 1p13. *American Journal of Human Genetics*, **53**, 220–233.

Pemble SE, Wardle AF and Taylor JB (1996) Glutathione S-transferase Kappa: characterization by the cloning of rat mitochondrial GST and identification of a human homologue. *Biochemical Journal*, **319**, 749–754.

Polyak K, Xia Y, Zweier JL, Kinzler KW and Vogelstein B (1997) A model for p53-induced apoptosis. *Nature (London)*, **389**, 300–305.

Prestera T, Holtzclaw WD, Zhang Y and Talalay P (1993) Chemical and molecular regulation of enzymes that detoxify carcinogens Proceedings of the National Academy of Sciences USA, **90**, 2965–2969.

Prough RA, Xiao GH, Pinaire JA and Falkner KC (1996) Hormonal regulation of xenobiotic drug metabolizing enzymes. FASEB Journal, **10**, 1369–1377.

Pulford DJ and Hayes JD (1996) Characterization of the rat glutathione S-transferase Yc_2 subunit gene, GSTA5: identification of a putative antioxidant responsive element in the 5′-flanking region of rat GSTA5 that may mediate chemoprotection against aflatoxin B_1. Biochemical Journal, **318**, 75–84.

Reinemer P, Dirr HW, Ladenstein R, Schäffer J, Gallay O and Huber R (1991) The three-dimensional structure of class-pi glutathione S-transferase in complex with glutathione sulfonate at 2.3 Å resolution. The EMBO Journal, **10**, 1997–2005.

Reiter R and Wendel A (1985) Selenium and drug metabolism-III. Regulation of glutathione-peroxidase and other hepatic enzyme modulations to dietary supplements. Biochemical Pharmacology, **34**, 2287–2290.

Rowe JD, Nieves E and Listowsky I (1997) Subunit diversity and tissue distribution of human glutathione S-transferases: interpretations based on electrospray ionization-MS and peptide sequence-specific antisera. Biochemical Journal, **325**, 481–486.

Rushmore TH and Pickett CB (1990) Transcriptional regulation of the rat glutathione S-transferase Ya subunit gene. Characterization of a xenobiotic-responsive element controlling inducible expression by phenolic antioxidants. The Journal of Biological Chemistry, **265**, 14648–14653.

Rushmore TH, Morton MR and Pickett CB (1991) The antioxidant responsive element. Activation of oxidative stress and identification of the DNA consensus sequence for functional activity. The Journal of Biological Chemistry, **266**, 11632–11639.

Sakai M, Okuda A and Muramatsu M (1988) Multiple regulatory elements and phorbol 12-O-tetradecanoate 13-acetate responsiveness of the rat placental glutathione transferase gene. Proceedings of the National Academy of Sciences, USA, **85**, 9456–9460.

Schmidt-Krey I, Mitsuoka K, Hirai T, Murata K, Cheng Y, Fujiyoshi Y, Morgenstern R and Herbert H (2000) The three-dimensional map of microsomal glutathione transferase 1 at 6 Å resolution. The EMBO Journal, **19**, 6311–6316

Scoggan KA, Jakobsson P-J and Ford-Hutchinson AW (1997) Production of leukotriene C_4 in different human tissues is attributable to distinct membrane bound biosynthetic enzymes. The Journal of Biological Chemistry, **272**, 10182–10187.

Sherratt PJ, Pulford DJ, Harrison DJ, Green T and Hayes JD (1997) Evidence that human class Theta glutathione S-transferase T1-1 can catalyse the activation of dichloromethane, a liver and lung carcinogen in the mouse. Comparison of the tissue distribution of GSTT1-1 with that of classes Alpha, Mu and Pi GST in human. Biochemical Journal, **326**, 837–846.

Sherratt PJ, Manson MM, Thomson AM, Hissink EAM, Neal GE, van Bladeren PJ, Green T and Hayes JD (1998) Increased bioactivation of dihaloalkanes in rat liver due to induction of class Theta glutathione S-transferase T1-1. Biochemical Journal, **335**, 619–630.

Singhal SS, Saxena M, Awasthi S, Ahmad H, Sharma R and Awasthi YC (1992) Gender-related differences in the expression and characteristics of glutathione S-transferases of human colon. Biochimica et Biophysica Acta, **1171**, 19–26.

Slater TF (1984) Free radical mechanism in human tissue injury. Biochemical Journal, **222**, 1–15.

Sreerama L, Rekha GK and Sladek NE (1995) Phenolic antioxidant-reduced overexpression of class-3 aldehyde dehydrogenase and oxazaphosphorine-specific resistance. Biochemical Pharmacology, **49**, 669–675.

Staffas L, Mankowitz L, Söderström M, Blanck A, Porsch-Hallström I, Sundberg C, Mannervik B, Olin B, Rydström J and DePierre JW (1992) Further characterization of hormonal regulation of glutathione transferase in rat liver and adrenal glands. Sex differences and demonstration that growth hormone regulates the hepatic levels. Biochemical Journal, **286**, 65–72.

Staffas L, Ellis EM, Hayes JD, Lundgren B, De Pierre JW and Mankowitz L (1998) Growth hormone- and testosterone-dependent regulation of glutathione transferase subunit A5 in rat liver. Biochemical Journal, **332**, 763–768.

Suzuki T, Imagawa M, Hirabayashi M, Yuki A, Hisatake K, Nomura K, Kitagawa T and Muramatsu M (1995) Identification of an enhancer responsible for tumour marker gene expression by means of transgenic rats. Cancer Research, **55**, 2651–2655.

Suzuki T, Morimura S, Diccianni MB, Yamada R, Hochi S-I, Hirabayashi M, Yuki A, Nomura K, Kitagawa T, Imagawa M and Muramatsu M (1996) Activation of glutathione transferase P gene expression by lead requires glutathione transferase P enhancer I. *The Journal of Biological Chemistry*, **271**, 1626–1632.

Sweeney C, McClure GY, Fares MY, Stone A, Coles BF, Thompson PA, Korourian S, Hutchins LF, Kadlubar FF and Ambrosone CB (2000) Association between survival after treatment for breast cancer and glutathione *S*-transferase P1 Ile[105]Val polymorphism. *Cancer Research*, **60**, 5621–5624.

Takahashi Y, Campbell EA, Hirata Y, Takayama T and Listowsky I (1993) A basis for differentiating among the multiple human Mu-glutathione S-transferases and molecular cloning of brain *GSTM5*. *The Journal of Biological Chemistry*, **268**, 8893–8898.

Temellini A, Castiglioni M, Giuliani L, Mussi A, Giulianotti PC, Pietrabissa A, Angeletti CA, Mosca F and Pacifici GM (1995) Glutathione conjugation with 1-chloro-2,4-dinitrobenzene (CDNB)-interindividual variability in human liver, lung, kidney and intestine. *International Journal of Clinical Pharmacology and Therapeutics*, **33**, 498–503.

Tew KD (1994) Glutathione-associated enzymes in anticancer drug resistance. *Cancer Research*, **54**, 4313–4320.

Trush MA and Kensler TW (1991) An overview of the relationship between oxidative stress and chemical carcinogenesis. *Free Radical Biology and Medicine*, **10**, 201–209.

Urade Y, Fujimoto N, Ujihara M and Hayaishi O (1987) Biochemical and immunological characterization of rat spleen prostaglandin D synthetase. *The Journal of Biological Chemistry*, **262**, 3820–3825.

Urade Y, Watanabe K and Hayaishi O (1995) Prostaglandin D, E, and F synthase. *Journal of Lipid Mediators in Cell Signalling*, **12**, 257–273.

van Bladeren PJ, Breimer DD, Rotteveel-Smijs GMT, de Jong RAW, Buijs W, van der Gen A and Mohn GR (1980) The role of glutathione conjugation in the mutagenicity of 1,2-dibromoethane. *Biochemical Pharmacology*, **29**, 2975–2982.

Verhoeven DTH, Verhagen H, Goldbohm RA, van den Brandt PA and van Poppel G (1997) A review of mechanisms underlying anticarcinogenicity by brassica vegetables. *Chemico-Biological Interactions*, **103**, 79–129.

Wassernam WW and Fahl WE (1997) Functional antioxidant responsive elements. *Proceedings of the National Academy of Sciences, USA*, **94**, 5361–5366.

Waxman DJ, Sundseth SS, Srivastava PK and Lapenson DP (1992) Gene-specific oligonucleotide probes for α, μ, π, and microsomal rat glutathione S-transferases: analysis of liver transferase expression and its modulation by hepatic enzyme inducers and platinum anticancer drugs. *Cancer Research*, **52**, 5797–5802.

Watabe T, Hiratsuka A, Ogura K, Okuda H and Nishiyame T (1996) Class Theta glutathione S-transferases in rats, mice and guinea pigs. In *Glutathione S-Transferases: Structure, Function and Clinical Implications*, Vermeulen NPE, Mulder GJ, Nieuwenhuyse H, Peters WHM and van Bladeren PJ (eds), Taylor and Francis, London, pp. 57–69.

Welsch DJ, Creely DP, Hauser SD, Mathis KJ, Krivi GG and Isakson PC (1994) Molecular cloning and expression of leukotriene-C_4 synthase. *Proceedings of the National Academy of Sciences, USA*, **91**, 9745–9749.

Xu SJ, Wang YP, Roe B and Pearson WR (1998) Characterization of the human class mu-glutathione S-transferase gene-cluster and the gstm1 deletion. *The Journal of Biological Chemistry*, **273**, 3517–3527.

Yin ZM, Ivanov VN, Habelhah H, Tew K and Ronai Z (2000) Glutathione S-transferase p elicits protection against H_2O_2-induced cell death via coordinated regulation of stress kinases. *Cancer Research*, **60**, 4053–4057.

10 Sulphotransferases

Hansruedi Glatt

German Institute of Human Nutrition (DIfE), Germany

Abbreviations

AAF, 2-acetylaminofluorene
AST, aryl sulphotransferase
BR-STL, brain sulphotransferase-like protein
DCNP, 2,6-dichloro-4-nitrophenol
DHEA, dehydroepiandrosterone
DHEA-S, dehydroepiandrosterone sulphate
E_1, oestrone
E_2, 17β-oestradiol
E_3, 16α,17β-oestriol
EST, oestrogen sulphotransferase
1-HEP, 1-(α-hydroxyethyl)pyrene
1-HMP, 1-hydroxymethylpyrene
HST, hydroxysteroid sulphotransferase
OH-AAF, N-hydroxy-2-acetylaminofluorene
PAP, 3′,5′-diphosphoadenosine
PAPS, 3′-phosphoadenosine-5′-phosphosulphate
PCP, pentachlorophenol
PST, phenol sulphotransferase
rT_3, reverse (3,5,5′) triiodothyronine
SMP-2, senescence marker protein 2
SNP, single-nucleotide polymorphism
SULT, member of the SULT gene/enzyme superfamily
T_3, 3,3′,5-triiodothyronine
T_4, thyroxine

Enzyme Systems that Metabolise Drugs and Other Xenobiotics. Edited by C. Ioannides.
© 2002 John Wiley & Sons Ltd

Introduction

REACTIONS, TERMINOLOGY

Sulphotransferases (EC 2.8.2), in the older literature also termed sulphokinases, transfer the sulpho (SO_3^-) moiety from a donor, usually 3'-phosphoadenosine-5'-phosphosulphate (PAPS), to a nucleophilic group of the acceptor molecule (Figure 10.1). The sulpho group has a pK_a value of smaller than 1.5 in most structures. The introduced negative charge affects important properties of the acceptor molecule, such as the interaction with receptors and transport proteins (including albumin), the water solubility and the penetration of cell membranes. Since the sulpho moiety as well as the sulphate, sulphamate and thiosulphate groups (formed by sulphonation of O, N and S atoms, respectively) are electron-withdrawing, certain conjugates are chemically reactive. They may spontaneously react with nucleophiles (such as DNA or water) or undergo elimination reactions. Therefore, sulphotransferases can activate various protoxicants and can also catalyse the dehydration and isomerisation of certain substrates.

Since sulphotransferases transfer the sulpho moiety (not the sulphate group), we term the reaction sulphonation rather than sulphation. The latter designation may be tolerable if the sulpho moiety is transferred to an oxygen atom, but is misleading in all other cases. For the products, the established names will be used, even if these are not correct.

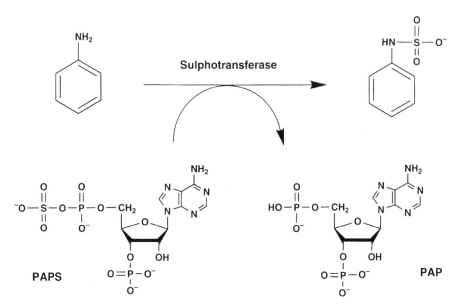

Figure 10.1 Scheme of the sulphotransferase-mediated sulpho transfer reaction shown for the substrate aniline. Enzyme, substrate and cofactor (PAPS) form a trimolecular complex. The sulpho group is directly transferred from PAPS to the substrate. In general, the reverse reaction is negligible.

The known sulphotransferases can be classified into three groups according to the principal acceptor molecules. Protein tyrosine sulphotransferases (Huttner 1987; Ouyang *et al.* 1998) and carbohydrate sulphotransferases (Bowman and Bertozzi 1999) sulphonate tyrosine residues of proteins and glycans attached to proteins and lipids, respectively. Since no xenobiotic-metabolising activities are known for these enzymes, only selected aspects are mentioned in the present review. The third group of enzymes sulphonates relatively small endogenous compounds (such as steroids and catecholamines) and xenobiotics. All enzymes of this group, investigated so far, belong to a common enzyme/gene superfamily. In the present review, the term SULT is used for any member of this superfamily, even if no sulphotransferase activity has been demonstrated so far, and the term sulphotransferase is used for any enzyme that transfers the sulpho group regardless of whether it belongs to the SULT superfamily. In particular, the SULT superfamily does not include the protein tyrosine sulphotransferases and carbohydrate sulphotransferases, as their structures and genetics are very distinct.

Various sulphoconjugates are substrates of sulphatases (also termed sulphohydrolases, EC 3.1.6). These enzymes transfer the sulpho group from the conjugate to water. They are presented in detail in Chapter 15 of this book.

HISTORY

In 1865, Baumann reported that urine of dogs, humans and horses contains sulphuric acid 'paired' with phenol, catechol and an indigo-forming compound (indoxyl) (Baumann 1865a,b). Heating in concentrated hydrochloric acid resulted in the hydrolysis of these conjugates to the phenols and inorganic sulphate. Application of phenol to the skin of a dog and feeding of catechol led to strong increases in the urinary levels of the corresponding sulphoconjugates (Baumann 1876). Half a century later, sulphoconjugates of the steroid hormones oestrone (E_1) (Schachter and Marrian 1938) and androsterone (Venning *et al.* 1942) were detected in mammalian urine. DeMeio and Tkacz (1950) were the first to demonstrate a sulphoconjugation in a subcellular preparation. Phenol was converted to phenyl sulphate in the presence of rat liver homogenate supplemented with α-ketoglutarate, adenylic acid and inorganic sulphate. Since all these components were necessary, it was concluded that oxidation reactions were needed to provide the energy for the conjugation reaction. Subsequently, this process could be subdivided into two steps, the activation of sulphate and the transfer of the sulpho group to the phenolic acceptor. Eventually, Robbins and Lipmann (1956, 1957) elucidated the structure of 'active sulphate', PAPS (Figure 10.1).

Phenol (or aryl) sulphotransferases (PSTs or ASTs) I, II and IV from rat liver were among the first sulphotransferases that were purified to apparent homogeneity (Sekura and Jakoby 1979, 1981). In particular, AST IV shows an extremely broad substrate tolerance. More than 100 substrates have been reported for this enzyme. In addition to many phenols, it sulphonates many other types of aryl compounds and even a few compounds that do not contain an aryl moiety, such as 2-nitropropane. AST IV is the sulphotransferase form that has been studied most thoroughly also in many other respects, such as kinetic parameters, mechanism of action and inhibition. However, later it was found that AST IV, purified from male rat liver, comprised two distinct homodimers (most likely rSULT1A1 and rSULT1C1, see below) and the corresponding heterodimer and that these enzymes differ in their substrate specificity (Kiehlbauch *et al.* 1995). Since differing purification methods and enzyme sources have been used, the exact nature of the enzyme(s) investigated in various studies is elusive. This example illustrates two common situations: (a) The name of an individual xenobiotic-metabolising enzyme may not exactly reflect its function and substrate specificity. (b) A purified enzyme that appeared homogeneous in state-of-the-art analyses may demonstrate a relevant heterogeneity with the refinement of the analytical methods.

Chatterjee *et al.* (1987) were the first to clone the cDNA of a member of the SULT superfamily. This cDNA was reported to encode rat senescence marker protein 2 (SMP-2). This protein was found in the liver of senescent male rats, but was absent in young adult males. Its homology to sulphotransferases was detected only later, when the cDNAs of known sulphotransferases were cloned (Ogura *et al.* 1990b). SMP-2 is a member of the SULT2A subfamily, which encodes hydroxysteroid sulphotransferases (HSTs). Since SMP-2 was never studied for sulphotransferase activity, we wanted to perform such analyses using cDNA-expressed SMP-2. However, we failed to isolate the cDNA from the liver of senescent rats, but always ended up with the cDNA of ST-60 (another member of the SULT2A subfamily). Moreover, the cDNA sequences of rat SMP-2 and a mouse HST (STa-2) published in the GenBank (accession numbers J02643 and L27121, respectively) were gradually corrected and differ now only in one nucleotide of the available sequence of 970 nucleotides, including the entire coding region. Therefore, it is suspected that the SMP-2 cDNA of Chatterjee *et al.* (1987) has been cloned from a mouse rather than a rat cDNA bank or from a contaminated rat cDNA bank, and that the expressed SMP-2 protein (i.e. mouse STa-2) was so similar in its electrophoretic mobility and immunoreactivity to rat HSTs that the mistake was not noticed. It is probable that the true rat SMP-2 protein is one of the known rat HSTs, probably STa and/or ST-60 (described below). This account serves to remind us that cloning and sequencing errors were common in the pioneering period of that technology, although the error rate can be kept very low nowadays.

In a recent review, Nagata and Yamazoe (2000) have listed the cDNAs of 10 human and 31 other mammalian *SULT* genes. Many of these cDNAs have been expressed individually in bacteria or in mammalian cells, allowing the characterisation of precisely defined enzyme proteins. The comparison of their activities and levels with those in tissues of the corresponding species is very useful in the elucidation of the enzymic basis of sulphonation.

Elisabeth and James Miller (1966) pioneered the concept that most carcinogens are electrophiles or are metabolised to electrophiles. A sulphuric acid ester, *N*-sulphooxy-AAF, was the first electrophilic metabolite of a carcinogen to be discovered (DeBaun *et al.* 1968; King and Phillips 1968). Bioactivation via sulphonation is not taken into account in standard *in vitro* mutagenicity test systems, because the target cells used do not express sulphotransferases and externally generated reactive sulphuric acid esters do not readily penetrate into the cells. This problem can be solved by the expression of sulphotransferases in target cells of test systems. Initial results from such systems indicate that a substantial number of compounds can be activated to mutagens via metabolic sulphonation (Glatt 2000a).

SCOPE OF THIS REVIEW

This chapter focuses on the metabolic and toxicological aspects of the xenobiotic-metabolising sulphotransferases with special consideration of new molecular-biological findings. The following articles may be useful to the reader looking for more detailed information on other aspects of sulphonation. A comprehensive review written by Mulder and Jakoby (1990) presents in detail the sulphonation pharmacokinetics in intact human and animal organisms, isolated organs and cell cultures; studies on enzyme kinetics and enzymic mechanisms using subcellular preparations and purified enzymes are also presented in depth. Other recent reviews are focused on the genetics of *SULTs* (Weinshilboum *et al.* 1997; Nagata and Yamazoe 2000; Glatt and Meinl 2001), the enzymic properties of individual SULT forms (Falany 1997), steroid sulphotransferases (Hobkirk 1993; Strott 1996), the regulation of the expression of SULTs (Runge-Morris 1997), the cofactor supply of sulphotransferases (Klaassen and Boles 1997) and the sulphotransferase-mediated activation of carcinogens and mutagens (Miller and Surh 1994; Glatt 1997, 2000a). Purification procedures for sulphotransferases and characteristics of purified enzymes have been reviewed by Singer (1985). Furthermore, two volumes of the journal *Chemico-Biological Interactions* (Vol. 92, 1994; Vol. 109, 1998) are dedicated to the sulphotransferases. Older reviews on sulphoconjugation (Dodgson and Rose 1970; Roy 1971; DeMeio 1975) are not only of historical interest, but contain numerous remarkable and puzzling findings that now can be readily followed up using the actual molecular biology knowledge and techniques.

Functions and effects of sulphonation

FACILITATION OF EXCRETION

Numerous xenobiotics are excreted as sulpho- and glucuronic acid conjugates in the urine or faeces. The negative charge introduced with these moieties usually increases the water solubility and inhibits the passive penetration of cell membranes by the conjugates. The transmembrane transport of conjugates requires special mediators, e.g. the ATP-dependent multidrug-resistance proteins (Jedlitschky *et al.* 1996; Walle *et al.* 1999). In polar cells, such as the hepatocytes, the multidrug-resistance proteins are not uniformly distributed on the plasma membrane (König *et al.* 1999). Therefore, selective transport to the bile duct or the basolateral site (and from there to the blood) is possible. According to a traditional rule, small molecules are preferentially excreted in the urine, whereas large molecules prefer the biliary route. The critical size varies among species. In the rat, molecules with a mass of less than 325 ± 50 usually are not excreted into the bile (Hirom *et al.* 1972). The corresponding value in the human is 500 to 700. The molecular mass of sulphoconjugates is lower by 98 units than that of the corresponding glucuronides; often a higher percent of the sulphoconjugate than of the glucuronide is excreted in the urine (Mulder and Bleeker 1975; Møller and Sheikh 1982). However, the size-related rules are simplistic. A better knowledge of the distribution and substrate specificity of the various anion transmembrane transporters will help to better predict the excretion routes of individual metabolites. Probenecid, an inhibitor of the active tubular transport of organic anions, decreased the renal clearance of certain sulphoconjugates, e.g. diflusinal sulphate (Macdonald *et al.* 1995) and *p*-hydroxytriamterene sulphate (Muirhead and Somogyi 1991).

Bile and urine are not the only excretion routes for conjugates. Enterocytes express transmembrane transporters on the luminal as well as the serosal surface and export conjugates in both directions (e.g. Sund and Lauterbach 1986, 1993; Walle *et al.* 1999). The human and the guinea pig, in contrast to the rat, express high levels of sulphotransferases in the intestinal mucosa (see below). Therefore, the direct excretion of sulphoconjugates into the gut lumen, without the loop via the liver, could be particularly important in the former species.

INACTIVATION OF BIOLOGICALLY ACTIVE MOLECULAR SPECIES

The predominant mechanism of pharmacological effects involves the non-covalent binding of the agent to specific receptors. The conjugation with the negatively charged sulpho group will abolish or drastically alter this interaction. Indeed, the sulphonated steroids do not show any relevant affinity to steroid receptors nor any hormonal activity as such (e.g. Hähnel *et al.* 1973). Likewise, catecholamine sulphates do not bind to adrenoreceptors (Lenz *et al.* 1991 and references cited therein) nor to dopamine receptors (Werle *et al.* 1988), and T_3 sulphate does not bind to the thyroid hormone receptor (Spaulding 1994). This metabolic inactivation of pharmacologically active species, complemented by the facilitation of the excretion, dominates the view of many pharmacologists on conjugation reactions so strongly that other important roles are often ignored.

FORMATION OF INACTIVE STORAGE AND TRANSPORT FORMS OF HORMONES

Dehydroepiandrosterone sulphate (DHEA-S) is the major circulating steroid in the human. In young adults, it reaches plasma levels of 10 μM, whereas unconjugated dehydroepiandrosterone (DHEA) is present at concentrations of about 10 nM (Baulieu 1996). Likewise, E_1 sulphate is the most abundant C18-steroid in human blood (Loriaux *et al.* 1971). Half-lives of several hours have been reported for DHEA-S and testosterone sulphate in human blood; their metabolic clearance rate was approximately 40 times lower than those of the respective free steroids (Wang *et al.* 1967). In rat and rabbit, this factor was only approximately 2. A high affinity to serum albumin (Puche and Nes 1962; Plager 1965) and the reabsorption from the renal tubules (Longcope 1995) are involved in the retention of steroid sulphates in the organism. Moreover, numerous tissues express steroid sulphatases (Chapter 15) that can release the free hormone from the conjugate (Martel *et al.* 1994; Dooley *et al.* 2000). Therefore, it is obvious that steroid sulphates are not metabolic end products destined only for excretion, but also serve as storage and transport forms of steroid hormones. DHEA-S and 16α-hydroxy-DHEA-S produced by the adrenal gland of the foetus are a major source of the maternal oestrogens during pregnancy in primates (Siiteri and MacDonald 1966; Parker *et al.* 1984; Kuss 1994); their conversion into the oestrogens occurs in the placenta. Manipulation of the activities of oestrogen sulphotransferase and sulphatase is considered a very promising approach for altering the oestrogen levels in target tissues such as breast tumours (Pasqualini and Chetrite 1999).

The much higher plasma levels and longer half-lives of steroid sulphates in the human, in comparison with rodents, indicates that this storage function may vary substantially between species.

Regulation via sulphonation/desulphonation may not be restricted to steroids, since mammalian sulphatases also can hydrolyse the sulphoconjugates of iodothyronines (Santini *et al.* 1992; Spaulding 1994; Richard *et al.* 2001) and dopamine (Yoshizumi *et al.* 1992).

Vitamin D_3 3β-sulphate has been detected in human milk (Boulch *et al.* 1982). Treatment of vitamin D-depleted mother rats with vitamin D_3 3β-sulphate significantly improved the biochemical plasma parameters of pups (Cancela *et al.* 1987), but was inefficient in the mother rat (Cancela *et al.* 1985). Since administration of unconjugated vitamin D_3 to the mother rat was more effective even for the pup than administration of vitamin D_3 3β-sulphate, the role of the sulphoconjugation of vitamin D_3 is not understood.

Futile cycling between the sulphonated and deconjugated forms has been observed for some xenobiotics such as 4-methylumbelliferone (Kauffman *et al.* 1991; Ratna *et al.* 1993).

DIRECTION OF METABOLIC PROCESSING OF ENDOGENOUS COMPOUNDS

Sulphonation strongly accelerates the deiodination of thyroxine (T_4) and $3,3',5$-triiodothyronine (T_3) by the type I deiodinase (Visser 1994). Interestingly, T_4 sulphate is deiodinated exclusively in the inner ring to the sulphate of the hormonally inactive $3,5,5'$-triiodothyronine (rT_3); the free prohormone T_4 is preferentially deiodinated in the outer ring, leading to the formation of the highly active hormone, T_3. Thus, sulphonation initiates the irreversible degradation of thyroid hormones by type I deiodinase. If type I deiodinase activity is low, as is the case in the foetus, T_3 sulphate may function as a reservoir from which active T_3 can be recovered by tissue sulphatase activity (Visser 1994).

In the same way, steroid sulphates do not require deconjugation in order to be further metabolised. They are substrates for various cytochromes P450 and dehydrogenases. For example, the following metabolic pathway has been established in the foetus: pregnenolone sulphate $\rightarrow$ 17-hydroxypregnenolone sulphate $\rightarrow$ DHEA-S $\rightarrow$ 16-hydroxy-DHEA-S (Pasqualini and Kinel 1985). Furthermore, testosterone sulphate can be converted directly to E_2 17-sulphate by human term placental microsomes (Satoh *et al.* 1992).

The loop sulphonation/further metabolism/desulphonation may lead to other products than the direct metabolism of the unconjugated species and avoids the formation of unconjugated intermediates.

BIOACTIVATION OF ENDOGENOUS MESSENGER MOLECULES

Not all biological effects of steroid hormones are mediated by the classical, intracellular steroid receptors. While sulphonation abolishes the interaction of steroids with steroid receptors, it appears to be important for other effects.

Various steroids, termed neurosteroids, are synthesised in the nervous system and exert a wide variety of diverse functions (Compagnone and Mellon 2000). Some of them are potent allosteric modulators of the γ-aminobutyric acid A (GABA$_A$) receptor (Paul and Purdy 1992; Park-Chung et al. 1999). Whereas various unconjugated steroids (including allopregnanolone and androsterone) are positive modulators, many steroid sulphates (such as DHEA-S and pregnenolone sulphate) negatively modulate the GABA$_A$ receptor. In addition, steroid sulphates can potentiate the activation of the N-methyl-D-aspartate (NMDA) receptor (Bowlby 1993) and are agonists of the σ1 receptor (Noda et al. 2000; Zou et al. 2000). These molecular interactions of the steroid sulphates with receptors have been shown to lead to improved learning, enhanced memory performance, antidepressant effects, reduced conditioned-fear stress as well as to convulsions. Such effects were observed after the intracerebro-ventricular (Flood et al. 1995; Kokate et al. 1999) and subcutaneous (Reddy et al. 1998; Noda et al. 2000) administration of steroid sulphates to laboratory animals. Likewise, the administration of a steroid sulphatase inhibitor to rats for 15 days markedly increased the ratio of the plasma levels of DHEA-S to DHEA, and diminished a scopolamine-induced amnesia (Rhodes et al. 1997).

DHEA at pharmacological levels is a peroxisome proliferator in rats and mice; this effect appears to be mediated by DHEA-S rather than the unconjugated DHEA (Yamada et al. 1994; Peters et al. 1996). It has also been suggested that certain steroid sulphates, e.g. 2-hydroxy-E$_2$ 17-sulphate, may act as antioxidants in the organism (Takanashi et al. 1995).

Cholesterol sulphate, formed in the differentiating keratinocytes, is essential for the keratinisation by activating the η isoform of protein kinase C (Kawabe et al. 1998). Also isoforms ε and ζ, and to a lower extent isoforms α and δ, are activated by cholesterol sulphate (Denning et al. 1995). Cholesterol sulphate is present at high levels in the acrosomes of spermatozoa and acts as a potent inhibitor of acrosin, a key proteolytic enzyme involved in the acrosome reaction; this inhibition appears to be reversed by sulphatases present in the female reproductive tract (Roberts 1987).

17,20β-Dihydroxy-4-pregnen-3-one 20-sulphate functions as a potent and specific olfactory stimulant with pheromonal actions in the goldfish (Sorensen et al. 1995).

Dopamine 4-sulphate demonstrated vasopressor activity in the peripheral and central nervous system, whereas dopamine 3-sulphate acted as a central depressor; these effects do not appear to be mediated via free dopamine formed by deconjugation (Minami et al. 1995a,b).

Various peptide hormones contain sulphonated tyrosine residues. In general, their sulphonation occurs by membrane-bound enzymes in the Golgi apparatus of the cells in which the peptides have been synthesised (Huttner 1987; Ouyang et al. 1998). However, there is evidence that some soluble enzymes are also capable of sulphonating certain peptide hormones. Hence, several peptides with an N-terminal tyrosine residue are substrates of rat AST IV (Sekura and Jakoby 1981). Furthermore, a soluble gastrin sulphotransferase has been detected in rat gastric mucosa, and sulphonated gastrin was a substrate of an aryl sulphatase (a commercial enzyme whose origin was not indicated) (Chen and Rosenquist 1990). Thus, reversible sulphonation may not only be used for the regulation of steroid and thyroid hormones, but also for the regulation of certain peptide hormones.

BIOACTIVATION OF PHARMACEUTICAL DRUGS

Minoxidil is an antihypertensive drug that also stimulates hair growth. In order to exert these effects, the N-oxide group of minoxidil has to be sulphonated (McCall *et al.* 1983; Buhl *et al.* 1990). The hypotensive effects of minoxidil were ablated in rats treated with the sulphotransferase inhibitor pentachlorophenol (PCP) and accentuated after sulphotransferase induction by the glucocorticoid triamcinolone acetonide (Duanmu *et al.* 2000). Minoxidil sulphate has the unique ability to non-enzymically sulphonate proteins, and it has been postulated that this mechanism is central to the pharmacological action of minoxidil (Groppi *et al.* 1990; Meisheri *et al.* 1991).

Triamterene, a potassium-sparing diuretic, is rapidly metabolised to *p*-sulphooxy-triamterene (Mutschler *et al.* 1983). This sulphoconjugate is pharmacologically active and may be the primary mediator of diuretic and natriuretic effects observed after the administration of triamterene (Leilich *et al.* 1980). *p*-Sulphooxytriamterene as well as minoxidil sulphate are zwitterions and, therefore, are more lipophilic than their cationic parent drugs.

The salidiuretic effect of cicletanine is not caused by the parent drug but by its sulphoconjugate (Garay *et al.* 1991). In the rat, (+)-cicletanine was sulphonated to a fivefold higher extent than its antipode, and the resulting sulphoconjugate was three to four times more potent than that of (−)-cicletanine in stimulating the renal sodium excretion and in inhibiting the sodium-dependent Cl^-/HCO_3^- exchanger of erythrocytes (Garay *et al.* 1995).

The 3-sulphate and, in particular, the 6-sulphate of morphine are potent analgesics if injected intracerebroventricularly in mice (Brown *et al.* 1985). Under these conditions, morphine 6-sulphate is about 30 times more potent than unconjugated morphine. Morphine 6-sulphate also shows a high analgesic activity after subcutaneous and intravenous administration to rodents (Preechagoon *et al.* 1998). The analgesic activity of morphine 6-sulphate (and morphine 6-glucuronide) appears to be mediated by another isoform of the MOR-1 receptor (a μ receptor) than that of unconjugated morphine (Zuckerman *et al.* 1999). Codeine 6-sulphate also has analgesic activity in the mouse, but leads to seizures at doses below full analgesic activity (Zuckerman *et al.* 1999). Enkephalins, which are endogenous ligands of the opioid receptors, contain an N-terminal tyrosine residue that can be enzymically sulphonated (Sekura and Jakoby 1981). Structural similarities between this tyrosine residue and the hydroxybenzo moiety of morphine and codeine are obvious.

UNSTABLE SULPHOCONJUGATES

Spontaneous hydrolysis and displacement reactions of sulphoconjugates

Sulphoconjugates of strongly acidic phenols are unstable. The O–S bond is cleaved and the sulpho group is transferred to water. In some cases these conjugates can also be used as sulpho donors by sulphotransferases (see below).

Also, many benzylic and allylic sulphoconjugates are readily hydrolysed. However, in this case the C–O bond rather than the O–S bond is cleaved. For example, incubation of $[^{18}O]$-1-hydroxymethylpyrene (1-HMP) (Figure 10.4, compound **I**) with a rat HST and PAPS led to the rapid formation of $[^{16}O]$-1-HMP (Landsiedel 1998). The intermediate formed, 1-sulphooxymethylpyrene, has been demonstrated, although it is short-lived in aqueous media ($t_{1/2} \sim 2.8$ min at 37°C). The heterolytic breakage of the C–O bond leads to the formation of a cation, which is stabilised by mesomerism. Such cations are strong electrophiles. In fact, 1-sulphooxymethylpyrene reacts in aqueous solution with nearly any nucleophile present (halogenide anions, thiols, amines, alcohols, amino acids) via an S_N1 mechanism (Landsiedel et al. 1996). Many of these displacement products are still reactive and eventually are converted to 1-HMP. In other cases, the displacement reactions of reactive sulphuric acid esters appear to follow an S_N2 mechanism. For example, the $t_{1/2}$ of 4-sulphooxycyclopenta[def]chrysene was strongly reduced in the presence of ethane thiolate (Landsiedel et al. 1996).

Acyl sulphates also are very unstable and alkylate the nucleophile 4-(p-nitrobenzyl)pyridine, but have not been reported as xenobiotic metabolites to date (van Breemen et al. 1985).

Sulphotransferase-mediated isomerisation reactions

If the benzylic/allylic hydroxyl group is attached to a chiral centre, its stereochemistry may be altered during a sulphonation/hydrolysis cycle. Thus, incubation of the individual enantiomers of 1-HEP with sulphotransferase in vitro led to their chiral inversion (Landsiedel et al. 1998). When an enantioselective sulphotransferase was used, such as human SULT1E1 (which has a 150-fold preference for (S)-1-HEP), a nearly complete chiral inversion was reached (reaction scheme in Figure 10.2).

Stiripentol is a chiral allylic alcohol that has been developed as an antiepileptic drug. After oral administration to rats, the (R)-enantiomer was extensively converted to its antipode, whereas little conversion was observed from the (S)-enantiomer (Zhang et al. 1994). This inversion was strongly reduced by PCP, an inhibitor of phenol sulphotransferases (see below), suggesting that the chiral inversion was mediated, at least in part, by enantioselective sulphonation. However, a final conclusion is not possible, since the inversion occurred only after oral but not after parenteral administration of the drug, and since a PAPS-dependent racemisation of (R)-stiripentol in hepatic cytosolic preparations was not observed (Zhang et al. 1994).

(E)-α-Hydroxytamoxifen was converted to its diastereomer, (Z)-α-hydroxytamoxifen, in the presence of a rat hydroxysteroid sulphotransferase (STa) and PAPS (Figure 10.4, compound IV) (Shibutani et al. 1998). It is probable that the Z/E-conversion occurred on the level of the carbo cation, when the double bond was less fixed due to mesomerism.

The mesomerism of benzylic and allylic carbenium ions and their nitrenium analogues not only may lead to cis/trans-isomerisations but also to positional isomerisations. For example, hydrolysis of 1'-sulphooxysafrole led not only to the formation of 1'-hydroxysafrole but also to 3'-hydroxyisosafrole (Figure 10.4, compounds II and III, respectively); likewise both alcohols were formed from 3'-sulphooxyisosafrole (Boberg et al. 1986). In the presence of rat liver cytosol, PAPS-dependent conversion of 3'-hydroxyisosafrole to 1'-hydroxysafrole was observed.

N-Sulphooxy-AAF, a reactive metabolite of the carcinogen AAF, has a half-life of approximately 4 s in aqueous solution at 37°C (Panda et al. 1989). The products formed include N-hydroxy-, 1-hydroxy-, 3-hydroxy-, 1-sulphooxy- and 3-sulphooxy-

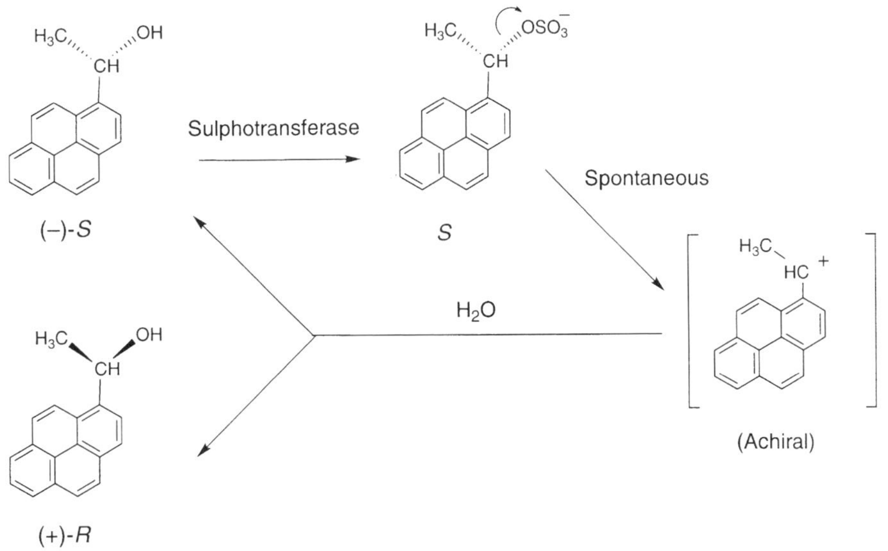

Figure 10.2 Sulphotransferase-mediated chiral inversion, shown for (-)-S-1-(α-hydroxyethyl) pyrene [(-)-S-1-HEP]. The sulphoconjugate of this substrate is unstable and hydrolyses via a carbonium intermediate to racemic 1-HEP. In the presence of a stereoselective sulphotransferase, such as human SULT1E1, a nearly complete inversion may occur after multiple cycles of sulphonation and hydrolysis (Landsiedel et al. 1998).

AAF (Panda et al. 1989; Smith et al. 1989; Kolanczyk et al. 1991; Novak and Rangappa 1992). Reaction of N-sulphooxy-AAF with glutathione yielded four conjugates which have been identified as 1-, 3-, 4- and 7-(glutathion-S-yl)-AAF (Beland et al. 1983). The formation of isomeric displacement products reflects the mesomeric properties of the compound. The isomerisation of N-sulphooxy-AAF to 1-sulphooxy- and 3-sulphooxy-AAF involves intramolecular substitution reactions.

Rat liver ASTs catalyse the Beckmann-like rearrangement of 9-fluorenone oxime to phenathridone (Mangold et al. 1986). While the sulphoconjugate of 9-fluorenone oxime was formed rapidly in the presence of ASTs, the subsequent conversion to phenathridone was slow and did not require the enzymes.

Sulphotransferase-mediated dehydration

The heterolytic cleavage of the sulphate group is not always associated with a substitution reaction (see above). Alternatively, a proton may be eliminated. In this case, the net effect of the sulphonation and the elimination reaction corresponds to a dehydration.

Cloning and sequencing of the cDNA of an insect retinol dehydratase indicated that it is a member of the SULT superfamily (Grün et al. 1996; Vakiani et al. 1998). Indeed, dehydration of retinol by this enzyme in subcellular systems required the presence of

PAPS, suggesting the reaction mechanism shown in Figure 10.3. The intermediate sulphate ester could not be demonstrated, probably because it is very short-lived.

Likewise, atropine (Figure 10.4, compound **V**) and scopolamine were dehydrated to apoatropine and aposcopolamine in the presence of guinea pig liver cytosol, ATP and sodium sulphate; the requirement of all three components strongly indicated that the reaction is mediated by a sulphotransferase (Wada *et al.* 1994). This notion was supported by the observation that DHEA, an alternate-substrate inhibitor of hydro-xysteroid sulphotransferases, decreased the dehydration of atropine *in vitro* and *in vivo* (Wada *et al.* 1994). The dehydration of atropine and scopolamine is particularly remarkable, as these compounds are neither benzylic nor allylic alcohols. Their hydroxyl group is localised in the β-position of a side chain. The authors of that study

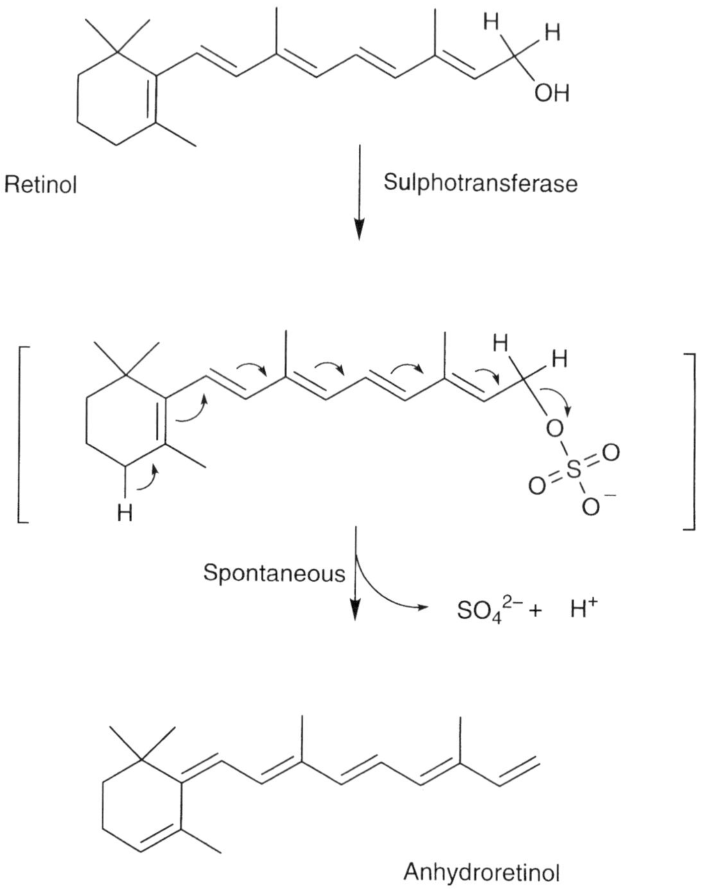

Figure 10.3 Sulphotransferase-mediated dehydration, shown for retinol. This reaction has been observed with an insect retinol dehydratase, which requires PAPS and is a member of the SULT superfamily (Vakiani *et al.* 1998).

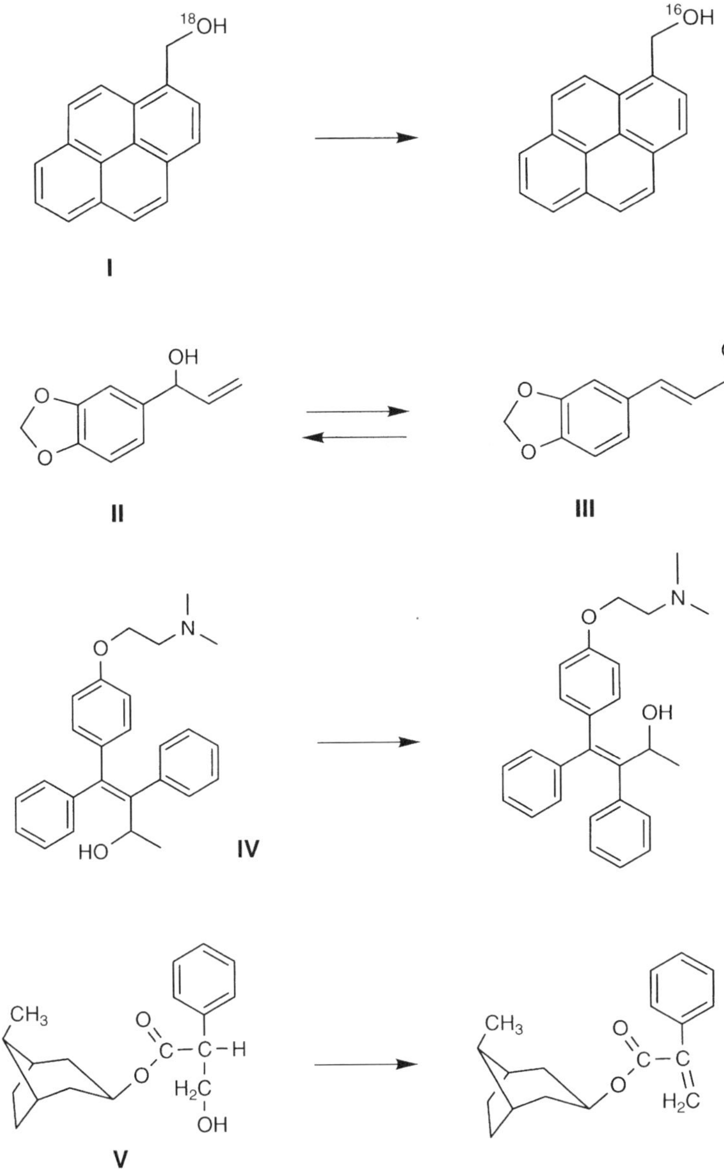

Figure 10.4 Sulphotransferase-mediated oxygen exchange in 1-hydroxymethylpyrene (1-HMP, **I**) (Landsiedel 1998), positional isomerisation of 1'-hydroxysafrole (**II**) and 3'-hydroxyisosafrole (**III**) (Boberg *et al.* 1986), *cis/trans* isomerisation of (*E*)-α-hydroxytamoxifen (**IV**) (Shibutani *et al.* 1998) and dehydration of atropine (**V**) (Wada *et al.* 1994).

suggested that the existence of a carboxyl group or a similar electron-attracting group at the benzylic position was pivotal to the dehydration reaction.

FORMATION OF REACTIVE SPECIES THAT COVALENTLY BIND TO MACROMOLECULES

As described above sulphate is a good leaving group in certain conjugates, leading to the formation of an electrophilic cation, which can react with water and numerous other nucleophiles. Besides small molecules, DNA and proteins are important nucleophilic reactants. N-Sulphooxy-AAF was the first electrophilic metabolite of a carcinogen to be discovered that reacted with cellular macromolecules (King and Phillips 1968; DeBaun et al. 1968). In the subsequent years, the research group of James and Elizabeth Miller demonstrated that various other carcinogens—including 4-aminoazobenzene, safrole, 2',3'-estragole, and 6-hydroxymethylbenzo[a]pyrene—are also metabolised to reactive species via sulphonation (reviewed by Miller and Surh 1994). They showed that (a) the authentic sulphuric acid esters formed adducts with macromolecules, (b) the bioactivation of the precursors by subcellular preparations required the presence of PAPS, (c) inhibitors of sulphotransferases reduced the formation of macromolecular adducts in vitro and in vivo and (d) brachymorphic mice, which are genetically deficient in the synthesis of PAPS (see below), are resistant to the adduct formation and liver tumorigenicity by these compounds.

We extended these studies by expressing individual rodent and human SULTs in target cells of standard mutagenicity tests (Ames's his⁻ Salmonella typhimurium strains and Chinese hamster V79 cells). These strains have been used for the detection of mutagens and the elucidation of activation pathways. Since we have reviewed this work recently (Glatt 1997, 2000a; Glatt et al. 2000), the findings are presented here only in a condensed form:

Using recombinant bacterial strains and mammalian cell lines, we have demonstrated SULT-dependent mutagenicity for approximately 100 compounds. Most of them belong to the following classes: benzylic alcohols, allylic alcohols, aromatic hydroxylamines, aromatic hydroxamic acids, secondary nitroalkanes and compounds that can be metabolised to any of these classes by other enzymes present in the test system.

For several compounds it was shown that activation within the target cell was important for the detection of a mutagenic effect; external activation using the same enzymes only led to a meagre effect in many cases.

Large differences in the substrate specificity towards these promutagens were observed between individual SULTs, e.g. between different forms from the same species, between orthologous enzymes of rat and human and between the allelic variants of human SULTs.

Nomenclature and substrate specificities of sulphotransferases

CLASSIFICATION SYSTEMS

After the metabolic sulphoconjugation was recognised as an enzymic reaction (DeMeio and Tkacz 1950), it was noticed that the activities towards different substrates can differ in tissue distribution, sex dependence, ontogenetic development, pH optima, thermostability and sensitivity towards inhibitors. This strongly indicated the existence of multiple forms of sulphotransferases and required a specification of the investigated enzyme activities. The substrates were obvious parameters for a specification, even at a stage when little was known about the number and characteristics of the different enzyme forms. Therefore, many substrate-based names were used operationally. The creation of a name does not imply that a single enzyme form (or a group of similar enzymes) causes the sulphonation of the namesake nor that these enzymes do not sulphonate other substrates. In fact, it is usually difficult to judge the identity or non-identity of enzymes purified in different laboratories using different enzyme sources, purification procedures and/or substrates. Also the possible role of contamination is difficult to assess. Using precisely defined, cDNA-expressed enzymes, it is now evident that many SULTs have very broad, overlapping substrate specificities, that various sulphotransferases named after different substrates are identical and that sometimes the same name has been used by different authors for clearly distinct enzyme forms.

Moreover, many SULTs were detected from their cDNA and/or genomic DNA sequences; as yet, only minimal information is available on the substrate specificity of these forms. This would result in additional problems for a substrate-based nomenclature. On the other hand, the cloned nucleic acid sequences of xenobiotic-metabolising sulphotransferases demonstrated substantial similarities in the deduced amino acid sequences and in the gene structures of these forms (Yamazoe *et al.* 1994; Weinshilboum *et al.* 1997; Nagata and Yamazoe 2000). These findings strongly suggest that all these forms are derived from a common ancestral gene and thus form an enzyme/gene superfamily. The participants of the 3rd International Sulphation Workshop, held in Drymen (Scotland) in 1996, agreed to use SULT as the name of the superfamily. A systematic nomenclature, based on the structural/genetic similarity, analogous to that of the cytochromes P450 (Nelson *et al.* 1996), is in preparation but not yet finalised. Members of each family (indicated by the number after 'SULT') and subfamily (capital letter after subfamily number) show at least 45% and 60% amino acid sequence identity, respectively. Although not formally ratified, this nomenclature is already widely used, at least for human SULTs. The systematic designations as well as traditional names for the human SULTs are listed in Table 10.1. Orthologous enzymes from different species will receive the same systematic designation. If the species is not clear from the context, a prefix is used, such as h for human, r for rat, m for mouse, or gp for guinea pig. Moreover, in agreement with the practice for other gene families, only the first letter is usually capitalised for mouse and *Drosophila* forms, whereas all letters are capitalised for other species. Italics can be used to indicate genes, whereas names in Roman letters normally represent gene products (RNA, cDNA, and protein).

Some confusion may arise from the existence of an older, preliminary nomenclature proposed by Yamazoe and co-workers (Yamazoe *et al.* 1994; Nagata and Yamazoe 2000). In this nomenclature, ST is used as the abbreviation for the superfamily, and no attempts are made to identify orthologous forms. Each member of a subfamily receives a separate number in the order of the detection. It is important to realise this difference in the nomenclature. For example, ST1A3 is identical with human SULT1A1 (P-PST),

Table 10.1 Classification and characteristics of human SULTs[a]

Systematic name	Other names, characteristics properties, selected references
SULT1A1	Other names: phenol-sulphating phenol sulphotransferase (P-PST, P-PST-1), thermostable phenol sulphotransferase (TS PST, TS PST 1), hippocampus phenol sulphotransferase (H-PST) (refers to a specific transcript), ST1A3, human aryl sulphotransferase (HAST) 1 and 2 (indicating two transcripts that differ in the first, non-translated exon). State of knowledge: excellent compared to other hSULTs; the enzyme has been purified from various human tissues. Gene: located on 16p12.1-11.2; various SNPs, some of them involving an amino acid exchange; the most prominent polymorphism is in codon 213 (Arg/His exchange) (Raftogianis et al. 1999) RNA: two alternative first exons (in the non-coding region). Protein: dimer; the electrophoretic mobility of the subunit (apparent $M_r = 32$ kDa) differs from that of all other known hSULTs Substrates: numerous phenols (Campbell et al. 1987) and other aryl compounds, and also some small molecules that do not contain an aryl moiety; including drugs and drug metabolites (paracetamol, minoxidil, 4-hydroxytamoxifen) and many promutagens (Glatt 2000a); 4-nitrophenol at a low concentration (~ 4 μM) is often used to selectively detect 1A1 activity ($K_m \sim 1$ μM) in tissues that co-express the other major PST (1A3, $K_m \sim 2$ mM); 1A1 is more sensitive to inhibition by DCNP and more thermostable than 1A3 (however, smaller differences in thermostability are also found between different 1A1 alloenzymes) (Raftogianis et al. 1999). Tissue distribution: 1A1 is the dominating PST in liver; it is also expressed at lower levels in nearly all investigated extrahepatic tissues, including platelets, which have been used extensively for phenotyping (Weinshilboum 1990).
SULT1A2	Other names: phenol-sulphating phenol sulphotransferase 2 (P-PST-2), thermostable phenol sulphotransferase (TS PST 2), ST1A2, HAST4 and HAST4v (representing alloenzymes). State of knowledge: detected via its cDNA, only minute knowledge of function and distribution in the organism. Gene: located on 16p12.2 (close to 1A1 and 1A3); various SNPs, some of them leading to an amino acid exchange (Raftogianis et al. 1999); the most prominent polymorphism involves codon 235 (Asn/Thr exchange) and usually is associated with polymorphisms in codon 7 (Ile/Thr) of 1A2 and in codon 213 (Arg/His) of the neighbouring 1A1 gene (Raftogianis et al. 1999; Engelke et al. 2000a). RNA: two alternative first exons (in the non-coding region); an incompletely spliced RNA has been detected in many tissues (Dooley et al. 2000). Protein: the wild-type alloenzymes of 1A1 and 1A2 differ in only six amino acid residues; the electrophoretic mobility of the subunit of 1A2 (apparent $M_r = 32.5$ kDa) differs from that of all other known hSULTs, including 1A1 (apparent $M_r = 32$ kDa). Substrates: 2-naphthol, minoxidil, 4 nitrophenol and 1-hydroxymethylpyrene are sulphonated by 1A2, although less efficiently than by 1A1; OH-AAF is much more efficiently activated by 1A2 than by any other hSULT; some aromatic hydroxylamines are also good substrates of 1A2 (Ozawa et al. 1994; Meinl et al. 2001a). Tissue distribution: 1A2 protein has not yet been demonstrated in any tissue, but is difficult to distinguish from 1A1 due to the high similarity (resolution in Western Blots is only possible if both forms are present at similar levels); 1A2

Table 10.1 (*continued*)

Systematic name	Other names, characteristics properties, selected references
	cDNA has been cloned from the liver; correctly spliced RNA has also been detected in the brain, lung, kidney and gut (Dooley *et al*. 2000).
SULT1A3	Other names: catecholamine-sulphating phenol sulphotransferase (M-PST), thermolabile phenol sulphotransferase (TL PST), human aryl sulphotransferase (HAST) 3 and 5 (indicating two transcripts that differ in the first, non-translated exon), placental oestrogen sulphotransferase (refers to a specific transcript; this name is misleading since E_1 and E_2 are only poor substrates of 1A3).

State of knowledge: excellent compared to other hSULTs; the enzyme has been purified from various human tissues.
Gene: located on 16p11.2.
RNA: three alternative first exons (in the non-coding region).
Protein: dimer; the electrophoretic mobility of the subunit ($M_r = 34$ kDa) is similar to that of 1C2 and 2A1; highly specific peptide antibodies detecting 1A3 are available (Rubin *et al*. 1996); 1A3 has been crystallised (Bidwell *et al*. 1999; Dajani *et al*. 1999b).
Substrates: although numerous phenols are substrates for 1A1 and 1A3, their K_m values are often very different between these enzymes (Brix *et al*. 1999; Dajani *et al*. 1999b); 1A3 displays high affinity for monocyclic phenols containing hydrogen bond donors, e.g. an additional amino group (Dajani *et al*. 1999b) (as is the case with catecholamines); various β receptor agonists (Hartman *et al*. 1998), minoxidil, paracetamol and oxamniquine (which is activated to a mutagen) (Glatt 2000a) are among the drugs that are substrates of 1A3; moderate activity has been reported with E_1 by some research groups (Bernier *et al*. 1994; Suiko *et al*. 2000) but not by others; dopamine at a low concentration (~ 4 µM) is often used to selectively detect 1A3 activity ($K_m \sim 1$ µM) in tissues that co-express the other major PST (1A1, $K_m \sim 100$ µM).
Tissue distribution: particularly high 1A3 protein and activity levels are found in jejunum and colon mucosa (Sundaram *et al*. 1989; Teubner *et al*. 1998; Eisenhofer *et al*. 1999); 1A3 is also expressed in many other tissues, including platelets, which have been used extensively for phenotyping (Weinshilboum 1990); expression is negligible in the liver (Heroux *et al*. 1989; Ozawa *et al*. 1998).

SULT1B1	Other names: thyroid hormone sulphotransferase (hST1B2), ST1B2.

State of knowledge: detected via its cDNA; protein detected in various tissues.
Gene: located on 4q13.1, close to *SULT1E1* and a *SULT1D* pseudogene (Meinl *et al*. 2001b); the sequence of Wang *et al*. (1998) encodes Glu in position 186, that of Fujita *et al*. (1997) encodes Gly in that position; we have sequenced the corresponding DNA region from 10 Chinese and 5 Caucasian subjects; all were homozygous for Glu (Meinl *et al*. 2001b).
Protein: electrophoretic mobility of the subunit (apparent $M_r = 33$ kDa) similar to that of 1C1 and 4A1.
Substrates: 4-nitrophenol, 1-naphthol and iodothyronines are normally used as substrates (Fujita *et al*. 1999; Wang *et al*. 1998); diethylstilboestrol; the promutagens 6-hydroxymethylbenzo[a]pyrene and 4-hydroxycyclopenta[*def*]chrysene are activated with high efficiency by this form (Glatt 2000a).

continued overleaf

Table 10.1 (*continued*)

Systematic name	Other names, characteristics properties, selected references
	Tissue distribution: the highest 1B1 protein levels have been detected in colon; 1B1 protein was also found in other parts of the gastrointestinal tract, liver and leukocytes (Teubner *et al.* 1998; Wang *et al.* 1998); high RNA levels were additionally detected in brain, ovary and kidney (Dooley *et al.* 2000).
SULT1C1	Other names: SULT1C sulphotransferase 1, ST1C2. The systematic name SULT1C1 may be changed to SULT1C4 because of the degree of the structural and functional similarities observed between various rodent and human SULT1C forms.
	State of knowledge: detected via genomic analysis and cDNA cloning (Her *et al.* 1997; Sakakibara *et al.* 1998a; Yoshinari *et al.* 1998a); only minute knowledge of function and distribution in the organism.
	Gene: located on 2q11.2 (Freimuth *et al.* 2000).
	RNA: An in-frame insert (leading to exon 3b instead of the regular exon 3a) has been found in some transcripts; however, no functional protein was produced from a cDNA containing exon 3b (Freimuth *et al.* 2000).
	Protein: the electrophoretic mobility of the subunit (apparent $M_r = 33.3$ kDa) is similar to that of 1B1, 1E1, 2A1 and 4A1 (33 to 33.5 kDa).
	Substrates: activity has been reported only with two substrates: 4-nitrophenol (requiring high substrate concentrations, 50 μM to 10 mM) and—in a single study (Yoshinari *et al.* 1998a)—with OH-AAF (50 μM); even with these substrates, the efficiency of 1C1 is very meagre compared to other hSULTs and rSULT1C1. For example, OH-AAF is not activated to a mutagen in *S. typhimurium* strains expressing 1C1, whereas 4 nM of OH-AAF was sufficient to double the number of revertants in the strain expressing 1A2 (Meinl *et al.* 2001a).
	Tissue distribution: 1C1 protein/activity has not yet been demonstrated in any tissue (the detection being complicated by the lack of specific antibodies and substrates); RNA has been detected in kidney, stomach, thyroid gland and foetal liver (Her *et al.* 1997), and also in ovary and some regions of the brain (Dooley *et al.* 2000).
SULT1C2	Other names: SULT1C sulphotransferase 2, ST1C3.
	State of knowledge: detected via genomic analysis and cDNA cloning (Sakakibara *et al.* 1998a); only minute knowledge of function and distribution in the organism.
	Gene: located on 2q11.2, close to *SULT1C1* (Freimuth *et al.* 2000) and other, not fully elucidated, *SULT1C* genes or pseudogenes.
	Protein: the electrophoretic mobility of the subunit (apparent $M_r = 34$ kDa) is similar to that of 1A3.
	Substrates: 4-nitrophenol, OH-AAF (Yoshinari *et al.* 1998a); E_1, bisphenol A, 4-octylphenol, 4-nonylphenol, diethylstilboestrol (Suiko *et al.* 2000); 1-hydroxymethylpyrene (Glatt *et al.* 2001); in all these studies, relatively high substrate concentrations were used.
	Tissue distribution: 1C2 protein/activity has not yet been demonstrated in any tissue (the detection being complicated by the lack of specific antibodies and substrates); the highest RNA levels were found in foetal lung and kidney; lower RNA levels were detected in foetal heart and in adult kidney, ovary and spinal cord (Sakakibara *et al.* 1998a).
SULT1E1	Other names: oestrogen sulphotransferase (EST, hEST-1), ST1E4.

Table 10.1 (*continued*)

Systematic name	Other names, characteristics properties, selected references
	State of knowledge: detected via its cDNA; subsequently, the protein was also purified from human liver (Forbes-Bamforth and Coughtrie 1994). Gene: 4q13.1 Protein: dimer; the electrophoretic mobility of the subunit (apparent $M_r = 33.5$ kDa) is similar to that of 1C1 and 2A1. Substrates: high affinity for E_1 and E_2 ($K_m = 5$ nM, Zhang *et al.* 1998); for the selective detection of 1E1-mediated activities, they have to be used at a low concentration (5 to 20 nM), which is the physiological range; at micromolar concentrations, several other hSULTs show substantial activity towards E_1 and E_2 (Falany 1997); high activity towards iodothyronines (Kester *et al.* 1999); activity also with pregnenolone, DHEA, diethylstilboestrol, 1-naphthol and naringenin (Falany 1997); very efficient, stereoselective activation of the promutagen (-)-1-(α-hydroxyethyl)pyrene (Hagen *et al.* 1998; Landsiedel *et al.* 1998). Tissue distribution: 1E1 protein was detected in liver (Forbes-Bamforth and Coughtrie 1994), endometrium (Falany *et al.* 1998), jejunum (Her *et al.* 1996) and mammary epithelial cells in primary culture (Falany and Falany 1996a); RNA was detected by Northern Blot analysis in adult adrenal gland, liver and small intestine and in foetal kidney, lung and liver (Her *et al.* 1996) and by reverse transcription/polymerase chain reaction analysis also in skin, brain, vagina (Dooley *et al.* 2000).
SULT2A1	Other names: dehydroepiandrosterone sulphotransferase (DHEA-ST) hydroxysteroid sulphotransferase (HST, HSST), alcohol/hydroxysteroid sulphotransferase (hST$_a$), ST2A3. State of knowledge: excellent compared to other hSULTs; the enzyme has been purified from liver and adrenal gland. Gene: located on 19q13.3; SNPs involving amino acid exchanges (Met57Thr, Glu186Val) have been observed (with frequencies of 0.027 and 0.038) (Wood *et al.* 1996); two other published sequences involve amino acid exchanges, Thr90Ser (Comer *et al.* 1993) and Leu159Val (Kong *et al.* 1992), but have not yet been corroborated in other investigations. Protein: dimer; the electrophoretic mobility of the subunit (apparent $M_r = 33.5$ kDa) is similar to that of some hSULT1 forms; however, antibodies with high selectivity for SULT2 forms are available. Substrates: the highest activity among the steroids has been found with DHEA; pregnenolone, testosterone, cortisol, and E_2 (Comer and Falany 1992; Forbes *et al.* 1995), bile acids (Radominska *et al.* 1990), cholesterol (Aksoy *et al.* 1994); digoxigenin (Schmoldt *et al.* 1992); 3α-hydroxycyproterone acetate and various benzylic alcohols are activated to mutagens by 2A1 (Glatt 2000a); activity has also been observed with 4-nitrophenol (Fujita *et al.* 1999), minoxidil (Kudlacek *et al.* 1997), iodothyronines (Li and Anderson 1999), E_1 and various xenoestrogens (Suiko *et al.* 2000) and OH-AAF (Lewis *et al.* 2000)—however, these substrates are conjugated much more efficiently by some SULT1 enzymes. Tissue distribution: in the adult, 2A1 protein has been detected in liver, adrenal gland and jejunum (in this order) but is absent or extremely low in other tissues investigated (Comer and Falany 1992; Her *et al.* 1996); the same expression pattern was observed on the RNA level, with the exception that moderate 2A1 RNA levels were detected , in addition, in ovary (Luu-The *et al.* 1995; Dooley *et al.* 2000); very high levels of SULT2A1 protein and activity have been

continued overleaf

Table 10.1 (*continued*)

Systematic name	Other names, characteristics properties, selected references
	detected in the foetal adrenal gland (Boström and Wengle 1967; Barker *et al.* 1994; Forbes *et al.* 1995).
SULT2B1	Other names: hydroxysteroid sulphotransferase State of knowledge: detected via its cDNA; only minute knowledge of function and distribution in the organism. Gene: 19q13.3 (close to *SULT2A1*). RNA: two alternative first exons (that contain coding sequences in contrast to the alternative first exons of the *SULT1A* genes) (Her *et al.* 1998). Protein: SULT2B1a and SULT2B1b differ in their *N*-terminal sequence and their electrophoretic mobility (apparent M_r of subunits: 39 and 41 kDa, respectively). Substrate: the only known substrate is DHEA. Tissue distribution: 2B1a and 2B1b proteins have not yet been detected in tissues; RNA was detected primarily in placenta, prostate and trachea (Her *et al.* 1998). Using reverse transcription/polymerase chain reaction, Dooley *et al.* (2000) detected high levels of 2B1a RNA in skin and high levels of 2B1b RNA in vagina, oral mucosa, colorectal mucosa and prostate.
SULT4A1	Other names: brain sulphotransferase-like cDNA/protein (BR-STL), ST5A2 State of knowledge: recently detected via its cDNA (Falany *et al.* 2000), function not known Gene: 22q13.2-.31 Protein: the electrophoretic mobility of the subunit (apparent M_r = 33 kDa) is similar to that of 1B1 and 1C1; the amino acid sequence of human 4A1 shows an unusually high degree of identity (98%) with the orthologous rat and mouse SULTs (that have identical amino acid sequences but markedly differ in their nucleotide sequences). Substrates: unknown. Tissue distribution: significant levels of 4A1 protein and RNA have been detected only in the brain.

[a] Genetic polymorphisms in *SULT* genes are mentioned under the keyword 'gene'. The electrophoretic mobilities were determined under uniform conditions in our laboratory using cDNA-expressed enzyme from *S. typhimurium*. SULT1A1, 1A3, 1B1, 1E1 and 2A1 showed the same migration as the cDNA-expressed enzymes; no standards from tissues were available for the other forms.

but not with human SULT1A3 (M-PST); in Yamazoe's system, M-PST is designated as ST1A5. The different names of the human SULTs are included in Table 10.1.

The genetic nomenclature differs in several principal aspects from the substrate-based nomenclature. Several RNAs may be formed from the same gene by differential splicing. The existence of multiple first exons has been observed for several human *SULT* genes (reviewed by Nagata and Yamazoe 2000). In most cases, the variable first exon is not translated. However, two different translation products (termed SULT2B1a and SULT2B1b) are formed from the human *SULT2B1* gene, as its first exon is variable and contains a part of the coding region (Her *et al.* 1998). Moreover, translation products may associate to homo- and hetero-oligomers and undergo other post-translational modifications (see below). Therefore, several distinct proteins may be formed from the same gene. Conversely, a protein may comprise products of more than one gene. Besides, sequence similarities between genes do not automatically imply functional similarity of the gene products. Although human BR-STL (brain sulphotransferase-like protein, SULT4A1) (Falany *et al.* 2000), rat BR-STL (Falany *et al.* 2000), rat SMP-2 (Chatterjee *et al.* 1987), and rat ST-60 (Watabe *et al.* 1994) represent members of the SULT superfamily, enzyme activities have not yet been reported for these forms. It remains to be investigated whether they act as sulphotransferases or have adopted other functions.

In the following sections, an overview of the different forms of sulphotransferases is given. Both classification systems will be used, as none of these systems is comprehensive. For purified sulphotransferases the substrate-based classification is primarily employed. With the genetic classification, substrate specificities were normally studied using cDNA-expressed enzymes. The focus will be on sulphotransferases of the rat (from which the largest number of enzymes have been purified) and the human as well as special forms from other species. Whereas all four purified human forms can be associated unambiguously with genetically defined forms, this is not yet possible for most enzymes of the rat and other laboratory animals. The human sulphotransferases are presented in detail in Table 10.1.

SUBSTRATE-BASED NAMES OF SULPHOTRANSFERASES

Phenol (aryl) sulphotransferases (PSTs, ASTs)

A large number of phenols are metabolically sulphonated. The tissue distribution of phenol sulphotransferase activity substantially differs from that of the various steroid sulphotransferases, as already reported by Boström and Wengle (1967). Such observations led to the postulation of special PSTs. As discussed above, ASTs I to IV were purified from rat liver. The substrate specificity of these forms, in particular AST IV, is very broad. More than 100 substrates of this enzyme have been reported in the literature. They comprise not only numerous phenols (Sekura and Jakoby 1981; Guo *et al.* 1994; Parker *et al.* 1994) but also various benzylic alcohols (Rao and Duffel 1991a; Glatt *et al.* 1995), aromatic amines (Ozawa *et al.* 1994; Guo *et al.* 1994; Marshall *et al.* 2000), aromatic hydroxylamines and hydroxamic acids (Duffel *et al.* 1992; Gilissen *et al.* 1992; King *et al.* 2000), aryl oximes (Mangold *et al.* 1993) and the *N*-oxide, minoxidil (Hirshey and Falany 1990). For this reason, the name AST is often used instead of PST. However, some alkyl alcohols, nitroalkanes and steroids are also substrates of AST IV, at least under certain experimental conditions (Sodum *et al.* 1994; Andrae *et al.* 1999; Marshall *et al.* 2000). AST IV corresponds to rSULT1A1 or, in some cases, a mixture of rSULT1A1 and rSULT1C1. The presence of additional forms in some preparations of AST IV purified from tissues cannot be ruled out. For AST I to III, the correlates in the sequence-based nomenclature are not yet known. One of them may be rSULT1B1. It is also possible that the cDNA of some of these forms has not yet been cloned.

In the human, two different PSTs could be distinguished and were eventually purified (see below). One of them prefers simple phenols (a property that has led to the designation 'P-PST'), is relatively thermostable ('TS PST') and is highly sensitive to inhibition by 2,6-dichloro-4-nitrophenol (DCNP, see below and Table 10.3); the other form has a high affinity for catecholamines (phenols with a positively charged group in the molecule) ('M-PST'), is relatively thermolabile ('TL PST'), and is inhibited only at relatively high concentrations of DCNP. The systematic names of P-PST and M-PST are hSULT1A1 and hSULT1A3, respectively. Several other human SULTs (1A2, 1B1, 1C1, 1C2, 1E1; see Table 10.1), detected via their cDNA, are also capable of sulphonating various phenols.

Dopamine sulphotransferases

Singer *et al.* (1988) have chromatographically resolved two dopamine sulphotransferases from rat liver cytosol. Dopamine sulphotransferase II (representing the second eluting peak) comprised 79% and 68% of hepatic dopamine sulphotransferase activity in untreated male and female rats, respectively, and appears to be identical to AST IV and minoxidil sulphotransferase 2 (see below). The relationship of dopamine sulphotransferase I with other purified and cDNA-expressed forms is not known.

In the human, dopamine is a diagnostic substrate for M-PST (SULT1A3) if used at low substrate concentrations. Human M-PST shows an apparent K_m value for dopamine of ~1 μM (Dajani *et al.* 1999a), much lower than rat liver cytosol (22.4 μM) and rat dopamine sulphotransferase II (47.5 μM) (Singer *et al.* 1988).

Dopamine is a good substrate for mouse St1d1, a recently detected member of a novel SULT subfamily (see below).

6-Hydroxymelatonin sulphotransferases

Melatonin, the main pineal gland hormone, is excreted mainly as 6-sulphooxymelatonin. Singer *et al.* (1995) have chromatographically resolved two 6-hydroxymelatonin sulphotransferases from rat liver cytosol. 6-Hydroxymelatonin sulphotransferase II comprised 80–90% of the recovered activity in untreated rats of both sexes. Although it co-eluted with dopamine sulphotransferase II/AST IV, it is not clear whether it is identical to these enzymes, as it has not been purified to homogeneity. The relationship of 6-hydroxymelatonin sulphotransferase I with other purified and cDNA-expressed forms is not known.

Minoxidil sulphotransferases

A minoxidil sulphotransferase from rat liver cytosol has been purified to apparent homogeneity (Hirshey and Falany 1990) and its cDNA has been cloned (Hirshey *et al.* 1992). The coding region of this cDNA is identical to that of PST-1 (ST1A1, rSULT1A1) (Ozawa *et al.* 1990).

Singer (1994) chromatographically resolved two minoxidil sulphotransferases from rat liver cytosol. Minoxidil sulphotransferases 1 and 2 comprised nearly 30% and 70% of the recovered enzyme activity from columns, and appear to be identical to dopamine sulphotransferases I and II, respectively. The author suspects that minoxidil sulphotransferase 2/dopamine sulphotransferase II is identical to AST IV.

Several human SULTs (e.g. 1A1, 1A2, 1E1, 2A1) display activity towards minoxidil

(Falany and Kerl 1990; Ozawa *et al.* 1994; Kudlacek *et al.* 1997; Anderson *et al.* 1998).

Oestrogen sulphotransferases (ESTs)

ESTs sulphonate the 3-hydroxy group of the physiological oestrogens E_1, E_2 and E_3. In general, but not always, the name is reserved for forms that have a high affinity for these substrates (apparent $K_m \ll 1$ µM) and are members of the SULT1E subfamily.

Although the physiological oestrogens are phenols, ESTs can be considered as a separate subgroup of enzymes, since they differ from PSTs in several properties. For example, E_1 and phenol sulphotransferase activities show different tissue distributions in the human; this is the case in the foetus (Wengle 1966) as well as in the adult (Boström and Wengle 1967). Likewise, the male/female ratio of E_3 sulphotransferase activity in adult rat liver was 21.3, whereas the same figure for 1-naphthol was only 2.2 (Borthwick *et al.* 1993). This observation led to the purification of a male-specific EST from rat liver (Borthwick *et al.* 1993). The purified enzyme is markedly unstable and has not been studied extensively. Bovine EST was the first mammalian sulphotransferase whose cDNA was cloned (Nash *et al.* 1988), apart from SMP-2. Mouse EST was the first sulphotransferase to be crystallised and subjected to X-ray analysis (Kakuta *et al.* 1997). In the guinea pig, four forms of EST exist that differ in their isoelectric point but have the same amino acid sequence (see below). All of these ESTs are members of the subfamily SULT1E.

Additional sulphotransferases, belonging to other subfamilies (e.g. human SULT1A1, 1A3, 1C2 and 2A1, Table 10.1) are capable of sulphonating physiological oestrogens and, therefore, have also been termed ESTs sporadically. While they can show substantial sulphonating activities towards E_1 and in particular towards E_2, high substrate concentrations (6 to 20 µM) are required for maximal activity, whereas maximal activity of human SULT1E1 is reached at 20 nM of these substrates (Falany 1997). Since human SULT1A1, 1A3 and/or 2A1 are expressed at much higher levels than SULT1E1 in most human tissues, they may dominate the sulphoconjugation of oestrogens in tissue preparations and cell cultures when high substrate concentrations are used. Nevertheless, these sulphotransferases are usually classified as PSTs and hydroxysteroid sulphotransferases (HSTs) rather than ESTs.

Several synthetic oestrogens are poor substrates of ESTs and are more efficiently conjugated by PSTs (Table 10.1) (Falany 1997; Suiko *et al.* 2000). This difference could be important since sulphotransferases appear to regulate the local oestrogen levels and ESTs and PSTs differ in their tissue distribution and regulation; therefore, synthetic oestrogens may be unable to exactly mimic physiological oestrogens (Kotov *et al.* 1999).

Hydroxysteroid, DHEA, cortisol, bile acid, cholesterol and alcohol sulpho-transferases

The term hydroxysteroid sulphotransferase (HST) is normally reserved for enzymes that sulphonate alcoholic steroids. Phenolic steroids are not considered to be characteristic substrates of HSTs, although E_2 is metabolised by some HSTs and has even been used in one study for monitoring the fractions in the purification of an HST (Lyon and Jakoby 1980). Whereas ESTs appear to sulphonate exclusively the phenolic 3-hydroxyl group of E_2, sulphonation of the alcoholic (17β) as well as the phenolic hydroxyl groups has been observed with certain HSTs. DHEA is a particularly good substrate of various HSTs and therefore has been used in many studies. The substrate specificity of HSTs is not limited to steroids; other alcohols and sometimes even alkyl amines (see below) are also included. All HSTs whose nucleotide sequences have been determined are members of the SULT2A or 2B subfamilies.

The complexity of HSTs and SULT2 forms appears to be low in the human but high in the rat. Only one HST has been purified from human tissues (human DHEA sulphotransferase, SULT2A1) (Falany et al. 1989). Its substrates include DHEA, cortisol, cholesterol and bile acids (Table 10.1). It appears to be responsible for all, or most, of the sulphotransferase activity towards hydroxysteroids in the liver and adrenal glands. Two other human HSTs have been identified via their cDNAs (SULT2B1a and 2B1b, Table 10.1). The cDNA-expressed enzymes sulphonate DHEA whereas other steroidal substrates have not yet been studied. The corresponding RNAs have not been detected in liver but, for example, in placenta, prostate and skin (Table 10.1).

The HSTs in rat liver have been studied extensively by Singer (1985). His group has separated three glucocorticoid sulphotransferases from female rat liver using anion exchange chromatography. These enzymes catalysed the sulphoconjugation of the 21-OH group of cortisol and corticosterone. One of these forms (STI) was female-specific, preferred DHEA over any other steroid substrate investigated, and showed a molecular weight of 160 000. STII and STIII were detected in the liver of both sexes, but their levels were higher in females. With STII, as with STI, DHEA was the preferred sulpho acceptor. STIII was a 66-kDa protein, composed of two electrophoretically identical subunits; glucocorticoids rather than DHEA served as the preferential substrates. Singer also studied the conversion of deoxycorticosterone to its 21-sulphate in rat liver. In males, this activity appeared to be due to STIII. However, in females, about half of the activity eluted between STI and STII, indicating the existence of additional HST(s), which preferentially metabolise mineralocorticoids. Bile acid sulphotransferase activity co-eluted with STI from the anion exchange column but unlike STI was detected in both sexes, thus demonstrating the existence of a further HST form, implying a minimum of five HSTs in rat liver.

Although E_2 was associated with ESTs rather than HSTs in later studies, Lyon and Jakoby (1980) used that substrate and butanol for the purification of an HST. They separated three peaks of activity towards both substrates by anion exchange chromatography using female rat liver cytosol; only two of these peaks were found in males. The female-specific form, termed HST 1, was purified to apparent homogeneity. The purified enzyme sulphonated various primary and secondary alcohols (including ethanol, 1-butanol, 1-hexanol, vitamin A, ascorbic acid, ephedrine and chloramphenicol), various alcoholic steroids and E_2. However, it was inactive towards E_1, 2-naphthol and 2-naphthylamine. The same research group used DHEA as the substrate during the purification of a second form (HST 2) from female rat liver (Marcus et al. 1980). Purified HST 2 conjugated various alcoholic steroids and E_2, but was inactive towards E_1, 2-naphthol, OH-AAF, deoxycorticosterone and cholesterol.

Barnes et al. (1989) separated three peaks of sulphotransferase activity towards a bile acid, glycolithocholic acid, by anion exchange chromatography using female rat liver cytosol. An apparently homogenous enzyme (BAST I) was purified from one of the peaks. Its substrates included DHEA, testosterone, cortisol and E_2 in addition to various bile acids. The authors suggested that BAST I is the same protein as HST 2 of Marcus et al. (1980).

Ogura et al. (1990a) studied the enzymes involved in the sulphoconjugation of the benzylic alcohol, 5-hydroxymethylchrysene. They separated three peaks of enzyme activity, all of which were associated with HST activity, from female rat liver by anion exchange chromatography. An apparently homogeneous enzyme, termed STa, was purified to apparent homogeneity from the major peak. Later, apparently homogeneous proteins, termed STb and STc, were also isolated from the other peaks (Watabe et al. 1994). All three proteins showed equal subunit size (30.5 kDa) and cross-reacted with an antibody raised against STa. The form STa was also purified by Czich et al. (1994), who used the activation of 1-HMP to a bacterial mutagen for monitoring the chromatographic fractions. In that study, STa contributed nearly 70% to the activation of this promutagen by hepatic cytosol of female rats.

Using an antibody raised against STa and synthetic oligonucleotide probes, several cDNAs were isolated from a female rat liver cDNA library (Ogura et al. 1989, 1990a; Watabe et al. 1994). They were termed ST-20, ST-21, ST-40, ST-41 and ST-60. The encoded subunits consist of 284 or 285 amino acid residues. ST-40 and ST-41 differ by three nucleotides in the coding region involving only one amino acid substitution. ST-20 and ST-21 differ by eight nucleotides in the coding sequence, leading to six amino acid substitutions. One of these amino acid exchanges is localised in the N-terminus. Takahashi et al. (1998) have separated sulphotransferases from rat liver, whose N-terminal sequences corresponded to those of ST-20 and ST-21. Therefore, the sequence differences between their cDNAs appear to be real and can hardly be attributed to a genetic polymorphism. Apart from these findings, the precise relationships between ST-40 and ST-41 and between ST-20 and ST-21 are not known. Otherwise, the amino acid sequence identity between ST-20/21, ST-40/41 and ST-60 varies from 83% to 91% The differences are spread over the entire sequences, indicating that separate genes encode these three forms. The N-terminal and two internal peptide fragments of STa were contained in the amino acid sequence deduced from the ST-40/41 cDNAs, but not in those of the ST-20/21 and ST-60 cDNAs. Thus, ST-40/41 appears to be the cDNA that encodes STa, as corroborated by comparison of chromatographic, electrophoretic and functional characteristics of natural STa and cDNA-expressed ST-40 or ST-41 (Watabe et al. 1994). The N-terminal 24 amino acid residues of BAST I of Barnes et al. (1989) show the highest similarity (two substitutions) with the deduced amino acid sequence of ST-40/41. In conclusion, the precise relationship among the various purified rat HSTs, among some rat HST cDNAs, as well as between purified enzymes and cDNAs is not clear with the exception that STa is encoded by ST-40/41.

Two distinct HSTs that demonstrate substrate specificity with respect to the orientation of the 3-hydroxyl group have been isolated from the guinea pig adrenal gland (Driscoll et al. 1993). One form has a strong preference for 3α-hydroxysteroids (such as allopregnenolone and androsterone); the other form displays a similarly strong preference for 3β-hydroxysteroids (such as pregnenolone, 17-hydroxypregnenolone and DHEA). The corresponding cDNAs have been cloned (Lee et al. 1994a; Luu et al.

1995; Dufort *et al.* 1996). A distinction between 3α-HSTs and 3β-HSTs has not yet been reported for any other species.

Amine sulphotransferases

Ramaswamy and Jakoby (1987) have purified from guinea pig liver a sulphotransferase that showed broad substrate specificity towards primary and secondary aryl and alkyl amines. Although highly purified, this enzyme also sulphonated phenols, hydroxysteroids and even E_1. Two sulphotransferases with high amine *N*-sulphonating activity were purified from male rabbit liver cytosols (Shiraga *et al.* 1999b). One of these forms, designated AST-RB1, efficiently catalysed the sulphonation of alicyclic, alkyl, and aryl amines, but showed negligible activities towards typical substrates of PSTs and HSTs (Yoshinari *et al.* 1998b; Shiraga *et al.* 1999b). The second form, AST-RB2, efficiently catalysed the sulphoconjugation of desipramine; it showed also high activity towards DHEA, but was inactive towards 1-naphthol (Yoshinari *et al.* 1998c; Shiraga *et al.* 1999b). Cloning of the cDNA of these forms indicated that AST-RB2 is a member of the SULT2 family; therefore, it was given an additional name, ST2A8 (Yoshinari *et al.* 1998c). AST-RB1 differed markedly in its amino acid sequence from all known SULTs and therefore was considered to be the first representative of a new family, which led to the designation ST3A1 (Yoshinari *et al.* 1998b). The sequence of an orthologous cDNA from the mouse has been deposited in the GenBank (AF026075). Members of this family have not yet been detected in any other species, nor are human or rat sulphotransferases known that show a high selectivity for amines as substrates. However, certain PSTs and HSTs of these species have *N*-sulphonating activities.

A rat enzyme that *N*-sulphonates 4-phenyl-1,2,3,6-tetrahydropyridine, *N*-deethanolated tiaramide and desipramine was purified to apparent homogeneity from female liver (Naritomi *et al.* 1994). It also showed high activity towards DHEA; its *N*-terminal amino acid sequence was very similar, but not identical, to those of known rat liver HSTs. Therefore, the purified enzyme appeared to be a new form of the HSTs. Rat SULT1A1 (AST IV), a typical PST, *N*-sulphonates the heterocyclic amines 2-amino-3-methylimidazo[4,5-*f*]quinoline (IQ) and 2-amino-3,8-dimethylimidazo[4,5-*f*]quinoxaline (MeIQx) (Ozawa *et al.* 1995a).

In human liver, the *N*-sulphonation of various alicyclic amines is mediated primarily by SULT2A1, which represents the prime HST in that tissue (Shiraga *et al.* 1999a). The heterocyclic amine 1,2,3,4-tetraisoquinoline is metabolised to its sulphamate by cytosolic fractions of human liver, duodenal mucosa and colon mucosa (Pacifici *et al.* 1997b). When hepatic cytosols from different subjects were compared, the rate of conjugation of 1,2,3,4-tetraisoquinoline correlated with that of testosterone, which is a diagnostic substrate of SULT2A1. In the colon, which does not express SULT2A1 (nor any other testosterone sulphotransferase activity), the conjugation of 1,2,3,4-tetraisoquinoline correlated with that of 4-nitrophenol, suggesting the involvement of a PST; among the three sulphotransferases detected in that tissue (SULT1A1, 1A3, 1B1) (Teubner *et al.* 1998), SULT1A1 shows clearly the highest activity towards 4-nitrophenol and therefore is likely to mediate the sulphonation of 1,2,3,4-tetraisoquinoline. Desipramine is *N*-sulphonated in the cytosolic fraction of human liver, lung, kidney, ileum, colon and platelets (Romiti *et al.* 1992). In platelets, only SULT1A1 and 1A3 appear to be present, suggesting that one of these forms has *N*-sulphonating activity towards desipramine. In human liver, SULT1A1 appears to be the major enzyme catalysing the sulphonation of 2-naphthylamine (Hernandez *et al.* 1991).

N-Hydroxylamine sulphotransferases (HASTs)

An *N*-hydroxy-2-acetylaminofluorene (OH-AAF) sulphotransferase was purified to apparent homogeneity from male rat liver by Wu and Straub (1976). This enzyme is expressed nearly exclusively in males. Using the same substrate, Gong *et al.* (1991) purified a male-dominant and a male-specific sulphotransferase from male rat liver to homogeneity. They were termed HAST-I and HAST-II, respectively. (Note that the research group of McManus uses the abbreviation HAST in a totally different sense, namely for human aryl sulphotransferase—see Table 10.1.) It is possible that HAST-I and/or II correspond to the OH-AAF sulphotransferase of Wu and Straub. All these purified forms also showed high PST activities but differed from the known rat PST forms (AST I, II, III, and IV). The cDNA of HAST-I was identified, cloned and designated as ST1C1 (Nagata *et al.* 1993).

It was previously thought that OH-AAF is a good substrate of AST IV (e.g. Gilissen *et al.* 1992). However, Kiehlbauch *et al.* (1995) demonstrated that an AST IV, purified from male rat liver, contains several distinct components; one component encoded by *rSULT1A1* catalysed the sulphoconjugation of phenols, whereas another component (most likely encoded by *rSULT1C1*) dominated the conjugation of OH–AAF. Nevertheless, AST IV from rat liver, whose homogeneity and identity with rSULT1A1 was corroborated by peptide analysis, efficiently sulphonated various unsubstituted and *N*-alkylated *N*-hydroxyanilines (King and Duffel 1997; King *et al.* 2000).

In the human, two SULT1C forms (1C1 and 1C2) have been detected. They show only very low activity towards OH-AAF and other aromatic hydroxylamines/amides (references in Table 10.1). hSULT1A2 and 1A1 are much more active against OH-AAF and 2-hydroxylamino-5-phenylpyridine than are hSULT1C1 and 1C2 (Meinl *et al.* 2001a).

Dopa/tyrosine and thyroid hormone sulphotransferases

Sakakibara *et al.* (1995) purified a sulphotransferase from rat liver using tyrosine as the substrate. The purified enzyme also metabolised L-dopa, D-dopa, thyroid hormones and 4-nitrophenol. The author called it dopa/tyrosine sulphotransferase. It is now classified in the SULT1B family (ST1B1 or rSULT1B1). This rat enzyme and its human orthologue (ST1B2 or hSULT1B1) have been referred to as thyroid hormone sulpho-transferases by Fujita *et al.* (1997; 1999). The high activity of ST1B2 (hSULT1B1) towards thyroid hormones has also been recognised by Wang *et al.* (1998). However, several other SULTs sulphonate thyroid hormones (for human SULTs see Table 10.1). High activity has been observed in particular with human EST (SULT1E1) (Kester *et al.* 1999). Moreover, in the rat, expression of EST but not of SULT1B1 is under the control of thyroid hormones.

Interestingly, as will be discussed later, the human SULT1A3 changed its substrate specificity and displayed high dopa/tyrosine sulphotransferase activity in the presence of Mn^{2+} ions.

Eicosanoids as substrates of sulphotransferases

Liu et al. (1999) have recently cloned the cDNA (clone 679153) of a mouse SULT that efficiently catalyses the sulphonation of various prostaglandins, thromboxane B_2, and leukotriene E_4. Dopamine, serotonin, 4-nitrophenol, 2-naphthol, 2-naphthylamine and minoxidil are other substrates of this enzyme, whereas none of the investigated steroids was sulphonated (Sakakibara et al. 1998a; Liu et al. 1999). It is a member of a novel SULT subfamily. Nagata and Yamazoe (2000) have given the name St1d1 to this form. The sequence of an orthologous rat cDNA has been deposited in the GenBank under the accession number U32372. A homologous human pseudogene is located in the SULT1E1/SULT1B1 cluster (Meinl et al. 2001b).

Protein tyrosine, tyrosine ester and gastrin sulphotransferases

Various plasma membrane and secretory proteins as well as peptide hormones contain sulphonated tyrosine residues. This modification is usually caused by membrane-bound protein tyrosine sulphotransferases, which differ genetically from the SULT superfamily (Huttner 1987; Ouyang et al. 1998). Nevertheless, xenobiotic-metabolising sulphotransferases were also studied for such activities. Rat AST IV sulphonates tyrosine methyl and ethyl esters as well as some peptides containing an N-terminal tyrosine residue (enkephalins and cholecystokinin heptapeptide) (Sekura and Jakoby 1981); free tyrosine and peptides with only internal tyrosine residues (cholecystokinin octapeptide, gastrin, angiotensin 1 and 2) were ineffective as sulpho acceptors. Due to this activity, rat AST IV has also been designated as tyrosine ester sulphotransferase (Chen et al. 1992). The same name has been given to a recently detected form (GenBank accession number U32372, see above). A soluble gastrin sulphotransferase has been detected in rat gastric mucosa (Chen and Rosenquist 1990). It is not yet known whether it is a member of the SULT superfamily.

Carbohydrate sulphotransferases

Carbohydrate sulphotransferases (Bowman and Bertozzi 1999) sulphonate glycans attached to proteins and lipids. They are localised in the Golgi membranes and in the serum. They are not members of the SULT superfamily but form a separate superfamily (http://www.gene.ucl.ac.uk/nomenclature/genefamily/sulfotrans.shtml). No xenobiotic-metabolising activity has been detected for these enzymes.

Arylsulphate sulphotransferases

Arylsulphate sulphotransferases transfer the sulpho group from arylsulphates to phenolic acceptors without the requirement of PAP (see below). They have been found in intestinal bacteria (Kim et al. 1986; Baek et al. 1998; Kwon et al. 1999), but not in mammals. They show high activity towards numerous phenolic substrates (Koizumi *et al.* 1991; Konishi-Imamura *et al.* 1991; Kim *et al.* 1992). An arylsulphate sulpho-transferase has also been detected in *Aspergillus orizae* (Burns *et al.* 1977).

SEQUENCE-BASED NAMES OF SULPHOTRANSFERASES

The classification principals for the sequence-based names have already been discussed. A complete list of the human members of the SULT family, together with their substrates, is given in Table 10.1. Therefore, only a few special forms require comment here.

Sakakibara *et al.* (1998a) have searched through a murine-expressed sequence tag bank and subsequently cloned two new SULT cDNAs. One of them (clone 679153 ST) is a member of the SULT1 family, but not of the established subfamilies (SULT1A, 1B, 1C and 1E). Therefore, Nagata and Yamazoe (2000) have classified it into a separate subfamily (ST1D) of their classification system. Its substrate specificity has already been described. The sequence of an orthologous rat cDNA (ST1D2, thyrosine–ester SULT) has been published in GenBank (accession number U32372). We have detected a human pseudogene belonging to the same subfamily (Meinl *et al.* 2001b).

The ST3 family (in Yamazoe's nomenclature) contains only two members, the rabbit amine ST3A1 and a mouse sequence, SULT-X2 (GenBank accession number AF026075). No further information (e.g. concerning enzymic activity) has been published for the latter form.

QUATERNARY STRUCTURE AND POST-TRANSLATIONAL MODIFICATIONS OF XENOBIOTIC-METABOLISING SULPHOTRANSFERASES

Four SULTs have been purified from human tissues: SULT1A1 from brain (Whittemore *et al.* 1986) and liver (Falany *et al.* 1990; Otterness *et al.* 1992), SULT1A3 from brain (Whittemore *et al.* 1985) and platelets (Heroux and Roth 1988), SULT2A1 from liver (Falany *et al.* 1989) and adrenal gland (Comer and Falany 1992), and SULT1E1 from liver (Forbes-Bamforth and Coughtrie 1994). All purified forms were homodimers, as deduced from size-exclusion chromatography under non-denaturing conditions and electrophoresis of the denatured proteins. The quaternary structure of other human SULTs has not been studied.

Many sulphotransferases from other species also are homodimers. However, ESTs purified from rat liver (Borthwick *et al.* 1993), mouse placenta (Hobkirk *et al.* 1985) and porcine endometrium (Brooks *et al.* 1987) exist as monomers. Substitution of two amino acid residues (Pro269Thr and Glu270Val) converts cDNA-expressed mouse EST to a dimer; conversely, the exchange of a single amino acid (Val269 Glu) was sufficient to transmute the normally dimeric human EST (SULT1E1) to a monomer (Petrotchenko *et al.* 2001).

Several rodent HSTs, including rat HST 1 (Lyon and Jakoby 1980), rat HST 2 (Marcus *et al.* 1980) and rat BAST I (Barnes *et al.* 1989), have a high tendency to aggregate; the enzymes found in freshly prepared subcellular preparations appear to consist of at least four subunits. Kiehlbauch *et al.* (1995) demonstrated that an AST IV purified from male rat liver comprised two distinct homodimers (most likely rSULT1A1 and 1C1) and the corresponding heterodimer. It is not known whether any functional character-istics are affected by heterodimer formation.

No post-translational modification has been reported for any human SULT. However, an internal disulphide bridge is readily formed between Cys66 and Cys232 in rSULT1A1. The reduced and oxidised enzyme forms differ substantially in their pH optima and substrate specificities (Marshall *et al.* 1997). Furthermore, two distinct forms of rSULT1A1 (α and β) can be isolated from rat liver as well as from a recombinant *Escherichia coli* strain (Yang *et al.* 1996). Form α has one molecule of PAP tightly bound per enzyme dimer and can use certain phenolic sulphoconjugates (e.g. 4-nitrophenyl sulphate) but not PAPS as sulpho donors. Form β does not contain a tightly bound PAP; it can utilise PAPS as well as 4-nitrophenyl sulphate as sulpho donor. The use of 4-nitrophenyl sulphate by form β, but not by form α, requires the addition of PAP to the incubation mixture. However, it is clear that the β-PAP complex is different from the α form, most likely due to a slowly occurring change in conformation. The purified α and β forms have been interconverted into the other form by extended incubation with PAP-degrading enzymes and a high concentration of PAP, respectively.

Likewise, guinea pig EST, purified from adrenal gland (and also from recombinant CHO cells), exists in multiple forms which differ in the isoelectric point but appear to have the same primary amino acid sequence (Lee *et al.* 1994b). The isoformation is caused, at least in part, by the binding of PAP. An acidic form containing an endogenous PAP molecule showed a particularly high affinity for 17β-E$_2$ but no enzyme activity. It has been suggested that this form functions *in vivo* as an oestrogen-binding protein rather than an enzyme. This notion is supported by the observation that about two-thirds of the total oestrogen (E$_2$ and E$_1$) content of the cytosolic fraction of guinea pig adrenal gland is in an unconjugated form, although this tissue shows a very high EST activity (Lee *et al.* 1994b).

Sulpho transfer reactions and sulpho donors

REACTION SCHEME AND MECHANISM

All enzymic sulphoconjugations investigated in mammalian organisms involve the transfer of the sulpho moiety (SO_3^-) from the cofactor PAPS to a nucleophilic site of the substrate (A) (Figure 10.1) or, in the reverse reaction, from a donor (AS) to PAP. Both substrates must be present on the enzyme active site simultaneously to yield a ternary complex. In general, PAPS is the leading substrate and PAP is the following product in the forward reaction (Whittemore *et al.* 1985, 1986). However, a random order of binding (and leaving) has also been reported for some sulphotransferase forms, e.g. rat AST IV (Duffel and Jakoby 1981), human EST (SULT1E1) (Zhang *et al.* 1998) and human bile salt sulphotransferase (SULT2A1) (Chen and Segel 1985).

In agreement with this reaction mechanism, PAP is an inhibitor of many sulpho-transferases (Table 10.2). It competes with PAPS. Sulphotransferases may be inhibited at high substrate concentrations, possibly via formation of the dead-end complex SULT-PAP-A. Marked substrate inhibition is commonly observed for substrates with a low K_m, such as 3-chlorophenol for rat AST I (Sekura and Jakoby 1979), 2-naphthol (Duffel and Jakoby 1981) and 4-nitrophenol (Hirshey and Falany 1990) for rSULT1A1

Table 10.2 K_m and K_i values of various sulphotransferases for PAPS and PAP, respectively[a]

Enzyme, enzyme source[b]	Sulpho acceptor	K_m (PAPS), μM	K_i (PAP), μM	References
hSULT2B1a, expressed in COS-1 cells (cytosol)	DHEA	0.0033		Her et al. (1998)
hSULT2A1 (DHEA sulphotransferase), expressed in COS-1 cells (cytosol)	DHEA	0.0067		Wood et al. (1996)
hSULT2A1, expressed in COS-1 cells (cytosol)	DHEA	0.01		Kudlacek et al. (1997)
hSULT2A1, expressed in COS-1 cells (cytosol)	T_3	0.12		Li and Anderson (1999)
hSULT2A1, expressed in COS-1 cells (cytosol)	Minoxidil	0.13		Kudlacek et al. (1997)
hSULT2A1 (bile salt sulphotransferase), purified from liver	Glycolithocholic acid	0.7	0.2	Chen and Segel (1985)
hSULT2A1 (DHEA sulphotransferase), purified from adrenal gland	DHEA	1.6		Comer and Falany (1992)
hSULT2A1 (HST), expressed in V79 cells (postmitochondrial supernatant)	DHEA	2.2		Forbes et al. (1995)
hSULT2A1 (ST2A3), expressed in E. coli (purified enzyme)	DHEA	4.81		Fujita et al. (1999)
hSULT1E1 (EST), expressed in E. coli (purified enzyme)	E_2	0.059	0.038[c]	Zhang et al. (1998)
5β-Scymnol sulphotransferase of Heterodontus portusjacksoni (partially purified enzyme)	5β-Scymnol	4.35	0.37	Pettigrew et al. (1998)
Rat AST I, purified from liver	2-Naphthol	6.5	0.89	Sekura and Jakoby (1979)
Rat AST II, purified from liver	2-Naphthol	12	1.2	Sekura and Jakoby (1979)
Rat HST 1, purified from liver	E_2	12	35	Lyon and Jakoby (1980)
Phenol (tyrosine) sulphotransferase of Euglena gracilis (purified enzyme)	Tyrosine	15	20	Saidha and Schiff (1994)
Rat AST IV (rSULT1A1), purified from liver	Tyrosine methyl ester	23	29 - 50[d]	Sekura and Jakoby (1981)
Rat HST 2, purified from liver	DHEA	47	14	Marcus et al. (1980)
Human liver cytosol (adult)	2-Naphthol	155		Pacifici et al. (1988)

[a] The K_m values presented are selected from nearly 100 values compiled from the literature. For hSULT2A1, figures from several studies are shown to illustrate the variation. In the other cases, data are preferentially presented from studies in which the K_i for PAP was also determined. In general, values represent apparent K_m and K_i.
[b] Name used by the authors and, for human enzymes, systematic name.
[c] K_m for reverse reaction.
[d] Under different conditions (pH).

(AST IV, minoxidil sulphotransferase) or E_2 for rat and human EST (Sugiyama et al. 1984; Zhang et al. 1998). In addition, allosteric inhibition of human EST by E_2 has been observed (Zhang et al. 1998). Therefore, it is important to vary the substrate concentration when new enzymes or substrates are investigated. Inhibition by the product AS rarely has been reported. DHEA-S (Lyon and Jakoby 1980) and glycolithocholic acid sulphate (Chen and Segel 1985) are relatively potent inhibitors of some HSTs. Furthermore, formation of the dead-end complex SULT-PAPS-AS has been observed for human EST and E_2 sulphate (Zhang et al. 1998).

The transfer of the sulpho moiety ($SO_3{}^-$) from the cofactor PAPS to most substrates is virtually irreversible. However, with a small number of acidic phenolic substrates [in particular, 4-nitrophenol ($pK_a = 7.2$), 2-chloro-4-nitrophenol and 3,5-dinitrophenol ($pK_a = 6.4$)], the reverse reaction (PAP + AS → PAPS + A) has been observed (Gregory and Lipmann 1957; Robbins and Lipmann 1957; Duffel and Jakoby 1981; Anhalt et al. 1982). Even in the case of 4-nitrophenyl sulphate and 3,5-dinitrophenyl sulphate, PAP and AS are thermodynamically strongly favoured over PAPS and A ($K_{app} =$ [AS] × [PAP] × [A]$^{-1}$ × [PAPS]$^{-1}$ = 26.4 and 4.1, respectively) (Gregory and Lipmann 1957; Robbins and Lipmann 1957). Therefore, the reverse reaction can occur in vitro, if only PAP and AS are added to the enzyme. It is likely that the reverse reaction is negligible in vivo due to the presence of PAPS and the normally very low levels of PAP.

A transfer of the sulpho moiety from one conjugate (e.g. 4-nitrophenyl sulphate) to a substrate (e.g. 2-naphthol) is possible (Sekura and Jakoby 1981; Yang et al. 1996; Marshall et al. 1997; Frame et al. 2000). However, this transfer is indirect and requires the presence of PAP. In some cases, PAP is tightly bound to the active centre of mammalian sulphotransferases; in these cases, the addition of external PAP is not required for enzyme activity.

In the absence of PAPS and PAP, rat AST IV catalysed the hydrolysis of 2-chloro-4-nitrophenyl sulphate, i.e. transferred its sulpho group to water (Duffel and Jakoby 1981). This reaction was inhibited in the presence of the sulphotransferase inhibitor PCP. Likewise, rat AST IV may hydrolyse PAPS in the absence of an other sulpho acceptor; this effect has been observed especially with the reduced form of the enzyme (Marshall et al. 2000) and under alkaline conditions (pH 8.0) (Lin and Yang 1998). However, such sulphatase activity of sulphotransferases has been observed only rarely and only under very special experimental conditions.

In bacteria, some sulphoconjugation reactions are not supported by PAPS, but require an arylsulphate as sulpho donor (Kim et al. 1994; Kwon et al. 1999). Studies with an arylsulphate sulphotransferase from Eubacterium A-44 demonstrated the formation of an intermediate in which the sulpho group is covalently bound to the enzyme (Kim et al. 1986). The sulpho acceptors in the enzyme are a tyrosine residue and probably, at other stages of the transfer reaction, a histidine residue. The sulpho group can then be transferred from the enzyme to a phenolic substrate. Thus, this bacterial aryl sulphotransferase operates via a ping-pong reaction mechanism.

PAPS, THE SULPHO DONOR

The sulpho donor PAPS was detected by Robbins and Lipmann (1956; 1957). It is a mixed anhydride of a phosphoric acid residue and sulphuric acid. Due to the high potential for hydrolysis of this bond (dG° ∼ −19 kcal/mol; cited in Zhang *et al.* 1998), PAPS has a high 'sulpho donor potential'. A similar situation is found in sulphoconjugates of acidic phenols (see above) and in sulphoconjugates of carboxylic acids (which apparently are not formed metabolically or are too unstable to be detected).

PAPS is normally present in rat liver at levels of 30–70 nmol/g tissue (Klaassen and Boles 1997); in some other studies somewhat higher values have been reported (e.g. 138 nmol/g tissue, Kim *et al.* 1995). In other tissues and species, lower levels (3.6 to 32.7 nmol/g tissue) have been detected (Brzeznicka *et al.* 1987; Klaassen and Boles 1997). Sulphonation is a high-affinity, low-capacity enzymic process in which the entire content of PAPS in the liver can be consumed in less than 1 min (Klaassen and Boles 1997). Therefore, the rate of PAPS synthesis can be pivotal for the rate of sulphonation reactions. Rates of biosynthesis of up to 100 nmol/min/g of liver have been observed (Pang *et al.* 1981). Nevertheless, compounds that are extensively sulphonated (e.g. paracetamol, salicylamide, phenol, 1-naphthol) can decrease tissue levels of PAPS and inorganic sulphate (Krijgsheld *et al.* 1981; Hjelle *et al.* 1985; Kim *et al.* 1995).

PAPS is formed in two steps from inorganic sulphate and ATP. In the first reaction, these substrates are converted by ATP sulphurylase (EC 2.7.7.4) to adenosine-5'-phosphosulphate (APS) and pyrophosphate. APS and ATP are then converted by APS kinase (EC 2.7.1.25) into PAPS and ADP. In the mammalian organism, ATP-sulphurylase and APS-kinase are fused to a common enzyme protein, termed PAPS synthetase (Lyle *et al.* 1995; Kurima *et al.* 1999). The apparent K_m for sulphate in the sulphonation of various substrates in isolated hepatocytes is 0.3 to 0.5 mM (Mulder and Jakoby 1990). Serum levels of inorganic sulphate in humans are approximately 0.3 mM and decrease after the administration of therapeutic doses of acetaminophen (paracetamol), a sulpho acceptor. In some cases, the availability of inorganic sulphate is limiting for the formation of PAPS and sulphoconjugation (Galinsky *et al.* 1979; Klaassen and Boles 1997). Structural analogues of sulphate (such as molybdate, chromate, perchlorate, and nitrate), adenosyl-trapping agents (ethionine), uncouplers of oxidative phosphorylation, inhibitors of the mitochondrial electron transport (menadione), some hepatotoxicants (cadmium) and some glutathione depletors (diethyl maleate) can decrease the rate of synthesis and the tissue levels of PAPS (reviewed by Klaassen and Boles 1997).

In the mouse and in the human, two different PAPS synthetases, encoded by different genes, have been detected (Kurima *et al.* 1998, 1999; Franzon *et al.* 1999; Xu *et al.* 2000). Brachymorphic mice have a genetic deficiency in the synthesis of PAPS, which leads to reduced PAPS levels and decreased rates of sulphonation of xenobiotics (e.g. 4-nitrophenol) *in vivo* (Lyman and Poland 1983). The deficiency is due to a missense mutation in PAPS synthetase 2; this mutation selectively destroys its APS kinase activity (Kurima *et al.* 1998). Various carcinogens, including OH-AAF (Lai *et al.* 1985), *N*-hydroxy-2-aminofluorene (Lai *et al.* 1987), 4-aminoazobenzene and *N,N*-dimethyl-4-aminoazobenzene (Delclos *et al.* 1984) and 1'-hydroxysafrole (Boberg *et al.* 1983) showed substantially reduced DNA adduct-forming and/or carcinogenic activities in brachymorphic mice.

Assays for sulphotransferase activities

COFACTOR SUPPLY

In early times, a cofactor-generating system (e.g. 10 mM $MgSO_4$ and 10 mM ATP) was often used if sulphotransferase activities were determined in subcellular prepara- tions. Under these conditions, the amount of sulphoconjugate formed may depend not only on the level of the sulphotransferase but also on that of the PAPS synthetase. For precise studies of the sulphotransferase reaction, the cofactor PAPS should be added directly. A wide range of apparent K_m values for PAPS has been reported for different sulphotransferases (0.0033 to 155 µM) and even for a single form, as illustrated in Table 10.2 for human SULT2A1 (0.0067 to 4.81 µM). Commercial PAPS preparations often contain substantial levels of its hydrolysis product, PAP (e.g. Lin and Yang 1998). Hydrolysis also occurs during its storage in aqueous solution under inappropriate conditions (we therefore store our stock solutions at −80°C). PAP is a potent competitive inhibitor of all investigated sulphotransferases; often, the K_i value for PAP is lower than the K_m value of the corresponding enzyme for PAPS (Table 10.2).

If sulphoconjugation is studied in intact cells, tissue sections or perfused organs, addition of PAPS to the medium is not required (and is useless). However, inorganic sulphate (in mM concentrations) is required to sustain the endogenous synthesis of PAPS used as sulpho donor.

ASSAYS USING [^{35}S]-LABELLED PAPS OR INORGANIC SULPHATE

The use of [^{35}S]-PAPS as a sulpho donor has led to convenient and sensitive assays. In general, unused [^{35}S]-PAPS and [^{35}S]-SO_4^{2-} are precipitated after the incubation as barium salts. Most sulphoconjugates are not precipitated by barium, and therefore, the radioactivity remaining in the supernatant after centrifugation can be used to quantify the sulphoconjugate formed (Wengle 1964; Foldes and Meek 1973). A few points have to be considered to avoid pitfalls with this assay:

(1) Due to problems with the background radioactivity and for cost reasons, the PAPS concentration normally used in these assays (0.1 to 1 µM) is low compared to the K_m values for PAPS reported for some enzymes (Table 10.2) and with regard to the concentrations used in some other assays (up to 1.8 mM, e.g. Marcus *et al.* 1980).
(2) Although most sulphoconjugates are soluble in the presence of barium, others are precipitated to various degrees; specifically, sulphoconjugates containing carboxyl groups may form insoluble barium salts (Foldes and Meek 1973).
(3) In the standard version of this assay, the sulphoconjugate formed is not further analysed; in practice, the product determined may result from endogenous substrates present in the enzyme preparation (Spencer 1960) or impurities in the investigated compound (e.g. 2-naphthol in some commercial batches of 2-naphthylamine) rather than the putative substrate; therefore controls (enzyme plus [^{35}S]-PAPS, without substrate) are pivotal and analysis of the product may be required in some cases.

(4) The assay is further complicated when enzyme inhibitors that are alternate substrates are studied.
(5) Unstable sulphoconjugates may not be detected.

The use of [^{35}S]-PAPS—with or without barium precipitation—can be combined with other analytical procedures, such as thin-layer chromatography (Sekura and Jakoby 1979) or high-pressure liquid chromatography. Similar analytical procedures can be used in intact cell systems if [^{35}S]-SO$_4^{2-}$ is used for the labelling of the products (Suiko *et al.* 2000).

ASSAYS USING RADIOLABELLED SULPHO ACCEPTORS

Most sulphoconjugates are much more water-soluble than the corresponding un-conjugated molecules. This property has been exploited in partition assays using radiolabelled substrates, in particular steroid hormones (Singer 1985).

If reactive, unstable conjugates are formed, they may be detected by trapping with an appropriate nucleophile, such as methionine (Wu and Straub 1976), guanosine (Lai *et al.* 1985; Surh *et al.* 1991), DNA (Ozawa *et al.* 1994, 1995a; Chou *et al.* 1995a,b), RNA (King and Phillips 1968; Boberg *et al.* 1983; Fennell *et al.* 1985), or methionine agarose beads (Ringer *et al.* 1990). This assay has been used for studying the sulphonation of various radiolabelled procarcinogens. It allows the determination of relative activities, but not of absolute rates. Other reactions of reactive sulphoconjugates also have been exploited for the quantification of the sulphonation reaction. For example, *N*-sulphooxy-AAF has been converted to AAF in the presence of dithiothreitol (Yamazoe *et al.* 1987).

FORMATION OF PAP

Since the sulphotransferase reaction leads to the stoichiometric formation of sulpho-conjugate and PAP, the quantification of the PAP formed may be used for monitoring the sulphoconjugation of any substrates, including those which form unstable conjugates (Duffel *et al.* 1989). The assay is primarily suited for purified enzymes because crude subcellular preparations often contain substantial PAP-degrading activity.

The availability of highly pure PAPS (free from PAP) is particularly important for the performance and sensitivity of this assay.

PHOTOMETRIC AND FLUORIMETRIC ASSAYS

4-Nitrophenyl sulphate and 4-nitrophenol, in the phenolate form (p$K_a \sim 7$), differ in their spectral absorption. This property has been exploited for the continuous measurement of sulphonation reactions (Yang *et al.* 1996).

In the 'physiological assay' (4-nitrophenol + PAPS → 4-nitrophenyl sulphate + PAP), the decrease in the absorption ($\lambda = 400$ nm) is monitored.

In the 'transfer assay', 4-nitrophenyl sulphate, in the presence of a small amount of PAP, is used as the sulpho donor for another substrate, usually 2-naphthol. The increase in absorption ($\lambda = 400$ nm) is monitored to assess the overall reaction:

4-nitrophenyl sulphate + 2-naphthol → 4-nitrophenol + 2-naphthyl sulphate. Frame *et al.* (2000) have developed a microtitre plate version of this assay.

Various phenolic sulphuric acid esters can be extracted as ion pairs with methylene blue into chloroform. Nose and Lipmann (1958) have used this property to estimate the amount of sulphoconjugates formed from the absorption of the co-extracted methylene blue. Although this assay is not particularly sensitive and accurate, it is simple and has been used, for example, for monitoring the fractions during the purification of enzymes (Sekura and Jakoby 1979, 1981).

In continuous fluorimetric assays, resorufin (Beckmann 1991) or 7-hydroxycoumarin (Leach *et al.* 1999) have been used as the substrate. Because the decrease in the fluorescence is recorded, low substrate concentrations (usually much below the K_m value) have to be employed.

DETECTION OF SULPHOCONJUGATES BY LIQUID CHROMATOGRAPHY/ MASS SPECTROMETRY

Reverse-phase high-pressure liquid chromatography coupled with negative-ion electrospray mass spectrometry is well suited for the identification and quantification of sulphoconjugates (Engst *et al.* 1997). (M-H)$^-$ can be monitored using the selected-ion recording technique. The selectivity and sensitivity can be increased by usage of the multiple-reaction-monitoring technique. Structural information about the sulphoconjugates studied can be obtained from the HSO_4^- (m/z = 97) and/or SO_3^- (m/z = 80) fragments produced by higher cone voltages or by collision-induced dissociation processes using tandem mass spectrometry. In particular, the HSO_4^- fragment is only formed from alcoholic sulphoconjugates, but not from phenolic sulphoconjugates due to the lack of an available hydrogen atom. This differentiation is useful for the identification of the site of sulphonation in substrates (e.g. E_2) that contain phenolic as well as alcoholic hydroxyl groups.

UNUSUAL ASSAYS

If the sulphoconjugates formed are unstable and hydrolyse immediately, special conditions have been used for measuring reaction rates, such as the chiral inversion of enantiomerically pure 1-(α-hydroxyethyl)pyrenes (1-HEP) (Landsiedel *et al.* 1998) (*cf.* Figure 10.2) or the conversion of [^{18}O]-1-HMP to [^{16}O]-1-HMP (*cf.* Figure 10.4, compound **I**) (Landsiedel 1998).

A bacterial mutagenicity assay, a modification of the Ames test, can detect extremely low sulphotransferase levels. Varying concentrations of rat liver cytosol, or fractions obtained during the purification of rat HSTs, were used as the activating system for a promutagenic substrate, 1-HMP (Czich *et al.* 1994; Glatt *et al.* 1994). The mutagenic response was linear over a more than 100-fold range of the amount of enzyme source used. For the detection of HST activity with the standard substrate, (radiolabelled) DHEA, a 100-fold amount of enzyme source was required over that used to detect a mutagenic response with 1-HMP.

Occurrence of sulphotransferases in different species

NON-MAMMALIAN SPECIES

PAPS-dependent sulphotransferases are known from diverse living creatures. The nodulation (Nod) H protein of the bacterium *Rhizobium meliloti* transfers the sulpho group from PAPS to the nodulation factor core structure (Schultze and Kondorosi 1998). However, sulpho transfer from PAPS to small acceptor molecules has so far been observed only in eukaryotic organisms. A sulphotransferase purified from the macroalgae *Porphyra yezoensis* uses dithiothreitol and other thiols as sulpho acceptors (Kanno *et al.* 1996). Several flavonol sulphotransferases with different regio specificities are known from *Flaveria* species, plants belonging to the family of Asteraceae (Marsolais and Varin 1998). An enzyme that catalyses the sulphonation of brassinosteroids and mammalian oestrogenic steroids has been found *in Brassica napus* (Rouleau *et al.* 1999). Choline sulphotransferases occur in the halophytic genus *Limonium* and other Plubaginaceae (Rivoal and Hanson 1994) and in some fungi (Orsi and Spencer 1964; Renosto and Segel 1977). A membrane-bound mitochondrial sulphotransferase that transfers the sulpho group of PAPS to L-tyrosine and other phenols has been purified from *Euglena gracilis* (Saidha and Schiff 1994). The filamentous fungus *Cunninghamella elegans* displays PST activities (Cerniglia *et al.* 1982; Zhang *et al.* 1996). In marine sponges and echinoderms, approximately 90 and 350 sulphonated compounds have been detected, respectively, but very little is known about their biosynthesis (Kornprobst *et al.* 1998). The retinol dehydratase of the insect *Spodoptera frugiperda* is a sulphotransferase (Vakiani *et al.* 1998). The schistosomacidal drugs hycanthone and oxamniquine require bioactivation by an enzyme, probably a sulphotransferase, of the parasite (Pica-Mattoccia *et al.* 1997). Sulphotransferases and/or sulphoconjugates have also been detected in cnidarians (Cormier *et al.* 1970), snails (Takimoto *et al.* 1987b), crustaceans (Elmamlouk and Gessner 1978; Sanborn and Malins 1980; Takimoto *et al.* 1987a; Swevers *et al.* 1991; Li and James 2000), onychophors (Jordan *et al.* 1970), insects (Khoo and Wong 1993; Ngah and Smith 1983), sea urchins (Malins and Roubal 1982), sharks (Cuevas *et al.* 1992; Pettigrew *et al.* 1998), bony fishes (Layiwola and Linnecar 1981; Layiwola *et al.* 1983; Watkins and Klaassen 1986; Kasokat *et al.* 1987; James *et al.* 1997; Coldham *et al.* 1998; Finnson and Eales 1998; Tong and James 2000), frogs (DeMeio 1945; Bridgwater and Ryan 1957; Beyer and Frank 1985; Frank and Beyer 1986; Gorge *et al.* 1987), reptiles (Smith 1968; Huf *et al.* 1987) and birds (Raud and Hobkirk 1968; Collett and Ungkitchanukit 1979; Dickstein *et al.* 1980; Gregus *et al.* 1983; Singer *et al.* 1985; Yang *et al.* 1986; Watkins and Klaassen 1986; Short *et al.* 1988).

DIFFERENCES IN SULPHOTRANSFERASES BETWEEN MAMMALIAN SPECIES

Numerous xenobiotics and their metabolites can be conjugated by UDP-glucuronosyltransferases as well as sulphotransferases. This functional redundancy may be one of the reasons for large species-dependent differences in the preferred conjugation reaction of a chemical.

The house cat and other felines appear to be unable to form glucuronides from various phenols and, therefore, excrete them chiefly as sulphoconjugates (Capel *et al.*

1974; Williams 1974; Mulder and Bleeker 1975; Caldwell 1982). The reverse situation is found in the pig and the opossum (Roy 1963; Capel *et al.* 1974; Williams 1974). These deficiencies do not affect all phenols to the same extent. For example, *in vivo*, substantial amounts of phenolphthalein are glucuronidated in the cat, and sulphonation is a significant elimination pathway of 1-naphthol in the pig.

Watkins and Klaassen (1986) have determined the sulphotransferase activity towards three substrates (2-naphthol, E_2, taurolithocholate) in liver cytosol of eleven species (rat, mouse, guinea pig, rabbit, cat, dog, quail, trout, cattle, sheep and pig). Each substrate was conjugated in each species. Particularly high and low activities (percentage of the respective activity in the rat) were observed with taurolithocholate in dog (10), rabbit (290) and sheep (300); with E_2 in trout (20), quail (320) and cat (620); and with 2-naphthol in trout (5), pig (10) and cattle (380).

Due to the existence of a large number of sulphotransferase forms and competing enzymes, it is difficult to predict the importance of the sulphonation in the metabolism of a compound in a specific species from the known structure–activity relationships, which are very limited, and/or from investigations using subcellular preparations. The situation is further complicated, as homologous sulphotransferases may strongly differ in their tissue distribution between species.

Distribution of sulphotransferases

TISSUE DISTRIBUTION

Methodological aspects

The expression of sulphotransferases can be studied on various levels, i.e. those of enzyme activity, enzyme protein (usually with an antibody) and RNA (Northern Blotting, dot blots, or reverse transcription/polymerase chain reaction). Each method and each probe have their own limitations in specificity and accuracy. Furthermore, a high level of a SULT RNA is not always associated with a high level of protein and enzyme activity. Since the available data are quite fragmentary, many findings about the tissue distribution of SULTs are preliminary.

Tissue distribution of human SULTs

All SULT forms investigated show tissue-selective expression. Data for the human forms are contained in Table 10.1. In this species, only SULT1A1 and 1A3 are found in many different tissues. SULT1A1 displays its highest expression in the liver, but it is also found at lower levels in most other tissues. Expression of SULT1A3 is particularly high in the gut, significant in most other tissues, but negligible in the liver. The high intestinal expression of SULT1A3 appears to reflect an enzymic 'gut–blood barrier' for detoxification of dietary biogenic amines and the delimiting effects of endogenous dopamine generated in the 'third catecholamine system' in the gut (nearly 50% of the dopamine production of the body) (Eisenhofer *et al.* 1999).

The other human SULTs show narrower tissue distributions and often are absent in the liver. The tissue distribution of certain SULTs changes during ontogeny. For example, the liver is the tissue with the highest DHEA sulphotransferase activity (primarily encoded by SULT2A1) in the adult; only half of this activity is detected in adult adrenal gland (Boström and Wengle 1967; Comer and Falany 1992). However, the activity in foetal adrenal gland is approximately thirty- and eight-fold higher than that in foetal and adult liver, respectively (Wengle 1966; Boström and Wengle 1967). Particularly high levels of SULT2A1 RNA have been detected in the foetal zone of the foetal human adrenal gland (Parker 1999). Other changes in the tissue distribution during ontogeny will be discussed later.

Tissue distribution of rat SULTs

The tissue distribution of SULTs in the rat strongly differs from that in the human and other species. In the rat, the expression of many SULTs is conspicuously focused on the liver (e.g. DeBaun et al. 1970; Wong and Yeo 1982; Glatt et al. 1990; Singer et al. 1995; Araki et al. 1997; Takahashi et al. 1998; Dunn II and Klaassen 1998). However, significant levels of minoxidil sulphotransferase (AST IV) have been detected in the outer root sheath of rat hair follicles (Dooley et al. 1991). Only recently, two rat SULT forms (SULT1C2 and 1C2A) were detected whose RNA is abundant in the kidney, at moderate levels in the stomach, very low in the liver and not detectable in the other investigated tissues, including intestine, brain and lung (Xiangrong et al. 2000).

It is important to realise that the enzyme level is not always the limiting factor for the sulphonation reactions occurring in the organism. For example, Mulder et al. (1984) studied the elimination of harmol in the rat and observed that its clearance rate via sulphonation substantially exceeded the hepatic blood flow rate, implying an important role of extrahepatic sulphonation. Likewise, phenol, at low levels, appears to be conjugated primarily in the lung, rather than in the liver, possibly because the perfusion rate of the lung is higher (nearly four-fold) than that of the liver (Cassidy and Houston 1980). Thus, even in the rat, the low extrahepatic levels of sulphotransferases can be sufficient for a substantial contribution to the sulphonation of certain substrates.

Distribution of SULTs in other species

Mouse olfactory mucosa shows higher sulphotransferase activity towards various phenolic odorants than hepatic tissue of this species (Tamura et al. 1997). At least part of this activity is mediated by mSult1c1, a form specifically expressed in the olfactory mucosa; no expression of this form was detected in any other tissue examined (liver, kidney, intestine, spleen, lung, brain, heart) (Tamura et al. 1998).

In the mouse, E_1 sulphotransferase activity is highest in the testis, followed by placenta and uterus, and very low in liver and adrenal gland (Hobkirk and Glasier 1992). However, this pattern may be changed in pathological states. For example, mouse EST RNA is dramatically induced in the livers of obese and diabetic C57BL/KsJ db/db mice, but its expression is unaffected in the testis of these animals (Song et al. 1995). In the guinea pig, the tissue distribution of E_1 sulphotransferase activity is very different from that in the mouse (Hobkirk and Glasier 1992). Highest levels are found in the adrenal gland and midgestational chorion, whereas activities are much lower in the liver and negligible in the gonads.

Substantial species differences in the tissue distribution of sulphotransferase activities were also observed with other substrates. Wong and Yeo (1982) studied the sulphonation of harmol and isoprenaline in cytosol preparations of four tissues (liver, kidney, small intestine, lung) from six species. In the rat and rabbit, both substrates were conjugated only in the liver. The dog showed activities in all tissues, with the highest levels occurring in liver and kidney. In the other species, the highest activities were detected in extrahepatic tissues: lung (monkey), kidney (mouse) and small intestine (guinea pig). Likewise, Schwenk and Locher (1985) observed higher 1-naphthol sulphotransferase activity in jejunal, ileal and colon cells than in hepatocytes of the guinea pig.

Thus, it appears that each SULT form shows a unique tissue distribution, and that orthologous SULT forms (as well as the activity towards a specific substrate) may fundamentally differ in their tissue distribution between species.

CELLULAR DISTRIBUTION

Sulphotransferases are not uniformly distributed among the different cell types of a tissue. For example, the levels of 1-HMP sulphotransferase activity in parenchymal, endothelial, and Kupffer cells of rat liver had a ratio of 216:40:1, respectively (Glatt et al. 1990). These differences are toxicologically important, as 1-HMP is activated to a DNA adduct-forming species by sulphotransferase; adduct levels in the different liver cell types of 1-HMP-treated rats were directly proportional to their respective 1-HMP sulphotransferase activities (Monnerjahn et al. 1993). Furthermore, sulphotransferases are not even distributed uniformly among the parenchymal liver cells, as demonstrated by immunohistochemical analysis of rat liver (Chen et al. 1995). In both sexes, centrilobular hepatocytes contained a higher level of AST IV than midzonal cells; even lower levels of AST IV were present in periportal cells (Chen et al. 1995). Rat STa (or sulphotransferases that cross-react with the antibody raised against STa) exhibited the opposite distribution. Similar results were obtained by Homma et al. (1997), who separated periportal and perivenous hepatocytes using the dual digitonin perfusion technique and then investigated enzyme activities in cytosol preparations of the harvested cells. Sulphotransferase activity towards 2-naphthol (a substrate of AST IV and other SULT1 forms) was approximately 1.5-fold higher in perivenous hepatocytes than in periportal cells. The activities towards DHEA and cortisol (substrates of STa and other SULT2 forms) were 1.6- to 5-fold higher in the periportal cells. However, chromatofocusing of cytosol of perivenous and periportal hepatocytes separated 2-naphthol sulphotransferase activity into three major fractions, which varied in their distribution between perivenous and periportal cells.

In adult male rat liver, only androgen-responsive hepatocytes located around the central vein contain immunoreactive EST protein and the corresponding RNA (Mancini et al. 1992).

Immunohistochemical analysis of human colon and ileum demonstrated high levels of SULT1A and 1B proteins in the differentiated enterocytes and negligible levels in the crypts, where the stem cells are localised (W. Teubner, M. Kretzschmar, C. N. Falany and H. R. Glatt, manuscript in preparation).

Antibodies raised against human SULT1A1 (P-PST) strongly stained neurones in human hippocampus and thalamus, whereas they showed a weak reaction or none at all with the neighbouring glial cells (Zou *et al.* 1990).

SUBCELLULAR LOCALISATION

Sulphotransferase activities towards xenobiotics nearly always were associated with the soluble fraction of tissue and cell homogenates. However, Fernando *et al.* (1993) reported on the presence of a PST activity (towards 1-naphthol, 2-naphthol, 3-nitrophenol and 4-nitrophenol) in microsomal preparations of bovine liver.

Immunohistochemical analyses usually indicate that the sulphotransferases are present primarily in the cytoplasm. However, high levels of EST protein have been detected in cell nuclei of the adrenal cortex of guinea pigs (Whitnall *et al.* 1993) and in a small number of hepatocytes in rats (Mancini *et al.* 1992). In rat brain, a higher specific DHEA sulphotransferase activity was observed in the nuclei fraction than in the cytosol fraction; however, the absolute level of these activities was very low (Rajkowski *et al.* 1997). Besides, rat SULT1C2 and 1C2A appear to be present primarily in the lysosomes in gastric mucosa as well as in transiently transfected baby hamster kidney (BHK) cells (Xiangrong *et al.* 2000).

Tyrosine protein sulphotransferases are localised in the Golgi membranes (Huttner 1987; Ouyang *et al.* 1998). Likewise, the Golgi membranes are the primary location of the carbohydrate sulphotransferases (Bowman and Bertozzi 1999). However, some of them have also been detected in serum (Huynh *et al.* 1999; Nadanaka *et al.* 1999).

Regulation of sulphotransferase expression

ONTOGENETIC DEVELOPMENT, AGE- AND SEX-DEPENDENCE

Human

High levels of sulphotransferases have been detected in various foetal tissues. HST activity and protein level in the adrenal gland (Wengle 1966; Barker *et al.* 1994), HST activity in the kidney (Sharp *et al.* 1993), PST and HST protein levels in the lung (Hume *et al.* 1996) and 2-naphthol sulphotransferase activity in lung, kidney and gut (Pacifici *et al.* 1988) are much higher in the foetus than in the adult. Ritodrine sulphotransferase activity in various tissues is also higher in the foetus than in the adult (Pacifici *et al.* 1993b). SULT1C2 RNA was detected in the lung and heart of the foetus but not of the adult; likewise, its level was higher in foetal than in adult kidney (Sakakibara *et al.* 1998b). Approximately two-fold higher platelet 4-nitrophenol and dopamine sulphotransferase activities were observed in newborns than in adults (Pacifici and Marchi 1993).

The situation is somewhat different in the liver. Whereas sulphotransferase activities towards numerous substrates have been detected in foetal liver, these were usually lower than the corresponding activities in the adult (Wengle 1966; Boström and Wengle 1967; Pacifici *et al.* 1988; Barker *et al.* 1994). However, the hepatic rates of the sulphonation of dopamine and ritodrine (β_2 agonist), which are mediated by SULT1A3, are higher in the foetus than in the adult (Cappiello *et al.* 1991; Pacifici *et al.* 1993b). In contrast, hepatic glucuronidation of ritodrine is hardly developed at the midgestational stage. In newborns, the ratio of urinary sulpho- and glucuronic acid conjugates of ritodrine was markedly higher than in the mothers (Brashear *et al.* 1988).

Age-dependent expression of human sulphotransferases has not been observed in a large number of studies, apart from the foetal and early postnatal periods, nor was sex dependence detected in most tissues and with most substrates. The hepatic sulphotransferase activities towards testosterone, budesonide and 1,2,3,4-tetrahydroi-soquinoline are slightly higher (18–28%) in males (Pacifici *et al.* 1997a,b). A pronounced sexual dimorphism in the platelet 4-nitrophenol sulphotransferase activity was detected in Finns but not in Italians; the median activity was nearly three-fold higher in Finnish men than in Finnish women and in Italians of either sex (Brittelli *et al.* 1999).

Rat

Little is known about the expression of sulphotransferases in the foetal rat. 2-Naphthol sulphotransferase activity was identified in several tissues on day 21 of gestation, although the activities in the extrahepatic tissues were low compared to those found in the human foetus (Pacifici *et al.* 1988). Between birth and 10 weeks of age, increases in PST activity were observed in liver and brain (approximately six-fold) but not in kidney (Maus *et al.* 1982). The hepatic level of SULT1B1 RNA was approximately four times higher at the age of 90 days than at birth in either sex; however, whereas the increase was continuous in females, it reached a peak in males between 30 and 45 days of age (Dunn II *et al.* 1999b). In the adult rat, SULT1B1 RNA and protein levels are similar in either sex.

In rat liver, the expression of all other SULT forms investigated is sex-dependent. SULT1A1 (AST IV) is moderately higher in males; EST (ST1E2) and SULT1C1 (ST1C1) are constitutively expressed only in males (Borthwick *et al.* 1993; Nagata *et al.* 1993; Liu and Klaassen 1996a). In contrast, expression of various HSTs is female-dominant or female-specific (Singer *et al.* 1976; Lyon and Jakoby 1980; Liu and Klaassen 1996b).

The hepatic RNA level of SULT1A1 increases several-fold between birth and puberty in either sex; in the same period, EST (ST1E2) and SULT1C1 RNA rise dramatically from a very low starting level but only in males (Liu and Klaassen 1996a). Hepatic cortisol sulphotransferase activity is very low at birth, develops in parallel in both sexes until 30 days after birth, then rises in females and drops in males until 50 days after birth (Singer *et al.* 1976). Among the different cortisol sulphotransferase forms, STII predominates in immature animals, STIII is the major form in adult males, and STI is essentially restricted to adult females (which also express high levels of STII and STIII). In senescent male rats, certain hepatic HSTs (initially termed SMP-2) are de-repressed (Chatterjee *et al.* 1987; Song *et al.* 1990). Liu and Klaassen (1996b) studied the ontogeny of the hepatic expression of the RNAs of the following HSTs: ST-20/21, ST-40/41 and ST-60. The level of ST-40/41 RNA was highest in immature animals of either sex. In mature males, only ST-20/21 RNA was found, whereas in adult females all three RNA types were detected. The ontogenetic development and the sexual dimorphisms associate STI protein with ST-60 RNA, STII with ST-40/41 (which is identical to STa), and STIII with ST-20/21. However, it is not known whether the associated forms are identical or only show similarities in their regulation.

Castration or oestrogen treatment of male rats leads to increases in various hepatic hydroxysteroid (cortisol, glycolithocholic acid) sulphotransferase activities and decreases in hepatic phenol (dopamine) sulphotransferase activity (Singer *et al.* 1976, 1988; Kirkpatrick *et al.* 1985). The anti-oestrogen tamoxifen, which is a partial agonist for some oestrogen receptors, also induces certain HSTs in male but not female rats (Kirkpatrick *et al.* 1985; Hellriegel *et al.* 1996; Nuwaysir *et al.* 1996). This induction is toxicologically significant, since tamoxifen is activated to a carcinogen by STa, an enzyme that is constitutively expressed only in females. This sex-dependent expression could explain why a short treatment with tamoxifen led to a high level of hepatic DNA adducts in female rats but only to a very low level of adducts in males (Davis *et al.* 2000). Treatment of rats with tamoxifen for a few weeks led to the induction of STa in males and, eventually, to the formation of similar levels of DNA adducts and to similar carcinogenic activities in both sexes (Davis *et al.* 2000).

Ovariectomy led to a decrease in hydroxysteroid (cortisol) sulphotransferase activity and an increase in phenol (dopamine) sulphotransferase activity in the liver, whereas treatment with testosterone had only minor effects on these activities (Singer *et al.* 1976, 1988).

The sex-dependent expression is primarily controlled by the pituitary gland via the pattern of growth hormone secretion (Gong *et al.* 1991, 1992; Borthwick *et al.* 1995b; Liu and Klaassen 1996a,b).

The expression of SULTs in extrahepatic tissues does not reflect the sexual dimorphism observed in the liver; in our experience, the extrahepatic expression is, in general, sex-independent.

Mouse

Similar to rats, substantial sex-dependent differences in hepatic sulphotransferase activities have also been found in mice. However, these dimorphisms differ fundamentally from those observed in the rat. In murine liver, most sulphotransferase activities are higher in females than in males. For example, hepatic T_3 sulphotransferase activity was five-fold higher in the females (Gong *et al.* 1992). Likewise, Borthwick *et al.* (1995a) found 13-, 5-, 60- and 4-fold higher hepatic E_1, E_3, DHEA and 1-naphthol sulphotransferase activities, respectively, in females than in males. Thus, hepatic EST expression is female-dominant in the mouse, but male-specific in the rat. The male mouse may not need EST in the liver, as it is highly expressed in the testis (Hobkirk and Glasier 1992), whereas no EST RNA is expressed in rat testis (Dunn II and Klaassen 1998).

DIFFERENTIATION OF CELLS AND TISSUES

In epidermal cells of murine skin, cholesterol sulphotransferase activity reached a sharp peak around day 16 of gestation; four months after birth, the activity decreased by a factor of 46 (Kagehara *et al.* 1994). Likewise, a strong increase in cholesterol sulphotransferase activity was observed during *in vitro* squamous differentiation of rabbit tracheal epithelial cells (Rearick *et al.* 1987b), human bronchial epithelial cells (Rearick *et al.* 1987a) and human epidermal keratinocytes (Jetten *et al.* 1989). As already described, cholesterol sulphate is an important factor in the squamous differentiation. Minoxidil sulphotransferase activity also increased during the differentiation of human keratinocytes, but with a time course differing from that of cholesterol sulphotransferase activity, suggesting the involvement of distinct enzyme forms (Johnson *et al.* 1992).

In the uterus of the guinea pig (Freeman *et al.* 1983) and the mouse (Hobkirk *et al.* 1983), EST activity was detected in the second half of gestation, but not in the non-pregnant state. Treatment of pigs with progesterone led to an induction of EST activity in the proliferative endometrium (Meyers *et al.* 1983; Brooks *et al.* 1987). Expression of SULT1E1 in the human endometrium is low during the luteal phase, but high during the follicular phase (Rubin *et al.* 1999). It was strongly reduced in women using oral contraceptives, and no expression was detected during early pregnancy. The levels of the other sulphotransferases that have been established in the human endometrium, SULT1A1 and 1A3, were altered only moderately during these physiological stages. Progesterone induced SULT1E1 protein and activity in endometrial adenoma cells in culture (Falany and Falany 1996b), and thus the progesterone level may be important for the endometrial SULT1E1 expression during the menstrual cycle and in women using contraceptives. However, other factors are required to explain the lack of SULT1E1 expression during early pregnancy.

Pseudopregnancy, which was induced by treatment with E_2 followed by chorionic gonadotropin, led to a 30-fold increase in endometrial cholesterol sulphotransferase activity in the rabbit (Momoeda *et al.* 1994).

After partial hepatectomy, a marked decrease in sulphotransferase activity towards OH-AAF (but not towards 4-nitrophenol) was observed in the liver of male rats during the first days (Gilissen and Meerman 1992). During the period of rapid liver growth, the level of various SULT RNAs was decreased (Dunn II *et al.* 1999a). Particularly marked losses were observed for the male-specific forms SULT1C1 and EST (SULT1E2) in males and for the female-dominant form ST-20/21 in females.

CORTICOID AND THYROID HORMONES

Adrenalectomy led to a decrease in hepatic dopamine sulphotransferase II (AST IV) in male rats (Singer *et al.* 1988). Under non-hypertensive regimens, glucocorticoids induce AST IV, repress ST1C1, and have only minor effects on HSTs and ESTs in the liver of male rats (Kirkpatrick *et al.* 1985; Runge-Morris *et al.* 1996; Liu and Klaassen 1996d; Duanmu *et al.* 2000). Dexamethasone induced PST activity strongly in the kidney (6.6-fold), moderately in the liver (1.3-fold) and not at all in the brain of male rats (Maus *et al.* 1982). In females, dexamethasone induced ST-40/41 (STa) in addition to ST1A1 (AST IV) (Liu and Klaassen 1996d).

At hypertensive regimens of cortisol (Singer *et al.* 1988), the induction of dopamine sulphotransferase (AST IV) in the male rat liver was enhanced and associated with a clear increase in cortisol sulphotransferase activity (in particular. form STIII) (Singer *et al.* 1977). However, these effects are not specific for cortisol, but also occur in hypertension produced in other ways (see below), e.g. induced by mineralocorticoids (Singer *et al*. 1977).

In the C57BL/Ks/J mouse (wild-type and *fat/fat* mutant), dexamethasone led to an approximately 10-fold induction of hepatic E_1 sulphotransferase activity (Leiter *et al*. 1999).

Thyroidectomy and drug-induced hypothyroidism led to a decrease in hepatic T_3 sulphotransferase activity in male rats but not in females (Gong *et al*. 1992; Kaptein *et al*. 1997). Administration of T_3 restored the activity. It also led to a 2-fold induction of T_3 sulphotransferase activity in hypophysectomised male and female rats (Gong *et al*. 1992). Treatment of hypophysectomised male rats with T_4 virtually abolished hepatic EST activity and immunoreactive protein, but had no effect on PST and HST activity and on protein levels (Borthwick *et al*. 1995b). Dunn II and Klaassen (2000) observed an increase in hepatic EST RNA in thyroidectomised rats of either sex and partial reversal of this effect by the infusion of T_3 and T_4. These findings indicate that EST is negatively regulated by thyroid hormones. Thyroidectomy had no effect on the hepatic levels of SULT1B1 and 1A1 RNA but altered the pattern of HST RNAs (with sex-dependent decreases in some forms and increases in other forms). Expression of the male-specific form SULT1C1, which exhibits T_3 sulphotransferase activity (Visser 1994), tended to be decreased in the liver of thyroidectomised rats (Dunn II and Klaassen 2000). All these effects of thyroidectomy were reversed, at least partially, by the infusion of thyroid hormone.

INFLUENCE OF PHYSIOLOGICAL STATE, DISEASE AND NUTRITIONAL FACTORS

Leiter *et al*. (1991) observed a dramatic increase in a high-affinity E_1 sulphotransferase activity and a concomitant strong decrease in DHEA sulphotransferase activity in the liver of genetically obese and diabetic female mice (*ob/ob* or *db/db*) compared to normal mice and *fat/fat* mice (which were obese and hyperinsulinaemic but not hyerglycaemic). Borthwick *et al*. (1995a) detected that the increase in E_1 and E_3 sulphotransferase activities is even stronger (nearly 100-fold) in male *ob/ob* mice than in females due to a lower constitutive activity in lean males. Interestingly, HST and PST activities were diminished in females but enhanced in males; these differential effects resulted in a strong mitigation of the sexual dimorphism in the *ob/ob* mouse compared to the controls. The increase in the EST activities was associated with the induction of a novel SULT RNA (Leiter and Chapman 1994) and a novel immunoreactive protein (Borthwick *et al*. 1995a), which were not detected in the liver of control animals of either sex.

Genetically, surgically as well as drug-induced hypertension leads to an induction of hepatic cortisol sulphotransferase activity, in particular of form STIII, in male rats (Turcotte and Silah 1970; Singer *et al*. 1977). Hypertension also induces hepatic dopamine sulphotransferase activity, primarily the form II (AST IV); conversely, anti-hypertensive treatments with spironolactone or hydralazine reduced dopamine sul-photransferase activity (Singer *et al*. 1988).

In Parkinson's disease, strong decreases in dopamine sulphotransferase activity were reported for various regions of the human brain, in particular for hypothalamus, frontal and temporal cortex, amygdaloid nucleus, and occipital and frontal cortex (Baran and Jellinger 1992).

Selenium deficiency led to a drastic decrease (by 92.7%) in the expression of EST (SULT1E) RNA in male rat liver (Yang and Christensen 1998). A moderate decrease (by 38%) was also observed for the 4-nitrophenol sulphotransferase activity, along with various other alterations in the xenobiotic-metabolising system, in male mice receiving a selenium-deficient diet (Reiter and Wendel 1984). The mechanism and functional significance of these down-regulations are not known.

Protein-free diet enhanced hepatic UDP-glucuronosyltransferase activities in immature and adult rats but did not affect the sulphotransferase activities measured with 4-nitrophenol and DHEA (Woodcock and Wood 1971). Food restriction may lead to moderate alterations in the levels of hepatic sulphotransferase activities and proteins (Witzmann et al. 1996; Kaptein et al. 1997).

XENOBIOTIC INDUCERS

Induction of sulphotransferases by hormones and hormonally-active drugs has been presented in the preceding sections. Classical inducers of cytochromes P450 (e.g. phenobarbital and 3-methylcholanthrene) often co-induce conjugating enzymes, such as glutathione transferases and UDP-glucuronosyltransferases. These treatments usually had little or no effect on the sulphotransferase activities studied (Gram et al. 1974; Nemoto and Takayama 1978; Thompson et al. 1982; Watkins 1991; further references in Mulder and Jakoby, 1990). Under some conditions, certain sulphotransferase activities were moderately induced by 3-methylcholanthrene. However, most of these studies were conducted using standard substrates whose conjugation is catalysed by one or usually several major sulphotransferase forms. Moreover, most investigations were restricted to the liver. In light of the molecular complexity of the sulphotransferases, a re-examination is required, since minor forms and extrahepatic enzymes could be pivotal for the metabolism of specific xenobiotics and in particular the activation of procarcinogens in target tissues.

Runge-Morris et al. (1998) observed that treatment of male rats with phenobarbital did not significantly affect the hepatic RNA levels of SULT1C1 and EST (SULT1E2), increased those of SULT1B1 (SULT-Dopa/tyrosine, 4.2-fold), ST-20/21 (SULT-20/21, 1.6-fold, starting from a low level) and ST-60 (SULT-60, 4.2-fold, starting from a low level), and decreased those of SULT1A1 (by a factor of 2.4) and ST-40/41 (SULT-40/41, by a factor of 3.3). Hellriegel et al. (1996) reported on moderate (statistically not verified) changes in the levels of several hepatic SULT RNAs in phenobarbital-treated rats of either sex. Garcia-Allan et al. (2000) searched for phenobarbital-modulated genes in mouse liver using the differential cDNA display technique. The most strongly elevated RNA, detected as cDNA, was shown to encode a sulphotransferase, SULT-N (St1d1), rather than a cytochrome P450, as might have been expected. St1d1 shows a unique catalytic activity for the sulphoconjugation of eicosanoids.

Phenobarbital treatment of pregnant rats produced a 5-fold increase in the hepatic bile salt sulphotransferase activity of the neonate; an even stronger increase in this activity (17-fold) was observed after intrauterine exposure to lithocholate or maternal bile duct ligation (Chen et al. 1982).

Administration of subcarcinogenic doses of AAF to male rats leads to a transient down-regulation of the hepatic OH-AAF sulphotransferase activity (attributed now to SULT1C1) (Ringer *et al.* 1990, 1994). Similar effects were detected with other genotoxic hepatocarcinogens, e.g. benzidine, aflatoxin B_1, ethionine and thiacetamide (Ringer *et al.* 1985).

Treatment of human hepatocytes in culture with rifampicin led to an induction of 17α-ethinyl-E_2 sulphotransferase activity (Li *et al.* 1999). Other typical inducers of cytochromes P450, such as 3-methylcholanthrene, phenobarbital, dexamethasone and omeprazole, did not induce 17α-ethinyl-E_2 sulphotransferase activity. Rifampicin also induced 4-nitrophenol sulphotransferase activity in rat and human hepatocytes in culture (Kern *et al.* 1997).

A number of other compounds that are not prototypic inducers of cytochromes P450 caused induction of sulphotransferases. For example, Singer *et al.* (1984) reported that PCP and acetylsalicylic acid preferentially induced glucocorticoid sulphotransferase form STIII, whereas propranolol, metyrapone and aminoglutethimide selectively enhanced the levels of forms STI and/or STII in male rat liver.

SULPHOTRANSFERASES IN PRENEOPLASTIC AND NEOPLASTIC LIVER TISSUE

Expression of immunoreactive HST proteins is decreased in preneoplastic, ATPase-deficient foci in female rat liver (Werle-Schneider *et al.* 1993). In male rats, decreased levels of OH-AAF sulphotransferase activity and immunoreactive AST IV protein (SULT1A1 and/or 1C1) were detected in hepatic nodules and tumours (Ringer *et al.* 1994; Malejka-Giganti *et al.* 1997). Thus, the regulation of sulphotransferases in preneoplastic lesions and tumours of the liver differs from that of other conjugating enzymes, which often are increased in the altered tissue (Bock *et al.* 1982; Buchmann *et al.* 1985).

SULPHOTRANSFERASES IN CELLS IN CULTURE

In rats, expression of sulphotransferases is sexually dimorphic and strongly focused on the liver. The expression of various sulphotransferases falls dramatically in primary cultures of hepatocytes from male and female rats. This has been demonstrated on the RNA level in particular for ST1C1, EST (ST1E2), ST-20/21, ST-40/41 (STa) and ST-60 (Liu *et al.* 1996). SULT1A1 (AST IV), a form that shows only minor sexual dimorphism, was less sensitive to this down-regulation, and its decrease could be reversed by the addition of dexamethasone to the culture medium. It is probable that this form is responsible for the residual phenol sulphotransferase activities observed in cultured hepatocytes (Grant and Hawksworth 1986; Kane *et al.* 1991; Utesch and Oesch 1992).

In humans, expression of sulphotransferases is sex-independent in general, and several forms display high expression in certain extrahepatic tissues. Preliminary findings suggest that the tissue-dependent expression of sulphotransferases is maintained, at least partially, in primary cultures and cell lines produced from human tissues (Rearick *et al.* 1987a; Johnson *et al.* 1992; Falany and Falany 1996b; Dooley *et al.* 2000).

In order to achieve maximal sulphonation activity in cells in culture, a relatively high concentration of inorganic sulphate in the medium is required (1 to 3 mM, depending on the cells used) (Schwarz and Schwenk 1984; Chen and Schwarz 1985; van de Poll *et al.* 1990).

Inhibitors and activators of sulphotransferases

SULPHOTRANSFERASE INHIBITORS AS DIAGNOSTIC PROBES—GENERAL ASPECTS

Specific inhibitors for an enzyme can be used to demonstrate its involvement in the metabolism or the activation/inactivation of a xenobiotic. In subcellular systems, the addition or omission of the cofactor PAPS can globally probe the role of sulphotransferases. Selective inhibitors for certain sulphotransferase forms (e.g. DCNP, PCP and DHEA) are occasionally used to specify the involved form(s). This specification is only possible to a limited extent for two reasons. (a) These inhibitors interact with the substrate-binding site; in analogy to their promiscuity for the substrates, different sulphotransferase forms show a broad overlap in their interaction with inhibitors, as illustrated in Table 10.3 for various human sulphotransferases and DCNP; (b) the knowledge about the inhibition of different sulphotransferase forms is insufficient; for example, PCP has only been tested with one human sulphotransferase and two rat sulphotransferases individually.

The examples presented in Tables 10.4 and 10.5 demonstrate that the sulphotransferase inhibitors PCP and DHEA can dramatically influence the metabolic fate and the pharmaco-toxicological effects of many xenobiotics in rodents. Strong modulating effects in animal experiments have also been observed with DCNP, e.g. with regard to the sulphoconjugation of harmol (Mulder and Scholtens 1977) and the activation of nitrotoluenes (Kedderis *et al.* 1984; Rickert *et al.* 1984), β-hydroxylated nitrosamines (Sterzel and Eisenbrand 1986; Kroeger-Koepke *et al.* 1992), 2-nitropropane (Sodum *et al.* 1994), N-hydroxy-4'-fluoro-4-acetylaminobiphenyl (van de Poll *et al.* 1989), tamoxifen (Randerath *et al.* 1994a) and safrole (Randerath *et al.* 1994a).

Table 10.3 Inhibition of individual human SULT forms by 2,6-dichloro-4-nitrophenol (DCNP)[a]

Enzyme form[b]	Substrate	IC_{50}, μM	References
SULT1A1*1	4-Nitrophenol, 4 μM	1.44	Raftogianis *et al.* (1999)
SULT1A2*1	4-Nitrophenol, 100 μM	6.94	Raftogianis *et al.* (1999)
SULT1A3	4-Nitrophenol, 3000 μM	86.9	Raftogianis *et al.* (1999)
SULT1B1 (ST1B2)	T_3, 60 μM	400	Fujita *et al.* (1999)
SULT1E1	T_3, 75 μM	30	Li and Anderson (1999)
SULT2A1 (DHEA ST)	DHEA, 5 μM	40	Aksoy *et al.* (1994)

[a] Since DCNP is a competitive inhibitor with respect to the sulpho acceptor, IC_{50} depends on the substrate concentration; the substrates were used at concentrations that lead to near-maximal activity. No data are available about the influence of DCNP on other human SULTs (1C1, 1C2, 2B1a, 2B1b, 4A1).
[b] The name used in the reference cited is given in parentheses if it is different from the systematic name.

Table 10.4 Modulation of the metabolism and/or pharmaco-toxicological effects of xenobiotics in animals by the sulphotransferase inhibitor PCP (examples)[a]

Agent	Animal model	Modulation of the fate/effect of the xenobiotic
Harmol	Male rat	Decrease in the biliary and urinary excretion of harmol sulphate (by up to 71%); increased excretion of harmol glucuronide (by up to 120%) (Meerman *et al.* 1983)
1-Naphthol	Male rat	Decreased urinary excretion of 1-naphthyl sulphate (by up to 48%); mitigated effect at high doses of 1-naphthol (that may overcome the competitive inhibition by PCP) (Boles and Klaassen 1999)
Toluene, benzyl alcohol, xylene and *o*-methylbenzyl alcohol	Male rat	Strong decrease in the urinary excretion of thio compounds (mercapturic acids) (van Doorn *et al.* 1981)
Benzyl acetate	Male rat	Abolition of the excretion of benzyl mercapturic acid (Chidgey *et al.* 1986)
(*R*)-Stiripentol	Male rat	Inhibition of the chiral inversion (Zhang *et al.* 1994)
Minoxidil	Male rat	Ablation of the hypotensive effect (Duanmu *et al.* 2000)
2-Acetylaminofluorene (AAF)	Male rat	Decreased excretion of glutathione conjugates (probably formed from a reactive sulphoconjugate) (by 67–75%); increase in the *N-O*-glucuronide (by 21–32%) (Meerman *et al.* 1983) Complete protection against the induction of liver tumours and resulting death (Ringer *et al.* 1988) Decrease in unscheduled DNA synthesis (determined in the isolated hepatocytes after exposure *in vivo*) (Monteith 1992) Protection against the loss of hepatic OH-AAF sulphotransferase activity (Ringer *et al.* 1985)
	Preweaning mouse	Decrease in hepatic DNA adducts (by 90%) and induced hepatomas (by 80–90%) (Lai *et al.* 1985, 1987)
N-Hydroxy-2-aminofluorene (OH-AAF)	Male rat	Decrease in hepatic RNA adducts (by 61%) and acetylated DNA adducts (by 70%); no influence on the level of deacetylated DNA adducts (Meerman *et al.* 1981). Decrease in the biliary excretion of 1- and 3-(glutathion-*S*-yl)-*N*-acetyl-2-aminofluorene (probably formed from the sulphoconjugate) (by 50%) (Meerman *et al.* 1982)
	Preweaning mouse	Decrease in hepatic DNA adducts (by 75–91%) and hepatomas (by ⩾ 80%) (Lai *et al.* 1987)
2,6-Dinitrotoluene	Male rat	Decrease in binding to hepatic macromolecules (by 65%) and DNA (by > 96%) (Kedderis *et al.* 1984)

continued overleaf

Table 10.4 (*continued*)

Agent	Animal model	Modulation of the fate/effect of the xenobiotic
Safrole	Male rat	Decrease in hepatic DNA adducts and in various cytogenetic effects in hepatocytes (determined in primary culture after the exposure *in vivo*) (Daimon *et al.* 1998)
	Female mouse	Decrease in hepatic DNA adducts (by 91%) (Randerath *et al.* 1994a)
1'-Hydroxysafrole	Preweaning mouse	Decrease in hepatic DNA, RNA and protein adducts (by 50–85%) and hepatomas (by ≥ 90% (Boberg *et al.* 1983)
	Partially hepatectomised rat	Abolition of the acute lethality and the initiation of (phenobarbital-promoted) enzyme-altered foci in the liver; decrease in hepatic DNA, RNA and protein adducts (by 74–89%); nearly complete abolition of the promotion of diethylnitrosamine-initiated enzyme-altered foci and the formation of neoplasms in the liver (Boberg *et al.* 1987)
Diethylstilboestrol	Female mouse	Decrease in hepatic DNA adducts (by 33–61%) (Moorthy *et al.* 1995)
Tamoxifen	Female mouse	Up to 7-fold intensification of the formation of total hepatic DNA adducts; 11-fold increase in polar adducts but 6-fold suppression of unpolar adducts in the liver; also changes in the pattern of DNA adduct in kidney but not in lung (Randerath *et al.* 1994a,b)
4-Hydroxytamoxifen	Female mouse	4-fold increase in hepatic DNA adducts (Randerath *et al.* 1994a)

[a] PCP also strongly decreased the formation of hepatic macromolecular adducts and/or the induction of tumours by 4-aminoazobenzene, *N*-methyl-4-aminoazobenzene, *N*,*N*-dimethyl-4-aminoazobenzene and 2-methyl-4-aminoazobenzene (Delclos *et al.* 1984), *N*-hydroxy-4'-fluoro-4-acetylaminobiphenyl (van de Poll *et al.* 1989), *N*-hydroxy-4-acetylaminobiphenyl (Lai *et al.* 1987), 2-nitrotoluene (Rickert *et al.* 1984) and 1'-hydroxy-2',3'-dehydroestragole (Fennell *et al.* 1985).

Table 10.5 Examples of modulation of the metabolism and/or pharmaco-toxicological effects of xenobiotics in animals by the sulphotransferase inhibitor DHEA

Agent	Animal model	Modulation of the fate/effect of the xenobiotic
Atropine	Guinea pig	Inhibition of dehydration (by 57%) (Wada *et al.* 1994)
6-Hydroxmethyl-benzo[a]pyrene	Preweaning female rat	Prevention of the formation of hepatic DNA adducts (Surh *et al.* 1989)
7-Hydroxmethyl-12-methylbenz[a]-anthracene	Preweaning female rat	Prevention of the formation of hepatic DNA adducts (Surh *et al.* 1987)
2-Nitropropane	Male rat	Marked decrease in various RNA and DNA modifications, e.g. 8-NH_2-dGua (by 92%) (Sodum *et al.* 1994)

2,6-DICHLORO-4-NITROPHENOL (DCNP) AND PENTACHLOROPHENOL (PCP)

Various phenols with electron-withdrawing substituents, such as Cl and NO_2, are poor substrates but potent competitive inhibitors of rat PSTs (Mulder and Scholtens 1977; Koster et al. 1979). Among them, DCNP and PCP show the highest inhibitory activity in vivo. Many other phenols that are inhibitors and competitive substrates in vitro did not significantly affect the sulphoconjugation in vivo (Koster et al. 1979).

In general, PCP is a more potent inhibitor than DCNP. Thus, partially purified rat AST IV was inhibited 97.7% and 51.3% by 1 μM PCP and DCNP, respectively (Sodum et al. 1994). Initial observations suggesting a relatively high toxicity of PCP were not confirmed in later studies, when purified batches of PCP were used. Indeed, PCP can be utilised for long-term inhibition of sulphotransferases in animal experiments (Meerman et al. 1983).

When PCP or DCNP are used as diagnostic inhibitors, it is important to take into account the very limited knowledge about the interaction of these inhibitors with individual forms of sulphotransferases. Rat AST IV and rSULT1C1 are the only rodent sulphotransferases that appear to have been studied as individual (purified or cDNA-expressed) enzymes with PCP. The inhibition of AST IV by PCP is competitive with regard to the sulpho acceptor ($K_i = 0.2$ μM); it could be defined as a dead-end inhibition, since no PCP sulphate is formed (Duffel and Jakoby 1981). rSULT1C1 (ST1C1) was inhibited completely in the presence of 1 μM PCP (Nagata et al. 1993). However, PCP was conjugated by a purified amine sulphotransferase from the guinea pig (Ramaswamy and Jakoby 1987). Likewise PCP, added to a tank with Zebra fishes, was converted to its sulphoconjugate to an extent of 31.4% (Kasokat et al. 1987). Sulphonation of PCP has also been observed in the sea urchin (Tjeerdema et al. 1994), abalone (Tjeerdema and Crosby 1992) and goldfish (Stehly and Hayton 1988).

Inhibition by DCNP has been studied with several individual human sulphotransferase forms (Table 10.3), but not with any highly purified or cDNA-expressed rodent forms. Inhibition of rat enzymes has been investigated in cytosol preparations of hepatic and extrahepatic tissues (e.g. Mulder and Scholtens 1977; Wong et al. 1993). DCNP is an alternate-substrate inhibitor of rat hepatic PST(s) in vitro ($K_m = 4.3$ μM) (Seah and Wong 1994).

Nevertheless, no sulphoconjugation of either DCNP or PCP has been observed in the rat in vivo. Both compounds can elevate hepatic PAPS levels, possibly via inhibition of the conjugation of endogenous substrates (Dills and Klaassen 1986). Through this mechanism, they may enhance the sulphonation of compounds that are conjugated by sulphotransferases resistant to these inhibitors, as has been shown for DHEA (Boles and Klaassen 1998a).

In subcellular preparations of mouse liver, PCP inhibited OH-AAF sulphotransferase activity ($IC_{50} < 1$ μM) without affecting the N-hydroxy-2-aminofluorene acetyltransferase, OH–AAF deacetylase and OH–AAF N,O-transacylase ($IC_{50} > 10$ to 100 μM) (Lai et al. 1985). Shinohara et al. (1986) reported that PCP, but not DCNP, inhibits the O-acetylation of various aromatic hydroxylamines in rat and hamster liver cytosols ($IC_{50} = 12$ to 25 μM). Flammang and Kadlubar (1986) found a moderate inhibition (~ 40%) of rat acetyltransferase and N,O-transacylase activities in the presence of 100 μM PCP. PCP also inhibited murine microsomal epoxide hydrolase ($IC_{50} = 35$ μM) and a glutathione transferase activity ($IC_{50} = 23.5$ μM) *in vitro* (Moorthy *et al.* 1995). Thus, inhibition of non-sulphotransferase enzymes requires substantially higher concentrations of PCP or DCNP than inhibition of certain PSTs. Under conditions that led to inhibition of sulphoconjugation, DCNP did not affect other conjugation reactions in the rat *in vivo* (Koster *et al.* 1979). Of course, only representative reactions could be studied.

Enzyme induction may occur, especially when inhibitors are administered to animals over long periods. Induction of hepatic CYP enzymes has been observed in rats treated for two weeks with PCP (Vizethum and Goerz 1979). Likewise, PCP induced hepatic glucocorticoid sulphotransferase (form STI) in male rats (Singer *et al.* 1984).

DHEA AND OTHER INHIBITORS OF HSTs

DHEA is an inhibitor (alternate substrate) of the 1-HMP, 7-hydroxymethyl-12-methyl-benz[*a*]anthracene and 6-hydroxymethylbenzo[*a*]pyrene sulphotransferase activities in rat hepatic cytosol (Surh *et al.* 1989, 1990, 1991). It has been used in animals to probe the role of HSTs in the metabolism and bioactivation of xenobiotics (Table 10.5).

DHEA-S is a product inhibitor of rat HST 1 ($K_i = 4$ μM) (Lyon and Jakoby 1980) and HST 2 ($K_i = 290$ μM) (Marcus *et al.* 1980). The inhibition is competitive with regard to the substrate DHEA. DHEA-S decreased the formation of ductal carcinomas by *N*-nitroso-bis(2-oxopropyl)amine in Syrian hamsters (Tsutsumi *et al.* 1995). Propylene glycol strongly reduced the DNA alkylation in the liver of rats treated with *N*-nitrosomethyl(2-hydroxyethyl)amine (Kroeger-Koepke *et al.* 1992). It has been postulated that these effects occurred via inhibition of HSTs.

When modulating effects of DHEA and DHEA-S are analysed, it is important to take into account that these compounds not only inhibit sulphotransferases but also exhibit numerous intrinsic activities on the organism and are the metabolic precursors of other steroid hormones. Furthermore, DHEA can diminish the expression of CYP1A1 protein via a stimulation of the degradation of CYP1A1 RNA (Ciolino and Yeh 1999).

MOLYBDATE

Molybdate lowers the serum levels of inorganic sulphate and the availability of PAPS. It is a relatively potent inhibitor of the sulphoconjugation of paracetamol (Oguro *et al.* 1994), harmol (Boles and Klaassen 1998b), 1-naphthol (Boles and Klaassen 1999) and DHEA (Boles and Klaassen 1998a) in the rat *in vivo*. Thus, it appears that sulphonation is decreased regardless of the enzyme form involved; this characteristic is consistent with the postulated mechanism of inhibition. However, the PAPS depletion is tissue-dependent. Molybdate strongly reduced the hepatic PAPS level, but showed negligible effects on the renal level in the rat (Oguro *et al.* 1994). Furthermore, the decrease in the sulphonation was stronger at high doses of the substrates (requiring a high expenditure of PAPS) than at low doses. The opposite was observed with PCP and DCNP (in agreement with the proposed mechanism, competitive inhibition of sulpho-transferases).

In the cases described above, molybdate did not inhibit metabolic pathways competing with sulphonation, in particular glucuronidation. In some cases, these pathways were even enhanced, probably because substrate is saved due to decreased sulphonation.

OTHER INHIBITORS

Numerous drugs (Singer *et al.* 1984; Bamforth *et al.* 1992; Harris *et al.* 1998; Vietri *et al.* 2000), food additives (Bamforth *et al.* 1993) and phytochemicals (Gibb *et al.* 1987; Harris and Waring 1996; Glatt 2000b; Otake *et al.* 2000) are potent inhibitors (and, in part, competitive substrates) of sulphotransferases *in vitro*. Various flavonoids inhibit hSULT1A1 with K_i values of < 1 µM (Ghazali and Waring 1999; De Santi *et al.* 2000). Several hydroxylated metabolites of polychlorinated biphenyls showed IC_{50} values of < 1 nM towards human EST; it has been suggested that this EST inhibition may be the cause of the oestrogenic activity of polychlorinated biphenyls observed in rodents (Kester *et al.* 2000). Tetraalkyl (*n*-propyl, *n*-butyl, *n*-pentyl) ammonium salts are potent inhibitors of DHEA and cortisol sulphotransferase activities ($IC_{50} = 15$ to 66 µM), but they do not affect the sulphonation of 2-naphthol in rat liver cytosol preparations (Matsui *et al.* 1995).

With some chiral compounds, one enantiomer is a substrate of a sulphotransferase whereas its antipode is an inhibitor (Rao and Duffel 1991a; Banoglu and Duffel 1997, 1999). Therefore, caution is required when testing racemic compounds as substrates or inhibitors of sulphotransferases.

Various primary alcohols are substrates of sulphotransferases, but can also be oxidised enzymically via aldehydes to carboxylic acids. The hydrated forms of the aldehydes and the carboxylic acids are nucleophiles and thus potential substrates of sulphotransferases. Whereas sulphonation of such compounds has not been observed to date, some aryl aldehydes and aryl carboxylic acids have displayed relatively strong competitive inhibition of rat AST IV (Rao and Duffel 1991b; Duffel and Zheng 1993).

Various aromatic hydroxylamines are substrates of sulphotransferases. During their incubation in enzyme assays, they may be autoxidised to potent irreversible sulphotransferase inhibitors (King and Duffel 1997).

METAL IONS, pH

Interestingly, the medium composition (salts, pH) not only affects the level of activity of sulphotransferases, but also can modify their substrate specificity. The pH effect is trivial for substrates whose protonation is altered in the investigated range, e.g. certain phenols and amines (Ramaswamy and Jakoby 1987). However, the pH optimum of rat AST IV can also vary between 5.5 and 9 for substrates whose protonation is not altered (Sekura and Jakoby 1981; Marshall *et al.* 1997). Some of these substrates were conjugated only within a small pH range that did not overlap with the appropriate pH range for another substrate. Likewise, the addition of $MnCl_2$ to human M-PST (SULT1A3) drastically affected its substrate and stereo selectivity towards tyrosine derivatives (Sakakibara *et al.* 1997). For example, the activity towards D-*p*-tyrosine was stimulated 130-fold in the presence of 10 mM $MnCl_2$, whereas the activity towards L-dopa was decreased by a factor of 3.7 under the same conditions.

SUICIDE SUBSTRATES

Incubation of purified rat AST IV (probably containing rSULT1C1) with [^{14}C]-OH-AAF and PAPS led to the radiolabelling of the enzyme (primarily at cysteine residues) and to a concurrent loss of OH-AAF sulphotransferase activity (Ringer et al. 1992). Omission of PAPS, addition of an enzyme inhibitor (PCP) and the presence of high levels of an exogenous nucleophile (methionine) abolished these effects, suggesting that they were caused by the product sulphooxy-AAF and that this product was released from the active centre before adduction. Therefore, this self-inactivation may be favoured when using purified enzyme. However, it may also be involved in the decrease in AST IV protein and enzyme activity observed in AAF-treated rats.

The PAPS-dependent inhibition by OH-AAF has to be distinguised from the PAPS-independent, autoxidation-mediated inhibition observed with its deacetylated congener, N-hydroxy-2-aminofluorene.

SULPHOTRANSFERASE INHIBITION BY DRUGS IN THE HUMAN

Data on drug–drug interactions in the human via inhibition of sulphotransferases are scarce. If paracetamol and salicylamide, two substrates of sulphotransferases, were administered together to human subjects, the sulphonation rate of each substrate was lower than when the compounds were administered individually (Levy and Yamada 1971). Dapsone significantly decreased the sulphonation of paracetamol in a clinical study; lamivudine had a similar effect in the cross-sectional part, but not in the longitudinal part, of the study (O'Neil et al. 1999).

Strain differences in animals and genetic polymorphisms in the human

STRAIN DIFFERENCES

Only scarce information is available about strain-dependent differences of sulphotransferases in laboratory animals. King and Olive (1975) reported that Fischer 344 rats express much higher hepatic OH-AAF sulphotransferase activity than Sprague–Dawley rats. In particular, Sprague–Dawley females were virtually deficient of this activity, whereas Fischer 344 females showed activities that were even higher than those in Sprague–Dawley males. Maus et al. (1982) determined the basal and dexamethasone-induced levels of PST activity (substrate 3-methoxy-4-hydroxyphenyl-glycol) in liver, kidney and brain of various rat strains; they did not find any major differences between the ten strains investigated (including Sprague–Dawley and Fischer 344). ACI rats exhibited lower hepatic T_3 sulphotransferase activity than Sprague–Dawley rats (Gong et al. 1992).

The AKR mouse showed higher N-sulphotransferase activities towards several substrates than three other strains (BALB/c, C57/6 and DBA/2) (Shiraga et al. 1995).

INTERINDIVIDUAL DIFFERENCES AND GENETIC POLYMORPHISMS IN HUMANS

STUDIES IN VIVO

Paracetamol, an analgesic, is chiefly excreted as sulpho and glucuronic acid conjugates in the urine. In several studies, the ratio of these conjugates in 8- or 24-hour urine showed an approximately 6-fold variation among healthy adult subjects, e.g. 0.25 to 1.41 (Critchley *et al.* 1986), 0.22 to 1.42 (Lee *et al.* 1992) and 0.26 to 1.33 (Esteban *et al.* 1996). It has not yet been elucidated to which extent this variation is due to differences in the levels and properties of sulphotransferases, UDP-glucuronosyltransferases or other factors (e.g. cofactor supply), and whether genetic or environmental factors are more important. On the one hand, the genetic contribution to the total interindividual variation appeared to be low in a study performed in mono- and di-zygotic twins (Nash *et al.* 1984). On the other hand, Chinese excreted a significantly higher percentage of the sulphoconjugate (35.9%) than Indians (28.9%) ; since both populations, university students in Singapore, had a similar life-style, the role of genetic factors is probable (Lee *et al.* 1992). Methyldopa (Campbell *et al.* 1985), salicylamide (Levy and Yamada 1971; Bonham Carter *et al.* 1983), ritodrine (Brashear *et al.* 1988) and diflunisal (Herman *et al.* 1994) are other drugs that have been used to study individual differences in the rate of excretion sulphoconjugates in the urine.

Phenotyping in tissue samples

The level of sulphotransferase activity and immunoreactive protein substantially varies among tissue samples obtained from different subjects (examples in Table 10.6). In liver, gut and lung samples from adult subjects, 4- to 12-fold variations are typical for sulphotransferase activities towards various substrates. Much larger variations have been observed in platelets, which express SULT1A1 and 1A3. Due to their accessibility, platelets have been used extensively for studying intra- and inter-individual variations of sulphotransferases (examples in Tables 10.6 and 10.8).

Platelet SULT1A1 activity (usually measured with 4-nitrophenol as a characteristic substrate) correlates with the activity in other tissues, e.g. brain, intestine and liver (reviewed by Weinshilboum 1990). It also correlates with the platelet level of immunoreactive SULT1A1 protein (Jones *et al.* 1993). Furthermore, high activity in platelets is associated with the high-thermostability phenotype and the *SULT1A1*1* genotype (see below), although this factor cannot explain all variations of the platelet SULT1A1 activity (Raftogianis *et al.* 1999).

The frequency distribution of platelet SULT1A3 activity (usually measured with dopamine as the diagnostic substrate) is normal in newborns but more complex in adults (Pacifici and Marchi 1993). No significant correlation was observed between the level of SULT1A3 activity in platelets and that in liver, small-intestinal mucosa and brain (reviewed by Weinshilboum 1990).

DHEA sulphotransferase activity in liver and duodenum is primarily mediated by SULT2A1. The frequency distribution of this activity in hepatic samples from 94 different subjects was bimodal with approximately 25% of the subjects included in the high-activity group (Aksoy *et al.* 1993). A bimodal distribution of the level of

Table 10.6 Variation of sulphotransferases activity in tissue samples from different subjects[a]

Substrate	Tissue	Number of subjects	Variation factor[b]	References
4-Nitrophenol	Platelets	905	55	Raftogianis et al. (1999)
4-Nitrophenol	Platelets	100	>100	Pacifici and Marchi (1993)
4-Nitrophenol	Liver	100	10	Pacifici et al. (1997c)
4-Nitrophenol	Lung	96	12	Pacifici et al. (1996)
4-Nitrophenol	Jejunum	64	11	Sundaram et al. (1989)
4-Nitrophenol	Duodenum	100	8	Pacifici et al. (1997c)
4-Nitrophenol	Colon	56	16	Pacifici et al. (1997b)
2-Naphthol	Platelets	174	>100	Frame et al. (1997)
2-Naphthol	Liver	42	4.8	Pacifici et al. (1988)
Dopamine	Platelets	100	>100	Pacifici and Marchi (1993)
Dopamine	Jejunum	64	11	Sundaram et al. (1989)
Dopamine	Lung	96	7	Pacifici et al. (1996)
DHEA	Liver	94	4.6	Aksoy et al. (1993)
Ritodrine	Liver	100	5.4	Pacifici et al. (1998)
Ritodrine	Duodenum	100	8	Pacifici et al. (1998)
(−)-Salbutamol	Liver	100	4.6	Pacifici et al. (1997c)
(−)-Salbutamol	Lung	96	6	Pacifici et al. (1996)
(−)-Salbutamol	Duodenum	100	5.6	Pacifici et al. (1997c)
Desipramine	Platelets	105	64	Romiti et al. (1992)
Desipramine	Liver	118	27	Romiti et al. (1992)
Minoxidil	Platelets	100	>100	Pacifici et al. (1993a)
Minoxidil	Liver	118	10	Pacifici et al. (1993a)

[a] Adult subjects.
[b] Ratio of highest to lowest activity observed (sometimes estimated from figures in the cited article).

immunoreactive SULT2A1 protein was also observed in jejunal tissue samples; however, in this case only 13% of the subjects were in the low SULT2A1 level group (Her et al. 1996).

Genetic polymorphisms

Single nucleotide polymorphisms (SNPs) have been detected in each of the ten human SULT genes known (reviewed by Glatt and Meinl 2001). The only other genetic polymorphisms reported involve single-base deletions in introns of SULT1A and SULT4A1. Most of the SNPs observed in SULTs do not affect the amino acid sequence, nor are any influences on the expression known. I present here only those polymorphisms that affect the amino acid sequence and have been investigated with regard to their frequency and/or functional aspects.

Raftogianis et al. (1997, 1999) have detected 15 different SULT1A1 alleles, which encode four different amino acid sequences. The frequencies of these alloenzymes in 150 random blood donors are shown in Table 10.7. Two further amino acid sequence variants of SULT1A1 have been reported by other groups (Table 10.7). Subjects with the SULT1A1*2/*2 genotype showed a 7.7-fold lower platelet enzyme activity than subjects with the SULT1A1*1/*1 genotype (Raftogianis et al. 1997). In the liver,

Table 10.7 Genetic polymorphisms of human *SULT1A1 and 1A2* involving amino acid substitutions

SULT form	Alloenzyme[a]	Amino acid substitution	Allele frequency (Caucasians)[b]
1A1	*1	(reference sequence)	0.674
	*2	Arg213His	0.313
	*3	Met223Val	0.010
	*4	Arg37Gly	0.003
	*V	Ala147Thr, Glu181Gly, Arg213His	0
	*VI	Pro90Leu, Val243Ala	0
1A2	*1	(reference sequence)	0.508
	*2	Ile7Thr, Asn235Thr	0.287
	*3	Pro19Leu	0.180
	*4	Ile7Thr, Arg184Cys, Asn235Thr	0.008
	*5	Ile7Thr	0.008
	*6	Asn235Thr	0.008

[a] The Arabic numerals of alleles/alloenzymes have been introduced by Raftogianis *et al.* (1997, 1999). Roman numerals are working designations for other variants; since they are only known from individual sequences (Jones *et al.* 1995a; Hwang *et al.* 1995), cloning/sequencing errors cannot be ruled out completely.
[b] Frequencies determined in US Caucasians, 150 random blood donors (1A1) or 61 liver samples (1A2) (Raftogianis *et al.* 1999). The frequencies of the 1A1 Arg213His and 1A2 Asn235Thr polymorphisms have also been determined in various other populations (reviewed by Glatt *et al.* 2001).

however, *SULT1A1*2* was not consistently associated with a low level of enzyme activity (Raftogianis *et al.* 1999), suggesting a tissue-dependent modulation of the SULT1A1 activity by the genotype and/or a contribution of other sulphotransferase forms to the investigated activity in the liver. Differences in enzyme kinetic parameters have been observed between cDNA-expressed SULT1A1 alloenzymes (Raftogianis *et al.* 1999). With all promutagens studied (1-HMP, *N*-hydroxy-2-amino-1-methyl-6-phenylimidazo[4,5-*b*]pyridine, OH-AAF, 2-nitropropane), SULT1A1*His^{213} (*2 and *V) alloenzymes were less active than the wild-type form (*1) (Glatt *et al.* 2001).

A total of 13 different *SULT1A2* alleles, encoding six alloenzymes, have been reported (Ozawa *et al.* 1995b; Zhu *et al.* 1996; Raftogianis *et al.* 1999). The two most common alloenzymes differ by two amino acid exchanges (Ile7Thr and Asn235Thr). In particular, the exchange in codon 235 appears to be functionally important; it strongly decreases the affinity for the standard substrate, 4-nitrophenol (Brix *et al.* 1998; Raftogianis *et al.* 1999; Meinl *et al.* 2001a). Expression of SULT1A2*1 and *5 (having Asn in position 235) in *S. typhimurium* led to a stronger activation of the promutagens OH-AAF, 2-hydroxylamino-5-phenylpyridine and 1-HMP than expression of SULT1A2*2 and *6 (Thr in position 235) (Meinl *et al.* 2001a).

The *SULT1A2* alleles are in a linkage disequilibrium with the alleles of the neighbouring *SULT1A1* gene (Raftogianis *et al.* 1999; Engelke *et al.* 2000a). Usually alleles that encode the high-activity enzyme variants (SULT1A1*Arg^{213} and 1A2*Asn^{235}) are associated with each other; accordingly, the alleles encoding the low-activity enzyme variants (SULT1A1*His^{213} and 1A2*Thr^{235}) form another common haplotype.

Several rare nucleotide exchanges in the *SULT2A1* gene (some leading to amino acid exchanges) were detected (Table 10.1), but could not be related to the levels of enzyme activity or immunoreactive protein in hepatic samples (Wood *et al.* 1996). Mutant proteins (Thr90Ser and/or Leu159Val) expressed in COS-1 cells did not differ in their K_m values for PAPS and DHEA from the wild-type protein.

For SULT1B1, two different cDNA sequences have been published, but not yet verified; they would imply an amino acid exchange, Glu186Gly (Table 10.1). Although the pattern of platelet SULT1A3 activity in families suggests that genetic polymorphisms affect this enzyme (Price *et al.* 1988), these have not yet been elucidated. Since the identical cDNA sequence was invariably isolated from human tissues in several laboratories, it is probable that critical polymorphism(s) are located outside the coding region or even outside the *SULT1A3* gene.

SULT genotypes and phenotypes in different ethnic groups

The percentage of sulphoconjugate among the urinary metabolites of paracetamol is higher in Chinese than in Indians (Lee *et al.* 1992). The mean platelet sulphotransfer-ase activities towards phenol, 4-nitrophenol and 2-naphthol (which are mediated by SULT1A1) are nearly 2-fold higher in African-Americans than in Caucasians, whereas the activities towards dopamine (mediated by SULT1A3) are similar in both ethnic groups (Anderson and Jackson 1984; Frame *et al.* 1997). Significant differences in platelet SULT1A1 as well as 1A3 activities were observed between Finns and Italians (Brittelli *et al.* 1999). A pronounced sexual dimorphism for the SULT1A1 activity was detected in Finns but not in Italians. The finding is unusual, since sexual dimorphisms of sulphotransferases have been detected only very rarely in the human. It is likely that a special life-style, rather than genetic factors, was involved in the aberrant platelet SULT1A1 activity in Finnish males. A role for non-genetic influences is also demon-strated by a study conducted in Italy: platelet SULT1A1 and 1A3 activities showed seasonal rhythms; they were $\sim$ 4-fold higher in summer than in winter (Marazziti *et al.* 1995).

Lower frequencies of *SULT1A1*His²¹³* alleles have been observed in Chinese (0.110) and Japanese (0.168) than in Nigerians (0.269) and various Caucasian popula-tions (0.311 to 0.365); *SULT1A1*Val²²³* (allele *3) occurs in Caucasians in the USA but has not been found in Caucasians in Germany nor in Chinese; it is rather common in African Americans (allele frequency 0.229) (reviewed by Glatt *et al.* 2001).

Associations between disease, age and SULT genotype or phenotype

In several studies, a decrease in platelet sulphotransferase (usually SULT1A1) activity was observed in patients suffering from migraine (Table 10.8). Aberrant platelet sulphotransferase activities were also detected in patients with certain psychiatric disorders and in colon tumour patients. Genotyping of *SULTs* is rather new and therefore has only been used in very few epidemiological studies (Table 10.8); associations between *SULT* genotype and longevity, obesity and certain cancer types have been reported, but require corroboration in prospective studies.

Table 10.8 SULT1A1 phenotype and genotype: association with diseases and other health-related parameters

Investigated disease/factor	Test parameter	Observation
Migraine	Sulphotransferase activities in platelets	Decreased 1A1 activity but normal 1A3 activity (Davis et al. 1987; Marazziti et al. 1996; Alam et al. 1997), especially in patients with diet-induced migraine (Littlewood et al. 1982) Divergent result: normal 1A1 activity but decreased 1A3 activity (Jones et al. 1995b)
	Urinary paracetamol metabolites, ratio of sulphoconjugate to glucuronide	Not significantly different from controls (Alam et al. 1997)
Psychiatric disorders	Sulphotransferase activities in platelets	1A1 and 1A3 activities elevated in patients with obsessive-compulsive disorder and manic patients; normal in dysthymic patients, bipolar depressives and patients with panic disorder; decreased in unipolar depressives (Marazziti et al. 1996)
Hyertension (in males)	SULT1A1 Arg213His and SULT1A2 Asn235Thr polymorphisms	Frequencies not significantly different from controls (Engelke et al. 2000b)
Obesity (in males)	SULT1A1 Arg213His and SULT1A2 Asn235Thr polymorphisms	Increased frequencies of 1A1*His[213] and 1A2*Thr[235] alleles (Engelke et al. 2000b)
Longevity	SULT1A1 Arg213His polymorphism	Decrease in the frequency of the 1A1*His[213] allele with increasing age (Coughtrie et al. 1999)
Colon cancer	2-Naphthol (SULT1A1) sulphotransferase activity in platelets	Increased frequency of the slow-activity phenotype (57% versus 40% in the controls) (Frame et al. 1997)
Prostate cancer	SULT1A1 Arg213His polymorphism	Frequencies not significantly different from controls (Steiner et al. 2000)
Breast cancer	SULT1A1 Arg213His polymorphism	Frequencies not significantly different from controls; however, association of the 1A1*Arg[213] allele(s) with early onset of breast tumour and with the likelihood of having other tumours in addition to breast cancer (Seth et al. 2000)

Competition of sulphotransferases with other enzymes

Nearly every substrate of a sulphotransferase is also metabolised by UDP-glucurono-syltransferases. The sulphotransferases show lower apparent K_m values for many phenolic substrates than the UDP-glucuronosyltransferases (Mulder and Meerman 1978). Besides, the supply of the cofactor PAPS becomes limiting more readily than that of UDP-glucuronic acid. Therefore, it is not surprising that a decrease in the ratio of sulphonated metabolites to glucuronidated metabolites in urine and/or bile was frequently observed when the dose of the xenobiotic was increased, e.g. for phenol, harmol, salicylamide, 4-nitrophenol, 1-naphthol and paracetamol in the rat (Mulder and Meerman 1978; Weitering *et al.* 1979; Hjelle *et al.* 1985; Kane *et al.* 1991; Kim *et al.* 1995) or perfused rat liver (Minck *et al.* 1973; Mulder *et al.* 1975); for paracetamol (Liu and Klaassen 1996c), *o*-phenylphenol (Bartels *et al.* 1998) and phenol (Kenyon *et al.* 1995) in the mouse; for 4-nitrophenol in the perfused foetal sheep liver (Ring *et al.* 1996); for phenol and 1-naphthol in various non-human primates (Mehta *et al.* 1978) and for paracetamol in the human (Clements *et al.* 1984). However, other dosage effects have also been observed, although much more rarely. The proportion of xamoterol excreted as sulphoconjugate remained constant over a 100-fold dose range in the dog (Groen *et al.* 1988). When the dose of diflusinal administered to humans was increased, the recovery of urinary diflunisal sulphate was increased (Loewen *et al.* 1988). In an *in-situ* intestinal loop preparation of the rat, glucuronidation rather than sulphonation became saturated at relatively low doses of various phenols, which were administered into the gut lumen (Goon and Klaassen 1990, 1991).

The trend to develop highly potent drugs that require a very low dosage may lead to an increase in the importance of the sulphonation pathway in pharmacology. Also in environmental toxicology, which usually deals with low exposure levels, the sulphonation pathway may be more important than in long-term carcinogenicity studies in animals, for example, where high doses are used. This discrepancy between human-relevant and animal-experimental situation is problematic because sulphonation involves a much higher risk of the formation of a reactive metabolite than does glucuronidation.

Competition between sulphotransferases and alcohol dehydrogenases is toxicologically significant with benzylic alcohols (L. Ma and H. R. Glatt, unpublished results). *N*-Acetylation and *N*-sulphonation are detoxification pathways for various aromatic amines, whereas *O*-acetylation and *O*-sulphonation of the corresponding hydroxylamines represent alternative activation pathways.

References

Aksoy IA, Sochorová V and Weinshilboum RM (1993) Human liver dehydroepiandrosterone sulfotransferase: nature and extent of individual variation. *Clinical Pharmacology and Therapeutics*, **54**, 498–506.

Aksoy IA, Wood TC and Weinshilboum R (1994) Human liver estrogen sulfotransferase: identification by cDNA cloning and expression. *Biochemical and Biophysical Research Communications*, **200**, 1621–1629.

Alam Z, Coombes N, Waring RH, Williams AC and Steventon GB (1997) Platelet sulphotransferase activity, plasma sulphate levels and sulphation capacity in patients with migraine and tension headache. *Cephalalgia*, **17**, 761–764.

Anderson RJ and Jackson BL (1984) Human platelet phenol sulfotransferase: stability of two forms of the enzyme with time and presence of a racial difference. *Clinica Chimica Acta*, **138**, 185–196.

Anderson RJ, Kudlacek PE and Clemens DL (1998) Sulfation of minoxidil by multiple human cytosolic sulfotransferases. *Chemico-Biological Interactions*, **109**, 53–67.

Andrae U, Kreis P, Coughtrie MWH, Pabel U, Meinl W, Bartsch I and Glatt HR (1999) Activation of propane 2-nitronate to a genotoxicant in V79-derived cell lines engineered for the expression of rat hepatic sulfotransferases. *Mutation Research*, **439**, 191–197.

Anhalt E, Holloway CJ, Brunner G and Trautschold I (1982) Mechanism of sulphate transfer from

4-nitrophenylsulphate to phenolic acceptors via liver cytosolic sulphotransferase. *Enzyme*, **27**, 171–178.

Araki Y, Sakakibara Y, Boggaram V, Katafuchi J, Suiko M, Nakajima H and Liu M-C (1997) Tissue-specific and developmental stage-dependent expression of a novel rat dopa/tyrosine sulfotransferase. *International Journal of Biochemistry & Cell Biology*, **29**, 801–806.

Baek M-C, Kwon A-R, Chung Y-J, Kim B-K and Choi E-C (1998) Distribution of bacteria with the arylsulfate sulfotransferase activity. *Archives of Pharmaceutical Research*, **21**, 475–477.

Bamforth KJ, Dalgliesh K and Coughtrie MWH (1992) Inhibition of human liver steroid sulfotransferase activities by drugs: a novel mechanism of drug toxicity. *European Journal of Pharmacology*, **228**, 15–21.

Bamforth KJ, Jones AL, Roberts RC and Coughtrie MWH (1993) Common food additives are potent inhibitors of human liver 17α-ethinyloestradiol and dopamine sulphotransferases. *Biochemical Pharmacology*, **46**, 1713–1720.

Banoglu E and Duffel MW (1997) Studies on the interactions of chiral secondary alcohols with rat hydroxysteroid sulfotransferase STa. *Drug Metabolism and Disposition*, **25**, 1304–1310.

Banoglu E and Duffel MW (1999) Importance of peri-interactions on the stereospecificity of rat hydroxysteroid sulfotransferase STa with 1-arylethanols. *Chemical Research in Toxicology*, **12**, 278–285.

Baran H and Jellinger K (1992) Human brain phenolsulfotransferase: regional distribution in Parkinson's disease. *Journal of Neural Transmission—Parkinson's Disease and Dementia Section*, **4**, 267–276.

Barker EV, Hume R, Hallas A and Coughtrie MWH (1994) Dehydroepiandrosterone sulfotransferase in the developing human fetus: quantitative biochemical and immunological characterization of the hepatic, renal, and adrenal enzymes. *Endocrinology*, **134**, 982–989.

Barnes S, Buchina ES, King RJ, McBurnett T and Taylor KB (1989) Bile acid sulfotransferase I from rat liver sulfates bile acids and 3-hydroxy steroids: purification, N-terminal amino acid sequence, and kinetic properties. *Journal of Lipid Research*, **30**, 529–540.

Bartels MJ, McNett DA, Timchalk C, Mendrala AL, Christenson WR, Sangha GK, Brzak KA and Shabrang SN (1998) Comparative metabolism of o-phenylphenol in mouse, rat and man. *Xenobiotica*, **28**, 579–594.

Baulieu EE (1996) Dehydroepiandrosterone (DHEA): a fountain of youth? *Journal of Clinical Endocrinology Metabolism*, **81**, 3147–3151.

Baumann E (1865a) Ueber gepaarte Schwefelsäuren im Harn. *Archiv für die gesammte Physiologie des Menschen und der Thiere*, **12**, 69–70.

Baumann E (1865b) Ueber das Vorkommen von Brenzcatechin im Harn. *Archiv für die gesammte Physiologie des Menschen und der Thiere*, **12**, 63–68.

Baumann E (1876) Ueber Sulfosäuren im Harn. *Berichte der Deutschen Chemischen Gesellschaft*, **9**, 54–58.

Beckmann JD (1991) Continuous fluorometric assay of phenol sulfotransferase. *Analytical Biochemistry*, **197**, 408–411.

Beland FA, Miller DW and Mitchum RK (1983) Synthesis of the ultimate hepatocarcinogen, 2-acetylaminofluorene N-sulphate. *Journal of the Chemical Society, Chemical Communications*, 30–31.

Bernier F, Solache IL, Labrie F and Luuthe V (1994) Cloning and expression of cDNA encoding human placental estrogen sulfotransferase. *Molecular and Cellular Endocrinology*, **99**, R11–R15.

Beyer J and Frank G (1985) Hydroxylation and conjugation of phenol by the frog *Rana temporaria*. *Xenobiotica*, **15**, 277–280.

Bidwell LM, McManus ME, Gaedigk A, Kakuta Y, Negishi M, Pedersen L and Martin JL (1999) Crystal structure of human catecholamine sulfotransferase. *Journal of Molecular Biology*, **293**, 521–530.

Boberg EW, Miller EC, Miller JA, Poland A and Liem A (1983) Strong evidence from studies with brachymorphic mice and pentachlorophenol that 1'-sulfoöxysafrole is the major ultimate electrophilic and carcinogenic metabolite of 1'-hydroxysafrole in mouse liver. *Cancer Research*, **43**, 5163–5173.

Boberg EW, Miller EC and Miller JA (1986) The metabolic sulfonation and side-chain oxidation of 3'-hydroxyisosafrole in the mouse and its inactivity as a hepatocarcinogen relative to 1'-hydroxysafrole. *Chemico-Biological Interactions*, **59**, 73–97.

Boberg EW, Liem A, Miller EC and Miller JA (1987) Inhibition by pentachlorophenol of the initiating and promoting activities of 1'-hydroxysafrole for the formation of enzyme-altered foci and tumors in rat liver. *Carcinogenesis*, **8**, 531–539.

Bock KW, Lilienblum W, Pfeil H and Eriksson LC (1982) Increased uridine diphosphate-glucuronyltransferase activity in preneoplastic liver nodules and Morris hepatomas. *Cancer Research*, **42**, 3747–3752.

Boles JW and Klaassen CD (1998a) Effects of molybdate and pentachlorophenol on the sulfation of dehydroepiandrosterone. *Toxicology and Applied Pharmacology*, **151**, 105–109.

Boles JW and Klaassen CD (1998b) Effects of molybdate on the sulfation of harmol and α-naphthol. *Toxicology*, **127**, 121–127.

Boles JW and Klaassen CD (1999) Effects of molybdate and pentachlorophenol on the sulfation of α-naphthol. *Toxicology Letters*, **106**, 1–8.

Bonham Carter SM, Rein G, Glover V, Sandler M and Caldwell J (1983) Human platelet phenolsulphotransferase M and P: substrate specificities and correlation with *in vivo* sulpho-conjugation of paracetamol and salicylamide. *British Journal of Clinical Pharmacology*, **15**, 323–330.

Borthwick EB, Burchell A and Coughtrie MWH (1993) Purification and immunochemical characterization of a male-specific rat liver oestrogen sulphotransferase. *Biochemical Journal*, **289**, 719–725.

Borthwick EB, Burchell A and Coughtrie MWH (1995a) Differential expression of hepatic oestrogen, phenol and dehydroepiandrosterone sulphotransferases in genetically obese dia-betic (*ob/ob*) male and female mice. *Journal of Endocrinology*, **144**, 31–37.

Borthwick EB, Voice MW, Burchell A and Coughtrie MWH (1995b) Effects of hypophysectomy and thyroxine on the expression of hepatic oestrogen, hydroxysteroid and phenol sulphotrans-ferases. *Biochemical Pharmacology*, **49**, 1381–1386.

Boström H and Wengle B (1967) Studies on ester sulphates: 23. Distribution of phenol and steroid sulphokinase in adult human tissues. *Acta Endocrinologica*, **56**, 691–704.

Boulch NL, Cancela L and Miravet L (1982) Cholecalciferol sulfate identification in human milk by HPLC. *Steroids*, **39**, 391–398.

Bowlby MR (1993) Pregnenolone sulfate potentiation of N-methyl-D-aspartate receptor channels in hippocampal neurons. *Molecular Pharmacology*, **43**, 813–819.

Bowman KG and Bertozzi CR (1999) Carbohydrate sulfotransferases: mediators of extracellular communication. *Chemistry & Biology*, **6**, R9–R22.

Brashear WT, Kuhnert BR and Wei R (1988) Maternal and neonatal urinary excretion of sulfate and glucuronide ritodrine conjugates. *Clinical Pharmacology & Therapeutics*, **44**, 634–641.

Bridgwater RJ and Ryan DA (1957) Sulphate conjugation with ranol and other steroid alcohols in liver homogenates from *Rana temporaria*. *Biochemical Journal*, **65**, 24P–25P.

Brittelli A, de Santi C, Raunio H, Pelkonen O, Rossi G and Pacifici GM (1999) Interethnic and interindividual variabilities of platelet sulfotransferases activity in Italians and Finns. *European Journal of Clinical Pharmacology*, **55**, 691–695.

Brix LA, Nicoll R, Zhu XY and McManus ME (1998) Structural and functional characterisation of human sulfotransferases. *Chemico-Biological Interactions*, **109**, 123–127.

Brix LA, Barnett AC, Duggleby RG, Leggett B and McManus ME (1999) Analysis of the substrate specificity of human sulfotransferases SULT1A1 and SULT1A3: site-directed mutagenesis and kinetic studies. *Biochemistry*, **38**, 10474–10479.

Brooks SC, Battelli MG and Corombos JD (1987) Endocrine steroid sulfotransferases: porcine endometrial estrogen sulfotransferase. *Journal of Steroid Biochemistry*, **26**, 285–290.

Brown CE, Roerig SC, Burger VT, Cody RB, Jr. and Fujimoto JM (1985) Analgesic potencies of morphine 3- and 6-sulfates after intracerebroventricular administration in mice: relationship to structural characteristics defined by mass spectrometry and nuclear magnetic resonance. *Journal of Pharmaceutical Sciences*, **74**, 821–824.

Brzeznicka EA, Hazelton GA and Klaassen CD (1987) Comparison of adenosine 3'-phosphate

5′-phosphosulfate concentrations in tissues from different laboratory animals. *Drug Metabolism and Disposition*, **15**, 133–135.

Buchmann A, Kuhlmann W, Schwarz M, Kunz W, Wolf CR, Moll E, Friedberg T and Oesch F (1985) Regulation and expression of four cytochrome P450 isoenzymes, NADPH–cytochrome P450 reductase, the glutathione transferases B and C and microsomal epoxide hydrolase in preneoplastic and neoplastic lesions in rat liver. *Carcinogenesis*, **6**, 513–521.

Buhl AE, Waldon DJ, Baker CA and Johnson GA (1990) Minoxidil sulfate is the active metabolite that stimulates hair follicles. *Journal of Investigative Dermatology*, **95**, 553–557.

Burns GR, Galanopoulou E and Wynn CH (1977) Kinetic studies of the phenol sulphate-phenol sulphotransferase of *Aspergillus oryzae*. *Biochemical Journal*, **167**, 223–227.

Caldwell J (1982) Conjugation reactions in foreign-compound metabolism: definition, consequences, and species variations. *Drug Metabolism Reviews*, **13**, 745–777.

Campbell NR, Sundaram RS, Werness PG, van Loon J and Weinshilboum RM (1985) Sulfate and methyldopa metabolism: metabolite patterns and platelet phenol sulfotransferase activity. *Clinical Pharmacology & Therapeutics*, **37**, 308–315.

Campbell NRC, van Loon JA and Weinshilboum RM (1987) Human liver phenol sulfotransferase: assay conditions, biochemical properties and partial purification of isozymes of the thermostable form. *Biochemical Pharmacology*, **36**, 1435–1446.

Cancela L, Marie PJ, Le Boulch N and Miravet L (1985) Vitamin D3 3β-sulfate has less biological activity than free vitamin D3 during pregnancy in rats. *Biology of the Neonate*, **48**, 274–284.

Cancela L, Marie PJ, Le Boulch N and Miravet L (1987) Lack of biological activity of vitamin D3 3β-sulfate during lactation in vitamin D-deficient rats. *Reproduction Nutrition Development*, **27**, 979–997.

Capel ID, Millburn P and Williams RT (1974) The conjugation of 1- and 2-naphthols and other phenols in the cat and pig. *Xenobiotica*, **4**, 601–615.

Cappiello M, Giuliani L, Rane A and Pacifici GM (1991) Dopamine sulphotransferase is better developed than *para*-nitrophenol sulphotransferase in the human fetus. *Developmental Pharmacology and Therapeutics*, **16**, 83–88.

Cassidy MK and Houston JB (1980) Phenol conjugation by lung *in vivo*. *Biochemical Pharmacology*, **29**, 471–474.

Cerniglia CE, Freeman JP and Mitchum RK (1982) Glucuronide and sulfate conjugation in the fungal metabolism of aromatic hydrocarbons. *Applied and Environmental Microbiology*, **43**, 1070–1075.

Chatterjee B, Majumdar D, Ozbilen O, Murty CVR and Roy AK (1987) Molecular cloning and characterization of cDNA for androgen-repressible rat liver protein, SMP-2. *Journal of Biological Chemistry*, **262**, 822–825.

Chen LJ and Rosenquist GL (1990) Enzymatic sulfation of gastrin in rat gastric mucosa. *Biochemical and Biophysical Research Communications*, **170**, 1170–1176.

Chen QY and Schwarz LR (1985) Sulfation in isolated kidney tubule fragments of rats: dependence on inorganic sulfate. *Biochemical Pharmacology*, **34**, 1363–1366.

Chen LJ and Segel IH (1985) Purification and characterization of bile salt sulfotransferase from human liver. *Archives of Biochemistry and Biophysics*, **241**, 371–379.

Chen LJ, Kane Bd, Bujanover Y and Thaler MM (1982) Development and regulation of bile salt sulfotransferase in rat liver. *Biochimica et Biophysica Acta*, **713**, 358–364.

Chen X, Yang YS, Zheng YQ, Martin BM, Duffel MW and Jakoby WB (1992) Tyrosine-ester sulfotransferase from rat liver: bacterial expression and identification. *Protein Expression and Purification*, **3**, 421–426.

Chen GP, Baron J and Duffel MW (1995) Enzyme- and sex-specific differences in the intralobular localizations and distributions of aryl sulfotransferase IV (tyrosine-ester sulfotransferase) and alcohol (hydroxysteroid) sulfotransferase a in rat liver. *Drug Metabolism and Disposition*, **23**, 1346–1353.

Chidgey MA, Kennedy JF and Caldwell J (1986) Studies on benzyl acetate: II. Use of specific metabolic inhibitors to define the pathway leading to the formation of benzylmercapturic acid in the rat. *Food and Chemical Toxicology*, **24**, 1267–1272.

Chou H-C, Lang NP and Kadlubar FF (1995a) Metabolic activation of N-hydroxy arylamines and N-hydroxy heterocyclic amines by human sulfotransferase(s). *Cancer Research*, **55**, 525–529.

Chou HC, Lang NP and Kadlubar FF (1995b) Metabolic activation of the *N*-hydroxy derivative of the carcinogen 4-aminobiphenyl by human tissue sulfotransferases. *Carcinogenesis,* **16,** 413–417.

Ciolino HP and Yeh GC (1999) The steroid hormone dehydroepiandrosterone inhibits CYP1A1 expression *in vitro* by a post-transcriptional mechanism. *Journal of Biological Chemistry,* **274,** 35186–35190.

Clements JA, Critchley JAJH and Prescott LF (1984) The role of sulphate conjugation in the metabolism and disposition of oral and intravenous paracetamol in man. *British Journal of Clinical Pharmacology,* **18,** 481–485.

Coldham NG, Sivapathasundaram S, Dave M, Ashfield LA, Pottinger TG, Goodall C and Sauer MJ (1998) Biotransformation, tissue distribution, and persistence of 4-nonylphenol residues in juvenile rainbow trout (*Oncorhynchus mykiss*). *Drug Metabolism and Disposition,* **26,** 347–354.

Collett RA and Ungkitchanukit A (1979) Phenol sulphotransferase in developing chick embryo. *Biochemical Society Transactions,* **7,** 132–134.

Comer KA and Falany CN (1992) Immunological characterization of dehydroepiandrosterone sulfotransferase from human liver and adrenal. *Molecular Pharmacology,* **41,** 645–651.

Comer KA, Falany JL and Falany CN (1993) Cloning and expression of human liver dehydro-epiandrosterone sulphotransferase. *Biochemical Journal,* **289,** 233–240.

Compagnone NA and Mellon SH (2000) Neurosteroids: biosynthesis and function of these novel neuromodulators. *Frontiers in Neuroendocrinology,* **21,** 1–56.

Cormier MJ, Hori K and Karkhanis YD (1970) Studies on the bioluminescence of *Renilla reniformis*: VII. Conversion of luciferin into luciferyl sulfate by luciferin sulfokinase. *Biochemistry,* **9,** 1184–1189.

Coughtrie MWH, Gilissen RAHJ, Shek B, Strange RC, Fryer AA, Jones PW and Bamber DE (1999) Phenol sulphotransferase SULT1A1 polymorphism: molecular diagnosis and allele frequencies in Caucasian and African populations. *Biochemical Journal,* **337,** 45–49.

Critchley JAJH, Nimmo GR, Gregson CA, Woolhouse NM and Prescott LF (1986) Inter-subject and ethnic differences in paracetamol metabolism. *British Journal of Clinical Pharmacology,* **22,** 649–657.

Cuevas ME, Miller W and Callard G (1992) Sulfoconjugation of steroids and the vascular pathway of communication in dogfish testis. *Journal of Experimental Zoology,* **264,** 119–129.

Czich A, Bartsch I, Dogra S, Hornhardt S and Glatt HR (1994) Stable heterologous expression of hydroxysteroid sulphotransferase in Chinese hamster V79 cells and their use for toxicological investigations. *Chemico-Biological Interactions,* **92,** 119–128.

Daimon H, Sawada S, Asakura S and Sagami F (1998) *In vivo* genotoxicity and DNA adduct levels in the liver of rats treated with safrole. *Carcinogenesis,* **19,** 141–146.

Dajani R, Sharp S, Graham S, Bethell SS, Cooke RM, Jamieson DJ and Coughtrie MWH (1999a) Kinetic properties of human dopamine sulfotransferase (SULT1A3) expressed in prokaryotic and eukaryotic systems: comparison with the recombinant enzyme purified from *Escherichia coli. Protein Expression and Purification,* **16,** 11–18.

Dajani R, Cleasby A, Neu M, Wonacott AJ, Jhoti H, Hood AM, Modi S, Hersey A, Taskinen J, Cooke RM, Manchee GR and Coughtrie MWH (1999b) X-Ray crystal structure of human dopamine sulfotransferase, SULT1A3: molecular modeling and quantitative structure-activity relationship analysis demonstrate a molecular basis for sulfotransferase substrate specificity. *Journal of Biological Chemistry,* **274,** 37862–37868.

Davis BA, Dawson B, Boulton AA, Yu PH and Durden DA (1987) Investigation of some biological trait markers in migraine: deuterated tyramine challenge test, monoamine oxidase, phenolsul-fotransferase and plasma and urinary biogenic amine and acid metabolite levels. *Headache,* **27,** 384–389.

Davis W, Hewer A, Rajkowski KM, Meinl W, Glatt HR and Phillips DH (2000) Sex differences in the activation of tamoxifen to DNA binding species in rat liver *in vivo* and in rat hepatocytes *in vitro*: role of sulfotransferase induction. *Cancer Research,* **60,** 2887–2891.

De Santi C, Pietrabissa A, Spisni R, Mosca F and Pacifici GM (2000) Sulphation of resveratrol, a natural product present in grapes and wine, in the human liver and duodenum. *Xenobiotica,* **30,** 609–617.

DeBaun JR, Rowley JY, Miller EC and Miller JA (1968) Sulfotransferase activation of N-hydroxy-2-acetylaminofluorene in rodent livers susceptible and resistant to this carcinogen. Proceedings of the Society for Experimental Biology and Medicine, 129, 268–273.

DeBaun JR, Miller EC and Miller JA (1970) N-Hydroxy-2-acetylaminofluorene sulfotransferase: its probable role in carcinogenesis and in protein-(methion-S-yl) binding in rat liver. Cancer Research, 30, 577–595.

Delclos KB, Tarpley WG, Miller EC and Miller JA (1984) 4-Aminoazobenzene and N,N-dimethyl-4-aminoazobenzene as equipotent hepatic carcinogens in male C57BL/6 × C3H/He F₁ mice and characterization of N-(deoxyguanosin-8-yl)-4-aminoazobenzene as the major persistent hepatic DNA-bound dye in these mice. Cancer Research, 44, 2540–2550.

DeMeio RH (1945) Phenol conjugation III. The type of conjugation in different species. Archives of Biochemistry, 7, 323–327.

DeMeio RH (1975) Sulfate activation and transfer. In Metabolism of Sulfur Compounds, Greenberg DM (ed.), Academic Press, New York, pp. 287–358.

DeMeio RH and Tkacz L (1950) Conjugation of phenol by rat liver homogenate. Archives of Biochemistry, 27, 242–243.

Denning MF, Kazanietz MG, Blumberg PM and Yuspa SH (1995) Cholesterol sulfate activates multiple protein kinase C isoenzymes and induces granular cell differentiation in cultured murine keratinocytes. Cell Growth & Differentiation, 6, 1619–1626.

Dickstein Y, Schwartz H, Gross J and Gordon A (1980) The metabolism of T4 and T3 in cultured chick-embryo heart cells. Molecular and Cellular Endocrinology, 20, 45–57.

Dills RL and Klaassen CD (1986) The effect of inhibitors of mitochondrial energy production on hepatic glutathione, UDP-glucuronic acid, and adenosine 3'-phosphate-5'-phosphosulfate concentrations. Drug Metabolism and Disposition, 14, 190–196.

Dodgson KS and Rose FA (1970) Sulfoconjugation and sulfohydrolysis. In Metabolic Conjugation and Metabolic Hydrolysis, Fishman WH (ed.), Academic Press, New York, pp. 239–325.

Dooley TP, Walker CJ, Hirshey SJ, Falany CN and Diani AR (1991) Localization of minoxidil sulfotransferase in rat liver and the outer root sheath of anagen pelage and vibrissa follicles. Journal of Investigative Dermatology, 96, 65–70.

Dooley TP, Haldeman-Cahill R, Joiner J and Wilborn TW (2000) Expression profiling of human sulfotransferase and sulfatase gene superfamilies in epithelial tissues and cultured cells. Biochemical and Biophysical Research Communications, 277, 236–245.

Driscoll WJ, Martin BM, Chen HC and Strott CA (1993) Isolation of two distinct 3-hydroxysteroid sulfotransferases from the guinea pig adrenal: evidence for 3α-hydroxy versus 3β-hydroxy stereospecificity. Journal of Biological Chemistry, 268, 23496–23503.

Duanmu Z, Dunbar J, Falany CN and Runge-Morris M (2000) Induction of rat hepatic aryl sulfotransferase (SULT1A1) gene expression by triamcinolone acetonide: impact on minoxidil-mediated hypotension. Toxicology and Applied Pharmacology, 164, 312–320.

Duffel MW and Jakoby WB (1981) On the mechanism of aryl sulfotransferase. Journal of Biological Chemistry, 256, 11123–11127.

Duffel MW and Zheng YQ (1993) Naphthaldehydes as reversible inhibitors of rat hepatic aryl sulfotransferase IV (tyrosine-ester sulfotransferase). Drug Metabolism and Disposition, 21, 400–402.

Duffel MW, Binder TP and Rao SI (1989) Assay of purified aryl sulfotransferase suitable for reactions yielding unstable sulfuric acid esters. Analytical Biochemistry, 183, 320–324.

Duffel MW, Modi RB and King RS (1992) Interactions of a primary N-hydroxy arylamine with rat hepatic aryl sulfotransferase IV. Drug Metabolism and Disposition, 20, 339–341.

Dufort I, Tremblay Y, Bélanger A, Labrie F and Luu-The V (1996) Isolation and characterization of a stereospecific 3β-hydroxysteroid sulfotransferase (pregnenolone sulfotransferase) cDNA. DNA and Cell Biology, 15, 481–487.

Dunn II RT and Klaassen CD (1998) Tissue-specific expression of rat sulfotransferase messenger RNAs. Drug Metabolism and Disposition, 26, 598–604.

Dunn II RT and Klaassen CD (2000) Thyroid hormone modulation of rat sulphotransferase mRNA expression. Xenobiotica, 30, 345–357.

Dunn II RT, Kolaja KL and Klaassen CD (1999a) Effect of partial hepatectomy on the expression of seven rat sulphotransferase mRNAs. Xenobiotica, 29, 583–593.

Dunn II RT, Gleason BA, Hartley DP and Klaassen CD (1999b) Postnatal ontogeny and hormonal regulation of sulfotransferase SULT1B1 in male and female rats. *Journal of Pharmacology and Experimental Therapeutics*, **290**, 319–324.

Eisenhofer G, Coughtrie MWH and Goldstein DS (1999) Dopamine sulphate: an enigma resolved. *Clinical and Experimental Pharmacology and Physiology*, **26**, S41–S53.

Elmamlouk TH and Gessner T (1978) Carbohydrate and sulfate conjugations of *p*-nitrophenol by hepatopancreas of *Homarus americanus*. *Comparative Biochemistry and Physiology C*, **61**, 363–367.

Engelke CE, Meinl W, Boeing H and Glatt HR (2000a) Association between functional genetic polymorphisms of human sulfotransferases 1A1 and 1A2. *Pharmacogenetics*, **10**, 163–169.

Engelke CEH, Meinl W, Boeing H and Glatt HR (2000b) Association between obesity and genetic polymorphisms of human sulfotransferases 1A1 and 1A2. *Kidney and Blood Pressure Research*, **23**, 64.

Engst W, Landsiedel R and Glatt HR (1997) Conjugates formed by recombinant rat and human sulfotransferases: analysis using electrospray LC-MS. In *Advances in Mass Spectrometry* Vol. 14, Proceedings of the 14th International Mass Spectrometry Conference, Tampere (Finland), Karjalainen EJ, Hesso AE, Jalonen JE and Karjalainen UPK (eds), CD-ROM.

Esteban A, Calvo R and Perez-Mateo M (1996) Paracetamol metabolism in two ethnically different Spanish populations. *European Journal of Drug Metabolism and Pharmacokinetics*, **21**, 233–239.

Falany CN (1997) Sulfation and sulfotransferases: 3. Enzymology of human cytosolic sulfotransferases. *FASEB Journal*, **11**, 206–216.

Falany CN and Kerl EA (1990) Sulfation of minoxidil by human liver phenol sulfotransferase. *Biochemical Pharmacology*, **40**, 1027–1032.

Falany CN, Vazquez ME and Kalb JM (1989) Purification and characterization of human liver dehydroepiandrosterone sulphotransferase. *Biochemical Journal*, **260**, 641–646.

Falany CN, Vazquez ME, Heroux JA and Roth JA (1990) Purification and characterization of human liver phenol-sulfating phenol sulfotransferase. *Archives of Biochemistry and Biophysics*, **278**, 312–318.

Falany JL and Falany CN (1996a) Expression of cytosolic sulfotransferases in normal mammary epithelial cells and breast cancer cell lines. *Cancer Research*, **56**, 1551–1555.

Falany JL and Falany CN (1996b) Regulation of estrogen sulfotransferase in human endometrial adenocarcinoma cells by progesterone. *Endocrinology*, **137**, 1395–1401.

Falany CN, Xie X, Wang J, Ferrer J and Falany JL (2000) Molecular cloning and expression of novel sulphotransferase-like cDNAs from human and rat brain. *Biochemical Journal*, **346**, 857–864.

Falany JL, Azziz R and Falany CN (1998) Identification and characterization of cytosolic sulfotransferases in normal human endometrium. *Chemico-Biological Interactions*, **109**, 329–339.

Fennell TR, Wiseman RW, Miller JA and Miller EC (1985) Major role of hepatic sulfotransferase activity in the metabolic activation, DNA adduct formation, and carcinogenicity of 1'-hydroxy-2',3'-dehydroestragole in infant male C57BL/6J × C3H/HeJ F_1 mice. *Cancer Research*, **45**, 5310–5320.

Fernando PHP, Sakakibara Y, Nakatsu S, Suiko M, Han JR and Liu MC (1993) Isolation and characterization of a novel microsomal membrane-bound phenol sulfotransferase from bovine liver. *Biochemistry and Molecular Biology International*, **30**, 433–441.

Finnson KW and Eales JG (1998) Sulfation of thyroid hormones by liver of rainbow trout, *Oncorhynchus mykiss*. *Comparative Biochemistry and Physiology C*, **120**, 415–420.

Flammang TJ and Kadlubar FF (1986) Acetyl coenzyme A-dependent metabolic activation of *N*-hydroxy-3,2'-dimethyl-4-aminobiphenyl and several carcinogenic *N*-hydroxy arylamines in relation to tissue and species differences, other acyl donors, and arylhydroxamic acid-dependent acyltransferases. *Carcinogenesis*, **7**, 919–926.

Flood JF, Morley JE and Roberts E (1995) Pregnenolone sulfate enhances post-training memory processes when injected in very low doses into limbic system structures: the amygdala is by far the most sensitive. *Proceedings of the National Academy of Sciences, USA*, **92**, 10806–10810.

Foldes A and Meek JL (1973) Rat brain phenolsulfotransferase: partial purification and some properties. *Biochimica et Biophysica Acta*, **327**, 365–374.

Forbes KJ, Hagen M, Glatt HR, Hume R and Coughtrie MWH (1995) Human fetal adrenal hydroxysteroid sulphotransferase: cDNA cloning, stable expression in V79 cells and functional characterisation of the expressed enzyme. *Molecular and Cellular Endocrinology*, **112**, 53–60.

Forbes-Bamforth KJ and Coughtrie MWH (1994) Identification of a new adult human liver sulfotransferase with specificity for endogenous and xenobiotic estrogens. *Biochemical and Biophysical Research Communications*, **198**, 707–711.

Frame LT, Gatlin TL, Kadlubar FF and Lang NP (1997) Metabolic differences and their impact on human disease: sulfotransferase and colorectal cancer. *Environmental Toxicology and Pharmacology*, **4**, 277–281.

Frame LT, Ozawa S, Nowell SA, Chou HC, DeLongchamp RR, Doerge DR, Lang NP and Kadlubar FF (2000) A simple colorimetric assay for phenotyping the major human thermostable phenol sulfotransferase (SULT1A1) using platelet cytosols. *Drug Metabolism and Disposition*, **28**, 1063–1068.

Frank G and Beyer J (1986) Metabolism of 3-nitrophenol by the frog *Rana temporaria*. *Xenobiotica*, **16**, 291–294.

Franzon VL, Gibson MA, Hatzinikolas G, Woollatt E, Sutherland GR and Cleary EG (1999) Molecular cloning of a novel human PAPS synthetase which is differentially expressed in metastatic and nonmetastatic colon carcinoma cells. *International Journal of Biochemistry & Cell Biology*, **31**, 613–626.

Freeman DJ, Saidi F and Hobkirk R (1983) Estrogen sulfotransferase activity in guinea pig uterus and chorion. *Journal of Steroid Biochemistry*, **18**, 23–27.

Freimuth RR, Raftogianis RB, Wood TC, Moon E, Kim U-J, Xu J, Siciliano MJ and Weinshilboum RM (2000) Human sulfotransferases SULT1C1 and SULT1C2: cDNA characterization, gene cloning, and chromosomal localization. *Genomics*, **65**, 157–165.

Fujita K, Nagata K, Ozawa S, Sasano H and Yamazoe Y (1997) Molecular cloning and characterization of rat ST1B1 and human ST1B2 cDNAs, encoding thyroid hormone sulfotransferases. *Journal of Biochemistry (Tokyo)*, **122**, 1052–1061.

Fujita K, Nagata K, Yamazaki T, Watanabe E, Shimada M and Yamazoe Y (1999) Enzymatic characterization of human cytosolic sulfotransferases: identification of ST1B2 as a thyroid hormone sulfotransferase. *Biological & Pharmaceutical Bulletin*, **22**, 446–452

Galinsky RE, Slattery JT and Levy G (1979) Effect of sodium sulfate on acetaminophen elimination by rats. *Journal of Pharmaceutical Sciences*, **68**, 803–805.

Garay RP, Nazaret C and Cragoe Jr. EJ (1991) Evidence for the *O*-sulfo derivative of MK-447 as active metabolite of MK-447. *European Journal of Pharmacology*, **274**, 175–180.

Garay RP, Rosati C, Fanous K, Allard M, Morin E, Lamiable D and Vistelle R (1995) Evidence for (+)-cicletanine sulfate as an active natriuretic metabolite of cicletanine in the rat. *European Journal of Pharmacology*, **274**, 175–180.

Garcia-Allan C, Lord PG, Loughlin JM, Orton TC and Sidaway JE (2000) Identification of phenobarbitone-modulated genes in mouse liver by differential display. *Journal of Biochemical and Molecular Toxicology*, **14**, 65–72.

Ghazali RA and Waring RH (1999) The effects of flavonoids on human phenolsulphotransferases: potential in drug metabolism and chemoprevention. *Life Sciences*, **65**, 1625–1632.

Gibb C, Glover V and Sandler M (1987) *In vitro* inhibition of phenolsulphotransferase by food and drink constituents. *Biochemical Pharmacology*, **36**, 2325–2330.

Gilissen RAHJ and Meerman JHN (1992) Bioactivation of the hepatocarcinogen *N*-hydroxy-2-acetylaminofluorene by sulfation in the rat liver changes during the cell cycle. *Life Sciences*, **51**, 1255–1260.

Gilissen RAHJ, Ringer DP, Stavenuiter HJFC, Mulder GJ and Meerman JHN (1992) Sulfation of hydroxylamines and hydroxamic acids in liver cytosol from male and female rats and purified aryl sulfotransferase IV. *Carcinogenesis*, **13**, 1699–1703.

Glatt HR (1997) Sulfation and sulfotransferases: 4. Bioactivation of mutagens via sulfation. *FASEB Journal*, **11**, 314–321.

Glatt HR (2000a) Sulfotransferases in the bioactivation of xenobiotics. *Chemico-Biological Interactions*, **129**, 141–170.

Glatt HR (2000b) An overview of bioactivation of chemical carcinogens. *Biochemical Society Transactions*, **28**, 1–6.

Glatt HR, Henschler R, Phillips DH, Blake JW, Steinberg P, Seidel A and Oesch F (1990) Sulfotransferase-mediated chlorination of 1-hydroxymethylpyrene to a mutagen capable of penetrating indicator cells. *Environmental Health Perspectives*, **88**, 43–48.

Glatt HR, Pauly K, Frank H, Seidel A, Oesch F, Harvey RG and Werle-Schneider G (1994) Substrate-dependent sex differences in the activation of benzylic alcohols to mutagens by hepatic sulfotransferases of the rat. *Carcinogenesis*, **15**, 2605–2611.

Glatt HR, Pauly K, Czich A, Falany JL and Falany CN (1995) Activation of benzylic alcohols to mutagens by rat and human sulfotransferases expressed in *Escherichia coli*. *European Journal of Pharmacology*, **293**, 173–181.

Glatt HR, Engelke CEH, Pabel U, Teubner W, Jones AL, Coughtrie MWH, Andrae U, Falany CN and Meinl W (2000) Sulfotransferases: genetics and role in toxicology. *Toxicology Letters*, **112–113**, 341–348.

Glatt HR, Boeing H, Engelke CEH, Kuhlow A, Ma L, Pabel U, Pomplun D, Teubner W and Meinl W (2001) Human cytosolic sulphotransferases: genetics, characteristics, toxicological aspects. *Mutation Research*, in press.

Glatt HR and Meinl W (2001) Sulfotransferases. In *Pharmacogenetics*, Roots I, Gonzalez FJ and Brockmöller J (eds), Springer-Verlag, Heidelberg, in press.

Gong DW, Ozawa S, Yamazoe Y and Kato R (1991) Purification of hepatic N-hydroxyarylamine sulfotransferases and their regulation by growth hormone and thyroid hormone in rats. *Journal of Biochemistry (Tokyo)*, **110**, 226–231.

Gong DW, Murayama N, Yamazoe Y and Kato R (1992) Hepatic triiodothyronine sulfation and its regulation by growth hormone and triiodothyronine in rats. *Journal of Biochemistry (Tokyo)*, **112**, 112–116.

Goon D and Klaassen CD (1990) Dose-dependent intestinal glucuronidation and sulfation of acetaminophen in the rat *in situ*. *Journal of Pharmacology and Experimental Therapeutics*, **252**, 201–207.

Goon D and Klaassen CD (1991) Intestinal biotransformation of harmol and 1-naphthol in the rat: further evidence of dose-dependent phase-II conjugation *in situ*. *Drug Metabolism and Disposition*, **19**, 340–347.

Gorge G, Beyer J and Urich K (1987) Excretion and metabolism of phenol, 4-nitrophenol and 2-methylphenol by the frogs *Rana temporaria* and *Xenopus laevis*. *Xenobiotica*, **17**, 1293–1298.

Gram TE, Litterst CL and Mimnaugh EG (1974) Enzymatic conjugation of foreign chemical compounds by rabbit lung and liver. *Drug Metabolism and Disposition*, **2**, 254–258.

Grant MH and Hawksworth GM (1986) The activity of UDP-glucuronyltransferase, sulphotransferase and glutathione-S-transferase in primary cultures of rat hepatocytes. *Biochemical Pharmacology*, **35**, 2979–2982.

Gregory JD and Lipmann F (1957) The transfer of sulfate among phenolic compounds with 3′,5′-diphosphoadenosine as coenzyme. *Journal of Biological Chemistry*, **229**, 1081–1090.

Gregus Z, Watkins JB, Thompson TN, Harvey MJ, Rozman K and Klaassen CD (1983) Hepatic phase I and phase II biotransformations in quail and trout: comparison to other species commonly used in toxicity testing. *Toxicology and Applied Pharmacology*, **67**, 430–441.

Groen K, Warrander A, Miles GS, Booth BS and Mulder GJ (1988) Sulphation and glucuronidation of xamoterol in the dog: dose dependence and site of sulphation. *Xenobiotica*, **18**, 511–518.

Groppi VE, Burnett B-A, Maggoria L, Bannow C, Zurcher-Neeley HA, Heinrikson RL, Schostarez HJ and Bienkowski MJ (1990) Protein sulfation: a unique mechanism of action of minoxidil sulfate. *Journal of Investigative Dermatology*, **94**, 532.

Grün F, Noy N, Hämmerling U and Buck J (1996) Purification, cloning, and bacterial expression of retinol dehydratase from *Spodoptera frugiperda*. *Journal of Biological Chemistry*, **271**, 16135–16138.

Guo WXA, Yang YS, Chen X, McPhie P and Jakoby WB (1994) Changes in substrate specificity of the recombinant form of phenol sulfotransferase IV (tyrosine-ester sulfotransferase). *Chemico-Biological Interactions*, **92**, 25–31.

Hagen M, Pabel U, Landsiedel R, Bartsch I, Falany CN and Glatt HR (1998) Expression of human estrogen sulfotransferase in *Salmonella typhimurium:* differences between hHST and hEST in the enantioselective activation of 1-hydroxyethylpyrene to a mutagen. *Chemico-Biological Interactions,* **109**, 249–253.

Hähnel R, Twaddle E and Ratajczak T (1973) The specificity of the estrogen receptor of human uterus. *Journal of Steroid Biochemistry,* **4**, 21–31.

Harris RM and Waring RH (1996) Dietary modulation of human platelet phenolsulphotransferase activity. *Xenobiotica,* **26**, 1241–1247.

Harris RM, Hawker RJ, Langman MJS, Singh S and Waring RH (1998) Inhibition of phenolsulphotransferase by salicylic acid: a possible mechanism by which aspirin may reduce carcinogenesis. *Gut,* **42**, 272–275.

Hartman AP, Wilson AA, Wilson HM, Aberg G, Falany CN and Walle T (1998) Enantioselective sulfation of β_2-receptor agonists by the human intestine and the recombinant M-form phenolsulfotransferase. *Chirality,* **10**, 800–803.

Hellriegel ET, Matwyshyn GA, Fei PW, Dragnev KH, Nims RW, Lubet RA and Kong ANT (1996) Regulation of gene expression of various phase I and phase II drug-metabolizing enzymes by tamoxifen in rat liver. *Biochemical Pharmacology,* **52**, 1561–1568.

Her C, Szumlanski C, Aksoy IA and Weinshilboum RM (1996) Human jejunal estrogen sulfotransferase and dehydroepiandrosterone sulfotransferase: immunochemical characterization of individual variation. *Drug Metabolism and Disposition,* **24**, 1328–1335.

Her C, Kaur GP, Athwal RS and Weinshilboum RM (1997) Human sulfotransferase SULT1C1: cDNA cloning, tissue-specific expression, and chromosomal localization. *Genomics,* **41**, 467–470.

Her C, Wood TC, Eichler EE, Mohrenweiser HW, Ramagli LS, Siciliano MJ and Weinshilboum RM (1998) Human hydroxysteroid sulfotransferase SULT2B1: two enzymes encoded by a single chromosome 19 gene. *Genomics,* **53**, 284–295.

Herman RJ, Loewen GR, Antosh DM, Taillon MR, Hussein S and Verbeeck RK (1994) Analysis of polymorphic variation in drug metabolism: III. Glucuronidation and sulfation of diflunisal in man. *Clinical and Investigative Medicine,* **17**, 297–307.

Hernandez JS, Powers SP and Weinshilboum RM (1991) Human liver arylamine *N*-sulfotransferase activity: thermostable phenol sulfotransferase catalyzes the *N*-sulfation of 2-naphthylamine. *Drug Metabolism and Disposition,* **19**, 1071–1079.

Heroux JA and Roth JA (1988) Physical characterization of a monoamine-sulfating form of phenol sulfotransferase from human platelets. *Molecular Pharmacology,* **34**, 194–199.

Heroux JA, Falany CN and Roth JA (1989) Immunological characterization of human phenol sulfotransferase. *Molecular Pharmacology,* **36**, 29–33.

Hirom PC, Millburn P, Smith RL and Williams RT (1972) Species variations in the threshold molecular-weight factor for the biliary excretion of organic anions. *Biochemical Journal,* **129**, 1071–1077.

Hirshey SJ and Falany CN (1990) Purification and characterization of rat liver minoxidil sulphotransferase. *Biochemical Journal,* **270**, 721–728.

Hirshey SJ, Dooley TP, Reardon IM, Heinrikson RL and Falany CN (1992) Sequence analysis, *in vitro* translation, and expression of the cDNA for rat liver minoxidil sulfotransferase. *Molecular Pharmacology,* **42**, 257–264.

Hjelle JJ, Hazelton GA and Klaassen CD (1985) Acetaminophen decreases adenosine 3'-phosphate 5'-phosphosulfate and uridine diphosphoglucuronic acid in rat liver. *Drug Metabolism and Disposition,* **13**, 35–41.

Hobkirk R (1993) Steroid sulfation: current concepts. *Trends in Endocrinology and Metabolism,* **4**, 69–74.

Hobkirk R and Glasier MA (1992) Estrogen sulfotransferase distribution in tissues of mouse and guinea pig: steroidal inhibition of the guinea pig enzyme. *Biochemistry and Cell Biology,* **70**, 712–715.

Hobkirk R, Cardy CA, Saidi F, Kennedy TG and Girard LR (1983) Development and characteristics of an oestrogen sulphotransferase in placenta and uterus of the pregnant mouse: comparison between mouse and rat. *Biochemical Journal,* **216**, 451–457.

Hobkirk R, Girard LR, Durham NJ and Khalil MW (1985) Behavior of mouse placental and uterine estrogen sulfotransferase during chromatography and other procedures. *Biochimica et Biophysica Acta,* **828**, 123–129.

Homma H, Tada M, Nakamura T, Yamagata S and Matsui M (1997) Heterogeneous zonal distribution of sulfotransferase isoenzymes in rat liver. *Archives of Biochemistry and Biophysics*, **339**, 235–241.

Huf PA, Bourne AR and Watson TG (1987) Identification of testosterone sulfate in the plasma of the male lizard *Tiliqua rugosa. General and Comparative Endocrinology*, **66**, 364–368.

Hume R, Barker EV and Coughtrie MWH (1996) Differential expression and immunohistochemical localisation of the phenol and hydroxysteroid sulphotransferase enzyme families in the developing lung. *Histochemistry and Cell Biology*, **105**, 147–152.

Huttner WB (1987) Protein tyrosine sulfation. *Trends in Biochemical Sciences*, **12**, 361–363.

Huynh QK, Shailubhai K, Boddupalli H, Yu HH, Broschat KO and Jacob GS (1999) Isolation and characterization from porcine serum of a soluble sulfotransferase responsible for 6-*O*-sulfation of the galactose residue in 2′-fucosyllactose: Implications in the synthesis of the ligand for L-selectin. *Glycoconjugate Journal*, **16**, 357–363.

Hwang SR, Kohn AB and Hook VYH (1995) Molecular cloning of an isoform of phenol sulfotransferase from human brain hippocampus. *Biochemical and Biophysical Research Communications*, **207**, 701–707.

James MO, Altman AH, Morris K, Kleinow KM and Tong Z (1997) Dietary modulation of phase1 and phase 2 activities with benzo[a]pyrene and related compounds in the intestine but not the liver of the channel catfish, *Ictalurus punctatus. Drug Metabolism and Disposition*, **25**, 346–354.

Jedlitschky G, Leier I, Buchholz U, Barnouin K, Kurz G and Keppler D (1996) Transport of glutathione, glucuronate, and sulfate conjugates by the MRP gene-encoded conjugate export pump. *Cancer Research*, **56**, 988–994.

Jetten AM, George MA, Nervi C, Boone LR and Rearick JI (1989) Increased cholesterol sulfate and cholesterol sulfotransferase activity in relation to the multi-step process of differentiation in human epidermal keratinocytes. *Journal of Investigative Dermatology*, **92**, 203–209.

Johnson GA, Baker CA and Knight KA (1992) Minoxidil sulfotransferase, a marker of human keratinocyte differentiation. *Journal of Investigative Dermatology*, **98**, 730–733.

Jones AL, Roberts RC and Coughtrie MWH (1993) The human phenolsulphotransferase polymorphism is determined by the level of expression of the enzyme protein. *Biochemical Journal*, **296**, 287–290.

Jones AL, Hagen M, Coughtrie MWH, Roberts RC and Glatt HR (1995a) Human platelet phenolsulphotransferases: cDNA cloning, stable expression in V79 cells and identification of a novel allelic variant of the phenol-sulfating form. *Biochemical and Biophysical Research Communications*, **208**, 855–862.

Jones AL, Roberts RC, Colvin DW, Rubin GL and Coughtrie MWH (1995b) Reduced platelet phenolsulphotransferase activity towards dopamine and 5-hydroxytryptamine in migraine. *European Journal of Clinical Pharmacology*, **49**, 109–114.

Jordan TW, McNaught RW and Smith JN (1970) Detoxications in *Peripatus:* sulphate, phosphate and histidine conjugations. *Biochemical Journal*, **118**, 1–8.

Kagehara M, Tachi M, Harii K and Iwamori M (1994) Programmed expression of cholesterol sulfotransferase and transglutaminase during epidermal differentiation of murine skin development. *Biochimica et Biophysica Acta*, **1215**, 183–189.

Kakuta Y, Pedersen LG, Carter CW, Negishi M and Pedersen LC (1997) Crystal structure of estrogen sulphotransferase. *Nature Structural Biology*, **4**, 904–908.

Kane RE, Tector J, Brems JJ, Li A and Kaminski D (1991) Sulfation and glucuronidation of acetaminophen by cultured hepatocytes reproducing *in vivo* sex-differences in conjugation on matrigel and type-1 collagen. *In Vitro Cellular & Developmental Biology*, **27**, 953–960.

Kanno N, Nagahisa E, Sato M and Sato Y (1996) Adenosine 5′-phosphosulfate sulfotransferase from the marine macroalga *Porphyra yezoensis* Ueda (Rhodophyta): stabilization, purification, and properties. *Planta*, **198**, 440–446.

Kaptein E, van Haasteren GAC, Linkels E, de Greef WJ and Visser TJ (1997) Characterization of iodothyronine sulfotransferase activity in rat liver. *Endocrinology*, **138**, 5136–5143.

Kasokat T, Nagel R and Urich K (1987) The metabolism of phenol and substituted phenols in zebra fish. *Xenobiotica*, **17**, 1215–1221.

Kauffman FC, Whittaker M, Anundi I and Thurman RG (1991) Futile cycling of a sulfate conjugate by isolated hepatocytes. *Molecular Pharmacology*, **39**, 414–420.

Kawabe S, Ikuta T, Ohba M, Chida K, Ueda E, Yamanishi K and Kuroki T (1998) Cholesterol sulfate activates transcription of transglutaminase 1 gene in normal human keratinocytes. *Journal of Investigative Dermatology*, **111**, 1098–1102.

Kedderis GL, Dyroff MC and Rickert DE (1984) Hepatic macromolecular covalent binding of the hepatocarcinogen 2,6-dinitrotoluene and its 2,4-isomer *in vivo:* modulation by the sulfotransferase inhibitors pentachlorophenol and 2,6-dichloro-4-nitrophenol. *Carcinogenesis*, **5**, 1199–1204.

Kenyon EM, Seeley ME, Janszen D and Medinsky MA (1995) Dose-, route-, and sex-dependent urinary excretion of phenol metabolites in B6C3F₁ mice. *Journal of Toxicology and Environmental Health*, **44**, 219–233.

Kern A, Bader A, Pichlmayr R and Sewing KF (1997) Drug metabolism in hepatocyte sandwich cultures of rats and humans. *Biochemical Pharmacology*, **54**, 761–772.

Kester MHA, van Dijk CH, Tibboel D, Hood AM, Rose NJM, Meinl W, Pabel U, Glatt HR, Falany CN, Coughtrie MWH and Visser TJ (1999) Sulfation of thyroid hormone by estrogen sulfotransferase. *Journal of Clinical Endocrinology and Metabolism*, **84**, 2577–2580.

Kester MHA, Bulduk S, Tibboel D, Meinl W, Glatt HR, Falany CN, M. W. H. Coughtrie, Bergman A, Safe SH, Kuiper GGJM, Schuur AG, Brouwer A and Visser TJ (2000) Potent inhibition of estrogen sulfotransferase by hydroxylated PCB metabolites: a novel pathway explaining the estrogenic activity of PCBs. *Endocrinology*, **141**, 1897–1900.

Khoo HG and Wong KP (1993) Sulphate conjugation of serotonin and *N*-acetylserotonin in the mosquito, *Aedes togoi*. *Insect Biochemistry and Molecular Biology*, **23**, 507–513.

Kiehlbauch CC, Lam YF and Ringer DP (1995) Homodimeric and heterodimeric aryl sulfotransferases catalyze the sulfuric acid esterification of *N*-hydroxy-2-acetylaminofluorene. *Journal of Biological Chemistry*, **270**, 18941–18947.

Kim D-H, Konishi L and Kobashi K (1986) Purification, characterization and reaction mechanism of novel arylsulfotransferase obtained from an anaerobic bacterium of human intestine. *Biochimica et Biophysica Acta*, **872**, 33–41.

Kim DH, Yoon HK, Koizumi M and Kobashi K (1992) Sulfation of phenolic antibiotics by sulfotransferase obtained from a human intestinal bacterium: note. *Chemical & Pharmaceutical Bulletin (Tokyo)*, **40**, 1056–1057.

Kim DH, Kim B, Kim HS, Sohng IS and Kobashi K (1994) Sulfation of parabens and tyrosylpeptides by bacterial arylsulfate sulfotransferases. *Biological & Pharmaceutical Bulletin*, **17**, 1326–1328.

Kim HJ, Cho JH and Klaassen CD (1995) Depletion of hepatic 3′-phosphoadenosine 5′-phosphosulfate (PAPS) and sulfate in rats by xenobiotics that are sulfated. *Journal of Pharmacology and Experimental Therapeutics*, **275**, 654–658.

King CM and Olive CW (1975) Comparative effects of strain, species, and sex on the acyltransferase- and sulfotransferase-catalyzed activations of *N*-hydroxy-*N*-2-fluorenylacetamide. *Cancer Research*, **35**, 906–912.

King CM and Phillips B (1968) Enzyme-catalyzed reactions of the carcinogen *N*-hydroxy-2-fluorenylacetamide with nucleic acid. *Science*, **159**, 1351–1353.

King RS and Duffel MW (1997) Oxidation-dependent inactivation of aryl sulfotransferase IV by primary N-hydroxy arylamines during *in vitro* assays. *Carcinogenesis*, **18**, 843–849.

King RS, Sharma V, Pedersen LC, Kakuta Y, Negishi M and Duffel MW (2000) Structure-function modeling of the interactions of *N*-alkyl-*N*-hydroxyanilines with rat hepatic aryl sulfotransferase IV. *Chemical Research in Toxicology*, **13**, 1251–1258.

Kirkpatrick RB, Wildermann NM and Killenberg PG (1985) Androgens and estrogens affect hepatic bile acid sulfotransferase in male rats. *American Journal of Physiology*, **248**, G639–G642.

Klaassen CD and Boles JW (1997) Sulfation and sulfotransferases: 5. The importance of 3′-phosphoadenosine 5′-phosphosulfate (PAPS) in the regulation of sulfation. *FASEB Journal*, **11**, 404–418.

Koizumi M, Akao T, Kadota S, Kikuchi T, Okuda T and Kobashi K (1991) Enzymatic sulfation of polyphenols related to tannins by arylsulfotransferase. *Chemical & Pharmaceutical Bulletin (Tokyo)*, **39**, 2638–2643.

Kokate TG, Juhng KN, Kirkby RD, Llamas J, Yamaguchi S and Rogawski MA (1999) Convulsant actions of the neurosteroid pregnenolone sulfate in mice. *Brain Research*, **831**, 119–124.

Kolanczyk RC, Rutks IR and Gutmann HR (1991) The catalytic effect of bovine serum albumin on the ortho rearrangement of the potential ultimate carcinogen, *N*-(sulfooxy)-2-(acetylamino)fluorene, generated enzymically from *N*-hydroxy-2-(acetylamino)fluorene and evidence for substrate specificity of the enzymic sulfonation of arylhydroxamic acids. *Chemical Research in Toxicology*, **4**, 187–194.

Kong ANT, Yang LD, Ma MH, Tao D and Bjornsson TD (1992) Molecular cloning of the alcohol/hydroxysteroid form (hST$_a$) of sulfotransferase from human liver. *Biochemical and Biophysical Research Communications*, **187**, 448–454.

König J, Nies AT, Cui YH, Leier I and Keppler D (1999) Conjugate export pumps of the multidrug resistance protein (MRP) family: localization, substrate specificity, and MRP2-mediated drug resistance. *Biochimica et Biophysica Acta*, **1461**, 377–394.

Konishi-Imamura L, Kim DH and Kobashi K (1991) Effect of enzymic sulfation on biochemical and pharmacological properties of catecholamines and tyrosine-containing peptides. *Chemical & Pharmaceutical Bulletin (Tokyo)*, **39**, 2994–2998.

Kornprobst JM, Sallenave C and Barnathan G (1998) Sulfated compounds from marine organisms. *Comparative Biochemistry and Physiology B*, **119**, 1–51.

Koster H, Scholtens E and Mulder GJ (1979) Inhibition of sulfation of phenols *in vivo* by 2,6-dichloro-4-nitrophenol: selectivity of its action in relation to other conjugations in the rat *in vivo*. *Medical Biology*, **57**, 340–344.

Kotov A, Falany JL, Wang J and Falany CN (1999) Regulation of estrogen activity by sulfation in human Ishikawa endometrial adenocarcinoma cells. *Journal of Steroid Biochemistry and Molecular Biology*, **68**, 137–144.

Krijgsheld KR, Scholtens E and Mulder GJ (1981) An evaluation of methods to decrease the availability of inorganic sulphate for sulphate conjugation in the rat *in vivo*. *Biochemical Pharmacology*, **30**, 1973–1979.

Kroeger-Koepke MB, Koepke SR, Hernandez L and Michejda CJ (1992) Activation of a β-hydroxyalkylnitrosamine to alkylating agents: evidence for the involvement of a sulfotransferase. *Cancer Research*, **52**, 3300–3305.

Kudlacek PE, Clemens DL, Halgard CM and Anderson RJ (1997) Characterization of recombinant human liver dehydroepiandrosterone sulfotransferase with minoxidil as the substrate. *Biochemical Pharmacology*, **53**, 215–221.

Kurima K, Warman ML, Krishnan S, Domowicz M, Krueger RC, Deyrup A and Schwartz NB (1998) A member of a family of sulfate-activating enzymes causes murine brachymorphism. *Proceedings of the National Academy of Sciences of the United States of America*, **95**, 8681–8685.

Kurima K, Singh B and Schwartz NB (1999) Genomic organization of the mouse and human genes encoding the ATP sulfurylase/adenosine 5′-phosphosulfate kinase isoform SK2. *Journal of Biological Chemistry*, **274**, 33306–33312.

Kuss E (1994) The fetoplacental unit of primates. *Experimental and Clinical Endocrinology*, **102**, 135–165.

Kwon AR, Oh TG, Kim DH and Choi EC (1999) Molecular cloning of the arylsulfate sulfotransferase gene and characterization of its product from *Enterobacter amnigenus* AR-37. *Protein Expression and Purification*, **17**, 366–372.

Lai C-C, Miller JA, Miller EC and Liem A (1985) *N*-Sulfoöxy-2-aminofluorene is the major ultimate electrophilic and carcinogenic metabolite of *N*-hydroxy-2-acetylaminofluorene in the livers of infant male C57BL/6J × C3H/HeJ F$_1$ (B6C3F$_1$) mice. *Carcinogenesis*, **6**, 1037–1045.

Lai C-C, Miller EC, Miller JA and Liem A (1987) Initiation of hepatocarcinogenesis in infant male B6C3F$_1$ mice by *N*-hydroxy-2-aminofluorene or *N*-hydroxy-2-acetylaminofluorene depends primarily on metabolism to *N*-sulfooxy-2-aminofluorene and formation of DNA-(deoxyguanosin-8-yl)-2-aminofluorene adducts. *Carcinogenesis*, **8**, 471–478.

Landsiedel R (1998) *Stoffwechsel und Mutagenität benzylischer Verbindungen*, Logos Verlag, Berlin.

Landsiedel R, Engst W, Scholtyssek M, Seidel A and Glatt HR (1996) Benzylic sulphuric acid esters react with diverse functional groups and often form secondary reactive species. *Polycyclic Aromatic Compounds*, **11**, 341–348.

Landsiedel R, Pabel U, Engst W, Ploschke J, Seidel A and Glatt HR (1998) Chiral inversion of 1-hydroxyethylpyrene enantiomers mediated by enantioselective sulfotransferases. *Biochemical and Biophysical Research Communications*, **247**, 181–185.

Layiwola PJ and Linnecar DFC (1981) The biotransformation of [^{14}C]-phenol in some freshwater fish. *Xenobiotica*, **11**, 167–171.

Layiwola PJ, Linnecar DFC and Knights B (1983) The biotransformation of three ^{14}C-labelled phenolic compounds in twelve species of freshwater fish. *Xenobiotica*, **13**, 107–113.

Leach M, Cameron E, Fite N, Stassinopoulos J, Palmreuter N and Beckmann JD (1999) Inhibition and binding studies of coenzyme A and bovine phenol sulfotransferase. *Biochemical and Biophysical Research Communications*, **261**, 815–819.

Lee HS, Ti TY, Koh YK and Prescott LF (1992) Paracetamol elimination in Chinese and Indians in Singapore. *European Journal of Clinical Pharmacology*, **43**, 81–84.

Lee YC, Park CS and Strott CA (1994a) Molecular cloning of a chiral-specific 3α-hydroxysteroid sulfotransferase. *Journal of Biological Chemistry*, **269**, 15838–15845.

Lee YC, Komatsu K, Driscoll WJ and Strott CA (1994b) Structural and functional characterization of estrogen sulfotransferase isoforms: distinct catalytic and high affinity binding activities. *Molecular Endocrinology*, **8**, 1627–1635.

Leilich G, Knauf H, Mutschler E and Völger K-D (1980) Influence of triamterene and hydroxytriamterene sulfuric acid ester on diuresis and saluresis in rats after oral and intravenous application. *Arzneimittelforschung/Drug Research*, **30**, 949–953.

Leiter EH and Chapman HD (1994) Obesity-induced diabetes (diabesity) in C57BL/KsJ mice produces aberrant trans-regulation of sex steroid sulfotransferase genes. *Journal of Clinical Investigation*, **93**, 2007–2013.

Leiter EH, Chapman HD and Falany CN (1991) Synergism of obesity genes with hepatic steroid sulfotransferases to mediate diabetes in mice. *Diabetes*, **40**, 1360–1363.

Leiter EH, Kintner J, Flurkey K, Beamer WG and Naggert JK (1999) Physiologic and endocrinologic characterization of male sex-biased diabetes in C57BLKS/J mice congenic for the fat mutation at the carboxypeptidase E locus. *Endocrine*, **10**, 57–66.

Lenz T, Werle E, Strobel G and Weicker H (1991) O-Methylated and sulfoconjugated catecholamines: differential activities at human platelet α2-adrenoceptors. *Canadian Journal of Physiology and Pharmacology*, **69**, 929–937.

Levy G and Yamada H (1971) Drug biotransformation interactions in man: 3. Acetaminophen and salicylamide. *Journal of Pharmaceutical Sciences*, **60**, 215–221.

Lewis AJ, Otake Y, Walle UK and Walle T (2000) Sulphonation of N-hydroxy-2-acetylaminofluorene by human dehydroepiandrosterone sulphotransferase. *Xenobiotica*, **30**, 253–261.

Li AP, Hartman NR, Lu C, Collins JM and Strong JM (1999) Effects of cytochrome P450 inducers on 17α-ethinyloestradiol (EE$_2$) conjugation by primary human hepatocytes. *British Journal of Clinical Pharmacology*, **48**, 733–742.

Li CL and James MO (2000) Oral bioavailability and pharmacokinetics of elimination of 9-hydroxybenzo[a]pyrene and its glucoside and sulfate conjugates after administration to male and female American lobsters, *Homarus americanus*. *Toxicological Sciences*, **57**, 75–86.

Li XY and Anderson RJ (1999) Sulfation of iodothyronines by recombinant human liver steroid sulfotransferases. *Biochemical and Biophysical Research Communications*, **263**, 632–639.

Lin E-S and Yang Y-S (1998) Colorimetric determination of the purity of 3'-phospho adenosine 5'-phosphosulfate and natural abundance of 3'-phospho adenosine 5'-phosphate at picomole quantities. *Analytical Biochemistry*, **264**, 111–117.

Littlewood J, Glover V, Sandler M, Petty R, Peatfield R and Rose FC (1982) Platelet phenolsulphotransferase deficiency in dietary migraine. *Lancet*, **1**, 983–986.

Liu L and Klaassen CD (1996a) Ontogeny and hormonal basis of male-dominant rat hepatic sulfotransferases. *Molecular Pharmacology*, **50**, 565–572.

Liu L and Klaassen CD (1996b) Ontogeny and hormonal basis of female-dominant rat hepatic sulfotransferases. *Journal of Pharmacology and Experimental Therapeutics*, **279**, 386–391.

Liu L and Klaassen CD (1996c) Different mechanism of saturation of acetaminophen sulfate conjugation in mice and rats. *Toxicology and Applied Pharmacology*, **139**, 128–134.

Liu L and Klaassen CD (1996d) Regulation of hepatic sulfotransferases by steroidal chemicals in rats. *Drug Metabolism and Disposition*, **24**, 854–858.

Liu L, Lecluyse EL, Liu J and Klaassen CD (1996) Sulfotransferase gene expression in primary cultures of rat hepatocytes. *Biochemical Pharmacology*, **52**, 1621–1630.

Liu MC, Sakakibara Y and Liu CC (1999) Bacterial expression, purification, and characterization of a novel mouse sulfotransferase that catalyzes the sulfation of eicosanoids. *Biochemical and Biophysical Research Communications*, **254**, 65–69.

Loewen GR, Herman RJ, Ross SG and Verbeeck RK (1988) Effect of dose on the glucuronidation and sulphation kinetics of diflunisal in man: single dose studies. *British Journal of Clinical Pharmacology*, **26**, 31–39.

Longcope C (1995) Metabolism of dehydroepiandrosterone. *Annals of the New York Academy of Sciences*, **774**, 143–148.

Loriaux D, Ruder H and Lipsett M (1971) The measurement of estrone sulfate in plasma. *Steroids*, **18**, 463–473.

Luu NX, Driscoll WJ, Martin BM and Strott CA (1995) Molecular cloning and expression of a guinea pig 3-hydroxysteroid sulfotransferase distinct from chiral-specific 3α-hydroxysteroid sulfotransferase. *Biochemical and Biophysical Research Communications*, **217**, 1078–1086.

Luu-The V, Dufort I, Paquet N, Reimnitz G and Labrie F (1995) Structural characterization and expression of the human dehydroepiandrosterone sulfotransferase gene. *DNA and Cell Biology*, **14**, 511–518.

Lyle S, Stanczak JD, Westley J and Schwartz NB (1995) Sulfate-activating enzymes in normal and brachymorphic mice: evidence for a channeling defect. *Biochemistry*, **34**, 940–945.

Lyman SD and Poland A (1983) Effect of the brachymorphic trait in mice on xenobiotic sulfate ester formation. *Biochemical Pharmacology*, **32**, 3345–3350.

Lyon ES and Jakoby WB (1980) The identity of alcohol sulfotransferases with hydroxysteroid sulfotransferases. *Archives of Biochemistry and Biophysics*, **202**, 474–481.

Macdonald JI, Wallace SM, Herman RJ and Verbeeck RK (1995) Effect of probenecid on the formation and elimination kinetics of the sulphate and glucuronide conjugates of diflunisal. *European Journal of Clinical Pharmacology*, **47**, 519–523.

Malejka-Giganti D, Ringer DP, Vijayaraghavan, Kielbauch CC and Kong J (1997) Aryl sulfotransferase IV deficiency in rat liver carcinogenesis initiated with diethylnitrosamine and promoted with N-2-fluorenenylacetamide or its C-9-oxidized metabolites. *Experimental and Molecular Pathology*, **64**, 63–77.

Malins DC and Roubal WT (1982) Aryl sulfate formation in sea urchin (*Strongylocentrotus droebachiensis*) ingesting marine algae (*Fucus distichus*) containing 2,6-dimethylnaphthalene. *Environmental Research*, **27**, 290–297.

Mancini MA, Song CS, Rao TR, Chatterjee B and Roy AK (1992) Spatio-temporal expression of estrogen sulfotransferase within the hepatic lobule of male rats: implication of *in situ* estrogen inactivation in androgen action. *Endocrinology*, **131**, 1541–1546.

Mangold JB, Mangold BLK and Spina A (1986) Rat liver sulfotransferase-catalyzed sulfation and rearrangement of 9-fluorenone oxime. *Biochimica et Biophysica Acta*, **874**, 37–43.

Mangold JB, McCann DJ and Spina A (1993) Aryl sulfotransferase-IV-catalyzed sulfation of aryl oximes: steric and substituent effects. *Biochimica et Biophysica Acta*, **1163**, 217–222.

Marazziti D, Palego L, Mazzanti C, Silvestri S and Cassano GB (1995) Human platelet sulfotransferase shows seasonal rhythms. *Chronobiology International*, **12**, 100–105.

Marazziti D, Palego L, Dell'Osso L, Batistini A, Cassano GB and Akiskal HS (1996) Platelet sulfotransferase in different psychiatric disorders. *Psychiatry Research*, **65**, 73–78.

Marcus CJ, Sekura RD and Jakoby WB (1980) A hydroxysteroid sulfotransferase from rat liver. *Analytical Biochemistry*, **107**, 296–304.

Marshall AD, Darbyshire JF, Hunter AP, McPhie P and Jakoby WB (1997) Control of activity through oxidative modification at the conserved residue Cys[66] of aryl sulfotransferase IV. *Journal of Biological Chemistry*, **272**, 9153–9160.

Marshall AD, McPhie P and Jakoby WB (2000) Redox control of aryl sulfotransferase specificity. *Archives of Biochemistry and Biophysics*, **382**, 95–104.

Marsolais F and Varin L (1998) Recent developments in the study of the structure-function relationship of flavonol sulfotransferases. *Chemico-Biological Interactions*, **109**, 117–122.

Martel C, Melner MH, Gagné D, Simard J and Labrie F (1994) Widespread tissue distribution of steroid sulfatase, 3β-hydroxysteroid dehydrogenase/Δ^5-Δ^4 isomerase (3β-HSD), 17β-HSD 5α-reductase and aromatase activities in the rhesus monkey. *Molecular and Cellular Endocrinology*, **104**, 103–111.

Matsui M, Takahashi M, Miwa Y, Motoyoshi Y and Homma H (1995) Structure-activity relationships of alkylamines that inhibit rat liver hydroxysteroid sulfotransferase activities *in vitro*. *Biochemical Pharmacology*, **49**, 739–741.

Maus TP, Pearson RK, Anderson RJ, Woodson LC, Reiter C and Weinshilboum RM (1982) Rat phenol sulfotransferase: assay procedure, developmental changes, and glucocorticoid regulation. *Biochemical Pharmacology*, **31**, 849–856.

McCall JM, Aiken JW, Chidester CG, DuCharme DW and Wendling MG (1983) Pyrimidine and triazine 3-oxide sulfates: a new family of vasodilators. *Journal of Medicinal Chemistry*, **26**, 1791–1793.

Meerman JHN, Beland FA and Mulder GJ (1981) Role of sulfation in the formation of DNA adducts from N-hydroxy-2-acetylaminofluorene in rat liver *in vivo*: inhibition of N-acetylated aminofluorene adduct formation by pentachlorophenol. *Carcinogenesis*, **2**, 413–416.

Meerman JHN, Beland FA, Ketterer B, Srai SKS, Bruins AP and Mulder GJ (1982) Identification of glutathione conjugates formed from N-hydroxy-2-acetylaminofluorene in the rat. *Chemico-Biological Interactions*, **39**, 149–168.

Meerman JHN, Sterenborg HM and Mulder GJ (1983) Use of pentachlorophenol as long-term inhibitor of sulfation of phenols and hydroxamic acids in the rat *in vivo*. *Biochemical Pharmacology*, **32**, 1587–1593.

Mehta R, Hirom PC and Millburn P (1978) The influence of dose on the pattern of conjugation of phenol and 1-naphthol in non-human primates. *Xenobiotica*, **8**, 445–452.

Meinl W, Meerman JHN and Glatt HR (2001a) *Salmonella typhimurium* strains expressing alloenzymes of human sulfotransferase 1A2 for the activation of promutagens. *Cancer Research*, submitted.

Meinl W and Glatt HR (2001b) Structure and localization of the human *SULT1B1* gene: neighborhood to *SULT1E1* and a *SULT1D* pseudogene. *Biochemical and Biophysical Research Communications*, submitted.

Meisheri KD, Oleynek JJ and Puddington L (1991) Role of protein sulfation in vasodilation induced by minoxidil sulfate, a K$^+$ channel opener. *Journal of Pharmacology and Experimental Therapeutics*, **258**, 1091–1097.

Meyers SA, Lozon MM, Corombos JD, Saunders DE, Hunter K, Christensen C and Brooks SC (1983) Induction of porcine uterine estrogen sulfotransferase activity by progesterone. *Biology of Reproduction*, **28**, 1119–1128.

Miller EC and Miller JA (1966) Mechanism of chemical carcinogenesis: nature of proximate carcinogens and interaction with macromolecules. *Pharmacological Reviews*, **18**, 805–838.

Miller JA and Surh Y-J (1994) Sulfonation in chemical carcinogenesis. In *Handbook of Pharmacology, Vol. 112, Conjugation-Deconjugation Reactions in Drug Metabolism and Toxicity*, Kauffman FC (ed.), Springer-Verlag, Berlin, pp. 429–458.

Minami M, Kawaguchi T, Takenaka K, Hamaue N, Endo T, Hirafuji M, Ohno K, Hagihara K and Parvez HS (1995a) Inotropic effects of dopamine-3-O-sulfate and dopamine-4-O-sulfate. *Biogenic Amines*, **11**, 469–477.

Minami M, Terado M, Kudo M, Endo T, Hamaue N, Hirafuji M, Ohno K, Itoh S, Yoshizawa I and Parvez HS (1995b) The hypotensive effects of intracerebroventricularly administered dopamine 3 O sulfate in anesthetized rats. *Biogenic Amines*, **11**, 487 504.

Minck K, Schupp RR, Illing HP, Kahl GF and Netter KJ (1973) Interrelationship between demethylation of *p*-nitroanisole and conjugation of *p*-nitrophenol in rat liver. *Naunyn–Schmiedeberg's Archives of Pharmacology*, **279**, 347–360.

Møller JV and Sheikh MI (1982) Renal organic anion transport system: pharmacological, physiological, and biochemical aspects. *Pharmacological Reviews*, **34**, 315–358.

Momoeda M, Cui YX, Sawada Y, Taketani Y, Mizuno M and Iwamori M (1994) Pseudopregnancy-dependent accumulation of cholesterol sulfate due to up-regulation of cholesterol sulfotransferase and concurrent down-regulation of cholesterol sulfate sulfatase in the uterine endometria of rabbits. *Journal of Biochemistry (Tokyo)*, **116**, 657–662.

Monnerjahn S, Seidel A, Steinberg P, Oesch F, Hinz M, Stezowsky JJ, Hewer A, Phillips DH and Glatt HR (1993) Formation of DNA adducts from 1-hydroxymethylpyrene in liver cells *in vivo* and *in vitro*. In *Postlabelling Methods for Detection of DNA Adducts*, Phillips DH, Castegnaro M and Bartsch H (eds), IARC, Lyon (France), pp. 189–193.

Monteith DK (1992) Inhibition of sulfotransferase affecting unscheduled DNA synthesis induced by 2-acetylaminofluorene: an *in vivo* and *in vitro* comparison. *Mutation Research*, **282**, 253–258.

Moorthy B, Liehr JG, Randerath E and Randerath K (1995) Evidence from ^{32}P-postlabeling and the use of pentachlorophenol for a novel metabolic activation pathway of diethylstilbestrol and its dimethyl ether in mouse liver: likely alpha-hydroxylation of ethyl group(s) followed by sulfate conjugation. *Carcinogenesis*, **16**, 2643–2648.

Muirhead MR and Somogyi AA (1991) Effect of H_2 antagonists on the differential secretion of triamterene and its sulfate conjugate metabolite by the isolated perfused rat kidney. *Drug Metabolism and Disposition*, **19**, 312–316.

Mulder GJ and Bleeker B (1975) UDP glucuronyltransferase and phenolsulfotransferase from rat liver *in vivo* and *in vitro* IV: species differences in harmol conjugation and elimination in bile and urine *in vivo*. *Biochemical Pharmacology*, **24**, 1481–1484.

Mulder GJ and Jakoby WB (1990) Sulfation. In *Conjugation Reactions in Drug Metabolism: An Integrated Approach*, Mulder GJ (ed.), Taylor and Francis, London, pp. 107–161.

Mulder GJ and Meerman JHN (1978) Glucuronidation and sulphation *in vivo* and *in vitro*: selective inhibition of sulphation by drugs and deficiency of inorganic sulphate. In *Conjugation Reactions in Drug Biotransformation*, Aitio A (ed.), Elsevier/North-Holland Biomedical Press, Amsterdam, pp. 389–397.

Mulder GJ and Scholtens E (1977) Phenol sulphotransferase and uridine diphosphate glucuronyltransferase from rat liver *in vivo* and *in vitro*: 2,6-dichloro-4-nitrophenol as selective inhibitor of sulphation. *Biochemical Journal*, **165**, 553–559.

Mulder GJ, Hayen-Keulemans K and Slutter NE (1975) UDP glucuronyltransferase and phenolsulfotransferase from rat liver *in vivo* and *in vitro*: characterization of conjugation and biliary excretion of harmol *in vivo* and in the perfused liver. *Biochemical Pharmacology*, **24**, 103–107.

Mulder GJ, Weitering JG, Scholtens E, Dawson JR and Pang KS (1984) Extrahepatic sulfation and glucuronidation in the rat *in vivo*: determination of the hepatic extraction ratio of harmol and the extrahepatic contribution to harmol conjugation. *Biochemical Pharmacology*, **33**, 3081–3087.

Mutschler E, Gilfrich H-J, Knauf K, Möhrke W and Völger K-D (1983) Pharmacokinetics of triamterene. *Clinical and Experimental Hypertension*, **A5**, 249–269.

Nadanaka S, Fujita M and Sugahara K (1999) Demonstration of a novel sulfotransferase in fetal bovine serum, which transfers sulfate to the C^6 position of the GalNAc residue in the sequence iduronic acid α1-3 GalNAc β1-4 iduronic acid in dermatan sulfate. *FEBS Letters*, **452**, 185–189.

Nagata K and Yamazoe Y (2000) Pharmacogenetics of sulfotransferase. *Annual Review of Pharmacology and Toxicology*, **40**, 159–176.

Nagata K, Ozawa S, Miyata M, Shimada M, Gong DW, Yamazoe Y and Kato R (1993) Isolation and expression of a cDNA encoding a male-specific rat sulfotransferase that catalyzes activation of N-hydroxy-2-acetylaminofluorene. *Journal of Biological Chemistry*, **268**, 24720–24725.

Naritomi Y, Niwa T, Shiraga T, Iwasaki K and Noda K (1994) Isolation and characterization of an alicyclic amine N-sulfotransferase from female rat liver. *Biological & Pharmaceutical Bulletin*, **17**, 1008–1011.

Nash RM, Stein L, Penno MB, Passananti GT and Vesell ES (1984) Sources of interindividual variations in acetaminophen and antipyrine metabolism. *Clinical Pharmacology and Therapeutics*, **36**, 417–430.

Nash AR, Glenn WK, Moore SS, Kerr J, Thompson AR and Thompson EOP (1988) Oestrogen sulfotransferase: molecular cloning and sequencing of cDNA for the bovine placental enzyme. *Australian Journal of Biological Sciences*, **41**, 507–516.

Nelson DR, Koymans L, Kamataki T, Stegeman JJ, Feyereisen R, Waxman DJ, Waterman MR, Gotoh O, Coon MJ, Estabrook RW, Gunsalus IC and Nebert DW (1996) P450 superfamily: update on new sequences, gene mapping, accession numbers and nomenclature. *Pharmacogenetics*, **6**, 1–42.

Nemoto N and Takayama S (1978) Inducibility of activation and/or detoxication enzymes in benzo[a]pyrene metabolism by 3-methylcholanthrene. *Toxicology Letters*, **1**, 247–252.

Ngah WZ and Smith JN (1983) Acidic conjugate of phenols in insects; glucoside phosphate and glucoside sulphate derivatives. *Xenobiotica*, **13**, 383–389.

Noda Y, Kamei H, Kamei Y, Nagai T, Nishida M and Nabeshima T (2000) Neurosteroids ameliorate conditioned fear stress: an association with sigma$_1$ receptors. *Neuropsychopharmacology*, **23**, 276–284.

Nose Y and Lipmann F (1958) Separation of steroid sulfokinases. *Journal of Biological Chemistry*, **233**, 1348–1351.

Novak M and Rangappa KS (1992) Nucleophilic substitution on the ultimate hepatocarcinogen N-(sulfonatooxy)-2-(acetylamino)fluorene by aromatic amines. *Journal of Organic Chemistry*, **57**, 1285–1290.

Nuwaysir EF, Daggett DA, Jordan VC and Pitot HC (1996) Phase II enzyme expression in rat liver in response to the antiestrogen tamoxifen. *Cancer Research*, **56**, 3704–3710.

O'Neil WM, Pezzullo JC, di Girolamo A, Tsoukas CM and Wainer IW (1999) Glucuronidation and sulphation of paracetamol in HIV-positive patients and patients with AIDS. *British Journal of Clinical Pharmacology*, **48**, 811–818.

Ogura K, Kajita J, Narihata H, Watabe T, Ozawa S, Nagata K, Yamazoe Y and Kato R (1989) Cloning and sequence analysis of a rat liver cDNA encoding hydroxysteroid sulfotransferase. *Biochemical and Biophysical Research Communications*, **165**, 168–174.

Ogura K, Sohtome T, Sugiyama A, Okuda H, Hiratsuka A and Watabe T (1990a) Rat liver cytosolic hydroxysteroid sulfotransferase (sulfotransferase a) catalyzing the formation of reactive sulfate esters from carcinogenic polycyclic hydroxymethylarenes. *Molecular Pharmacology*, **37**, 848–854.

Ogura K, Kajita J, Narihata H, Watabe T, Ozawa S, Nagata K, Yamazoe Y and Kato R (1990b) cDNA cloning of the hydroxysteroid sulfotransferase STa sharing a strong homology in amino acid sequence with the senescence marker protein SMP-2 in rat livers. *Biochemical and Biophysical Research Communications*, **166**, 1494–1500.

Oguro T, Gregus Z, Madhu C, Liu L and Klaassen CD (1994) Molybdate depletes hepatic 3-phosphoadenosine 5-phosphosulfate and impairs the sulfation of acetaminophen in rats. *Journal of Pharmacology and Experimental Therapeutics*, **270**, 1145–1151.

Orsi BA and Spencer B (1964) Choline sulphokinase (sulphotransferase). *Journal of Biochemistry (Tokyo)*, **56**, 81–91.

Otake Y, Nolan AL, Walle UK and Walle T (2000) Quercetin and resveratrol potently reduce estrogen sulfotransferase activity in normal human mammary epithelial cells. *Journal of Steroid Biochemistry and Molecular Biology*, **73**, 265–270.

Otterness DM, Wieben ED, Wood TC, Watson RWG, Madden BJ, McCormick DJ and Weinshilboum RM (1992) Human liver dehydroepiandrosterone sulfotransferase: molecular cloning and expression of cDNA. *Molecular Pharmacology*, **41**, 865–872.

Ouyang YB, Lane WS and Moore KL (1998) Tyrosylprotein sulfotransferase: Purification and molecular cloning of an enzyme that catalyzes tyrosine O-sulfation, a common posttranslational modification of eukaryotic proteins. *Proceedings of the National Academy of Sciences of the United States of America*, **95**, 2896–2901.

Ozawa S, Nagata K, Gong DW, Yamazoe Y and Kato R (1990) Nucleotide sequence of a full-length cDNA (PST-1) for aryl sulfotransferase from rat liver. *Nucleic Acids Research*, **18**, 4001.

Ozawa S, Chou H-C, Kadlubar FF, Nagata K, Yamazoe Y and Kato R (1994) Activation of 2-hydroxyamino-1-methyl-6-phenylimidazo[4,5-b]pyridine by cDNA-expressed human and rat arylsulfotransferases. *Japanese Journal of Cancer Research*, **85**, 1220–1228.

Ozawa S, Nagata K, Yamazoe Y and Kato R (1995a) Formation of 2-amino-3-methylimidazo[4,5-f]quinoline- and 2-amino-3,8-dimethylimidazo[4,5-f]quinoxaline-sulfamates by cDNA-expressed mammalian phenol sulfotransferases. *Japanese Journal of Cancer Research*, **86**, 264–269.

Ozawa S, Nagata K, Shimada M, Ueda M, Tsuzuki T, Yamazoe Y and Kato R (1995b) Primary structures and properties of two related forms of aryl sulfotransferases in human liver. *Pharmacogenetics*, **5**, S135–S140.

Ozawa S, Tang YM, Yamazoe Y, Kato R, Lang NP and Kadlubar FF (1998) Genetic polymorphisms in human liver phenol sulfotransferases involved in the bioactivation of *N*-hydroxy derivatives of carcinogenic arylamines and heterocyclic amines. *Chemico-Biological Interactions*, **109**, 237–248.

Pacifici GM and Marchi G (1993) Interindividual variability of phenol- and catechol-sulphotransferases in platelets from adults and newborns. *British Journal of Clinical Pharmacology*, **36**, 593–597.

Pacifici GM, Franchi M, Colizzi C, Giuliani L and Rane A (1988) Sulfotransferase in humans: development and tissue distribution. *Pharmacology*, **36**, 411–419.

Pacifici GM, Bigotti R, Marchi G and Giuliani L (1993a) Minoxidil sulphation in human liver and platelets—a study of interindividual variability. *European Journal of Clinical Pharmacology*, **45**, 337–341.

Pacifici GM, Kubrich M, Giuliani L, Devries M and Rane A (1993b) Sulphation and glucuronidation of ritodrine in human foetal and adult tissues. *European Journal of Clinical Pharmacology*, **44**, 259–264.

Pacifici GM, De Santi C, Mussi A and Ageletti CA (1996) Interindividual variability in the rate of salbutamol sulphation in the human lung. *European Journal of Clinical Pharmacology*, **49**, 299–303.

Pacifici GM, Gucci A and Giuliani L (1997a) Testosterone sulphation and glucuronidation in the human liver: interindividual variability. *European Journal of Drug Metabolism and Pharmacokinetics*, **22**, 253–258.

Pacifici GM, Dalessandro C, Gucci A and Giuliani L (1997b) Sulphation of the heterocyclic amine 1,2,3,4-tetrahydroisoquinoline in the human liver and intestinal mucosa: Interindividual variability. *Archives of Toxicology*, **71**, 477–481.

Pacifici GM, Giulianetti B, Quilici MC, Spisni R, Nervi M, Giuliani L and Gomeni R (1997c) (-)-Salbutamol sulphation in the human liver and duodenal mucosa: interindividual variability. *Xenobiotica*, **27**, 279–286.

Pacifici GM, Quilici MC, Giulianetti B, Spisni R, Nervi M, Giuliani L and Gomeni R (1998) Ritodrine sulphation in the human liver and duodenal mucosa: interindividual variability. *European Journal of Drug Metabolism and Pharmacokinetics*, **23**, 67–74.

Panda M, Novak M and Magonski J (1989) Hydrolysis kinetics of the ultimate hepatacarcinogen *N*-(sulfonatooxy)-2-(acetylamino)fluorene: detection of long-lived hydrolysis intermediates. *Journal of the American Chemical Society*, **111**, 4524–4525.

Pang KS, Koster H, Halsema IC, Scholtens E and Mulder GJ (1981) Aberrant pharmacokinetics of harmol in the perfused rat liver preparation: sulfate and glucuronide conjugations. *Journal of Pharmacology and Experimental Therapeutics*, **219**, 134–140.

Park-Chung M, Malayev A, Purdy RH, Gibbs TT and Farb DH (1999) Sulfated and unsulfated steroids modulate γ-aminobutyric acid A receptor function through distinct sites. *Brain Research*, **830**, 72–87.

Parker CR, Jr (1999) Dehydroepiandrosterone and dehydroepiandrosterone sulfate production in the human adrenal during development and aging. *Steroids*, **64**, 640–647.

Parker CR, Jr, Leveno K, Carr BR, Hauth J and MacDonald PC (1984) Umbilical cord plasma levels of dehydroepiandrosterone sulfate during human gestation. *Clinical Journal of Endocrinology and Metabolism*, **54**, 1216–1220.

Parker MH, McCann DJ and Mangold JB (1994) Sulfation of di- and tricyclic phenols by rat liver aryl sulfotransferase isozymes. *Archives of Biochemistry and Biophysics*, **310**, 325–331.

Pasqualini JR and Chetrite GS (1999) Estrone sulfatase versus estrone sulfotransferase in human breast cancer: potential clinical applications. *Journal of Steroid Biochemistry and Molecular Biology*, **69**, 287–292.

Pasqualini JR and Kinel F (1985) *Hormones and the Fetus*, Pergamon Press, Oxford, pp. 73–334.

Paul SM and Purdy RH (1992) Neuroactive steroids. *FASEB Journal*, **6**, 2311–2322.

Peters JM, Zhou YC, Ram PA, Lee SST, Gonzalez FJ and Waxman DJ (1996) Peroxisome proliferator-activated receptor α required for gene induction by dehydroepiandrosterone-3β-sulfate. *Molecular Pharmacology*, **50**, 67–74.

Petrotchenko EV, Pedersen LC, Borcher CH, Tomer KB and Negishi M (2001) The dimerization motif of cytosolic sulfotransferases. *FEBS Letters*, **490**, 39–43.

Pettigrew NE, Wright PFA and Macrides TA (1998) 5β-Scymnol sulfotransferase isolated from the tissues of an Australian shark species. *Comparative Biochemistry and Physiology B*, **121**, 299–307.

Pica-Mattoccia L, Novi A and Cioli D (1997) Enzymatic basis for the lack of oxamniquine activity in *Schistosoma haematobium* infections. *Parasitology Research*, **83**, 687–689.

Plager JE (1965) The binding of androsterone sulfate, etiocholanolone sulfate, and dehydroepiandrosterone sulfate by human plasma protein. *Journal of Clinical Investigation*, **44**, 1234–1239.

Preechagoon D, Smith MT and Prankerd RJ (1998) Investigation of the antinociceptive efficacy and relative potency of extended duration injectable 3-acylmorphine-6-sulfate prodrugs in rats. *International Journal of Pharmaceutics*, **163**, 191–201.

Price RA, Cox NJ, Spielman RS, Van Loon JA, Maidak BL and Weinshilboum RM (1988) Inheritance of human platelet thermolabile phenol sulfotransferase (TL PST) activity. *Genetic Epidemiology*, **5**, 1–15.

Puche RC and Nes WR (1962) Binding of dehydroepiandrosterone sulfate to serum albumin. *Endocrinology*, **70**, 857–863.

Radominska A, Comer KA, Zimniak P, Falany J, Iscan M and Falany CN (1990) Human liver steroid sulphotransferase sulphates bile acids. *Biochemical Journal*, **272**, 597–604.

Raftogianis RB, Wood TC, Otterness DM, van Loon JA and Weinshilboum RM (1997) Phenol sulfotransferase pharmacogenetics in humans: association of common SULT1A1 alleles with TS PST phenotype. *Biochemical and Biophysical Research Communications*, **239**, 298–304.

Raftogianis RB, Wood TC and Weinshilboum RM (1999) Human phenol sulfotransferases SULT1A2 and SULT1A1: genetic polymorphisms, allozyme properties, and human liver genotype-phenotype correlations. *Biochemical Pharmacology*, **58**, 605–616.

Rajkowski KM, Robel P and Baulieu EE (1997) Hydroxysteroid sulfotransferase activity in the rat brain and liver as a function of age and sex. *Steroids*, **62**, 427–436.

Ramaswamy SG and Jakoby WB (1987) Amino N-sulfotransferase. *Journal of Biological Chemistry*, **262**, 10039–10043.

Randerath K, Moorthy B, Mabon N and Sriram P (1994a) Tamoxifen: evidence by ^{32}P-postlabeling and use of metabolic inhibitors for two distinct pathways leading to mouse hepatic DNA adduct formation and identification of 4-hydroxytamoxifen as a proximate metabolite. *Carcinogenesis*, **15**, 2087–2094.

Randerath K, Bi J, Mabon N, Sriram P and Moorthy B (1994b) Strong intensification of mouse hepatic tamoxifen DNA adduct formation by pretreatment with the sulfotransferase inhibitor and ubiquitous environmental pollutant pentachlorophenol. *Carcinogenesis*, **15**, 797–800.

Rao SI and Duffel MW (1991a) Benzylic alcohols as stereospecific substrates and inhibitors for aryl sulfotransferase. *Chirality*, **3**, 104–111.

Rao SI and Duffel MW (1991b) Inhibition of aryl sulfotransferase by carboxylic acids. *Drug Metabolism and Disposition*, **19**, 543–545.

Ratna S, Chiba M, Bandyopadhyay L and Pang KS (1993) Futile cycling between 4-methylumbelliferone and its conjugates in perfused rat liver. *Hepatology*, **17**, 838–853.

Raud HR and Hobkirk R (1968) *In vitro* biosynthesis of steroid sulfates by cell-free preparations from tissues of the laying hen. *Canadian Journal of Biochemistry*, **46**, 749–757.

Rearick JI, Hesterberg TW and Jetten AM (1987a) Human bronchial epithelial cells synthesize cholesterol sulfate during squamous differentiation *in vitro*. *Journal of Cellular Physiology*, **133**, 573–578.

Rearick JI, Albro PW and Jetten AM (1987b) Increase in cholesterol sulfotransferase activity during *in vitro* squamous differentiation of rabbit tracheal epithelial cells and its inhibition by retinoic acid. *Journal of Biological Chemistry*, **262**, 13069–13074.

Reddy DS, Kaur G and Kulkarni SK (1998) Sigma (σ_1) receptor mediated anti-depressant-like effects of neurosteroids in the Porsolt forced swim test. *Neuroreport*, **9**, 3069–3073.

Reiter R and Wendel A (1984) Selenium and drug metabolism: II. Independence of glutathione peroxidase and reversibility of hepatic enzyme modulations in deficient mice. *Biochemical Pharmacology*, **33**, 1923–1928.

Renosto F and Segel IH (1977) Choline sulfokinase of *Penicillium chrysogenum:* partial purification and kinetic mechanism. *Archives of Biochemistry and Biophysics*, **180**, 416–428.

Rhodes ME, Li P-K, Burke AM and Johnson DA (1997) Enhanced plasma DHEAS, brain acetylcholine and memory mediated by steroid sulfatase inhibition. *Brain Research*, **773**, 28–32.

Richard K, Hume R, Kaptein E, Visser TJ and Coughtrie MWH (2001) Sulfation of thyroid hormone and dopamine during human development: ontogeny of phenol sulfotransferases and arylsulfatase in liver, lung and brain. *Journal of Clinical Endocrinology and Metabolism*, **86**, 2734–2742.

Rickert DE, Long RM, Dyroff MC and Kedderis GL (1984) Hepatic macromolecular covalent binding of mononitrotoluenes in Fischer 344 rats. *Chemico-Biological Interactions*, **52**, 131–139.

Ring JA, Ghabrial H, Ching MS, Shulkes A, Smallwood RA and Morgan DJ (1996) Conjugation of *p*-nitrophenol by isolated perfused fetal sheep liver. *Drug Metabolism and Disposition*, **24**, 1378–1384.

Ringer DP, Norton TR and Kizer DE (1985) Effect of sulfotransferase inhibitors on the 2-acetylaminofluorene-mediated lowering of rat liver *N*-hydroxy-2-acetylaminofluorene sulfotransferase activity. *Biochemical Pharmacology*, **34**, 3380–3383.

Ringer DP, Norton TR, Cox B and Howell BA (1988) Changes in rat liver *N*-hydroxy-2-acetylaminofluorene aryl sulfotransferase activity at early and late stages of hepatocarcinogenesis resulting from dietary administration of 2-acetylaminofluorene. *Cancer Letters*, **40**, 247–255.

Ringer DP, Norton TR and Howell BA (1990) 2-Acetylaminofluorene-mediated alteration in the level of liver aryl sulfotransferase IV during rat hepatocarcinogenesis. *Cancer Research*, **50**, 5301–5307.

Ringer DP, Norton TR and Self RR (1992) Reaction product inactivation of aryl sulfotransferase IV following electrophilic substitution by the sulfuric acid ester of *N*-hydroxy-2-acetylaminofluorene. *Carcinogenesis*, **13**, 107–112.

Ringer DP, Howell BA, Norton TR, Woulfe GW, Duffel MW and Kosanke SD (1994) Evidence of two separate mechanisms for the decrease in aryl sulfotransferase activity in rat liver during early stages of 2-acetylaminofluorene-induced hepatocarcinogenesis. *Molecular Carcinogenesis*, **9**, 2–9.

Rivoal J and Hanson AD (1994) Choline-*O*-sulfate biosynthesis in plants: identification and partial characterization of a salinity-inducible choline sulfotransferase from species of *Limonium* (Plumbaginaceae). *Plant Physiology*, **106**, 1187–1193.

Robbins PW and Lipmann F (1956) Identification of enzymically active sulfate as adenosine-3′-phosphate-5′-phosphosulfate. *Journal of the American Chemical Society*, **78**, 2652–2653.

Robbins PW and Lipmann F (1957) Isolation and identification of active sulfate. *Journal of Biological Chemistry*, **229**, 837–851.

Roberts KD (1987) Sterol sulfates in the epididymis; synthesis and possible function in the reproductive process. *Journal of Steroid Biochemistry*, **27**, 337–341.

Romiti P, Giuliani L and Pacifici GM (1992) Interindividual variability in the *N*-sulphation of desipramine in human liver and platelets. *British Journal of Clinical Pharmacology*, **33**, 17–23.

Rouleau M, Marsolais F, Richard M, Nicolle L, Voigt B, Adam G and Varin L (1999) Inactivation of brassinosteroid biological activity by a salicylate-inducible steroid sulfotransferase from *Brassica napus*. *Journal of Biological Chemistry*, **274**, 20925–20930.

Roy AB (1963) The arylsulphatases and some related enzymes in livers of some lower vertebrates. *Australian Journal of Experimental Biology and Medical Sciences*, **148**, 270–276.

Roy AB (1971) Sulphate conjugation enzymes. In *Concepts in Biochemical Pharmacology (Part 2)*, Brodie BB and Gillette JR (eds), Springer-Verlag, Berlin, pp. 536–563.

Rubin GL, Sharp S, Jones AL, Glatt HR, Mills JA and Coughtrie MWH (1996) Design, production and characterization of antibodies discriminating between the phenol- and monoamine-sulphating forms of human phenol sulphotransferase. *Xenobiotica*, **26**, 1113–1119.

Rubin GL, Harrold AJ, Mills JA, Falany CN and Coughtrie MWH (1999) Regulation of sulphotransferase expression in the endometrium during the menstrual cycle, by oral contraceptives and during early pregnancy. *Molecular Human Reproduction*, **5**, 995–1002.

Runge-Morris MA (1997) Sulfation and sulfotransferases: 2. Regulation of expression of the rodent cytosolic sulfotransferases. *FASEB Journal*, **11**, 109–117.

Runge-Morris M, Rose K and Kocarek TA (1996) Regulation of rat hepatic sulfotransferase gene expression by glucocorticoid hormones. *Drug Metabolism and Disposition*, **24**, 1095–1101.

Runge-Morris M, Rose K, Falany CN and Kocarek TA (1998) Differential regulation of individual sulfotransferase isoforms by phenobarbital in male rat liver. *Drug Metabolism and Disposition*, **26**, 795–801.

Saidha T and Schiff JA (1994) Purification and properties of a phenol sulphotransferase from *Euglena* using L-tyrosine as substrate. *Biochemical Journal*, **298**, 45–50.

Sakakibara Y, Takami Y, Zwieb C, Nakayama T, Suiko M, Nakajima H and Liu MC (1995) Purification, characterization, and molecular cloning of a novel rat liver dopa/tyrosine sulfotransferase. *Journal of Biological Chemistry*, **270**, 30470–30478.

Sakakibara Y, Katafuchi J, Takami Y, Nakayama T, Suiko M, Nakajima H and Liu M-C (1997) Manganese-dependent dopa/tyrosine sulfation in HepG2 human hepatoma cells: novel dopa/tyrosine sulfotransferase activities associated with the human monoamine-form phenol sulfotransferase. *Biochimica et Biophysica Acta*, **1355**, 102–106.

Sakakibara Y, Yanagisawa K, Takami Y, Nakayama T, Suiko M and Liu M-C (1998a) Molecular cloning, expression, and functional characterization of novel mouse sulfotransferases. *Biochemical and Biophysical Research Communications*, **247**, 681–686.

Sakakibara Y, Yanagisawa K, Katafuchi J, Ringer DP, Takami Y, Nakayama T, Suiko M and Liu MC (1998b) Molecular cloning, expression, and characterization of novel human SULT1C sulfotransferases that catalyze the sulfonation of N-hydroxy-2-acetylaminofluorene. *Journal of Biological Chemistry*, **273**, 33929–33935.

Sanborn HR and Malins DC (1980) The disposition of aromatic hydrocarbons in adult spot shrimp (*Pandalus platyceros*) and the formation of metabolites of naphthalene in adult and larval spot shrimp. *Xenobiotica*, **10**, 193–200.

Santini F, Chopra IJ, Wu SY, Solomon DH and Teco GNC (1992) Metabolism of 3,5,3'-triiodothyronine sulfate by tissues of the fetal rat: a consideration of the role of desulfation of 3,5,3'-triiodothyronine sulfate as a source of T_3. *Pediatric Research*, **31**, 541–544.

Satoh T, Watanabe K, Takanashi K, Itoh S, Takagi H and Yoshizawa I (1992) Evidence of direct conversion of testosterone sulfate to estradiol 17-sulfate by human placental microsomes. *Journal of Pharmacobio-Dynamics*, **15**, 427–436.

Schachter B and Marrian GF (1938) The isolation of estrone sulfate from the urine of pregnant mares. *Journal of Biological Chemistry*, **126**, 663–669.

Schmoldt A, Blomer I and Johannes A (1992) Hydroxysteroid sulfotransferase and a specific UDP-glucuronosyltransferase are involved in the metabolism of digitoxin in man. *Naunyn–Schmiedeberg's Archives of Pharmacology*, **346**, 226–233.

Schultze M and Kondorosi A (1998) Regulation of symbiotic root nodule development. *Annual Review of Genetics*, **32**, 33–57.

Schwarz LR and Schwenk M (1984) Sulfation in isolated enterocytes of guinea pig: dependence on inorganic sulfate. *Biochemical Pharmacology*, **33**, 3353–3356.

Schwenk M and Locher M (1985) 1-Naphthol conjugation in isolated cells from liver, jejunum, ileum, colon and kidney of the guinea pig. *Biochemical Pharmacology*, **34**, 697–701.

Seah VMY and Wong KP (1994) 2,6-Dichloro-4-nitrophenol (DCNP), an alternate-substrate inhibitor of phenolsulfotransferase. *Biochemical Pharmacology*, **47**, 1743–1749.

Sekura RD and Jakoby WB (1979) Phenol sulfotransferases. *Journal of Biological Chemistry*, **254**, 5658–5663.

Sekura RD and Jakoby WB (1981) Aryl sulfotransferase IV from rat liver. *Archives of Biochemistry and Biophysics*, **211**, 352–359.

Seth P, Lunetta KL, Bell DW, Gray H, Nasser SM, Rhei E, Kaelin CM, Iglehart DJ, Marks JR,

Garber JE, Haber DA and Polyak K (2000) Phenol sulfotransferases: hormonal regulation, polymorphism, and age of onset of breast cancer. *Cancer Research*, **60**, 6859–6863.

Sharp S, Barker EV, Coughtrie MWH, Lowenstein PR and Hume R (1993) Immunochemical characterisation of a dehydroepiandrosterone sulfotransferase in rats and humans. *European Journal of Biochemistry*, **211**, 539–548.

Shibutani S, Dasaradhi L, Terashima I, Banoglu E and Duffel MW (1998) α-Hydroxytamoxifen is a substrate of hydroxysteroid (alcohol) sulfotransferase, resulting in tamoxifen DNA adducts. *Cancer Research*, **58**, 647–653.

Shinohara A, Saito K, Yamazoe Y, Kamataki T and Kato R (1986) Inhibition of acetyl-coenzyme A dependent activation of N-hydroxyarylamines by phenolic compounds, pentachlorophenol and 1-nitro-2-naphthol. *Chemico-Biological Interactions*, **60**, 275–285.

Shiraga T, Iwasaki K, Takeshita K, Matsuda H, Niwa T, Tozuka Z, Hata T and Guengerich FP (1995) Species- and gender-related differences in amine, alcohol and phenol sulphoconjugations. *Xenobiotica*, **25**, 1063–1071.

Shiraga T, Hata T, Yamazoe Y, Ohno Y and Iwasaki K (1999a) N-Sulphoconjugation of amines by human cytosolic hydroxysteroid sulphotransferase. *Xenobiotica*, **29**, 341–347.

Shiraga T, Iwasaki K, Hata T, Yoshinari K, Nagata K, Yamazoe Y and Ohno Y (1999b) Purification and characterization of two amine N-sulfotransferases, AST-RB1 (ST3A1) and AST-RB2 (ST2A8), from liver cytosols of male rabbits. *Archives of Biochemistry and Biophysics*, **362**, 265–274.

Short CR, Flory W, Hsieh LC, Aranas T, Ou SP and Weissinger J (1988) Comparison of hepatic drug metabolizing enzyme activities in several agricultural species. *Comparative Biochemistry and Physiology C*, **91**, 419–424.

Siiteri PK and MacDonald PC (1966) Placental estrogen biosynthesis during human pregnancy. *Journal of Clinical Endocrinology and Metabolism*, **26**, 751–761.

Singer SS (1994) The same enzymes catalyze sulfation of minoxidil, minoxidil analogs and catecholamines. *Chemico-Biological Interactions*, **92**, 33–45.

Singer SS (1985) Preparation and characterization of the different kinds of sulfotransferases. In *Methodological Aspects of Drug Metabolizing Enzymes,* Zakim D and Vessey DA (eds), John Wiley, New York, pp. 95–159.

Singer SS, Giera D, Johnson J and Sylvester S (1976) Enzymatic sulfation of steroids: I. The enzymic basis for the sex difference in cortisol sulfation by rat liver preparations. *Endocrinology*, **99**, 963–974.

Singer SS, Hess E and Sylvester S (1977) Hepatic cortisol sulfotransferase activity in several types of experimental hypertensions in male rats. *Biochemical Pharmacology*, **26**, 1033–1038.

Singer SS, Ansel AZ, van Brunt N, Torres J and Galaska EG (1984) Enzymatic sulfation of steroids: XX. Effects of ten drugs on the hepatic glucocorticoid sulfotransferase activity of rats *in vitro* and *in vivo*. *Biochemical Pharmacology*, **33**, 3485–3490.

Singer SS, Galaska EG, Feeser TA, Benak RL, Ansel AZ and Moloney A (1985) Enzymatic sulfation of steroids: XIX. Cortisol sulfotransferase activity, glucocorticoid sulfotransferases, and tyrosine aminotransferase induction in chicken, gerbil, and hamster liver. *Canadian Journal of Biochemistry and Cell Biology*, **63**, 23–32.

Singer SS, Palmert MR, Redman MD, Leahy DM, Feeser TC, Lucarelli MJ, Volkwein LS and Bruns M (1988) Hepatic dopamine sulfotransferases in untreated rats and in rats subjected to endocrine or hypertension-related treatments. *Hepatology*, **8**, 1511–1520.

Singer SS, Hagedorn JE, Smith DM and Williams JL (1995) The enzymic basis for the rat liver 6-hydroxymelatonin sulfotransferase activity. *Journal of Pineal Research*, **18**, 49–55.

Smith JN (1968) The comparative metabolism of xenobiotics. *Advances in Comparative Physiology and Biochemistry*, **3**, 173–232.

Smith BA, Gutmann HR and Springfield JR (1989) Catalytic effect of serum albumin on the o-rearrangement of N-sulfooxy-2-acetylaminofluorene, a potential hepatocarcinogen in the rat, to nonmutagenic sulfuric acid esters of o-amidofluorenols. *Biochemical Pharmacology*, **38**, 3987–3994.

Sodum RS, Sohn OS, Nie G and Fiala ES (1994) Activation of the liver carcinogen 2-nitropropane by aryl sulfotransferase. *Chemical Research in Toxicology*, **7**, 344–351.

Song C-S, Kim JM, Roy AK and Chatterjee B (1990) Structure and regulation of the senescence marker protein 2 gene promoter. *Biochemistry*, **29**, 542–551.

Song WC, Moore R, McLachlan JA and Negishi M (1995) Molecular characterization of a testis-specific estrogen sulfotransferase and aberrant liver expression in obese and diabetogenic C57BL/KsJ-*db/db* mice. *Endocrinology*, **136**, 2477–2484.

Sorensen PW, Scott AP, Stacey NE and Bowdin L (1995) Sulfated 17,20β-dihydroxy-4-pregnen-3-one functions as a potent and specific olfactory stimulant with pheromonal actions in the goldfish. *General and Comparative Endocrinology*, **100**, 128–142.

Spaulding SW (1994) Bioactivities of conjugated iodothyronines. In *Thyroid Hormone Metabolism: Molecular Biology and Alternate Pathways*, Wu SY and Visser TJ (eds), CRC Press, Boca Raton, FL, pp. 139–153.

Spencer B (1960) Endogenous sulphate acceptors in rat liver. *Biochemical Journal*, **77**, 294–304.

Stehly GR and Hayton WL (1988) Detection of pentachlorophenol and its glucuronide and sulfate conjugates in fish bile and exposure water. *Journal of Environmental Science and Health B*, **23**, 355–366.

Steiner M, Bastian M, Schulz WA, Pulte T, Franke KH, Rohring A, Wolff JM, Seiter H and Schuff-Werner P (2000) Phenol sulphotransferase SULT1A1 polymorphism in prostate cancer: lack of association. *Archives of Toxicology*, **74**, 222–225.

Sterzel W and Eisenbrand G (1986) *N*-Nitrosodiethanolamine is activated in the rat to an ultimate genotoxic metabolite by sulfotransferase. *Journal of Cancer Research and Clinical Oncology*, **111**, 20–24.

Strott CA (1996) Steroid sulfotransferases. *Endocrine Reviews*, **17**, 670–697.

Sugiyama Y, Stolz A, Sugimoto M, Kuhlenkamp J, Yamada T and Kaplowitz N (1984) Identification and partial purification of a unique phenolic steroid sulphotransferase in rat liver cytosol. *Biochemical Journal*, **224**, 947–953.

Suiko M, Sakakibara Y and Liu M-C (2000) Sulfation of environmental estrogen-like chemicals by human cytosolic sulfotransferases. *Biochemical and Biophysical Research Communications*, **267**, 80–84.

Sund RB and Lauterbach F (1986) Drug metabolism and metabolite transport in the small and large intestine: experiments with 1-naphthol and phenolphthalein by luminal and contraluminal administration in the isolated guinea pig mucosa. *Acta Pharmacologica et Toxicologica*, **58**, 74–83.

Sund RB and Lauterbach F (1993) 2-Naphthol metabolism and metabolite transport in the isolated guinea pig mucosa: further evidence for compartmentation of intestinal drug metabolism. *Pharmacology & Toxicology*, **72**, 84–89.

Sundaram RS, Szumlanski C, Otterness D, Van Loon JA and Weinshilboum RD (1989) Human intestinal phenol sulfotransferase: assay conditions, activity levels and partial purification of the thermolabile form. *Drug Metabolism and Disposition*, **17**, 255–264.

Surh Y-J, Lai C-C, Miller JA and Miller EC (1987) Hepatic DNA and RNA adduct formation from the carcinogen 7-hydroxymethyl-12-methylbenz[*a*]anthracene and its electrophilic sulfuric acid ester metabolite in preweaning rats and mice. *Biochemical and Biophysical Research Communications*, **144**, 576–582.

Surh Y-J, Liem A, Miller EC and Miller JA (1989) Metabolic activation of the carcinogen 6-hydroxymethylbenzo[*a*]pyrene: formation of an electrophilic sulfuric acid ester and benzylic DNA adducts in rat liver *in vivo* and in reactions *in vitro*. *Carcinogenesis*, **10**, 1519–1528.

Surh Y-J, Blomquist JC and Miller JA (1990) Activation of 1-hydroxymethylpyrene to an electrophilic and mutagenic metabolite by rat hepatic sulfotransferase activity. In *Biological Reactive Intermediates IV: Molecular and Cellular Effects and Their Impact on Human Health*, Witmer CM, Snyder RR, Jollow DJ, Kalf GF, Kocsis JJ and Sipes IG (eds), Plenum Press, New York, pp. 383–391.

Surh Y-J, Liem A, Miller EC and Miller JA (1991) Age- and sex-related differences in activation of the carcinogen 7-hydroxymethyl-12-methylbenz[*a*]anthracene to an electrophilic sulfuric acid ester metabolite in rats: possible involvement of hydroxysteroid sulfotransferase activity. *Biochemical Pharmacology*, **41**, 213–221.

Swevers L, Lambert JG, Novak F, Paesen G and De Loof A (1991) Lack of essential enzymes for the biosynthesis of C19 and C18 steroids in gonads of the migratory locust, Locusta migratoria. General and Comparative Endocrinology, **84**, 237–248.

Takahashi M, Tamura H, Kondo S, Kobayashi K and Matsui M (1998) Identification of three hydroxysteroid sulfotransferase isoenzymes in the rat liver. Biological & Pharmaceutical Bulletin, **21**, 10–15.

Takanashi K, Watanabe K and Yoshizawa I (1995) On the inhibitory effect of C_{17}-sulfoconjugated catechol estrogens upon lipid peroxidation of rat liver microsomes. Biological & Pharmaceutical Bulletin, **18**, 1120–1125.

Takimoto Y, Ohshima M and Miyamoto J (1987a) Comparative metabolism of fenitrothion in aquatic organisms: III. Metabolism in the crustaceans, Daphnia pulex and Palaemon paucidens. Ecotoxicology and Environmental Safety, **13**, 126–134.

Tamura H, Miyawaki A, Inoh N, Harada Y, Mikoshiba K and Matsui M (1997) High sulfotransferase activity for phenolic aromatic odorants present in the mouse olfactory organ. Chemico-Biological Interactions, **104**, 1–9.

Takimoto Y, Ohshima M and Miyamoto J (1987b) Comparative metabolism of fenitrothion in aquatic organisms: II. Metabolism in the freshwater snails, Cipangopaludina japonica and Physa acuta. Ecotoxicology and Environmental Safety, **13**, 118–125.

Tamura H, Harada Y, Miyawaki A, Mikoshiba K and Matsui M (1998) Molecular cloning and expression of a cDNA encoding an olfactory-specific mouse phenol sulphotransferase. Biochemical Journal, **331**, 953–958.

Teubner W, Pabel U, Meinl W, Coughtrie MWH, Falany CN, Kretzschmar M, Seidel A and Glatt HR (1998) Characterisation of sulfotransferases in human colon mucosa and their expression in Salmonella typhimurium for the study of the activation of promutagens. Naunyn–Schmiedeberg's Archives of Pharmacology, **357**, R 135.

Thompson TN, Watkins JB, Gregus Z and Klaassen CD (1982) Effect of microsomal enzyme inducers on the soluble enzymes of hepatic phase II biotransformation. Toxicology and Applied Pharmacology, **66**, 400–408.

Tjeerdema RS and Crosby DG (1992) Disposition and biotransformation of pentachlorophenol in the red abalone (Haliotis rufescens). Xenobiotica, **22**, 681–690.

Tjeerdema RS, Lukrich KL and Stevens EM (1994) Toxicokinetics and biotransformation of pentachlorophenol in the sea urchin (Strongylocentrotus purpuratus). Xenobiotica, **24**, 749–757.

Tong Z and James MO (2000) Purification and characterization of hepatic and intestinal phenol sulfotransferase with high affinity for benzo[a]pyrene phenols from channel catfish, Ictalurus punctatus. Archives of Biochemistry and Biophysics, **376**, 409–419.

Tsutsumi M, Noguchi O, Okita S, Horiguchi K, Kobayashi E, Tamura K, Tsujiuchi T, Denda A, Konishi Y, Iimura K and Mori Y (1995) Inhibitory effects of sulfation inhibitors on initiation of pancreatic ductal carcinogenesis by N-nitrosobis(2-oxopropyl)amine in hamsters. Carcinogenesis, **16**, 457–459.

Turcotte G and Silah JG (1970) Corticosterone sulfation by livers of normal and hypertensive rats. Endocrinology, **87**, 723–729.

Utesch D and Oesch F (1992) Phenol sulfotransferase activity in rat liver parenchymal cells cultured on collagen gels. Drug Metabolism and Disposition, **20**, 614–615.

Vakiani E, Luz JG and Buck J (1998) Substrate specificity and kinetic mechanism of the insect sulfotransferase, retinol dehydratase. Journal of Biological Chemistry, **273**, 35381–35387.

van Breemen RB, Fenselau CC and Dulik DM (1985) Activated phase II metabolites: comparison of alkylation by 1-O-acyl glucuronides and acyl sulfates. In Biological Reactive Intermediates III: Mechanisms of Action in Animal Models and Human Disease, Kocsis JJ, Jollow DJ, Witmer CM, Nelson JO and Snyder R (eds), Plenum Press, New York, pp. 423–429.

van de Poll MLM, Tijdens RB, Vondracek P, Bruins AP, Meijer DKF and Meerman JHN (1989) The role of sulfation in the metabolic activation of N-hydroxy-4'-fluoro-4-acetylaminobiphenyl. Carcinogenesis, **10**, 2285–2291.

van de Poll ML, Venizelos V and Meerman JH (1990) Sulfation-dependent formation of N-acetylated and deacetylated DNA adducts of N-hydroxy-4-acetylaminobiphenyl in male rat liver in vivo and in isolated hepatocytes. Carcinogenesis, **11**, 1775–1781.

van Doorn R, Leijdekkers CM, Bos RP, Brouns RME and Henderson PT (1981) Alcohol and sulphate intermediates in the metabolism of toluene and xylenes to mercapturic acids. *Journal of Applied Toxicology*, **1**, 236–242.

Venning EH, Hoffman MM and Browne JSL (1942) Isolation of androsterone sulfate. *Journal of Biological Chemistry*, **146**, 369–379.

Vietri M, De Santi C, Pietrabissa A, Mosca F and Pacifici GM (2000) Fenamates and the potent inhibition of human liver phenol sulphotransferase. *Xenobiotica*, **30**, 111–116.

Visser TJ (1994) Role of sulfation in thyroid hormone metabolism. *Chemico-Biological Interactions*, **92**, 293–303.

Vizethum W and Goerz G (1979) Induction of the hepatic microsomal and nuclear cytochrome P450 system by hexachlorobenzene, pentachlorophenol and trichlorophenol. *Chemico-Biological Interactions*, **28**, 291–299.

Wada S, Shimizudani T, Yamada H, Oguri K and Yoshimura H (1994) Sulphotransferase-dependent dehydration of atropine and scopolamine in guinea pig. *Xenobiotica*, **24**, 853–861.

Walle UK, Galijatovic A and Walle T (1999) Transport of the flavonoid chrysin and its conjugated metabolites by the human intestinal cell line Caco-2. *Biochemical Pharmacology*, **58**, 431–438.

Wang DY, Bulbrook RD, Sneddon A and Hamilton T (1967) The metabolic clearance rates of dehydroepiandrosterone, testosterone and their sulphate esters in man, rat and rabbit. *Journal of Endocrinology*, **38**, 307–318.

Wang J, Falany JL and Falany CN (1998) Expression and characterization of a novel thyroid hormone-sulfating form of cytosolic sulfotransferase from human liver. *Molecular Pharmacology*, **53**, 274–282.

Watabe T, Ogura K, Satsukawa M, Okuda H and Hiratsuka A (1994) Molecular cloning and functions of rat liver hydroxysteroid sulfotransferases catalysing covalent binding of carcinogenic polycyclic arylmethanols to DNA. *Chemico-Biological Interactions*, **92**, 87–105.

Watkins JB (1991) Effect of microsomal enzyme inducing agents on hepatic biotransformation in cotton rats (*Sigmodon hispidus*): comparison to that in Sprague–Dawley rats. *Comparative Biochemistry and Physiology C*, **98**, 433–439.

Watkins JBd and Klaassen CD (1986) Xenobiotic biotransformation in livestock: comparison to other species commonly used in toxicity testing. *Journal of Animal Science*, **63**, 933–942.

Weinshilboum R (1990) Sulfotransferase pharmacogenetics. *Pharmacology & Therapeutics*, **45**, 93–107.

Weinshilboum RM, Otterness DM, Aksoy IA, Wood TC, Her C and Raftogianis RB (1997) Sulfotransferase molecular biology. 1. cDNAs and genes. *FASEB Journal*, **11**, 3–14.

Weitering JG, Krijgsheld KR and Mulder GJ (1979) The availability of inorganic sulphate as a rate limiting factor in the sulphate conjugation of xenobiotics in the rat. Sulphation and glucuronidation of phenol. *Biochemical Pharmacology*, **28**, 757–762.

Wengle B (1964) Studies on ester sulphates: XVI. Use of ^{35}S-labelled inorganic sulphate for quantitative studies on sulphate conjugation in liver extracts. *Acta Chemica Scandinavica*, **18**, 65–76.

Wengle B (1966) Distribution of some steroid sulphokinases in foetal human tissues. *Acta Endocrinologica*, **52**, 607–618.

Werle E, Lenz T, Strobel G and Weicker H (1988) 3- and 4-*O*-sulfoconjugated and methylated dopamine: highly reduced binding affinity to dopamine D_2 receptors in rat striatal membranes. *Naunyn–Schmiedeberg's Archives of Pharmacology*, **338**, 28–34.

Werle-Schneider G, Schwarz M and Glatt HR (1993) Development of hydroxysteroid sulfotransferase-deficient lesions during hepatocarcinogenesis in rats. *Carcinogenesis*, **14**, 2267–2270.

Whitnall MH, Driscoll WJ, Lee YC and Strott CA (1993) Estrogen and hydroxysteroid sulfotransferases in guinea pig adrenal cortex: cellular and subcellular distributions. *Endocrinology*, **133**, 2284–2291.

Whittemore RM, Pearce LB and Roth JA (1985) Purification and kinetic characterization of a dopamine-sulfating form of phenol sulfotransferase from human brain. *Biochemistry*, **24**, 2477–2482.

Whittemore RM, Pearce LB and Roth JA (1986) Purification and kinetic characterization of a phenol-sulfating form of phenol sulfotransferase from human brain. *Archives of Biochemistry and Biophysics*, **249**, 464–471.

Williams RT (1974) Inter-species variations in the metabolism of xenobiotics. *Biochemical Society Transactions*, **2**, 359–377.

Witzmann F, Coughtrie M, Fultz C and Lipscomb J (1996) Effect of structurally diverse peroxisome proliferators on rat hepatic sulfotransferase. *Chemico-Biological Interactions*, **99**, 73–84.

Wong KO, Tan AYH, Lim BG and Wong KP (1993) Sulphate conjugation of minoxidil in rat skin. *Biochemical Pharmacology*, **45**, 1180–182.

Wong KP and Yeo T (1982) Importance of extrahepatic sulphate conjugation. *Biochemical Pharmacology*, **31**, 4001–4003.

Wood TC, Her C, Aksoy I, Otterness DM and Weinshilboum RM (1996) Human dehydroepiandrosterone sulfotransferase pharmacogenetics: quantitative western analysis and gene sequence polymorphisms. *Journal of Steroid Biochemistry and Molecular Biology*, **59**, 467–478.

Woodcock BG and Wood GC (1971) Effect of protein-free diet on UDP-glucuronyltransferase and sulphotransferase activities in rat liver. *Biochemical Pharmacology*, **20**, 2703–2713.

Wu S-CG and Straub KD (1976) Purification and characterization of N-hydroxy-2-acetylaminofluorene sulfotransferase from rat liver. *Journal of Biological Chemistry*, **251**, 6529–6536.

Xiangrong L, Jöhnk C, Hartmann D, Schestag F, Krömer W and Gieselmann V (2000) Enzymatic properties, tissue-specific expression, and lysosomal location of two highly homologous rat SULT1C2 sulfotransferases. *Biochemical and Biophysical Research Communications*, **272**, 242–250.

Xu ZH, Otterness DM, Freimuth RR, Carlini EJ, Wood TC, Mitchell S, Moon E, Kim UJ, Xu JP, Siciliano MJ and Weinshilboum RM (2000) Human 3'-phosphoadenosine 5'-phosphosulfate synthetase 1 (PAPSS1) and PAPSS2: gene cloning, characterization and chromosomal localization. *Biochemical and Biophysical Research Communications*, **268**, 437–444.

Yamada J, Sakuma M, Ikeda T and Suga T (1994) Activation of dehydroepiandrosterone as a peroxisome proliferator by sulfate conjugation. *Archives of Biochemistry and Biophysics*, **313**, 379–381.

Yamazoe Y, Manabe S, Murayama N and Kato R (1987) Regulation of hepatic sulfotransferase catalyzing the activation of N-hydroxyarylamide and N-hydroxyarylamine by growth hormone. *Molecular Pharmacology*, **32**, 536–541.

Yamazoe Y, Nagata K, Ozawa S and Kato R (1994) Structural similarity and diversity of sulfotransferases. *Chemico-Biological Interactions*, **92**, 107–117.

Yang HY, Namkung MJ, Nelson WL and Juchau MR (1986) Phase II biotransformation of carcinogens/atherogens in cultured aortic tissues and cells: I. Sulfation of 3-hydroxybenzo[a]-pyrene. *Drug Metabolism and Disposition*, **14**, 287–92.

Yang QF and Christensen MJ (1998) Selenium regulates gene expression for estrogen sulfotransferase and α2U-globulin in rat liver. *Journal of Steroid Biochemistry and Molecular Biology*, **64**, 239–244.

Yang Y-S, Marshall AD, McPhie P, Guo W-XA, Xie X, Chen X and Jakoby WB (1996) Two phenol sulfotransferase species from one cDNA: nature of the differences. *Protein Expression and Purification*, **8**, 423–429.

Yoshinari K, Nagata K, Shimada M and Yamazoe Y (1998a) Molecular characterization of ST1C1-related human sulfotransferase. *Carcinogenesis*, **19**, 951–953.

Yoshinari K, Nagata K, Ogino M, Fujita K, Shiraga T, Iwasaki K, Hata T and Yamazoe Y (1998b) Molecular cloning and expression of an amine sulfotransferase cDNA: a new gene family of cytosolic sulfotransferases in mammals. *Journal of Biochemistry (Tokyo)*, **123**, 479–486.

Yoshinari K, Nagata K, Shiraga T, Iwasaki K, Hata T, Ogino M, Ueda R, Fujita K, Shimada M and Yamazoe Y (1998c) Molecular cloning, expression, and enzymic characterization of rabbit hydroxysteroid sulfotransferase AST-RB2 (ST2A8). *Journal of Biochemistry (Tokyo)*, **123**, 740–746.

Yoshizumi M, Ohuchi T, Masuda Y, Katoh I and Oka M (1992) Deconjugating activity for sulfoconjugated dopamine in homogenates of organs from dogs. *Biochemical Pharmacology*, **44**, 2263–2265.

Zhang DL, Yang YF, Leakey JEA and Cerniglia CE (1996) Phase I and phase II enzymes produced by *Cunninghamella elegans* for the metabolism of xenobiotics. *FEMS Microbiology Letters*, **138**, 221–226.

Zhang HP, Varmalova O, Vargas FM, Falany CN and Leyh TS (1998) Sulfuryl transfer: the catalytic mechanism of human estrogen sulfotransferase. *Journal of Biological Chemistry*, **273**, 10888–10892.

Zhang KY, Tang CY, Rashed M, Cui DH, Tombret F, Botte H, Lepage F, Levy RH and Baillie TA (1994) Metabolic chiral inversion of stiripentol in the rat: 1. Mechanistic studies. *Drug Metabolism and Disposition*, **22**, 544–553.

Zhu XY, Veronese ME, Iocco P and McManus ME (1996) cDNA cloning and expression of a new form of human aryl sulfotransferase. *International Journal of Biochemistry & Cell Biology*, **28**, 565–571.

Zou JY, Pentney R and Roth JA (1990) Immunohistochemical detection of phenol sulfotransferase-containing neurons in human brain. *Journal of Neurochemistry*, **55**, 1154–1158.

Zou LB, Yamada K, Sasa M, Nakata Y and Nabeshima T (2000) Effects of σ_1 receptor agonist SA4503 and neuroactive steroids on performance in a radial arm maze task in rats. *Neuropharmacology*, **39**, 1617–1627.

Zuckerman A, Bolan E, de Paulis T, Schmidt D, Spector S and Pasternak GW (1999) Pharmacological characterization of morphine-6-sulfate and codeine-6-sulfate. *Brain Research*, **842**, 1–5.

11 Arylamine Acetyltransferases

Gerald N. Levy and Wendell W. Weber

University of Michigan Medical School, Ann Arbor, Michigan, USA

Arylamine *N*-acetyltransferase is an enzyme activity widely distributed across species and throughout many tissues. Mammalian arylamine *N*-acetyltransferase enzymes (NAT) EC 2.3.1.5 are cytosolic in location and have molecular weights of about 31 000–33 000 daltons. They catalyse three reactions involving the transfer of an acetyl moiety (reviewed in Weber and Hein 1985; Weber 1987; Hein 1988). The reaction for which the enzyme is named is the transfer of the acetyl group from acetyl CoA (AcCoA) to an arylamine to produce an arylacetamide. The reaction occurs as two half-reactions: first, the acetyl is transferred to an evolutionarily conserved cysteine residue of the NAT protein forming an acetyl-cysteinyl-NAT catalytic intermediate (Andres *et al.* 1988), and second, the acetyl is transferred to the amino nitrogen of an acceptor arylamine. The reaction mechanism is classified as ping-pong Bi Bi as the first product (CoA) is released before the second reactant (arylamine) is bound (see Figure 11.1). Amino acid sequences for NATs of many species show conservation of cysteines at positions 44, 68, and 223. In particular the sequence surrounding C68 (C69 in *Salmonella*) is highly conserved and this therefore is believed to be the active site for acetyl transfer (Watanabe *et al.* 1992).

In addition to *N*-acetylation, NAT can also catalyse the *O*-acetylation of arylhydroxylamines. This reaction is formally similar to *N*-acetylation except that the NAT enzyme, acetylated by AcCoA, donates the acetyl group to the oxygen of an arylhydroxylamine to form an acetoxyarylamine.

A third significant reaction catalysed by NAT is independent of AcCoA and involves transfer of acetyl from the nitrogen of an arylhydroxamic acid to the oxygen, producing an acetoxyarylamine. This reaction referred to as *N*,*O*-transacetylation or AHAT (arylhydroxamic acid acetyl transfer) produces the acetyl-enzyme intermediate

$$\text{AcCoA} + \text{NAT} \Leftrightarrow \text{Ac-NAT} + \text{CoA}$$
$$\text{Ac-NAT} + \text{Ar-NH}_2 \Leftrightarrow \text{Ar-NH-Ac} + \text{NAT}$$

Figure 11.1 The *N*-acetyltransfer reaction comprises two half reactions. Initially the enzyme (NAT) is acetylated by AcCoA. The second step is the transfer of the acetyl group to the acceptor arylamine, releasing the enzyme protein for further catalysis.

Enzyme Systems that Metabolise Drugs and Other Xenobiotics. Edited by C. Ioannides.
© 2002 John Wiley & Sons Ltd

$$\text{NAT:} \qquad \text{Ar-NH}_2 \xrightarrow{\text{AcCoA}} \text{Ar-NH-COCH}_3$$
$$\qquad\qquad\quad \text{amine} \qquad\qquad\quad \text{acetamide}$$

$$\text{OAT:} \qquad \text{Ar-NHOH} \xrightarrow{\text{AcCoA}} \text{Ar-NHO-COCH}_3$$
$$\qquad\qquad \text{hydroxylamine} \qquad\quad \text{acetoxyamine}$$

$$\text{AHAT:} \qquad \text{Ar-NOH-COCH}_3 \longrightarrow \text{Ar-NHO-COCH}_3$$
$$\qquad\qquad\quad \text{hydroxamic acid} \qquad\qquad \text{acetoxyamine}$$

Figure 11.2 The three acetyl transfer reactions catalyzed by arylamine acetyltransferase. NAT: N-acetyltransfer. OAT: O-acetyltransfer. AHAT: arylhydroxamic acid acetyltransfer.

from the arylhydroxamic acid rather than from AcCoA. The N to O transfer of the acetyl group can be either intra- or inter-molecular depending on if the acetyl donor and acceptor are the same or different molecules. The three activities of NAT are illustrated in Figure 11.2.

Role of NAT in arylamine metabolism and toxicity

NAT, by virtue of its ability to catalyse O-acetylation and AHAT, in addition to N-acetylation, can both metabolically activate and detoxify arylamines. As a phase II conjugating enzyme, NAT produces water-soluble amides from less water-soluble, lipophilic amines, thereby hastening elimination of these potentially toxic compounds. However, further activation of hydroxylamines to acetoxyamines by the OAT and AHAT activities leads to the formation of arylnitrenium ions which combine with proteins and nucleic acids. DNA adduct formation with arylamine-derived nitrenium ions is a mechanism of DNA mutation and initiation of carcinogenesis.

An arylamine faces two primary metabolic fates: it can be conjugated by various phase II enzymes to a water-soluble, excretable product or it can undergo oxidation. Oxidation can be catalysed by a number of enzymes or enzyme families including the flavin monooxygenases (chapter 3), prostaglandin H synthases (chapter 6) and other peroxidases, and cytochromes P450. In the liver, the prevalent arylamine oxidative enzymes appear to be cytochromes P450, and for arylamines CYP1A2 has been found to be a major enzyme of oxidation. Here too there is a fork in the metabolic path, for oxidation of a carbon of the aromatic ring will usually, but not always (Harris *et al.* 1989), lead to a non-toxic compound that can be conjugated and excreted. In contrast, oxidation of the amino nitrogen forms a hydroxylamine which, although often toxic itself, can be further activated by NAT leading to acetoxyamine-derived nitrenium ions. The factors that determine which path an arylamine follows, activation or detoxification, are not completely known. It is reasonable to believe that the relative expression and activities of the oxidative and conjugating enzymes, the selectivity of the competing enzymes for the specific arylamine substrate, the concentration of the arylamine, and the intracellular environment are among the factors which influence arylamine metabolism.

NAT in drug metabolism

The role of NAT in detoxification of drugs can be illustrated by a few examples involving commonly used therapeutics. Early interest in NAT was sparked by the introduction of isoniazid (INH) as an antitubercular agent. Isoniazid is an example of an arylhydrazine (isonicotinic acid hydrazide) which is a substrate for NAT. N-Acetylation is the primary pathway of isoniazid metabolism. Although many metabolites are produced during INH metabolism, acetylisoniazid is the most prominent product initially formed. Acetylisoniazid is further metabolised to isonicotinic acid, which is conjugated with glycine and excreted, and to monoacetylhydrazine. Monoacetylhydrazine can follow any of three metabolic pathways: 1- hydrazone formation, 2- further acetylation to diacetylhydrazine, or 3- oxidation. The first two paths produce non-toxic excretable products while the third, probably catalysed by a P450 enzyme, leads to electrophilic reactive intermediates capable of inducing hepatotoxicity (Weber 1991).

Another example of the competition between detoxification by acetylation and activation by oxidation is seen in metabolism of the antiarrhythmic procainamide. N-Acetylation is the primary detoxification pathway of procainamide, producing N-acetylprocainamide and some N-acetyl-p-aminobenzoic acid and desethyl-N-acetylprocainamide (Weber 1987). Procainamide can also undergo oxidation to a hydroxylamine and subsequently to the nitroso and nitro derivatives. The catalytic activity for the oxidation is thought to be a hepatic cytochrome P450 protein (probably CYP2D6; Lessard et al. 1999), but more importantly for toxicities such as drug-induced lupus, myeloperoxidase catalyses the reaction in leukocytes. Activation of procainamide and similar drugs by leukocytes provides active metabolites direct access to the immune system and leads to various expressions of immunotoxicity (Uetrecht 1989).

One of the most important occurrences of the competition between detoxifying N-acetylation and activating oxidation is in the metabolism of sulphonamide derivatives. These antibacterial agents, such as sulphamethoxazole, are used as primary therapy of infection in transplantation and AIDS-related complications such as pneumonia caused by Pneumocystis carinii. For sulphonamides, as for the other examples, NAT catalyses formation of the nontoxic N-acetyl derivative while oxidation produces a reactive hydroxylamine. The hydroxylamine can be further oxidized to a nitroso compound leading to hypersensitivity, toxicity, and immune suppression. For sulphamethoxazole, hydroxylamine formation is catalysed primarily by CYP2C9 in the liver. The further oxidation to the nitroso compound can occur spontaneously or by CYP catalysis. The very high incidence of sulpha drug toxicity among AIDS patients is apparently due to the AIDS-induced depletion of $CD4^+$ white blood cells, leaving an excess of $CD8^+$ white cells which are very susceptible to the toxic effects of sulphamethoxazole hydroxylamine (Hess et al. 1999).

NAT in metabolism of carcinogens

In addition to detoxification of arylamine and arylhydrazine drugs, NAT is significant in the metabolism of carcinogenic arylamines which may be encountered in the environment. Here again there is competition between conjugation and oxidation pathways. As mentioned earlier, conjugation of arylamines by acetylation or other phase II enzymes will lead towards excretion while oxidation at the amino nitrogen will form a reactive intermediate which can be further activated by the O-acetylation capacity of NAT. Arylamine carcinogens can be detoxified not only by conjugation as the initial step in metabolism but also by oxidation at a ring carbon followed by glucuronide formation and excretion. It is oxidation at the amino nitrogen that truly commits the amine towards activation. Differences in the level of NAT activity can play a large part in determining the fate of arylamine carcinogens; high NAT activity will send more of the dose towards detoxification while low NAT activity will allow for more N-oxidation to occur. Acetylarylamines (arylacetamides) produced by NAT can also be oxidised to arylhydroxamic acids and then converted to acetoxyarylamines by AHAT, but this appears to be a less important pathway than OAT. It is also known that microsomal de-acetylases can transform the arylacetamide back to arylamine and allow the oxidative enzymes 'another chance' at activation. Genetic factors which are involved in determining the level of NAT activity will be discussed later in the section on Pharmacogenetics.

The high activity of NAT is not uniformly protective against carcinogenic arylamines because the OAT activity parallels NAT activity. Some amines which are poor substrates for N-acetylation become, after oxidation, good substrates for O-acetylation to more reactive forms. This is readily illustrated by the processing of several of the carcinogenic and mutagenic heterocyclic amines produced by cooking of protein-containing food. 2-Amino-3-methylimidazo[4,5-f]quinoline (IQ) is not a good substrate for N-acetylation by NAT, but following oxidation by CYP1A2 in the liver (or prostaglandin H synthase 2 extrahepatically; (Liu and Levy 1998)), the heterocyclic hydroxylamine is O-acetylated by NAT producing a more reactive acetoxyheterocyclic amine that forms adducts with DNA (Probst et al. 1992).

Substrates

Arylamine N-acetyltransferase carries out the acetylation of arylamines and arylhydrazines. The latter category includes isoniazid, hydralazine and phenelzine, while the former includes procainamide, sulphamethazine, aminoglutethimide, dapsone, proguanil, and p-aminobenzoic and p-aminosalicylic acids. Substrates for NAT also include compounds which are biotransformed to arylamines by other steps in their metabolism. In this group are acebutolol, nitrazepam, sulphasalazine and caffeine (Weber 1987).

In addition to therapeutic agents, NAT is a part of the metabolic pathways for environmental arylamines such as β-naphthylamine, benzidine, 4-aminobiphenyl and 2-aminofluorene. Nitro arenes which are reduced to arylamines must also be considered potential substrates for NAT. Heterocyclic arylamines, including those produced by pyrolysis during the cooking of food, may also be NAT substrates. The structure and properties of the specific substrate determine how likely it is for N-acetylation to occur. For some compounds, such as the heterocyclic arylamines, N-acetylation is slight to non-existent, but O-acetylation of the hydroxylamine is significant.

In humans, NAT is encoded by two independently expressed loci termed NAT1 and NAT2, which are responsible for the corresponding cytosolic enzymes NAT1 and NAT2. The reactions catalysed by the two NATs are identical, but the enzymes show somewhat different substrate selectivity. Multiple NATs had been postulated for many years (Jenne 1965), but the actual separation of the two human NATs was not achieved until 1989 (Grant et al. 1989). Substrate selectivity profiles of the NATs have been studied using partially purified liver cytosolic fractions (Grant et al. 1991) and with enzyme expressed in bacteria (Dupret and Grant 1992; Hein et al. 1993), in COS cells (Grant et al. 1991; Minchin et al. 1992) and in CHO cells (Ohsako and Deguchi 1990). Results of all systems are in general agreement indicating that for N-acetylation NAT1 is significantly more active than NAT2 with p-aminobenzoic acid and p-aminosalicylic acid, somewhat more active with 2-aminofluorene, 4-aminobiphenyl, β-naphthylamine, and 5-aminosalicylic acid (5-ASA), but less active with sulpha-methazine. Both enzymes are nearly equally active with procainamide. O-acetylation activities of the expressed enzymes were examined with the hydroxylamines of 2-aminofluorene and 4-aminobiphenyl. The activities of the two enzymes were similar with NAT1 having slightly greater activity. For the few heterocyclic hydroxylamines tested, the activity observed was very low, but N-OH-IQ appeared somewhat selective for NAT1 (summarised from data compiled by Vatsis and Weber 1997).

While the involvement of NAT in the metabolism of drugs and other xenobiotics has been known for decades, the function of NAT in the transformation of endogenous compounds is not clear. Since NAT is found in a wide range of organisms, from bacteria to primates, and since it is rather highly conserved throughout evolution, the expectation is that NAT must have some essential role other than protecting organisms from man-made chemicals. A possible endogenous substrate for NAT was recently demonstrated to be a metabolic product of folate metabolism, p-aminobenzoylgluta-mate (p-ABG) (Minchin 1995). p-ABG is a selective substrate for human NAT1 and mouse NAT2. Circumstantial evidence including the role of folate in protecting the developing foetus from neural tube defects and the pre-natal expression of NAT, suggests that a relationship of NAT activity and neural tube defects might be worthy of investigation. In support of a role for NAT in foetal development, Stanley et al. (1998) using immunochemical detection found NAT in developing neuronal tissue of embryonic mice. The NAT protein was expressed intensely in the neural tube around the time of closure. NAT was also detected in the developing heart and gut. mRNA for NAT has been found in mouse embryos during gestation and in the early postnatal period (Mitchell et al. 1999). NAT activity and mRNA were measured in new-born and early postnatal mouse liver and kidney by Estrada et al. (2000).

Interestingly, some earlier work had attributed determination of teratogen-induced and spontaneous cleft palate in mice to a specific region of chromosome 8 which included the Nat genes (Liu and Erickson 1986; Karolyi et al. 1990). At the time, it was suggested that since mice with the slow acetylator allele (now called Nat2*9) were more susceptible to cleft palate than mice with the Nat2*8 rapid allele, NAT might be a marker for the gene(s) responsible for cleft palate in mice (Karolyi et al.

1990). However, now that a role for NAT in folate metabolism has been observed (Minchin 1995), and the expression of NAT in the area of the closing neural tube in embryonic mice detected (Stanley *et al.* 1998), it is quite conceivable that NAT may function as more than just a marker for susceptibility to cleft palate in mice.

Tissue distribution

NAT activity is found in a wide range of human tissues. Although no systematic assay of enzyme distribution has been carried out in humans as has been done for mouse NAT (Chung *et al.* 1993), numerous individual tissues have been examined. Human NAT1 and NAT2 differ in their tissue distribution in that NAT2 is primarily hepatic and NAT1 is primarily extrahepatic. Even though both NATs have been partially purified from liver (Grant *et al.* 1989), NAT1 is the primary activity in mononuclear blood cells (Cribb *et al.* 1991), colon epithelium (Rodriguez *et al.* 1993) and throughout the intestine (Hickman *et al.* 1998), placenta (Smelt *et al.* 1997), bladder urothelium (Kloth *et al.* 1994) and mammary epithelium (Sadrieh *et al.* 1996). NAT2 activity is either low or absent in these extrahepatic tissues.

The different tissue distribution of the human NATs together with their different substrate selectivity suggest that not all NAT substrates will be acetylated by first pass metabolism. Substances which are good NAT2 substrates may be expected to be metabolised by hepatic NAT2, while NAT1 selective substrates may be partially metabolised by NAT1 activity in the liver, but also by NAT1 in the various tissues. Also one should not neglect NAT1 activity in colon as a possible site of pre-hepatic xenobiotic activation and/or detoxification.

Oxidative activation of arylamines by hepatic CYP1A2 and other enzymes produces hydroxylamines, some of which are stable enough to be transported through blood to the tissues. Additionally, the hydroxylamines can be conjugated such as to *N*-glucuronides which are later hydrolysed by the low pH in bladder or by bacterial glucuronidases in the gut to release the hydroxylamine. In the various target tissues, the hydroxylamines may be further activated by the *O*-acetylation activity of NAT1 to cause local toxicity. The relative activities of NAT1 and NAT2 as well as the relative activities of oxidative enzymes are important factors in determining the fate of NAT substrates. This is particularly true for those substrates, such as the carcinogenic arylamines, that are relatively good substrates for both NATs.

Species distribution

N-Acetyltransferases are widely distributed among animal species. Enzymes with homology to human NAT occur in mammals including monkeys (Goedde *et al.* 1967), baboons (Radtke *et al.* 1979), common rodents such as mice (Tannen and Weber 1979; Levy *et al.* 1992), rats (Tannen and Weber 1979), and hamsters (Hein *et al.* 1982), rabbits (Frymoyer and Jacox 1963), and livestock animals (Watkins and Klaassen 1986). NAT is also found in quail (Watkins and Klaassen 1986), pigeons (Andres *et al.* 1983) and chickens (Deguchi *et al.* 1988). Arylamine *N*-acetyltransferase activity has been found in trout (Watkins and Klaassen 1986), frogs (Ho *et al.* 1996), and nematodes (Chung *et al.* 1996). *N*-Acetyltransferases have been reported in insects where the enzyme is important in sclerotisation of the insect cuticle (Karlson and Sekeris 1962), puparium formation (Sekeris 1964), and tanning of oviposited eggs (Li and Nappi 1992). In insects, NAT is the major route for metabolism of biogenic amines (Dewhurst *et al.* 1972). The enzyme from housefly has been purified and found to have a molecular weight of approximately 27 000 daltons, but terminal and partial internal sequence analysis did not show any similarity to the mammalian or avian NAT enzymes (Whittaker and Goosey 1993).

Bacteria have *N*- and *O*-acetylating activity. The best studied bacterial NAT is from *Salmonella typhimurium* which is used in the Ames assay for testing the mutagenicity of compounds. Many mutagenic and carcinogenic compounds activated by mammalian microsomes show increased mutagenicity in *Salmonella* strains overexpresssing NAT compared with those lacking NAT (Watanabe *et al.* 1987). The NAT gene coding for this enzyme has been cloned and the protein sequenced. The enzyme shows sequence homology to the mammalian NATs, has a similar molecular weight of about 33 000 daltons, and has a similar catalytic mechanism involving the conserved cysteine residue at position 69 (Watanabe *et al.* 1992). The *N*-acetylation substrate selectivity of the enzyme suggests it should be classified as an NAT2 and has been officially designated NAT2 10 (Vatsis *et al.* 1995).

Among mammals there are some notable exceptions to the seemingly universal distribution of NAT. While most species that have been studied have two or more isozymes of NAT, the domestic dog and other canids lack cytosolic NAT. This deficiency has been known for some time (Marshall 1954) and has more recently been shown to be caused by the absence of NAT genes (Trepanier *et al.* 1997). The only other mammal reported to lack cytosolic NAT activity is the shrew *Suncus murinus* (Nakura *et al.* 1994), for which an activity assay rather than molecular techniques were used to seek NAT.

Interestingly, although dogs lack cytosolic NAT, they are capable of acetylation, deacetylation, and transacylation of amines and their derivatives (Sone *et al.* 1991). The enzymes which catalyse these reactions include at least three microsomal activities. One of these enzymes which is responsible for most of the *N,O*-acyltransferase capacity has been purified to a mannose-rich glycoprotein of molecular weight 58 000 daltons. The enzyme was partially sequenced and the *N*-terminal area showed strong homology to amidase/carboxylesterase of rabbit, hamster, and rat. It was concluded that the metabolism of arylamines in dog is attributable to microsomal carboxylesterases rather than to intestinal flora (Sone *et al.* 1994).

The domestic cat and wild felids have cytosolic NAT, but only a single isoform, corresponding to NAT1, has been found. Although the activity of feline NAT for *p*-aminobenzoic acid is lower than in other species, it is present as an enzyme with sequence homology to mammalian NAT1. No NAT2 activity with the NAT2 selective substrate sulphamethazine could be demonstrated in cat hepatic cytosol (Trepanier *et al.* 1998). The lack of NAT2 activity agrees with earlier studies which could not detect *N*-acetylation of isoniazid or sulphamethazine by cat liver cytosol. Use of molecular techniques showed the presence of only the *NAT1* gene in felids and the absence of any other sequences with homology to NAT (Trepanier *et al.* 1998).

The consequences of lacking cytosolic NAT activity in canids (and shrews) and the lack of NAT2 activity in felids are not apparent. Although both dogs and cats may develop signs of toxicity when given sulphonamides, and the toxicities are similar to those seen more commonly in humans deficient in NAT2 activity, it is not possible to relate the idiosyncratic response of the animals to their lack of NAT activity (Trepanier *et al.* 1998).

Regulation of NAT activity and expression

There are large gaps in our knowledge of the regulation of NAT expression in humans. Most of the studies which investigated variation in NAT activity occurred before the age of molecular biology, heterologous expression systems, and the appreciation of transcription elements. These early studies, for the most part, examined *in vivo* differences in drug acetylation (most often isoniazid) by administering the drug and measuring metabolites in blood or urine. The method does not give direct information about NAT activity as several possible influences on the drug, in addition to NAT expression, will affect these ratios. Absorption, distribution, metabolism and elimination will all have significant effects on the ratio of acetylated to non-acetylated drug measured. Even within the process of metabolism, we have seen that there is competition for the substrate, and N-acetylation rates may be altered by changes in, for example, cytochrome P450 activities. Another difficulty with the whole-body approach to determining the extent of NAT expression is that the liver is the primary site of *in vivo* drug acetylation and, therefore, liver mass has a great influence on observed acetylation. The many studies of NAT activity and age in children through the elderly, once corrected for changes in liver mass, do not convincingly demonstrate an age-related effect beyond the first year of life (Weber 1987).

Malfunction of specific organs can give the appearance of a change in NAT activity when drug/metabolite ratios are measured in blood or urine. In severe renal disease, the acetylated metabolite or the parent drug may be relatively retained resulting in an altered ratio in urine leading to the incorrect conclusion of altered NAT activity. For patients with acute or chronic liver disease, the acetylation of drugs may appear decreased. This effect is due to having fewer active hepatocytes in the affected liver and not to any intrinsic change in NAT activity.

During absorption, there are also a number of problems with the *in vivo* approach. As an example, absorption of isoniazid is reduced by high carbohydrate meals (Mannisto *et al.* 1982), antacids (Hurwitz and Schlozman 1974) and laxatives. In contrast, insulin increases isoniazid uptake (Danyz and Wisniewski 1970).

Various drugs may appear to alter the rate and extent of NAT reactions. Obviously, administration of drugs which are NAT substrates will compete for the enzyme and reduce acetylation of the test drug. PAS, procainamide, PABA, chlorpromazine, and phenylramidol will all increase the concentration of co-administered isoniazid and prolong its half-life (Weber and Hein 1979). Ethanol has been observed to increase the acetylation rate and decrease the biological half-life of many NAT substrates including isoniazid (Lester 1964), sulphamethazine (Olsen and Morland, 1978) and procainamide (Olsen and Morland 1982). The exact mechanism for the ethanol effect is not proven, but experiments indicate that ethanol increases the acetyl CoA level in liver and drives the NAT reaction in the direction of acetylation (Olsen and Morland 1982).

Inhibitors of NAT include compounds which bind the essential sulphydryl group of cysteine 68: metal ions, p-chloromercuribenzoate, N-ethylmaleimide, phenylglyoxal, diethylpyrocarbonate among others (Andres et al. 1988). As mentioned previously, other NAT substrates are competitive inhibitors. Folate and the folate mimic, methotrexate, are inhibitors of NAT (Andres et al. 1983). N-Acetyltransferases are also inhibited by CoA-SH, which can be a problem in in vitro assays of NAT activity if the assay does not include an AcCoA recycling system (Andres et al. 1985). Two other inhibitors of NAT are pentachlorophenol and paracetamol (acetaminophen). Pentachlorophenol has been used to inhibit sulphotransferases in metabolic investigations. However, the compound is also a good inhibitor of NAT and thus can not be used to implicate sulphotransferase in a metabolic pathway as acetyltransferases will also be inhibited by this phenolic compound (Shinohara et al. 1986). Paracetamol is not a substrate for NAT but is a product analogue which shows inhibition of NAT in vivo and in vitro. The in vivo inhibition can be a problem in patients using paracetamol, as NAT inhibition by the achievable serum concentration of the drug can give the appearance of decreased acetylation activity and lead to higher and more persistent levels of NAT substrates such as sulpha drugs which may cause toxic reactions (Rothen et al. 1998).

Pharmacogenetics of NATs

Throughout this chapter we have referred to various effects caused by differences in NAT activity. The most common cause of altered NAT activity is the genetic polymorphism of the NAT enzymes. Discovery of the polymorphism occurred with the introduction of isoniazid. The definition of rapid and slow acetylators as well as phenotyping assays to distinguish them developed subsequent to clinical observation of interindividual variations in response to the drug (reviewed in Weber 1987). Discovery of the acetylation polymorphism was followed by many genetic studies of the mode of inheritance of NAT activity and the frequency of rapid or slow acetylation in different populations. Associations between acetylator phenotype and various disease states were investigated and several were found (Weber and Hein 1985; Weber 1987). Among the most important associations observed was that between slow acetylation and occupational bladder cancer (Cartwright et al. 1982). There followed much work examining the distribution of the different acetylator phenotypes in patients with various cancers as compared to healthy controls. An association was found between rapid acetylators and increased likelihood of colorectal cancer (Lang et al. 1986; Ilett et al. 1987). Other cancers were found to be more or less common in rapid or slow acetylators in some studies, but often other studies could not confirm these findings. The importance of NAT activity as a risk factor for cancers is still under active investigation.

Two events took place that dramatically changed NAT research. The first was the separation and characterisation of the second NAT isozyme (ironically called NAT1) (Grant et al. 1989, 1991). The second event was the blossoming of molecular biology and molecular genetics. For nearly four decades after the discovery of the isoniazid-sulphamethazine polymorphism only NAT2 was known. After the discovery of NAT1, the isozymes were often referred to as polymorphic (NAT2) and monomorphic (NAT1)

NATs. We know now that both forms are polymorphic when the proper substrates are tested, and that both enzymes are coded for by multi-allelic loci, illustrating investigators' proclivity to find what is being sought.

The human NAT genes are located on chromosome 8 in the region 8p22 as is a pseudogene, NATP (Matas et al. 1997). The order of the loci are (from telomeric end towards the centromere) NAT1, NATP, NAT2 with the loci within 1000 kb of each other (Thygesen et al. 1999). The NAT1 and NAT2 loci have intronless open reading frames of 870 bases coding for proteins of molecular mass of about 33 500 daltons. More than 20 allelic variants of NAT1 have been identified, distinguished from each other and from the 'wild type' NAT1*4 by mutations in the coding and/or 3'-UTR (Vatsis et al. 1995). While many of the substitutions in the coding region cause amino acid changes in the NAT1 protein or introduce stop signals, others are silent but may influence the rate of transcription or translation or affect the stability of mRNA. Several of the mutations in the 3'-UTR are hypothesized to alter the function or efficiency of the polyadenylation signal and alter the mRNA stability (Vatsis and Weber 1993; deLeón et al. 2000).

Human NAT2 also has in excess of 20 alleles differing from the 'wild type' NAT2*4. The known mutations of NAT2 are in the coding region and include silent mutations as well as those which cause amino acid substitutions. Currently much more information is available on the distribution and frequencies of NAT2 alleles in different populations and about the phenotype-genotype correlations than is available for NAT1. Both in vivo and in vitro studies as well as several heterologous expression experiments have shown that almost all the naturally occurring mutations in NAT2*4 lead to enzymes with reduced acetylation activity. Because of this, individuals with 2 mutated NAT2 genes are slow acetylators, those with one mutated allele are heterozygous rapid, and those with two wild type NAT2*4 alleles are homozygous rapid acetylators. In many phenotyping assays, the homozygous and heterozygous rapid acetylators can be distinguished resulting in a trimodal distribution of the population. The quantitative aspects of the relative 'slowness' of slow alleles have not been well studied.

The frequency of the slow NAT2 alleles shows surprising variability in various populations. Japanese populations are about 90% rapid acetylators (Vatsis and Weber 1997) while the Moroccan population was found to be 90% slow acetylators (Karim et al. 1981). For mixed Caucasian populations the slow and rapid phenotypes are almost equally represented. The different mutations causing the slow phenotype also show ethnic differences, as illustrated in Table 11.1. In many populations, three or four

Table 11.1 Approximate frequency (in %) of NAT2* alleles in various populations

Allele	Caucasian	Afro-American	Hispanic-American	Asian (Japan)
*4	25	34	42	70
*5B	40	30	23	1
*6A	30	25	17	20
*14A, *14B		10		
*7A, *7B			17	7

mutations account for the vast majority of slow acetylators. Although other, much rarer slow alleles may occur, their frequencies are so small that, for most purposes, they can be ignored. A listing of reported *NAT1* and *NAT2* alleles is maintained at http://www.louisville.edu/medschool/pharmacology/NAT.html.

Our knowledge of *NAT1* allelic frequency and distribution is still sparse. The original report of structural heterogeneity of *NAT1* identified the wild type (*NAT1*4*) and two mutant alleles (*NAT1*10* and *NAT1*11*) (Vatsis and Weber 1993). This report also demonstrated Mendelian inheritance of the variants and independent expression of NAT1 and NAT2. In the small population examined, 42% of *NAT1* alleles of unrelated individuals were variants. Later reports found 28.5% of alleles from 280 subjects were variants (Hughes *et al.* 1998) and, in a large study, only a 5% mutation frequency was found in over 1800 subjects (Lin *et al.* 1998), but this study did not test for *NAT1*10*. A number of other studies of *NAT1* allelic frequency have been done, but several considered only three or four alleles. While this approach is acceptable in populations where the major alleles accounting for >95% of all alleles are known (as discussed in the previous paragraph), it is not reliable when examining new populations for a different gene. When examining an untested population it is a better strategy to genotype for as many alleles as practical. Dhaini and Levy (2000) found a combined frequency of 23.8% for the *NAT1*14A* and *14B* alleles and 6% for previously unreported alleles in a Lebanese population living in Dearborn, Michigan. This result is considerably higher than the previously reported *NAT1*14* frequencies of 1.3–3% in Caucasians of European origin (Lin *et al.* 1998; Butcher *et al.* 1998) and exemplifies that different ethnic groups may vary very greatly in *NAT1* allelic distribution. This result also demonstrates that had the authors not specifically tested for this allele, most of the *NAT1*14* individuals would have been incorrectly classified as *NAT1*4* or *NAT1*10*, giving greatly misleading information about the population.

The *NAT1*3* allele occurs with a high frequency in Asians as shown by a study of different ethnic groups living in Singapore. The population was genotyped for alleles *NAT1*3*, *4*, *10*, and *11*. Among the 122 Malay and 181 Chinese in the study, the distribution was 30–35% *NAT1*3*, 30–34% *NAT1*4*, and 30–39% *NAT1*10*. Among the 140 Indians, *NAT1*4* was increased to 51% and *NAT1*10* reduced to 17%. In all three ethnic groups *NAT1*11* occurred at a 2% frequency (Zhao *et al.* 1998).

The activity of the various *NAT1* alleles, that is, the phenotype–genotype correlation for NAT1, is still being discovered. Before the identification of NAT1 was achieved, it was noted that large variation in PAS and PABA acetylation occurred between individuals. An 80-fold variation in PAS acetylation in 131 human blood samples (Motulsky and Steinman 1962; Evans 1963), a 4-fold variation in PABA acetylation by seven human liver biopsy samples (Glowinski *et al.* 1978), a 90-fold variation in V_{max} values for PABA acetylation by 39 (Grant *et al.* 1991) and 8 (Cribb *et al.* 1993) liver samples, as well as several other reports, illustrated the variability in human NAT1 activity.

Further evidence for an NAT1 activity polymorphism has come from a study of the urinary N-Ac-PAS to PAS ratio of 130 subjects given PAS (Grant *et al.* 1992) and a measurement of whole blood lysate PABA acetylation activity from 200 subjects (Weber and Vatsis 1993). Both studies showed a significant variability in activity and a strong tendency towards a bimodal distribution. More recently, Butcher *et al.* (1998)

Table 11.2 Frequency of NAT1 alleles in 314 Germans

NAT1 Allele	%
*3	3.0
*4	70.9
*10	20.1
*11	3.3
*14	2.2
*15	0.5

Data from Bruhn *et al.* (1999)

reported that lysed white cells from 85 individuals showed a bimodal distribution of PABA-NAT activity with roughly 8% of the subjects being slow (NAT1) acetylators. Genotyping showed that the seven slow individuals had either $G^{560}A$ (*NAT1*14*) or $C^{190}T$ (*NAT1*17*) mutations together with an allele lacking these mutations. Both mutations resulted in a V_{max} approximately half that of the wild type.

Hughes *et al.* (1998) examined the ratio of N-Ac-PAS to PAS in urine and plasma of 144 subjects given PAS. A greater than 65-fold variation in the urinary ratio and a greater than 5.6-fold variation in the plasma ratio was found. *In vitro* determination of PAS acetylation (whole blood lysates), cloning, sequencing, and expression in *E. coli* were carried out for a subset of eight individuals. No significant differences were found between *NAT1*4* and *NAT1*11* in terms of activity, apparent K_m or V_{max}. However, *NAT1*14* ($G^{560}A$: $Arg^{187}Gln$) had a 15-fold higher apparent K_m and 4-fold lower V_{max} than *NAT1*4*. *NAT1*15* ($C^{559}T$: $Arg^{187}stop$) had no detectable activity and thus non-measureable K_m and V_{max} values. A single individual in the study with the *NAT1*14/*15* genotype had an extremely low N-Ac-PAS to PAS ratio, while subjects with *NAT1*4/*14*, *10/*14*, *4/*15*, and *10/*15* had urinary ratios not markedly different from the wild type or group means. This suggests that presence of a single allele with normal activity (*NAT1*4* or *NAT1*10*) can compensate for the low or absent activity of the *NAT1*14* or *NAT1*15* alleles, at least *in vivo*.

A group of 314 German volunteers were genotyped for *NAT1*3, *4, *10, *11, *14*, and *15* by Bruhn *et al.* (1999) and a subset of 105 were phenotyped *in vitro* (whole blood lysate) with PABA. Allelic frequencies are shown in Table 11.2. There was no functional difference between alleles *NAT1*3, *4*, or *10*. Carriers of *NAT1*11*, an allele with mutations in the 5′ and 3′ regions as well as in the coding region (Val^{149}Ile, Ser^{214}Ala), had reduced enzyme activity, those with *NAT1*14* had a further reduction in activity, and an individual with the *NAT1*15/*15* genotype had no detectable PABA-NAT activity. These results demonstrate the variable effects of *NAT1** mutations on NAT1 activity: a range of activity is possible depending on the specific allele(s) present. However, none of the studies has yet dealt with the question of variability in gene expression and the factors which may influence this important determinant of activity.

Summary

N-Acetyltransferase activity, along with O-acetyltransferase and N,O-acyltransferase, are important determinants of the fate of numerous environmental and pharmaceutical arylmines. Disposition of these aromatic nitrogen compounds has very significant effects on their interactions with organisms. The NATs influence activation and detoxification of arylamine carcinogens and the efficacy and toxicity of therapeutic agents. It is also possible that NAT is important in normal foetal development.

Any compound with an arylamine moiety, or that can be biotransformed to expose such moeity, must be considered a potential substrate for acetyl transferases. The very significant differences in human NAT activity resulting from genetic polymorphisms in both NAT1 and NAT2 can render substrates toxic in certain individuals while non-toxic in others. The combination of oxidative and acetyltransfer activities of an individual needs to be considered in determining the safety and efficacy of arylamine drugs, both in drug trials and in prescribing doses of approved pharmaceutical agents. Knowledge of the frequency and activity of the major NAT alleles in the ethnic group of the individual given these agents can help to guide the selection of a safe and effective dosage.

References

Andres HH, Klem AJ, Szabo SM and Weber WW (1985) New spectrophotometric and radio-chemical assays for acetyl-CoA: arylamine N-acetyltransferase applicable to a variety of arylamines. *Analytical Biochemistry*, **145**, 367–375.

Andres HH, Klem AJ, Schopfer LM, Harrison JK and Weber WW (1988) On the active site of liver Acetyl-CoA Arylamine N-Acetyltransferase from Rapid Acetylator Rabbits (III/J). *Journal of Biological Chemistry*, **263**, 7521–7527.

Andres HH, Kolb HJ, Schreiber RJ and Weis L (1983) Characterization of the active site, substrate specificity and kinetic properties of acetyl CoA: arylamine N-acetyltransferase from pigeon liver. *Biochimica Biophysica Acta*, **746**, 192–203.

Bruhn C, Brockmöller J, Cascorbi I, Roots I and Borchert H-H (1999) Correlation between genotype and phenotype of the human N-acetyltransferase type 1 (NAT1). *Biochemical Pharmacology*, **58**, 1759–1764.

Butcher NJ, Ilett KF and Minchin RF (1998) Functional polymorphism of the human arylamine N-acetyltransferase type 1 gene caused by $C^{190}T$ and $G^{560}A$ mutations. *Pharmacogenetics*, **8**, 67–72.

Cartwright RA, Glasham RW, Roger HJ, Ahmad RA, Hall DB, Higgins E and Kahn MA (1982) The role of N-acetyltransferase phenotypes in bladder carcinogenesis. A pharmacogenetics epi-demiological approach to bladder cancer. *Lancet*, **2**, 842–846.

Chung JG, Levy GN and Weber WW (1993) Distribution of 2-aminofluorene and p-aminoben-zoic acid N-acetyltransferase activity in tissues of C57BL/6J rapid and B6.A-Nat^S slow acetylator congenic mice. *Drug Metabolism and Disposition*, **21**, 1057–1063.

Chung JG, Kuo HM, Lin TH, Ho CC, Lee JH, Lai JM, Levy GN and Weber WW (1996) Evidence for arylamine N-acetyltransferase in the nematode Anisakis simplex. *Cancer Letters*, **106**, 1–8.

Cribb AE, Grant DM, Miller MA and Spielberg SP (1991) Expression of monomorphic arylamine N-acetyltransferase (NAT1) in human leukocytes. *Journal of Pharmacology and Experimental Therapeutics*, **259**, 1241–1246.

Cribb AE, Nakamura H, Grant DM, Miller MA and Spielberg SP (1993) Role of polymorphic and monomorphic human arylamine N-acetyltransferases in determining sulfamethoxazole meta-bolism. *Biochemical Pharmacology*, **45**, 1277–1282.

Danysz A and Wisniewski K (1970) Control of drug transport through cell membranes. *Materia Medica Polona*, **2**, 35–44.

Deguchi T, Sakamoto Y, Sasaki Y and Uyemura K (1988) Arylamine N-acetyltransferase from chicken liver: 1. Monoclonal antibodies, immunoaffinity purification, and amino acid sequences. *Journal of Biological Chemistry*, **263**, 7528–7533.

deLeón JH, Vatsis KP and Weber WW (2000) Characterization of naturally occurring and recombinant human N-acetyltransferase variants encoded by *NAT1**. *Molecular Pharmacology*, in press.

Dewhurst SA, Croker SG, Ikeda K and McCaman RE (1972) Metabolism of biogenic amines in *Drosophila* nervous tissue. *Comparative Biochemistry and Physiology*, **43B**, 975–981.

Dhaini HR and Levy GN (2000) Arylamine N-acetyltransferase 1 (*NAT1*) genotypes in a Lebanese population. *Pharmacogenetics*, **10**, 79–83.

Dupret J-M and Grant DM (1992) Site-directed mutagenesis of recombinant human arylamine N-acetyltransferase expressed in *Escherichia coli*: Evidence for direct involvement of Cys^{68} in the catalytic mechanism of polymorphic human NAT2. *Journal of Biological Chemistry*, **267**, 7381–7385.

Estrada L, Kanelakis KC, Levy GN and Weber WW (2000) Tissue- and gender-specific expression of N-acetyltransferase 2 (*NAT2**) during development of the outbred mouse strain CD-1. *Drug Metabolism and Disposition*, **28**, 139–146.

Evans DAP (1963) Pharmacogenetics. *American Journal of Medicine*, **34**, 639–662.

Frymoyer JW and Jacox RF (1963) Investigation of the genetic control of sulfadiazine and isoniazid metabolism in the rabbit. *Journal of Laboratory and Clinical Medicine*, **62**, 891–904.

Glowinski IB, Radtke HE and Weber WW (1978) Genetic variation in N-acetylation of carcinogenic arylamines by human and rabbit liver. *Molecular Pharmacology*, **14**, 940–949.

Goedde BW, Schloot W and Valesky A (1967) Individual different response to drugs: Characterization of an INH-acetylating system in Rhesus monkeys. *Biochemical Pharmacology*, **16**, 1793–1799.

Grant DM, Blum M, Beer M and Meyer UA (1991) Monomorphic and polymorphic human N-acetyltransferases: A comparison of liver isozymes and expressed products of two clones genes. *Molecular Pharmacology*, **39**, 184–191.

Grant DM, Lottspeich F and Meyer UA (1989) Evidence for two closely related isozymes of arylamine N-acetyltransferase in human liver. *FEBS Letters*, **244**, 203–207.

Grant DM, Vohra P, Avis Y and Ima A (1992) Detection of a new polymorphism of human arylamine N-acetyltransferase NAT1 using *p*-aminosalicylic acid as an in vivo probe. *Journal of Basic and Clinical Physiology and Pharmacology*, **3**(Suppl.), 244.

Harris C, Stark KL, Luchtel DL and Juchau MR (1989) Abnormal neurulation induced by 7-hydroxy-2-acetylaminofluorene and acetaminophen: Evidence for catechol metabolites as proximate dysmorphogens. *Toxicology and Applied Pharmacology*, **101**, 432–446.

Hein DW (1988) Acetylation genotype and arylamine-induced carcinogenesis. *Biochimica and Biophysica Acta*, **948**, 37–66.

Hein DW, Doll MA, Rustan TD, Gray K, Feng Y, Ferguson RJ and Grant DM (1993) Metabolic activation and deactivation of arylamine carcinogens by recombinant human NAT1 and polymorphic NAT2 acetyltransferases. *Carcinogenesis*, **14**, 1633–1638.

Hein DW, Omichinski JG, Brewer JA and Weber WW (1982) A unique pharmacogenetic expression of the N-acetylation polymorphism in the inbred hamster. *Journal of Pharmacology and Experimental Therapeutics*, **220**, 8–15.

Hess DA, Sisson ME, Suria H, Wijsman J, Puvanesasingham R, Madrenas J and Rieder MJ (1999) Cytotoxicity of sulfonamide reactive metabolites: apoptosis and selective toxicity of $CD8^+$ cells by the hydroxylamine of sulfamethoxazole. *FASEB Journal*, **13**, 1688–1698.

Hickman D, Pope J, Patil SD, Fakis G, Smelt V, Stanley LA Payton M, Unadkat JD and Sim E (1998) Expression of arylamine N-acetyltransferase in human intestine. *Gut*, **42**, 402–409.

Ho CC, Lin TH, Lai YS, Chung JG, Levy GN and Weber (1996) Kinetics of acetyl coenzyme A:arylamine N-acetyltransferase from rapid and slow acetylator frog tissues. *Drug Metabolism and Disposition*, **24**, 137–141.

Hughes NC, Janezic SA, McQueen KL, Jewett MAS, Castranio T, Bell DA and Grant DM (1998) Identification and characterization of variant alleles of human acetyltransferase NAT1 with defective function using *p*-aminosalicylate as an *in-vivo* and *in-vitro* probe. *Pharmacogenetics*, **8**, 55–66.

Hurwitz A and Schlozman DL (1974) Effects of antacids on gastrointestinal absorption of isoniazid in rat and man. *Annual Review of Respiratory Diseases*, **109**, 41–47.

Ilett KP, David BM, Detchon P, Castleden WM and Kwa R (1987) Acetylation phenotype in colorectal carcinoma. *Cancer Research*, **47**, 1466–1469.

Jenne JW (1965) Partial purification and properties of the isoniazid transacetylase of human liver: Its relationship to the acetylation of p-aminosalicylic acid. *Journal of Clinical Investigation*, **44**, 1992–2002.

Karim AKMB, Elfellah MS and Evans DAP (1981) Human acetylator polymorphism: estimate of allele frequency in Libya and details of global distribution. *Journal of Medical Genetics*, **18**, 325–333.

Karlson P and Sekeris CE (1962) N-Acetyl-dopamine as sclerotizing agent of the insect cuticle. *Nature*, **195**, 183–184.

Karolyi J, Erickson RP, Liu S and Killewald L (1990) Major effects on teratogen-induced facial clefting in mice determined by a single genetic region. *Genetics*, **126**, 201–205.

Kloth MT, Gee RL, Messing EM and Swaminathan S (1994) Expression of N-acetyltransferase (NAT) in cultured human uroephelial cells. *Carcinogenesis*, **15**, 2781–2787.

Lang NP, Chu DZJ, Hunter CF, Kendall DC, Flemmang TJ and Kadlubar FF (1986) Role of aromatic amine acetyltransferase in human colorectal cancer. *Archives of Surgery*, **121**, 1259–1261.

Lessard E, Hamelin BA, Labbe L, O'Hara G, Belanger PM and Turgeon J (1999) Involvement of CYP2D6 activity in the N-oxidation of procainamide in man. *Pharmacogenetics*, **9**, 683–696.

Lester D (1964) The acetylation of isoniazid in alcoholics. *Quarterly Journal of Studies on Alcohol*, **25**, 541–543.

Levy GN, Martell KJ, deLeón JH and Weber WW (1992) Metabolic, molecular genetic and toxicological aspects of the acetylation polymorphism in inbred mice. *Pharmacogenetics*, **2**, 197–206.

Li JY and Nappi AJ (1992) N-Acetyltransferase activity during ovarium development in the mosquito *Aedes aegypti* following blood feeding. *Insect Biochemistry and Molecular Biology*, **22**, 44–54,

Lin HJ, Probst-Hensch NM, Hughes NC, Sakamoto GT, Louie AD, Kau IH, Lin BK, Lee DB, Lin J, Frankl HD, Lee ER, Hardy S, Grant DM and Haile RW (1998) Variants of N-acetyltransferase NAT1 and a case-control study of colorectal adenomas. *Pharmacogenetics*, **8**, 269–281.

Liu S and Erickson RP (1986) Genetic differences among the A/J × C57BL/6J recombinant inbred mouse lines and their degree of association with glucocoticoid-induced cleft palate. *Genetics*, **111**, 745–754.

Liu Y and Levy GN (1998) Activation of heterocyclic amines by combinations of prostaglandin H synthase-1 and -2 with N-acetyltransferase 1 and 2. *Cancer Letters*, **133**, 115–123.

Mannisto P, Mantayla R, Klinge E, Nykanen S, Koponen A and Lamminisivu U (1982) Influence of various diets on the bioavailability of isoniazid. *Journal of Antimicrobial Therapy*, **10**, 427–434.

Marshall EK (1954) Acetylation of sulfonamides in the dog. *Journal of Biological Chemistry*, **211**, 499–503.

Matas N, Thygesen P, Stacey M, Risch A and Sim E (1997) Mapping AAC1, AAC2 and AACP, the genes for arylamine N-acetyltransferases, carcinogen metabolizing enzymes on human chromosome 8p22, a region frequently deleted in tumors. *Cytogenetics and Cell Genetics*, **77**, 290–295.

Minchin RF (1995) Acetylation of p-aminobenzoylglutamine, a folic acid catabolite, by recombinant human N-acetyltransferase and U937 cells. *Biochemical Journal*, **307**, 1–3.

Minchin RF, Reeves PT, Teitel CH, McManus ME, Mojarrabi B, Ilett KF and Kadlubar FF (1992) N- and O-Acetylation of aromatic and heterocyclic amine carcinogens by human mono-morphic and polymorphic acetyltransferases expressed in COS-1 cells. *Biochemical and Biophysical Research Communications*, **185**, 839–844.

Mitchell MK, Futscher BW and McQueen CA (1999) Developmental expression of N-acetyl-transferases in C57BL/6J mice. *Drug Metabolism and Disposition*, **27**, 261–264.

Motulsky AG and Steinman L (1962) Arylamine acetylation in human red cells. *Journal of Clinical Investigation*, **41**, 1387.

Nakura H, Itoh S, Kusano H, Ishizone H and Kamataki T (1994) Activities of drug-metabolizing enzymes in the liver of *Suncus murinus*: Possible lack of *N*-acetyltransferase activity in liver cytosol. *Pharmacology*, **48**, 201–204.

Ohsako S and Deguchi T (1990) Cloning and expression of cDNAs for polymorphic and monomorphic arylamine *N*-acetyltransferases from human liver. *Journal of Biological Chemistry*, **265**, 4630–4634.

Olsen H and Morland J (1978) Ethanol-induced increase in drug acetylation in man and isolated rat liver cells. *British Medical Journal*, **2**, 1260–1262.

Olsen H and Morland J (1982) Ethanol-induced increase in procainamide acetylation in man. *British Journal of Clinical Pharmacology*, **13**, 203–208.

Probst MR, Blum M, Fasshauer I, D'Orazio D, Meyer UA and Wild D (1992) The role of human acetylation polymorphism in the metabolic activation of the food carcinogen 2-amino-3-methylimidazo-[4,5-*f*]quinoline (IQ). *Carcinogenesis*, **13**, 1713–1717.

Radtke HE, Brenner WB and Weber WW (1979) *N*-Acetylation of drugs: Search for the INH acetylation polymorphism in baboons. *Drug Metabolism and Disposition*, **7**, 194–195.

Rodriguez JW, Kirlin WG, Ferguson RJ, Doll MA, Gray K, Rustin TD, Lee ME, Kemp K, Urso P and Hein DW (1993) Human acetylator genotype: relationship to colorectal cancer incidence and arylamine *N*-acetyltransferase expression on colon cytosol. *Archives of Toxicology*, **67**, 445–452.

Rothen J-P, Haefeli WE, Meyer UA, Todesco L and Wenk M (1998) Acetaminophen is an inhibitor of hepatic *N*-acetyltransferase 2 *in vitro* and *in vivo*. *Pharmacogenetics*, **8**, 553–559.

Sadrieh N, Davis CD and Snyderwine EG (1996) *N*-Acetyltransferase expression and metabolic activation of food-derived heterocyclic amines in human mammary gland. *Cancer Research*, **56**, 2683–2687.

Sekeris CE (1964) Sclerotization of the blowfly imago. *Science*, **144**, 419–420.

Shinohara A, Saito K, Yamazoe Y, Kamataki T and Kato R (1986) Inhibition of acetyl-coenzyme A dependent activation of *N*-hydroxyarylamines by phenolic compounds. Pentachlorophenol and 1-nitro-2-naphthol. *Chemico-Biological Interactions*, **60**, 275–285.

Smelt VA, Mardon HJ, Redman CWG, and Sim E (1997) Acetylation of arylamines by the placenta. *European Journal of Drug Metabolism and Pharmacokinetics*, **22**, 403–408.

Sone T, Zukowski K, Land SJ, King CM and Wang CY (1991) Acetylation of 2-aminofluorene derivatives by dog liver microsomes. *Carcinogenesis*, **12**, 1887–1891.

Sone T, Zukowski K, Land SJ, King CM, Martin BM, Pohl LR and Wang CY (1994) Characteristics of a purified dog hepatic microsomal *N*,*O*-acyltransferase. *Carcinogenesis*, **15**, 595–599.

Stanley LA, Copp AJ, Pope J, Rolls S, Smelt V. Perry VH and Sim E (1998) Immunochemical detection of arylamine *N*-acetyltransferase during mouse embryonic development and in adult mouse brain. *Teratology*, **58**, 174–182.

Tannen RH and Weber WW (1979) Rodent models of the human isoniazid acetylator polymorphism. *Drug Metabolism and Disposition*, **7**, 274–279.

Thygesen P, Risch A, Stacey M, Fakis G, Takle L, Knowles M and Sim E (1999) Genes for human arylamine *N*-acetyltransferase in relation to loss of the short arm of chromosome 8 in bladder cancer. *Pharmacogenetics*, **9**, 1–8.

Trepanier LA, Cribb AE, Spielberg SP and Ray K (1998) Deficiency of cytosolic arylamine *N*-acetylation in the domestic cat and wild felids caused by the presence of a single *NAT1*-like gene. *Pharmacogenetics*, **8**, 169–179.

Trepanier LA, Ray K, Winard NJ, Spielberg SP and Cribb AE (1997) Cytosolic arylamine *N*-acetyltransferase (*NAT*) deficiency in the dog and other canids due to an absence of NAT genes. *Biochemical Pharmacology*, **54**, 73–80.

Uetrecht JP (1989) Idiosyncratic drug reactions: Possible role of reactive metabolites generated by leukocytes. *Pharmacological Research*, **6**, 265–273.

Vatsis KP and Weber WW (1993) Structural heterogeneity of Caucasian *N*-acetyltransferase at the *NAT1* gene locus. *Archives of Biochemistry and Biophysics*, **301**, 71–76.

Vatsis KP and Weber WW (1997) Acetyltansferases. In *Comprehensive Toxicology*, Sipes IG, McQueen CA and Gandolfi AJ (eds), Pergamon, New York, Vol. 3, pp. 385–399.

Vatsis KP, Weber WW, Bell DA, Dupret JM, Evans DAP, Grant DM, Hein DW, Lin HJ, Meyer UA, Relling MV, Sim E, Suzuki T and Yamazoe Y (1995) Nomenclature for *N*-acetyltransferases. *Pharmacogenetics*, **5**, 1–17.

Watanabe M, Nohmi T and Ishidate M (1987) New tester strains of Salmonella typhimurium highly sensitive to mutagenic nitroarenes. *Biochemical and Biophysical Research Communications*, **147**, 974–979.

Watanabe M, Sofuni T and Nohmi T (1992) Involvement of Cys 69 residue in the catalytic mechanism of *N*-hydroxyarylamine *O*-acetyltransferase of *Salmonella typhimurium*. Sequence similarity at the amino acid level suggests a common catalytic mechanism of acetyltransferase for *S. typhimurium* and higher organisms. *Journal of Biological Chemistry*, **267**, 8429–8436.

Watkins JB and Klaassen CD (1986) Xenobiotic biotransformation in livestock: Comparison to other species commonly used in toxicity testing. *Journal of Animal Science*, **63**, 933–942.

Weber WW (1987) *The Acetylator Genes and Drug Response*, Oxford University Press, New York.

Weber WW (1991) Isoniazid. In *Therapeutic Drugs*, Dollery C (ed.), Churchill Livingstone, Edinburgh, pp. 191–197.

Weber WW and Hein DW (1979) Clinical pharmacokinetics of isoniazid. *Clinical Pharmacokinetics*, **4**, 401–422.

Weber WW and Hein DW (1985) *N*-Acetylation pharmacogenetics. *Pharmacological Reviews*, **37**, 25–79.

Weber WW and Vatsis KP (1993) Individual variability in *p*-aminobenzoic acid *N*-acetylation by human *N*-acetyltransferase (NAT1) of peripheral blood. *Pharmacogenetics*, **3**, 209–212.

Whittaker DP and Goosey MW (1993) Purification and properties of the enzyme arylamine *N*-acetyltransferase from the housefly *Musca domestica*. *Biochemical Journal*, **295**, 149–154.

Zhao B, Lee EJD, Yeoh PN and Gong NH (1998) Detection of mutations and polymorphisms of *N*-acetyltransferase 1 gene in Indian, Malay and Chinese populations. *Pharmacogenetics*, **8**, 299–304.

12 Mammalian Xenobiotic Epoxide Hydrolases

Michael Arand and Franz Oesch

Institute of Toxicology, University of Mainz, Germany.

General characteristics

FUNCTION

Epoxide hydrolases (EH, E.C.3.3.2.3) hydrolyse oxiranes, a specific class of cyclic ethers (Oesch 1973). The common feature of these compounds, their three-membered ring system, is under high tension, due to the unusually small bonding angles. Together with the polarisation of the C–O bonds, this leads to an enhanced chemical reactivity that is further modulated by the substitution pattern at the epoxide ring. Epoxides act as electrophiles with the reactive centre being one of the ring carbon atoms. As a general rule, asymmetric substitution at the ring enhances the reactivity. Important targets for epoxides in living organisms are nucleophilic sites in biomacro-molecules, in particular proteins and nucleic acids. Chemical attack of these leads to cytotoxic and genotoxic effects. In particular, the modification of DNA bases can result in inheritable changes and such changes may ultimately give rise to carcinogen-esis (Miller and Miller 1981). The primary function of xenobiotic epoxide hydrolases is to defeat such hazardous effects of epoxides. Epoxides can enter the body pre-formed or may arise from the metabolism of xenobiotic and, in some cases, of endogenous compounds (Figure 12.1).

In contrast to the great number of enzymes that can metabolise arenes or alkenes to epoxides, there are at present only two distinct mammalian xenobiotic epoxide hydrolases known (Oesch and Bentley 1976; Ota and Hammock 1980; Guenthner et al. 1981; Thomas et al. 1990; Hammock et al. 1997; Armstrong 1999). These two, the membrane-bound microsomal epoxide hydrolase (mEH) and the soluble epoxide hydrolase (sEH), will be described in detail in this chapter. Three more EHs have been identified in mammals that all have a narrow substrate specificity for epoxides formed from endogenous precursors, namely the leukotriene A4 hydrolase (Haeggstrom et al. 1990), the cholesterol epoxide hydrolase (Levin et al. 1983; Oesch et al. 1984) and

Enzyme Systems that Metabolise Drugs and Other Xenobiotics. Edited by C. Ioannides.
© 2002 John Wiley & Sons Ltd

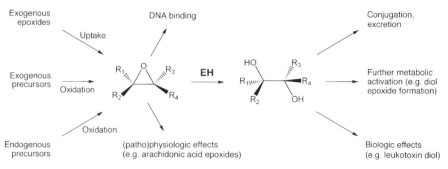

Figure 12.1 Role of epoxide hydrolases in the metabolism of exogenous and endogenous epoxides. EH: epoxide hydrolase.

the hepoxilin epoxide hydrolase (Pace-Asciak and Lee 1989). These will not be discussed here, since they have little, if any, impact on xenobiotic metabolism.

In general, xenobiotic epoxide hydrolases serve the above-described detoxification function, yet, as always, there are some exceptions to this rule (Bentley *et al.* 1977). A prominent example for this is the metabolic activation of polycyclic aromatic hydrocarbons (PAH) to the corresponding dihydrodiol epoxides, the ultimate carcinogenic metabolites of this class of compounds (Holder *et al.* 1974; Sims *et al.* 1974) (Figure 12.2). PAH with a so-called bay region are metabolically activated in a first step by, e.g. CYP (cytochrome P450), to pre-bay epoxides. These genotoxic metabolites can rapidly rearrange to the corresponding, much less toxic, phenols and thus undergo spontaneous detoxification. Likewise, enzymic cleavage by epoxide hydrolases to the corresponding vicinal dihydrodiols leads to *per se* inactive products. However, these diols are again substrates for a variety of CYP and COX isoenzymes which finally generate the highly reactive dihydrodiol epoxides. These compounds neither undergo rearrangement to phenols (they are alkene, not arene oxides), nor are they substrates (or in some cases only extremely poor substrates) for epoxide hydrolases. Their detoxification by glutathione conjugation (Jernstrom *et al.* 1992), the last line of defence, is obviously not sufficient to protect the organism from the potent carcinogenic effect of these metabolites. The central importance of mEH for this activation pathway has finally been proven using mEH *knockout* mice (Miyata *et al.* 1999). In contrast to their wild-type relatives, these animals were highly resistant to the carcinogenic effect of 7,12-dimethylbenz[*a*]anthracene in the mouse skin tumorigenesis test.

A second, recently discovered and somehow unexpected activation pathway driven by epoxide hydrolases is the formation of toxic vicinal diols from the epoxides of unsaturated fatty acids. Leukotoxin, the epoxide of linolenic acid, has earlier been reported to be the chemical mediator in multiple organ failure and adult respiratory distress syndrom (ARDS) (Ozawa *et al.* 1991). It now appears that the diol rather than the epoxide seems to be the causative agent (Moghaddam *et al.* 1997): in mice, sEH is the major leukotoxin-metabolising EH. Pretreatment of mice with an sEH inhibitor significantly increased the tolerance of the animals to the toxic effects of leukotoxin.

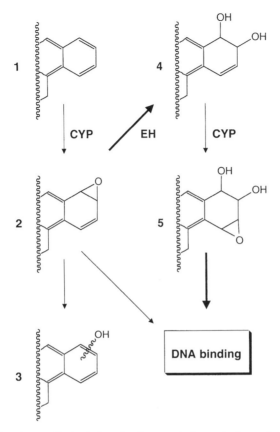

Figure 12.2 Role of epoxide hydrolase in the metabolic activation of polycyclic aromatic hydrocarbons (PAH) to diol epoxides.
1 = parent PAH; 2 = PAH pre-bay epoxide; 3 = PAH phenol (two regioisomers possible); 4 = PAH dihydrodiol; 5 = PAH dihydrodiol epoxide. CYP, cytochrome P_{450}: EH, epoxide hydrolase.

Likewise, recombinant expression of sEH enhanced leukotoxin cytotoxicity in a number of different cell systems.

In summary, the vast majority of substrates are chemically inactivated and thus detoxified by EH but in some specific cases, epoxide hydrolysis can directly or indirectly increase the toxicity of the respective substrate.

PHYLOGENETICS

The mEH was among the first xenobiotic metabolising enzymes to be cloned (Gonzalez and Kasper 1981) and characterised in terms of amino acid sequence (Heinemann and Ozols 1984), yet little immediate progress resulted from these early findings. At the end of the 1980s, a bacterial enzyme with marginal sequence similarity to the mEH was discovered, a haloalkane dehalogenase (Janssen *et al.* 1989), but this possible relationship was largely ignored. This changed dramatically when the molecular characterisation of sEH was reported (Beetham *et al.* 1993; Grant *et al.* 1993; Knehr *et al.* 1993). At that time, direct comparison of the two epoxide hydrolase sequences did not show any convincing relationship between the two, yet the sEH, like the mEH before, showed marginal but significant similarity to the bacterial haloalkane dehalogenase (Arand *et al.* 1994), of which the three-dimensional structure had just been determined (Franken *et al.* 1991; Verschueren *et al.* 1993). The dehalogenase had been identified as a member of the α/β hydrolase fold family of enzymes (Ollis *et al.* 1992), its most famous relative thus being the acetylcholine esterase. The fact that the overall protein fold is conserved in this family of enzymes, despite the lack of evident sequence similarity, strongly suggested that proteins related to the dehalogenase by sequence similarity should have the same three-dimensional structure, and therefore EHs should also be members of the α/β hydrolase fold enzyme family (Arand *et al.* 1994; Lacourciere and Armstrong 1994; Pries *et al.* 1994). Final proof for this has recently been provided by X-ray analysis of a variety of EH structures (Argiriadi *et al.* 1999; Nardini *et al.* 1999; Zou *et al.* 2000). The generic EH structure derived from this is the following (Figure 12.3).

The central domain of EH is the α/β hydrolase fold that is composed of a central β-

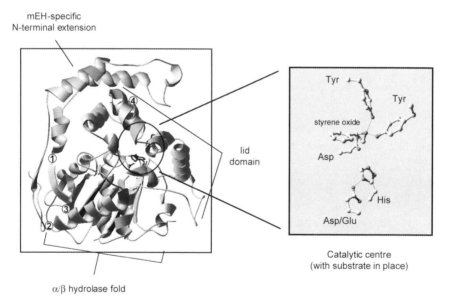

Figure 12.3 Three-dimensional structure of epoxide hydrolases.
The left part of the figure shows a ribbon model of the EH structure while the right side shows a magnification of the active centre, with styrene oxide as a generic EH substrate in place. Note that the *N*-terminal extension is a unique feature of the mEH-like epoxide hydrolases. The numbered circles in the ribbon representation denote the position of the following individual polymorphisms of EHs that are discussed later in this chapter. $1 = \text{Tyr}_{113}\text{His}$ (mEH); $2 = \text{Arg}_{139}\text{His}$ (mEH); $3 = \text{Arg}_{287}\text{Gln}$ (sEH); $4 = \text{Arg}_{402}\text{Arg/Arg}$ (insertion;sEH).

sheet flanked by α-helices. On top of this fold sits a so-called lid. The catalytic site is situated at the interface of these two structural entities, with a catalytic triad (Ollis et al. 1992) being anchored in the α/β hydrolase fold and two catalytic tyrosines hanging from the lid into the substrate binding pocket. One very important aspect of this discovery was its impact on the understanding of the catalytic mechanism of enzymic epoxide hydrolysis, (see below). Furthermore, the comparison between the EHs and the dehalogenase revealed a number of typical signature sequences that define a subgroup in the family of α/β hydrolase fold enzymes. These signatures can be used to scan the available biological databases for the identification of other potential epoxide hydrolases. A phylogenetic tree of a selection of sequences retrieved this way is shown in Figure 12.4.

The first lesson to be learned from this multiple-sequence comparison is that mEH

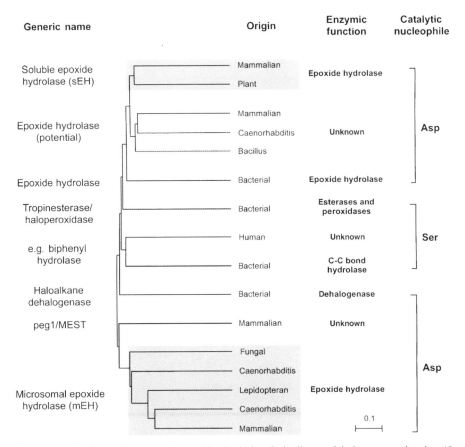

Figure 12.4 Phylogenetic tree of epoxide hydrolase/haloalkane dehalogenase-related α/β hydrolase fold enzymes.

Comparison has been performed using CLUSTAL-X (Thompson et al. 1997). For further details of the analysis see Arand et al. (1999a).

and sEH are, indeed, at the opposite ends of this comparison, and they must have evolved from their common ancestor several billion years ago. The mammalian mEH has apparent orthologues in insects (Wojtasek and Prestwich 1996), nematodes (Wilson *et al.* 1994) and in fungi (Arand *et al.* 1999a), while sEH orthologous enzymes have been identified in plants (Kiyosue *et al.* 1994; Stapleton *et al.* 1994). A second finding is the broad variety of different enzymes in this family tree, ranging from epoxide hydrolases over esterases to C–C bond hydrolases. As will be detailed later, mammalian xenobiotic epoxide hydrolases have an aspartic acid residue in their active site serving as the catalytic nucleophile. Unexpectedly, a number of enzymes in the alignment revealed a serine in this position, namely the esterases and C–C bond hydrolases, and this results in a dilemma in terms of enzyme nomenclature: while these serine-nucleophile enzymes are functionally related to esterases, they are included in the epoxide hydrolase/dehalogenase-like α/β hydrolase fold enzymes on the basis of their structure, and it will be difficult to establish a widely accepted nomenclature system as is now available for CYP (Nelson *et al.* 1996), UGT (Mackenzie *et al.* 1997) and GST (Hayes and Pulford 1995) enzymes. So far, it has been proposed to name mEH HYL1, mammalian sEH HYL2 and plant sEH HYL3 (Beetham *et al.* 1995), yet this attempt must be regarded as preliminary since the other enzymes related to EH should be incorporated into this nomenclature system.

MECHANISM

The mechanism of enzymic epoxide hydrolysis has been subject of intense investigation since the 1970s (DuBois *et al.* 1978), and its understanding offers a clue to the incredible yet hidden efficacy of EHs. It was a kind of mystery how a single enzyme, the mEH, could—on the one hand—have an enormously broad substrate specificity while—on the other—displaying an apparently high affinity to different substrates, sufficient to detoxify these at low concentrations.

The first indication of an unusual mode of action was provided by the notion that a single round of substrate turnover in the presence of heavy water led to the incorporation of ^{18}O into the enzyme rather than into the reaction product (Lacourciere and Armstrong 1993), an observation incompatible with the previously favoured direct hydrolysis of epoxides by EH (Armstrong 1987). The authors reasoned that the formation of an enzyme-substrate ester intermediate must have taken place, a deduction that was further substantiated by the above-described sequence comparison (Arand *et al.* 1994) between EH and other enzymes for which the formation of similar ester intermediates in the course of their enzymic reaction had already been shown. Biochemical (Pinot *et al.* 1995; Arand *et al.* 1996, 1999b; Rink *et al.* 1997, 2000; Laughlin *et al.* 1998; Tzeng *et al.* 1998; Yamada *et al.* 2000) and structural (Argiriadi *et al.* 1999; Nardini *et al.* 1999; Zou *et al.* 2000) analyses of EHs then led to a detailed understanding of the process (Figure 12.5).

The key event in the initial substrate recognition of epoxides by EH seems to be the trapping of the epoxide oxygen by two tyrosine residues via hydrogen bonding in the active site of the enzyme. Also contributing to this initial, reversible binding may be some hydrophobic interactions between the lipophilic side chain of the substrate and the surface of the substrate access tunnel of the enzyme. Here, also, some constraints

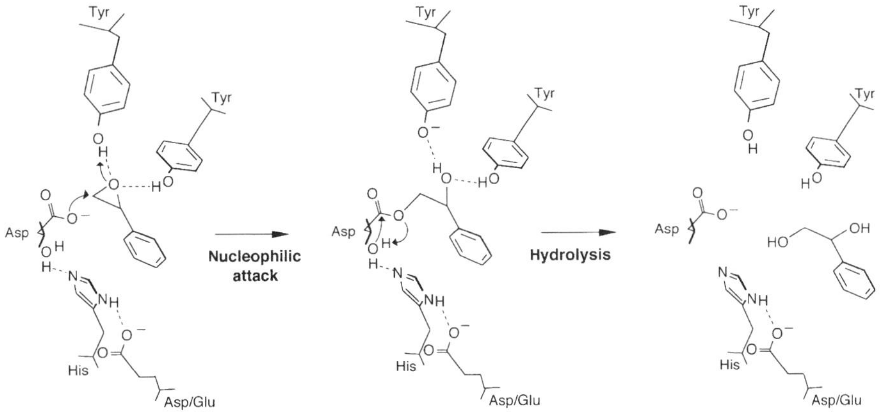

Figure 12.5 Catalytic mechanism of enzymic epoxide hydrolysis.

do apply. Due to the position of the active site residues at the end of its substrate access tunnel (Zou *et al.* 2000), mEH may be unable to hydrolyse *trans*-substituted epoxides, while the sEH, with its active site residues sitting on the side of a bent narrow tunnel (Argiriadi *et al.* 1999), does interact with these *trans*-substituted epoxides but cannot breakdown particularly bulky substrates.

The first chemical reaction step is the crucial one for substrate inactivation: the hydrogen bonding of the epoxide oxygen positions a ring carbon favourable for nucleophilic attack by the catalytic nucleophile of the EH catalytic triad, an aspartic acid residue. In a push–pull mechanism, this aspartic acid forms an ester bond with the ring carbon under scission of the respective C–O bond in the epoxide ring. Simultaneously, the oxygen is saturated by a proton from one of the two tyrosines. The resulting enzyme-substrate ester intermediate is subsequently hydrolysed by the water-activating charge relay system of the catalytic triad, composed of a histidine and an acidic residue, an aspartic acid in the case of sEH and a glutamic acid in the case of mEH.

An important observation was that the first step of this reaction, the ester formation, proceeds by orders of magnitudes faster than the second, hydrolytic step (Tzeng *et al.* 1996). First, this explains a likely underestimation of the detoxication efficacy of EH as illustrated in Figure 12.6, if the product formation is used as the measure for this. Second, it explains the apparent contradiction between broad substrate specificity and high substrate affinity. The relationship between the real affinity of the substrate to the enzyme, characterised by the dissociation constant K_D, and the apparent affinity measured as the Michaelis–Menten constant K_M is dependent on the rate constants of the nucleophilic attack (k_1) and the hydrolysis (k_2) as follows:

$$K_M = K_D \times \frac{k_2}{k_1 + k_2} \tag{12.1}$$

If, as in the present case, k_1 is orders of magnitudes higher than k_2, this equation essentially reduces to

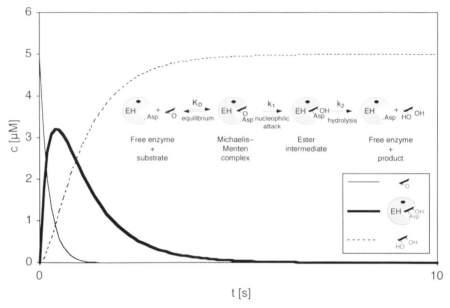

Figure 12.6 Detoxification kinetics of enzymic epoxide hydrolysis.

Displayed is the result of a computer simulation of styrene epoxide hydrolysis by human mEH. The concentrations of epoxide (narrow line), diol (broken line) and ester intermediate (bold line) are plotted over time. Under the chosen conditions (enzyme in excess over its substrate, as is probably true for mEH in most real life settings), the decline of the substrate concentration (i.e. of the toxic challenge) proceeds significantly faster than the increase in product formation, due to the intermediate accumulation of the enzyme-substrate ester. Thus, considering rate of detoxification by taking the initial rate of product formation as the measure results in a strong underestimation of the efficacy of the enzyme. For further discussion see Oesch *et al.* (2000).

$$K_M = K_D \times \frac{k_2}{k_1} \tag{12.2}$$

which can be further transformed to

$$\frac{K_M}{K_D} = \frac{k_2}{k_1} \tag{12.3}$$

Thus, the apparent high affinity of mEH for many substrates is actually based on a comparatively low real affinity and reflects the great difference between k_1 and k_2.

From the above it is evident that the enzymic epoxide hydrolysis is not optimised for product release. EH turnover rates are in the order of a few substrate molecules per second at best (Thomas *et al.* 1990). In contrast, acetylcholine esterase, a structural and functional relative (see above), achieves a turnover rate of 25,000 s^{-1}, despite the fact that the second step for both enzymic reactions is chemically practically the same, i.e. hydrolysis of the ester intermediate. We speculate that optimisation of the hydrolytic step could not be more successful in the case of EHs, since the ester intermediates of different substrates may have different spatial location in the active site. We suggest that the epoxide side chain will put some constraint on the position of the attacked epoxide ring carbon which can most likely be compensated by an inferred flexibility of the aspartic acid that acts as the nucleophile. Since ester formation is the detoxification step, selection pressure is in favour to speed up this step rather than the subsequent hydrolysis. Indeed, it was possible to increase k_2 of mEH for a variety of different substrates by introducing a single-point mutation (Arand *et al.* 1999b), but the consequences for the turnover of other substrates have not been thoroughly investigated.

Microsomal epoxide hydrolase

STRUCTURAL CHARACTERISTICS

Mammalian mEH has a molecular mass of 51 kDa, corresponding to 455 amino acid residues (Porter *et al.* 1986). It is attached to the ER membrane by a single *N*-terminal membrane anchor (Friedberg *et al.* 1994). This anchor is connected to the generic α/β hydrolase fold by a stretch of about 100 amino acid residues that wraps around the molecule in a single large meander (see Figure 12.3), thereby apparently clamping together the α/β hydrolase fold and the lid on its top (Zou *et al.* 2000). This very compact structure may be a reason for an observed comparatively high resistance of mEH against thermal inactivation and proteolytic digestion. The quaternary structure of mammalian mEH is presently unknown. It has been speculated that the enzyme associates with the CYP and CYP reductase to a multienzyme complex, yet the few experimental approaches to prove this have not been conclusive (Oesch and Daly 1972; Etter *et al.* 1991). Since EH substrates are, in general, highly lipophilic, it is conceivable that the entry to the EH active site is directly connected to the lipid bilayer of the membrane, so that lipophilic compounds that would travel along the lipid phase could directly enter into it. A similar topology has just been reported for the xenobiotic-metabolising CYP2C5 (Williams *et al.* 2000). Such a scenario, with the two-dimensional membrane being the universal and efficient adaptor, would facilitate the interaction between EH and any CYP, without the need of direct interaction between the proteins.

After solubilisation from the membrane with detergents, EH is in the state of a homo-oligomer of an apparent molecular weight of 700-800 kDa (Guengerich and Davidson 1982). A soluble enzyme related to the mammalian mEH has been identified in *Aspergillus niger* (Morisseau *et al.* 1999a). This enzyme has been cloned (Arand *et al.* 1999a) and crystallised (Zou *et al.* 2000), and proved to be a homodimer in solution and in crystal form. The interaction surface between the two subunits involves the lid and the *N*-terminal meander and is apparently well conserved between fungal and mammalian enzymes. Thus, mEH may also exist as a homodimer in the membrane, possibly with its active site bent towards the lipid bilayer.

METABOLIC FUNCTION

Microsomal epoxide hydrolase is believed to be the major xenobiotic metabolising EH (Armstrong 1987). It has an extremely broad range of substrates, examples of which are shown in Figure 12.7. In general, an mEH substrate should be an epoxide (so far, no exceptions have been identified), should be hydrophobic in nature and should be mono-,1,1-di- or 1,2-*cis*-disubstituted.

Bulky substrates, such as benzo[a]pyrene-4,5-oxide are as well accepted as slim compounds, e.g. octene-1,2-oxide. Since epoxides are potentially hazardous, not very many therapeutically used drugs undergo this metabolic pathway and thus, implications of mEH in clinical drug metabolism are, from a quantitative standpoint, not so numerous as those associated with CYP, glucuronide and sulphate conjugating enzymes, yet toxicologically especially important. One example is the anticonvulsant carbamazepine, a major metabolite of which is the 10,11-oxide (Eichelbaum *et al.* 1979). This symmetric epoxide is not very reactive and did not act as a mutagen in the Ames test (Glatt *et al.* 1983). Nevertheless, it has been suggested to be the cause of the adverse drug effects of carbamazepine after *in vivo* inhibition of mEH by co-medication with valpromide (see below) (Meijer *et al.* 1984).

Some industrial compounds are metabolically activated to epoxides. A prominent case is styrene, of which more than 90% of a given dose is converted to the 7,8-epoxide in the human body (Jenkins Sumner and Fennell 1994). It is at the same time an impressive example of the detoxification efficacy of mEH. Despite the fact that styrene oxide is a proven carcinogen (Ponomarkov *et al.* 1984), styrene itself is orders

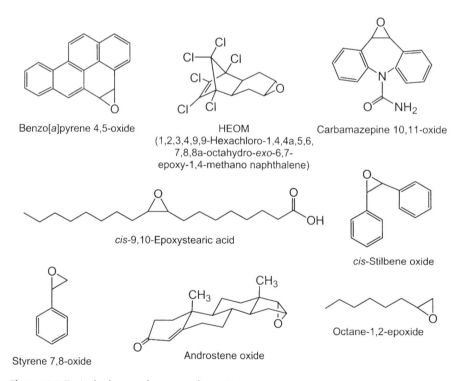

Benzo[a]pyrene 4,5-oxide

HEOM
(1,2,3,4,9,9-Hexachloro-1,4,4a,5,6,
7,8,8a-octahydro-*exo*-6,7-
epoxy-1,4-methano naphthalene)

Carbamazepine 10,11-oxide

cis-9,10-Epoxystearic acid

cis-Stilbene oxide

Styrene 7,8-oxide

Androstene oxide

Octane-1,2-epoxide

Figure 12.7 Typical substrates for mammalian mEH.

of magnitudes less harmful since the metabolically formed styrene-7,8-oxide is almost immediately hydrolysed by mEH, probably because the liver is the major site for both, formation as well as breakdown of the epoxide. Thus, it has been observed in styrene-exposed workers that the biomarkers of exposure to the reactive metabolite styrene oxide hardly correlate with the level of styrene exposure, but show significant correlation with exposure to exogenous styrene oxide that was present in roughly 1,000-fold lower concentration in the ambient air as compared to the styrene (Rappaport *et al.* 1996). This is in line with the observation that recombinant cell lines that express human mEH at a level comparable to that observed in human liver can tolerate up to a definable threshold a high concentration of styrene, without showing detectable signs of genotoxic damage (Herrero *et al.* 1997).

A number of possible endogenous functions have been attributed to mEH, the significance of which is not perfectly clear:

(1) The formation of $16\alpha,17\alpha$-epoxides first from oestradiol (Breuer and Knuppen 1961) and later from androsterone was reported (Disse *et al.* 1980), and it was found thereafter that these were good substrates for mEH (Vogel-Bindel *et al.* 1982). Furthermore, the adrenal gland was reported to contain exceptionally high amounts of mEH (Papadopoulos *et al.* 1994). Very recently, high expression of mEH was observed in the corpus luteum, and a decrease of oestradiol production was observed on treatment with the mEH inhibitor 1,1,1-trichloro-2,3-propene oxide under conditions where the aromatase activity remained unaffected (Hattori *et al.* 2000). Finally, mEH has been identified as one component of a so-called anti-oestrogen binding site (AEBS) (Mesange *et al.* 1998).
(2) Similarly, mEH has been proposed to be a component of the vitamin K epoxide reductase (VKOR) (Guenthner *et al.* 1998).
(3) A highly controversial issue that splits the EH community in believers and disbelievers is its possible role in the membrane transport of bile acids (Alves *et al.* 1993), which is discussed in the context of subcellular localisation of the enzyme in the next section.

SUBCELLULAR LOCALISATION

The major location of mEH within the cell is the ER membrane. The above-mentioned bile acid carrier function implies a location of mEH on the plasma membrane that has, indeed, been claimed in a few reports (Alves *et al.* 1993; von Dippe *et al.* 1993, 1996; Zhu *et al.* 1999). However, many attempts by other researchers, including ourselves, to reproduce these findings have reportedly failed (Waechter *et al.* 1982; Craft *et al.* 1990; Honscha *et al.* 1995; Friedberg *et al.* 1996; Holler *et al.* 1997), and we, therefore, find the present proof for the mEH being a genuine plasma membrane constituent not unambiguously conclusive. The topology of mEH within the ER membrane has been a subject of intense research for some time (Porter *et al.* 1986; Craft *et al.* 1990). It finally turned out that mEH is attached to the membrane with a single *N*-terminal anchor (Friedberg *et al.* 1994) and that at least the mammalian enzyme is oriented towards the cell cytosol (Holler *et al.* 1997), as are the CYP enzymes, and not towards the ER lumen, as are the UGTs (glucuronosyl transferases; Figure 12.8).

TISSUE DISTRIBUTION

Early studies have identified EH activity in almost every tissue that was analysed, which led to the statement that mEH is apparently ubiquitously expressed in rat organs (Oesch *et al.* 1977a). These findings, referring to the distribution of enzyme activity among organs, were later refined to tissue compartments and cell types, using a variety of techniques such as immunohistochemistry or cell sorting (Bentley *et al.* 1979; Wolf *et al.* 1984; Guenthner and Karnezis 1986; Steinberg *et al.* 1987; Bogdanffy 1990; Farin and Omiecinski 1993; Backman *et al.* 1999; Hattori *et al.* 2000; Kessler *et al.* 2000). Indeed, a large variety of cell types express mEH to appreciable levels, but in many others mEH expression is below the level of detection. This should be borne in mind when talking about the apparently ubiquitous mEH expression. Expression of mEH is usually highest in the liver, followed by testes, adrenal gland, lung, kidney and intestine (in the mouse, interestingly, higher in testis than in the liver) (Thomas *et al.* 1990; Hammock *et al.* 1997). However, this order may vary from species to species. In humans, for instance, a particularly high mEH content has been reported for the adrenal gland (Papadopoulos *et al.* 1994). In rats, three different transcripts for mEH have been described, that are divergent in their 5'-non-coding region but code for identical polypeptides (Honscha *et al.* 1991). It was concluded, that at least three alternative non-coding first exons exist in the rat mEH gene which would allow for three independant regions of transcriptional control. Five such non-coding first exons

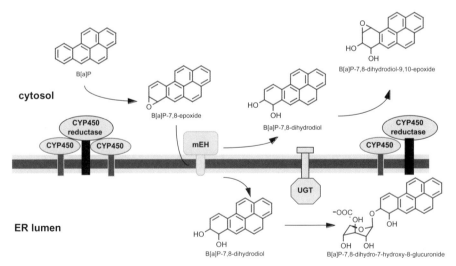

Figure 12.8 Membrane topology of mEH in comparison to other ER-resident xenobiotic-metabolising enzymes.
Part of the metabolism of benzo[a]pyrene is incorporated into the figure as an example of the metabolic cooperation of the different enzymes.

have recently been described for the human mEH gene and their tissue-specific expression was reported (Gaedigk *et al.* 1997). Thus, one reason for the wide distribution of mEH in mammalian tissues is obviously due to the existence of multiple promoters in the mEH gene.

SPECIES DIFFERENCES

Marked species differences in the expression level of mEH between mammals exist. Under certain circumstances, these may be crucial for differential toxic effects in different species (Oesch *et al.* 1977b). Of the classic laboratory animals, the mouse is low in mEH, with about 0.1–0.2% of the microsomal protein being mEH. In contrast, human liver mEH constitutes well above 1% of the microsomal protein, while in the rat it is between 0.5% and 1%. This is possibly one reason why high doses of styrene, above a threshold of about 300 ppm in the ambient air, led to a strong increase in styrene oxide blood levels in mice while this was not observed with rats under similar conditions (Kessler *et al.* 1992).

INDUCIBILITY BY FOREIGN COMPOUNDS

Microsomal EH is—despite its already high concentration in liver—inducible by a large variety of different compounds in laboratory animals. In view of the above-described enzymic mechanism this is beneficial even if the enzyme is already in apparent excess over its substrate, because the steady-state level of its substrates is, in any case, inversely correlated with the mEH concentration. 2-Acetylaminofluorene (Astrom and DePierre 1981) and *trans*-stilbene oxide (Schmassmann and Oesch 1978) are among the most potent inducers, leading to up to a 7-fold increase of enzyme activity in rat liver. Other inducers include phenobarbital (Oesch *et al.* 1971a), imidazole derivatives (Kim *et al.* 1995), lead acetate (Sheehan *et al.* 1991) and peroxisome proliferators (Oesch and Arand 1994). The existence of several independent promoters in the mEH gene (see above) that are differentially regulated complicates the analysis of transcriptional regulation since the different inducers will most likely act on different transcriptional units of the gene.

INHIBITORS

The first mEH inhibitors that were identified (Oesch *et al.* 1971b) can, on the basis of current understanding, all be regarded as substrates with a low k_2, i.e. a low K_M and a low V_{max} (see above). Of these, 1,1,1-trichloro-2,3-propene oxide (TCPO) (Figure 12.9) has been the most widely used. Later, valpromide was identified as the first non-substrate inhibitor of mEH, on the basis of its interference with the carbamazepine metabolism (see above) (Meijer *et al.* 1984). The amide group seems to mimic the epoxide ring, most likely in that the amide carbonyl hydrogen bonds to the tyrosines while the amino group hydrogen bonds to the nucleophilic aspartate. The hydrophobic side chain of valpromide is obviously well suited to fit the substrate access tunnel. Obviously, valpromide has a much lower K_D with mEH than most of the substrates, thus making up for the lack of covalent binding to the enzyme. The advantage of using

1,1,1-Trichloro-2,3-propene oxide (TCPO)   Valpromide

Figure 12.9 Prototypes of mEH inhibitors.

a non-substrate inhibitor is obvious: as long as the inhibitor is itself a substrate for the enzyme it can be consumed over time and thus lose its inhibitory potency. Indeed, TCPO is no safe inhibitor if used over a longer period of time in the presence of substantial amounts of mEH.

GENETIC POLYMORPHISMS

Two genetic polymorphisms affecting the primary sequence of the mEH protein in humans have been described (Hassett *et al.* 1994), and quite a number of epidemiological studies has monitored the prevalence of the different alleles in different subgroups of the population, especially with respect to disease susceptibility (McGlynn *et al.* 1995; Lancaster *et al.* 1996; Benhamou *et al.* 1998). The two polymorphisms represent single amino acid exchanges, namely $Tyr_{113}His$ or $Arg_{139}His$, none of which apparently affects the enzyme kinetics or substrate selectivity. This is not surprising, since both polymorphic sites lie on the surface of the protein (see Figure 12.3), far away from the catalytic centre. However, both polymorphisms appear to moderately affect the protein stability, resulting—at best—in a 2-fold difference in the enzyme tissue concentration. This is far from the observed maximum interindividual difference in enzymic activity reported in human liver (Mertes *et al.* 1985; Hassett *et al.* 1997), and thus is unlikely to be a major contributor to this variability. Likewise, polymorphisms (single nucleotide polymorphisms; SNPs) in the promoter region of the human mEH gene have been identified (Raaka *et al.* 1998) that may have a minor influence ($\pm30\%$) on the transcription efficacy of the gene, which is, at best, a minor contribution to overall variability. However, since only one of the at least five promoter regions of the human gene (see above) has been addressed in this study, there is a good chance that more relevant polymorphisms in the regulatory regions of the gene await detection.

Soluble epoxide hydrolase

STRUCTURAL CHARACTERISTICS

Mammalian sEH is a homodimer in solution, with a subunit molecular mass of 62 kDa, corresponding to 554 amino acid residues (Beetham *et al.* 1993; Grant *et al.* 1993; Knehr *et al.* 1993). The generic EH α/β hydrolase fold is built by the C-terminal 320 amino acid residues while the N-terminal 220 amino acid residues comprise a second domain, harbouring a potential second catalytic site, the function of which is as yet unknown. From X-ray analysis of mouse sEH (Argiriadi *et al.* 1999), one important function deduced for the *N*-terminal domain was to stabilise the overall structure, since the *N*-terminal domain of subunit A largely interacts with the *C*-terminal domain of subunit B and vice versa.

METABOLIC FUNCTION

The soluble epoxide hydrolase complements the mEH in the metabolism of xenobiotic epoxides in that it is capable of hydrolysing 1,2-*trans*-substituted oxiranes (Ota and Hammock 1980). Typical examples of this group of compounds are *trans*-stilbene oxide and *trans*-ethylstyrene oxide. Treatment with the latter compound leads to sister chromatide exchange in human lymphocytes (Krämer *et al.* 1991). In these cells the individual susceptibility to this is negatively correlated with the expression level of sEH. As a general rule, bulky substrates are not accepted, but a limited number of PAH epoxides are converted to diols by sEH (Figure 12.10) (Oesch and Golan 1980). Nevertheless, the clear domain of sEH are epoxides derived from fatty acids. Substrates within this group range from arachidonic acid epoxides (Zeldin *et al.* 1993) over

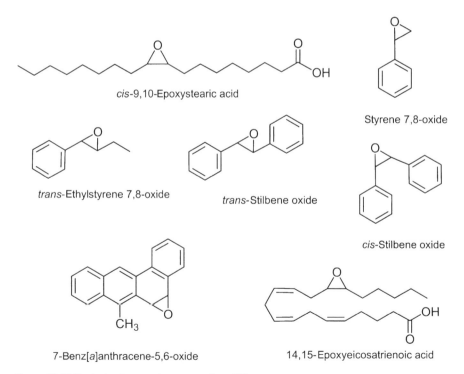

cis-9,10-Epoxystearic acid

Styrene 7,8-oxide

trans-Ethylstyrene 7,8-oxide

trans-Stilbene oxide

cis-Stilbene oxide

7-Benz[*a*]anthracene-5,6-oxide

14,15-Epoxyeicosatrienoic acid

Figure 12.10 Typical substrates for mammalian sEH.

leukotoxin (Moghaddam *et al.* 1997) to diepoxides (Nourooz-Zadeh *et al.* 1992). The observation that some of these substrates and/or their corresponding diols apparently have (patho)physiological functions points towards a major endogenous role for sEH. In line with this, the sEH knockout mouse has a reduced blood pressure, indicating a modulating function of sEH in blood pressure regulation (Sinal *et al.* 2000).

A particularly interesting aspect of sEH is the potential second catalytic centre. Sequence similarity analysis has grouped the *N*-terminal domain into a large family of hydrolytic enzymes, including dehalogenases (different from the haloalkane dehalogenase) and phosphatases (Koonin and Tatusov 1994). The fact that a putative substrate binding cavity as well as the catalytic residues are conserved in the *N*-terminal sEH domain suggests that it probably has a second catalytic activity.

SUBCELLULAR LOCALISATION

As indicated by its former name cytosolic EH, sEH has first been identified in the cell cytosol (Gill *et al.* 1974). Later, a similar enzyme was found in the matrix of peroxisomes (Waechter *et al.* 1983), organelles that are separated from the cytosol by a single membrane and harbour a substantial number of different metabolic pathways, such as formation and degradation of long-chain and branched fatty acids or the degradation of urate (Lazarow and Fujiki 1985). Several of these pathways lead to the stoichiometric formation of hydrogen peroxide as a by-product, hence the name peroxisome. The presence of sEH in peroxisomes may protect the cell from secondary oxidation products generated by hydrogen peroxide and by lipid peroxidation initiated by it, but the true function is as yet unclear. Comparison of the biochemical characteristics of cytosolic and peroxisomal sEH did not reveal any significant difference between the two (Meijer and DePierre 1988; Chang and Gill 1991). Sequence analysis of the *C*-terminal of sEH appeared to explain the situation in that an imperfect carboxy terminal peroxisome targeting signal (PTS I) was identified in the rat sEH sequence (Arand *et al.* 1991), and it was concluded that the lack of perfection resulted in a reduced translocation efficacy into the peroxisomal matrix, thus leading to an unusual bi-compartmental localisation of the same enzyme. This interpretation has recently been challenged by the observation that the native sEH does not translocate into peroxisomes after recombinant expression in mammalian cells (Mullen *et al.* 1999) while a mutant with an $Ile_{554}Leu$ substitution, that restores the perfect PTS I, is exclusively localised in peroxisomes under otherwise identical conditions.

TISSUE DISTRIBUTION

The major location of sEH in most species is the liver, followed by kidney, heart, brain, lung, testes, spleen and lymphocytes (Gill and Hammock 1980; Seidegard *et al.* 1984; Schladt *et al.* 1986). At least in the organs with higher expression levels, this seems to correlate with the expression of peroxisome proliferator-activated receptor α (PPARα) (Issemann and Green 1990), the transcription factor important for the regulation of sEH expression (see below).

SPECIES DIFFERENCES

Species differences in sEH are much more pronounced than they are for mEH (Hammock *et al.* 1997). A 100-fold difference exists between the expression level of sEH in rat and mouse liver. Rat is particularly low in sEH. The sEH expression in human liver is intermediate, about 10-fold below the mouse expression level and 10-fold above the rat expression level. This suggests a significant difference between the above species with respect to the adverse effects of *trans*-substituted epoxides.

INDUCIBILITY BY FOREIGN COMPOUNDS

In contrast to mEH, sEH is not inducible by administration of classical inducers of xenobiotic metabolising enzymes. The only group of compounds known so far that enhances sEH expression are the peroxisome proliferators (Waechter *et al.* 1984). It was shown that in rodents, sEH expression is co-ordinately regulated with that of the enzymes involved in peroxisomal β-oxidation of fatty acids (Schladt *et al.* 1987), which are transcriptionally regulated by the PPARα (Issemann and Green 1990). A functionally active PPARα-responsive element that mediates this effect has, indeed, been identified in the rat sEH gene (Hinz W, Oesch F, and Arand M, unpublished observations).

INHIBITORS

The first established inhibitors of sEH were chalcone oxide derivatives (Mullin and Hammock 1982), the 4-fluorochalcone oxide possibly being the most important representative (Figure 12.11). Like the epoxide-derived inhibitors for mEH, these compounds are essentially low k_2 substrates for sEH (Morisseau *et al.* 1998). Very recently, alkyl urea derivatives have evolved as a novel, especially potent group of non-substrate competitive inhibitors of sEH (Morisseau *et al.* 1999b). X-ray analysis of the enzyme-inhibitor complex indicates a molecular interaction that resembles the one speculated about above for the mEH valpromide interaction (Argiriadi *et al.* 2000). The urea carbonyl seems to hydrogen bond to the active site tyrosines, while a nitrogen-bound proton hydrogen bonds to the catalytic nucleophile. These class of inhibitors possess a surprisingly low K_i—in view of the fact that they do not covalently bind to the enzyme—that is, in the nanomolar range. Thus, these compounds represent promising candidates for a possible therapeutic interaction with sEH, e.g. in prevention of multiple organ failure (see above).

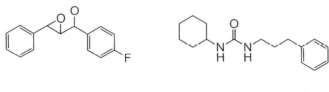

4-Fluorochalcone oxide   *N*-Cyclohexyl-*N*'-(3-phenylpropyl)urea

Figure 12.11 Prototypes of sEH inhibitors.

GENETIC POLYMORPHISMS

Very recently, two polymorphisms affecting the protein sequence have been described for sEH (Sandberg *et al.* 2000). Of these, an Arg_{287} Gln exchange that results in a surface modification of the α/β hydrolase fold domain at the dimerisation interphase seemed to have little effect on enzymic activity and protein stability, while an insertion of an additional arginine in position 402/403 seemed to decrease both, specific enzymic activity as well as protein stability. The latter change that affects a loop structure in the lid domain was proposed by the authors to slightly influence the geometry of the active centre of the enzyme. The exact prevalence of this polymorphism as well as its possible consequences for human health remain to be established.

References

Alves C, von Dippe P, Amoui M and Levy D (1993) Bile acid transport into hepatocyte smooth endoplasmic reticulum vesicles is mediated by microsomal epoxide hydrolase, a membrane protein exhibiting two distinct topological orientations. *Journal of Biological Chemistry*, **268**, 20148–20155.

Arand M, Knehr M, Thomas H, Zeller HD and Oesch F (1991) An impaired peroxisomal targeting sequence leading to an unusual bicompartmental distribution of cytosolic epoxide hydrolase. *FEBS Letters*, **294** 19–22.

Arand M, Grant DF, Beetham JK, Friedberg T, Oesch F and Hammock BD (1994) Sequence similarity of mammalian epoxide hydrolases to the bacterial haloalkane dehalogenase and other related proteins—Implication for the potential catalytic mechanism of enzymic epoxide hydrolysis. *FEBS Letters*, **338**, 251–256.

Arand M, Wagner H and Oesch F (1996) Asp[333], Asp[495], and His[523] form the catalytic triad of rat soluble epoxide hydrolase. *Journal of Biological Chemistry*, **271**, 4223–4229.

Arand M, Hemmer H, Dürk H, Baratti J, Archelas A, Furstoss R and Oesch F (1999a) Cloning and molecular characterisation of a soluble epoxide hydrolase from Aspergillus niger that is related to mammalian microsomal epoxide hydrolase. *Biochemical Journal*, **344**, 273–280.

Arand M, Müller F, Mecky A, Hinz W, Urban P, Pompon D, Kellner R and Oesch F (1999b) Catalytic triad of microsomal epoxide hydrolase: replacement of Glu404 with Asp leads to a strongly increased turnover rate. *Biochemical Journal*, **337**, 37–43.

Argiriadi MA, Morisseau C, Hammock BD and Christianson DW (1999) Detoxification of environmental mutagens and carcinogens: structure, mechanism, and evolution of liver epoxide hydrolase. *Proceedings of the National Academy of Sciences USA*, **96**, 10637–10642.

Argiriadi MA, Morisseau C, Goodrow MH, Dowdy DL, Hammock BD and Christianson DW (2000) Binding of alkylurea inhibitors to epoxide hydrolase implicates active site tyrosines in substrate activation. *Journal of Biological Chemistry*, **275**, 15265–15270.

Armstrong RN (1987) Enzyme-catalyzed detoxication reactions: Mechanisms and stereochemistry. *CRC Critical Reviews in Biochemistry*, **22**, 39–88.

Armstrong RN (1999) Kinetic and chemical mechanism of epoxide hydrolase. *Drug Metabolism Reviews*, **3**, 71–86.

Astrom A and DePierre JW (1981) Characterisation of the induction of drug-metabolising enzymes by 2-acetylaminofluorene. *Biochimica et Biophysica Acta*, **673**, 225–233.

Backman JT, Siegle I, Zanger UM and Fritz P (1999) Immunohistochemical detection of microsomal epoxide hydrolase in human synovial tissue. *Histochemical Journal*, **31**, 645–649.

Beetham JK, Tian TG and Hammock BD (1993) cDNA cloning and expression of a soluble epoxide hydrolase from human liver. *Archives of Biochemistry and Biophysics*, **305**, 197–201.

Beetham JK, Grant D, Arand M, Garbarino J, Kiyosue T, Pinot F, Oesch F, Belknap WR, Shinozaki K and Hammock BD (1995) Gene evolution of epoxide hydrolases and recommended nomenclature. *DNA and Cell Biology*, **14**, 61–71.

Benhamou S, Reinikainen M, Bouchardy C, Dayer P and Hirvonen A (1998) Association between lung cancer and microsomal epoxide hydrolase genotypes. *Cancer Research*, **58**, 5291–5293.

Bentley P, Oesch F and Glatt HR (1977) Dual role of epoxide hydratase in both activation and inactivation. *Archives of Toxicology*, **39**, 65–75.

Bentley P, Waechter F, Oesch F and Staubli W (1979) Immunochemical localisation of epoxide hydratase in rat liver: effects of 2-acetylaminofluorene. *Biochemical Biophysical Research Communications*, **91**, 1101–1108.

Bogdanffy MS (1990) Biotransformation enzymes in the rodent nasal mucosa: the value of a histochemical approach. *Environmental Health Perspectives*, **85**, 177–186.

Breuer H and Knuppen R (1961) The formation of $16\alpha,17\alpha$-epoxy-oestratriene-3-ol by rat liver tissue. *Biochimica et Biophysica Acta*, **49**, 620–621.

Chang C and Gill SS (1991) Purification and characterisation of an epoxide hydrolase from the peroxisomal fraction of mouse liver. *Archives of Biochemistry and Biophysics*, **285**, 276–284.

Craft JA, Baird S, Lamont M and Burchell B (1990) Membrane topology of epoxide hydrolase. *Biochimica et Biophysica Acta*, **1046**, 32–39.

Disse B, Siekmann L and Breuer H (1980) Biosynthesis of $16\alpha,17\alpha$-epoxy-4-androstene-3-one in rat liver microsomes. *Acta Endocrinologica*, **95**, 58–66.

DuBois GC, Appella E, Levin W, Lu AY and Jerina DM (1978) Hepatic microsomal epoxide hydrase. Involvement of a histidine at the active site suggests a nucleophilic mechanism. *Journal of Biological Chemistry*, **253**, 2932–2939.

Eichelbaum M, Kothe KW, Hoffman F and von Unruh GE (1979) Kinetics and metabolism of carbamazepine during combined antiepileptic drug therapy. *Clinical Pharmacology and Therapeutics*, **26**, 366–371.

Etter H-U, Richter C, Ohta Y, Winterhalter KH, Sasabe H and Kawato S (1991) Rotation and interaction with epoxide hydrase of cytochrome P-450 in proteoliposomes. *Journal of Biological Chemistry*, **266**, 18600–18605.

Farin FM and Omiecinski CJ (1993) Regiospecific expression of cytochrome P-450s and microsomal epoxide hydrolase in human brain tissue. *Journal of Toxicology and Environmental Health*, **40**, 317–335.

Franken SM, Rozeboom HJ, Kalk KH and Dijkstra BW (1991) Crystal structure of haloalkane dehalogenase: an enzyme to detoxify halogenated alkanes. *EMBO Journal*, **10**, 1297–1302.

Friedberg T, Holler R, Löllmann B, Arand M and Oesch F (1996) The catalytic activity of the endoplasmic reticulum-resident protein microsomal epoxide hydrolase towards carcinogens is retained on inversion of its membrane topology. *Biochemical Journal*, **319**, 131–136.

Friedberg T, Löllmann B, Becker R, Holler R and Oesch F (1994) The microsomal epoxide hydrolase has a single membrane signal anchor sequence which is dispensable for the catalytic activity of this protein. *Biochemical Journal*, **303**, 967–972.

Gaedigk A, Leeder JS and Grant DM (1997) Tissue-specific expression and alternative splicing of human microsomal epoxide hydrolase. *DNA and Cell Biology*, **16**, 1257–1266.

Gill SS and Hammock BD (1980) Distribution and properties of a mammalian soluble epoxide hydrase. *Biochemical Pharmacology*, **29**, 389–395.

Gill SS, Hammock BD and Casida JE (1974) Mammalian metabolism and environmental degradation of the juvenoid, 1-(4'-ethylphenoxy)-3,7-dimethyl-6,7-epoxy-*trans*-2-octene. *Journal of Agricultural and Food Chemistry*, **22**, 386–395.

Glatt H, Jung R and Oesch F (1983) Bacterial mutagenicity investigation of epoxides: drugs, drug metabolites, steroids and pesticides. *Mutation Research*, **111**, 99–118.

Gonzalez FJ and Kasper CB (1981) Cloning of epoxide hydratase complementary DNA. *Journal of Biological Chemistry*, **256**, 4697–4700.

Grant DF, Storms DH and Hammock BD (1993) Molecular cloning and expression of murine liver soluble epoxide hydrolase. *Journal of Biological Chemistry*, **268**, 17628–17633.

Guengerich FP and Davidson NK (1982) Interaction of epoxide hydrolase with itself and other microsomal proteins. *Archives of Biochemistry and Biophysics*, **215**, 462–477.

Guenthner TM and Karnezis TA (1986) Immunochemical characterisation of human lung epoxide hydrolases. *Journal of Biochemical Toxicology*, **1**, 67–81.

Guenthner TM, Hammock BD, Vogel U and Oesch F (1981) Cytosolic and microsomal epoxide hydrolases are immunologically distinguishable from each other in the rat and mouse. *Journal of Biological Chemistry*, **256**, 3163–3166.

Guenthner TM, Cai D and Wallin R (1998) Co-purification of microsomal epoxide hydrolase with the warfarin-sensitive vitamin K1 oxide reductase of the vitamin K cycle. *Biochemical Pharmacology*, **55**, 169–175.

Haeggstrom JZ, Wetterholm A, Vallee BL and Samuelsson B (1990) Leukotriene A4 hydrolase: an epoxide hydrolase with peptidase activity. *Biochemical Biophysical Research Communications*, **173**, 431–437.

Hammock BD, Storms DH and Grant DF, 1997, Epoxide hydrolases. In *Biotransformation*, Guengerich FP (ed.), Elsevier Sciences, New York, pp. 283–305.

Hassett C, Aicher L, Sidhu JS and Omiecinski CJ (1994) Human microsomal epoxide hydrolase: genetic polymorphism and functional expression in vitro of amino acid variants. *Human Molecular Genetics*, **3**, 421–428.

Hassett C, Lin J, Carty CL, Laurenzana EM and Omiecinski CJ (1997) Human hepatic microsomal epoxide hydrolase: comparative analysis of polymorphic expression. *Archives of Biochemistry and Biophysics*, **337**, 275–283.

Hattori N, Fujiwara H, Maeda M, Fujii S and Ueda M (2000) Epoxide hydrolase affects oestrogen production in the human ovary. *Endocrinology*, **141**, 3353–3365.

Hayes JD and Pulford DJ (1995) The glutathione S-transferase supergene family: regulation of GST and the contribution of the isoenzymes to cancer chemoprotection and drug resistance. *Critical Reviews in Biochemistry and Molecular Biology*, **30**, 445–600.

Heinemann FS and Ozols J (1984) The covalent structure of microsomal epoxide hydrolase. II. The complete amino acid sequence. *Journal of Biological Chemistry*, **259**, 797–804.

Herrero ME, Arand M, Hengstler JG and Oesch F (1997) Recombinant expression of human microsomal epoxide hydrolase protects V79 Chinese hamster cells from styrene oxide—but not from ethylene oxide-induced DNA strand breaks. *Environmental Molecular Mutagenesis*, **30**, 429–439.

Holder G, Yagi H, Dansette P, Jerina DM, Levin W, Lu AY and Conney AH (1974) Effects of inducers and epoxide hydrase on the metabolism of benzo(a)pyrene by liver microsomes and a reconstituted system: analysis by high pressure liquid chromatography. *Proceedings of the National Academy of Sciences, USA* **71**, 4356–4360.

Holler R, Arand M, Mecky A, Oesch F and Friedberg T (1997) The membrane anchor of microsomal epoxide hydrolase from human, rat, and rabbit displays an unexpected membrane topology. *Biochemical Biophysical Research Communications*, **236**, 754–759.

Honscha W, Oesch F and Friedberg T (1991) Tissue-specific expression and differential inducibility of several microsomal epoxide hydrolase mRNAs which are formed by alternative splicing. *Archives of Biochemistry and Biophysics*, **287**, 380–385.

Honscha W, Platte HD, Oesch F and Friedberg T (1995) Relationship between the microsomal epoxide hydrolase and the hepatocellular transport of bile acids and xenobiotics. *Biochemical Journal*, **311**, 975–979.

Issemann I and Green S (1990) Activation of a member of the steroid hormone receptor superfamily by peroxisome proliferators. *Nature*, **347**, 645–650.

Janssen DB, Fries F, van der Ploeg J, Kazemier B, Terpstra P and Witholt B (1989) Cloning of 1,2-dichloroethane degradation genes of *Xanthobacter autotrophicus* GJ10 and expression and sequencing of the *dhlA* gene. *Journal of Bacteriology*, **171**, 6791–6799.

Jenkins Sumner S and Fennell TR (1994) Review on the metabolic fate of styrene. *Critical Reviews in Toxicology*, **24**, S11–S33.

Jernstrom B, Seidel A, Funk M, Oesch F and Mannervik B (1992) Glutathione conjugation of trans-3,4-dihydroxy 1,2-epoxy 1,2,3,4-tetrahydrobenzo[c]phenanthrene isomers by human glutathione transferases. *Carcinogenesis*, **13**, 1549–1555.

Kessler W, Jiang X and Filser JG (1992) Pharmakokinetik von Styrol-7,8-Oxid bei Maus und Ratte. In *Arbeitsmedizinische Aspekte der Arbeits(-zeit)organisation—Skeletterkrankungen und Beruf—Arbeitsmedizinisches Kolloquium der gewerblichen Berufsgenossenschaften*, Kreutz R and Piekarski C (eds), Gentner Verlag, Stuttgart, pp. 622–626.

Kessler R, Hamou MF, Albertoni M, de Tribolet N, Arand M and Van Meir EG (2000)

Identification of the putative brain tumor antigen BF7/GE2 as the (de)toxifying enzyme microsomal epoxide hydrolase. *Cancer Research*, **60**, 1403–1409.

Kim SG, Cho JY and Jung KH (1995) Differential expression of rat microsomal epoxide hydrolase gene by imidazole and triazole antimycotic agents. *Drug Metabolism and Disposition*, **23**, 460–464.

Kiyosue T, Beetham JK, Pinot F, Hammock BD, Yamaguchi-Shinozaki K and Shinozaki K (1994) Characterisation of an *Arabidopsis* cDNA for a soluble epoxide hydrolase gene that is inducible by auxin and water stress. *Plant Journal*, **6**, 259–269.

Knehr M, Thomas H, Arand M, Gebel T, Zeller HD and Oesch F (1993) Isolation and characterisation of a cDNA encoding rat liver cytosolic epoxide hydrolase and its functional expression in Escherichia-Coli. *Journal of Biological Chemistry*, **268**, 17623–17627.

Koonin EV and Tatusov RL (1994) Computer analysis of bacterial haloacid dehalogenases defines a large superfamily of hydrolases with diverse specificity—application of an iterative approach to database search. *Journal of Molecular Biology*, **244**, 125–132.

Krämer A, Frank H, Setiabudi F, Oesch F and Glatt H (1991) Influence of the level of cytosolic epoxide hydrolase on the induction of sister chromatid exchanges by trans-beta-ethylstyrene 7,8-oxide in human lymphocytes. *Biochemical Pharmacology*, **42**, 2147–2152.

Lacourciere GM and Armstrong RN (1993) The catalytic mechanism of microsomal epoxide hydrolase involves an ester intermediate. *Journal of the American Chemical Society*, **115**, 10466–10467.

Lacourciere GM and Armstrong RN (1994) Microsomal and soluble epoxide hydrolases are members of the same family of C–X bond hydrolase enzymes. *Chemical Research in Toxicology*, **7**, 121–124.

Lancaster JM, Brownlee HA, Bell DA, Futreal PA, Marks JR, Berchuck A, Wiseman RW and Taylor JA (1996) Microsomal epoxide hydrolase polymorphism as a risk factor for ovarian cancer. *Molecular Carcinogenesis*, **17**, 160–162.

Laughlin LT, Tzeng H-F, Lin S and Armstrong RN (1998) Mechanism of microsomal epoxide hydrolase. Semifunctional site-specific mutants affecting the alkylation half-reaction. *Biochemistry*, **37**, 2897–2904.

Lazarow PB and Fujiki Y (1985) Biogenesis of peroxisomes. *Annual Review of Cell Biology*, **1**, 489–530.

Levin W, Michaud DP, Thomas PE and Jerina DM (1983) Distinct rat hepatic microsomal epoxide hydrolases catalyze the hydration of cholesterol 5,6 alpha-oxide and certain xenobiotic alkene and arene oxides. *Archives of Biochemistry and Biophysics*, **220**, 485–494.

Mackenzie PI, Owens IS, Burchell B, Bock KW, Bairoch A, Belanger A, Fournel-Gigleux S, Green M, Hum DW, Iyanagi T, Lancet D, Louisot P, Magdalou J, Chowdhury JR, Ritter JK, Schachter H, Tephly TR, Tipton KF and Nebert DW (1997) The UDP glycosyltransferase gene superfamily: recommended nomenclature update based on evolutionary divergence. *Pharmacogenetics*, **7**, 255–269.

McGlynn KA, Rosvold EA, Lustbader ED, Hu Y, Clapper ML, Zhou T, Wild CP, Xia XL, Baffoe-Bonnie A, Ofori-Adjei D, Chen GC, London WT, Shen FM and Buetow K (1995) Susceptibility to hepatocellular carcinoma is associated with genetic variation in the enzymic detoxification of aflatoxin B1. *Proceedings of the National Academy of Sciences USA*, **92**, 2384–2387.

Meijer J and DePierre JW (1988) Immunoaffinity purification and comparison of epoxide hydrolases from liver cytosol and peroxisomes of untreated and clofibrate-treated mice. *Archives of Toxicology*, **Suppl. 12**, 283–287.

Meijer JW, Binnie CD, Debets RM, van Parys JA and de Beer-Pawlikowski NK (1984) Possible hazard of valpromide-carbamazepine combination therapy in epilepsy. *Lancet*, **8380**, 802.

Mertes I, Fleischmann R and Oesch F (1985) Interindividual variations in the activities of cytosolic and microsomal epoxide hydrolase in human liver. *Carcinogenesis*, **6**, 219–223.

Mesange F, Sebbar M, Kedjouar B, Capdevielle J, Guillemot JC, Ferrara P, Bayard F, Delarue F, Faye JC and Poirot M (1998) Microsomal epoxide hydrolase of rat liver is a subunit of the anti-ooestrogen-binding site. *Biochemical Journal*, **334**, 107–112.

Miller EC and Miller JA (1981) Mechanisms of chemical carcinogenesis. *Cancer*, **47**, 1055–1064.

Miyata M, Kudo G, Lee YH, Yang TJ, Gelboin HV, Fernandez-Salguero P, Kimura S and Gonzalez FJ (1999) Targeted disruption of the microsomal epoxide hydrolase gene. Microsomal epoxide hydrolase is required for the carcinogenic activity of 7,12-dimethylbenz[a]anthracene. *Journal of Biological Chemistry*, **274**, 23963–23968.

Moghaddam MF, Grant DF, Cheek JM, Greene JF, Williamson KC and Hammock BD (1997) Bioactivation of leukotoxins to their toxic diols by epoxide hydrolase. *Nature Medicine*, **3**, 562–566.

Morisseau C, Du G, Newman JW and Hammock BD (1998) Mechanism of mammalian soluble epoxide hydrolase inhibition by chalcone oxide derivatives. *Archives of Biochemistry and Biophysics*, **356**, 214–228.

Morisseau C, Archelas A, Guitton C, Faucher D, Furstoss R and Baratti JC (1999a) Purification and characterisation of a highly enantioselective epoxide hydrolase from Aspergillus niger. *European Journal of Biochemistry*, **263**, 386–395.

Morisseau C, Goodrow MH, Dowdy D, Zheng J, Greene JF, Sanborn JR and Hammock BD (1999b) Potent urea and carbamate inhibitors of soluble epoxide hydrolases. *Proceedings of the National Academy of Sciences, USA*, **96**, 8849–8854.

Mullen RT, Trelease RN, Duerk H, Arand M, Hammock BD, Oesch F and Grant DF (1999) Differential subcellular localisation of endogenous and transfected soluble epoxide hydrolase in mammalian cells: evidence for isozyme variants. *FEBS Letters*, **445**, 301–305.

Mullin CA and Hammock BD (1982) Chalcone oxides—Potent selective inhibitors of cytosolic epoxide hydrolase. *Archives of Biochemistry and Biophysics*, **216**, 423–439.

Nardini M, Ridder IS, Rozeboom HJ, Kalk KH, Rink R, Janssen DB and Dijkstra BW (1999) The x-ray structure of epoxide hydrolase from Agrobacterium radiobacter AD1. An enzyme to detoxify harmful epoxides. *Journal of Biological Chemistry*, **274**, 14579–14586.

Nelson DR, Koymans L, Kamataki T, Stegeman JJ, Feyereisen R, Waxman DJ, Waterman MR, Gotoh O, Coon MJ, Estabrook RW, Gunsalus IC and Nebert DW (1996) P450 superfamily: update on new sequences, gene mapping, accession numbers and nomenclature. *Pharmacogenetics*, **6**, 1–42.

Nourooz-Zadeh J, Uematsu T, Borhan B, Kurth MJ and Hammock BD (1992) Characterisation of the epoxide hydrolase-catalyzed hydration products from 9,10:12,13-diepoxy stearic esters. *Archives of Biochemistry and Biophysics*, **294**, 675–685.

Oesch F (1973) Mammalian epoxide hydrases: Inducible enzymes catalysing the inactivation of carcinogenic and cytotoxic metabolites derived from aromatic and olefinic compounds. *Xenobiotica*, **3**, 305–340.

Oesch F and Arand M, (1994), Induction of drug-metabolising enzymes by short/intermediate-term exposure to peroxisome proliferators: a synopsis. In *Peroxisome Proliferators: Unique Inducers of Drug-Metabolizing Enzymes*, Moody DE (ed.), CRC Press, Boca Raton, FL, pp. 161–174.

Oesch F and Bentley P (1976) Antibodies against homogeneous epoxide hydratase provide evidence for a single enzyme hydrating styrene oxide and benz(a)pyrene 4,5-oxide. *Nature*, **259**, 53–55.

Oesch F and Daly J (1972) Conversion of naphthalene to trans -naphthalene dihydrodiol: Evidence for the presence of a coupled aryl monooxygenase-epoxide hydrase system in hepatic microsomes. *Biochemical Biophysical Research Communications*, **46**, 1713–1720.

Oesch F and Golan M (1980) Specificity of mouse liver cytosolic epoxide hydrolase for K-region epoxides derived from polycyclic aromatic hydrocarbons. *Cancer Letters*, **9**, 169–175.

Oesch F, Jerina DM and Daly J (1971a) A radiometric assay for hepatic epoxide hydrase activity with 7-3H-styrene oxide. *Biochimica et Biophysica Acta*, **227**, 685–691.

Oesch F, Kaubisch N, Jerina DM and Daly J (1971b) Hepatic epoxide hydrase: Structure-activity relationships for substrates and inhibitors. *Biochemistry*, **10**, 4858-4866.

Oesch F, Glatt HR and Schmassmann HU (1977a) The apparent ubiquity of epoxide hydratase in rat organs. *Biochemical Pharmacology*, **26**, 603–607.

Oesch F, Raphael D, Schwind H and Glatt HR (1977b) Species differences in activating and inactivating enzymes related to the control of mutagenic metabolites. *Archives of Toxicology*, **39**, 97–108.

Oesch F, Timms CW, Walker CH, Guenthner TM, Sparrow A, Watabe T and Wolf CR (1984)

Existence of multiple forms of microsomal epoxide hydrolases with radically different substrate specificities. *Carcinogenesis*, **5**, 7–9.

Oesch F, Herrero ME, Hengstler JG, Lohmann M and Arand M (2000) Metabolic detoxification: implications for thresholds. *Toxicological Pathology*, **28**, 382–387

Ollis DL, Cheah E, Cygler M, Dijkstra B, Frolow F, Franken SM, Harel M, Remington SJ, Silman I, Schrag J, Sussman JL, Verschueren KHG and Goldman A (1992) The α/β hydrolase fold. *Protein Engineering*, **5**, 197–211.

Ota K and Hammock BD (1980) Cytosolic and microsomal epoxide hydrolases: Differential properties in mammalian liver. *Science*, **207**, 1479–1481.

Ozawa T, Hayakawa M, Kosaka K, Sugiyama S, Ogawa T, Yokoo K, Aoyama H and Izawa Y (1991) Leukotoxin, 9,10-epoxy-12-octadecenoate, as a burn toxin causing adult respiratory distress syndrome. *Advances in Prostaglandin, Thromboxane and Leukotoxin Research*, **21B**, 569–572.

Pace-Asciak CR and Lee WS (1989) Purification of hepoxilin epoxide hydrolase from rat liver. *Journal of Biological Chemistry*, **264**, 9310–9313.

Papadopoulos D, Jornvall H, Rydstrom J and Depierre JW (1994) Purification and initial characterisation of microsomal epoxide hydrolase from the human adrenal gland. *Biochimica et Biophysica Acta*, **1206**, 253–262.

Pinot F, Grant DF, Beetham JK, Parker AG, Borhan B, Landt S, Jones AD and Hammock BD (1995) Molecular and biochemical evidence for the involvement of the Asp-333- His-523 pair in the catalytic mechanism of soluble epoxide hydrolase. *Journal of Biological Chemistry*, **270**, 7968–7974.

Ponomarkov V, Cabral JR, Wahrendorf J and Galendo D (1984) A carcinogenicity study of styrene-7,8-oxide in rats. *Cancer Letters*, **24**, 95–101.

Porter TD, Beck TW and Kasper CB (1986) Complementary DNA and amino acid sequence of rat liver microsomal, xenobiotic epoxide hydrolase. *Archives of Biochemistry and Biophysics*, **248**, 121–129.

Pries F, Kingma J, Pentenga M, Vanpouderoyen G, Jeronimusstratingh CM, Bruins AP and Janssen DB (1994) Site-Directed mutagenesis and oxygen isotope incorporation studies of the nucleophilic aspartate of haloalkane dehalogenase. *Biochemistry*, **33**, 1242–1247.

Raaka S, Hassett C and Omiencinski CJ (1998) Human microsomal epoxide hydrolase: 5'-flanking region genetic polymorphisms. *Carcinogenesis*, **19**, 387–393.

Rappaport SM, Yeowell-O'Connell K, Bodell W, Yager JW and Symanski E (1996) An investigation of multiple biomarkers among workers exposed to styrene and styrene-7,8-oxide. *Cancer Research*, **56**, 5410–5416.

Rink R, Fennema M, Smids M, Dehmel U and Janssen DB (1997) Primary structure and catalytic mechanism of the epoxide hydrolase from *Agrobacterium radiobacter* AD1. *Journal of Biological Chemistry*, **272**, 14650–14657.

Rink R, Kingma J, Lutje Spelberg JH and Janssen DB (2000) Tyrosine residues serve as proton donor in the catalytic mechanism of epoxide hydrolase from Agrobacterium radiobacter. *Biochemistry*, **39**, 5600–5613.

Sandberg M, Hassett C, Adman ET, Meijer J and Omiecinski CJ (2000) Identification and functional characterisation of human soluble epoxide hydrolase genetic polymorphisms. *Journal of Biological Chemistry*, **275**, 28873–28881.

Schladt L, Woerner W, Setiabudi F and Oesch F (1986) Distribution and inducibility of cytosolic epoxide hydrolase in male Sprague-Dawley rats. *Biochemical Pharmacology*, **35**, 3309–3316.

Schladt L, Hartmann R, Timms C, Strolin-Benedetti M, Dostert P, Wörner W and Oesch F (1987) Concomitant induction of cytosolic but not microsomal epoxide hydrolase with peroxisomal β-oxidation by various hypolipidemic compounds. *Biochemical Pharmacology*, **36**, 345–351.

Schmassmann HU and Oesch F (1978) *Trans*-stilbene oxide: A selective inducer of rat liver epoxide hydratase. *Molecular Pharmacology*, **14**, 834–847.

Seidegard J, DePierre JW and Pero RW (1984) Measurement and characterisation of membrane-bound and soluble epoxide hydrolase activities in resting mononuclear leukocytes from human blood. *Cancer Research*, **44**, 3654–3660.

Sheehan JE, Pitot HC and Kasper CB (1991) Transcriptional regulation and localisation of the tissue-specific induction of epoxide hydrolase by lead acetate in rat kidney. *Journal of Biological Chemistry*, **266**, 5122–5127.

Sims P, Grover PL, Swaisland A, Pal K and Hewer A (1974) Metabolic activation of benzo(a)-pyrene proceeds by a diol-epoxide. *Nature*, **252**, 326–328.

Sinal CJ, Miyata M, Tohkin M, Nagata K, Bend JR and Gonzalez FJ (2000) Targeted disruption of soluble epoxide hydrolase reveals a role in blood pressure regulation. *Journal of Biological Chemistry*, in press.

Stapleton A, Beetham JK, Pinot F, Garbarino JE, Rockhold DR, Friedman M, Hammock BD and Belknap WR (1994) Cloning and expression of soluble epoxide hydrolase from potatoe. *Plant Journal*, **6**, 251–258.

Steinberg P, Lafranconi WM, Wolf CR, Waxman DJ, Oesch F and Friedberg T (1987) Xenobiotic metabolising enzymes are not restricted to parenchymal cells in rat liver. *Molecular Pharmacology*, **32**, 463–470.

Thomas H, Timms CW and Oesch F (1990), Epoxide hydrolases: Molecular properties, induction, polymorphisms and function. In *Frontiers in Biotransformation*, Ruckpaul K and Rein H (eds), Akademie-Verlag, Berlin, pp. 278–337.

Thompson JD, Gibson TJ, Plewniak F, Jeanmougin F and Higgins DG (1997) The CLUSTAL_X windows interface: flexible strategies for multiple sequence alignment aided by quality analysis tools. *Nucleic Acids Research*, **25**, 4876–4882.

Tzeng H-F, Laughlin LT, Lin S and Armstrong RN (1996) The catalytic mechanism of microsomal epoxide hydrolase involves reversible formation and rate-limiting hydrolysis of the alkyl-enzyme intermediate. *Journal of the American Chemical Society*, **118**, 9436–9437.

Tzeng H-F, Laughlin LT and Armstrong RN (1998) Semifunctional site-specific mutants affecting the hydrolytic half-reaction of microsomal epoxide hydrolase. *Biochemistry*, **37**, 2905–2911.

Verschueren KHG, Seljée F, Rozeboom HJ, Kalk KH and Dijkstra BW (1993) Crystallographic analysis of the catalytic mechanism of haloalkane dehalogenase. *Nature*, **363**, 693–698.

Vogel-Bindel U, Bentley P and Oesch F (1982) Endogenous role of microsomal epoxide hydrolase: ontogenesis, induction, inhibition, tissue distribution, immunological behaviour and purification of microsomal epoxide hydrolase with 16α,17α-epoxy-androstene3-one as substrate. *European Journal of Biochemistry*, **126**, 425–431.

von Dippe P, Amoui M, Alves C and Levy D (1993) Na(+)-dependent bile acid transport by hepatocytes is mediated by a protein similar to microsomal epoxide hydrolase. *American Journal of Physiology*, **264**, G528–534.

von Dippe P, Amoui M, Stellwagen RH and Levy D (1996) The functional expression of sodium-dependent bile acid transport in Madin-Darby canine kidney cells transfected with the cDNA for microsomal epoxide hydrolase. *Journal of Biological Chemistry*, **27**, 18176–18180.

Waechter F, Bentley P, Germann M, Oesch F and Stäubli W (1982) Immuno-electron-microscopic studies on the subcellular distribution of rat liver epoxide hydrolase and the effect of phenobarbitone and 2-acetamidofluorene treatment. *Biochemical Journal*, **202**, 677–686.

Waechter F, Bentley P, Bieri F, Stäubli W, Völkl A and Fahimi HD (1983) Epoxide hydrolase activity in isolated peroxisomes of mouse liver. *FEBS Letters*, **158**, 225–228.

Waechter F, Bieri F, Stäubli W and Bentley P (1984) Induction of cytosolic and microsomal epoxide hydrolases by the hypolipidaemic compound nafenopin in the mouse liver. *Biochemical Pharmacology*, **33**, 31–34.

Williams PA, Cosme J, Sridhar V, Johnson EF and McRee DE (2000) Mammalian microsomal cytochrome P450 monooxygenase: structural adaptations for membrane binding and functional diversity. *Molecular Cell*, **5**, 121–131.

Wilson R, Ainscough R, Anderson K, Baynes C, Berks M, Bonfield J, Burton J, Connell M, Copsey T, Cooper J, Coulson A, Craxton M, Dear S, Du Z, Durbin R, Favello A, Fulton L, Gardner A, Green P, Hawkins T, Hillier L, Jier M, Johnston L, Jones M, Kershaw J, Kirsten J, Laister N, Latreille P, Lightning J, Lloyd C, McMurray A, Mortimore B, O'Callaghan M, Parsons J, Percy C, Rifken L, Roopra A, Saunders D, Shownkeen R, Smaldon N, Smith A, Sonnhammer E, Staden R, Sulston J, Thierry-Mieg J, Thomas K, Vaudin M, Vaughan K, Waterston R, Watson A, Weinstock L, Wilkinson-Sproat J and Wohldman P (1994) 2.2 Mb of contiguous nucleotide sequence from chromosome III of C. elegans. *Nature*, **368**, 32–38.

Wojtasek H and Prestwich GD (1996) An insect juvenile hormone-specific epoxide hydrolase is related to vertebrate microsomal epoxide hydrolase. *Biochemical Biophysical Research Communications*, **220**, 323–329.

Wolf CR, Buchmann A, Friedberg T, Moll E, Kuhlmann WD, Kunz HW and Oesch F (1984) Dynamics of the localisation of drug metabolising enzymes in tissues and cells. *Biochemical Society Transactions*, **12**, 60–62.

Yamada T, Morisseau C, Maxwell JE, Argiriadi MA, Christianson DW and Hammock BD (2000) Biochemical evidence for the involvement of tyrosine in epoxide activation during the catalytic cycle of epoxide hydrolase. *Journal of Biological Chemistry*, **275**, 23082–23088.

Zeldin DC, Kobayashi J, Falck JR, Winder BS, Hammock BD, Snapper JR and Capdevila JH (1993) Regiofacial and enantiofacial selectivity of epoxyeicosatrienoic acid hydration by cytosolic epoxide hydrolase. *Journal of Biological Chemistry*, **268**, 6402–6407.

Zhu Q, von Dippe P, Xing W and Levy D (1999) Membrane topology and cell surface targeting of microsomal epoxide hydrolase. Evidence for multiple topological orientations. *Journal of Biological Chemistry*, **274**, 27898–27904.

Zou J, Hallberg BM, Bergfors T, Oesch F, Arand M, Mowbray SL and Jones TA (2000) Structure of Aspergillus niger epoxide hydrolase at 1.8 A resolution: implications for the structure and function of the mammalian microsomal class of epoxide hydrolases. *Structure with Folding & Design*, **8**, 111–122.

13 Methyltransferases

C. R. Creveling

NIDDK, NIH, USA

Introduction

The transfer of methyl groups from *S*-adenosyl-L-methionine (AdoMet) to methyl acceptor substrates is one of the most extensive reactions in nature. The consequences of the methyl-transfer reaction are attested to by the great diversity of the methyl-acceptor substrates found in biological systems. These methyl-transfer reactions almost exclusively utilise *S*-adenosyl-L-methionine (AdoMet) as the methyl-donor co-substrate. The transfer of the methyl group occurs to a sulphur-, nitrogen-, or oxygen-nucleophile. The conjugation reactions include, *O*-, *N*- and *S*-methyl-transferases. At present more that 100 methyltransferases have been identified. These methyltransferases catalyse the methylation of a diverse group of small molecules, either as drugs, xenobiotics and endogenous hormones and neurotransmitters, as well as macromolecules including proteins, lipids (Hirata, 1982), RNA and DNA. A series of functionally significant, genetic polymorphisms have been discovered for many methyltransferases, some of which have critical importance in clinical therapeutics. The structure and function of selected methyltransferases is described including catechol-*O*-methyltransferase (COMT), phenethanolamine-*N*-methyltransferase (PNMT), histamine-N-methyltransferase (HNMT), and the cytosolic-*S*-methyltransferases, thiopurine-*S*-methyltransferase (TPMT) and thioether-*S*-methyltransferase (TEMT) and finally a membrane bound thiomethyltransferase (TMT).

The co-substrate: *S*-Adenosylmethionine

Any consideration of methyltransferases must begin with an understanding of the nature of the co-substrate AdoMet. AdoMet is the methyl donor of virtually all methyltransferases. Since its discovery in 1953 (Cantoni 1953), AdoMet has been demonstrated in an increasing number of transmethylation reactions, many critical for the survival of cells. AdoMet is formed by the condensation of ATP and L-methionine catalysed by ATP:L-methionine *S*-adenosyltransferase (EC 2.5.1.6) (MAT). MAT is both a synthetase and a tripolyphosphatase resulting in a complete but asymmetric depho-

Enzyme Systems that Metabolise Drugs and Other Xenobiotics. Edited by C. Ioannides.
© 2002 John Wiley & Sons Ltd

sphorylation of ATP, followed by the release of inorganic pyrophosphate as well as phosphate. The enzyme requires divalent cations for its activity, and is activated by monovalent cations (Green 1969; Lombardini and Talalay 1971). The energy from the dephosphorylation of ATP is transferred to the energy-rich sulphonium complex that contains the reactive methyl group. The product synthesised by the enzymic condensation, AdoMet, has (S) configuration at the sulphonium centre (Conforth *et al.* 1977; de la Haba *et al.* 1959). The same diastereomer with *S*-configuration is required for all the methyl-transfer reactions (de la Haba *et al.* 1959; Zapia *et al.* 1969; Borchardt *et al.* 1976a). Permitted structural changes in AdoMet are limited. Any change in the base, sugar or the amino acid moiety causes a dramatic increase in the K_m values for the methyltransferases (Borchardt *et al.* 1976b).

Upon donating its methyl group, AdoMet is converted to *S*-adenosylhomocysteine (AdoHcy) which then is hydrolysed to adenosine and homocysteine. Homocysteine can undergo remethylation to methionine and re-enter the trans-methylation cycle. It should be noted that AdoHcy, which is a potent, competitive inhibitor of the methyltransferases, has no effect on the MAT reaction. The role of MAT in the level and control of AdoMet synthesis is complex. There are two major mammalian forms of MAT, which differ in both tissue distribution and kinetic properties. The products AdoMet, pyrophosphate, and phosphate regulate the reaction, but the nature of the regulation, feedback inhibition or stimulation varies depending upon the MAT isoenzyme (Kotb and Kredich 1990). For a recent review of the structure and function of MAT see Kotb and Geller (1993).

Methionine, the primary initiator and source of AdoMet, is evident from a rapid increase in the levels of AdoMet in various tissues after administration of methionine, either orally or intraperitoneally (Regina *et al.* 1993). It should be noted that MAT activity in various organs of rats does not correlate well with the levels of AdoMet. While the specific activity of MAT (pmole product/min/mg protein) ranged from a low value of 40 in the heart and brain of rats to a high value of 7700 in the liver, an over 180-fold increase, the concentrations of AdoMet ranged from a low of 25 (nmoles/g of tissue) in the brain to a high of 68 in the liver (less than 3-fold increase) (Eloranta 1977). Similarly, despite 13-fold higher specific activity of MAT in the pancreas over that in the heart, the levels of AdoMet in both the tissues were similar. Thus, the complex nature of the MAT isoforms has important roles in controlling the levels of AdoMet in different tissues and organs (Kotb and Geller, 1993). Furthermore, drugs or other xenobiotics that can be methylated by AdoMet-dependent methyltransferases have been shown to affect the levels of AdoMet in various tissues (Fuller *et al.* 1983; Borchardt *et al.* 1976b). While the decreased levels in AdoMet are generally transient, the administration of high levels of quercetin or fisetin, both excellent substrates for COMT, resulted in a decrease of kidney AdoMet levels by 25%. As a consequence of both the reduction in AdoMet levels and the increase in adenosine homocysteine levels, an inhibition of the *O*-methylation of catecholoestrogens could be demonstrated (Zhu and Liehr 1996).

An overview: *O*-methylation

A major AdoMet-dependent methyltransferase is the *O*-methylation reaction. This reaction includes the catechol-*O*-methyltransferase catalysed *O*-methylation of the catechols dopamine, norepinephrine, epinephrine, L-DOPA and the catecholoestrogens. Another enzyme, hydroxyindole O-methyltransferase (EC 2.1.1.4), catalyses the O-methylation of the phenolic group of N-acetylserotonin to form melatonin (Axelrod and Weissbach 1961). An important O-methylation reaction is catalysed by protein-carboxy O-methyl-transferase (EC.2.1.1.24). This reaction catalyses the esterification of aspartic and glutamic acid residues on many proteins. Methylation and demethylation may serve as signals to reversibly affect the three-dimensional configuration of proteins and thus their function (Kim *et al.* 1982). Since COMT is the first and, at present the best, described O-methyltransferase a more complete review of this methyltransferase is given below.

CATECHOL-O-METHYLTRANSFERASE

Catechol-O-methyltransferase (EC 2.1.1.6) (COMT) was first described by Axelrod and Tomchick (1958). At present the enzyme has been cloned, the genetic locus determined, crystallised, the molecular structure of the two isoforms determined, the nature of the active site formulated, effective inhibitors developed, and the tissue localisation, both in the brain and in peripheral tissues, has been described. These developments have been the subject of several extensive reviews (Guldberg and Marsden 1975; Creveling and Hartman 1982; Thakker and Creveling 1990; Creveling and Thakker 1994; Lundstrom *et al.* 1995; Mannisto and Kaakkola 1999).

STRUCTURE OF COMT

COMT is derived from a single gene localised on chromosome 22, band q11.2 (Grossman *et al.* 1992) which codes for both the soluble (S-COMT) and the membrane-bound (MB-COMT) form of the enzyme. S-COMT contains 221 amino acids with a molecular mass of 24.4 kDa. MB-COMT contains an additional 50 amino acids with a molecular mass of 30 kDa. Of the extra amino acids, 20 function as a hydrophobic membrane anchor (Lundstrom *et al.* 1995; Salminen *et al.* 1990). S-COMT is quantitatively the predominant form in peripheral tissues while MB-COMT accounts for 70% in the brain (Tenhunen *et al.* 1994). The MB-COMT is bound by the hydrophobic anchor to the cytoplasmic side of intracellular membranes with the remainder of the enzyme suspended in the cytoplasm (Bertocci *et al.* 1991).

A detailed description of the atomic structure of S-COMT from rat has been derived from the crystallised enzyme at a 1.7 to 2.0-Å resolution (Vidgren and Ovaska 1997; Vidgren *et al.* 1994). COMT has a single domain of eight α-helices around a central β-sheet. The active site is located in a groove on the outer surface of the enzyme. The binding site for the methionine of AdoMet is deeper within the protein and is bound first followed by Mg^{2+}, which promotes the ionisation of one of the hydroxyl groups of the catechol. Mg^{2+} forms an octahedral coordination structure with aspartic acids (Asp141 and 169), asparagine (Asn170), a water molecule and the hydroxyls of the catechol substrate. This configuration controls the orientation of the catechol. At the surface of the protein, leading to the active site is a hydrophobic slot composed of 'gatekeeper' residues Trp38, Trp143 and Pro174. This configuration interacts with the side chains of various substrates and keeps the catechol ring in a planar position. A lysine (Lys144), acting as a general catalytic base, accepts a proton from the ionised hydroxyl followed by transfer of the methyl group from AdoMet to that hydroxyl group. The methyl transfer results from a direct nucleophilic attack by one of the hydroxyl groups of the catechol on the methyl carbon of AdoMet in a tight SN2 transition state (Woodard *et al.* 1980).

THE SUBSTRATES

COMT catalyses transfer of the methyl group from AdoMet to either the *meta-* or the *para*-hydroxyl group of virtually all substituted catechol derivatives to form a mono-methyl ether. The enzyme accommodates catechol substrates with positively charged, negatively charged or neutral substituents (Creveling *et al.* 1970, 1972). Recent studies using recombinant enzyme isoforms have clearly demonstrated that the kinetic differences are due to interactions of the substrate sidechains with COMT (Lotta *et al.* 1992). The preference for *para-O*-methylation is exhibited by catechols that contain electronegative substituents like nitro, cyano or fluoro groups. These derivatives are poor substrates but can be potent inhibitors (Thakker *et al.* 1986; Backstrom *et al.* 1989). The *meta/para* ratios of substituted catechols are a consequence of their ability to bind in two dissimilar orientations in the active site (Lan and Bruice 1998).

The endogenous substrates include: the catecholamines, dopamine, norepinephrine and epinephrine and their metabolites which retain the catechol moiety; L-dopa; 2- and 4-hydroxycatechol oestrogens and oestrones (Cavalieri and Rogan 2000; Raftogianis *et al.* 2000) and the dihydroxyindolic intermediates in melanin formation (Smit *et al.* 2000) and the alkaloids, hydroxyisoquinolines and apomorphine. Many pharmacologic agents are substrates for COMT including isoprenaline, α-methyldopa, carbidopa, benserazide, β-lactam antibiotics with a 3,4-dihydroxybenzoyl-function-ality, dihydroxyphenylserine and dobutamine. Even drug molecules that contain a single phenolic moiety, which are not subject to *O*-methylation, can become sub-strates for COMT after hydroxylation *ortho* to the phenolic group by cytochrome P450-dependent hydroxylases. Successive oxidation followed by *O*-methylation re-sults in the metabolism of many phenolic steroids, phenethylamine, and hydroxyin-doles (Thakker and Creveling 1990).

Most xenobiotics ranging from dihydroxybenzene to complex polycyclic catechols of plant origin are COMT substrates. Of particular interest are flavonoids containing a catechol moiety. Several polyphenols, water-soluble bearing catechol groups from tea, have been shown to be excellent substrates form COMT (Zhu *et al.* 2000). Of particular interest are flavonoids such as quercetin and fisetin, which are rapidly *O*-methylated by COMT. Such compounds have been shown to be mutagenic *in vivo* by the Ames test and other indicators. However, *in vivo*, these compounds are so rapidly *O*-methylated by COMT and other conjugation reactions that they do not enter a mutagenic pathway (Zhu *et al.* 1994).

THE FUNCTIONS OF COMT IN VIVO

The functions of COMT *in vivo* have become more apparent with the expanding knowledge of the specific cellular localisation of the enzyme (Inoue *et al.* 1977;

Creveling and Hartman 1982; Creveling 1984, 1988; Karhunen *et al.* 1994, 1995).

Several general aspects about COMT can now be appreciated. First, while there is a wide variation in the level of COMT activity in various tissues, among species and strains, the individual level of activity in most tissues and strains show great similarities. In most cases COMT activity increases rapidly from relatively low levels at birth to a characteristic level early in life and remains essentially constant throughout the adult life (Goldstein *et al.* 1980). Second, in certain tissue sites such as the epithelial lining of the uterus and in the ductal epithelium of breast, the level of COMT activity undergoes marked increases in response to pregnancy, lactation and oestrus (Inoue and Creveling 1991, 1995; Amin *et al.* 1983; Creveling 1984).

Elevated levels of COMT are present in breast adenocarcinomas in women, mouse and rat (Assicot *et al.* 1977; Amin *et al.* 1983; Hoffman *et al.* 1979), and in β-islet insulinomas of rat and hamster. COMT activity is relatively constant and characteristic of the individual at most sites in the adult animal. However, in certain cell types, the epithelial cell of uterus and breast, COMT activity increases in response to hormonal or physiological cues (Inoue and Creveling 1995). These observations strongly suggest that, in addition to the now classical function of COMT in the inactivation of circulating of catecholamines in the liver, originally described by Axelrod and his coworkers (Axelrod and Tomchick 1958), COMT appears to have a much wider role in the control of the level and distribution of substances bearing the reactive catechol moiety.

COMT is an important determinant in the effective use of L-DOPA for the symptomatic therapy of Parkinson's disease. With the discovery of dopaminergic cell loss, the accompanying decrease in striatal dopamine and the introduction of L-DOPA as an effective means for the restoration of central dopamine stores, there has been increasing attention directed towards the COMT-catalysed formation of 3-O-methyl-DOPA (OMD). OMD is a major metabolite of L-DOPA and its formation is increased with higher levels of L-DOPA achieved in the presence of the peripheral decarboxylase inhibitor, carbidopa. OMD formation after the oral administration L-DOPA is of special significance due to the relatively high levels of both S-COMT and MB-COMT in the gut wall (Nissenin *et al.* 1988). OMD competes with L-DOPA with greater affinity for the neutral amino acid transport system in both gut and at the blood-brain barrier, and furthermore OMD has a longer biological half-life than L-DOPA. A new class of inhibitors has been developed which are selective and potent inhibitors of COMT. One derivative, OR-462 [(3-(3,4-dihydroxy-5-nitrobenzylidene)-2,4-pentanedione], effectively inhibits COMT activity in the gut wall for up to 10 hours when given orally to rats. (Nissenin *et al.* 1988). When given with L-DOPA, it produces a long lasting inhibition of OMD formation both peripherally and in the striatum. With this inhibitor, equivalent levels of striatal dopamine were achieved with one-forth as much L-DOPA compared to controls given L-DOPA alone. An extensive review of the current studies on the biochemistry, toxicology and therapeutic applications of this group of nitrocatechols, has been published. This review includes the status of the available therapeutic agents, Tolcapone, Netcapone and Entacpone, and their use in the clinical treatment of Parkinson's disease (Mannisto and Kaakkola 2000).

THE CELLULAR DISTRIBUTION OF COMT

The distribution of COMT in brain is divided between the high-affinity, MB-COMT and the cytosolic S-COMT. MB-COMT accounts for approximately 70% of brain COMT. A specific polyvalent antisera to both MB- and S-COMT has been used for cell-specific immunochemical localisation of COMT (Inoue et al. 1977; Grossman et al. 1985). COMT is present in the cytoplasm of the ciliated, cuboidal cells of the ventricular ependyma along the borders of the lateral, 3rd and 4th ventricles of the rat brain, perhaps as a barrier between the CSF and the brain parenchyma (Kaplan et al. 1979, 1981a). In the brain parenchyma proper, COMT is found primarily in glial elements, oligodendrocytes and fibrous astrocytes. Spatz et al. (1986) showed that the endothelium of cerebral capillaries, arterioles, and larger vessels contained COMT. Thus COMT, in conjunction with MAO, may provide cerebral capillaries with an enzymic barrier for the passage of catechols. In the cerebellum, in addition to glial elements, the cell bodies of Bergmann cells, adjacent to S-COMT-negative Purkinje cells, contained S-COMT as did the Bergmann fibres ascending through the molecular layer to the pial surface (Kaplan et al. 1981b).

Immunological localisation of COMT in the uterus, oviduct, placenta, mammary gland, and vas deferens and seminal vesicle has led to a greater appreciation of the role of COMT in the reproductive process (Inoue et al. 1977; Amin et al. 1983; Inoue and Creveling 1986, 1991, 1995).

Of interest is the presence of elevated levels of COMT in mammary glands, human breast tumours and the apparent positive relationship between the COMT activity and the grade of malignancy in primary carcinomas (Assicot et al. 1977; Creveling and Inoue, 1994), and the demonstration of de novo synthesis of catecholsteroids in breast tumours (Hoffman et al. 1979; Raftogianis et al. 2000).

POLYMORPHISM OF COMT

The levels of COMT activity were shown to be controlled by a common genetic polymorphism over 20 years ago (Weinshilboum and Raymond 1977). The phenotypic trait of low COMT was detected as present in approximately 25% of a caucasian population. Subsequent molecular pharmacogenetic studies have identified a single nucleotide polymorphism in the COMT gene that results in a Val 108/Met substitution in S-COMT. This amino acid substitution results in a enzyme with low activity. The frequency occurrence of each allele is approximately 50%. The variation in these alleles has been determined in a variety of racial groups and various pathologies (Weinshilboum et al. 1999). For example, of great interest is the possible relationship of the low activity form of COMT and the pathophysiology of breast cancer (Lavigne et al. 1977; Thompson et al. 1998; Millikan et al. 1998; Thompson and Ambrosone 2000; Yager 2000).

CONCLUSION

In conclusion, it should be emphasised that methyltransferases and the methyl-donor cosubstrate, SAM, play a pivotal role in diverse biological systems. In this chapter, the role of one such methyltransferase, COMT has been discussed in considerable detail. It appears that the role of COMT is more extensive than just the inactivation of catechol xenobiotics, circulating catecholamines, and catecholamine neurotransmitters. The function of COMT in the reproductive system, the presence of sexual dimorphism with regard to COMT, the physiological and neoplastic alterations in the activity of COMT, and the extensive localisation of COMT, clearly point to a significant role in the inactivation of catechol oestrogens, as a barrier for the passage of catechols between tissue compartments, and the control of other, as yet unrecognised, catechol-mediated functions. The recent development by Orion Pharmaceutica of Finland of a series of derivatives of 3-nitrocatechol appears to have provided the research community with a selection of specific, long-lasting, essentially irreversible inhibitors of COMT.

N-Methyltransferases: an overview

N-Methylation is a prominent pathway for the metabolism of many endogenous hormones and neurotransmitters. It is an important metabolic step in the transformation of molecules containing primary, secondary, or tertiary amino groups. As is the case with *O*-methylation, *N*-methylation also occurs by the transfer of a methyl group from AdoMet to nucleophilic amino groups. The resultant products are the *N*-methylated metabolites and AdoHcy. Amine-*N*-methyltransferase catalyses the *N*-methylation of amines with a very wide variety of structures. Despite its broad substrate selectivity, it does not catalyse such important molecules as histamine and norepinephrine. Interestingly, the highly specific enzyme, phenethanolamine-*N*-methyltransferase, a biosynthetic enzyme, is responsible for the *N*-methylation of norepinephrine to form epinephrine. Equally specific is histamine-*N*-methyltransferase, which catalyses the *N*-methylation of histamine. It is of interest that histamine may exhibit genetic polymorphism. There are many *N*-methyltransferases in the liver and gastrointestinal mucosa which *N*-methylate both exogenous compounds as well as endogenous compounds. Nicotinamide-*N*-methyltransferase activity exhibits wide variations in both human and animal liver and is of interest in that it may have genetic polymorphisms. The *N*-methylation reaction plays an important role in the metabolic inactivation of drugs since a large number of drug molecules contain an amino functionality. It should be noted that in selected cases, *N*-methylation may result in the formation of active metabolites. This is best illustrated by the conversion of apomorphine to the pharmacologically active morphine. *N*-methylation of 4-amino-azobenzene followed by biological oxidation results in the formation of the *N*-hydroxyarylamine that is a proximate carcinogen (Ziegler *et al.* 1988).

PHENETHANOLAMINE-*N*-METHYLTRANSFERASE (PNMT)

PNMT is the terminal enzyme in the biosynthetic pathway for the catecholamines and catalyses the *N*-methylation of norepinephrine to yield epinephrine. It is located primarily in the adrenal medulla but smaller amounts are also present in small intensity fluorescent cells of sympathetic ganglia and sensory nuclei of the vagus nerve. Peripheral PNMT is responsive to glucocorticoids. In the brain, PNMT is found in small cell groups in the medulla oblongata, hypothalamus, amygdala, and in retinal and amacrine cells of the retina. PNMT is highly specific for phenethanolamines and does not accept phenethylamines not bearing a β-hydroxyl group (Rafferty and Grunewald 1982). The endogenous substrates are limited to norepinephrine and epinephrine. Curiously, certain compounds in which the aromatic ring is replaced by cyclohex-3-enyl, cyclohexyl or cyclooctyl rings are good substrates for PNMT. An extended series of investigations seeking effective substrates or inhibitors have been published. The goal of these investigations is through inhibition of central PNMT to determine the possible function of the enzyme in brain (Liang *et al.* 1982).

The nucleotide sequence and the deduced amino acid sequence of bovine PNMT was reported and a full-length clone isolated using PNMT mRNA (Joh *et al.* 1983; Baetge *et al.* 1986). Subsequently, the PNMT-coding sequence and amino acid sequence for human PNMT were published (Kane *et al.* 1988). The human enzyme consists of 282 amino acids with a molecular mass of 30.9 kDa. The gene for human PNMT is assigned to a single gene on chromosome 17. The protein has binding sites for glucocorticoid regulatory elements. The central PNMT is not regulated by glucocorticoids. It should be noted that the limited amounts and substrate specificity makes it unlikely that PNMT plays an important role in the *N*-methylation of drugs or dietary xenobiotics.

HISTAMINE-*N*-METHYLTRANSFERASE (HNMT)

HNMT is a cytosolic, monomeric, AdoMet-dependent enzyme. A cDNA clone of 1.3 kb has been derived from rat kidney consisting of a coding region of 876 nucleotides and an amino acid sequence of 292 residues for a molecular mass of 34 kDa. Expression in *E. coli* yielded a catalytically active transferase (Takamura *et al.* 1992). Northern Blot analysis established that HNMT is widely expressed in human tissues. HNMT is approximately 34 kb in length and mapped to human chromosome 2 (Aksoy *et al.* 1996). HNMT is a very specific methyltransferase and will *N*-methylate only histamine derivatives in which positions 1, 2, and 3 are unsubstituted (Ansher and Jakoby 1990). Further substrates must have a positive charge on the side chain. Histamine derivatives with a negative charge on the side chain are not substrates. The methyl group transfer from AdoMet to histamine is by direct transfer to the histamine nitrogen and is accompanied by inversion of the configuration (Asano *et al.* 1984).

Histamine is exclusively metabolised in the brain by transfer of the methylgroup from AdoMet to form *N*-methylhistamine, followed by oxidative deamination by monoamine oxidase B. The enzyme is localised primarily in neurons and the vascular walls of vessels in the brain. The localisation of HNMT in the blood vessel walls suggests that vascular levels of histamine and histamine released from mast cells associated with blood vessels are metabolised locally. Within the brain, the cell bodies of histaminergic neurons are present in the tuberomannillary nucleus of the posterior hypothalamus. Fibres from this nucleus are distributed widely throughout the brain in a manner similar to the distribution of catecholamine fibres from the locus ceruleus (Watanabe *et al.* 1984). In peripheral sites, primarily the bronchi, HNMT has been shown to be a major metabolic pathway. This localisation may be important since histamine is an important factor in asthma and allergies (Wasserman 1983).

S-Methylation: an overview

Unlike O- and N-methylation reactions which have many biosynthetic roles as well as interactions with xenobiotics, the AdoMet-dependent S-methylases appear to be associated primarily with the metabolic inactivation function. The cytosolic sulpho-transferases catalyse sulphoconjugation of relatively small lipophilic endogenous compounds and xenobiotics (see Chapter 10). Recent reviews suggest that there are at least forty-four cytosolic sulphotransferases, which have been identified in mammalian tissues. The identification of the sulphotransferases is based upon their amino acid sequences, and the enzymes constitute five different families which are localised on at least five different chromosomal sites. Most sulphotransferases are active in the sulphation of various xenobiotics and well as such endogenous compounds like oestrogens, corticoids, and thyroxin. An understanding of sulphotransferases is com-plicated both by the overlapping nature of the substrate specificity and a non-uniform nomenclature. The relationship between the cytosolic and membrane bound sulpho-transferases is at present unclear. The biochemistry and pharmacogenetics have been extensively reviewed (Weinshilboum 1989a, 1989b) and more recently by Nagata and Yamazoe (2000).

THIOPURINE-S-METHYLTRANSFERASE (TPMT)

A major group of cytosolic sulphotransferases catalyses the methylation sulphydryl groups to form methyl thioether products. A characteristic thiopurinemethyltransferase (TPMT, 2.1.1.67) catalyses the methylation of the sulphur atom in aromatic and heterocyclic thiols such as 6-mercaptopurine. TPMT exhibits genetic polymorphism where approximately 5–10% people are heterozygotic at the TPMT gene locus and have intermediate enzyme activity, while less than 1% inherit two mutant TPMT alleles and are TPMT deficient. The role of TPMT is clinically important in the treatment of acute lymphoblastic leukaemia with 6-mercaptopurine. Patients with homozygous and heterozygous deficiency for TPMT have extreme sensitivity to 6-mercaptopurine due to the accumulation of thioguanine nucleotides. Unless such patients are treated with lower doses of 6-mercaptopurine, they develop profound haematopoietic toxicity that can be fatal. Thus determination of the patients TPMT status is essential (Relling et al. 1999).

THIOETHER-S-METHYLTRANSFERASE (TEMT)

An interesting cytosolic mammalian S-methylase, thioether S-methyltransferase (TEMT, 2.1.1.96), was discovered in a search for the biosynthetic source of a urinary meta-bolite of selenium, trimethyl selenonium ion. The enzyme was originally known as a selenoether methyltransferase (Mozier and Hoffman 1988). The purified enzyme has a molecular mass of 28 kDa and has a broad substrate specificity, methylating the sulphur atom as well as selenium and telluride atoms in a variety of thio ethers, and thus was termed TEMT.

THIOMETHYLTRANSFERASE (TMT)

A microsomal, membrane-bound thiomethyltransferase (TMT, EC 2.1.1.9) TMT has a broad specificity towards thiol-containing compounds and catalyses the methylation of sulphur atoms in many drugs like captopril, thiopurines and cephalosporins, thiols like D-penicillamine and other aliphatic sulphydryl compounds like 2-mercaptoethanol. The distribution of TMT activity is highest in intestinal mucosa suggesting a primary role in the detoxification of hydrogen sulphide formed by anaerobes in the intestinal tract (Weisinger and Jakoby 1980). It should be noted that many substrates for thiol methylation are generated by initial conjugation with glutathione followed by enzymic degradation to the cysteine conjugates, and finally cleavage by various β-lyases (see Chapter 1). S-Methylation of the thiol metabolites generated by β-lyase action is known as the thiomethyl shunt. This pathway diverts conjugates to the formation and excretion of mercapturic acids (Stevens and Bakke 1990).

Conclusions

Our knowledge of the methyltransferases has expanded rapidly over the last three decades so that at present at least 100 AdoMet-dependent methyltransferases have been described. One can expect that the variety and functions of O-, N- and S-methyltransferases will continue to be discovered. The appreciation of the complexity of the genetic diversity resulting in the polymorphisms, from the apparent simplicity of COMT to the growing number of forms of S-methyltransferases, is now clearly recognised. Many of these studies have made serious additions to clinical therapeutics. The role of the level of TPMT in the toxicity of 6-mercaptopurine in the treatment of acute lymphoblastic leukaemia is a clear example of the knowledge of the polymorphism of this enzyme. Studies of the risk factor for breast cancer in women with the low activity form of COMT, while not yet fully understood are an active area of research. The methods of molecular biology, the cloning of cDNAs and genes have enhanced the definitive discovery of genetic polymorphisms. With the availability of access to the human genome, one challenge for the research community will be to discover signature sequences and polymorphisms for methyltransferases. Knowledge of the genetic variation in methyltransferases will lead to an enhanced understanding of individual variation and the pathophysiology of disease, and guides for clinical therapeutics.

References

Aksoy S, Raftogianis R and Weinshilboum RM (1996) Human histamine N-methyltransferase gene: Structural characterization and chromosomal localisation. *Biochemical Biophysical Research Communications*, **219**, 548–554.

Amin AM, Creveling CR and Lowe MC (1983) Immunocytochemical localization of catechol-O-methyltransferase in normal and cancerous breast tissues of mouse and rat. *Journal of the National Cancer Institute*, **70**, 337–342.

Ansher SS and Jacoby WB (1990) N-Methylation. In *Conjugation Reactions in Drug Metabolism*, Mulder GJ (ed.), Taylor and Francis, London, pp. 233–245.

Asano Y, Woodward WB, Houck DR and Floss HG (1984) Stereochemical course of the transmethylation catalyzed by histamine N-methyltransmethylase. *Archives of Biochemistry and Biophysics*, **231**, 253–256.

Assicot M, Contesso G, and Bohuon C (1977) Catechol-O-methyltransferase in human breast cancers. *European Journal of Cancer*, **13**, 961–966.

Axelrod J and Tomchick R (1958) Enzymatic O-methylation of epinephrine and other catechols. *Journal of Biological Chemistry*, **233**, 702–705.

Axelrod J and Weissbach H (1961) Purification and properties of hydroxyindole-O-methyltransferase. *Journal of Biological Chemistry*, **236**, 211–213.

Backstrom R, Honkanen E, Pippuri A, Kairisalo P, Pystynen J, Heinola K, Nissinen E, Linden I-B, Mannisto PT, Kaakkloa S and Pohto P (1989) Synthesis of some novel potent and selective catechol-O-methyltransferase inhibitors. *Journal of Medicinal Chemistry*, **32**, 841–846.

Baetge EE, Suh YH and Joh TH (1986) Complete nucleotide and deduced amino acid sequence of bovine phenethanolamine-N-methyltransferase: partial amino acid homology with rat tyrosine hydroxylase. *Proceedings of the National Academy of Sciences, USA*, **83**, 5454–5458.

Bertocci B, Miggiano V, Da Prada M, Dembic Z, Lahm HW and Malherbe P (1991) Human catechol-O-methyltransferase: Cloning and expression of the membrane-associated form. *Proceedings of the National Academy of Sciences, USA*, **88**, 1416–1420.

Borchardt RT, Wu YS, Huber JA and Wycpalek AF (1976a) Potential inhibitors of S-adenosylmethionine dependent methyltransferases. V. The role of the asymetric sulphonium pole in the enzymatic binding of S-adenosyl-L-methionine. *Journal of Medicinal Chemistry*, **19**, 1104–1110.

Borchardt RT, Huber JA, Wu YS (1976b) Potential inhibitors of S-adenosylmethionine dependent methyltransferases IV. Further modification of the amino acid and base portions of S-adenosylmethionine. *Journal of Medicinal Chemistry*, **19**, 1094–1099.

Cantoni GL (1953) S-Asdenosyl-L-methionine: a new intermediate formed enzymatically form L-methionine and adenosinetriphosphate. *Journal of Biolological Chemistry*, **204**, 403–416.

Cavaleri E and Rogan E (2000) The key role of catecholestrogen-3,4-quinones, in tumor initiation, In *Role of Catechol Quinone Species in Cellular Toxicity*. Creveling CR (ed.) F. P. Graham, Johnson City, TN pp. 247–260.

Conforth JW, Reichard SA, Talalay P, Correll HL and Glusker JP (1977) Determination of the absolute configuration at the sulphonium center of S-adenosylmethionine: Correlation with the absolute configuration of the diastereomeric S-carboxymethyl-(S)-methionine salts. *Journal of the American Chemical Society*, **99**, 7292–7300.

Creveling CR (1984) The functional significance of the cellular localization of catechol-O-methyltransferase. In *The Role of Catecholamines and Other Neurotransmitters in Stress*. Usdin E and Kvetnansky R (eds), Gordon and Brench, New York, pp. 447–455.

Creveling CR (1988) On the nature and function of catechol-O-methyltransferase. In *Perspectives in Psychopharmacology: A Collection of Papers in Honor of Earl Usdin*. Barchas J and Bunny W (eds), Alan R. Liss, New York, pp. 55–64.

Creveling CR and Hartman BK (1982) Relationships between the cellular localization and the physiological function of catechol-O-methyltransferase. In *Biochemistry of S-Adenosylmethionine and Related Compounds*, Usdin E, Borchardt RT and Creveling CR (eds), Macmillan Press, London. pp. 479–486.

Creveling CR and Inoue K (1994) Catechol-O-methyltransferase: Factors relating to the carcinogenic potential of catecholestrogens. *Polycyclic Aromatic Compounds*, **6**, 253–259.

Creveling CR and Thakker DR (1994) O-, N-, and S-Methyltransferases. In *Handbook of Experimental Pharmacology*, Kaufmann FC (ed.), Springer-Verlag, Berlin, Vol. 112, pp. 189–216.

Creveling CR, Dalgard N, Shimizu H and Daly JW (1970) Catechol-O-methyltransferase III. m- and p-O-methylation of catecholamines and their metabolites. *Molecular Pharmacology*, **6**, 691–696.

Creveling CR, Morris N, Shimizu H, Ong HH and Daly JW (1972) Catechol-O-methyltransferase IV. Factors affecting m- and p-methylation of substituted catechols. *Molecular Pharmacology*, **8**, 398–409.

De La Haba G, Jamieson GA, Mudd SH and Richards HH (1959) S-Adenosylmethionine the relationship of configuration at the sulphonium center to enzymatic activity. *Journal of the American Chemical Society*, **81**, 3975–3980.

Eloranta TO (1977) Tissue distribution of S-adenosylmethionine and S-adenosylhomocysteine in the rat. *Biochemical Journal*, **166**, 521–529.

Fuller RW, Perry KW and Hemrick-Luecke SK (1983) Tropolone antagonism of the L-dopa-induced elevation of S-adenosylhomocysteine: S-Adenosylmethionine ratio but not depletion of adrenaline in rat hypothalamus. *Journal of Pharmacy and Pharmacolology*, **36**, 419–429.

Guldberg HC and Marsden CA (1975) Catechol-O-methyltransferase: Pharmacological aspects and physiological role. *Pharmacological Reviews*, **27**, 135–206.

Goldstein DJ, Weinshilboum RM, Dunnette JH and Creveling CR (1980) Developmental patterns of catechol-O-methyltransferase in genetically different rat strains: enzymatic and immuno-chemical studies. *Journal of Neurochemistry*, **34**, 153–162.

Green R (1969) Kinetic studies of the mechanism of S-adenosylmethionine synthetase from yeast. *Biochemistry*, **8**, 2255–2265.

Grossman MH, Creveling C R, Rybczynski R, Braverman M, Isersky C and Breakfield XO (1985) Soluble and particulate forms of catechol-O-methyltransferase distinguished by gel electrophoresis and immune fixation. *Journal of Neurochemistry*, **44**, 421–432.

Grossman MH, Emanuel BS and Budarf ML (1992) Chromosomal mapping of the human catechol-O-methyltransferase gene to 22q11.1-q11.2. *Genomics*, **12**, 822–825.

Grunewald GL, Grindel JM, Vincek WC and Borchardt RT (1975) Importance of aromatic ring in adrenergic amines: Nonaromatic analogues of phenethanolamine as substrates for phenethanolamine N-methyltransferase. *Molecular Pharmacology*, **11**, 694–699.

Hirata F (1982) Overviews on phopholipid methylation. In *Biochemistry of S-Adenosylmethionine and Related Compounds*, Usdin E, Borchardt RT, Creveling CR (eds), Macmillan, London, pp. 109–117.

Hoffman JL, Paul SM and Axelrod J (1979) Catecholestrogens: synthesis and metabolism by human breast tumors in vitro. *Cancer Research*, **39**, 4584–4587.

Inoue K and Creveling CR (1986) Immunocytochemical localization of catechol-O-methyltransferase in the oviduct and in macrophages in corpora lutea of rat. *Cell and Tissue Research*, **245**, 623–628.

Inoue K and Creveling CR (1991) Induction of catechol-O-methyltransferase in the luminal epithelium of rat uterus by progesterone. *Journal of Histochemistry and Cytochemistry*, **39**, 823–828.

Inoue K and Creveling CR (1995) Induction of catechol-O-methyltransferase in the luminal epithelium of rat uterus by progesterone: Inhibition by RU-386. *Drug Metabolism and Disposition*, **23**, 430–432.

Inoue K, Tice LW and Creveling CR (1977) Immunological localization of catechol-O-methyltransferase. In *Structure and Function of Monoamine Enzymes*, Usdin E, Weiner N, and Youdim MBH (eds), Marcel Dekker, New York, pp. 835–859.

Joh TH, Baetge EE , Ross MD and Feis DJ (1983) Evidence for the existence of homologous gene coding region for the catecholamine biosynthetic enzymes. *Cold Spring Harbor Symposium on Quantitative Biology*, **48**, 327–335.

Karhunen T, Tilgmann C, Ulmanen I, Julkunen I and Panula P (1994) Distribution of catechol-O-methyltransferase enzyme in rat tissues. *Journal of Histochemistry and Cytochemistry*, **42**, 1079–1090.

Karhunen T, Tilgmann C, Ulmanen I and Panula P (1995) Catechol-O-methyltransferase(COMT) in rat brain: Immunoelectron microscopic study with an antiserum against rat recombinant COMT protein. *Neuroscience Letters*, **13**, 825–834.

Kaneda N, Icinose H, Kobayashi K, Oka K, Kishi F, Nakazawa Y, Fujita K and Nagatsu T (1988) Molecular cloning of cDNA and chromosomal assignment of the gene for human phenylethanolamine N-methyltransferase, the enzyme for epinephrine biosynthesis. *Journal of Biological Chemistry*, **263**, 7672–7677.

Kaplan GP, Hartman BK and Creveling CR (1979) Immunohistochemical demonstration of catechol-O-methyltransferase in mammalian brain. *Brain Research*, **167**, 241–250.

Kaplan GP, Hartman BK and Creveling CR (1981a) Localization of catechol-O-methyltransferase in the leptomeninges, choroid plexus and ciliary epithelium: Implications for the separation of central and peripheral catechols. *Brain Research*, **204**, 323–335.

Kaplan GP, Hartman BK and Creveling CR (1981b) Immunocytochemical localization of catechol-O-methyltransferase in the circumventricular organ: Potential for variation in the blood brain barrier to native catechols. *Brain Research*, **229**, 323–335.

Kim S, Ro J-Y, Manna C and Glusko, VG (1982) Enzymatic carboxyl methyl esterification of proteins: Studies on sickle erythrocyte membranes. In *Biochemistry of S-Adenosylmethionine and Related Compounds*, Usdin E, Borchardt RT, Creveling CR (eds). Macmillan, London, pp. 39–48.

Kotb M and Geller AM (1993) Methionine adenosyltransferase: structure and function. *Pharmacology and Therapeutics*, **59**, 125–143.

Kotb M and Kredich NM (1990) Regulation of human lymphocyte S-adenosylmethionine synthetase by product inhibition. *Biochimica et Biophysica Acta*, **1039**, 253–260.

Lan EY and Bruice TC (1998) Importance of correlated motions in forming highly reactive near attack formations in catechol-O-methyltransferase. *Journal of the American Chemical Society*, **120**, 12387–12394.

Lavigne JA, Helzlsouer KJ, Huarg HY, Strickland PT and Bell DA (1977) An association between the allele coding for a low activity variant of catechol-O-methyltransferase and the risk for breast cancer. *Cancer Research*, **57**, 5493–5497.

Liang NY, Tessel RE, Borchardt RT, Vincek WC and Grunewald GL (1982) The use of phenethanol N-methyltransferase inhibitors in elucidating the role of brain epinephrine in the control of cardiovascular function. In *Biochemistry of S-Adenosylmethionine and Related Compounds*, Usdin E, Borchardt RT, Creveling CR (eds), Macmillan Press Ltd, London, pp. 457–460.

Lombardini JB and Talalay P (1971) Formation, functions and regulatory importance of S-adenosyl-L-methionine. *Advances in Enzyme Regulation*, **9**, 349–384.

Lotta T, Takinen J, Backstrom R and Nissinen E (1992) PLS modeling of structure-activity relationships of catechol-O-methyltransferase inhibitors. *Journal of Computor Aided Molecular Description*, **6**, 253–257.

Lundstrom K, Tehunen J, Tilgmann C, Karhunen T, Panula P and Ulmanen I (1995) Cloning, expression and structure of catechol-O-methyltransferase. *Biochimica et Biophysica Acta*, **125**, 1–10.

Mannisto PT and Kaakkola S (1999) Catechol-O-methyltransferase (COMT): biochemistry, molecular biology, pharmacology, and clinical efficacy of the new selective COMT inhibitors. *Pharmacological Reviews*, **512**, 593–628.

Millikan RC, Pittman GS, Tse CK, Duell E, Newman B and Savitz D (1998) Catechol-O-methyltransferase and breast cancer risk. *Carcinogenesis*, **19**, 1943–1947.

Mozier NM and Hoffman JL (1988) S-adenosyl-L-methionine:thioether S-methyltransferase, a new enzyme in sulphur and selenium metabolism. *Journal of Biolological Chemistry*, **263**, 4527–4531.

Nagata K and Yamazoe Y (2000) Pharmacogenetics of sulphotransferase. *Annual Review of Pharmacology and Toxicology*, **40**, 159–176.

Nishibori M, Tahara A, Sawada K, Sakiyama J, Nakaya N and Saeki K (2000) Neuronal and vascular localization of histamine N-methyltransferase in the bovine central nervous system. *European Journal of Neuroscience*, **12**, 415–424.

Nissinen E, Linden I-B, Schultz E and Kaakkola S (1988) Catechol-O-methyltransferase activity in human and rat small intestine. *Life Science*, **42**, 2609–2614.

Raftogianis R, Creveling CR, Weinshilboum R, Weisz J (2000) Estrogen metabolism by conjugation. *Journal of the National Cancer Institute*, **27**, 113–124.

Rafferty MF and Grunewald GL (1982) The remarkable substrate activity for phenethanolamine-N-methyltransferase of some configurationally defined adrenergic compounds lacking a side chain hydroxyl group: Conformationally defined adrenergic agents 6. *Molecular Pharmacology*, **22**, 127–132.

Regina M, Korhonen VP, Smith TK, Alakeuijala L and Eloranta TO (1993) Methionine toxicity in the rat in relation to hepatic accumulation of S-adenosylmethionine: Prevention by dietary stimulation of the hepatic transulphuration pathway. *Archives of Biochemistry and Biophysics*, **300**, 598–607.

Relling MV, Hancock ML, Rivera GK, Sandlund JT, Riberio RC, Kryneski EY, Pui C-H and Evans WE (1999) Mercaptopurine therapy intolerance and heterozygosity at the thiopurine S-methyltransferase gene locus. *Journal of the National Cancer Institute*, **91**, 2001–2008.

Salminen M, Lundstrum K, Tilgmann C, Savolainen R, Kalkkinen N and Ulmanen I (1990) Molecular cloning and characterization of rat liver catechol-O-methyltransferase. *Gene*, **93**, 241–247.

Smit NPM, Pavel S and Riley PA (2000) Mechanisms of control of the cytotoxicity of orthoquinone intermediates of melanogenesis. In *Role of Catechol Quinone Species in Cellular Toxicity*, Creveling CR (ed.), F. P. Graham, Johnson City, TN, pp. 191–245.

Spatz M, Kaneda N, Sumi C, Nagatsu I, Creveling CR and Nagatsu T (1986) Presence of catechol-O-methyltransferase activity in separately cultured cerebromicrovascular endothelial and smooth muscle cells. *Brain Research*, **381**, 363–367.

Stevens JL and Bakke JE (1990) S-Methylation. In *Conjugation Reactions in Drug Metabolism*, Mulder GJ (ed.), Taylor and Francis, New York, pp. 251–265.

Takamura M, Tanaka T, Taguchi Y, Imamura I, Miszuguchi H, Kuroda M, Fukui H, Yamatodani A and Wada H (1992) Histamine N-methyltransferase from rat kidney: cloning nucleotide sequence, and expression in E. coli cells. *Journal of Biological Chemistry*, **267**, 15687–15691.

Thakker DR and Creveling CR (1990) O-Methylation, *In Conjugation Reactions in Drug Metabolism*, Mulder GJ (ed.), Taylor and Francis, London, pp. 193–231.

Thakker DR, Boehlert C, Kirk KL, Antkowiak R and Creveling CR (1986) Regioselectivity of catechol-O-methyltransferase. *Journal of Biological Chemistry*, **261**, 178–184.

Thompson PA and Ambrosone C (2000) Molecular epidemiology of genetic polymorphisms in estrogen metabolizing enzymes in human breast cancer. *Journal of the National Cancer Institute*, **27**, 125–134.

Thompson PA, Shields PG, Freudenheim JL, Stone A and Vena JE (1998) Genetic polymorphism in catechol-O-methyltransferase, menopausal status, and breast cancer risk. *Cancer Research*, **58**, 2107–2110.

Tenhunen J, Salminen M, Lundstrom K, Kiviluoto T, Savolainen R and Ulmanen I (1994) Genomic organization of the human catecol O-methyltransferase gene and its expression from two distinct promoters. *European Journal of Biochemistry*, **223**, 1049–1059.

Vidgren J and Ovaska M (1997) Structural aspects in the inhibitor design of catechol-O-methyltransferase, In *Structure-based Drug Design*, Veerapandian P (ed.), Marcel Dekker, New York, pp. 343–363.

Vidgren J, Svensson LA and Liljas A (1994) Crystal structure of catechol-O-methyltransferase, *Nature*, **368**, 354–358.

Vidgren J, Ovaska M, Tenhunen K, Tilgmann C, Lotta T, and Mannisto PT (1999) Catechol-O-methyltransferase In *Structure and Function of AdoMet-dependant Methyltransferases*, Cheng X and Blumenthal RM (eds), World Scientific, Singapore, pp. 55–91.

Wasserman SI (1983) Mediators of immediate hypersensitivity (1983) *Journal of Allergy Clinical Immunology*, **72**, 101–105.

Watanabe T, Taguchi N, Shiosaka S, Tanaka J, Kubota H, Terao Y, Tohyama M and Wada H (1984) Distribution of the histaminergic neuron system of rat: a fluorescent immunohistochemical analysis with histidine decarboxylase as a marker. *Brain Research*, **295**, 13–25.

Weinshilboum RM and Raymond FA (1977) Inheritance of low erythrocyte catechol-O-methyltransferase activity in man. *American Journal of Human Genetics*, **29**, 125–135.

Weinshilboum R (1989a) Thiol S-methyltransferases. I. Biochemistry. In *Sulfur-containing Drugs and Related Organic Compounds*, Damani LA (ed.), Ellis Horwood, Chichester, Vol. 2, Part A, pp. 121–142.

Weinshilboum R (1989b) Thiol S-methyltransferases. II. Pharmacogenetics In *Sulfur-containing Drugs and Related Organic Compounds*. Damani LA (ed.), Ellis Horwood, Chichester, Vol. 2, Part A, pp. 144–157.

Weinshilboum RM, Otterness DM and Szumlanski CL (1999) Methylation pharmacogenetics: catechol-O-methyltransferase, thiopurine methyltransferase, and histamine N-methyltransferase. *Annual Reviews of Pharmacology and Toxicology*, **39**, 19–52.

Weisinger RA and Jacoby WB (1980) S-Methylation: thiol S-methyltransferases. In *Enzymatic Basis of Detoxification*, Jacoby WB (ed.), Academic Press, New York, Vol. 2, pp. 131–145.

Woodward RW, Tsai MD, Floss HG, Gooks PA and Coward JK (1980) Stereochemical course of the transmethylation catalyzed by catechol-O-methyltransferase. *Journal of Biological Chemistry*, **255**, 9124–9127.

Yager JD (2000) Endogenous estrogens as carcinogens through metabolic activation. *Journal of the National Cancer Institute*, **27**, 67–73.

Zapia V, Zydeck-Cwich CR and Schlenk F (1969) The specificity of S-adenosylmethionine derivatives in methyltransfer reactions. *Journal of Biological Chemistry*, **214**, 4499–4509.

Zhu BT and Liehr JG (1996) Inhibition of Catechol O-methyltransferase-catalyzed O-methylation of 2- and 4-hydroxyestradiol by quercetin. *Journal of Biological Chemistry*, **271**, 1357–1363.

Zhu BT, Ezell EL and Liehr JG (1994) Catechol-O-methyltransferase-catalyzed rapid O-methylation of mutagenic flavonoids. *Journal of Biolological Chemistry*, **269**, 292–299.

Zhu BT, Patel UK, Cai MX, Lee AJ and Conney AH (2000) Rapid conversion of tea catechins to monomethyl products by rat liver cytosol. *Analytical Biochemistry*, in press.

Ziegler DM, Ansher SS, Nagata T, Kadluber FF and Jacoby WB (1988) N-Methylation: Potential mechanism for metabolic activation of carcinogenic primary arylamines. *Proceedings of the National Academy of Sciences, USA*, **85**, 2514–2517.

14 The Amino Acid Conjugations

G. B. Steventon and A. J. Hutt

King's College London, UK

Introduction

In 1829 Liebig isolated a compound from horse urine which he called hippuric acid (Greek: acid from horse urine) and was able to show that the material contained both a benzoyl group and nitrogen. Keller isolated the same compound from his own urine, in 1842, following self-administration of benzoic acid and the structure of hippuric acid was determined by Dessaignes, three years later, when he found both benzoic acid and glycine on treatment of the material with inorganic acid (Conti and Bickel 1977). Thus conjugation of benzoic acid with glycine to yield hippuric acid or benzoylglycine (Figure 14.1) is generally accepted to be the first xenobiotic transformation reaction to be discovered (Smith and Williams 1970; Conti and Bickel 1977). The other major metabolic transformation of xenobiotic carboxylic acids, namely conjugation with glucuronic acid, was also discovered in the nineteenth century (Williams 1959; Conti and Bickel 1977).

The significance of xenobiotic metabolism and particularly that of the xenobiotic carboxylic acids to the initial development of biochemistry cannot be overemphasised. For example, the application of higher homologues of both benzoic acid and the phenylacetic acids ultimately resulted in the elucidation of the β-oxidation pathway of fatty acid metabolism (Dakin 1922), and glycine was shown to be a constituent of hippuric acid before it was found in glycocholic acid, a bile acid (Young 1977).

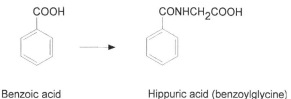

Benzoic acid   Hippuric acid (benzoylglycine)

Figure 14.1 Conjugation of benzoic acid to hippuric acid: the first reaction of drug metabolism.

Enzyme Systems that Metabolise Drugs and Other Xenobiotics. Edited by C. Ioannides.
© 2002 John Wiley & Sons Ltd

Since these initial observations, conjugation of xenobiotic carboxylic acids with endogenous amino acids has been shown to be an important pathway in the biotransformation of a number of compounds in a variety of species. The reaction involves the formation of an amide or peptide bond between the carboxyl group of the xenobiotic acid and the amino group of the endogenous compound. The conjugation reaction is generally accepted to be a two-step process involving initial activation of the carboxyl group to yield a reactive acyl-CoA thioester (Figure 14.2, equations 1 and 2), followed by acyl transfer to the amino group of an amino acid (Figure 14.2, equation 3) (Killenberg and Webster 1980; Caldwell 1982). Thus selectivity, or specificity, may be exerted at either the activation and/or acyl-transfer steps.

As pointed out above, the two major metabolic options of carboxylic acids have been known for well over a century and it was only relatively recently that a number of alternative metabolic pathways have been elucidated and their potential toxicological significance appreciated (Figure 14.3) (Hutson 1982; Caldwell 1984, 1985; Fears 1985). Conjugation with glucuronic acid is discussed elsewhere in this book (Chapters 5 and 8) and will not be examined in any detail here. The formation of acyl-coenzyme A (acyl-CoA) thioester intermediates (Figure 14.2, equations 1 and 2) is of significance in both the metabolism of xenobiotic acids and in intermediary biochemistry, and the alternative pathways associated with carboxylic acids are those of the acyl-CoA thioester intermediates. It is therefore appropriate to briefly provide an overview of the fate of these reactive intermediates.

1. $Ar\text{-}COOH + ATP \rightarrow Ar\text{-}CO{\sim}AMP + PPi + H_2O$

2. $Ar\text{-}CO{\sim}AMP + CoA\text{-}SH \rightarrow Ar\text{-}CO{\sim}S\text{-}CoA + AMP$

3. $Ar\text{-}CO{\sim}S\text{-}CoA + NH_2CH_2COOH \rightarrow Ar\text{-}CO\text{-}NHCH_2COOH + CoA\text{-}SH$

Figure 14.2. Reaction sequence of amino acid conjugation.

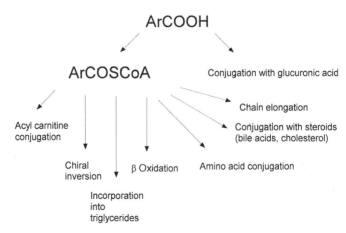

Figure 14.3 Biotransformation of xeno/endobiotic carboxylic acids.

Reactions of Acyl-coenzyme A thioesters

With respect to the topic of this chapter, the transfer of the acyl moiety to the amino group of an amino acid is obviously the most significant of the reactions outlined in Figure 14.3. However, a number of alternative pathways are possible and what may appear as relatively minor changes in the structure of the xenobiotic acid may result in significant alterations in the product ultimately formed.

Carnitine, an essential cofactor required for the transport of long-chain fatty acids into mitochondria, has been shown to yield conjugates with xenobiotic carboxylic acids following acyl transfer onto the secondary alcohol group. Thus cyclopropane carboxylic acid (Quistad et al. 1978a–c, 1986), pivalic acid (Vickers et al. 1985; Totsuka et al. 1992; Mizojiri et al. 1995) and valproic acid (Millington et al. 1985) are all excreted as carnitine conjugates following administration either as such, or as metabolic precursors, to both animals and man. In the case of cyclopropane carboxylic acid, the corresponding glycine conjugate is also excreted (Quistad et al. 1978a) and Quistad et al. (1986) were able to detect very small (ca 0.04% of the dose) quantities of benzoylcarnitine following administration of benzoic acid to the rat. Kanazu and Yamaguchi (1997) have carried out a comparative in vitro study to examine the relative extent of carnitine and glycine conjugation using rat hepatocytes and kidney slices. Cyclopropane and cyclobutane carboxylic acids were found to be the best substrates for carnitine conjugation in both tissues, both compounds also yielding glycine conjugates.

Acyl-transfer to oxygen may also result in the formation of sterol esters. For example, the pyrethroid insecticide fluvalinate undergoes ester hydrolysis to yield an anilino acid derivative which undergoes conjugation with bile acids in rats, chickens and cows (Quistad et al. 1982). Similarly, the hypolipidaemic drug, CCD (1-(4-carboxy-phenoxy)-10-(4-chlorophenoxy)decane) forms a cholesteryl ester conjugate in the rat (Fears et al. 1982).

Acyl-transfer to carbon results in the addition of either single or multiple acetate units and elongation of the carbon chain (Caldwell and Marsh 1983). For example, 5-(4-chlorobut-1-yl)picolinic acid undergoes addition of a two-carbon unit to yield products corresponding to the β-ketoacid, α,β-unsaturated acid and the corresponding saturated analogue (Miyazaki et al. 1976). These metabolites were found in the urine of both animals and man following drug administration. Similarly, 3-hydroxy-3-phenylpropionic acid has been identified in horse urine following administration of benzoic acid (Marsh et al. 1982).

The active acyl group may become involved with the intermediates of lipid biosynthesis (Caldwell 1984) and undergo acyl transfer to oxygen to yield hybrid triacylglycerols (Hutson 1982; Fears 1985) or alternatively may be incorporated into triglycerides following chain elongation (Hutson 1982; Caldwell 1985). The pharmacological and toxicological significance of these alternative pathways, the majority of which are quantitatively minor, have been discussed elsewhere (Fears 1985; Hutson et al. 1985).

In the case of the 2-arylpropionic acid non-steroidal anti-inflammatory drugs (NSAIDs), e.g. ibuprofen and fenoprofen, the formation of the acyl-CoA thioesters is of pharmacological significance. These agents are used as racemic mixtures even though their main pharmacological activity, inhibition of cyclooxygenase resides in the enantiomers of the *S*-configuration. However, the *R*-enantiomers of a number of these agents form the corresponding CoA thioesters which subsequently undergo inversion of chirality of the propionic acid moiety, followed by hydrolysis to yield the active enantiomer (Hutt and Caldwell 1983; Caldwell *et al.* 1988). Recent evidence, associated with the formation of amino acid conjugates, has indicated that the reaction, in some species, may be highly stereoselective rather than stereospecific (see below).

Amino acids and conjugation

As pointed out above, conjugation of benzoic acid with glycine is generally accepted to be the first reaction of drug metabolism to be discovered but it was not until the 1980s that Marsh *et al.* (1981) demonstrated the formation of hippuric acid following the administration of benzoic acid to the horse. Since the initial observation, a number of alternative amino acids have been shown to be involved and the history of the amino acid conjugations is summarised in Table 14.1.

The amino acid utilised for conjugation is highly dependent on both the structure of the xenobiotic carboxylic acid and the animal species under investigation. The most frequently observed amino acid conjugates are those with glycine, which is utilised by the majority of animal species for the conjugation of a wide variety of carboxylic acids including aliphatic, aromatic, heteroaromatic and phenylacetic acid derivatives.

The first example of an alternative to glycine conjugation was reported by Jaffe, in 1877, who found that benzoic acid underwent conjugation with ornithine in the hen. Ornithine conjugation, unlike the other amino acid conjugations, involves the acylation of both amino groups, thus in the case of benzoic acid the product is N^2, N^5-dibenzoylornithine or ornithuric acid. Conjugation with ornithine has been found to occur in other avian and in some reptile species (Smith 1958), and appears to be associated with uricotelic species (Killenberg and Webster 1980), species which excrete uric acid as the major nitrogenous waste product of amino acid metabolism, a

Table 14.1 Discovery of the amino acid conjugations

Amino acid	Carboxylic acid	Species	References
Glycine	Benzoic acid	Man	Keller (1842)
Ornithine	Benzoic acid	Hen	Jaffe (1877)
Glutamine	Phenylacetic acid	Man	Thierfelder and Sherwin (1914)
Serine	Xanthurenic acid	Rat	Rothstein and Greenberg (1957)
Glutamic acid	4-Nitrobenzoic acid	Spider	Smith (1962)
Arginine	4-Nitrobenzoic acid	Spider	Smith (1962)
Histidine	Benzoic acid	*Peripatus*	Jordan *et al.* (1970)
Taurine	Phenylacetic acid	Pigeon	James *et al.* (1971)
Alanine	4,4'-Dichlorodiphenylacetic acid	Mouse	Wallcave *et al.* (1974)
Aspartic acid	2,4'-Dichlorodiphenylacetic acid	Hamster	Reif and Sinsheimer (1975)
	2,4'-Dichlorodiphenylacetic acid	Rat	Reif and Sinsheimer (1975)

process characteristic of terrestrial species which develop within a shell where the nitrogenous waste products are stored in an insoluble form. Ornithine conjugation is not a general reaction of all avian species. For example, benzoic acid yields ornithine conjugates in domestic fowl (Galliformes), ducks and geese (Anseriformes), and hippuric acid in pigeons and doves (Columbiformes), whereas parrots yield neither amino acid conjugate (Baldwin *et al.* 1960). The formation of an *N*-acetylornithine conjugate of 3-phenoxybenzoic acid has been reported in the chicken (Huckle *et al.* 1982).

L-Glutamine conjugation, first reported to be the route of biotransformation of phenylacetic acid in humans (Thierfelder and Sherwin 1914), appears to be restricted in the main to arylacetic acids, e.g. phenylacetic acid and related compounds in mammals. The conjugation of phenylacetic acid and related compounds (e.g. 4-chlorophenylacetic acid and indol-3-ylacetic acid) with L-glutamine was believed to be confined to the anthropoid apes, Old and New World monkeys and humans (James *et al.* 1972a,b; Bridges *et al.* 1974). However, both phenylacetic acid and 4-chlorophenylacetic acid yield small quantities of the corresponding glutamine conjugates in the ferret (Hirom *et al.* 1977; Idle *et al.* 1978) and 2-naphthylacetic acid undergoes extensive conjugation with L-glutamine in the ferret, rabbit and rat (Emudianughe *et al.* 1977, 1978). Formation of L-glutamine conjugates of benzoic acid derivatives in the house fly and arachnids has also been reported (Smith 1962; Hitchcock and Smith 1964; Esaac and Casida 1968). With respect to drug metabolites, diphenylmethoxyacetic acid, a metabolite of diphenhydramine, yields an L-glutamine conjugate following drug administration to the Rhesus monkey (Drach and Howell 1968; Drach *et al.* 1970). In addition, α-fluorovalproic acid and the corresponding fluorinated 2-propyl-4-pentenoic acid oxidation product undergo glutamine conjugation in rats and mice (Tang and Abbott 1997; Tang *et al.* 1997).

Taurine, 1-aminoethane sulphonic acid, while not strictly an amino acid, yields conjugates with xenobiotic carboxylic acids and taurine conjugation is generally classified as an amino acid conjugation. The reaction appears to be primarily associated with carnivorous species, e.g. dog (Jordan and Rance 1974; Sakai *et al.* 1984) and ferret (Idle *et al.* 1978), and some aquatic species (James and Bend 1976; James 1982) but is known to occur also in the rat (Emudianughe *et al.* 1978; Egger *et al.* 1982; Peffer *et al.* 1987), horse (Marsh *et al.* 1981) and to a minor extent in humans (Shirley *et al.* 1994). In terms of xenobiotic acids, taurine conjugation was thought to be restricted to aryl- and aryloxyacetic acids but some aromatic, e.g. 3-phenoxybenzoic acid (Hutson and Casida 1978; Huckle *et al.* 1981c) and aliphatic acids, e.g. trimoprostil (Kolis *et al.* 1986), have been shown to undergo this conjugation.

The most commonly encountered alternative amino acids to glycine are L-glutamine, L-ornithine and taurine but there are, however, several examples of other amino acids involved in conjugation reactions but these appear to be restricted both in terms of their species occurrence and the substrate utilised (Table 14.2) (Caldwell *et al.* 1980; Quistad 1986). In addition, a small number of dipeptide conjugates have also been reported and these are summarised in Table 14.2.

Polyglutamyl conjugates of methotrexate, a folic acid antagonist, have been reported in several species including humans (Baugh *et al.* 1973; Shin *et al.* 1974;

Table 14.2 Atypical amino acids and dipeptides used in conjugation reactions

Amino acid or dipeptide	Carboxylic acid	Species	References
Alanine	4,4'-Dichlorodiphenylacetic acid	Mouse	Wallcave et al. (1974)
	Piperonylic acid	Hamster	Gingell (1976)
		Housefly	Esaac and Casida (1968)
Aspartic acid	2,4-Dichlorodiphenylacetic acid	Rat	Reif and Sinsheimer (1975)
Serine	4,8'-Dihydroxyquinaldic acid	Rat	Rothstein and Greenberg (1957)
	2,4'-Dichlorodiphenylacetic acid	Rat	Feil et al. (1973)
			Reif and Sinsheimer (1975)
	4,4'-Dichlorodiphenylacetic acid	Mouse	Gingell (1976)
	Piperonylic acid	Housefly	Esaac and Casida (1968, 1969)
Histidine	Benzoic acid	Peripatus	Jordan et al. (1970)
Glutamic acid	Benzoic acid	Indian fruit bat	Idle et al. (1975)
		African bat	Collins et al. (1977)
	trans 3-(2,2-Dichlorovinyl)-2,2-dimethylcyclopropane carboxylic acid	Cow	Gaughan et al. (1977)
	Piperonylic acid	Housefly	Esaac and Casida (1969)
	3-Phenoxybenzoic acid	Cow	Gaughan et al. (1977)
Arginine	Benzoic acid	Scorpion	Hitchcock and Smith (1966)
	p-Aminobenzoic acid	House spider	Smith (1962)
		Millipede	Hitchcock and Smith (1964)
		Arachnids	Hitchcock and Smith (1964)
Glycyltaurine	Quinaldic acid	Cat	Kaihara and Price (1961)
			Morris and Price (1963)
Glycylglycine	Quinaldic acid	Cat	Kaihara and Price (1965)
Aspartic acid and serine[a]	4,4'-Dichlorodiphenylacetic acid	Rat	Pinto et al. (1965)
Glycylvaline	3-Phenoxybenzoic acid	Mallard duck	Huckle et al. (1981a)

[a] Sequence unknown

Israili *et al.* 1977). The significance of this biotransformation is unclear but these conjugates may contribute to the biological activity of the drug (Montgomery *et al.* 1979).

Structure–metabolism relationships

The biotransformation of xenobiotic carboxylic acids is dependent on the size and the nature of the substituents surrounding the carboxyl group. The presence and nature of other functional groups within the molecule and their possible biotransformation will also influence the ultimate fate of the acid. To date there have been few systematic attempts to examine the structure–metabolism relationships of amino acid conjugation. The majority of the literature resulting from an examination of the metabolic fate of a particular xenobiotic carboxylic acid in a range of animal species, an approach which is handicapped by both species variation in the amino acid used for the conjugation and alternative, possibly competing, metabolic pathways.

Structure–metabolism relationships for amino acid conjugation need to be carried out with some care as selectivity, or specificity, may be exerted at either of the two steps of the reaction, i.e. formation of the acyl-CoA thioester or in the acyl-transfer. In *in vivo* studies of the two major metabolic options available to carboxylic acids, i.e. conjugation with either an amino acid or glucuronic acid, the ultimate fate of the compound will depend on a number of factors including the administered dose, availability of the conjugating agent, diet and species under examination (Hutt and Caldwell 1990).

Aliphatic and alicyclic carboxylic acids

There are few examples of exogenous aliphatic carboxylic acids, or 3-aryl substituted acids containing three carbon atoms in a straight chain, undergoing amino acid conjugation, presumably due to their facile β-oxidation. However, in addition to the metabolism of xenobiotic carboxylic acids, amino acid conjugation is an important metabolic pathway for the elimination of endobiotic acids that may accumulate in a number of metabolic diseases which result in acidaemia, e.g. isovaleric acid undergoes conjugation with glycine (Tanaka and Isselbacher 1967). In cases of medium-chain acyl-CoA dehydrogenase deficiency, 3-phenylpropionic acid, a compound that normally undergoes β-oxidation to benzoic acid followed by excretion as hippuric acid, is excreted as 3-phenylpropionylglycine (Bennett *et al*. 1992). In such cases, unusual amino acids, not normally associated with amino acid conjugation, may be utilised, e.g. sarcosine (Lehnert 1983). Several investigations using purified enzyme systems for the formation of both CoA thioesters and *N*-acyltransferase activity have indicated that short-chain aliphatic acids may form amino acid conjugates *in vitro* (Mahler *et al*. 1953; Schachter and Taggart 1954; Nandi *et al*. 1979).

Several 3-arylpropionic acid derivatives have been reported to yield amino acid conjugates, the carboxylic acid metabolites of the anti-histamines brompheniramine and chlorpheniramine have been found to yield amino acid conjugates in the urine of both humans and dogs (Bruce *et al*. 1968). Glycine conjugates have similarly been reported for the unsaturated acid, cinnamic acid and the related compounds β-methylcinnamic and 3,4-dimethoxycinnamic acids (Williams 1963; Solheim and Scheline 1976; Hoskins 1984).

The metabolism of the anticonvulsant agent valproic acid involves cytochrome P450-mediated formation of 2-propyl-4-pentenoic acid which is implicated in the hepatotoxicity of the drug via β-oxidation to yield a reactive metabolite. Valproic acid also undergoes glycine conjugation in rats but to a minor extent, less than 1% of the dose. The α-fluorinated derivatives of both valproic acid and 2-propyl-4-pentenoic acid have been synthesised in order to prevent β-oxidation and hence hepatotoxicity. As a result of the structural modification, neither compound undergoes β-oxidation and the metabolic pathway switches to amino acid conjugation. Both yield L-glutamine conjugates in the rat and mouse (Tang and Abbott 1997; Tang *et al*. 1997).

The formation of taurine conjugates of relatively long chain aliphatic acids has been reported. For example the metabolism of prostaglandin E_2(PGE$_2$) in rat hepatocytes yields taurine conjugates of the major metabolites arising from β-oxidation dinor and tetranor prostaglandin E_1 (PGE_1) and dinor-PGE_2 (Hankin *et al.* 1997). Whether these metabolites are produced *in vivo* is yet to be investigated. However, taurine conjugates of the PGE_2 analogue trimoprostil and its β-oxidation products have been identified in rat bile following administration of the parent drug (Kolis *et al.* 1986). Similarly both *all-trans* and 9-*cis* retinoic acids have been reported to undergo a variety of biochemical transformations in the rat resulting in the formation of highly polar metabolites identified as taurine conjugates (Skare *et al.* 1982; Shirley *et al.* 1996).

Glycine conjugation of isopropoxyacetic acid has been found in both rat and dog following the administration of isopropyl oxitol (Hutson and Pickering 1971). While the glycine conjugate of cyclopropylcarboxylic acid, a metabolite of the miticide, hexadecyclopane carboxylate, has been reported in the urine of rats, cows and dogs (Quistad *et al.* 1978a–c). The glutamic acid conjugate of *trans*-3-(2, 2-dichlorovinyl)-2, 2-dimethylcyclopropane carboxylic acid has been found in the cow. Both geometrical isomers have been reported to give rise to glycine, serine and glutamic acid conjugates in insects (Unai and Casida 1977). The formation of glycine conjugates of cyclohexanoic acid derivatives, hexahydrohippuric acid and 3,4,5,6-tetrahydrohippuric acid, together with hippuric acid, have been reported in rats and perfused rat liver following the administration of cyclohexanoic acid and shikimic acid (Brewster *et al.* 1977a,b, 1978). Hexahydrohippurate has also been reported to occur in the urine of cattle, horses and elephants (Balba and Evans 1977).

Aromatic carboxylic acids

Benzoic and heterocyclic aromatic acids are mainly conjugated with glycine in mammalian species, other amino acid conjugates are also generated but their occurrence is restricted in terms of species. Between 1944 and 1955, Bray and co-workers examined the fate, in terms of conjugation with either glycine or glucuronic acid, of a variety of substituted benzoic acids in the rabbit (summarised in Williams 1959), and this group of compounds is the most extensively examined in terms of quantitative structure–metabolism relationships. Hansch *et al.* (1968) using Bray's data for the *para*-substituted compounds established a parabolic relationship between the logarithm of the percentage of the dose undergoing conjugation with glycine (Log MR) and log P:

$$\text{Log MR} = -0.665(\text{Log P})^2 + 3.153\text{Log P} - 1.763$$
$$r = 0.916, \ s = 0.187, \ n = 8$$

Subsequently, examination of the data obtained for the *ortho*-substituted compounds indicated that the steric bulk of the substituent, as measured by Taft's steric parameter, E_s, was the determining factor (Caldwell *et al.* 1980), the extent of glycine conjugation expressed as a percentage of the dose decreasing as steric bulk of the *ortho* substituents increased, which Caldwell *et al.* (1980) interpreted as steric hindrance of the *ortho*-substituent for the formation of the CoA thioester.

Kasuya *et al.* (1990, 1991) examined structure–metabolism relationships for the glycine conjugation of a series of benzoic acids using liver and kidney mitochondria from rat and mouse. Physicochemical parameters found to be of significance for the rate of glycine conjugation were Van der Waals volume and the calculated logarithm of the octanol/water partition coefficient (CLOGP). These workers also re-examined Bray's data for *para*-substituted benzoic acid conjugation in the rabbit, and similar to their *in vitro* observations both Van der Waals volume and CLOGP were found to be significant parameters (Kasuya *et al.* 1990). Overall conjugation with glycine was found to increase with lipophilicity and decrease with steric bulk of substituents in the 3- and 4-positions of the aromatic ring. More recent investigations have examined the overall fate of benzoic acids, i.e. excretion either unchanged or as the glycine or glucuronide conjugates, following administration to rabbits (again using Bray's data) and rats (Ghauri *et al.* 1992; Cupid *et al.* 1996, 1999). Using computational chemistry and multivariate statistical methods, relationships were derived which allow prediction of the urinary excretion of both conjugates together with the unchanged acid. Interestingly, the urinary excretion of benzoylglycines in the rat was found to be dependent on molecular weight and the energy of the highest occupied molecular orbital (HOMO) of the acid (Cupid *et al.* 1999), whereas in the case of the rabbit the most significant parameters were CLOGP, the molar refractivity, the partial charge on C_1 of the aromatic ring and the second principal ellipsoid axis, several of which relate to molecular size and shape (Cupid *et al.* 1996; 1999). Similar to the *in vitro* study by Kasuya *et al.* (1990), the most important physicochemical property was CLOGP (Cupid *et al.* 1996). Such species differences in the structure–metabolism relationships of a relatively simple series of compounds serve to illustrate the complexity of the reaction sequence involved in amino acid conjugation.

Arylacetic acids

The amino acid utilised for conjugation of arylacetic acids varies between species, rodents such as rats producing predominantly glycine conjugates whereas primates, including humans, utilise L-glutamine (Table 14.3). Substitution of these compounds at the α-carbon atom, also has a marked effect on their metabolic fate. For example, 2-phenylpropionic and diphenylacetic acids undergo glucuronidation rather than amino acid conjugation in both animals and humans (Dixon *et al.* 1977a,b).

Table 14.3 Variation in the amino acid conjugation (% dose) of arylacetic acids with structure and species

Species	Rat	Man	Rhesus monkey	Capuchin monkey
Amino acid	Glycine	Glutamine	Glutamine	Glutamine
Phenylacetic acid	99	93	32	64
4-Chlorophenylacetic acid	92	90	40	14
4-Nitrophenylacetic acid	61	–	–	–
1-Naphthylacetic acid	15	0	0	5
2-Naphthylacetic acid	8.6	–	–	–

Data from Caldwell *et al.* (1980), Dixon *et al.* (1977c) and Emudianaghe *et al.* (1978).

The conjugation of the regioisomers 1- and 2-naphthylacetic acid is of interest as the pattern of conjugation both in terms of the pathway, glucuronidation versus amino acid conjugation and the amino acid utilised varies considerably. The major metabolic pathway of the 1-isomer in human, rat and rabbit is glucuronidation, the cat yielding both glycine and taurine conjugates while the glycine conjugation of the compound is relatively low (15% of dose) compared to the substituted phenylacetic acids (61-99%) (Dixon *et al.* 1977c). In contrast the 2-isomer undergoes conjugation with three amino acids simultaneously, namely glycine, L-glutamine and taurine, in addition to glucuronic acid conjugation in the rat, rabbit and ferret (Emudianughe *et al.* 1978), the overall urinary recovery of the 2-isomer being reduced in comparison to 1-naphthylacetic acid but the total undergoing amino acid conjugation being greater. This difference in amino acid conjugation may be due in part to steric hindrance in the case of 1-naphthylacetic acid for the formation of either the CoA thioester or *N*-acyltransferase(s). Thus it would appear that the 2-naphthylacetic acid readily undergoes activation but that selectivity is exerted in the acyl transfer step. These data suggest that 2-naphthylacetic acid may be a useful probe compound for investigating these mechanisms.

The fate of 2-phenylpropionic acid is of interest as this compound undergoes chiral inversion in the rat (Fournel and Caldwell 1986), the initial step of which is formation of a Coenzyme A thioester. As amino acid conjugation does not take place in this species, the selectivity presumably occurs at the level of *N*-acyltransferase. In recent years a number of 2-arylpropionic acid NSAIDs, together with 2-phenylpropionic acid, have been reported to undergo conjugation with taurine, mainly in the dog (Sakai *et al.* 1984; Mori *et al.* 1985; Asami *et al.* 1995; Konishi *et al.* 1999) but also in the rat and mouse (Mohri *et al.* 1998; Egger *et al.* 1982), and to a minor extent in humans (Shirley *et al.* 1994). In addition, 2-phenylpropionic acid has been reported to yield a glycine conjugate following administration to dogs (Tanaka *et al.* 1992). In the case of some compounds, the stereochemistry of the amino acid conjugates has been investigated and there are indications that the *S*-enantiomers of the 2-arylpropionic acids may also form CoA thioesters, but to a much smaller extent than their *R*-antipodes, and that the conjugation reaction may also be stereoselective and vary with species (Tanaka *et al.* 1992; Shirley *et al.* 1994; Mohri *et al.* 1998; Konishi *et al.* 1999).

Aryloxyacetic acids

Aryloxyacetic acid derivatives, e.g. the herbicides 2,4-dichlorophenoxyacetic acid (2, 4-D) and 2,4,5-trichlorophenoxyacetic acid (2,4,5-T) together with the structurally related hypolipidaemic agent, clofibric acid, are known to undergo taurine conjugation in carnivorous, marine and other species (James and Bend 1976; James 1982; Emundianughe *et al.* 1983). 3,4-Dichlorobenzyloxyacetic acid, an agent with potential for the treatment of sickle cell anaemia, has been shown to undergo extensive taurine conjugation (60% of dose) in the rat (Peffer *et al.* 1987). This compound may be useful as a probe for taurine conjugation in other species. Both 2,4-D and 2,4,5-T have been reported to yield small quantities of amino acid conjugates following administration to rats (Grunow and Bohme 1974). An explanation for this observation, in terms of the affinity of the corresponding CoA thioesters for the glycine N-acyltransferase(s) has been postulated (see below; Kelley and Vessey 1986).

Biochemical and molecular mechanisms of amino acid conjugation

ACYL-CoA SYNTHETASES

As pointed out above, the initial reaction in amino acid conjugation is the formation of an acyl CoA thioester mediated by an acyl CoA synthetase or ATP-dependent acid: CoA ligases. These synthetases are divided into three ATP- and one GTP-dependent systems, thus: short-chain/acetyl-CoA synthetase/acetate: CoA ligase (AMP), EC 6.2.1.1; medium-chain/butyryl-CoA synthetase/medium-chain fatty acid: CoA ligase (AMP), EC 6.2.1.2; long-chain fatty acyl-CoA synthetase/acyl-CoA synthetase/long-chain fatty acid: CoA ligase, EC 6.2.1.3 and medium-chain fatty acid: CoA ligase (GDP), EC 6.2.1.10.

Of these four systems, the medium-chain CoA synthetase (EC 6.2.1.2) is principally associated with the activation of benzoic acids and phenylacetic acids whereas the long-chain CoA synthetases (EC 6.2.1.3) are primarily involved in the activation of the 2-arylpropionic acids (Sevoz et al. 2000).

The medium-chain CoA synthetase was initially purified by Mahler et al. (1953) from bovine hepatic mitochondria. The enzyme showed broad substrate specificity for straight chain aliphatic acids (C_4-C_{12}) with optimal activity at C_7. In addition, benzoic and phenylacetic acids, together with several branched chain aliphatic carboxylic acids were found to be activated by this enzyme. However, salicylic and p-aminosalicylic acids were found not to be substrates (Schachter and Taggart 1954). Killenberg et al. (1971), again using the same enzyme source, purified two medium-chain CoA synthetases, one being able to activate salicylic and p-aminosalicylic acids while the other could not. Differences in enzyme stability were also reported, the salicylate CoA synthetase being less stable than the non-salicylate CoA synthetase. By the mid-1970s, three soluble CoA synthetases, medium-chain acyl-CoA synthetase, a salicylate CoA synthetase and a propionyl-CoA synthetase, had been purified from guinea pig liver mitochondria (Groot and Scheek 1976).

More recently, Kasuya et al. (1996) reported the purification and characterisation of a medium-chain acyl-CoA synthetase from bovine hepatic mitochondria. Enzyme specificity was examined using aliphatic acids (C_3-C_{10}), substituted benzoic acids and 1- and 2-naphthylacetic acid. Optimal activity was found with hexanoic acid, but benzoic acid derivatives with large alkyl and alkoxy groups in the para- or meta-positions were also highly active whereas ortho-substituted derivatives exhibited no activity (Kasuya et al. 1996). Such data corresponded to in vivo observations where the extent of glycine conjugation decreased with increased steric bulk of the ortho-substituent (Caldwell et al. 1980). Also of interest is that both regioisomers of naphthylacetic acid had activities similar to that of benzoic acid. The molecular mass of the Kasuya-enzyme was determined to be 65 kDa, whereas the mass of the Mahler-enzyme was estimated to be between 30 and 60 kDa (Mahler et al. 1953) and that of a similar enzyme isolated from rat liver mitochondria was 47 kDa (Groot and Scheek

1976). However, stability may be a problem and storage may result in either dissociation and/or degradation (Kasuya et al. 1996).

The rat possesses five long-chain acyl-CoA synthetases (ACS1-ACS5) (Sevoz et al. 2000). Each ACS appears to have a marked tissue distribution and completely different regulation from those of the others (Suzuki et al. 1995). Rat ACS1 is found predominantly in the liver, heart and adipose tissue (Suzuki et al. 1991), while ACS2 and ACS3 are the major forms in the CNS (Fujino and Yamamoto 1992, Fujino et al. 1996). The remaining enzymes, ACS4 and ACS5, are highly expressed in steroidogenic tissues and the small intestine (Kang et al. 1997; Oikawa et al. 1998). These regulator and tissue distribution differences may reflect the biological roles of these enzymes with regard to their function in fatty acid metabolism (Oikawa et al. 1998). Using rat recombinant ACS1 and ACS2 expressed in E. coli and $(-)$-(R)-ibuprofen or $(-)$-(R)-fenoprofen, Sevoz et al. (2000) reported that ACS1 appeared to be the major enzyme involved in the first step of the chiral inversion of the 2-arylpropionic acids in vitro.

ACYL-CoA: AMINO ACID N-ACYLTRANSFERASE

The transfer of the acyl group from CoA thioester to the amino group of the amino acid is catalysed by an N-acyltransferase. The first example of this type of enzyme to be partially purified was the glycine N-acyltransferase (EC 2.3.1.13) from bovine hepatic mitochondria (Schachter and Taggart 1954). The enzyme was found to show absolute specificity with respect to the amino acid but to catalyse the transfer of a variety of both aliphatic (C_2-C_{10}) and aromatic acyl groups. Moldave and Meister (1957) partially purified a glutamine N-phenylacetyltransferase (EC 2.3.1.14) from the cytosolic fractions of human liver and kidney. This enzyme was found to catalyse both phenacylation of glutamine and the benzoylation of glycine, the latter reaction being carried out at a considerably slower rate than the former. The kidney enzyme appeared to have a higher specific activity than the hepatic enzyme (Moldave and Meister 1957). Glycine N-acyltransferases have also been purified from human and bovine hepatic mitochondrial preparations (Tishler and Goldman 1970; Forman et al. 1971). Both enzyme preparations were found to transfer salicyl and benzoyl acyl groups from their corresponding CoA thioesters to produce salicyluric and hippuric acid respectively.

Webster et al. (1976) isolated and purified two acyl-CoA: amino acid N-acyltransferases, with a molecular mass approximately 24 kDa, from Rhesus monkey and human hepatic mitochondrial preparations. Both enzymes were shown to exhibit acyl-acceptor specificity with either glycine or L-glutamine. The 'glycine' conjugating enzyme showed acyl-donor specificity for benzoyl and salicyl Co-A while the 'glutamine' conjugating enzyme used either phenylacetyl or indolylacetyl-CoA (Webster et al. 1976). The presence of only one glutamine N-acyltransferase was indicated by the nearly constant ratio of phenylacetyl and indolylacetyl transferase activities during purification of the enzyme isolated from Rhesus monkey tissues.

Webster et al. (1976) also reported that the amino acid N-acyltransferase activity of both enzymes was inhibited by the acyl donors for the other enzyme. Nandi et al. (1979) found similar results following isolation of two enzymes from bovine hepatic mitochondria. However, in addition they observed that glycine was the preferred acyl acceptor for both enzymes, with L-glutamine and L-asparagine being weak acyl acceptors. The molecular mass of these two enzymes of approximately 33 kDa was in agreement with those of Lau *et al.* (1977) and Forman *et al.* (1971) of 36 and 32 kDa respectively.

The specificity of the two bovine transferases to the CoA thioesters of 2,4-D, 2,4,5-T and phenoxyacetic acid has been examined using benzoyl- and phenylacetyl-CoAs as reference standards for activity (Kelly and Vessey 1986). Phenoxyacetyl-CoA and 2,4,5-T-CoA were found to be substrates for the phenylacetyl- and benzoyl-transferases respectively, whereas 2,4-D-CoA was a substrate for both enzymes. Both enzymes showed a high affinity for the herbicide-CoA thioesters but the reaction rates were low, which was found to be due to increased K_m values for glycine in comparison to the normal substrates (Kelly and Vessey 1986).

Organ location

Although the liver is a major site of amino acid conjugation, the kidney has been known to be involved in this biotransformation since 1870. Quick (1931) extended the early work of Schmeideberg in the 1870s, in showing that in the dog hippuric acid biosynthesis was effected by the kidney and not the liver. Amino acid conjugation has been reported at very low levels of enzymic activity in rat intestinal slices and everted sections (Strahl and Barr 1971). The formation of *p*-aminohippuric acid from *p*-aminobenzoic acid has been reported for rat and guinea pig duodenum homogenates (Irjala 1972). Rabbit small intestine extracts possessed very low levels of glycine *N*-acyltransferase activity. Lung tissue, however, was found to contain no enzymatic activity at all (James and Bend 1978). A similar pattern of results were reported in that human brain, lung, intestine and heart possessed little or no ability to form hippuric acid from benzoic acid (Caldwell *et al.* 1976).

The relative contributions of the kidney and liver to the formation of amino acid conjugates is both species and substrate dependent. The formation of glycine conjugates from *p*-aminobenzoic acid, benzoic acid and salicylic acid was undertaken in tissue slices, homogenates and mitochondria from hepatic and renal sources in the rat, guinea pig, cat and dog (Irjala 1972). The conjugation of *p*-aminobenzoic acid was greater in renal tissue slices than the hepatic tissue in all four species investigated. Benzoic acid conjugation was greater in renal than hepatic slices from cats and dogs, but no significant differences were seen in the rat and guinea pig. With salicylic acid as substrate, the renal slices showed greater activity than the liver in rat, guinea pig and dog. When homogenates were used instead of the tissue slices, a different picture was seen. Both *p*-aminobenzoic acid and benzoic acid conjugation were greater in the liver than kidney in rat and guinea pig. However, this observation was reversed in the cat and dog. Using salicylic acid as substrate, little activity was found in rat hepatic homogenates whereas using guinea pig tissue, hepatic activity was greater than renal, and in the dog the reverse was observed. Finally, using mitochondrial preparations from both renal and hepatic sources in the rat, hepatic mitochondria showed greater activity in glycine conjugation of *p*-aminobenzoic acid than renal mitochondria.

The conjugation of benzoic acid with glycine has been investigated using human hepatic and renal cortex homogenates (Temellini *et al.* 1993). The kidney cortex was found to have the higher activity of the two organs but that the activity was normally distributed in both organs. The enzymic activity of both hepatic and renal preparations from rat, mouse, hamster, gerbil and ferret have been investigated using 3-phenoxybenzoic acid as substrate (Huckle *et al.* 1981b). These investigators examined the overall conjugation reaction, CoA thioester formation and glycine *N*-acyltransferase activity. Examination of the overall reaction indicated that in the ferret and mouse, renal activity was the greatest, the activities were similar in hamster and gerbil renal and hepatic tissue, and that the rat had greater activity in the hepatic tissue. When the activation of the substrate with CoA was investigated, similar results were observed. However, when the glycine *N*-acyltransferase activity was investigated the rat, ferret and gerbil showed higher activity in renal tissue than the liver, whereas the opposite was true for the hamster and mouse (Huckle *et al.* 1981b). It was also found that acyl-CoA formation was 10 to 300-fold slower than glycine *N*-acyltransferase activity, thus making the CoA thioester formation the rate-limiting step in glycine amino acid conjugation (Forman *et al.* 1971, Huckle *et al.* 1981c).

References

Asami M, Takasaki W, Iwabuchi H, Haruyama H, Wachi K, Terada A and Tanaka Y (1995) Stereospecific taurine conjugation of the trans-OH metabolite (active metabolite) of CS-670, a new 2-arylpropionic acid nonsteroidal anti-inflammatory drug in dogs. *Biological and Pharmaceutical Bulletin*, **18**, 1584–1589.

Balba MT and Evans WC (1977) The origin of hexahydrohippurate (cyclohexanoylglycine) in the urine of herbivores. *Biochemical Society Transactions*, **5**, 300–302.

Baldwin DC, Robinson D and Williams RT (1960) Studies in detoxication 82. The fate of benzoic acid in some domestic and other birds. *Biochemical Journal*, **76**, 595–600.

Baugh CM, Krumdieck CL and Nair MG (1973) Polygammaglutamyl metabolites of methotrexate. *Biochemical and Biophysical Research Communications*, **52**, 27–34.

Bennett MJ, Bhala A, Poirier SF, Ragui MC, Willi SM and Hale DE (1992) When do gut flora in the newborn produce 3-phenylpropionic acid? Implications for the early diagnosis of medium-chain acyl-CoA dehydrogenase deficiency. *Clinical Chemistry*, **38**, 278–281.

Brewster D, Jones RS and Parke DV (1977a) The metabolism of cyclohexane carboxylate in the rat. *Biochemical Journal*, **164**, 595–600.

Brewster D, Jones RS and Parke DV (1977b) The metabolism of cyclohexane carboxylic acid in the isolated perfused rat liver. *Xenobiotica*, **7**, 601–609.

Brewster D, Jones RS and Parke DV (1978) The metabolism of shikimate in the rat. *Biochemical Journal*, **170**, 257–264.

Bridges JW, Evans ME, Idle JR, Millburn P, Osiyemi FO, Smith RL and Williams RT (1974) The conjugation of indolylacetic acid in man, monkeys and other species. *Xenobiotica*, **4**, 645–652.

Bruce RB, Turnbull LB, Newman JH and Pitts JE (1968) Metabolism of brompheniramine. *Journal of Medicinal Chemistry*, **11**, 1031–1034.

Caldwell J (1982) Conjugation of xenobiotic carboxylic acids. In *Metabolic Basis of Detoxication Metabolism of Functional Groups*, Jakoby WB, Bend JR and Caldwell J (eds), Academic Press, New York, pp. 271–290.

Caldwell J (1984) Xenobiotic acyl-coenzymes A: critical intermediates in the biochemical pharmacology and toxicology of carboxylic acids. *Biochemical Society Transactions*, **12**, 9–11.

Caldwell J (1985) Novel xenobiotic-lipid conjugates. *Biochemical Society Transactions*, **13**, 852–854.

Caldwell J and Marsh MV (1983) Interrelationships between xenobiotic metabolism and lipid biosynthesis. *Biochemical Pharmacology*, **32**, 1667–1672.

Caldwell J, Idle JR, and Smith RL (1980) The amino acid conjugations. In *Extrahepatic Metabolism of Drugs and Other Foreign Compounds*, Gram TE (ed.), SP Medical and Scientific Books, New York, pp. 453–477.

Caldwell J, Hutt AJ and Fournel-Gigleux S (1988) The metabolic chiral inversion and dispositional enantioselectivity of the 2-arylpropionic acids and their biological consequences. *Biochemical Pharmacology*, **37**, 105–114.

Caldwell J, Moffatt JR and Smith RL (1976) Post-mortem survival of hippuric acid formation in rat and human cadaver tissue samples. *Xenobiotica*, **6**, 275–280.

Collins MW, French MR, Hirom PC, Idle JR, Bassir O and Williams RT (1977) The conjugation of benzoic acid in the African bat, *Epomops franqueti*. *Comparative Biochemistry and Physiology*, **56C**, 103–104.

Conti A and Bickel MH (1977) History of drug metabolism: discoveries of the major pathways in the 19th century. *Drug Metabolism Reviews*, **6**, 1–50.

Cupid BC, Beddell CR, Lindon JC, Wilson ID and Nicholson JK (1996) Quantitative structure–metabolism relationships for substituted benzoic acids in the rabbit: prediction of urinary excretion of glycine and glucuronide conjugates. *Xenobiotica*, **26**, 157–176.

Cupid BC, Holmes E, Wilson ID Lindon JC and Nicholson JK (1999) Quantitative structure–metabolism relationships (QSMR) using computational chemistry: pattern recognition analysis and statistical prediction of phase II conjugation reactions of substituted benzoic acids in rat. *Xenobiotica*, **29**, 27–42.

Dakin HD (1922) *Oxidations and Reductions in the Animal Body*, 2nd edition, Longmans, Green, London, pp. 26–47.

Dixon PAF, Caldwell J and Smith RL (1977a) Metabolism of arylacetic acids 2. The fate of [^{14}C]-hydratropic acid and its variation with species. *Xenobiotica*, **7**, 707–715.

Dixon PAF, Caldwell J and Smith RL (1977b) Metabolism of arylacetic acids 3. The metabolic fate of diphenylacetic acid and its variation with species and dose. *Xenobiotica*, **7**, 717–725.

Dixon PAF, Caldwell J and Smith RL (1977c) Metabolism of arylacetic acids 1. The fate of 1-naphthylacetic acid and its variation with species and dose. *Xenobiotica*, **7**, 695–706.

Drach JC and Howell JP (1968) Identification of diphenhydramine urinary metabolites in the rhesus monkey. *Biochemical Pharmacology*, **17**, 2125–2136.

Drach JC, Howell JP, Borondy PE and Glazko AJ (1970) Species differences in the metabolism of diphenhydramine (benadryl). *Proceedings of the Society for Experimental Biology and Medicine*, **135**, 849–853.

Egger H, Bartlett F, Yuan HP and Karliner J (1982) Metabolism of pirprofen in man, monkey, rat and mouse. *Drug Metabolism and Disposition*, **10**, 529–536.

Emudianughe TS, Caldwell J and Smith RL (1977) Structure–metabolism relationships of arylacetic acids: the metabolic fate of naphth-2-ylacetic acid *in vivo*. *Biochemical Society Transactions*, **5**, 1006–1007.

Emudianughe TS, Caldwell J, Dixon PAF and Smith RL (1978) Studies on the metabolism of arylacetic acids 5. The metabolic fate of 2-naphthylacetic acid in the rat, rabbit and ferret. *Xenobiotica*, **8**, 525–534.

Emudianughe TS, Caldwell J, Sinclair KA and Smith RL (1983) Species differences in the metabolic conjugation of clofibric acid and clofibrate in laboratory animals and man. *Drug Metabolism and Disposition*, **11**, 97–102.

Esaac EG and Casida JE (1968) Piperonyl acid conjugates with alanine, glutamate, glutamine, glycine, and serine in living houseflies. *Journal of Insect Physiology*, **14**, 913–925.

Esaac EG and Casida JE (1969) Metabolism in relation to mode of action of methylenedioxyphenyl synergists in houseflies. *Journal of Agricultural and Food Chemistry*, **17**, 539–550.

Fears R (1985) Lipophilic xenobiotic conjugates: the pharmacological and toxicological consequences of the participation of drugs and other foreign compounds as substrates in lipid biosynthesis. *Progress in Lipid Research*, **24**, 177–195.

Fears R, Baggaley KH, Walker P and Hindley RM (1982) Xenobiotic cholesteryl ester formation. *Xenobiotica*, **12**, 427–433.

Feil VJ, Lamoureux CJH, Styrvoky E, Zaylskie RG, Thacker EJ and Holman GM (1973) Metabolism of o, p'-DDT in rats. *Journal of Agricultural and Food Chemistry*, **21**, 1072–1078.

Forman WB, Davidson ED and Webster LT (1971) Enzymatic conversion of salicylate to salicylurate. *Molecular Pharmacology*, **7**, 247–259.

Fournel S and Caldwell J (1986) The metabolic chiral inversion of 2-phenylpropionic acid in rat, mouse and rabbit. *Biochemical Pharmacology*, **35**, 4153–4159.

Fujino T and Yamamoto T (1992) Cloning and functional expression of a novel long-chain acyl-CoA synthetase expressed in the brain. *Journal of Biochemistry*, **111**, 77–81.

Fujino T, Kang MJ, Suzuki H, Iijima H and Yamamoto T (1996) Molecular characterization and expression of rat acyl-CoA synthetase 3. *Journal of Biological Chemistry*, **271**, 16748–16752.

Gaughan LC, Unai T, and Casida JE (1977) Permethrin metabolism in rats and cows and in bean and cotton plants. In *Synthetic Pyrethroids*, ACS Symposia Series No. 42 Elloit M (ed.), pp. 186–193.

Ghauri FYK, Blackledge CA, Glen RC, Lindon JC, Bedell CR, Wilson ID and Nicholson JK (1992) Quantitative structure metabolism relationships for substituted benzoic acids in the rat: computational chemistry, NMR spectroscopy and pattern recognition studies. *Biochemical Pharmacology*, **44**, 1935–1946.

Gingell R (1976) Metabolism of ^{14}C-DDT in the mouse and hamster. *Xenobiotica*, **6**, 15–20.

Groot PHE and Scheek LM (1976) Acyl-CoA synthetases in guinea pig liver mitochondria. Purification and characterisation of a distinct propionyl-CoA synthetase. *Biochimica et Biophysica Acta*, **441**, 260–267.

Grunow W and Bohme C (1974) Metabolism of 2,4,5-T and 2,4-D in rats and mice. *Archives of Toxicology*, **32**, 217–225.

Hankin JA, Wheelan P and Murphy RC (1997) Identification of novel metabolites of prostagladin E-2 formed from isolated rat hepatocytes. *Archives of Biochemistry and Biophysics*, **340**, 317–330.

Hansch C, Lien EJ and Helmer F (1968) Structure-activity correlations in the metabolism of drugs. *Archives of Biochemistry and Biophysics*, **128**, 319–330.

Hirom PC, Idle JR, Millburn P and Williams RT (1977) Glutamine conjugation of phenylacetic acid in the ferret. *Biochemical Society Transactions*, **5**, 1033–1036.

Hitchcock M and Smith JN (1964) Comparative detoxication 13. Detoxication of aromatic acids in arachnids: arginine, glutamic acid and glutamine conjugations. *Biochemical Journal*, **93**, 392–400.

Hitchcock M and Smith JN (1966) The detoxication of aromatic acids by invertebrates: detection of agmatine conjugates in scorpions. *Biochemical Journal*, **98**, 736–741.

Hoskins JA (1984) The occurrence, metabolism and toxicity of cinnamic acid and related compounds. *Journal of Applied Toxicology*, **4**, 283–292.

Huckle KR, Tait GH, Millburn P and Hutson DH (1981a) Species variations in the renal and hepatic conjugation of 3-phenoxybenzoic acid with glycine. *Xenobiotica*, **11**, 635–644.

Huckle KR, Climie IJG, Hutson DH and Millburn P (1981b) Dipeptide conjugation of 3-phenoxybenzoic acid in the mallard duck. *Drug Metabolism and Disposition*, **9**, 147–149.

Huckle KR, Hutson DH and Millburn P (1981c) Species differences in the metabolism of 3-phenoxybenzoic acid. *Drug Metabolism and Disposition*, **9**, 352–359.

Huckle KR, Stoydin G, Hutson DH and Millburn P (1982) Formation of an N-acetylornithine conjugate of 3-phenoxybenzoic acid in the chicken. *Drug Metabolism and Disposition*, **10**, 523–528.

Hutson DH (1982) Formation of lipophilic conjugates of pesticides and other xenobiotic compounds. In *Progress in Pesticide Biochemistry*, Volume 2, Hutson DH and Roberts TR (eds), John Wiley, Chichester, pp. 171–184.

Hutson DH and Casida JE (1978) Taurine conjugation in metabolism of 3-phenoxybenzoic acid and the pyrethroid insecticide cypermethrin in mouse. *Xenobiotica*, **8**, 565–571.

Hutson DH and Pickering BA (1971) The metabolism of isopropyl oxitol in rat and dog. *Xenobiotica*, **1**, 105–119.

Hutson DH, Dodds PF and Logan CJ (1985) The significance of xenobiotic-lipid conjugation. *Biochemical Society Transactions*, **13**, 854–856.

Hutt AJ and Caldwell J (1983) The metabolic chiral inversion of 2-arylpropionic acids- a novel route with pharmacological consequences. *Journal of Pharmacy and Pharmacology*, **35**, 693–704.

Hutt AJ and Caldwell J (1990) Amino acid conjugation. In *Conjugation Reactions in Drug Metabolism. An Integrated Approach*, Mulder GJ (ed.), Taylor and Francis, London, pp. 273–305.

Idle JR, Millburn P and Williams RT (1975) Benzoylglutamic acid, a metabolite of benzoic acid in Indian Fruit bats. *FEBS Letters*, **59**, 234–236.

Idle JR, Millburn P and Williams RT (1978) Taurine conjugates as metabolites of arylacetic acids in the ferret. *Xenobiotica*, **8**, 253–264.

Irjala K (1972) Synthesis of p-aminohippuric, hippuric, and salicyluric acids in experimental animals and man. *Annals Academiae Scientiarum Fennicae, AV*, **154**, 1–40.

Israili ZH, Dayton PG and Kiechiel JR (1977) Novel routes of drug metabolism: a survey. *Drug Metabolism and Disposition*, **5**, 411–415.

Jaffe M (1877) Uber das Verhalten das Benzoesaure im Organismus der Vogel. *Berichte des Deutschen Chemischen Gesellschaft*, **10**, 1925–1930.

James MO (1982) Disposition and taurine conjugation of 2,4-dichlorophenoxyacetic acid, 2,4,5-trichlorophenoxyacetic acid, bis(4-chlorophenyl) acetic acid, and phenylacetic acid in the spiny lobster, *Panulirus argus. Drug Metabolism and Disposition*, **10**, 516–522.

James MO and Bend JR (1976) Taurine conjugation of 2,4-dichlorophenoxyacetic acid and phenylacetic acid in two marine species. *Xenobiotica*, **6**, 393–398.

James MO and Bend JR (1978) A radiochemical assay for glycine N-acyltransferase activity. Some properties of the enzyme in rat and rabbit. *Biochemical Journal*, **172**, 285–291.

James MO, Smith RL and Williams RT (1971) Conjugates of phenylacetic acid with taurine and other amino acids in various species. *Biochemical Journal*, **124**, 15P–16P.

James MO, Smith RL, Williams RT and Reidenberg M (1972a) The conjugation of phenylacetic acid in man, sub-human primates and some non-human primate species. *Proceedings of the Royal Society, London B*, **182**, 25–35.

James MO, Smith RL and Williams RT (1972b) The conjugation of 4-chloro- and 4-nitro-phenylacetic acids in man, monkey and rat. *Xenobiotica*, **2**, 499–506.

Jordan BJ and Rance MJ (1974) Taurine conjugation of fenclofenac in the dog. *Journal of Pharmacy and Pharmacology*, **26**, 359–361.

Jordan TW, McNaught RW and Smith JN (1970) Detoxications in Peripatus. Sulphate, Phosphate and Histidine conjugations. *Biochemical Journal*, **118**, 1–8.

Kaihara M and Price JM (1961) Quinaldylglycyltaurine: a urinary metabolite of quinaldic acid and kynurnic acid in the rat. *Journal of Biological Chemistry*, **236**, 508–511.

Kaihara M and Price JM (1965) The metabolism of quinaldylglycylglycine, a urinary metabolite of quinaldic acid in the cat. *Journal of Biological Chemistry*, **240**, 454–456.

Kanazu T and Yamaguchi T (1997) Comparison of *in vitro* carnitine and glycine conjugation with branched-side chain and cyclic side chain carboxylic acids in rats. *Drug Metabolism and Disposition*, **25**, 149–153.

Kang MJ, Fujino T, Sasano H, Minekura H, Yabuki N, Nagura H, Iijima H and Yamamoto T (1997) A novel arachidonate-preferring acyl-CoA synthetase is present in steroidogenic cells of the rat adrenal, ovary, and testis. *Proceedings of the National Academy of Sciences, USA*, **94**, 2880–2884.

Kasuya F, Igarashi K and Fukui M (1990) Glycine conjugation of the substituted benzoic acids *in vitro*: structure–metabolism relationship study. *Journal of Pharmacobiology and Dynamics*, **13**, 432–440.

Kasuya F, Igarashi K and Fukui M (1991) Glycine conjugation of the substituted benzoic acids in mice: structure–metabolism relationship Study II. *Journal of Pharmacodynamics*, **14**, 671–677.

Kasuya F, Igarashi K and Fukui M (1996) Participation of the medium chain Acyl-CoA synthetase in glycine conjugation of the benzoic acid derivatives with the electron-donating groups. *Biochemical Pharmacology*, **51**, 805–809.

Keller W (1842) Ueber Verwandlung der Benzoesaure in Hippursaure. *Justus Liebigs Annal Chemistry*, **43**, 108–111.

Kelley M and Vessey DA (1986) Interaction of 2,4-dichlorophenoxyacetate (2,4-D) and 2,4,5-

trichlorophenoxyacetate (2,4,5-T) with the acyl-CoA: amino acid N-acyltransferase enzymes of bovine liver mitochondria. *Biochemical Pharmacology*, **35**, 289–295.

Killenberg PG and Webster LT (1980) Conjugation by peptide bond formation. In *Enzymatic Basis of Detoxification* Volume 2, Jakoby WB (ed.), Academic Press, New York, pp. 142–167.

Killenberg PG Davidson ED and Webster LT (1971) Evidence for a medium-chain fatty acid: coenzyme A ligase (adenosine monophosphate) that activates salicylate. *Molecular Pharmacology*, **7**, 260–268.

Kolis SJ, Postma EJ, Williams TH and Sasso GJ (1986) Identification of trimoprostil metabolites excreted in rat bile formed by oxidation and taurine conjugation. *Drug Metabolism and Disposition*, **14**, 465–470.

Konishi T, Nishikawa H, Kitamura S and Tatsumi K (1999) In vivo studies on chiral inversion and amino acid conjugation of 2-[4-(3-methyl-2-thienyl)phenyl]propionic acid in rats and dog. *Drug Metabolism and Disposition*, **27**, 158–160.

Lau EP, Haley BE and Barden R (1977) Photoaffinity labeling of acyl-coenzyme A glycine N-acyltransferase with p-azidobenzoyl-coenzyme A. *Biochemistry*, **16**, 2581–2585.

Lehnert W (1983) N-Isovalerylalanine and N-isovalerylsarcosine: two new minor metabolites in isovaleric acidemia. *Clinica Chimica Acta*, **134**, 207–212.

Mahler HR, Wakil SJ and Bock RM (1953) Studies on fatty acid oxidation 1. Enzymatic activation of fatty acids. *Journal of Biological Chemistry*, **204**, 453–468.

Marsh MV, Caldwell J, Smith RL, Hormer MW, Houghton E and Moss MS (1981) Metabolic conjugation of some carboxylic acids in the horse. *Xenobiotica*, **11**, 655–663.

Marsh MV, Caldwell J, Hutt AJ, Smith RL, Horner MW, Houghton E and Moss MS (1982) 3-Hydroxy- and 3-Keto-3-phenylpropionic acids: novel metabolites of benzoic acid in horse urine. *Biochemical Pharmacology*, **31**, 3320–3325.

Millington DS, Bohman TP, Roe CR, Yergey AL and Liberato DJ (1985) Valproylcarnitine: a novel drug metabolite identified by fast atom bombardment and thermospray liquid chromatography-mass spectrometry. *Clinical Chimica Acta*, **145**, 69–75.

Miyazaki H, Takayama H, Minatogawa Y and Miyano K (1976) A noval metabolic pathway in the metabolism of 5-(4'-chloro-n-butyl) picolinic acid. *Biomedical Mass Spectroscopy*, **3**, 140–145.

Mizojiri K, Futaguchi S, Norikura R, Katsuyama Y, Nagasaki T, Yoshimori T and Nakanishi H (1995) Disposition of S-1108, a new oral cephem antibiotic and the metabolic fate of pivalic acid liberated from (pivaloyl-^{14}C) S-1108 in rat and dogs. *Antimicrobial Agents and Chemotherapy*, **39**, 1445–1453.

Moldave K and Meister A (1957) Synthesis of phenylacetylglutamine by human tissue. *Journal of Biological Chemistry*, **229**, 463–476.

Montgomery JA, Johnston TP and Shealy YF (1979) Drugs for neoplastic diseases. In *Burger's Medicinal Chemistry*, 4th edition part II, Wolff ME (ed.), John Wiley, New York, pp. 595–670.

Mohri K, Okada K and Benet LZ (1998) Stereoselective metabolism of benoxaprofen in rats. Biliary excretion of benoxaprofen taurine conjugate and glucuronide. *Drug Metabolism and Disposition*, **26**, 332–337.

Mori Y, Kuroda N, Sakai Y, Yokoya F, Toyoshi K and Baba, S (1985) Species differences in the metabolism of suprofen in laboratory animals and man. *Drug Metabolism and Disposition*, **13**, 239–245.

Morris JE and Price JM (1963) Origin of the taurine moiety of quinaldylglycyltaurine. *Journal of Biological Chemistry*, **238**, 3963–3965.

Nandi DL, Lucas SV and Webster LT (1979) Benzoyl-coenzyme A: glycine N-acyltransferase and phenylacetyl-coenzyme A: glycine N-acyltransferase from bovine liver mitochondria. Purification and characterization. *Journal of Biological Chemistry*, **254**, 7230–7237.

Oikawa E, Iijima H, Suzuki T, Sasano H, Sato H, Kamataki A, Nagura H, Kang MJ, Fujino T, Suzuki H and Yamamoto T (1998) A novel acyl-CoA synthetase, ACS5, expressed in intestinal epithelial cells and proliferating preadipocytes. *Journal of Biochemistry*, **124**, 679–685.

Peffer RC, Abraham DJ, Zemaitis MA, Wang LK and Alvin JD (1987) 3,4-Dichlorobenzyloxyacetic acid is extensively metabolized to a taurine conjugate in rats. *Drug Metabolism and Disposition*, **15**, 305–311.

Pinto JD, Camien MN and Dunn MS (1965) Metabolic fate of p,p'-DDT(1,1,1-trichloro-2,2-bis(p-chlorophenyl)ethane) in rats. *Journal of Biological Chemistry*, **240**, 2148–2154.

Quick AJ (1931) The conjugation of benzoic acid in man. *Journal of Biological Chemistry*, **92**, 65–85.

Quistad GB (1986) In *Xenobiotic Conjugation Chemistry*, Paulson GD, Caldwell J, Hutson DH and Menn JJ (eds), ACS Symposium Series No. 299, ACS, Washington, pp. 221–241.

Quistad GB, Staiger LE and Schooley DA (1978a) Environmental degradation of the miticide cycloprate (hexadecyl cyclopropane carboxylate).1. Rat metabolism. *Journal of Agricultural and Food Chemistry*, **26**, 60–66.

Quistad GB, Staiger LE and Schooley DA (1978b) Environmental degradation of the miticide cycloprate (hexadecyl cyclopropane carboxylate).3. Bovine metabolism. *Journal of Agricultural and Food Chemistry*, **26**, 71–75.

Quistad GB, Staiger LE and Schooley DA (1978c) Environmental degradation of the miticide cycloprate (hexadecyl cyclopropane carboxylate).4. Beagle Dog Metabolism. *Journal of Agricultural and Food Chemistry*, **26**, 76–80.

Quistad GB, Staiger LE and Schooley DA (1982) Xenobiotic conjugation: a novel role for bile acids. *Nature*, **296**, 462–464.

Quistad GB, Staiger LE and Schooley DA (1986) The role of carnitine in the conjugation of acidic xenobiotics. *Drug Metabolism and Disposition*, **14**, 521–525.

Reif VD and Sinsheimer JE (1975) Metabolism of 1-(o-chlorophenyl)-1-(p-chlorophenyl)-2,2-dichloroethane (o, p-DDD) in rats. *Drug Metabolism and Disposition*, **3**, 15–25.

Rothstein M and Greenberg DM (1957) Studies on the metabolism of xanthurenic acid-4-C^{14}. *Archives of Biochemistry and Biophysics*, **68** 206–214.

Sakai Y, Mori Y, Toyoshi K, Horie M and Baba S (1984) Metabolism of suprofen in the dog. *Drug Metabolism and Disposition*, **12**, 795–797.

Schachter D and Taggart JV (1954) Glycine N-acylase: purification and properties. *Journal of Biological Chemistry*, **208**, 263–275.

Sevoz C, Benoit E and Buronfosse T (2000) Thioesterification of the 2-arylpropionic acids by recombinant acyl-coenzyme A synthetases (ACS1 and ACS2). *Drug Metabolism and Disposition*, **28**, 398–402.

Shin YS, Buehring KU and Stokstad ELR (1974) The metabolism of methotrexate in *Lactobacillus casei* and rat liver and the influence of methotrexate on metabolism of folic acid. *Journal of Biological Chemistry*, **249**, 5772–5777.

Shirley MA, Guan X, Kaiser DG, Halstead GW and Baillie TA (1994) Taurine conjugation of ibuprofen in humans and in the rat liver *in vitro*. Relationship to metabolic chiral inversion. *Journal of Pharmacology and Experimental Therapeutics*, **269**, 1166–1175.

Shirley MA, Bennani YL, Boehm MF, Breau AP, Pathirana C and Ulm EH (1996) Oxidative and reductive metabolism of 9-*cis*-retinoic acid in the rat. Identification of the 13, 14-dihydro-9-*cis*-retinoic acid and its taurine conjugate. *Drug Metabolism and Disposition*, **24**, 293–302.

Skare KL, Schnoes HK and DeLuca HF (1982) Biliary metabolites of *all-trans*-retinoic acid in the rat: isolation and identification of a novel polar metabolite. *Biochemistry*, **21**, 3308–3317.

Smith JN (1958) Comparative detoxication 5. Conjugation of aromatic acids in reptiles: formation of ornithuric acid, hippuric acid and glucuronides. *Biochemical Journal*, **69**, 509–516.

Smith JN (1962) Detoxications of aromatic acids with glutamic acid and arginine in spiders. *Nature*, **195**, 399–400.

Smith RL and Williams RT (1970) History of the discovery of the conjugation mechanisms. In *Metabolic Conjugation and Metabolic Hydrolysis*, Volume 1, Fishman WH (ed.), Academic Press, London, pp. 1–19.

Solheim E and Scheline RR (1976) Metabolism of alkenebenzene derivatives in the rat II. Eugenol and isoeugenol methylethers. *Xenobiotica*, **6**, 137–150.

Strahl NR and Barr WH (1971) Intestinal drug absorption and metabolism. 3. Glycine conjugation and accumulation of benzoic acid in rat intestinal tissue. *Journal of Pharmaceutical Sciences*, **60**, 278–281.

Suzuki H, Kawarabaysai Y, Kondo J, Abe T, Nishikawa K, Hashimoto T and Yamamoto T (1991) Structure and regulation of rat long-chain acyl-CoA synthetase. *Journal of Biological Chemistry*, **265**, 8681–8685.

Suzuki H, Watanabe M, Fujino T and Yamamoto T (1995) Multiple promoters in rat acyl-CoA synthetase gene mediate differential expression of multiple transcripts with 5′-end heterogeneity. *Journal of Biological Chemistry*, **270**, 9676–9682.

Tanaka K and Isselbacher K (1967) The isolation and identification of N-isovalerylglycine from urine of patients with isovaleric acidemia. *Journal of Biological Chemistry*, **242**, 2966–2972.

Tanaka Y, Shimomura Y, Hirota T, Nozaki A, Ebata M, Takasaki W, Shigehara E, Hayashi R and Caldwell J (1992) Formation of glycine conjugate and (-)-(*R*)-enantiomer from (+)-(*S*)-2-phenylpropionic acid suggesting the formation of the CoA thioester intermediate of (+)-(*S*)-enantiomer in dogs. *Chirality*, **4**, 342–348.

Tang W and Abbott FS (1997) A comparative investigation of 2-propyl-4-pentenoic acid (4-ene VPA) and its α-fluorinated analogue: phase II metabolism and pharmacokinetics. *Drug Metabolism and Disposition*, **25**, 219–227.

Tang W, Palaty J and Abbott FS (1997) Time course of α-fluorinated valproic acid in mouse brain and serum and its effect on synaptosomal γ-aminobutyric acid levels in comparison to valproic acid. *Journal of Pharmacology and Experimental Therapeutics*, **282**, 1163–1172.

Temellini A, Mogavero S, Giulianotti PC, Pietrabissa A, Mosca F and Pacifici GM (1993) Conjugation of benzoic acid with glycine in human liver and kidney. *Xenobiotica*, **23**, 1427–1433.

Thierfelder H and Sherwin CO (1914) Phenylacetyl-glutamin ein stoffwechselprodukt des menschlichen. Korpers nach Eingabe von Phenyl-essigsaure. *Berichte des Deutschen Chemischen Gesellschaft*, **47**, 2630–2634.

Tishler SL and Goldman P (1970) Properties and reactions of salicyl-coenzyme A. *Biochemical Pharmacology*, **19**, 143–150.

Totsuka K, Shimizu K, Konishi M and Yamamoto S (1992) Metabolism of S-1108, a new cephem antibiotic and metabolic profiles of its metabolites in humans. *Antimicrobial Agents and Chemotherapy*, **36**, 757–761.

Unai T and Casida JE (1977) In *Synthetic Pyrethroids*, Elliot M (ed.), ACS Symposium No. 42, ACS, Washington, pp. 194–200.

Vickers S, Duncan CAH, White SD, Ramjit HG, Smith JL, Walker RW, Flynn H and Arison BH (1985) Carnitine and glucuronic acid conjugates of pivalic acid. *Xenobiotica*, **15**, 453–458.

Wallcave L, Bronczyk S and Gingell R (1974) Excreted metabolites of 1,1,1-trichloro-2,2-*bis*(*p*-chlorophenyl)ethane in the mouse and hamster. *Journal of Agricultural and Food Chemistry*, **22**, 904–908.

Webster LT, Siddiqui UA, Lucas SV, Strong JM and Mieyal JJ (1976) Identification of separate acyl-CoA:glycine and acyl-CoA:L-glutamine N-acyltransferase activities in mitochondrial fractions from liver of rhesus monkey and man. *Journal of Biological Chemistry*, **251**, 3352–3358.

Williams RT (1959) *Detoxication Mechanisms*, Chapman & Hall, London, pp. 348–367.

Williams RT (1963) In *Biogenesis of Natural Compounds*, 2nd edition, Bernfeld P (ed.), Pergamon Press, Oxford, pp. 427–474.

Young L (1977) The metabolism of foreign compounds. History and Development. In *Drug Metabolism from Microbe to Man*, Parke DV and Smith RL (eds), Taylor and Francis, London, pp. 1–11.

15 Deconjugating Enzymes: Sulphatases and Glucuronidases

Christiane Kunert-Keil, Christoph A. Ritter, Heyo K. Kroemer and Bernhard Sperker

Institut für Pharmakologie, Ernst Moritz Arndt Universität Greifswald, Greifswald, Germany

Introduction

The activity of deconjugating enzymes may have a significant impact on the pharmacokinetics of various xenobiotics. Release of the parent compound from a conjugate modifies the concentration and hence the action of a drug. Since conjugation of a drug during phase II metabolism can result in either activation or inactivation, enzymic deconjugation may diminish or augment the efficacy of such a compound. Two of the most important conjugation reactions are glucuronidation and sulphation which are mediated by glucuronosyltransferases and sulphotransferases, respectively. The possible contribution of deglucuronidation and desulphation reactions to the disposition of xenobiotics by β-glucuronidases and sulphatases has not been systematically evaluated. However, these enzymes are well characterised on a biochemical level since genetic deficiency of sulphatases and β-glucuronidase results in a variety of lysosomal storage diseases. This chapter summarises data available on enzyme function regarding metabolism and bioactivation of xenobiotics as well as the underlying mechanisms of action, regulation, tissue and species distribution and genetics.

Sulphatases

OVERVIEW

The sulphatase enzyme family is a group of proteins which catalyse the hydrolysis of sulphate esters from various compounds like steroids, glycosaminoglycans or glycolipids. The physiological role of most sulphatases is lysosomal catabolism of complex

Enzyme Systems that Metabolise Drugs and Other Xenobiotics. Edited by C. Ioannides.
© 2002 John Wiley & Sons Ltd

carbohydrates and lipids. In contrast, sulphatases which are located within the endoplasmic reticulum can be involved in synthetic pathways like that of steroid hormones (Parenti et al. 1997; Coughtrie et al. 1998). Beside the above-mentioned natural substrates, several sulphate esters of different other compounds like thyroxine, catecholamines, oligo-and mono-saccharides are hydrolysed by sulphatases (Gorham and Cantz 1978, Roy 1979; Roy and Turner 1982; Kung et al. 1988; Strobel et al. 1990). Members of the subgroup of arylsulphatases (EC 3.1.6.1) are able to hydrolyse sulphate esters of aromatic compounds like p-nitrocatechol, p-nitrophenol or 4-methylumbelliferone which are used as probes for determination of sulphatase activity (Roy 1979). Hydrolysis of sulphate esters by sulphatases is mediated by a catalytic site containing the unusual amino acid derivative Cα-formylglycine (2-amino-3-oxopro-pionic acid) which is generated by oxidation of the thiol group of a cysteine that is conserved among all known eukaryotic sulphatases (von Figura et al. 1998). Lack of this protein modification in humans results in multiple sulphatase deficiency, a rare lysosomal storage disorder characterised by severely decreased activity of all known sulphatases (Schmidt et al. 1995).

Sulphatases are found in lower and higher eukaryotes as well as in prokaryotes (von Figura et al. 1998; Schirmer and Kolter 1998). For example, sulphatases have been identified in algae, sea urchins, as well as in mammalian species like rodents, cats, dogs and humans with their amino acid sequences being highly conserved. In addition, the active site of bacterial sulphatases displays a high degree of similarity to eukaryotic sulphatases including the presence of Cα-formylglycine which points to similar catalytic mechanisms and to an evolutionary conservation of the sulphatase gene family.

To date, the role of sulphatases in the metabolism and bioactivation of drugs and other xenobiotics is not clear. However, there are several reports on the influence of mammalian sulphatases on the metabolism of thyroxine and steroids (Kung et al. 1988; Pasqualini et al. 1996a). In addition, some data are available on the role of bacterial sulphatases in the enterohepatic circulation of thyroxine, bile acids and steroid hormones (Huijghebaert et al. 1984; Hazenberg et al. 1988; Robben et al. 1988). Inhibition of sulphatases by xenobiotics might have consequences for mammalian development causing symptoms comparable to the genetic deficiency of sulpha-tases as demonstrated for inhibition of arylsulphatase E by warfarin (Franco et al. 1995).

ARYLSULPHATASES

Arylsulphatase A

Arylsulphatase A (ARSA; cerebroside 3-sulphatase; EC 3.1.6.8) is a lysosomal enzyme which catalyses the desulphation of cerebroside 3-sulphate and other sulphated glycolipids (Roy 1979; Parenti et al. 1997) as well as a wide variety of other substrates like 12-(1-pyrene)dodecanoyl cerebroside sulphate, adrenaline-4-sulphate, noradren-aline-4-sulphate, dopamine-4-sulphate, L-tyrosine O-sulphate, ascorbate 2-sulphate, monosaccharide sulphates, some steroid sulphates, adenosine 3',5'-monophosphate, nitrocatechol sulphate, nitroquinol sulphate, 2-nitropyridinyl 3-sulphate and 4-methylumbelliferyl sulphate with K_m values ranging from 1 to 220 mM (Fluharty *et al.* 1979; Roy and Turner 1982; Marchesini *et al.* 1989; Roy and Mantle 1989; Strobel *et al.* 1990). Among the sulphatase family, ARSA is one of the most extensively studied. However, the role of ARSA in the metabolism and bioactivation of xenobiotics has not been investigated in a systematic manner. To our knowledge there are no data on desulphation of pharmacologically relevant substrates. The hydrolysis of sulphate esters occurs via cleavage of the O–S bond (Roy and Mantle 1989). *In vivo*, the 3-sulphate moiety is hydrolysed by ARSA only if cerebroside 3-sulphate is complexed with the activator protein saposin B (Bierfreund *et al.* 2000). The structure of the active site of the enzyme has been elucidated and a catalytic mechanism involving a trans-esterification step has been proposed (Lukatela *et al.* 1998; Waldow *et al.* 1999). Recent crystal structure data point to an ARSA homooctamer composed of a tetramer of dimers (Lukatela *et al.* 1998). However, the quarternary structure seems to be dependent on enzyme concentration and pH (Roy 1979). The pH optimum has been described to be 4.5 or 5.6 depending on the substrate used (Roy and Mantle 1989) and the enzyme is inhibited by various compounds such as sulphate anions, thiosulphate, potassium ferrate, borate, phosphate, pyrophosphate, arsenate, selenate, silver nitrate and 2-hydroxy-dopamine, (Mercelis *et al.* 1979; Roy 1979; Laidler and Steczko 1986; Roy and Mantle 1989; Cawley and Shickley 1992).

ARSA has been characterised in different mammalian species like man, ox, horse, pig, sheep, dog, cat, rabbit, rat and mouse (Waheed and van Etten 1985, Roy and Mantle 1989; Kreysing *et al.* 1994). The enzyme has been found in most tissues investigated so far like liver, kidney, brain, placenta, spleen, testis and oviduct (Waheed and van Etten 1985; Kihara *et al.* 1986; van der Pal *et al.* 1991; Vitaioli *et al.* 1996).

Genetic deficiency of the human enzyme leads to lysosomal accumulation of cerebroside sulphate in the CNS resulting in a disorder called Metachromatic Leukodystrophy (MLD; Barth *et al.* 1994). Pseudodeficiencies of ARSA have been shown to be present in 10–20% of healthy individuals and are due to allelic mutations which result in low ARSA activities without exerting clinical evidence of disease (Thomas 1994).

ARSA has been shown to be regulated by a variety of stimuli. For example, activity of ARSA is inhibited by physiological concentrations of cortisol in myelinogenic cell cultures from embryonic mouse brain (Stephens and Pieringer 1984), whereas oestrogens seem to induce activity of ARSA in rabbit oviduct (Vitaioli *et al.* 1996). In addition, regulation of ARSA activity during menstrual cycle and by endogenous gonadal steroid hormones has been reported in humans (Oner *et al.* 1994; Kamei *et al.* 1997). Suramin has been demonstrated to decrease ARSA activity in mice but to increase it in rats (Marjomaki and Salminen 1986).

Throughout development of rat and mouse brain, and during growth and ageing of human liver cell lines activity of ARSA was increased (Le Gall *et al.* 1979; Bird *et al.* 1981; van der Pal *et al.* 1991). During galactose-induced cataract development in rats ARSA activity has been shown to increase (Harries *et al.* 1985). ARSA activity increases during development of granulomas after infection of mice with *Schistosoma mansoni* (Higuchi *et al.* 1984). In lesions of endodontic origin ARSA activity is elevated compared to healthy periodontal ligament (Aqrabawi *et al.* 1993). Urinary excretion of ARSA activity was reduced in malnourished children with mild vitamin A deficiency whereas activity was increased in cases of severe vitamin A deficiency (Latif *et al.* 1979).

Arylsulphatase B

Arylsulphatase B (ARSB; *N*-acetylgalactosamine 4-sulphatase; EC 3.1.6.12) is a lysosomal enzyme catalysing the hydrolysis of *N*-acetyl-*D*-galactosamine 4-sulphate moieties from complex molecules like dermatan and chondroitin sulphate as well as from UDP-*N*-acetylgalactosamine 4-sulphate (Fluharty *et al.* 1975; Gorham and Cantz 1978; Habuchi *et al.* 1979). As reported for ARSA, ARSB is able to hydrolyse substrates like catecholamine sulphates, oligosaccharide substrates, nitrocatechol sulphate or 4-methylumbelliferyl sulphate (Daniel 1978; Gibson *et al.* 1987; Strobel *et al.* 1990). A role of ARSB in the hydrolysis of pharmacologically relevant sulphates in *vivo* has not been described so far.

The crystal structure of ARSB has been demonstrated, indicating a monomeric protein consisting of two domains with the active site localised on the larger domain containing cysteine modified to Cα-formylglycine and a calcium ion (Bond *et al.* 1997). Although the geometry and the functional amino acids of the active site of ARSB have been shown to be identical to ARSA, the enzymic mechanism proposed is somewhat different (Bond *et al.* 1997; Lukatela *et al.* 1998). Enzymic activity of ARSB has been shown to be inhibited by compounds like ascorbic acid, sulphidopeptide leukotrienes, metal ions and iodoacetate (Agogbua and Wynn 1976; Weller *et al.* 1986; Selvidge and Verlangieri 1991).

ARSB has been purified or cloned from various mammalian species like man, ox, horse, dog, cat, rabbit, rat and mouse (Murata *et al.* 1975; Wojczyk 1986; Thompson and Daniel 1988; Peters *et al.* 1990; Jackson *et al.* 1992). A wide variety of tissues like liver, kidney, brain, lung, spleen and placenta has been shown to contain ARSB activity (Shapira and Nadler 1975; Daniel 1978; Gibson *et al.* 1987).

Deficiency of ARSB results in the rare lysosomal storage disorder Maroteaux–Lamy disease or mucopolysaccharidosis type VI which is characterised by progressive organ dysfunction with clinical symptoms like skeletal abnormalities and cardiac as well as respiratory failure (Neufeld and Muenzer 1995). About 50 different mutations have been identified pointing to a broad molecular heterogeneity of ARSB deficiency (Isbrandt *et al.* 1994; Villani *et al.* 1999; Wu *et al.* 2000).

A wide variation of ARSB expression among different murine tissues and inbred mouse strains has been described (Daniel 1978, 1987). In rats, ARSB activity of hepatocytes, Kupffer and endothelial cells increases with age (Ferland *et al.* 1990) which is consistent with the data of Le Gall *et al.* (1979) demonstrating a significantly higher ARSB activity in senescent human liver cell lines as compared to actively growing cells. In addition, glycosylation of rat liver ARSB changes depending on age (Przybylo and Litynska 2000).

ARSB has been reported to be inducible by amyloid β-peptide and hydrogen peroxide, compounds which play a central role in the pathogenesis of Alzheimer's disease. Cells resistant to amyloid β express higher levels of ARSB (Li *et al.* 1999). ARSB activity has been detected in lesions of endodontic origin whereas no activity could be found in the respective control tissue (Aqrabawi *et al.* 1993). Guinea pig natural killer cells (Kurloff cells) show an increased activity of anionic isoforms ARSB during development of acute lymphoblastic leukaemia (Taouji *et al.* 1996). In chronic myelogenous leukaemia, activity of phosphorylated forms of ARSB in leukocytes was increased (Uehara *et al.* 1983). Augmented phosphorylation of ARSB seems to result in elevated enzymic activity in tumours (Gasa *et al.* 1987).

Arylsulphatase C

Arylsulphatase C (ARSC, steroid sulphatase, EC 3.1.6.2) is a microsomal, membrane-bound enzyme, which cleaves the sulphate moiety of several endogenous 3-hydroxy-steroid sulphates like oestrone sulphate, dehydroepiandrosterone sulphate, pregnenolone sulphate, cholesterol sulphate and testosterone sulphate. Several nonsteroidal compounds are known to be cleaved like the endogenous 3,5,3'-triiodothyronine sulphate and the synthetic compounds *p*-nitrophenyl sulphate and 4-methylumbelliferyl sulphate. K_m values for the endogenous steroid sulphates range from 1 to 70 µM whereas synthetic compounds are desulphated with K_m values between 1 and 5 mM (Hobkirk 1985). For desulphation of 3,5,3'-triiodothyronine sulphate the K_m value was 390 µM (Kung *et al.* 1988). After solubilisation of the enzyme molecular weights from 330 kDa to 533 kDa depending on source and solubilisation method were found. The number of subunits varies between 3 and 8 monomers with molecular weights ranging from 72 to 78 kDa. The pH optimum of the enzyme appears to be substrate-dependent. Oestrone sulphate sulphatase activity exhibited an alkaline pH optimum approximating pH 8.0–8.6 regardless of source. However, sulphatase activity for cholesterol sulphate showed two optima at pH 5.0 and 7.5. The enzyme is inhibited by a variety of steroid sulphates and free steroids. Competitive inhibitors are oestrogens, *p*-nitrophenyl sulphate and nitrophenol. Inhibition by dehydroepiandrosterone sulphate, pregnenolone sulphate, testosterone sulphate, cholesterol sulphate and androgens is noncompetitive. Inorganic anions such as sulphate, sulphite, fluoride, phosphate and cyanide anions moderately inhibit sulphatase, whereas borate is a strong inhibitor (Daniel 1985; Hobkirk 1985). Arylsulphatase C actually consists of two biochemically distinct isozymes, 's' (slow) and 'f' (fast), identified by their electrophoretic mobility. Only the s form of arylsulphatase C exhibits steroid sulphatase activity whereas the f form hydrolyses substrates other than 3-hydroxysteroid sulphates (Shankaran *et al.* 1991). While less information on the f form of arylsulphatase C is available, steroid sulphatase has been investigated extensively.

ARSC is present in different mammalian species like man, sheep, guinea pig, rabbit, hamster, rat and mouse (Daniel 1985; Hobkirk 1985). The enzyme has been found in most tissues investigated so far. However, distribution of the two isozymes is not identical. In man the s form is present in spleen, thyroid, heart, skeletal muscle, placenta and adrenal tissue, while the f form is absent or poorly active. In ovary, testis, intestinal and lung tissue both isozymes are present and are usually of equal amount. In contrast, kidney, liver and pancreatic tissue exhibit only the f form of ARSC (Munroe and Chang 1987).

Genetic deficiency of the human enzyme occurs in 1 of 2000 to 6000 newborns (Bradshaw and Carr 1986). In more than 85% of patients with steroid sulphatase deficiency complete or partial deletions of the sulphatase gene is responsible for the absence of enzymic activity (Hernandez-Martin *et al.* 1999). However, several unique base pair substitutions have been characterised (Alperin and Shapiro 1997; Oyama *et al.* 2000). Steroid sulphatase deficiency syndrome is an X-linked inherited metabolic disease which is characterised by decreased maternal oestriol production due to deficient placental sulphatase activity during foetal life and postnatally by ichthyosis. Placental steroid sulphatase deficiency is manifested by low oestriol levels in urine and plasma, delay in the onset of labour, uterine resistance to oxytocin 1 and increased stillbirth frequency. The liveborn infants are, however, clinically normal at birth. Male infants with steroid sulphatase deficiency develop X-linked ichthyosis early in the first year of life. Prominent skin peeling with a reptilian appearance of the skin is the major clinical feature. Accumulation of undegraded cholesterol sulphate is thought to be responsible for scale-formation in steroid sulphate deficiency. In addition, corneal opacities, cryptorchidism and pyloric stenosis are associated with X-linked ichthyosis (Bradshaw and Carr 1986).

Expression of steroid sulphatase has been reported to depend on age, sex and presence of disease. During mouse ontogeny, expression of steroid sulphatase mRNA was observed in restricted areas of the liver, in cartilage of many tissues, in spleen and skin. Steroid sulphatase mRNA is expressed in the embryonic mouse cortex, hindbrain and thalamus during the last third of gestation (Compagnone *et al.* 1997). At birth, steroid sulphatase levels in brain are clearly higher than those in adult mice (Mortaud *et al.* 1996). The role of pre- and post-pubertal stage and sex on the steroid sulphatase activity has been investigated in human leukocytes. Pre- and post-pubertal females presented a higher sulphatase activity than the comparable male group. Enzymic activity in prepubertal subjects was higher than in postpubertal individuals (Cuevas-Covarrubias *et al.* 1993). Investigations in breast tissue from women with breast fibroadenomas have shown significantly increased steroid sulphatase levels in the fibroadenoma tissue as compared to normal tissue (Pasqualini *et al.* 1997). In addition, human serum albumin has a marked stimulatory effect on steroid sulphatase activity, which is almost completely inhibited by basic fibroblast growth factor in malignant breast tissue *in vitro* (Purohit *et al.* 1999). The cytokines interleukin-6 and tumour necrosis factor α both stimulated steroid sulphatase activity in MCF-7 human breast cancer cells (Purohit *et al.* 1996a). In post-menopausal women the concentration of various oestrogens in cancer tissues is much higher than those found in plasma, suggesting an accumulation of these substances in tumour tissue (Pasqualini *et al.* 1996b). As oestrogens, especially oestradiol, are known to be involved in both the aetiology and maintenance of growth of breast cancer, many steroid sulphatase inhibitors have been developed during the past few years. Oestradiol derivatives include amino oestrones, oestradiol amides and oestrone sulphamates with IC_{50} values of 10 μM, 80 nM and 0.5 nM, respectively (Selcer *et al.* 1996; Woo *et al.* 1998; Boivin *et al.* 1999). Oestrone-3-O-sulphamate is known to be the most potent steroid sulphatase inhibitor, but exhibits strong oestrogenic activity. Structural modifications lead to a variety of potent steroid sulphatase inhibitors lacking any oestrogenicity (Li *et al.* 1998; Purohit *et al.* 1998; Ciobanu *et al.* 1999). In hormone-dependent and hormone-independent mammary cancer cell lines, the androgen danazol and the progestin medrogestone have been observed to inhibit steroid sulphatase activity

(Nguyen *et al.* 1993; Chetrite *et al.* 1999). Several non-steroidal, non-oestrogenic steroid sulphatase inhibitors have been developed including substituted tyramine and coumarin sulphamates (Purohit *et al.* 1996b; Selcer *et al.* 1997). Finally, dietary compounds like natural flavonoids, especially quercetin and naringenin, as well as sulphoconjugates of daidzein, an isoflavone found in leguminosae, are potent inhibitors of steroid sulphatase and thus may offer the potential for breast cancer prevention therapy (Huang *et al.* 1997; Wong and Keung 1997).

Other arylsulphatases

Arylsulphatases D (ARSD), E (ARSE) and F (ARSF) are non-lysosomal enzymes localised in the endoplasmic reticulum (ARSD, ARSF) or the golgi apparatus (ARSE), respectively (Parenti *et al.* 1997). These enzymes have been first identified on genomic level by their high DNA sequence homology to the sulphatase gene family (Franco *et al.* 1995; Puca *et al.* 1997). The physiological role of these enzymes as well as their natural substrates have not yet been identified. However, ARSE and ARSF have been demonstrated to be heat-labile and to hydrolyse the synthetic substrate 4-methylumbelliferyl sulphate with the maximal activity of ARSF being at a pH of 8 (Franco *et al.* 1995; Puca *et al.* 1997). ARSE activity is inhibited by warfarin which seems to be responsible for warfarin embryopathy (Franco *et al.* 1995).

To date, ARSD, ARSE and ARSF homologues have not been identified in species other than man. ARSD mRNA has been detected in various tissues as pancreas, kidney, liver, lung, placenta, brain and heart whereas ARSE mRNA seems to be exclusively expressed in pancreas, liver and kidney (Franco *et al.* 1995).

Mutations in the ARSE gene have been shown to cause an X-linked recessive defect of bone and cartilage development called chondrodysplasia punctata which is characterised by abnormal calcium deposition in regions of enchondral bone formation and shares striking phenotypic similarities with warfarin embryopathy. However, the physiological role of ARSE in bone development remains unknown (Franco *et al.* 1995). Up to now less than ten different point mutations in the ARSE gene have been identified, four of them resulting in a complete loss of enzymic activity (Franco *et al.* 1995; Daniele *et al.* 1998; Sheffield *et al.* 1998; Dahl *et al.* 1999).

OTHER SULPHATASES

A wide variety of other non-arylsulphatases has been described, some of them being associated with genetic deficiencies resulting in lysosomal storage diseases. However, their role in the metabolism and bioactivation of xenobiotics has not been investigated.

N-Acetylgalactosamine-6-sulphatase (Chondroitinsulphatase, Galactose-6-sulphatase, EC 3.1.6.4) hydrolyses *N*-acetyl-*D*-galactosamine 6-sulphate and *D*-galactose 6-sulphate residues from chondroitin sulphate and keratan sulphate, respectively (Habuchi *et al.* 1979; Nakanishi *et al.* 1979). In addition, several other substrates like sulphated tetrasaccharide or UDP-*N*-acetylgalactosamine 6-sulphate have been identified (Singh *et al.* 1976; Nakanishi *et al.* 1979). Compounds like chondroitin 6-sulphate, keratan sulphate, heparin, heparan sulphate, hyaluronic acid, sulphated pentasaccharides and several inorganic ions including sulphate anions have been shown to inhibit the enzyme (Glössl *et al.* 1979; Lim and Horowitz 1981). *N*-Acetylgalactosamine-6-sulphatase activity has been identified in different tissues of man, dog, rat and in bacteria (Hayashi 1978; Glössl *et al.* 1979; Habuchi *et al.* 1979; Salyers and O'Brien 1980). Human deficiency of this enzyme leads to Morquio A syndrome (Mucopolysaccharidosis IV A) and is based on a great number of different mutations reflecting an excessive allelic heterogeneity (Singh *et al.* 1976; Bunge *et al.* 1997; Tomatsu *et al.* 1998).

Iduronate-2-sulphatase (Chondroitinsulphatase, EC 3.1.6.13) is a lysosomal enzyme acting on L-iduronate 2-sulphate units of dermatan sulphate, heparan sulphate and heparin (Yutaka *et al.* 1982) and is inhibited by compounds like inorganic and organic sulphates, heparan sulphate, chondroitin 4- and 6-sulphate, suramin and phosphate (Constantopoulos *et al.* 1980; Lissens *et al.* 1984). The enzyme exists as multiple molecular forms and has been identified in various human tissues as well as in rat liver (Constantopoulos *et al.* 1980; Di Natale and Ronsisvalle 1981; Archer *et al.* 1982; Yutaka *et al.* 1982). In addition, the cDNA of murine iduronate-2-sulphatase has been cloned (Daniele *et al.* 1993). Deficiency of human iduronate-2-sulphatase, which can be caused by several different types of gene mutations (Karsten *et al.* 1998, 1999), leads to lysosomal accumulation of heparan and dermatan sulphate resulting in a clinical disorder called Hunter syndrome (Mucopolysaccharidosis II), one of the most common mucopolysaccharidoses (Neufeld and Muenzer 1995)

N-Acetylglucosamine-6-sulphatase (Glucosamine-6-sulphatase, Chondroitinsulphatase, EC 3.1.6.14) catalyses the hydrolysis of *N*-acetyl-*D*-glucosamine 6-sulphate units from heparan and keratan sulphate (Freeman and Hopwood 1987). Additionally, other substrates like glucose 6-sulphate have been identified (Freeman and Hopwood 1987; Freeman *et al.* 1987). *N*-Acetylglucosamine-6-sulphatase has been found in different human tissues like placenta, liver, heart, spleen and kidney (Freeman *et al.* 1987). Deficiency of this enzyme has been reported to result in mucopolysaccharidosis III D (Sanfilippo D), the rarest of the mucopolysaccharidoses known (Neufeld and Muenzer 1995).

N-Sulphoglucosamine-3-sulphatase (Glucosamine-3-sulphatase, Chondroitinsulphatase, EC 3.1.6.15) acts on *N*-2-sulpho-*D*-glucosamine 3-sulphate residues of heparin as well as on methyl-2-deoxy-2-sulphamino-alpha-*D*-glucopyranoside 3-sulphate. The enzyme is inhibited by sulphate and phosphate anions (Leder 1980). In addition to the human enzyme, *N*-sulphoglucosamine-3-sulphatase activity has been observed in bacteria (Bruce *et al.* 1985).

Glucuronate-2-sulphatase (EC 3.1.6.18) hydrolyses the 2-sulphate groups of the 2-O-sulpho-*D*-glucuronate residues of glycosaminoglycans like chondroitin sulphate or heparan sulphate (Shaklee *et al.* 1985). The enzyme is stimulated by Cu^{2+} and Zn^{2+} and inhibited by compounds like sulphate anions, sodium hydrogen phosphate, EDTA, glucuronic acid 2,5-anhydro-*D*-mannose 6-sulphate and glucuronic acid 2-sulphate-2,5-anhydro-*D*-mannose 6-sulphate (Freeman and Hopwood 1989). Glucuronate-2-sulphatase has been purified from chicken embryo and from human skin fibroblasts as well as from liver, lung and kidney (Shaklee *et al.* 1985; Freeman and Hopwood 1989).

BACTERIAL SULPHATASES

Sulphatase activities have been identified in rat and human intestinal bacteria like *Peptococcus, Chlostridium, Lactobacillus, Eubacterium, Peptostreptococcus* and *Bacteroides* (Hazenberg *et al.* 1988; Van Eldere *et al.* 1988, 1991). Sulphates of endogenous and xenobiotic compounds have been shown to be hydrolysed by these enzymes pointing to an enterohepatic circulation of these substances. For example, *Lactobacillus* and *Eubacterium* strains as well as *Peptostreptococcus productus* isolated from rat and human faecal suspensions, respectively, hydrolysed different iodothyronine sulphates (Hazenberg *et al.* 1988). Reabsorption of 3,3',5-triiodothyronine released from its sulphate was significantly reduced in germ-free as compared to conventional rats (Rutgers *et al.* 1989). Contamination of germ-free rats with sulphatase-producing bacteria like *Chlostridium sp.* resulted in enhanced circulation of sulphated bile salts (Robben *et al.* 1988). As shown by Huijgebaert *et al.* (1984), oestrone sulphate is absorbed from the small intestine after deconjugation by the intestinal microflora. Reduced levels of contraceptive steroids in women treated with antibiotics have been attributed to impairment of the intestinal flora which resulted in interruption of the enterohepatic circulation (Orme and Back 1979). Bacterial strains from species like *Clostridium, Lactobacillus, Eubacterium,* and *Peptococcus* have been identified as producers of steroid sulphatase activity with a broad substrate specificity. For example, oestrone 3-sulphate, β-oestradiol-3-sulphate, the 3β-sulphates of 5α-androstane-17-one and 5β-androstane-17-one, the 3β-sulphates of 5α-androstane-17-one, Δ^5-androstane-17-one, 5α-pregnane-20-one and Δ^5-pregnene-17-one, the 3α and 3β-sulphates of 5α and 5β-bile acids, as well as *p*-nitrocatechol sulphate, *p*-nitrophenyl sulphate and phenolphthalein disulphate have been shown to be hydrolysed by bacterial steroid sulphatases (Van Eldere *et al.* 1987, 1988). Three types of steroid sulphatases have been characterised in *Peptococcus niger* H4, two of them being inducible by their substrates with the exception of bile acid sulphates. The K_m values vary from 180 to 643 µM and their activity is competitively inhibited by various substrates. In addition, sulphite, sulphate, taurine, cyanide and fluoride have been demonstrated to inhibit bacterial steroid sulphatase activity under certain conditions (Van Eldere *et al.* 1991).

An example of activation of xenobiotics by bacterial sulphatases is conversion of cyclamate (cyclohexylamine *N*-sulphonate) to the bladder carcinogen cyclohexylamine in the intestine (Goldin 1990). Further examples for desulphation by intestinal bacteria have been reviewed by Scheline (1973).

Glucuronidases

OVERVIEW

β-Glucuronidase (β gluc; EC 3.2.1.31) is one of the most extensively studied glycosidases active in the metabolic hydrolysis of conjugated compounds. The activity of β-gluc has been measured in tissue extracts of mammalians and other vertebrates, molluscs and bacteria (Fishman 1974).

β-Gluc is an exoglycosidase that cleaves β-D-glucuronic acid residues from the nonreducing termini of glycosaminoglycans such as chondroitin sulphate and hyaluronic acid, and is an essential enzyme for the normal restructuring and turnover of these extracellular matrix components (Paigen 1989). It has been suggested that β-gluc plays a role in the enzymic hydrolysis of glucuronides of endogenous compounds and xenobiotics in humans because it is capable of hydrolysing glucuronide conjugates *in vitro* (Schöllhammer *et al.* 1975; Kauffman 1994; Sperker *et al.* 1997). This could have clinical consequences, especially if glucuronides are accumulated during long-term therapy as described for oxazepam, imipramine or propranolol (Walle *et al.* 1979; Sisenwine *et al.* 1982; Sutfin *et al.* 1988). Even if glucuronides are not accumulated, the enzyme may alter the net rate of glucuronide formation because it can catalyse the reverse reaction.

In addition, many substances that undergo glucuronidation and secretion into the bile are hydrolysed through the action of bacterial and perhaps of enteric β-gluc and subsequently reabsorbed (Rollins and Klaassen 1979). This enterohepatic circulation can influence disposition and action of xenobiotics, endogenous compounds and toxic chemicals. The extended enterohepatic circulation of morphine is an example for such an effect (Plaa 1975). Furthermore, the enzyme plays an important role in the bioactivation of certain prodrugs (Sinhababu and Thakker 1996). Several glucuronide prodrugs of anticancer agents, from which the active drug is released by the action of β-gluc, are currently under development (Bosslet *et al.* 1994; Houba *et al.* 1999; Guerquin-Kern *et al.* 2000). The aim of the approach is to selectively activate the prodrug at the tumour site by utilising β-gluc.

In contrast to β-gluc, there are no reports on the relevance of other glucuronidases, such as hyaluronidase, glycyrrhizinate β-glucuronidase, glucuronosyl-disulphogluco-samine glucuronidase and α-glucuronidase for drug metabolism.

MAMMALIAN β-GLUCURONIDASES

Human β-glucuronidase

β-Gluc (exo-β-D-glucuronidase, β-glucuronide glucuronosylhydrolase) is an acid lysosomal hydrolase expressed at variable levels by virtually every cell in the body. The main metabolic role of β-gluc is thought to be glycosaminoglycan degradation in lysosomes by removing terminal β-glucuronosyl residues from dermatan sulphate and heparan sulphate (Stahl and Fishman 1984; Paigen 1989). The classical substrates of β-gluc are β-D-glucopyranosiduronic acids containing an aglycone from one of the following groups:

(1) drugs and other xenobiotics
(2) steroids
(3) endogenous non-steroid compounds

Release of the parent compound from glucuronide conjugates has been described for xenobiotics such as clofibric acid (Meffin *et al.* 1983), lorazepam (Herman *et al.* 1989), glycyrrhizin (Kanaoka *et al.* 1986), paracetamol (Bohnenstengel *et al.* 1999), doxycycline (Pedersen and Miller 1980), ranitidine (Miller 1984) and diflunisal (Brunelle and Verbeeck 1996). Three synthetic substrates are commonly used for determination of β-gluc activity: 4-nitrophenyl-β-D-glucuronide and phenolphthalein-

β-glucuronide for colorimetric tests and 4-methylumbelliferyl-β-D-glucuronide (MUG) for a more sensitive fluorometric assay (Fishman *et al.* 1967; Szasz 1967; Sperker *et al.* 1996). For *in situ* localisation of β-gluc by enzyme histochemical methods a number of substrates like 8-hydroxy-quinoline-β-D-glucuronide, naphthol AS-BI β-D-glucuronide (6-bromo-2'-hydroxy-3-naphthoyl-o-anisidine β-D-glucuronide) and 5-bromo-4-chloro-3-indolyl-β-D-glucuronide (X-gluc) have been described (Fishman and Baker 1956; Hayashi *et al.* 1964).

The function of β-gluc is dependent on several factors, such as pH and Ca^{2+} concentration at the enzyme site (Wakabayashi 1970; Foster and Conigrave 1999). The human enzyme has a pH optimum of 4 to 5 and at physiological tissue pH about 10% of the maximum activity of the enzyme is retained (Wakabayashi 1970).

β-Gluc has been shown to catalyse glycoside bond hydrolysis, presumably via a covalent glucuronyl-enzyme intermediate (Wong *et al.* 1998). Recently, the active site responsible for this mechanism and the crystal structure of the enzyme have been characterised (Jain *et al.* 1996; Islam *et al.* 1999a). Human β-gluc is a tetrameric glycoprotein of about 310 to 380 kDa with a dihedral symmetry resulting from disulphide-linked dimers consisting of two identical monomers. After cleavage of the signal peptide, the unglycosylated protein contains 629 amino acids and is modified during or after the transport of the enzyme to lysosomes (Oshima *et al.* 1987; Tanaka *et al.* 1992; Islam *et al.* 1993). Following glycosylation and *C*-terminal processing within the endoplasmic reticulum and the Golgi complex, the enzyme is directed to lysosomes via the mannose 6-phosphate receptor (Shipley *et al.* 1993). Recent data from Islam *et al.* (1999b) point to a dual localisation of the human enzyme in lysosomes as well as in the endoplasmic reticulum. This seems to be due to binding of the enzyme to a human functional homologue of the murine esterase egasyn.

β-Gluc is present in most cell types such as macrophages and most other blood cells except erythrocytes, in organs like liver, spleen, kidney, intestine, lung, muscle, reproductive organs, lymph nodes and pancreas as well as in body fluids including bile, intestinal juice, urine, seminal plasma and serum (Platt and Platt 1970; Wakabayashi 1970; Gupta and Singh 1983; Heinert *et al.* 1983; Paigen 1989; Sperker *et al.* 2000). A large interindividual variability in its activity and expression has been described for tissues like liver and kidney as well as in serum (Fishman *et al.* 1967; Corrales-Hernandez *et al.* 1988; Sperker *et al.* 1997). β-Gluc activity in plasma is about 500–1000 times lower than in liver or kidney (Sperker, unpublished data).

The autosomal recessive deficiency of β-gluc activity results in progressive accumulation of undegraded glycosaminoglycans in the lysosomal compartment of many tissues and organs, leading to the lysosomal storage disease mucopolysaccharidosis type VII (MPS VII). Lysosomal accumulation of glycosaminoglycans impairs normal function of several tissues, resulting in clinical abnormalities which include glycosaminoglycan excretion, bone deformities, growth and mental retardation, hepatosplenomegaly, corneal clouding and abdominal organ enlargement (Sly *et al.* 1973). About 50 cases with heterogeneous phenotypes have been reported. From these cases, more than 20 different point mutations in the coding region of the β-gluc gene have been identified (Tomatsu *et al.* 1990, 1991; Vervoort *et al.* 1996). In addition, several pseudogenes and a pseudodeficiency allele have been described (Shipley *et al.* 1993; Vervoort *et al.* 1995, 1998; Speleman *et al.* 1996). Since analysis of the 5' flanking region of the human β-gluc gene revealed properties commonly associated with 'housekeeping' genes the human enzyme was assumed to be unregulated (Shipley *et al.* 1991). In contrast, Sperker *et al.* (2001) have found that the calcium ionophore A23187 and the calcium ATPase inhibitor thapsigargin downregulate β-gluc expression in HepG2 cells, respectively. Several other reports have demonstrated that endogenous compounds or xenobiotics affect β-gluc release and activity. Triggiani *et al.* (2000) demonstrated that secretory phospholipase A2 and p-aminophenyl-mannopyranoside, a mannose-receptor-ligand, release β-gluc from human macrophages. 17β-Oestradiol decreases secreted levels of β-gluc in a dose-dependent manner (Kremer *et al.* 1995) and cyclosporin increases serum β-gluc after treatment for 8 to 16 weeks (Falkenbach *et al.* 1993).

Furthermore, clinical relevant concentrations of benzodiazepines, such as midazolam and flunitrazepam increased the activity of β-gluc released from polymorphonuclear neutrophil leukocytes (Krumholz *et al.* 2000). Interestingly, increased tissue and serum β-gluc activity has been observed in certain disease states like cancer, inflammatory joint and urinary tract diseases (active pyelonephritis, acute renal necrosis), dermatosis, some hepatic diseases, Type 1 diabetes mellitus, multiple sclerosis, periodontal disease and AIDS (Stephens *et al.* 1975; Camisa *et al.* 1988; Saha *et al.* 1991; Ohta *et al.* 1992; Waters *et al.* 1992; Boyer and Tannock 1993; Goi *et al.* 1993; Zenser *et al.* 1999; Layik *et al.* 2000). For example, β-gluc activity has been reported to be higher in breast, prostate, kidney and lung tumours as compared to peritumoral tissues (Albin *et al.* 1993; Mürdter *et al.* 1997). A significantly lower concentration of β-gluc has been observed in epithelial cells of the renal cortex under pathological conditions (Heinert *et al.* 1983). In addition, cataractous lenses exhibited decreased β-gluc activity compared to normal lenses (Kamai 1995). Serum levels of β-gluc are 16-fold higher in HIV-infected patients than in healthy individuals (Saha *et al.* 1991).

Plasma levels of β-gluc in males from 24 to 60 years were significantly higher (about 30%) than in females of the same age (Lombardo *et al.* 1981). Additionally, plasma level of β-gluc was highest in the umbilical cord blood, then dropped slowly to reach the absolute minimum at the age of 10–14 years. Thereafter the enzymic activity increased reaching a maximum around 20–24 years followed by a decrease (30–34 years) and a second slow increase up to the old age. Similarly, age-dependent variations of the specific β-gluc activities have been described for different tissues. For example, the level of total β-gluc activity in childhood liver was higher than in foetal liver (Minami *et al.* 1979). Platt (1970) has demonstrated that the enzyme activity increased significantly in liver and adrenal glands from birth to 80 years. The increase in β-gluc activity with age was also shown in the retinal pigment epithelium (Verdugo and Ray 1997). In contrast, activity in the kidney decreased from birth to adult age (Platt 1970).

In addition to glycosaminoglycan degradation, human β-gluc plays a role in the deconjugation of some other endogenous substances. For example, deconjugation of glucuronides of bilirubin in human bile and development of cholelithiasis is dependent on biliary β-gluc activity (Ho *et al.* 1986). Neonatal intestinal β-gluc antagonises the net clearance of bilirubin by increasing enterohepatic circulation and hence causing increased serum bilirubin levels (Poland and Odell 1971). In addition, intestinal β-glucuronidase originating from breast milk may cause hyperbilirubinaemia in infants of mothers with diabetes mellitus (Sirota et al. 1992).

Drug glucuronides can be susceptible to enterohepatic circulation since many glucuronides are secreted into bile. In the intestine, the glucuronides may be hydrolysed by the action of endogenous (biliary or enteric) and bacterial β-gluc and subsequently reabsorbed. This mechanism has been suggested for a variety of compounds like digitoxin, some benzodiazepines, morphine and indomethacin (Caldwell and Greenberger 1971; Rollins and Klaassen 1979; Herman et al. 1989). However, the relative influence of enteric β-gluc on enterohepatic circulation is not clear. Brunelle and Verbeeck (1996) could show that the glucuronidation rate of diflunisal is affected by the microsomal β-gluc activity.

Cleavage of glucuronides by β-gluc also plays a role in drug toxicity and chemical carcinogenesis. It has been suggested that non-toxic metabolites of aromatic hydrocarbons are exported from the liver as glucuronides which are subsequently hydrolysed to mutagenic metabolites in peripheral tissues. For example, it has been demonstrated that β-gluc deconjugates benzidine- and 4-aminobiphenyl-glucuronides as well as 3-benzo(α)pyrenyl glucuronide resulting in the release of the carcinogenic aromatic compounds. These aromatic amines may cause bladder cancer in humans (Moore et al. 1982; Zenser et al. 1999). Furthermore it has been suggested that inhibitors of β-gluc could be used to reduce the rate of deglucuronidation and to protect against carcinogenesis. Such inhibitors could be ascorbic acid (Young et al. 1990), silymarin and silybin (Kim et al. 1994) as well as cortisone and gold-l-complexes (Chang et al. 1993). In human liver and kidney homogenates, D-glucaro-1,4-lactone, glycyrrhizin, oestradiol 3-glucuronide and paracetamol glucuronide competitively inhibit the β-gluc-mediated cleavage of 4-methylumbelliferyl-D-glucuronide (Ho et al. 1985; Sperker et al. 1997).

An important approach in the alleviation of toxic adverse effects in cancer chemotherapy is the development of prodrugs of anticancer agents which are less reactive or less cytotoxic than the respective parent compounds. A desired objective of prodrug therapy is the selective delivery of the cytotoxic agent to the tumour tissue, which can be achieved by antibody-directed enzyme prodrug therapy (ADEPT) or by prodrug therapy based on elevated tumour activities of bioactivating enzymes. In both approaches, β-gluc can be used as the activating enzyme. Recently, fusion proteins consisting of a human β-gluc moiety and humanised antibody fragment directed against tumour specific antigens have been designed for use in ADEPT (Bosslet et al. 1992; Haisma et al. 1998). Furthermore, it was demonstrated that endogenous β-gluc mediates tumour-selective release of doxorubicin from a glucuronide prodrug (HMR1826) at the tumour site, using an isolated, perfused human lung tumour model (Mürdter et al. 1997). A reason for the tumour selective activation of the glucuronide prodrug is the localisation of β-gluc. In contrast to normal tissue, tumoral β-gluc is localised extracellularly probably due to secretion by inflammatory cells and disintegrating tumour cells (Bosslet et al. 1998). Weyel et al. (2000) have shown that transduction of human tumour cells with a secreted form of human β-gluc resulted in conversion of HMR1826 to doxorubicin and in enhanced tumour cell killing. Recently, an improved tumour targeting of daunorubicin and 5-fluorouracil in nude mice bearing human cancer xenografts has been demonstrated by using the respective glucuronide prodrugs (Houba *et al.* 1999; Guerquin-Kern *et al.* 2000).

Rodent β-glucuronidase

Rodent β-gluc is an unusual protein in the class of hydrolases because it is localised not only in lysosomes but also in the endoplasmic reticulum (ER) of liver parenchymal cells, kidney and lung, but not in spleen, brain, heart, erythrocytes, testis and skin (Fishman *et al.* 1967; Himeno *et al.* 1976; Lusis and Paigen 1977). Liver lysosomal and microsomal β-gluc are products of a single structural gene. Both enzymes are catalytically identical and show the same immune reactivity. They are similar in molecular weight, but differ in both sugar and amino acid composition (Wang and Touster 1975; Himeno *et al.* 1976; Owens and Stahl 1976; Lusis and Paigen 1977). The ER localisation of β-gluc is due to its interaction with the esterase egasyn (Lusis and Paigen 1977; Medda *et al.* 1986, 1987). This binding results in the retention of a fraction of β-gluc in the lumen of the ER I (Zhen *et al.* 1993). Inhibitors of egasyn-esterase activity like organophosphates caused rapid dissociation of egasyn-microsomal β-gluc complex as a result of a massive increase of microsomal β-gluc in plasma (Medda *et al.* 1987). Furthermore, mouse strains lacking egasyn also lack ER glucuronidase (Lusis and Paigen 1977).

Murine β-gluc gene complex (*Gus*) has been studied extensively because it provides a useful system for understanding mammalian gene regulation. Three common alleles of *Gus-s* ($-s^a$, $-s^b$ and $-s^h$) specify allozymes which differ in electrophoretic mobility and heat stability (Paigen 1961; Swank and Paigen 1973; Lalley and Shows 1974). Specific alleles of each of three GUS regulatory elements (*Gus-r, Gus-t and Gus-u*) are associated with specific alleles of *Gus-s*.

The combination of *Gus-s* with the regulatory elements define three common haplotypes (Gus^a, Gus^b, Gus^h) with different effects on the expression of GUS. *Gus-r* controls the androgen responsiveness of kidney GUS mRNA (Palmer *et al.* 1983). The second element, *Gus-u*, controls the levels of constitutive GUS synthesis in all tissues (Lusis 1983). In addition, the third regulatory element, *Gus-t*, exerts an additional temporal control over GUS synthesis in certain tissues (Meredith and Ganschow 1978; Lusis 1983). Mouse and rat β-gluc are identical in length and share 88.2% sequence identity. Mouse and human β-gluc are 75.3% identical (Funkenstein *et al.* 1988). Like the human enzyme, rodent β-gluc is a tetramer of identical subunits formed from a single gene product (Delvin and Granetto 1970).

Heringova *et al.* (1965) found that rat β-gluc activity in the intestinal mucosa is age-dependent. Relatively high activity was measured during the first two weeks of life and decreased to adult levels at the end of the third week. Activity of β-gluc in murine lungs increased markedly with advancing age (Traurig 1976). Significant age-dependent decrease of the total activity of β-gluc has been found in kidney, rib cartilage and skin whereas significant age-dependent increase of this enzyme activity has been demonstrated in the spleen and liver of rats (Lindner *et al.* 1986). In the studies by Watson *et al.* (1985), the enzyme has been shown to be both developmentally and hormonally regulated in rats and mice. In the mouse kidney, β-gluc is subject to induction by androgens, which results in an 120-fold elevation of mRNA concentration (Palmer *et al.* 1983). In female rat preputial glands β-gluc is regulated by oestrogen (Levy *et al.* 1958). The administration of oestradiol produced an increase of β-gluc activity (Briggs 1973; Gallagher and Sloane 1984) and glycyrrhizin significantly inhibited an increase of uterine β-gluc activity by oestradiol-17β (Kumagai *et al.* 1967). Furthermore, low levels of vitamin A caused an increase in β-gluc activity in rats and mice (Kostulak 1974; Rundell *et al.* 1974). In contrast, cortisone acetate (Baglioni *et al.* 1978), cortisone (Koldovsky and Palmieri 1971; Horowitz *et al.* 1978), thyroxine (Horowitz *et al.* 1978), ganoderenic acid A (Kim *et al.* 1999) and dexamethasone (Hicks *et al.* 1994) decreased β-gluc activity in rats.

Hepatic microsomal β-gluc can influence biliary excretion and hepatic elimination of bilirubin-IXα as well as other endogenous and exogenous compounds. Whiting *et al.* (1993) found an increased excretion of glucuronides of bilirubin-IXα in a strain of mice lacking hepatic microsomal β-gluc. In rat liver microsomes, the production of diflunisal glucuronide has been shown to increase when β-gluc is inhibited by D-saccharic acid 1,4-lactone (Brunelle and Verbeeck 1993). Furthermore, a mechanism of enterohepatic circulation in rodents has been suggested for digitalis glycosides (Caldwell and Greenberger 1971) and lorazepam (Ruelius 1978).

β-Glucuronidases in other species

In addition to humans and rodents, the enzyme has been characterised in many mammalian species like dog, monkey, cow, pig and cat. The optimal pH for β-gluc activity in the retinal pigment epithelium was found to range from 4.0 to 4.5. Except for the very unstable bovine enzyme, β-gluc activity displays a very high resistance to heat inactivation at pH 5.0 (Ray *et al.* 1997). Canine and feline β-gluc show a high degree of homology to the human enzyme (Fyfe *et al.* 1999; Ray *et al.* 1999). Schuchman et al. (1989) reported that about 65% of total β-gluc in canine liver is membrane associated and can be solubilised by detergent.

β-Gluc from bovine liver and porcine kidney catalysed the hydrolysis of some glucopyranosiduronamides and glucopyranosides as well as of oestrone-3-glucuronide (Gowers and Breuer 1980; Parker *et al.* 1981). Recently, some new compounds with β-gluc inhibitory activity have been synthesised and tested on bovine liver homogenate, such as glucuronic acid-type 1-*N*-iminosugar of D-arabinose (Igarashi *et al.* 1996), D-glucaro-δ-lactam, an oxidation product of nojirimycin (Niwa *et al.* 1972), nojirimycin A (Tsuruoka *et al.* 1996) and sistatin B analogues (Satoh *et al.* 1996).

Both, dogs and cats are useful animal models of human diseases, such as pancreatitis and mucopolysaccharidosis type VII (Dlugosz *et al.* 1977; Haskins *et al.* 1991; Gitzelmann *et al.* 1994). Dogs with acute experimental pancreatitis showed an increase of free β-gluc activity in pancreas homogenates. This increased β-gluc level was antagonized by glucagon (Dlugosz *et al.* 1977). Mucopolysaccharidosis type VII (MPD VII) was first described in a mixed-breed dog which was homozygous for a single G to A missense mutation in the β-gluc gene (Haskins *et al.* 1984, 1991). Ray *et al.* (1998) could demonstrate that the mutation reduces the canine β-gluc activity by more than 100-fold. Pathological lesions of MPS VII in dog and man are similar (Haskins *et al.* 1991; Neufeld and Muenzer 1995). In 1994, Gitzelmann *et al.* have described a case of feline MPS VII in a domestic male cat. Like dogs, a single G to A missense mutation of β-gluc cDNA reduced β-gluc activity to background levels.

BACTERIAL β-GLUCURONIDASE

Most *E. coli* strains, including many pathogenic serogroups, produce β-gluc (GUS) and are positive with the 4-methylumbelliferyl-β-D-glucuronide assay (Feng and Hartman 1982; Hartman 1989). In addition, Ralovich *et al.* (1991) found, that other gram-negative bacteria such as *Salmonella*, *Shigella* and *Yersinia* strains show a positive β-gluc reaction. β-Gluc is the first enzyme of the hexuronide-hexuronate pathway (Ashwell 1962). *E. coli* β-gluc is a very stable enzyme, with a molecular weight of 290 kDa and is composed of four subunits (Kim *et al.* 1995). The enzyme displays a broad pH optimum in the neutral range with its activity being decreased by 50% at pH 4.3 and 8.5. *E. coli* β-gluc is resistant to thermal inactivation at 50°C (Ho and Ho 1985; Jefferson *et al.* 1986).

Bacterial β-gluc plays an important role in the enterohepatic circulation of drugs and endogenous compounds. This enterohepatic circulation is likely to be clinically significant because the alteration of circulation can considerably decrease or increase the clearance of some drugs like digitalis glycosides, morphine, steroids, indomethacin, amphetamine, several antibiotics, doxycycline and some benzodiazepines and nonsteroidal anti-inflammatory agents (Plaa 1975; Rollins and Klaassen 1979). Moreover, clearance of progesterone has been reported to be increased if the enterohepatic circulation is interrupted by *D*-saccharic acid 1,4-lactone which has been shown to be a potent inhibitor of β-gluc (Marselos *et al.* 1975). In addition, it has been reported that lorazepam is extensively conjugated resulting in the 3-*O*-phenolic glucuronide which undergoes enterohepatic circulation (Ruelius 1978). The clearance of the drug increased by about 25% when its circulation is blocked by neomycin, an antibiotic effective against β-gluc producing bacteria (Herman *et al.* 1989). Arylamine *N*-glucuronides were found to be susceptible to hydrolysis by *E. coli* β-gluc, suggesting the release of carcinogenic arylamines (Lilienblum and Bock 1984). Furthermore, it has been shown that glucuronides of 1-nitropyrene-metabolites secreted into the bile can be hydrolysed in the intestine by bacterial β-gluc to potent mutagenic aglycones (Morotomi *et al.* 1985). In fact, as many carcinogens are detoxified by glucuronidation, it has been proposed that inhibitors of β-gluc could protect against carcinogenesis by reducing the rate of deglucuronidation. For example, *D*-glucarate and *D*-saccharic acid 1,4-lactone have been shown to inhibit benzo(α)pyrene-induced mammary, colonic and pulmonary tumorigenesis (Kinoshita and Gelboin 1978; Walaszek *et al.* 1984; Walaszek *et al.* 1990). Intestinal toxicity of the anticancer agent irinotecan is due to biliary excretion of the glucuronic acid conjugate of its active metabolite SN-38 and subsequent deglucuronidation by bacterial β-gluc. Administration of the β-gluc inhibitor baicalin has been shown to protect against intestinal toxicity of irinotecan metabolites in rats (Takasuna *et al.* 1995, 1996). Other inhibitors of bacterial β-gluc are glycyrrhizin (Takasuna *et al.* 1995), as well as silymarin and silybin (Kim *et al.* 1994).

E. coli β-gluc expression is negatively controlled by the products of the two regulatory genes *uidR* and *uxuR* (Novel and Novel 1976) and inducible by methyl-β-

D-glucuronate and alkaline pH (Novel *et al.* 1974; Kim et al. 1992). Results from the studies of Caldini *et al.* (1999) have shown that starch metabolism is involved in *β*-gluc induction in *E. coli.*

OTHER GLUCURONIDASES

Heparanase is the dominant endoglucuronidase in mammalian tissues. The enzyme is capable of cleaving heparan sulphate glycosaminoglycans to short 5–6 kDa fragments (Gallagher *et al.* 1988). Heparanase was found to cleave the single *β-D*-glucuronidic linkage of a heparin octasaccharide with *O*-sulphate groups being essential for substrate recognition by heparanase (Pikas *et al.* 1998). Heparanase has been implicated in metastasis, because non-metastatic murine T lymphoma and melanoma cells transfected with the heparanase gene acquired a highly metastatic phenotype *in vivo* (Vlodavsky *et al.* 2000). A 4-fold increase in serum heparanase activity was found in patients with tissue metastases (Nakajima *et al.* 1988). In addition, these authors observed that serum heparanase levels in rats were 17-fold increased after injection of highly metastatic adenocarcinoma cells. Furthermore, elevated levels of heparanase were detected in tumour biopsies of cancer patients (Vlodavsky *et al.* 1999). Calcium spirulan, a sulphated polysaccharide chelating calcium from a blue-green alga, significantly inhibited degradation of heparan sulphate by heparanase and significantly reduced experimental lung metastasis (Mishima *et al.* 1998). Moreover, the anti-metastatic effect of non-anticoagulant species of heparin and of certain sulphated polysaccharides like laminarin and the phosphorothioate homopolymer of cytidine was attributed to their heparanase-inhibiting activity (Bitan *et al.* 1995; Miao *et al.* 1999). Furthermore, heparanase is involved in tumour angiogenesis both directly, by promoting invasion of endothelial cells and indirectly, by releasing heparan-bound basic fibroblast growth factor (Vlodavsky *et al.* 2000). Parish *et al.* (1999) found that sulphated oligosaccharides are potent inhibitors of heparanase activity and *in vitro* angiogenesis.

Human heparanase is a 50 kDa enzyme (Toyoshima and Nakajima 1999) and its amino acid sequence is highly homologous to mouse and rat heparanase (Hulett *et al.* 1999; Dong *et al.* 2000). The intracellular heparanase from mouse melanoma cells is similar to the human platelet enzyme in terms of pH optimum and pI value, but appears to be bigger in size (Graham and Underwood 1996). Kussie *et al.* (1999) have shown high expression of heparanase in placenta and spleen. Furthermore, heparanase activity has been described in a number of tissues and cell types including rat liver (Gallagher *et al.* 1988), human placenta (Klein and von Figura 1979), human platelets (Hoogewerf *et al.* 1995), cultured human skin fibroblasts (Klein and von Figura 1976), human neutrophils (Matzner *et al.* 1985), rat T-lymphocytes (Naparstek *et al.* 1984), murine B-lymphocytes (Laskov *et al.* 1991) and human monocytes (Bartlett *et al.* 1995).

Bilirubin monoglucuronide transglucuronidase (bilirubin-glucuronoside glucuronosyltransferase, EC 2.4.1.95) catalyses a two- step degradation reaction of bilirubin mono- and di-glucuronides: (1) hexosyl group transfer and (2) transglucuronidation. The enzyme (16 kDa) with a pH optimum of 6.6 and a K_m value of 32–34 μM was found in rat liver and is localised in plasma membrane and microsomes (Jansen *et al.* 1977; Chowdhury *et al.* 1979; Chowdhury and Arias 1981).

α-Glucuronidase (EC 3.2.1.139) is a xylanolytic enzyme produced by a wide variety of fungi, such as *Trichoderma reesei* (Siika-aho *et al.* 1994), *Thermoascus aurantiacus* (Khandke *et al.* 1989), *Agaricus bisporus* (Puls *et al.* 1987) and *Aspergillus niger* (Uchida *et al.* 1992). Furthermore, α-glucuronidase activity has been identified in bacteria, e.g. *Streptomyces* (MacKenzie *et al.* 1987), *Fibrobacter* (Smith and Forsberg 1991),*Clostridium* (Trudeau *et al.* 1992; Bronnenmeier *et al.* 1995) and *Thermoanaer-obacterium* strains (Bronnenmeier *et al.* 1995).

Hyaluronidase (EC 3.2.1.36) catalyses *O*-glycosyl bond hydrolysis of 1,3-linkages between β-*D*-glucuronate and *N*-acetyl-*D*-glucosamine residues in hyaluronic acid and tetrasaccharides. The activity of this enzyme is inhibited by butane-2,3-dione and phenylglyoxal. The enzyme is localised in the salivary glands of *Hirudo medicinalis* (Linker *et al.* 1960; Hipkin *et al.* 1989).

Glycyrrhizinate β-glucuronidase (EC 3.2.1.128) from *Aspergillus niger* and *Eubac-terium sp.* is specific for the hydrolysis of *O*-glycosyl compounds, such as the triterpenoid glycoside glycyrrhizinate from roots of *Glycyrrhiza sp.* The enzyme with a pH optimum of 4.1 to 4.5 and a temperature optimum of 45°C has a molecular weight of 15 kDa and is localised in the cytosol (Muro *et al.* 1986; Akao *et al.* 1987; Sasaki *et al.* 1988).

Glucuronosyl-disulphoglucosamine glucuronidase (EC 3.2.1.56) with a pH opti-mum of 6.5 and a temperature optimum of 30°C was isolated from cells of *Flavobac-terium heparinum* and catalysed the hydrolysis of *O*-glycosyl bonds from desulphated disaccharides and heparin (Dietrich 1969; Dietrich *et al.* 1973; Hovingh and Linker 1977).

Conclusions

Sulphatases are widely distributed in various animal species and tissues. Whereas less information is available on their role in metabolising drugs and other xenobiotics, the main function appears to be metabolism of endogenous compounds. These include glycosaminoglycans, glycolipids, steroid as well as thyroid hormones and catechola-mines. Activation of steroid hormones by steroid sulphatase plays an important role in the pathogenesis of several tumours like breast or prostate cancer. Therefore inhibitors of steroid sulphatase are of great interest for prevention and therapy of these tumours. Bacterial sulphatase activity modulates bioavailability of substances that undergo enterohepatic circulation, especially thyroid and steroid hormones used in replace-ment therapy. In addition, transformation of substances into carcinogens via desulpha-tion has been reported.

β-Glucuronidase is the most important enzyme among the glucuronidases known. Localisation of the enzyme differs among species. Whereas human β-glucuronidase is located mainly in lysosomes, remarkable amounts of rodent β-glucuronidase have been detected in the endoplasmic reticulum. Thus the extent of net glucuronidation of xenobiotics in rodents depends on the level of hepatic microsomal β-glucuronidase. After glucuronidation, several drugs and endogenous compounds are known to be eliminated via the bile. Bacterial expression of β-glucuronidase in the intestine results in deglucuronidation and reabsorption of these compounds. Interference with this enterohepatic circulation sets the stage for a wealth of drug interactions. Finally, drugs have been developed which are selectively bioactivated by β-glucuronidase for the targeted treatment of cancer, thereby reducing adverse side effects.

References

Agogbua S and Wynn C (1976) Purification and properties of arylsulphatase B of human liver. *Biochemical Journal*, **153**, 415–421.

Akao T, Mibu K, Erabi T, Hattori M, Namba T and Kobashi K (1987) Non-enzymatic reduction of sennidins and sennosides by reduced flavin. *Chemical and Pharmaceutical Bulletin*, **35**, 1998–2003.

Albin N, Massaad L, Toussaint C, Mathieu M, Morizet J, Parise C, Gouyette A and Chabot G (1993) Main drug-metabolizing enzyme systems in human breast tumors and peritumoral tissues. *Cancer Research*, **53**, 3541–3546.

Alperin E and Shapiro L (1997) Characterization of point mutations in patients with X-linked ichthyosis. Effects on the structure and function of the steroid sulfatase protein. *Journal of Biological Chemistry*, **272**, 20756–20763.

Aqrabawi J, Schilder I I, Toselli P and Franzblau C (1993) Biochemical and histochemical analysis of the enzyme arylsulfatase in human lesions of endodontic origin. *Journal of Endodontics*, **19**, 335–338.

Archer I, Harper P and Wusteman F (1982) Multiple forms of iduronate 2-sulphate sulphatase in human tissues and body fluids. *Biochimica et Biophysica Acta*, **708**, 134–140.

Ashwell G (1962) Enzymes of glucuronic and galacturonic acid metabolism in bacteria. *Methods in Enzymology*, **5**, 190–208.

Baglioni T, Locatelli A, Sartorelli P and Simonic T (1978) Influence of age and cortisone treatment on two intestinal lysosomal hydrolases in young rats. *Italian Journal of Biochemistry*, **27**, 1–10.

Barth M, Fensom A and Harris A (1994) The arylsulphatase A gene and molecular genetics of metachromatic leucodystrophy. *Journal of Medical Genetics*, **31**, 663–660.

Bartlett MR, Underwood PA and Parish CR (1995) Comparative analysis of the ability of leucocytes, endothelial cells and platelets to degrade the subendohelial basement membrane: evidence for cytokine dependence and detection of a novel sulfatase. *Immunology and Cell Biology*, **73**, 113–124.

Bierfreund U, Kolter T and Sandhoff K (2000) Sphingolipid hydrolases and activator proteins. *Methods in Enzymology*, **311**, 255–276.

Bird T, Farrell D and Stranahan S (1981) Genetic control of developmental patterns of cerebral enzyme activities: further differences between C3H and ICR strains of mice. *Neurochemical Research*, **6**, 863–871.

Bitan M, Mohsen M, Levi E, Wygoda M, Miao H, Lider O, Svahn C, Erke H, Ishai-Michaeli R, Bar-Shavit R, Vlodavsky I and Peretz T (1995) Structural requirements for inhibition of melanoma lung colonization by heparanase inhibiting species of heparin. *Israel Journal of Medical Sciences*, **31**, 106–118.

Bohnenstengel F, Kroemer H and Sperker B (1999) In vitro cleavage of paracetamol glucuronide by human liver and kidney β-glucuronidase: determination of paracetamol by capillary electrophoresis. *Journal of Chromatography B*, **721**, 295–299.

Boivin R, Labrie F and Poirier D (1999) 17alpha-Alkan (or alkyn) amide derivatives of estradiol as inhibitors of steroid-sulfatase activity. *Steroids*, **64**, 825–833.

Bond C, Clements P, Ashby S, Collyer C, Harrop S, Hopwood J and Guss J (1997) Structure of a human lysosomal sulfatase. *Structure*, **5**, 277–289.

Bosslet K, Czech J, Lorenz P, Sedlacek H, Schuermann M and Seemann G (1992) Molecular and functional characterisation of a fusion protein suited for tumour specific prodrug activation. *The British Journal of Cancer*, **65**, 234–8.

Bosslet K, Czech J and Hoffmann D (1994) Tumor-selective prodrug activation by fusion protein-mediated catalysis. *Cancer Research*, **54**, 2151–2159.

Bosslet K, Straub R, Blumrich M, Czech J, Gerken M, Sperker B, Kroemer H, Gesson J, Koch M and Monneret C (1998) Elucidation of the mechanism enabling tumor selective prodrug monotherapy. *Cancer Research*, **58**, 1195–1201.

Boyer M and Tannock I (1993) Lysosomes, lysosomal enzymes, and cancer. *Advances in Cancer Research*, **60**, 269–291.

Bradshaw K and Carr B (1986) Placental sulfatase deficiency: maternal and foetal expression of steroid sulfatase deficiency and X-linked ichthyosis. *Obstetrical and Gynecological Survey*, **41**, 401–413.

Briggs M (1973) Lysosomal enzyme activation by steroid hormones in vivo. *Journal of Steroid Biochemistry*, **4**, 341–347.

Bronnenmeier K, Meissner H, Stocker S and Staudenbauer W (1995) alpha-*D*-glucuronidase from the xylanolytic thermophiles *Clostridium stercorarium* and *Thermoanaerobacterium saccharolyticum*. *Microbiology*, **141**, 2033–2040.

Bruce J, McLean M, Long W and Williamson F (1985) Flavobacterium heparinum 3-O-sulphatase for N-substituted glucosamine 3-O-sulphate. *European Journal of Biochemistry*, **148**, 359–365.

Brunelle F and Verbeeck R (1993) Glucuronidation of diflunisal by rat liver microsomes – effect of microsomal β-glucuronidase activity. *Biochemical Pharmacology*, **46**, 1953–1958.

Brunelle F and Verbeeck R (1996) Glucuronidation of diflunisal in liver and kidney microsomes of rat and man. *Xenobiotica*, **26**, 123–131.

Bunge S, Kleijer W, Tylki-Szymanska A, Steglich C, Beck M, Tomatsu S, Fukuda S, Poorthuis B, Czartoryska B, Orii T and Gal A (1997) Identification of 31 novel mutations in the N-acetylgalactosamine-6-sulfatase gene reveals excessive allelic heterogeneity among patients with Morquio A syndrome. *Human Mutation*, **10**, 223–232.

Caldini G, Strappini F, Trotta F and Cenci G (1999) Implications of α-amylase production and β-glucuronidase expression in *Escherichia coli* strains. *Microbios*, **99**, 123–130.

Caldwell J and Greenberger N (1971) Interruption of the enterohepatic circulation of digitoxin by cholestyramine.I. Protection against lethal digitoxin intoxication. *Journal of Clinical Investigation*, **50**, 2626–2637.

Camisa C, Hessel A, Rossana C and Parks A (1988) Autosomal dominant keratoderma, ichthyosiform dermatosis and elevated serum β-glucuronidase. *Dermatologica*, **177**, 341–347.

Cawley TJ and Shickley T (1992) The potential neurotoxin 2–OH-dopamine is an inhibitor of arylsulfatase. *Annals of the New York Academy of Sciences*, **648**, 256–259.

Chang H, Friedman M, Hileman D and Parish E (1993) Inhibition of human synovial β-glucuronidase by steroidal compounds – short communication. *Journal of Enzyme Inhibition*, **6**, 331–335.

Chetrite G, Ebert C, Wright F, Philippe A and Pasqualini J (1999) Control of sulfatase and sulfotransferase activities by medrogestone in the hormone-dependent MCF-7 and T-47D human breast cancer cell lines. *Journal of Steroid Biochemistry and Molecular Biology*, **70**, 39–45.

Chowdhury J and Arias I (1981) Dismutation of bilirubin monoglucuronide. *Methods in Enzymology*, **77**, 192–197.

Chowdhury J, Chowdhury N, Bhargava M and Arias I (1979) Purification and partial characterization of rat liver bilirubin glucuronoside glucuronosyltransferase. *Journal of Biological Chemistry*, **254**, 8336–8339.

Ciobanu L, Boivin R, Luu-The V, Labrie F and Poirier D (1999) Potent inhibition of steroid sulfatase activity by 3-O-sulfamate 17alpha-benzyl (or 4'-tert-butylbenzyl)estra-1,3,5(10)-trienes: combination of two substituents at positions C3 and C17alpha of estradiol. *Journal of Medical Chemistry*, **42**, 2280–2286.

Compagnone N, Salido E, Shapiro LJ and Mellon SH (1997) Expression of steroid sulfatase during embryogenesis. *Endocrinology*, **138**, 4768–4773.

Constantopoulos G, Rees S, Cragg B, Barranger J and Brady R (1980) Experimental animal model for mucopolysaccharidosis: suramin-induced glycosaminoglycan and sphingolipid accumulation in the rat. *Proceedings of the National Academy of Sciences, USA*, **77**, 3700–3704.

Corrales Hernandez J, Gonzalez Buitrago J, Pastor Encinas I, Garcia Diez L and Miralles J (1988) Androgen environment and β-glucuronidase activity in the human kidney. *Archives of Andrology*, **20**, 185–191.

Coughtrie M, Sharp S, Maxwell K and Innes N (1998) Biology and function of the reversible sulfation pathway catalysed by human sulfotransferases and sulfatases. *Chemico-Biological Interactions*, **109**, 3–27.

Cuevas-Covarrubias S, Juarez-Oropeza M, Miranda-Zamora R and Diaz-Zagoya J (1993) Comparative analysis of human steroid sulfatase activity in prepubertal and postpubertal males and females. *Biochemistry and Molecular Biology International*, **3**, 691–695.

Dahl H, Osborn A, Hutchison W, Thorburn D and Sheffield L (1999) Late diagnosis of maternal PKU in a family segregating an arylsulfatase E mutation causing symmetrical chondrodysplasia punctata. *Molecular Genetics and Metabolism*, **68**, 503–506.

Daniel W (1978) Murine arylsulfatase B: regulation of As-1 expression in different tissues. *Journal of Heredity*, **69**, 244–250.

Daniel W (1985) Arylsulfatase C and the steroid sulfatases. *Isozymes* , **12**, 189–228.

Daniel W (1987) Arylsulfatase B synthesis and clearance in inbred mouse strains. *Experientia*, **43**, 1209–1211.

Daniele A, Faust C, Herman G, Di Natale P and Ballabio A (1993) Cloning and characterization of the cDNA for the murine iduronate sulfatase gene. *Genomics*, **16**, 755–757.

Daniele A, Parenti G, d'Addio M, Andria G, Ballabio A and Meroni G (1998) Biochemical characterization of arylsulfatase E and functional analysis of mutations found in patients with X-linked chondrodysplasia punctata. *American Journal of Human Genetics*, **62**, 562–572.

Delvin E and Granetto R (1970) The purification of lysosomal rat-liver β-glucuronidase. *Biochimica et Biophysica Acta*, **220**, 93–100.

Di Natale P and Ronsisvalle L (1981) Identification and partial characterization of two enzyme forms of iduronate sulfatase from human placenta. *Biochimica et Biophysica Acta*, **661**, 106–111.

Dietrich C (1969) Studies on the induction of heparin-degrading enzymes in Flavobacterium heparinum. *Biochemistry*, **8**, 3342–3347.

Dietrich C, Silva M and Michelacci Y (1973) Sequential degradation of heparin in Flavobacterium heparinum. Purification and properties of five enzymes involved in heparin degradation. *Journal of Biological Chemistry*, **248**, 6408–6415.

Dlugosz J, Gabryelewicz A and Baiko K (1977) The lysosomal hydrolases in acute experimental pancreatitis in dogs treated with glucagon. *Acta Hepato-Gastroenterologica*, **24**, 44–51.

Dong J, Kukula A, Toyoshima M and Nakajima M (2000) Genomic organization and chromosome localization of the newly identified human heparanase gene. *Gene*, **253**, 171–178.

Falkenbach A, Wigand R, Unkelbach U, Jörgens K, Matinovic A, Scheuermann E, Seiffert U and Kaltwasser J (1993) Cyclosporin treatment in rheumatoid arthritis is associated with an increases serum activity of β-glucuronidase. *Scandinavian Journal of Rheumatology*, **22**, 83–85.

Feng P and Hartman P (1982) Fluorogenic assay for immediate confirmation of *Escherichia coli*. *Applied and Environmental Microbiology*, **43**, 1320–1329.

Ferland G, Perea A, Audet M and Tuchweber B (1990) Characterization of liver lysosomal enzyme activity in hepatocytes, Kupffer and endothelial cells during aging: effect of dietary restriction. *Mechanisms of Ageing and Development*, **56**, 143–154.

Fishman W (1974) β-Glucuronidase. In *Methods of Enzymatic Analysis,* Bergmeyer HU (ed.), Academic Press, New York, pp. 929–943.

Fishman W and Baker J (1956) Cellular localization of β-glucuronidase in benign and malignant tissue. *Journal of Histochemistry and Cytochemistry*, **4**, 412–414.

Fishman W, Kato K, Anstiss C and Green S (1967) Human serum β-glucuronidase; its measurement and some of its properties. *Clinica Chimica Acta*, **15**, 435–447.

Fluharty A, Stevens R, Fung D, Peak S and Kihara H (1975) Uridine diphospho-N-acetylgalactosamine-4-sulfate sulfohydrolase activity of human arylsulfatase B and its deficiency in the Maroteaux-Lamy syndrome. *Biochemical and Biophysical Research Communication*, **64**, 955–962.

Fluharty A, Stevens R, Goldstein E and Kihara H (1979) The activity of arylsulfatase A and B on tyrosine O-sulfates. *Biochimica et Biophysica Acta*, **566**, 321–326.

Foster F and Conigrave A (1999) Genistein inhibits lysosomal enzyme release by suppressing Ca2+ influx in HL-60 granulocytes. *Cell Calcium*, **25**, 69–76.

Franco B, Meroni G, Parenti G, Levilliers J, Bernard L, Gebbia M, Cox L, Maroteaux P, Sheffield L, Rappold G, Andria G, Petit C and Ballabio A (1995) A cluster of sulfatase genes on Xp22.3: mutations in chondrodysplasia punctata (CDPX) and implications for warfarine embryopathy. *Cell*, **81**, 15–25.

Freeman C and Hopwood J (1987) Human liver N-acetylglucosamine-6-sulphate sulphatase. Catalytic properties. *Biochemical Journal*, **246**, 355–365.

Freeman C and Hopwood J (1989) Human liver glucuronate 2-sulphatase. Purification, characterization and catalytic properties. *Biochemical Journal*, **259**, 209–216.

Freeman C, Clements P and Hopwood J (1987) Human liver N-acetylglucosamine-6-sulphate sulphatase. Purification and characterization. *Biochemical Journal*, **246**, 347–354.

Funkenstein B, Leary S, Stein J and Catterall J (1988) Genomic organization and sequence of the Gus-s$^\alpha$ allele of the murine β-glucuronidase gene. *Mollecular and Cellular Biology*, **8**, 1160–1168.

Fyfe J, Kunzhals R, Lassaline M, Henthorn P, Alur P, Wang P, Wolfe J, Giger U, Haskins M, Patterson D, Sun H, Jain S and Yuhki N (1999) Molecular basis of feline β-glucuronidase deficiency: an animal model of Mucopolysaccharidosis VII. *Genomics*, **58**, 121-128.

Gallagher L and Sloane B (1984) Effect of estrogen on lysosomal enzyme activities in rat heart. *Proceedings of the Society for Experimental Biology and Medicine*, **177**, 428–433.

Gallagher J, Walker A, Lyon M and Evans W (1988) Heparan sulphate-degrading endoglycosidase in liver plasma membranes. *Biochemical Journal*, **250**, 719–726.

Gasa S, Balbaa M, Nakamura M, Yonemori H and Makita A (1987) Phosphorylation of human lysosomal arylsulfatase B by cAMP-dependent protein kinase. Different sites of phosphorylation between normal and cancer tissues. *Journal of Biological Chemistry*, **262**, 1230–1238.

Gibson G, Saccone G, Brooks D, Clements P and Hopwood J (1987) Human N-acetylgalactosamine-4-sulphate sulphatase. Purification, monoclonal antibody production and native and subunit Mr values. *Biochemical Journal*, **248**, 755–764.

Gitzelmann R, Bosshard N, Supertifurga A, Spycher M, Briner J, Wiesmann U, Lutz H and Litschi B (1994) Feline mucopolysaccharidosis VII due to β-glucuronidase deficiency. *Veterinary Pathology*, **31**, 435–443.

Glössl J, Truppe W and Kresse H (1979) Purification and properties of N-acetylgalactosamine 6-sulphate sulphatase from human placenta. *Biochemical Journal*, **181**, 37–46.

Goi G, Caputo D, Bairati C, Lombardo A, Burlina A, Ferrate P, Cazzullo C and Tettamanti G (1993) Enzymes of lysosomal origin in the cerebrospinal fluid and plasma of patients with multiple sclerosis. *European Neurology*, **3**, 1–4.

Goldin B (1990) Intestinal microflora: metabolism of drugs and carcinogens. *Annals of Internal Medicine*, **22**, 43–48.

Gorham S and Cantz M (1978) Arylsulphatase B, an exo-sulphatase for chondroitin 4-sulphate tetrasaccharide. *Hoppe–Seyler's Zeitschrift Für Physiologische Chemie*, **359**, 1811–1814.

Gowers H and Breuer H (1980) Purification and properties of a β-glucuronidase from pig kidney. *Journal of Steroid Biochemistry*, **13**, 1021–1027.

Graham L and Underwood P (1996) Comparison of the heparanase enzymes from mouse melanoma cells, mouse macrophages, and human platelets. *Biochemistry and Molecular Biology International*, **39**, 563–571.

Guerquin-Kern J, Volk A, Chenu E, Lougerstay-Madec R, Monneret C, Florent J, Carrez D and Croisy A (2000) Direct in vivo observation of 5-fluorouracil release from a prodrug in human tumors heterotransplanted in nude mice: a magnetic resonance study. *NMR in Biomedicine*, **13**, 306–310.

Gupta G and Singh G (1983) Immunological specificity of β-glucuronidase from human seminal plasma and its immunolocalization in the human reproductive tract. *American Journal of Reproductive Immunology*, **4**, 122–126.

Habuchi H, Tsuji M, Nakanishi Y and Suzuki S (1979) Separation and properties of five glycosaminoglycan sulfatases from rat skin. *Journal of Biological Chemistry*, **254**, 7570–7578.

Haisma HJ, Sernee M, Hooijberg E, Brakenhoff R, van den Meulen-Muileman I, Pinedo H and Boven E (1998) Construction and characterization of a fusion protein of single-chain anti-CD20 antibody and human β-glucuronidase for antibody-directed enzyme prodrug therapy. *Blood*, **92**, 184–190.

Harries W, Tsui J and Unakar N (1985) Ultrastructural cytochemistry: effect of Sorbinil on arylsulfatases in cataractous lenses. *Current Eye Research*, **4**, 657–666.

Hartman P (1989) The MUG (glucuronidase) test for *Escherichia coli* in food and water. In *Rapid Methods and Automation in Microbiology and Immunology*, Balows A, Tilton RC and Turano A (eds), Brixia Academic Press, Brescia, pp. 290–308.

Haskins M, Desnick R, DiFerrante N, Jezyk P and Patterson D (1984) β-Glucuronidase deficiency in a dog: A model of human mucopolysaccharidosis VII. *Pediatric Research*, **18**, 980–984.

Haskins M, Aguirre G, Jezyk P, Schuchman E, Desnick R and Patterson D (1991) Mucopolysaccharidosis type VII (Sly syndrome). Beta-glucuronidase-deficient mucopolysaccharidosis in the dog. *American Journal of Pathology*, **138**, 1553–1555.

Hayashi S (1978) Study on the degradation of glycosaminoglycans by canine liver lysosomal enzymes. II. The contributions of hyaluronidase, β-glucuronidase, sulfatase, and β-N-acetyl-hexosaminidase in the case of chondroitin 4-sulfate. *Journal of Biochemistry*, **83**, 149–157.

Hayashi M, Nakajima Y and Fishman W (1964) The cytologic demonstration of β-glucuronidase employing naphtol AS-BI glucuronide and hexazonium pararosanilin; a preliminary report. *Journal of Histochemistry and Cytochemistry*, **12**, 293–297.

Hazenberg M, de Herder W and Visser T (1988) Hydrolysis of iodothyronine conjugates by intestinal bacteria. *FEMS Microbiological Reviews*, **54**, 9–16.

Heinert G, Scherberich J, Mondorf W and Weber W (1983) Quantitative enzymatic and immunologic computer-assisted histophotometry of human kidney tissue following neoplastic and other clinically significant alterations. *European Urology*, **9**, 235–241.

Heringova A, Jirosova V and Koldovsky O (1965) Postnatal development of β-glucuronidase in the jejunum and ileum of rats. *Canadian Journal of Biochemistry*, **43**, 173–178.

Herman R, Van Pham J and Szakacs C (1989) Disposition of lorazepam in human beings: Enterohepatic recirculation and first-pass effect. *Clinical Pharmacology & Therapeutics*, **46**, 18–25.

Hernandez-Martin A, Gonzalez-Sarmiento R and De Unamuno P (1999) X-linked ichthyosis: an update. *British Journal of Dermatology*, **141**, 617–627.

Hicks J, Duran-Reyes G and Diaz-Flores M (1994) Effect of dexamethasone as an inhibitor of implantation and embryo development in rat; lysosomal role. *Contraception*, **50**, 581–589.

Higuchi M, Ito Y, Fukuyama K and Epstein W (1984) Biochemical characterization of arylsulfatases detected in granulomatous inflammation. *Experimental and Molecular Pathology*, **40**, 70–78.

Himeno Mnishimura Y, Tsuji H and Kato K (1976) Purification and characterization of microsomal and lysosomal β-glucuronidase from rat liver by use of immunoaffinity chromatography. *European Journal of Biochemistry*, **70**, 349–359.

Hipkin J, Gacesa P, Olavesen A and Sawyer R (1989) Functional arginine residues in leech hyaluronidase (Orgelase). *Biochemical Society Transactions*, **17**, 784.

Ho Y and Ho K (1985) Differential quantitation of urinary β-glucuronidase of human and bacterial origins. *Journal of Urology*, **134**, 1227–1230.

Ho K, Hsu S, Chen J and Ho L (1986) Human biliary β-glucuronidase: correlation of its activity with deconjugation of bilirubin in the bile. *European Journal of Clinical Investigation*, **16**, 361–367.

Hobkirk R (1985) Steroid sulfotransferases and steroid sulfate sulfatases: characteristics and biological roles. *Canadian Journal of Biochemistry and Cell Biology*, **63**, 1127–1144.

Hoogewerf A, Leone J, Reardon I, Howe W, Asa D, Heinrikson R and Ledbetter S (1995) CXC chemokines connective tissue activating peptide-III and neutrophil activating peptide-2 are heparin/heparan sulfate degrading enzymes. *Journal of Biological Chemistry*, **270**, 3268–3277.

Horowitz C, Comer S, Lau H and Koldovsky O (1978) Effect of cortisone and thyroxine on acid β-glycosidases in the liver and kidney of suckling and adult rats. *Hormone and Metabolic Research*, **10**, 531–538.

Houba P, Boven E, van der Meulen-Muileman I, Leenders R, Scheeren J, Pinedo H and Haisma H (1999) Distribution and pharmacokinetics of the prodrug daunorubicin GA3 in nude mice bearing human ovarian cancer xenografts. *Biochemical Pharmacology*, **57**, 673–680.

Hovingh P and Linker A (1977) Specificity of flavobacterial glycuronidases acting on disaccharides derived from glycosaminoglycans. *Biochemical Journal*, **165**, 287-293.

Huang Z, Fasco M and Kaminsky L (1997) Inhibition of oestrone sulfatase in human liver microsomes by quercetin and other flavonoids. *Journal of Steroid Biochemistry and Molecular Biology*, **63**, 9–15.

Huijgebaert S, Sim S, Back D and Eyssen H (1984) Distribution of estrone sulfatase activity in the intestine of germfree and conventional rats. *Journal of Steroid Biochemistry*, **20**, 1175–1179.

Hulett M, Freeman C, Hamdorf B, Baker R, Harris M and Parish C (1999) Cloning of mammalian heparanase, an important enzyme in tumor invasion and metastasis. *Nature Medicine*, **5**, 803–809.

Igarashi Y, Ichikawa M and Ichikawa Y (1996) Synthesis of a potent inhibitor of β-glucuronidase. *Tetrahedron Letters*, **37**, 2707–2708.

Isbrandt D, Arlt G, Brooks D, Hopwood J, von Figura K and Peters C (1994) Mucopolysaccharidosis VI (Maroteaux-Lamy syndrome): six unique arylsulfatase B gene alleles causing variable disease phenotypes. *American Journal of Human Genetics* , **54**, 454–463.

Islam M, Grubb J and Sly W (1993) C-terminal processing of human β-glucuronidase – the propeptide is required for full expression of catalytic activity, intracellular retention, and proper phosphorylation. *Journal of Biological Chemistry*, **268**, 22627–22633.

Islam M, Tomatsu S, Shah G, Grubb J, Jain S and Sly W (1999a) Active site residues of human β-glucuronidase. Evidence for Glu(540) as the nucleophile and Glu(451) as the acid-base residue. *Journal of Biological Chemistry*, **274**, 23451–23455.

Islam M, Waheed A, Shah G, Tomatsu S and Sly W (1999b) Human egasyn binds β-glucuronidase but neither the esterase active site of egasyn nor the C terminus of β-glucuronidase is involved in their interaction. *Archives of Biochemistry and Biophysics*, **372**, 53–61.

Jackson C, Yuhki N, Desnick R, Haskins M, O'Brien S and Schuchman E (1992) Feline arylsulfatase B (ARSB): isolation and expression of the cDNA, comparison with human ARSB, and gene localization to feline chromosome A1. *Genomics*, **14**, 403–411.

Jain S, Drendel W, Chen Z, Mathews F, Sly W and Grubb J (1996) Structure of human β-glucuronidase reveals candidate lysosomal targeting and active-site motifs. *Nature Structural Biology*, **3**, 375–381.

Jansen P, Chowdhury J, Fischberg E and Arias I (1977) Enzymatic conversion of bilirubin monoglucuronide to diglucuronide by rat liver plasma membranes. *Journal of Biological Chemistry*, **252**, 2710–2716.

Jefferson R, Burgess S and Hirsh D (1986) β-Glucuronidase from Escherichia coli as a gene-fusion marker. *Proceedings of the National Academy of Sciences, USA* **83**, 8447–8451.

Kamai A (1995) Variation in the glycosidase activity of human lens during aging and advance of senile cataract. *Biological & Pharmaceutical Bulletin*, **18**, 1450–1453.

Kamei K, Kubushiro K, Fujii T, Tsukazaki K, Nozawa S and Iwamori M (1997) Menstrual cycle-associated regulation of anabolic and catabolic enzymes causes luteal phase-characteristic expression of sulfatide in human endometrium. *American Journal of Obstetrics and Gynecology*, **176**, 142–149.

Kanaoka M, Yano S, Kato H and Nakada T (1986) Synthesis and separation of 18β-glycyrrhetyl monoglucuronide from serum of a patient with glycyrrhizin-induced pseudo-aldosteronism. *Chemical & Pharmaceutical Bulletin*, **34**, 4978–4983.

Karsten S, Voskoboeva E, Carlberg B, Kleijer W, Tsnnesen T, Pettersson U and Bondeson M (1998) Identification of 9 novel IDS gene mutations in 19 unrelated Hunter syndrome (mucopolysaccharidosis Type II) patients. Mutations in brief no. 202. Online. *Human Mutation*, **12**, 433.

Karsten S, Voskoboeva E, Krasnopolskaja X and Bondeson M (1999) Novel type of genetic rearrangement in the iduronate-2-sulfatase (IDS) gene involving deletion, duplications, and inversions. *Human Mutation*, **14**, 471–476.

Kauffman FC (1994) Regulation of drug conjugate production by futile cycling in intact cells. In

Conjugation–Deconjugation Reactions in Drug Metabolism and Toxicity, Kauffman FC (ed.), Springer-Verlag, Berlin, pp. 245–255.

Khandke K, Vithayayathil P and Murthy S (1989) Purification and characterization of an alpha-D-glucuronidase from a thermophilic fungus, *Thermoascus aurantiacus. Archives of Biochemistry and Biophysics*, **274**, 511–517.

Kihara H, Meek W and Fluharty A (1986) Attenuated activities and structural alterations of arylsulfatase A in tissues from subjects with pseudo arylsulfatase A deficiency. *Human Genetics*, **74**, 59–62.

Kim D, Kang H, Kim S and Kobashi K (1992) pH-inducible β-glucosidase and β-glucuronidase of intestinal bacteria. *Chemical & Pharmaceutical Bulletin*, **40**, 1667–1699.

Kim D, Jin Y, Park J and Kobashi K (1994) Silymarin and its components are inhibitors of β-glucuronidase. *Biological & Pharmaceutical Bulletin*, **17**, 443–445.

Kim D, Jin Y, Jung E, Han M and Kobashi K (1995) Purification and characterization of β-glucuronidase from Escherichia coli HGU-3, a human intestinal bacterium. *Biological & Pharmaceutical Bulletin*, **18**, 1184–1188.

Kim D, Shim S, Kim N and Jang I (1999) Beta-glucuronidase-inhibitory activity and hepatoprotective effect of Ganoderma lucidum. *Biological & Pharmaceutical Bulletin*, **22**, 162–164.

Kinoshita N and Gelboin H (1978) Beta-glucuronidase catalyzed hydrolysis of Benzo[a]pyrene-3-glucuronide and binding to DNA. *Science*, **199**, 307–309.

Klein U and Von Figura K (1976) Partial purification and characterization of heparan sulfate specific endoglucuronidase. *Biochemical and Biophysical Research Communication*, **73**, 569–576.

Klein U and Von Figura K (1979) Substrate specificity of a heparan sulfate-degrading endoglucuronidase from human placenta. *Hoppe–Seylers Zeitschrift Für Physiologische Chemie*, **360**, 1465–1471.

Koldovsky O and Palmieri M (1971) Cortisone-evoked decrease of acid β-galactosidase, β-glucuronidase, N-Acetyl-β-glucosaminidase and arylsulphatase in the ileum of suckling rats. *Biochemical Journal*, **125**, 697–701.

Kostulak A (1974) Histochemical studies on the salivary glands on white rats in vitamin A deficiency. *Folia Morphologica* , **33**, 29–35.

Kremer M, Judd J, Rifkin B, Auszmann J and Oursler M (1995) Estrogen modulation of osteoclast lysosomal enzyme secretion. *Journal of Cellular Biochemistry*, **57**, 271–279.

Kreysing J, Polten A, Hess B, von Figura K, Menz K, Steiner F and Gieselmann V (1994) Structure of the mouse arylsulfatase A gene and cDNA. *Genomics*, **19**, 249–256.

Krumholz W, Weidenbusch H and Menges T (2000) The influence of midazolam and flunitrazepam on the liberation of lysosome and β-glucuronidase from neutrophile granulozytes in vitro. *Anästhesiologie, Intensivmedizin, Notfallmedizin, Schmerztherapie*, **35**, 316–318.

Kumagai A, Nishino K, Shimomura A, Kin T and Yamamura Y (1967) Effect of glycyrrhizin on estrogen action. *Endocrinology*, **14**, 34–38.

Kung M, Spaulding S and Roth J (1988) Desulfation of 3,5,3'-triiodothyronine sulfate by microsomes from human and rat tissues. *Endocrinology*, **122**, 1195–1200.

Kussie P, Hulmes J, Ludwig D, Patel S, Navarro E, Seddon A, Giorgio N and Bohlen P (1999) Cloning and functional expression of a human heparanase gene. *Biochemical and Biophysical Research Communication*, **261**, 183–187.

Laidler MP and Steczko J (1986) Catalytic and immunochemical properties of arylsulphatase A from urine, modified by potassium ferrate. *Acta Biochimica Polonica*, **33**, 101–108.

Lalley P and Shows T (1974) Lysosomal and microsomal glucuronidase: genetic variant alters electrophoretic mobility of both hydrolases. *Science*, **185**, 442–444.

Laskov R, Michaeli R, Sharir H, Yefenof E and Vlodavsky I (1991) Production of heparanase by normal and neoplastic murine B-lymphocytes. *International Journal of Cancer*, **47**, 92–90.

Latif K, Amla I and Rao P (1979) Urinary excretion of arylsulfatases in malnourished/vitamin A deficient children. *Clinica Chimica Acta*, **96**, 131–138.

Layik M, Yamalik N, Caglayan F, Kilinc K, Etikan I and Eratalay K (2000) Analysis of human gingival tissue and gingival crevicular fluid β-glucuronidase activity in specific peridontal diseases. *Journal of Periodontology*, **71**, 618–624.

Le Gall J, Khoi T, Glaise D, Le Treut A, Brissot P and Guillouzo A (1979) Lysosomal enzyme activities during ageing of adult human liver cell lines. *Mechanisms of Ageing and Development*, **11**, 287–293.

Leder I (1980) A novel 3-O sulfatase from human urine acting on methyl-2-deoxy-2-sulfamino-alpha-D-glucopyranoside 3-sulfate. *Biochemical and Biophysical Research Communication*, **94**, 1183–1189.

Levy G, McAllan A and Marsh C (1958) Purification of b-glucuronidase from the preputial gland of the female rat. *Biochemical Journal*, **69**, 22–27.

Li P, Chu G, Guo J, Peters A and Selcer K (1998) Development of potent non-estrogenic estrone sulfatase inhibitors. *Steroids*, **63**, 425–432.

Li Y, Xu C and Schubert D (1999) The up-regulation of endosomal-lysosomal components in amyloid β-resistant cells. *Journal of Neurochemistry*, **73**, 1477–1482.

Lilienblum W and Bock K (1984) N-glucuronide formation of carcinogenic aromatic amines in rat and human liver microsomes. *Biochemical Pharmacology*, **33**, 2041–2046.

Lim C and Horowitz A (1981) Purification and properties of human N-acetylgalactosamine-6-sulfate sulfatase. *Biochimica et Biophysica Acta*, **657**, 344–355.

Lindner J, Schönrock P, Nüßgen A and Schmiegelow P (1986) Altersabhängige Veränderungen des Zellgehaltes (DNA) und der Glykosaminoglykan-abbauenden Enzyme (β–Glucuronidase, β-N-Acetylglucosaminidase) bindegewebiger und parenchymatöser Organe der Ratte durch 6-Methylprednisolon (Enzyminduktion, Adaptation, Reifungsbeschleunigung und mögliche Altersbeeinflussung). *Zeitschrift für Gerontologie*, **19**, 190–205.

Linker A, Meyer K and Hoffman P (1960) The production of hyaluronate oligosaccharides by leech hyaluronidase and alkali. *Journal of Biological Chemistry*, **235**, 924–927.

Lissens W, Zenati A and Liebaers I (1984) Polyclonal antibodies against iduronate 2-sulphate sulphatase from human urine. *Biochimica et Biophysica Acta*, **801**, 365–371.

Lombardo A, Goi G, Marchesini S, Caimi L, Moro M and Tettamanti G (1981) Influence of age and sex on five human plasma lysosomal enzymes assayed by automated procedures. *Clinica Chimica Acta*, **113**, 141–152.

Lukatela G, Krauss N, Theis K, Selmer T, Gieselmann V, von Figura K and Saenger W (1998) Crystal structure of human arylsulfatase A: the aldehyde function and the metal ion at the active site suggest a novel mechanism for sulfate ester hydrolysis. *Biochemistry*, **37**, 3654–3664.

Lusis A (1983) Preparation of microsomal β-glucuronidase and its membrane anchor protein, egasyn. *Methods in Enzymology*, **96**, 557–565.

Lusis AJ and Paigen K (1977) Mechanisms involved in the intracellular localization of mouse glucuronidase. *Isozymes*, **2**, 63–106.

MacKenzie C, Bilous D, Schneider H and Johnson K (1987) Induction of cellulolytic and xylanolytic enzyme systems in *Streptomyces* spp. *Applied and Environmental Microbiology*, **53**, 2835–2839.

Marchesini S, Viani P, Cestaro B and Gatt S (1989) Synthesis of pyrene derivatives of cerebroside sulfate and their use for determining arylsulfatase A activity. *Biochimica et Biophysica Acta*, **1002**, 14–19.

Marjomaki V and Salminen A (1986) Morphological and enzymic heterogeneity of sumarin-induced lysosomal storage disease in some tissues of mice and rats. *Experimental and Molecular Pathology*, **45**, 76–83.

Marselos M, Dutton G and Hänninen O (1975) Evidence that D-glucaro-1,4-lactone shortens the pharmacological action of drugs being disposed via the bile as glucuronides. *Biochemical Pharmacology*, **24**, 1855–1858.

Matzner Y, Bar-Ner M, Yahalom J, Ishai-Michaeli R, Fuks Z and Vlodavsky I (1985) Degadation of heparan sulfate in the subendothelial extracellular matrix by a readily released heparanase from human neutrophils. Possible role in invasion through basement membranes. *Journal of Clinical Investigation*, **76**, 1306–1313.

Medda S, von Deimling O and Swank R (1986) Identity of esterase-22 and egasyn, the protein which complexes with microsomal β-glucuronidase. *Biochemical Genetics*, **24**, 229–243.

Medda S, Takeuchi K, Devore Carter D, von Deimling O, Heymann E and Swank R (1987) An accessory protein identical to mouse egasyn is complexed with rat microsomal β-glucuronidase and is identical to rat esterase-3. *Journal of Biological Chemistry*, **262**, 7248–7253.

Meffin P, Zilm D and Veenendaal J (1983) Reduced clofibric acid clearance in renal dysfunction is due to a futile cycle. *Journal of Pharmacology and Experimental Therapeutics*, **227**, 732–738.

Mercelis R, Van Elsen A and Leroy J (1979) Arylsulphatases A and B in human diploid fibroblasts: differential assay with 4-methylumbelliferylsulphate and $AgNO_3$. *Clinica Chimica Acta*, **93**, 85–92.

Meredith S and Ganschow R (1978) Apparent trans control of murine β-glucuronidase synthesis by a temporal genetic element. *Genetics*, **90**, 725–734.

Miao H, Elkin M, Aingorn E, Ishai-Michaeli R, Stein C and Vlodavsky I (1999) Inhibition of heparanase activity and tumor metastasis by laminarin sulfate and synthetic phosphorothioate oligodeoxynucleotides. *International Journal of Cancer*, **83**, 424–431.

Miller R (1984) Pharmacokinetics and bioavailability of ranitidine in humans. *Journal of Pharmaceutical Sciences*, **73**, 1376–1379.

Minami R, Sato S, Kudoh T, Oyanagi K and Nakao T (1979) Age-dependent variations of lysosomal enzymes in human liver. *The Tohoku Journal of Experimental Medicine*, **129**, 65–70.

Mishima T, Murata J, Toyoshima M, Fujii H, Nakajima M, Hayashi T, Kato T and Saiki I (1998) Inhibition of tumor invasion and metastasis by calcium spirulan (Ca-SP), a novel sulfated polysaccharide derived from a blue-green alga, *Spirulina platensis*. *Clinical & Experimental Metastasis*, **16**, 541–550.

Moore B, Hicks R, Knowles M and Redgrave S (1982) Metabolism and binding of benzo(a)pyrene and 2-acetylaminofluorene by short-term organ cultures of human and rat bladder. *Cancer Research*, **42**, 642–648.

Morotomi M, Nanno M, Watnabe T, Sakurai T and Mutai M (1985) Mutagenic activation of biliary metabolites of 1-nitropyrene by intestinal microflora. *Mutation Research*, **149**, 171–178.

Mortaud S, Donsez-Darcel E, Roubertoux P and Degrelle H (1996) Murine steroid sulfatase gene expression in the brain during postnatal development and adulthood. *Neuroscience Letters*, **215**, 145–148.

Munroe DG and Chang PL (1987) Tissue-specific expression of human arylsulfatase-C isozymes and steroid sulfatase. *American Journal of Human Genetics*, **40**, 102–114.

Murata F, Nagata T and Spicer S (1975) Fine structural localization of arylsulfatase B activity in the rabbit blood platelets. *Histochemistry*, **44**, 307–312.

Mürdter T, Sperker B, Kivistö K, McClellan M, Fritz P, Friedel G, Linder A, Bosslet K, Toomes H, Dierkesmann R and Kroemer H (1997) Enhanced uptake of doxorubicin into bronchial carcinoma: β-glucuronidase mediates release of doxorubicin from a glucuronide prodrug (HMR 1826) at the tumor site. *Cancer Research*, **57**, 2440–2445.

Muro T, Kuramoto T, Imoto K and Okada S (1986) Purification and some properties of glycyrrhizinic acid hydrolase from *Aspergillus niger* GRM3. *Agricultural and Biological Chemistry*, **50**, 687–692.

Nakajima M, Irimura T and Nicolson G (1988) Heparanases and tumor metastasis. *Journal of Cellular Biochemistry*, **36**, 157–167.

Nakanishi Y, Tsuji M, Habuchi H and Suzuki S (1979) Isolation of UDP-N-acetylgalactosamine-6-sulfate sulfatase from quail oviduct and its action on chondroitin sulfate. *Biochemical and Biophysical Research Communication*, **89**, 863–870.

Naparstek Y, Cohen I, Fuks Z and Vlodavsky I (1984) Activated T lymphocytes produce a matrix-degrading heparan sulphate endoglycosidase. *Nature*, **310**, 241–244.

Neufeld E and Muenzer J (1995) The mucopolysaccharidoses. In *The Metabolic and Molecular Bases of Inherited Disease*, Scriver C, Beaudet A, Sly W and Valle D (eds), McGraw-Hill, New York, pp. 2465–2494.

Nguyen B, Ferme I, Chetrite G and Pasqualini J (1993) Action of danazol on the conversion of estrone sulfate to estradiol and on the sulfatase activity in the MCF-7, T-47D and MDA-MB-231 human mammary cancer cells. *Journal of Steroid Biochemistry and Molecular Biology*, **46**, 17–23.

Niwa T, Tsuruoka T, Inoue S, Naito Y and Koeda T (1972) A new potent-glucuronidase inhibitor, D-glucaro-δ-lactam derived from nojirimycin. *Journal of Biochemistry*, **72**, 207–211.

Novel M and Novel G (1976) Regulation of β-glucuronidase synthesis in Escherichia coli K-12: Constitutive mutants specifically derepressed for uidA expression. *Journal of Bacteriology*, **127**, 406–417.

Novel G, Didier-Fichet M and Stoeber F (1974) Inducibility of β-glucuronidase in wild-type and hexuronate-negative mutants of *Escherichia coli* K-12. *Journal of Bacteriology*, **120**, 89–95.

Ohta H, Ono M, Sekiya C and Namiki M (1992) Serum immunoreactive β-glucuronidase determined by an enzyme-linked immunosorbent assay in patients with hepatic diseases. *Clinica Chimica Acta*, **208**, 9–21.

Oner P, Bekpinar S, Cinar F and Argun A (1994) Relationship of some endogenous sex steroid hormones to leukocyte arylsulphatase A activities in pre- and postmenopausal healthy women. *Hormone and Metabolic Research*, **26**, 301–304.

Orme M and Back D (1979) Therapy with oral contraceptive steroids and antibiotics. *Journal of Antimicrobial Chemotherapy*, **5**, 124–126.

Oshima A, Kyle J, Miller R, Hoffmann J, Powell P, Grubb J, Sly W, Tropak M, Guise K and Gravel R (1987) Cloning, sequencing, and expression of cDNA for human β-glucuronidase. *Proceedings of the National Academy of Sciences, USA*, **84**, 685–689.

Owens J and Stahl P (1976) Purification and characterization of rat liver microsomal β-glucuronidase. *Biochimica et Biophysica Acta*, **438**, 474–486.

Oyama N, Satoh M, Iwatsuki K and Kaneko F (2000) Novel point mutations in the steroid sulfatase gene in patients with X-linked ichthyosis: transfection analysis using the mutated genes. *Journal of Investigative Dermatology*, **114**, 1195–1199.

Paigen K (1961) The effect of mutation on the intracellular location of β-glucuronidase. *Experimental Cell Research*, **25**, 286–301.

Paigen K (1989) Mammalian β-glucuronidase: genetics, molecular biology, and cell biology. *Progress in Nucleic Acid Research and Molecular Biology*, **37**, 155–205.

Palmer R, Gallagher P, Boyko W and Ganschow R (1983) Genetic control of levels of murine kidney glucuronidase mRNA in response to androgen. *Proceedings of the National Academy of Sciences, USA*, **80**, 7596–7600.

Parenti G, Meroni G and Ballabio A (1997) The sulfatase gene family. *Current Opinion in Genetics & Development*, **7**, 386–391.

Parish C, Freeman C, Brown K, Francis D and Cowden W (1999) Identification of sulfated oligosaccharide-based inhibitors of tumor growth and metastasis using novel in vitro assays for angiogenesis and heparanase activity. *Cancer Research*, **59**, 3433–3441.

Parker A, Maw B and Fedor L (1981) The β-glucuronidase catalysed hydrolysis of a glucupyranosiduronamide and a glucopyranoside: evidence for the oxocarbonium ion mechanism for bovine liver β-glucuronidase. *Biochemical and Biophysical Research Communication*, **103**, 1390–1394.

Pasqualini J, Chetrite G and Nestour E (1996a) Control and expression of oestrone sulphatase activities in human breast cancer. *Journal of Endocrinology*, **150 Suppl.**, 99–105.

Pasqualini J, Chetrite G, Blacker C, Feinstein M, Delalonde L, Talbi M and Maloche C (1996b) Concentrations of estrone, estradiol, and estrone sulfate and evaluation of sulfatase and aromatase activities in pre- and postmenopausal breast cancer patients. *Journal of Clinical Endocrinology and Metabolism*, **81**, 1460–1464.

Pasqualini J, Cortes-Prieto J, Chetrite G, Talbi M and Ruiz A (1997) Concentrations of estrone, estradiol and their sulfates, and evaluation of sulfatase and aromatase activities in patients with breast fibroadenoma. *International Journal of Cancer*, **70**, 639–643.

Pedersen P and Miller R (1980) Pharmacokinetics of doxycycline reabsorption. *Journal of Pharmaceutical Sciences*, **69**, 204–207.

Peters C, Schmidt B, Rommerskirch W, Rupp K, Zühlsdorf M, Vingron M, Meyer HE, Pohlmann R and von Figura K (1990) Phylogenetic conservation of arylsulfatases. cDNA cloning and expression of human arylsulfatase B. *Journal of Biological Chemistry*, **265**, 3374–3381.

Pikas D, Li J, Vlodavsky I and Lindahl U (1998) Substrate specificity of heparanase from human hepatome and platelets. *Journal of Biological Chemistry*, **273**, 18770–18777.

Plaa GL (1975) The enterohepatic circulation. In *Handbook of Experimental Pharmacology*, Gilette JR and Mitchell JR (eds), Springer-Verlag, Berlin, pp. 130–149.

Platt D (1970) Altersabhängige Aktivitätsänderungen lysosomaler Enzyme (Glycosaminoglycano-Hydrolasen) in Serum und Organen des Menschen. *Deutsche Medizinische Wochenschrift*, **12**, 634–637.

Platt D and Platt M (1970) Biochemischer und histochemischer Nachweis der b-Glukuronidase und b-Azetylglukosaminidase in normalen und pathologisch veränderten Lymphknoten. *Blut*, **22**, 12–18.

Poland R and Odell G (1971) Physiologic jaundice: the enterohepatic circulation of bilirubin. *New England Journal of Medicine*, **284**, 1–6.

Przybylo M and Litynska A (2000) Changes in glycosylation of rat liver arylsulfatase B in relation to age. *Mechanisms of Ageing and Development*, **113**, 193–203.

Puca AA, Zollo M, Repetto M, Andolfi G, Guffanti A, Simon G, Ballabio A and Franco B (1997) Identification by shotgun sequencing, genomic organization, and functional analysis of a fourth arylsulfatase gene (ARSF) from the Xp22.3 region. *Genomics*, **42**, 192–199.

Puls J, Schmidt O and Granzow C (1987) alpha-Glucuronidase in two microbial xylanolytic systems. *Enzyme and Microbial Technology*, **9**, 83–88.

Purohit A, Wang D, Ghilchik M and Reed M (1996a) Regulation of aromatase and sulphatase in breast tumour cells. *Journal of Endocrinology*, **150**, 65–71.

Purohit A, Woo L, Singh A, Winterborn C, Potter B and Reed M (1996b) In vivo activity of 4-methylcoumarin-7-O-sulfamate, a nonsteroidal, nonoestrogenic steroid sulfatase inhibitor. *Cancer Research*, **56**, 4950–4955.

Purohit A, Vernon K, Hummelinck A, Woo L, Hejaz H, Potter B and Reed M (1998) The development of A-ring modified analogues of oestrone-3-O-sulphamate as potent steroid sulphatase inhibitors with reduced oestrogenicity. *Journal of Steroid Biochemistry and Molecular Biology*, **64**, 269–275.

Purohit A, Singh A and Reed MJ (1999) Regulation of steroid sulphatase and oestradiol 17 β-hydroxysteroid dehydrogenase in breast cancer. *Biochemical Society Transactions*, **27**, 323–327.

Ralovich B, Ibrahim G, Fabian A and Herpay M (1991) Beta-D-glucuronidase (BDG) activity of gram-negative bacteria. *Acta Microbiologica Hungarica*, **38**, 283–291.

Ray J, Bouvet A, DeSanto C, Fyfe JC, Xu D, Wolfe JH, Aguirre GD, Patterson DF, Haskins ME and Henthorn PS (1998) Cloning of the canine β-glucuronidase cDNA, mutation identification in canine MPS VII, and retroviral vector-mediated correction of MPS VII cells. *Genomics*, **48**, 248–253.

Ray J, Wu Y and Aguirre G (1997) Characterization of β-glucuronidase in the retinal pigment epithelium. *Current Eye Research*, **16**, 131–143.

Ray J, Scarpino V, Laing C and Haskins M (1999) Biochemical basis of the β-glucuronidase gene defect causing canine mucopolysaccharidosis VII. *The Journal of Heredity*, **90**, 119–123.

Robben J, Caenepeel P, Van Eldere J and Eyssen H (1988) Effects of intestinal microbial bile salt sulfatase activity on bile salt kinetics in gnotobiotic rats. *Gastroenterology*, **94**, 494–502.

Rollins D and Klaassen C (1979) Biliary excretion of drugs in man. *Clinical Pharmacokinetics*, **4**, 368–379.

Roy AB (1979) Sulphatase A: an arylsulphatase and a glycosulphatase. *Ciba Foundation Symposium*, **72**, 177–190.

Roy AB and Mantle TJ (1989) The anomalous kinetics of sulphatase A. *Biochemical Journal*, **261**, 689–697.

Roy A and Turner J (1982) The sulphatase of ox liver. XXIV. The glycosulphatase activity of sulphatase a. *Biochimica et Biophysica Acta*, **704**, 366–373.

Ruelius H (1978) Comparative metabolism of lorazepam in man and four animal species. *Journal of Clinical Psychiatry*, **39**, 11–15.

Rundell J, Sato T, Wetzelberger E, Ueda H and Brandes D (1974) Lysosomal enzyme release by vitamin A in L1210 leukemia cells. *Journal of the National Cancer Institute*, **52**, 1237–1244.

Rutgers M, Heusdens F, Bonthuis F, de Herder W, Hazenberg M and Visser T (1989) Enterohepatic circulation of triiodothyronine (T3) in rats: importance of the microflora for the liberation and reabsorption of T3 from biliary T3 conjugates. *Endocrinology*, **125**, 2822–2830.

Saha A, Glew R, Kotler D and Omene J (1991) Elevated serum β-glucuronidase activity in acquired immunodeficiency syndrome. *Clinica Chimica Acta*, **199**, 311–316.

Salyers A and O'Brien M (1980) Cellular location of enzymes involved in chondroitin sulfate breakdown by Bacteroides thetaiotaomicron. *Journal of Bacteriology*, **143**, 772–780.

Sasaki Y, Morita T, Kuramoto T, Mizutani K, Ikeda R and Tanaka O (1988) Substrate specificity of glycyrrhizinic acid hydrolase. *Agricultural and Biological Chemistry*, **52**, 207–210.

Satoh T, Nishimura Y, Kondo S and Takeuchi T (1996) Synthesis and antimetastatic activity of 6-trichloroacetamido and 6-guanidino analogues of sistatin B. *Journal of Antibiotics*, **49**, 321–325.

Scheline R (1973) Metabolism of foreign compounds by gastrointestinal microorganisms. *Pharmacological Reviews*, **25**, 451–523.

Schirmer A and Kolter R (1998) Computational analysis of bacterial sulfatases and their modifying enzymes. *Chemico-Biological Interactions*, **5**, R181–R186.

Schmidt B, Selmer T, Ingendoh A and von Figura K (1995) A novel amino acid modification in sulfatases that is defective in multiple sulfatase deficiency. *Cell*, **82**, 271–278.

Schöllhammer I, Poll D and Bickel M (1975) Liver microsomal β-glucuronidase and UDP-glucuronosyltransferase. *Enzyme*, **20**, 269–276.

Schuchman E, Toroyan T, Haskins M and Desnick R (1989) Characterization of the defective β-glucuronidase activity in canine mucopolysaccharidosis type VII. *Enzyme*, **42**, 174–180.

Selcer K, Jagannathan S, Rhodes M and Li P (1996) Inhibition of placental estrone sulfatase activity and MCF-7 breast cancer cell proliferation by estrone-3-amino derivatives. *Journal of Steroid Biochemistry and Molecular Biology*, **59**, 83–91.

Selcer K, Hegde P and Li P (1997) Inhibition of estrone sulfatase and proliferation of human breast cancer cells by nonsteroidal (p-O-sulfamoyl)-N-alkanoyl tyramines. *Cancer Research*, **57**, 702–707.

Selvidge L and Verlangieri A (1991) Inhibition of arylsulfatase B by ascorbic acid. *Research Communications in Chemical Pathology and Pharmacology*, **73**, 253–256.

Shaklee P, Glaser J and Conrad H (1985) A sulfatase specific for glucuronic acid 2-sulfate residues in glycosaminoglycans. *Journal of Biological Chemistry*, **260**, 9146–9149.

Shankaran R, Ameen M, Daniel W, Davidson R and Chang P (1991) Characterization of arylsulfatase C isozymes from human liver and placenta. *Biochimica et Biophysica Acta*, **1078**, 251–257.

Shapira E and Nadler H (1975) Purification and some properties of soluble human liver arylsulfatases. *Archives of Biochemistry and Biophysics*, **170**, 179–187.

Sheffield L, Osborn A, Hutchison W, Sillence D, Forrest S, White S and Dahl H (1998) Segregation of mutations in arylsulphatase E and correlation with the clinical presentation of chondrodysplasia punctata. *Journal of Medical Genetics* , **35**, 1004–1008.

Shipley JM, Miller RD, Wu BM, Grubb JH, Christensen SG, Kyle JW and Sly WS (1991) Analysis of the 5′ flanking region of the human β-glucuronidase gene. *Genomics*, **10**, 1009–1018.

Shipley J, Klinkenberg M, Wu B, Bachinsky D, Grubb J and Sly W (1993) Mutational analysis of a patient with mucopolysaccharidosis type VII, and identification of pseudogenes. *American Journal of Human Genetics*, **52**, 517–526.

Siika-aho M, Twnkanen M, Buchert J, Puls J and Viikari L (1994) An alpha-glucuronidase from *Trichoderma reesei* RUTC-30. *Enzyme and Microbial Technology*, **16**, 813–819.

Singh J, Di Ferrante N, Niebes P and Tavella D (1976) N-acetylgalactosamine-6-sulfate sulfatase in man. Absence of the enzyme in Morquio disease. *Journal of Clinical Investigation*, **57**, 1036–1040.

Sinhababu AK and Thakker DR (1996) Prodrugs of anticancer agents. *Advanced Drug Delivery Reviews*, **19**, 241–273.

Sirota L, Ferrera M, Lerer N and Dulitzky F (1992) Beta glucuronidase and hyperbilirubinaemia in breast fed infants of diabetic mothers. *Archives of Disease in Childhood*, **67**, 120–121.

Sisenwine S, Tio C, Hadley F, Lui A, Kimmel H and Ruelius H (1982) Species-related differences in the stereoselective glucuronidation of oxazepam. *Drug Metabolism and Disposition*, **10**, 605–608.

Sly W, Quinton B, McAlister W and Rimoin D (1973) Beta glucuronidase deficiency: report of clinical, radiologic, and biochemical features of a new mucopolysaccharidosis. *Journal of Pediatry*, **82**, 249–257.

Smith D and Forsberg C (1991) alpha-Glucuronidase and other hemicellulase activities of

Fibrobacter succinogenes S85 grown on crystalline cellulose or ball-milled barley straw. *Applied and Environmental Microbiology*, **57**, 3552–3557.

Speleman F, Vervoort R, Vanroy N, Liebaers I, Sly W and Lissens W (1996) Localization by fluorescence in situ hybridization of the human functional β-glucuronidase gene (GUSB) to 7q11.21->q11.22 and two pseudogenes to 5p13 and 5q13. *Cytogenetics and Cell Genetics*, **72**, 53–55.

Sperker B, Schick M and Kroemer HK (1996) High-performance liquid chromatographic quantification of 4-methylumbelliferyl-β-D-glucuronide as a probe for human β-glucuronidase activity in tissue homogenates. *Journal of Chromatography B*, **685**, 181–184.

Sperker B, Muerdter TE, Schick M, Eckhardt K, Bosslet K and Kroemer HK (1997) Interindividual variability in expression and activity of human β-glucuronidase in liver and kidney: consequences for drug metabolism. *Journal of Pharmacology and Experimental Therapeutics*, **281**, 914–920.

Sperker B, Werner U, Mündter T, Tekkaya C, Fritz P, Wacke R, Adam U, Gerken M, Drewelow B and Kroemer H (2000) Expression and function of β-glucuronidase in pancreatic cancer: potential role in drug targeting. *Naunyn–Schmiedeberg's Archives of Pharmacology*, **362**, 110–115.

Sperker B, Tomkiewicz C, Burk O, Barouki R and Kroemer H (2001) Regulation of human β-glucuronidase by A23187 and thapsigargin in the hepatoma cell line HepG2. *Molecular Pharmacology*, **59**, 177–182.

Stahl PD and Fischman WH (1984) Beta-D-Glucuronidase. In *Methods in Enzymatic Analysis*, Bergmeier J and Graßl M (eds), Verlag Chemie, Weinheim, pp. 246–256.

Stephens JL and Pieringer RA (1984) Regulation of arylsulfatase A and sulphogalactolipid turnover by cortisol in myelinogenic cultures of cells dissociated from embryonic mouse brain. *Biochemical Journal*, **219**, 689–697.

Stephens R, Ghosh P, Taylor T, Gale C, Swann J, Robinson R and Webb J (1975) The origins and relative distribution of polysaccharidases in rheumatoid and osteoarthritic fluids. *Journal of Rheumatology*, **2**, 393–400.

Strobel G, Werle E and Weicker H (1990) Isomer specific kinetics of dopamine β-hydroxylase and arylsulphatase towards catecholamine sulfates. *Biochemistry International*, **20**, 343–351.

Sutfin TA, Perini GI, Molnar G and Jusko WJ (1988) Multiple-dose pharmacokinetics of impramine and its major active and conjugated metabolites in depressed patients. *Journal of Clinical Pharmacology*, **8**, 48–53.

Swank R and Paigen K (1973) Genetic control of glucuronidase induction in mice. *Journal of Molecular Biology*, **81**, 225–243.

Szasz G (1967) Die Bestimmung der Beta-Glukuronidase Aktivität im Serum mit p-Nitrophenyl-glucuronid. *Clinica Chimica Acta*, **15**, 275–282.

Takasuna K, Kasai Y, Kitano Y, Mori K, Kobayashi R, Hagiwara T, Kakihata K, Hirohashi M, Nomura M, Nagai E and Kamataki T (1995) Protective effects of kampo medicines and baicalin against intestinal toxicity of a new anticancer camptothecin derivative, irinotecan hydrochloride (CPT-11), in rats. *Japanese Journal of Cancer Research*, **86**, 978–984.

Takasuna K, Hagiwara T, Hirohashi M, Kato M, Nomura M, Nagai E, Yokoi T and Kamataki T (1996) Involvement of β-glucuronidase in intestinal microflora in the intestinal toxicity of the antitumor camptothecin derivative irinotecan hydrochloride (CPT-11) in rats. *Cancer Research*, **56**, 3752–3757.

Tanaka J, Gasa S, Sakurada K, Miyazaki T, Kasai M and Makita A (1992) Characterization of the subunits and sugar moiety of human placental and leukemic β-glucuronidase. *Biological Chemistry Hoppe–Seyler*, **373**, 57–62.

Taouji S, Debout C and Izard J (1996) Arylsulfatase B in Kurloff cells: increased activity of anionic isoforms in guinea pig acute lymphoblastic leukemia. *Leukemia Research*, **20**, 259–264.

Thomas GH (1994) 'Pseudodeficiencies' of lysosomal hydrolases. *American Journal of Human Genetics*, **54**, 934–940.

Thompson D and Daniel W (1988) Comparative biochemistry of mammalian arylsulfatases A and B. *Comparative Biochemistry and Physiology*, **90**, 823–831.

Tomatsu S, Sukegawa K, Ikedo Y, Fukuda S, Yamada Y, Sasaki T, Okamoto H, Kuwabara T and

Orii T (1990) Molecular basis of mucopolysaccharidosis type VII: replacement of Ala619 in β-glucuronidase with Val. *Gene*, **89**, 283–287.

Tomatsu S, Fukuda S, Sukegawa K, Ikedo Y, Yamada S, Yamada Y, Sasaki T, Okamoto H, Kuwahara T and Yamaguchi S (1991) Mucopolysaccharidosis type VII: characterization of mutations and molecular heterogeneity. *American Journal of Human Genetics*, **48**, 89–96.

Tomatsu S, Fukuda S, Cooper A, Wraith J, Yamagishi A, Kato Z, Yamada N, Isogai K, Sukegawa K, Suzuki Y, Shimozawa N, Kondo N and Orii T (1998) Fifteen polymorphisms in the N-acetylgalactosamine-6-sulfate sulfatase (GALNS) gene: diagnostic implications in Morquio disease. *Human Mutation* , **Suppl 1**, 42–46.

Toyoshima M and Nakajima M (1999) Human heparanase. Purification, characterization, cloning and expression. *Journal of Biological Chemistry*, **274**, 24153–24160.

Traurig H (1976) Lysosomal acid hydrolase activities in the lungs of foetal, neonatal, adult and senile mice. *Gerontology*, **22**, 419–427.

Triggiani M, Granata F, Oriente A, De Marino V, Gentile M, Calabrese C, Palumbo C and Marone G (2000) Secretory phospholipase A2 induce β-glucuronidase release and IL-6 production from human lung macrophages. *Journal of Immunology*, **164**, 4908–4915.

Trudeau D, Bernier R, Gannon D and Forsberg C (1992) Isolation of *Clostridium acetobutylicum* strains and the preliminary investigation of the hemicellulolytic activities of isolate 3BYR. *Canadian Journal of Microbiology*, **38**, 1120–1127.

Tsuruoka T, Fukuyasu H, Ishii M, Ususi T, Shibahara S and Inouye S (1996) Inhibition of mouse tumor metastasis with nojirimycin-related compounds. *Journal of Antibiotics*, **49**, 155–161.

Uchida H, Nanri T, Kawabata Y, Kusakabe L and Murakami K (1992) Purification and characterization of intracellular alpha-glucuronidase from *Aspergillus niger 5-16*. *Bioscience, Biotechnology and Biochemistry*, **56**, 1608–1615.

Uehara Y, Gasa S, Makita A, Sakurada K and Miyazaki T (1983) Lysosomal arylsulfatases of human leukocytes: increment of phosphorylated B variants in chronic myelogenous leukemia. *Cancer Research*, **43**, 5618–5622.

van der Pal RHM, Klein W, van Golde LMG and Lopez-Cardoso M (1991) Developmental profiles of arylsulfatases A and B in rat cerebral cortex and spinal cord. *Biochimica et Biophysica Acta*, **1081**, 315–320.

van Eldere JR, de Pauw G and Eyssen HJ (1987) Steroid sulfatase activity in a peptococcus niger strain from the human intestinal microflora. *Applied and Environmental Microbiology*, **53**, 1655–1660.

van Eldere J, Robben J, de Pauw G, Merckx R and Eyssen H (1988) Isolation and identification of intestinal steroid-desulfating bacteria from rats and humans. *Applied and Environmental Microbiology*, **54**, 2112–2117.

van Eldere J, Parmentier G, Asselberghs S and Eyssen H (1991) Partial characterization of the steroidsulfatases in peptococcus niger H4. *Applied and Environmental Microbiology*, **57**, 69–76.

Verdugo M and Ray J (1997) Age-related increase in activity of specific lysosomal enzymes in the human retinal pigment epithelium. *Experimental Eye Research*, **65**, 231–240.

Vervoort R, Islam MR, Sly W, Chabas A, Wevers R, Dejong J, Liebaers I and Lissens W (1995) A pseudodeficiency allele (d152n) of the human β-glucuronidase gene. *American Journal of Human Genetics*, **57**, 798–804.

Vervoort R, Islam M, Sly W, Zabot M, Kleijer W, Chabas A, Fensom A, Young E, Liebaers I and Lissens W (1996) Molecular analysis of patients with β-glucuronidase deficiency presenting as hydrops foetalis or as early mucopolysaccharidosis VII. *American Journal of Human Genetics*, **58**, 457–471.

Vervoort R, Gitzelmann R, Bosshard N, Maire I, Liebaers I and Lissens W (1998) Low β-glucuronidase enzyme activity and mutations in the human β-glucuronidase gene in mild mucopolysaccharidosis type VII, pseudodeficiency and a heterozygote. *Human Genetics*, **102**, 69–78.

Villani G, Balzano N, Vitale D, Saviano M, Pavone V and Di Natale P (1999) Maroteaux-lamy syndrome: five novel mutations and their structural localization. *Biochimica et Biophysica Acta*, **1453**, 185–192.

Vitaioli L, Gobbetti A and Baldoni E (1996) Arylsulphatase A activity and sulphatide concentration in the female rabbit oviduct are under physiological hormonal influence. *Histochemical Journal*, **28**, 149-156.

Vlodavsky I, Friedmann Y, Elkin M, Aingorn H, Atzmon R, Ishai-Michaeli R, Bitan M, Pappo O, Peretz T, Michal I, Spector L and Pecker I (1999) Mammalian heparanase: gene cloning, expression and function in tumor progression and metastasis. *Nature Medicine*, **5**, 793–802.

Vlodavsky I, Elkin M, Pappo O, Aingorn H, Atzmon R, Ishai-Michaeli R, Aviv A, Pecker I and Friedmann Y (2000) Mammalian heparanase as mediator of tumor metastasis and angiogenesis. *Journal of the Israel Medical Association*, **2 Suppl.**, 37–45.

von Figura K, Schmidt B, Selmer T and Dierks T (1998) A novel protein modification generating an aldehyde group in sulfatases: its role in catalysis and disease. *Bioessays*, **20**, 505–510.

Waheed A and van Etten RL (1985) Purification of mammalian arylsulfatase A enzymes by subunit affinity chromatography. *International Journal of Peptide and Protein Research*, **26**, 362–372.

Wakabayashi M (1970) Beta-glucuronidases in metabolic hydrolysis. In *Metabolic Conjugation and Metabolic Hydrolysis*, Fishman WH (ed), Academic Press, New York, pp. 519–602.

Walaszek Z, Hanausek-Walaszek M and Webb TE (1984) Inhibition of 7,12-dimethylbenzynthracene-induced rat mammary tumorigenesis by 2,5-di-O-acetyl-D-glucaro-1,4:6,3-dilactone, an in vivo β-glucuronidase inhibitor. *Carcinogenesis*, **5**, 767–772.

Walaszek Z, Hanausek M, Sherman U and Adams AK (1990) Antiproliferative effect of dietary glucarate on the Sprague–Dawley rat mammary gland. *Cancer Letters*, **49**, 51–57.

Waldow A, Schmidt B, Dierks T, von Bulow R and von Figura K (1999) Amino acid residues forming the active site of arylsulfatase A. Role in catalytic activity and substrate binding. *Journal of Biological Chemistry*, **274**, 12284–12288.

Walle T, Conradi E, Walle U and Graffney T (1979) Propranolol glucuronide accumulation during long-term therapy: A proposed storage mechanism for propranolol. *Clinical Pharmacology and Therapeutics*, **26**, 686–695.

Wang C and Touster O (1975) Turnover studies on proteins of rat liver lysosomes. *Journal of Biological Chemistry*, **250**, 4896–4902.

Waters P, Flynn M, Corrall R and Pennock C (1992) Increases in plasma lysosomal enzymes in type 1 (insulin-dependent) diabetes mellitus: relationship to diabetic complications and glycaemic control. *Diabetologia*, **35**, 991–995.

Watson G, Felder M, Rabinow L, Moore K, Labarca C, Tietze C, Van der Molen G, Bracey L, Brabant M, Cai J and Paigen K (1985) Properties of rat and mouse β-glucuronidase mRNA and cDNA, including evidence for sequence polymorphism and genetic regulation of mRNA levels. *Gene*, **36**, 15–25.

Weller P, Corey E, Austen K and Lewis R (1986) Inhibition of homogeneous human eosinophil arylsulfatase B by sulfidopeptide leukotrienes. *Journal of Biological Chemistry*, **261**, 1737–1744.

Weyel D, Sedlacek H, Müller R and Brüsselbach S (2000) Secreted human β-glucuronidase: a novel tool for gene-directed enzyme prodrug therapy. *Gene Therapy*, **7**, 224–231.

Whiting JF, Narciso JP, Chapman V, Ransil BJ, Swank RT and Gollan JL (1993) Deconjugation of bilirubin-IX alpha glucuronides—a physiologic role of hepatic microsomal β-glucuronidase. *Journal of Biological Chemistry*, **268**, 23197–23201.

Wojczyk B (1986) Lysosomal arylsulfatases A and B from horse blood leukocytes: purification and physico-chemical properties. *Biology of the Cell*, **57**, 147-152.

Wong C and Keung W (1997) Daidzein sulfoconjugates are potent inhibitors of sterol sulfatase (EC 3.1.6.2). *Biochemical and Biophysical Research Communication*, **233**, 579–583.

Wong C, He S, Grubb J, Sly W and Withers S (1998) Identification of Glu-540 as the catalytic nucleophile of human β-glucuronidase using electrospray mass spectrometry. *Journal of Biological Chemistry*, **273**, 34057–34062.

Woo LW, Howarth NM, Purohit A, Hejaz HA, Reed MJ and Potter BV (1998) Steroidal and nonsteroidal sulfamates as potent inhibitors of steroid sulfatase. *Journal of Medicinal Chemistry*, **41**, 1068–83.

Wu J, Yang C, Lee C, Chang J and Tsai F (2000) A novel mutation (Q239R) identified in a Taiwan Chinese patient with type VI mucopolysaccharidosis (Maroteaux-Lamy syndrome). *Human Mutation*, **15**, 389–390.

Young J, Kenyon E and Calabrese E (1990) Inhibition of β-glucuronidase in human urine by ascorbic acid. *Human & Experimental Toxicology*, **9**, 165–170.

Yutaka T, Fluharty A, Stevens R and Kihara H (1982) Purification and some properties of human liver iduronate sulfatase. *Journal of Biochemistry*, **91**, 433–441.

Zenser T, Lakshmi V and Davis B (1999) Human and *Escherichia coli* β-glucuronidase hydrolysis of glucuronide conjugates of benzidine and 4-aminobiphenyl, and their hydroxy metabolites. *Drug Metabolism and Disposition*, **27**, 1064–1067.

Zhen L, Baumann H, Novak E and Swank R (1993) The signal for retention of the egasyn-glucuronidase complex within the endoplasmic reticulum. *Archives of Biochemistry and Biophysics*, **304**, 402–414.

16 Nitroreductases and Azoreductases

Shmuel Zbaida

Schering-Plough Research Institute, USA

Introduction

Reductive drug metabolism is the least studied biotransformation pathway in comparison to enzymic oxidation, hydrolysis and conjugation. There are numerous thiol, flavin and cytochrome P-450 enzymes that await individual recognition as well as enzymes with metal centres that may serve as electron sources. Reductive drug metabolism is a means by which drug detoxification or activation, that elevates toxicity and mutagenicity, may take place.

Nitro substituents are found in compounds such as industrial solvents, insecticides, food preservatives and other xenobiotics. Azo compounds are widely used as colourants in food, cosmetics, pharmaceuticals and other industries. The azo dyes' colour is derived from conjugated double bonds of aromatic residues through an azo linkage, permitting release of energy in the visible spectrum. The last 10 years of publications associated with the metabolism of nitro compounds and azo dyes, and its implication for both activation and detoxification are discussed in this chapter.

Nitroreductases

NITROREDUCTASES AND ENZYME MULTIPLICITY

Nitro moietices ($-NO_2$) can be reduced by three sets of two electron reductions to nitroso ($-NO$), hydroxylamine ($-NHOH$) and finally primary amine ($-NH_2$) (Figure 16.1) (McLane *et al.* 1983). Nitroreductase activity is associated with a diversity of enzymes, including xanthine oxidase (De Castro *et al.* 1990; Nakao *et al.* 1991),

$$RNO_2 \longrightarrow RNO \longrightarrow RNHOH \longrightarrow RNH_2$$

Figure 16.1 Reduction of nitro compounds to primary amine metabolites requiring a total of six electrons. Each reductive step requires two electrons.

Enzyme Systems that Metabolise Drugs and Other Xenobiotics. Edited by C. Ioannides.
© 2002 John Wiley & Sons Ltd

aldehyde oxidase (Belisario *et al.* 1990; Nakao *et al.* 1991), DT-diaphorase (Belisario *et al.* 1990), flavoprotein-enzymes (Bueding and Jolliffe 1946), NADPH-cytochrome P-450 reductase alone (Person *et al.* 1991) and/or in combination with cytochrome P-450s (Belisario *et al.* 1990, 1991). Consequently, multiple mechanisms are involved in nitroreduction activity. The main forms of cytochrome P-450 (CYP) which exhibit nitroreductase activity are associated with CYP1A and CYP3A subfamilies (Berson *et al.* 1993), primarily CYP1A2 (Chae *et al.* 1993; Lehman-McKeeman *et al.* 1997a), CYP3A4 (Chae *et al.* 1993; Seree *et al.* 1993) and CYP2B10 (Lehman-McKeeman *et al.* 1997a) isoforms.

The literature of the past 10 years discusses a variety of nitroreductase biotransformation pathways. Recent examples, in addition to those previously discussed in an excellent review by McLane *et al.* (1983), include studies on the anaerobic reductive metabolism of the hair dye constituent, 2-nitro-1,4-diaminobenzene (2-nitro-*p*-phenylenediamine), in rat liver microsomal and cytosolic fractions (Nakao *et al.* 1991). In the microsomal preparation, both air and carbon monoxide inhibit the nitroreduction of this compound while NADPH is more effective than NADH as an electron donor. This is consistent with the involvement of cytochrome P-450 in microsomal nitroreductase activity. Addition of FMN, together with NADPH and NADH, also elevates nitroreductase activity with 2-nitro-1,4-diaminobenzene. However, activity in the cytosolic fraction is attributed to xanthine oxidase, aldehyde oxidase and other enzymes.

De Castro *et al.* (1990) studied Nifurtimox nitroreductase activity in the microsomal, mitochondrial and cytosolic fractions of rat adrenals. Enzymic activity of all fractions was inhibited by oxygen. Carbon monoxide inhibited only 10% of the microsomal activity but did not affect the mitochondrial activity. Reduction of Nifurtimox in both the microsomal and mitochondrial fractions required NADPH. However, reduction of this drug in the cytosolic fraction required the addition of hypoxanthine and was inhibited by allopurinol, suggesting that xanthine oxidase was involved in its nitroreduction.

Belisario *et al.* (1990) studied the biotransformation of nitrofluoranthenes, mutagenic and carcinogenic environmental pollutants. These compounds undergo both oxidation and reduction by rat liver subcellular fractions. Under aerobic conditions only ring hydroxylation occurred, whereas under anaerobic conditions, reduction of the nitrofluoranthenes occurred in both cytosolic and microsomal fractions. Based on the cofactors required for the cytosolic nitroreduction, the authors attributed the activity of this fraction to DT-diaphorase, aldehyde oxidase and other unknown enzymes. Nitroreductase activity of the microsomal fraction was attributed by cytochrome P-450.

In another study (Belisario *et al.* 1991) the ring-oxidation and nitroreduction of 1-nitropyrene and 1,6-dinitropyrene were examined in human hepatoma cells, HepG2, following induction with either phenobarbital or 3-methylcholanthrene. 3-Methylcholanthrene selectively induced ring-oxidation of 1-nitropyrene, whereas phenobarbital stimulated its nitroreduction. Phenobarbital-inducible nitroreduction was consistent with cytochrome P-450, as indicated by the requirement of NADPH and the inhibition of nitroreduction activity by α-naphthoflavone and carbon monoxide.

The nitroreduction of the antiandrogen nilutamide was studied using rat liver microsomes and NADPH (Person *et al.* 1991). Under anaerobic conditions, a nitro anion-free radical was detected by electron spin resonance (EPR) spectroscopy while in air this radical reacts with oxygen, forming a reactive oxygen species.

Since this reaction was not inhibited by SKF 525-A or carbon monoxde (cytochrome P-450 inhibitors) but was inhibited by methylene blue and 2'-adenosine monophosphate (NADPH-cytochrome P-450 reductase inhibitors), the authors concluded that nilutamide was reduced by NADPH-cytochrome P-450 reductase alone.

Oxidations versus nitroreduction biotransformation pathways were investigated to determine the possible toxicity of nitro-aromatic compounds. Nitroreduction can biologically activate nitro compounds leading to carcinogenecity or mutagenicity. In other cases, nitroreduction can serve as a detoxification pathway.

Silvers *et al.* (1994) examined the ratio of cytochrome P-450-mediated C-oxidation to nitroreduction of 1-nitropyrene in HepG2 cells. Addition of 3-methylcholanthrene to the cells increased the ratio of C-oxidation to nitroreduction, which was accompanied by a decrease in the formation of 1-nitropyrene adduct *via* nitroreduction. These results suggest that the cytochrome P-450-mediated C-oxidation is not an activation pathway in HepG2 cells, and may explain the weak carcinogenecity of 1-nitropyrene where cytochrome P-450-mediated biotransformation predominates.

Studies of lung tumours in rats induced by 3,9-dinitrofluoranthrene (3,9-DNF) and 3-nitrofluoranthrene (3-NF) were investiaged by Mitchel *et al.* (1993). The two compounds exhibited different carcinogenicity. The former was a potent pulmonary carcinogen while the latter exhibited weak carcinogenic activity. It was found that there was greater ring hydroxylation of 3-NF than of 3,9-DNF, where nitroreduction was the major pathway. Thus, a higher ratio of nitroreduction to ring hydroxylation accounted for lower carcinogenicity of 3,9-DNF.

Studies were also directed toward identification of cytochrome P-450 (CYP) isoforms involved in the biological activation of nitrocompounds. Chae *et al.* (1993) demonstrated that human liver and lung are capable of metabolising 6-nitrochrysene to known potent carcinogenic metabolites *via* ring oxidation and nitroreduction. Rates of phenacetin O-deethylation (CYP1A2) and nifedipine oxidation (CYP3A4) were correlated with the rates of formation of *trans*-1,2-dihydro-1,2-dihydroxy-6-nitrochrysene (oxidation) and 6-aminochrysene (nitroreduction), respectively. This suggested that both CYP1A2 and CYP3A4 were involved in the biotransformation of 6-nitrochrysene. The involvement of CYP3A4 in the nitroreduction of clonazepam was also reported by Seree *et al.* (1993). The authors synthesised an oligonucleotide specific for the CYP3A4 protein gene and used it for hybridisation on total RNA from human liver samples. The 2.2 kb transcript correlated ($r = 0.61$) with the intensity of clonazepam nitroreduction in human liver microsomes.

Li *et al.* (1995) evaluated the biotransformation pathways catalysed by CYP3A4 in human liver preparations (microsomes, cultured hepatocytes) and in yeast cells genetically engineered to express CYP3A4. These authors demonstrated that CYP3A4 pathways included both oxidation reactions (N-oxidation, C-oxidation, N-dealkylation and O-dealkylation) as well as nitroreduction.

Musk xylene, 1,3,5-trinitro-2-t-butylxylene, a synthetic nitromusk perfume ingredient that induces cytochrome P-450 enzymes, undergoes nitroreduction by intestinal flora to yield aromatic amine metabolites (Lehman-McKeeman *et al.* 1997a,b). The authors examined the potential capability of the monoamine metabolites to induce CYP2B10 and CYP1A2 using Northern Blot analyses. mRNAs for both cytochromes were induced, but that for CYP1A2 only to a slight extent.

In vitro metabolic activation of flutamide, a nitroaromatic antiandrogen that produces hepatitis, was studied in male rat liver microsomes (Berson *et al.* 1993). The authors did not detect the presence of one-electron reduction intermediate of flutamide by electron spin resonance. However, flutamide formed reactive intermediate(s) that covalently bound to microsomal proteins. Formation of microsomal protein-flutamide adducts required NADPH and oxygen, and was suppressed by cytochrome P-450 inhibitors, e.g. SKF 525-A, piperonyl butoxide and troleandomycin (inhibitor of the CYP3A subfamily). The formation of the covalent adduct was enhanced considerably following pretreatment of the rats with dexamethasone (inducer of CYP3A subfamily) and moderately following pretreatment with β-naphthoflavone (inducer of the CYP1A subfamily). The formation of the covalent adduct(s) was enhanced with yeast microsomes expressing human CYP1A1, CYP1A2 or CYP3A4. Covalent binding was inhibited significantly by anti-CYP3A immunoglobulin G and moderately with anti-CYP1A immunoglobulin G. Thus, the authors concluded that CYP1A and CYP3A subfamilies mediated the formation of the covalent-adduct of flutamide.

NITROREDUCTASE ACTIVITY IN DIFFERENT SPECIES AND ORGANS

Nitroreductase activity was found in mammalian liver and adrenal cortex, and channel catfish (*Ictalurus punctatus*) (Washburn and Di Giulio 1988), as well as in different bacteria, such as *Bacillus licheniformis* isolated from industrial waste containing high concentrations of 5-nitro-1,2,4-triazol-3-one (Le Campion *et al.* 1999).

MECHANISM OF ENZYMATIC NITROREDUCTION

Due to the involvement of various enzymes in different tissues there is no comprehensive mechanism for enzymatic nitroreduction. Two general types of nitroreductase mechanisms are described:

(1) Oxygen-sensitive biotransformation pathway. This nitroreductase mechanism requires the formation of one-electron free-radical intermediate, which is stable enough to be detected by EPR and to react with oxygen (aerobic reduction) with the formation of the superoxide anion radical (Person *et al.* 1991; Washburn and Di Giulio 1998). This requires two separate one-electron reductions but not a single two-electron reduction.

Studies by Mason (Orna and Mason 1989; Mason 1997) also suggested a nitroreductase mechanism involving a one-electron reduction and formation of a free-radical intermediate at the initial step. These nitro free-radicals reductive intermediates were found to be unstable in air (Mason 1997; Orna and Mason 1989). A linear correlation between $\log V_{max}/K_m$ and one-electron reduction potentials of nitro compounds found by these researchers clearly supported such a mechanism.

(2) Oxygen-insensitive biotransformation pathway. Reduction of 5-nitrofuran is rela-

tively oxygen insensitive (Holtzman *et al.* 1981). This type of mechanism probably requires either a single two-electron step reduction or a stepwise mechanism of two one-electron reductions, with the formation of a one-electron free-radical intermediate, temporarily stable in air (oxygen), that 'immediately' proceeds to an additional one-electron reduction.

UTILISATION OF NITRO COMPOUNDS AS PRODRUGS

The possible use of nitro compounds as prodrugs activated by bacterial aerobic nitroreductase is currently being investigated. This prospect was discussed in a review article by Patterson and Raleigh (1998) on reductive drug metabolism, which includes a section on nitroreductase biotransformation pathways.

Azoreductases

Azo compounds (Ar-N = N-Ar) are widely used as colourants in the pharmaceutical, food, textile and printing industries (Catino and Farris 1978). The reduction of azo dyes to primary amines is catalysed by mammalian liver microsomal cytochrome P-450 (Fugita and Peisach 1978; Hernandez *et al.* 1967), cytosolic enzymes (Huang *et al.* 1979) and colonic bacteria (Walker 1970; Scheline 1980). Chung *et al.* (1992) discuss reductive cleavage of a large number of azo dyes to primary amines by intestinal microflora (Figure 16.2).

Azoreductase activity was also detected *in vitro* (diffusion cells) following percutaneous absorption and metabolism of three azo dyes, phenylazophenol, phenylazo-2-naphthol and 5-(phenylazo)-6-hydroxynaphthalene-2-sulphonic acid, in mouse, pig and human skin (Collier *et al.* 1983). Thus, dermal risk assessments from exposures to pharmaceutical preparations should be considered.

The NAD(P)H:quinone oxidoreductase, also known as DT-diaphorase, is a flavoprotein that catalyses the two-electron reduction of quinones, quinone imines and azo-compounds (Belinsky and Jaiswal 1993). Thus, this enzyme forms no free-radical intermediates resulting from one-electron reductions and consequently, no reactive oxygen intermediates are formed. In contrast, reduction by cytochrome P-450 is associated with one electron reduction (Porter and Coon 1991).

Metabolism of azo dyes may lead to either mutagenic or toxic effects or to detoxification. Sandhu and Chipman (1991) studied the oxidation by cytochrome P-450, particularly P-448 (CYP1), which resulted in the metabolic activation of chrysoidine azo dyes. However, addition of FMN inhibited mutagenicity, suggesting that reduction of the azo dyes resulted in detoxification.

$$Ar_1\text{-}N{=}N\text{-}Ar_2 \longrightarrow Ar_1\text{-}NH\text{-}NH\text{-}Ar_2 \longrightarrow Ar_1\text{-}NH_2 + Ar_2\text{-}NH_2$$

1   **2**   **3**

Figure 16.2 Reductive cleavage of azo dyes (**1**) to primary amine metabolites (**3**) requiring electrons. Two electrons are required for the reduction of an azo compound (**1**) to hydrazo intermediate (**2**) and an additional two electrons are required for the reduction of hydrazo compound (**2**) to primary amine metabolites (**3**).

Purified rabbit liver aldehyde oxidase readily reduced water-soluble azo dyes. However, lipophilic azodyes were poor substrates or not reduced at all by this enzyme but were readily reduced by microsomal cytochrome P-450 enzymes (Stoddart and Levine 1992).

Prediction of the azoreduction biotransformation pathway in liver microsomes is possible based on electronic aspects and structure-activity relationships. Highlights of these findings, which were previously presented in a detailed review article (Zbaida 1995), are summarised below:

- Microsomal reduction of azo dyes requires polar electron-donating substituents (Hammett σ constant of -0.37 or lower) such as hydroxyl, amino, methylamino or dimethylamino moieties *ortho* or *para* to the azo linkages for binding to enzyme (Levine and Zbaida 1988; Zbaida *et al.* 1989). These substituents are essential for both binding and reduction of azo dyes. Type I and type II binding-spectra were observed for the reactive azo dyes but not for the poorly reactive compounds (Levine and Zbaida 1988), suggesting that the electron-donating substituents are essential for both binding and reduction of azo dyes.
- High substrate reactivity was observed for 4-dimethylamino-, 4-methylamino- and 4-amino-azobenzene and for 4-hydroxyazobenzene. In contrast, substituents that nullify the non-bonding electron donation (Hammett σ constant higher than -0.37), such as the benzoylamide derivative of 4-aminoazobenzene, unsubstituted azobenzene or non-polar electron donating substituents such as 4-isopropylazobenzene, exhibited only negligible rates of azoreduction. These results suggest that polar electron-donating substituents are obligatory for microsomal azoreduction (Levine and Zbaida 1988; Zbaida *et al.* 1988, 1989).
- Two classes of azo dyes structurally related to 4-dimethylaminoazobenzene, based on the sensitivity of their microsomal azoreduction to oxygen and carbon-monoxide, were identified: I-substrates, *insensitive* to both oxygen and carbon-monoxide, and the S-substrates, sensitive to both oxygen and carbon-monoxide (Levine and Zbaida 1991; Zbaida and Levine 1990a).
- The two classes of dyes differ in their chemical structure (Figure 16.3). The I-substrates contain *only* polar electron donating substituents, whereas the S-substrates contain *both* electron-donating and electron-withdrawing substituents [such as carboxylic, sulphonic or arsenic (AsO_3H_2) acid residues on the opposite phenyl ring] (Zbaida and Levine 1990a). The combination of both electron-donating and electron-withdrawing substituents in the S-substrates promotes electronic resonance. Substituents on azo dyes are the key in defining the mechanism of their microsomal reduction. The electron-donating substituents are mainly associated with binding to cytochrome P-450, whereas the combination of additional electron-withdrawing substituents on the opposite ring (S-substrates) alters the charge and the redox potentials, and consequently changes their rates and mechanisms of microsomal reduction. The combination of electron-donating and/or -withdrawing groups alters the relative stability of their one-electron-reduced radical intermediate towards oxygen and their tendency to interact with ferrous cytochrome P-450. This results in a different sensitivity of their microsomal azoreduction towards oxygen and carbon monoxide.

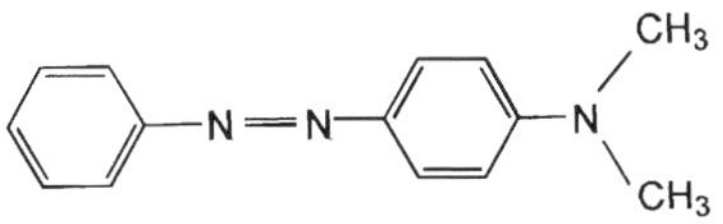

p-Dimethylaminoazobenzene

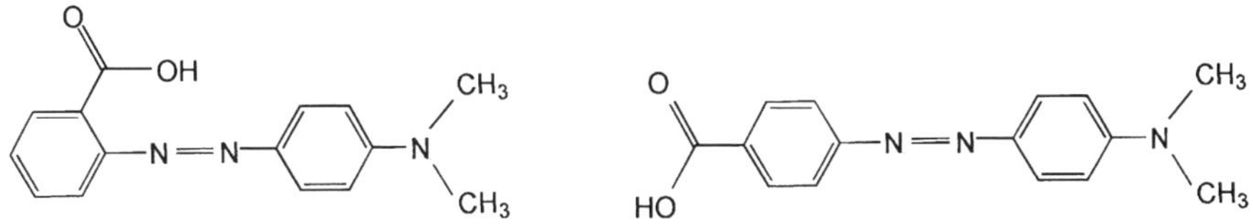

o-Methyl red

p-Methyl red

Figure 16.3 Chemical structures of a representative *l-substrate* (*p*-dimethylaminoazobenzene) containing *only* electron-donating substituent and two representative *S-substrates* (*o-* and *p-*methyl red) containing *both* electron-donating and electron-withdrawing substituents *ortho* or *para* to the azo linkage.

- Although a total of four electrons are required for the reduction of an azo linkage to primary amines, microsomal reduction of azo dyes probably involves a two-electron reduction cycle leading to a hydrazo intermediate (Zbaida *et al.* 1989). Non-substrate dyes (such as azobenzene) yield a stable hydrazo intermediate but non-reactive enzymically. Electron-donating groups *ortho* or *para* to the azo linkage are responsible for the instability of the hydrazo compounds (Figure 16.4). Non-bonding electrons from the electronegative substituent can produce a negative charge on the hydrazo nitrogen. Dissociable hydrogen from the substituent could then migrate to the negatively charged nitrogen leading to bond cleavage (Zbaida *et al.* 1989).
- The azoreduction of I-substrates is selectively induced by clofibrate, whereas the

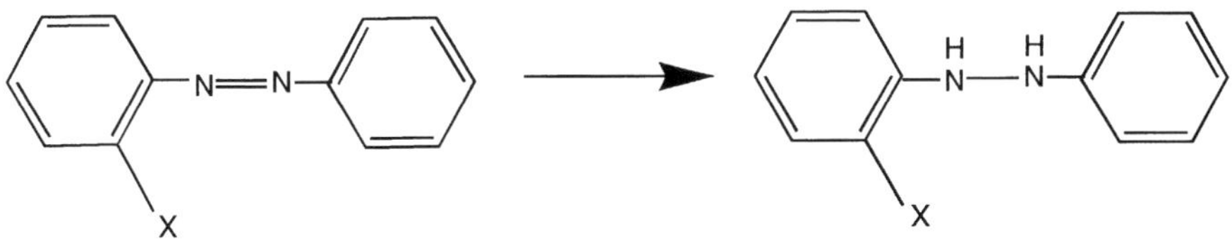

Azo compounds

Hydrazo intermediates

X = Electron-donating groups

Figure 16.4 Azo dyes and hydrazo intermediates substituted with polar electron-donating substituents.

azoreduction of S-substrates is selectively induced by phenobarbital, β-naphthoflavone, isosafrole and pregnenolone-16α-carbonitrile as well as clofibrate (Levine et al. 1992; Zbaida and Levine 1990a,b).

- The S-substrates are more readily reduced both enzymically and chemically (by sodium dithionite) than I-substrates (Zbaida and Levine 1990a; 1992b).

- Inhibition of microsomal azoreduction by CN^- also distinguishes the two classes of dyes. Reduction of I-substrates is, on average, more sensitive to CN^- than is reduction of S-substrates, possibly due to the alteration of the enzyme redox potentials by CN^- (Zbaida and Levine 1992a; Zbaida et al. 1992).

- The electrode potential at which a chemical undergoes reduction or oxidation can be detected by cyclic voltammetry. Electrochemical techniques are suitable means of studying oxidation-reduction potentials. The electrode potential at which a chemical undergoes reduction or oxidation can be rapidly detected by cyclic voltammetry. Electrochemical reductions and oxidations generate intermediates coupled to the electrode surface reaction, which may serve as models for those biologically formed. Enzymic reduction of substrate and nonsubstrate dyes was performed anaerobically in anhydrous dimethylformamide (Zbaida and Levine 1991a). All azo dye substrates exhibit two negative and one positive redox potential, as measured anaerobically by cyclic voltammetry. The negative potentials reflect one- and two-electron reduced intermediates while the positive potential is associated with the electron transfer from the microsomal cytochrome P-450 to the dyes. Nonsubstrate azo dyes did not exhibit positive anodic potentials. The S-substrate dyes exhibited less negative oxidation-reduction potentials (Levine et al. 1992; Zbaida and Levine 1991a). The sensitivity of the S-substrates to oxygen is well demonstrated by the cyclic voltammetry. The one-electron reduced intermediates of S-substrates, which are stable under nitrogen, immediately quenched in air, while those of I-substrates were relatively stable in air. The I-substrates exhibit on average potentials which are approximately 0.6 Volt more negative than those for S-substrates (Levine et al. 1992; Zbaida and Levine 1991a). Although microsomal azoreduction of S-substrates is sensitive to both oxygen and carbon monoxide, the mechanisms of the two inhibitions are totally different. Carbon monoxide interacts with a reduced form of cytochrome P-450 (Ortiz de Montellano and Reich 1986; Stanford et al. 1980) whereas oxygen reacts with the one-electron-reduced intermediate of the dye forming the superoxide anion radical (Peterson et al. 1988).

- These two classes of dyes also differ in their Hammett sigma values. The I-substrates exhibit negative values while the S-substrates exhibit a combination of negative and positive values on opposite rings. However, an overall value of -0.37 or lower is obligatory for microsomal azoreductases (binding) (Zbaida et al. 1994). Chemical modification which alters the magnitude (from negative to positive) of the Hammett substituent constants of the prime substituents also alters sensitivity to both oxygen and carbon monoxide. This implies that catalytic recognition of azo dyes by hepatic microsomes is regulated by charge and redox potentials (Zbaida et al. 1994).

- A linear correlation between V_{max} and K_m was observed for I-substrates, suggesting inverse relationships between binding affinity and rate of reduction (Zbaida and Levine 1990a). This relationship was not obvious for the S-substrate dyes. A lower K_m was observed for all dyes (both I- and S-substrates) bearing greater number of heteroatoms with non-bonding electrons in either electron-donating or -withdraw-ing groups (Levine and Zbaida 1992; Levine *et al.* 1992; Zbaida and Levine 1991b, 1992b).

- More basic dyes, which contain a higher density of non-bonding electrons, showed an inverse correlation with both K_m and V_{max} (Levine and Zbaida 1992; Levine *et al.* 1992; Zbaida and Levine 1991b; 1992b).
- NMR (nuclear magnetic resonance) studies revealed minor differences in the chemical shifts of protons attached to the phenyl ring substituted with electron-donating substituents. However, there are significant differences of the aromatic protons on the opposite prime ring, which also distinguish I- and S-substrates (Zbaida *et al.* 1992).
- This implies that the mechanism of microsomal azoreduction is critically dependent on charge and redox potentials of the dyes (Zbaida *et al.* 1994).
- Surprisingly, no significant studies of microsomal azoreduction were performed in the last 7 years. The commercial availability of cDNA expressed human cytochrome P-450s (Supersomes), chemical inhibitors and antibodies against specific forms of cytochrome P-450s and the considerable progress in molecular biology (Dachs *et al.* 1997) in recent years can provide an excellent opportunity to identify the form(s) of cytochrome P-450 involved in microsomal azo reduction. It is important to characterise the forms that metabolise I-substrate *versus* those that metabolise the S-substrate dyes. This may provide a better understanding of the enzymic mechanism of action of various forms of cytochrome P-450s.

Molecular biological approaches should also be considered for nitroreduction pathways. The nitroreductase from *E. coli* B has been studied for its use in antibody-directed enzyme prodrug therapy (ADEPT) (Dachs *et al.* 1997). Alternate prodrugs for use with *E. coli* nitroreductase in suicide gene approaches to cancer-therapy have been investigated (Bailey *et al.* 1996).

References

Bailey SM, Knox RJ, Hobbs SM, Jenkins TC, Mauger AB, Melton RG, Burke PJ, Connors TA and Hart IR (1996) Investigation of alternative prodrugs for use with *Escherichia-Coli* nitroreductase in suicide gene approaches to cancer-therapy. *Gene Therapy*, **3**, 1143–1150.

Belinsky M and Jaiswal AK (1993) NAD(P)H: quinone oxidoreductase 1 (DT-diaphorase) expression in normal and tumor tissues. *Cancer and Metastasis Reviews*, **12**, 103–117.

Belisario MA, Pecce R, Della Morte R, Arena AR, Cecinato A, Ciccioli P and Staiano N (1990) Characterization of oxidative and reductive metabolism *in vitro* of nitrofluoranthenes by rat liver enzymes. *Carcinogenesis*, **11**, 213–218.

Belisario MA, Arena AR, Pecce R, Borgia R, Staiano N and De Lorenzo F (1991) Effect of enzyme inducers on metabolism of 1-nitropyrene in human hepatoma cell line HepG2. *Chemico-Biological Interactions*, **78**, 253–268.

Berson A, Wolf C, Chachaty C, Fisch C, Fau D, Eugene D, Loeper J, Gauthier JC, Beaune P, Pompon D, Maurel P and Pessayre D (1993) Metabolic-activation of the nitroaromatic antiandrogen flutamide by rat and human cytochromes P-450, including forms belonging to the 3A and 1A subfamilies. *Journal of Pharmacology and Experimental Therapeutics*, **265**, 366–372.

Bueding E and Jolliffe N (1946) Metabolism of trinitrotoluene (TNT) in vitro. *Journal of Pharmacology and Experimental Therapeutics*, **88**, 300–312.

Catino SC and Farris RE (1978) Azo dyes. In *Kirk–Othmer Encyclopedia of Chemical Toxicology*,

Mark HF, Othmer DF, Overberger CG and Seaborg GT (eds.), Wiley, New York, Vol. 3, (3rd edition), pp. 387–433.

Chae YH, Yun CH, Guengerich FP, Kadlubar FF and El-Bayoumy K (1993) Roles of human hepatic and pulmonary cytochrome P450 enzymes in the metabolism of the environmental carcinogen 6-nitrochrysene. *Cancer Research*, **53**, 2028–2034.

Chung K-T, Stevens Jr. SE and Cerniglia CE (1992) The reduction of azo dyes by the intestinal microflora. *Critical Reviews in Microbiology*, **18**, 175–190.

Collier SW, Storm JE and Bronaugh RL (1993) Reduction of azo dyes during *in vitro* percutaneous absorption. *Toxicology and Applied Pharmacology*, **118**, 73–79.

Dachs GU, Dougherty GJ, Stratford IJ and Chaplin DJ (1997) Targeting gene therapy to cancer, a review. *Oncology Research*, **9**, 313–325.

De Castro CR, De Toranzo EG, Carbone M and Castro JA (1990) Ultrastructural effects of Nifurtimox on rat adrenal cortex related to reductive biotransformation. *Experimental and Molecular Pathology*, **52**, 98–108.

Fugita S and Peisach J (1978) Liver microsomal cytochrome P-450 and azo reductase activity. *Journal of Biological Chemistry*, **253**, 4512–4513.

Hernandez PH, Gillette JR and Mazel H (1967) Studies on the mechanism of mammalian hepatic azoreductase. I. Azoreductase activity of reduced nicotinamide adenine dinucleotide phosphate cytochrome c reductase. *Biochemical Pharmacology*, **16**, 1859–1875.

Holtzman JL, Crankshaw DL, Peterson FJ and Polnaszek CF (1981) The kinetics of the aerobic reduction of Nitrofurantoin by NADPH-cytochrome P-450(c) reductase. *Molecular Pharmacology*, **20**, 669–673.

Huang MT, Miwa GT, Cronheim N and Lu AYH (1979) Rat liver cytosolic azoreductase. Electron transport properties and the mechanism of dicumarol inhibition of the purified enzyme. *Journal of Biological Chemistry*, **254**, 11223–11227.

Le Campion L, Delaforge M, Noel JP and Ouazzani J (1999) Metabolism of ^{14}C-labelled 5-nitro-1,2,4-triazol-3-one (NTO): comparison between rat liver microsomes and bacterial metabolic pathways. *Journal of Molecular Catalysis—B Enzymatic*, **5**, 395–402.

Lehman-McKeeman LD, Stuard SB, Caudill D and Johnson DR (1997a) Induction of mouse cytochrome P-450 2B enzymes by amine metabolites of musk xylene; contribution of microsomal enzyme induction to the hepatocarcinogenicity of musk xylene. *Molecular Carcinogenesis*, **20**, 308–316.

Lehman-McKeeman LD, Johnson DR and Caudill D (1997b) Induction and inhibition of mouse cytochrome P-450 2B enzymes by musk xylene. *Toxicology and Applied Pharmacology*, **142**, 169–177.

Levine WG and Zbaida S (1988) Microsomal azoreductase mechanism studied with compounds structurally related to dimethylaminoazobenzene (DAB). II International ISSEX Meeting—ISSX, *Xenobiotic Metabolism and Disposition*, Kobe, Japan, Abstract No II–403-P3.

Levine WG and Zbaida S (1991) Two classes of azo dye reductase activity associated with rat liver microsomal cytochrome P-450. In *Advances in Experimental Medicine and Biology*, Vol. 283, *Biological Reactive Intermediates IV, Molecular and Cellular Effects and their Impact on Human Health*, Witmer CW, Snyder RR, Jollow DJ, Kalf GF, Kocsis JJ and Sipes IG (eds), Plenum Press, New York, pp. 315–321.

Levine WG and Zbaida S (1992) Chemical shifts and Hammett substituent constants in the mechanism of microsomal azoreductase. *FASEB Journal*, Abstract 3640.

Levine WG, Stoddard A and Zbaida S (1992) Multiple mechanisms in hepatic microsomal azoreduction. *Xenobiotica*, **22**, 1111–1120.

Li AP, Kaminski DL and Rasmussen A (1995) Substrates of human hepatic cytochrome P-450 3A4. *Toxicology*, **104**, 1–8.

Mason RP (1997) Physical chemical determinants of xenobiotic free-radical generation-the Marcus theory of electron transfer. In *Free Radical Toxicology*, Wallace KD (ed.), Taylor & Francis, London, pp. 15–24.

McLane KE, Fisher J and Ramakrishnan K (1983) Reductive drug metabolism. *Drug Metabolism Reviews*, **14**, 741–799.

Mitchell CE, Bechtold WE and Belinsky SA (1993) Metabolism of nitrofluoranthenes by rat lung subcellular fractions. *Carcinogenesis*, **14**, 1161–1166.

Nakao M, Goto Y, Hiratsuka A and Watabe T (1991) Reductive metabolism of nitro-para-phenylenediamine by rat liver. *Chemical and Pharmaceutical Bulletin*, **39**, 177–180.

Orna MV and Mason RP (1989) Correlation of kinetic parameters of nitroreductase enzymes with redox properties of nitroaromatic compounds. *Journal of Biological Chemistry*, **264**, 12379–12384.

Ortiz de Montellano PR and Reich NO (1986) Inhibition of cytochrome P-450 enzymes. In *Cytochrome P-450 Structure Mechanism and Biochemistry*, Ortiz de Montellano PR (ed.), Plenum Press, New York, pp. 273–314.

Patterson LH and Raleigh SM (1998) Reductive metabolism: its application in prodrug activation. *Biomedical Health Research*, **25**, 72–79.

Person A, Wolf C, Berger V, Fau D, Chachaty C and Fromenty B (1991) Generation of free-radicals during the reductive metabolism of the nitroaromatic compound, nilutamide. *Journal of Pharmacology and Experimental Therapeutics*, **257**, 714–719.

Peterson FJ, Holtzman JL, Crankshaw and Mason RP (1988) Two sites for azo reduction in the monooxygenase system. *Molecular Pharmacology*, **34**, 597–603.

Porter TD and Coon MJ (1991) Multiplicity of isoforms, substrates, and catalytic and regulatory mechanisms. *Journal of Biological Chemistry*, **266**, 13469–13472.

Sandhu P and Chipman JK (1991) Role of oxidation and azo reduction in the activation of chrysoidine dyes to genotoxic products. *Progress in Pharmacology and Clinical Pharmacology*, **8**, 319–325.

Scheline RR (1980) Drug metabolism by the gastrointestinal microflora. *Monographs in Pharmacology and Physiology*, **5**, 551–580.

Seree EJ, Pisano PJ, Placidi M, Rahmani R and Barra YA (1993) Identification of the human and animal hepatic cytochromes P-450 involved in clonazepam metabolism. *Fundamental and Clinical Pharmacology*, **7**, 69–75.

Silvers KJ, Eddy EP, McCoy EC, Rosenkranz HS and Howard PC (1994) Pathways for mutagenesis of 1-nitropyrene and dinitropyrenes in the human hepatoma cell line HepG2. *Environmental Health Perspectives*, **102 (Suppl. 6)**, 195–200.

Stanford MA, Swartz JC, Phillips TE and Hoffman BM (1980) Electronic control of ferroporphyrin ligand-binding kinetics. *Journal of American Chemical Society*, **102**, 4492–4499.

Stoddart AM and Levine WG (1992) Azoreductase activity by purified rabbit liver aldehyde oxidase. *Biochemical Pharmacology*, **43**, 2227–2235.

Walker R (1970) The metabolism of azo compounds. A review of the literature. *Food and Cosmetic Toxicology*, **8**, 656–676.

Washburn PC and Di Giulio RT (1988) Nitrofurantoin-stimulated superoxide production by channel catfish (*Ictalurus punctatus*) hepatic microsomal and soluble fractions. *Toxicology and Applied Pharmacology*, **95**, 363–377.

Zbaida S (1995) The mechanism of microsomal azoreduction: predictions on electronic aspects of structure–activity relationships. *Drug Metabolism Reviews*, **27**, 497–516.

Zbaida S and Levine WG (1990a) Characteristics of two classes of azo dye reductase activity associated with rat liver microsomal cytochrome P-450. *Biochemical Pharmacology*, **40**, 2415–2423.

Zbaida S and Levine WG (1990b) Sensitivity of azo dye reduction to carbon monoxide and oxygen. A probe for two different microsomal reduction pathways. *FASEB Journal*, Abstract 2743.

Zbaida S and Levine WG (1991a) A novel application of cyclic voltammetry for direct investigation of metabolic intermediates in microsomal azo reduction. *Chemical Research in Toxicology*, **4**, 82–88.

Zbaida S and Levine WG (1991b) The role of electronic factors in binding and reduction of azo dyes by hepatic microsomes. *FASEB Journal*, Abstract 2745.

Zbaida S and Levine WG (1992a) Microsomal azoreduction. Inhibition by CO, cyanide and azide. *FASEB Journal*, Abstract 5267.

Zbaida S and Levine WG (1992b) Role of electronic factors in binding and reduction of dyes by hepatic microsomes. *Journal of Pharmacology and Experimental Therapeutics*, **260**, 554–561.

Zbaida S, Stoddart AM and Levine WG (1988) Studies on the mechanism of azo dye carcinogen reduction by rat liver microsomes. *FASEB Journal*, **2**, Abstract 4412.

Zbaida S, Stoddart AM and Levine WG (1989) Studies on the mechanism of reduction of azo dye carcinogens by rat liver microsomal cytochrome P-450. *Chemico-Biological Interactions*, **69**, 61–71.

Zbaida S, Brewer CF and Levine WG (1992) Substrates for microsomal azoreductase. Hammett substituent effects, NMR studies, and response to inhibitors. *Drug Metabolism and Disposition*, **20**, 902–908.

Zbaida S, Brewer CF and Levine WG (1994) Hepatic microsomal azoreductase activity. Reactivity of azo dye substrates is determined by their electron densites and redox potentials. *Drug Metabolism and Disposition*, **22**, 412–418.

Index

Pages in bold denote principal subject areas that are discussed in depth.

Acebutolol 444
Acetaminophen, see under paracetamol
2-Acetylaminofluorene 23, 362–363, 399, 401, 471
N-Acetylases (N-Acetyltransferases) **441–457**
   catalytic activity 441–442
   genetics 29, 445–446
   in carcinogen metabolism 200, 297, 444
   in drug metabolism 175, 443
   in endogenous metabolism 445
   in the metabolism and bioactivation of arylamines 442–443
   inhibition 448–449
   isozymes 445, 449–450
   pharmacogenetics 449–452
   regulation 448–449
   species differences 446–448
   subcellular localisation 441
   substrates 444–446
   tissue distribution 446
Acetylation
   N-acetylation 4, 5, 16, 17
   O-acetylation 24
N-Acetylbenzidine 193, 196, 197, 201, 202
Acetylcholine esterase 466
N-Acetylcysteine 23
N-Acetyldopamine 241, 242
Acetyl salicylic acid 192, 202, 214, 260, 267, 399
Acridine 152
Acrolein 258, 325, 331
Acrylonitrile 247–248, 249
Acyl-CoA synthetases 511–512
Acyl-CoA:amino acid N-acyltransferase 512–513
S-Adenosyl methionine 17
Adrenaline (epinephrine) 98, 102, 125, 129, 204, 241, 242, 487, 488, 491, 492
Aflatoxin B₁ 5–6, 193, 200, 246–247, 324, 338, 399

Aflatoxin B₁-8,9-epoxide 322
Agaricus bisporus 1, 538
Agmatine 134
Ah (aryl hydrocarbon) receptor 37, 213, 282, 285, 302, 305, 340
AHAT (aryl hydoxyamic acid acetyl transfer) 441–442, 444
Albendazol 73
Alcohol dehydrogenase 159
Aldehyde dehydrogenases 95–96, 109, 110, 112, 113, 124, 131, 133,159
Aldehyde oxidases see under molybdenum hydroxylases
Aldehyde oxidoreductase 148
Aldehyde reductases 95–96, 109, 110, 112, 113, 124, 131
Aldrin 237, 248, 250
Aliphatic hydroxylation 7,8
Alizarin 293
Allopregnenolone 377
Allopurinol 165, 166–168, 175, 176, 177
Allylamine 129, 130
Alprenalol 108, 109
Alzheimer's disease 524
Amiflamine 99
Amifostine 131–132
Amine oxidases **95–136**
   catalytic activity 95–96
   diamine oxidase **132–134**
   monoamine oxidases **96–125**
   polyamine oxidase **134–136**
   semicarbazide-sensitive **125–132**
Amino acid conjugations **501–520**
   acyl CoA synthetases 511 512
   acyl–CoA:amino acid N-acyltransferase 512–513
   acyl transfer 503–504, 514
   mechanism 502, 504–506
   species differences 513–514
   tissue distribution 513–514

Amino acid conjugations (*cont.*)
  with aliphatic and alicyclic carboxylic acids
    507
  with aromatic carboxylic acids 508–509
  with arylacetic acids 509–510
  with aryloxyacetic acids 510–511
Aminoacetone 129
4-Aminoazobenzene 366, 385, 402, 491, 560
p-Aminobenzoylglutamate 445
2-Aminobiphenyl 28
4-Aminobiphenyl 10, 28, 237, 245, 282, 297,
    444, 445
Aminocarb 249, 250
2-Aminofluorene 19, 196, 197, 201, 204, 246,
    445
Aminoglutethimide 399, 444
Aminoguanidine 132
Aminopeptidase M 321
p-Aminophenol 265
Aminopyrine 193, 197–198, 237, 249–251,
    253
5-Aminosalicylic acid 7, 14, 445
1-Aminotriazole 202
Amitriptyline 104, 252, 253, 293
Amitrol 204
Amphetamine 76, 99, 106, 107, 536
Amsacrine 166, 167
Amyloid -peptide 524
Amytal 167
Androgens 288, 294, 295
Androsterone 294, 295, 377
  epoxides 468, 469
ANFT (2-amino-4-(5-nitro-2-furyl)thiazole) 204
Aniline 204
Anthraquinones 292, 294
Antimycin A 167
Antioxidant response elements (AREs) 303, 304
Apomorphine 488, 491
Apoptosis 324
Arabidopsis thaliana 35
Arachidonic acid 190, 203, 210, 213, 214,
    231–232, 247, 292, 473
Arginine decarboxylase 134
ARNT (Ah receptor nuclear transporter) 37, 302
Aromatic amines (arylamines) 7, 17, 25, 85,
    196, 242, 246, 282, 297, 373, 378
  N-acetylation **441–457**
Aromatic-L-amino-acid decarboxylase 103,
    107
Aromatic hydroxylation 5
Aryl-1,3-dithiolane 83
Aspergillus flavus 246
Aspergillus niger 467
Aspergillus orizae 380
Aspergillus parasiticus 246
Ascorbic acid 204, 241, 242, 376, 533

Aspirin see acetyl salicylic acid
Atherosclerosis 236
Atrazine 324
Atropine 364–366, 402
Autoantibodies 20
AVS (2(R,S)-1,2-bis(nicotinamido)propane) 150,
    151
Azathioprine 168
Azoreductases 163, **559–563**
  induction 561
  mechanism 559

Baicalein 261,
Baicalin 536
Bay X1005 261
BCNU (1,3-*bis*(2-chloroethyl)-1-nitrosourea)
    325
Benoxaprofen 298
Benserazide 488
Benzaldehyde 159, 160, 164, 173
Benzanthracene 342
7-Benz[a]anthracene-5,6-oxide 473
Benzidine 193, 196, 197, 201, 236, 237, 239,
    240, 246, 257, 293, 399, 444
Benzoic acid 17
Benzo[a]pyrene 26, 200, 213, **242–245**, 260,
    294, **295–297**, 324, 470
Benzo[e]pyrene 213
Benzo[a]pyrene 4,5-oxide 468
Benzoquinolines 151-152
Benzoquinone 167
Benzydamine 82
Benzyl isothiocyanate 322, 327, 331, 341, 342
Benzylamine 97, 98, 105–107, 125, 126, 127,
    128
Bile acids 371, 376–377, 383, 503, 522
Bilirubin conjugation 284, 287–288, 291–292,
    297, 305
Bioactivation **17–26**
Biochanin A 290
Biphenyl 5, 6
Bladder cancer 449
BOF-4242 168
Breast cancer 336, 344, 411, 490, 526
Brofaromine 99, 104
Brominidine 151, 154
Bromobenzene 19, 21–22
6-Bromomethyl-(9H)-purine 164
Bromosulphophthalein 320
Brompheniramine 507
Budenoside 394
Buprenorphine 292, 293
Butadiene 326
Butylated hydroxyanisole (BHA) 260, 303, 341
Butylated hydroxytoluene (BHT) 262

t-Butylhydroquinone 303, 304, 340
BW A4C 261

Cadaverine 134
Caenorhabditis elegans 35, 284
Caffeine 82, 175–176, 444
Calreticulin 71
Captopril 494
CAR (constitutively active receptor) 37
Carbamazepine 10,11-oxide 468
Carbazeran 151, 153, 154, 173
Carbidopa 488, 489
Carbon tetrachloride 9, 19
Carboxylesterases 447
Carnitene 503
-Carotene 264
Caroxazone 99
Catalase 27
Catechol-O-methyltransferase 110
Catecholoestrogens 199, 288, 294, 295, 487, 488
CCD ((1-(4-carboxy-phenoxy)-10-(4-chlorophenoxy)decane) 503
CDNB (1-chloro-2,4-dinitrobenzene) 320, 331, 332
Chlorambucil 9, 15, 325
Chloramphenicol 7, 13, 301, 376
Chlordimeform 250
Chloroperoxidase 35
Chlorpheniramine 507
Chlorpromazine 80, 166, 167, 239, 253, 254, 257, 448
Cholesterol sulphonation 360, 376, 395–396
Choline 76
Chondrodysplasia punctata 527
Cicletadine 361
Cimetidine 73, 81, 82, 167
Cimoxatone 99
Cinchonidine 152
Cinchonine 152
Cinnamic acids 507
Ciprofibrate ethyl ester 211, 212
Cirrhosis 304
Citalopram 111–114, 160, 161
Clofibrate 211, 212, 297, 302, 510, 530, 561, 562
Clomipramine 252, 253
Clonazepam 557
Clorgyline 95, 96, 97, 99, 101, 104, 112, 116, 118, 121
Clozapine 81, 82, 293
Codeine 293
  6-sulphate 361
Colon cancer 411
Colorectal cancer 449

Conjugation
  with aminoacids 13–14, 17, **501–520**
  with glucuronic acid 2, 3, 9, 23, **281–318**
    in Gilbert's syndrome 24
    in the transport of reactive intermediates 25
  with glutathione 11–13, 21–22, 23, 265–267, **319–352**
  with sulphate 2, 3, 9, 11, 23, **353–439**
    in bioactivation 24
Corticosterone 102
Cortisol 371, 376, 377, 392, 394, 396, 397, 405
Cotinine 159
Coumarins 294, 340, 341
Creatine 129
Crigler-Najjar syndrome 24, 284, 287
Crystalluria 4
Cumene hydroperoxide 331, 332, 334
Curcumin 240
Cycasin 21
Cyclamate 529
Cyclohexanone monooxygenase 68, 84
N-Cyclohexenyl-N-(3-phenylpropyl)urea 475
Cyclophosphamide 267, 325
Cyclosporin 532
Cyproheptadine 292, 293
4-S-Cysteaminylphenol 130, 131
Cysteinylglycine dipeptidase 321
Cytochrome P450 29, **33–55**, 467
  carbon hydroxylation 43
  catalytic cycle **41–43**
  catalytic activity **51–53**
  cooperativity 54
  CYP1A1 36, 37, 39, 303, 304, 404, 558
  CYP1A2 23, 29, 34, 36, 37, 39, 115, 175, 442, 444, 446, 556, 557, 558
  CYP1B1 36, 37, 39
  CYP2A6 175
  CYP2B 37, 38, 39, 52, 84, 556, 558
  CYP2C5 40, 41, 467
  CYP2C6 84
  CYP2C9 443
  CYP2C11 38, 52
  CYP2C19 74
  CYP2D6 35, 53, 74, 443
  CYP2E1 24, 38, 50, 51, 52, 76, 159
  CYP3A 165
  CYP3A4 34, 36, 39, 53, 73, 556, 557, 558
  CYP4A 39
  CYP19 53
  CYP101 40, 41, 42, 50
  CYP102 41
  CYP107 40
  CYP108 40
  dehydrogenation 43–44

Cytochrome (*cont.*)
  effect of disease 304
  heteroatom dealkylation 45–46
  heteroatom oxygenation 45
  in endogenous metabolism 34, 232, 359
  in xenobiotic metabolism 25, **33–34**, **43–51**,
      107, 112, 114, 119, 152, 154, 158, 159,
      160, 161, 163, 173, 200, 202, 243, 245,
      246, 247, 248, 251, 252, 254, 267, 321,
      442–443, 444, 446, 460, 488, 507,
      556, 557, 559
  induction 24, 29, 39, 76, 404, 557–558
  inhibition **53–55**, 259, 404
  nomenclature **34–35**
  one electron oxidation 44
  one electron reduction 44
  oxidation of aromatic rings 47–48
  oxidation of olefins and acetylenes 46–47
  polymorphisms 74
  regulation **36–39**
  structure **39–41**
  tissue distribution 24, 35–36
Cytochrome P450 reductase 160, 467, 556
Cytokines 208, 209, 295, 304, 526

DACA (*N*-[(2-dimethylamino)ethyl]acridine-
      4-carboxamide) 151, 152, 153
Daidzein 527
Danazol 526
Dapsone 406, 444
Daunorubicin 533–534
DCNB (1,2-dichloro-4-nitrobenzene) 320, 331,
      334
DDT (dichlorodiphenyltrichloroethane) 324
N-Deacetyl ketoconazole 82
Dealkylation 7, 9
Debrisoquine 4-hydroxylation 74
Deconjugating enzymes **521–554**
  glucuronidases **529–539**
  sulphatases **521–529**
Dehydroepiandrosterone 358, 360, 371, 372,
      376, 377, 378, 383, 392, 393, 404, 405,
      407, 408
Deprenyl 95, 96, 97, 99, 101, 102, 104, 112,
      113, 116, 119, 121
Desipramine 252, 253, 378–379, 408
Desulphovibrio gigas 148
Dexamethasone 102, 210, 214, 259, 302,
      340–342, 396, 535
Diabetes 258, 397, 532, 533
Diamine oxidase 125, 132–134
o-Dianisidine 257
DT-Diaphorase (quinone oxidoreductase) 26,
      27, 160, 166, 296, 303, 556, 559
Diazepam 2, 3

Dibromoethane 322, 326, 327, 331
2,6-Dichloro-4-nitrophenol (DCNP) 400,
      403–404
Dichloromethane 325–326, 327, 331
2,4-Dichlorophenoxy acetic acid (2,4-D)
      510–511, 513
Diclofenac 192, 214
Dicrotophos 250
Dieldrin 248
Diethylmaleate 341, 385
Diethylstilboestrol 193, 195, 201, 204, 253,
      254, 369, 370
N, *N*-Diethyltryptamine 106
Diflunisal 407, 412, 530
Digitalis glycosides 536
Dihomo–linolenate 203
Dihydralazine 20
Dihydropyrimidine dehydrogenase 153
5,6-Dihydroxyindole 241
5,6-Dihydroxyindole-2-carboxylic acid 241
1,25-Dihydroxyvitamin D_3 210, 259, 359
Diltiazem 252, 253
Dimethadione 202
Dimethyl aminoazobenzenes 163, 385, 402,
      560, 561
7,12-Dimethylbenz[a]anthracene 259, 343,
      460
1,2-Dimethylhydrazine 247
Dimethylnitrosamine 19, 106, 115
N, *N*-Dimethyltryptamine 111
3,9-Dinitrofluoranthene 557
1,6-Dinitropyrene 556
Diosgenin 293
Dipeptidase 5
Diphenylisobenzofuran 193
Disulfiram 76, 167
Dobutamine 488
Docarpamine 110, 111
Docosahexaenoic acid 203
L-Dopa 239 , 241, 242, 257, 379, 487, 488,
      489
Dopamine 97, 98, 102, 110, 111, 126, 167,
      241, 242, 257–258, 359, 374, 487, 488
  sulphate conjugation 360, 374, 380, 393,
      396–397, 408, 410
Dopamine–hydroxylase 105
Doxepin 252, 253
Doxocycline 530, 536
Doxorubicin 533
Drosophila melanogaster 35, 284, 287, 289
Dubin-Johnson syndrome 301

Eicosapentanoic acid 203
Eicosatetraenoic acid 212
Electrophile response elements (EpREs) 303

Enterohepatic circulation 5, 301, 530, 532, 533, 536
Ephedrine 376
(-)-Epicatechin 17
Epinephrine see under adrenaline
Epinine 111, 112
Epoxide hydrolases **459–483**
  catalytic activity 14–17, 459–461
  forms 459–460
  in the bioactivation of chemicals 460–461
  inhibition 404
  mechanism 464–467
  microsomal **467–472**
    induction 471
    inhibition 471–472
    polymorphisms 472
    species differences 471
    structure 467
    subcellular localisation 469–470
    substrates 467–469
    tissue distribution 470–471
  phylogenetics 461–464
  soluble **472–47**
    induction 475
    inhibition 475
    polymorphisms 476
    species differences 475
    structure 472–473
    subcellular localisation 474
    substrates 473–474
    tissue distribution 474
  structure 461–464
Epoxides
  formation 5–6, 46, 47–48, 459
  metabolism 14, 17
1,2-Epoxy-3-(p-nitrophenoxy)propane 320
cis-9,10-Epoxystearic acid 468, 473
Escherichia coli 382
Esculetin 261
Esterases 111
Estragole 1, 366
Ethacrynic acid 265, 266, 331
Ethionine 399
7-Ethoxycoumarin 2, 3, 7
Ethoxyquin 341
trans-Ethyl styrene 7,8-oxide 473
N-Ethylmaleimide 320, 449
2-Ethynyloestradiol 39, 292, 293
5-Ethynyluracil 153
ETYA (5,11,181,14-eicositetraynoic acid) 240, 261
Eugenol 288, 293, 295

Famciclovir 151, 155, 176
Famphur 250

Fenbendazole 10
Fenoprofen 503–504, 512
First-pass effect 36
Fisetin 486, 488
FLAP (five lipoxygenase activating protein) 236, 321, 323, 332, 337
Flavanols 2
Flavin monooxygenases 29, **67–86**
  biochemical properties **77–78**
  catalytic cycle 79–80
  FMO1 69, 70, 71, 72, 73, 74, 79, 80, 82, 84
  FMO2 69, 70, 71, 74, 77, 80, 84
  FMO3 69, 70, 71, 73, 74, 76, 80, 81, 82, 84
  FMO4 69, 73, 81
  FMO5 69, 73, 81, 85
  FMO6 69, 81
  forms 68
  genetics 69–70, **73–76**
  in bioactivation 25, 84–85, 246
  in pregnancy 74
  in xenobiotic metabolism 67–68, 160
  induction 74
  inhibition 76
  nomenclature 68–69
  ontogenic expression 73
  polymorphisms 74–75, 76
  sex differences 81
  species differences 74, 81
  stereoselective metabolism 81, **82–84**
  substrates 80–81
  thermal inactivation 77
  tissue distribution 73–74
Flavonoids 215, 265, 292, 294, 405, 488, 527
Flosequinan 164
Flunitrazepam 532
5-Fluoro-4-pyrimidinone 153–154
4-Fluorochalcone oxide 475
5-Fluorouracil 533–534
Flutamide 558
Fluvanilate 503
Formaldehyde dehydrogenase 112, 129
Formetanate 250

Garlic 2
Gilbert's syndrome 24, 288
Glucocorticoids 209–210, 214, 340–342, 396
Glucuronidases 25, **529–539**
  bacterial-glucuronidase 536–537, 538
  catalytic activity 529–530
  -glucuronidase 538
  glucuronosyl-diphosphglucosamine glucuronidase 538
  glycyrrhizinate -glucuronidase 538
  heparanase 537
  human -glucuronidase 530–534

Glucuronidases (cont.)
  hyaluronidase 538
  rodent -glucuronidase 534–536
Glucuronyl transferases **281-318**
  catalytic activity 282–284
  cofactor 300
  effect of disease 304–305
  effect on nutrition 398
  in the formation of reactive intermediates
      297–298
  genetics 285–286
  in carcinogen metabolism **295–297**
  induction 301–303, 398–399
  isoforms 285–286
  membrane topology 299–301
  nomenclature 284–285
  polymorphisms 287–289
  regulation **299–305**
  species differences 289, 302, **305–307**, 389
  structure 286–287
  substrates 290–298, 509, 510
  tissue distribution 281, 289–290
  transport of glucuronides 24–25, 281–282,
      300–301
GLUT 4 (Glucose transporter) 126
-Glutamyltransferase 5, 321
-Glutamyltranspeptidase 24
Glutathione 4, 5, 11, 21, 23, 78, 85, 236–237,
    241–242 see also under glutathione S-
    Transferases
Glutathione peroxidase 26, 27
Glutathione S-transferases **319–352**
  binding of non-substrate ligands 323
  catalytic activity 5, 11–12, 319, **321–323**
  effect of selenium deficiency 339–340
  genetics **329–336**
  hormonal regulation 337–339
  in endogenous metabolism 329
  in xenobiotic bioactivation **325–328**
  in xenobiotic detoxication 13, 296, 324–
      325, 460
  induction 303, 340–343
  inhibition 260, 404
  isomerisation activity 322–323
  peroxidase activity 321–322
  polymorphisms 334
  protein-protein interactions 324
  sex differences 337–339
  subcellular localisation 319–320
  superfamilies 319–320
  tissue distribution 336–337
P-Glycoprotein 73
Glycyrrhizin 530, 535, 536
Gonadotropins 209
GRE (glucocorticoid receptor element) 38
Guaiacol 257

Gunn rat 296
Gut microflora
  in xenobiotic metabolism 5
  in bioactivation 21

Halothane 7, 12, 20
Harmaline 99
Harmine 99
Harmol 391, 392, 400, 401, 404, 411
HEOM (1,2,3,4,9,9-Hexachloro-
    1,4,4a,5,6,7,8,8a-octahydro-exo-6,7-
    epoxy1,4-methano naphthalene 468
Heparanase 537
Hepoxins 233
Heterocyclic amines 1, 17, 298, 378
Hexachlorobutadiene 327
Hexestrol 204
HIF (hypoxia inducible factor) 1, 37
Hippuric acid 17, 501, 508
Histamine 103, 105, 132, 134
HMN (3-hydroxymethyl-1-{[3-(5-nitro-
    2-furyl)allylidene]amino}hydantoin) 204
HNF (hepatic nuclear factor) 37
Homocysteine 486
Homovanillamine 160
Hordenine 106
Hordinine 111
Horseradish peroxidase 48, 84, 195, 196, 197,
    204, 247
Hyaluronidase 538
Hydralazine 132, 167, 397, 444
Hydrazines 1, 17, 163
Hydrogen peroxide 26, 27, 77, 78
Hydrolysis 9, 15
Hydroquinone 204
7-Hydroxy-DACA (7-hydroxy-
    N[(2-dimethylamino)ethylacridin-
    6-caboxamide 167
1(-Hydroxyethyl)pyrene 362, 363, 371
5-Hydroxyindoleacetaldehyde 160
4-Hydroxyionone 241
Hydroxyl radical 18, 20, 26, 27
1-Hydroxymethylpyrene 365, 368, 370, 392,
    404, 409
4-Hydroxynonenal 325, 331, 334, 342–343
5-Hydroxytryptamine (serotonin) 97, 98, 102,
    103, 108, 126, 294 380
Hyodeoxycholic acid 288, 295
Hyperbilirubinaemia 287–288, 301, 533
Hypertension 397, 411
Hypoxanthine 151, 155, 175

Ibopamine 111, 112
Ibuprofen 192, 214, 503–504, 512

Ichthyosis 526
Iminotecan 24
Imipramine 80, 99, 252, 253, 292, 293, 307, 378, 530
Indole acetic acid 204
Indole-3-aldehyde 159
Indole-3-carbinol 76, 82, 340, 341
Indomethacin 192, 202, 214, 259, 260, 536
Interferon 171
Iodide 204
Iodothyronins 359, 369
-Ionone 241
Iproniazid 99, 107
IQ (2-amino-3-methylimidazo[4,5- f]quinoline 193, 196, 197, 200, 201, 202, 204–205, 378, 444
Irinotecan 288, 292, 301, 536
Isocarboxazid 99
Isomescaline 128
Isoniazid 9, 15, 443, 444, 447, 448, 449
Isoprenaline (isoproterenol) 253, 254, 392, 488
Isoprene 326
2-Isopropyl-MPTP 98
Isoproterenol see under isoprenaline
Isoquinoline 150, 151
Isosafrole 562
Isothiocyanates 321
Isotretinoin 202
Isovanillin 160, 167

Ketoconazole 53
Ketotifene 292
Kojic amine 106, 130
Kynuramine 98

Leopardus wiedii 306
Leukotoxin 460–461, 474
Leukotrienes 233, 320, 321, 329, 330, 337, 380
Lilly 51641 99
Lilly 54761 99
Lindane 324
Lipoic acid 204
Lipoxins 233
Lipoxygenases 231–279, 321
    drug-chemical interactions 257–258
    forms 23
    in glutathione conjugation 265–267
    in xenobiotic metabolism 236, 237–256
    induction 258–259
    inhibition 215, 259–265
    mechanism 233–236
    physiological function 203, 231–236
    subcellular localisation 236

    substrates 240–256
    tissue distribution 233, 235
Lorazepam 530, 536
-Lyase 4, 5, 12, 16, 326, 494

Maleylacetoacetate 322, 323, 329, 332, 334
Maleylacetoacetate isomerase 329
MD 220661 130
MD 240928 123–124
MD 240931 123–124
MD 780236 123–124, 130
MDL 27695 134–135
MDL 72145 126
MDL 72274A 127
MDL 72527 135
MDL 72727 134
MDL 72974A 127
Mechanism-based inhibitors 115–117
Medrogestone 526
MeIQx (2-amino-3,8-dimethylimidazo[4, 5- f]quinoxaline 23, 378
Melanin 242, 488
Melatonin 28, 374, 487
Menadione 155, 166, 167, 385
1-Menaphthyl sulphate 320, 331, 334
Mephalan 325
Mephenytoin 4-hydroxylation 74
6-Mercaptopurine 151, 156, 157, 167, 175, 493
Mercapturates 4, 5, 11, 16, 21, 321, 494
Mescaline 106, 107, 128
Metachromatic Leukodystrophy 523
Methadone 167
Methamphetamine 76
Methimazole 82
Methimidazole 85
L-Methionine S-adenosyltransferase (MAT) 485–486
Methotrexate 151, 153, 155, 173, 449, 505
3-Methoxytropolone 261
-Methyldopa 488
Methyl isocyanate 327
2-Methyl-MPTP 119
1-Methyl-4-phenyl-2,3-dihydropyridinium (MPDP$^+$) 158
1-Methyl-4-phenyl-1,2,3,6-tetrahydropyridine (MPTP) 85, 117–120
Methyl parathion 324–325
Methyl red 561
Methylamine 129
N-Methylaniline 193
Methylation 17
Methylazoxymethanol 21
3-Methylcholanthrene 74, 284, 302, 323, 342, 398, 556

-Methyldopa 107, 241, 407
-Methyldopamine 106, 107
N^r-Methylhistamine 98
3-Methylindole 202–203
N^l-Methylnicotinamide 157–158, 173
Methylphenylsulphide 194, 204
Methylsalicylate 293
Methyltransferases **485–499**
  catalytic activity 5, 16, 17, 485
  catechol-O-methyltransferase (COMT)
    487–491
    cellular distribution 490
    effect of disease 490
    forms 487
    hormonal regulation 489
    ontogenic expression 489
    polymorphisms 490
    structure 487–488
    substrates 488
    tissue distribution 490
  N-methyltransferases 491–492
    histamine-N-methyltransferase (HNMT)
      492
    phenetholamine-N-methyltransferase
      491–492
  O-methyltransferases **486–491**
  S-methyltransferases 493–494
    thioether-S-methyltransferase (TEMT) 493
    thiomethyltransferase (TMT) 494
    thiopurine-S-methyltransferase (TPMT) 493
  co-substrate 485–486
4-Methylumbelliferone 293, 306, 359, 522
Metyrapone 150–151, 259, 399
Midazolam 532
Migraine 411
Milacemide 98, 120–121, 128, 136
Minoxidil 361, 368, 373, 374–375, 380, 383,
  391, 396, 401, 408
Misonidazole 2, 4
Moclobemide 99, 104, 125
Molybdate 404–405
Molybdenum cofactor deficiency 177
Molybdenum hydroxylases **147–187**
  ethnic differences 176
  function 95–96, 112, 113,147–149, 555,
    556
  genetics 173, 176–177
  hormonal regulation 173–176
  inhibition 114,165–168
  ontogenic expression 176
  mechanism 148-149
  sex differences 173–174, 175
  species differences **172–173**
  strain differences 173
  structure 148
  substrates **150–165**, 560

  tissue distribution **168–172**
Molybdopterin 177
Monoacetylspermidine 134, 135
Monoacetylspermine 134, 135
Monoamine oxidases **95–125**
  cellular localisattion 100
  inhibitors 95, 97–100, 115–125
  isoenzymes 96–97
  physiological function 102–103
  regulation 101–102
  species differences 96–97
  structure 96–97
  substrates 85, 97–100, **103–115**, 160
  tissue distribution 100–101, 104
  turnover 102
Monoethylhexyl phthalate 211, 212
Monomethylhydrazine 127
Morphine 289, 293, 295, 301, 491, 530, 536
  3-glucuronide 305, 306, 361
  6-glucuronide 290, 305, 306, 361
MPTP (1-methyl-4-phenyl-1,2,3,6-
  tetrahydropyridine) 98
Mucopolysaccharidosis
  II (Hunter disease) 528
  III (Sanfilippo D) 528
  IV A (Morquio A syndrome) 528
  VII 535–536
Multidrug resistance proteins (MRPs) 281–282,
  321, 325, 329, 357
Multiple sulphatase deficiency 522
Musk xylene 557–558
Mycophenolic acid 293, 294
Mycotoxins 1
Myeloperoxidase 443

Nalorphine 292
Naloxone 293
Naphthalene 17
-Naphthoflavone 302, 303, 340, 341, 562
1-Naphthylamine 293, 294
2-Naphthylamine 204, 282, 293, 294, 297,
  380, 444, 445
Naringenin 293, 527
NDGA (nordihydroguairetic acid) 261
Neurosteroids 360
NF-B (transcription factor-kappaB) 208, 213,
  215
Nicotine 10, 81, 159
Nifurtimox 556
NIH shift 47
Nilutamide 556
Nitrazepam 444
Nitrenium ion 24, 25
Nitric acid synthase 35
4-Nitrobenzene sulphonamide 161

6-Nitrochrysene 557
3-Nitrofluoranthene 557
Nitrofurazone 162
Nitrogen oxidation 7, 10
1-Nitropyrene 556, 557
4-Nitroquinoline N-oxide 162
Nitroreductases **555–559**
  catalytic activity 555–556
  mechanism 558–559
Nitrosamines 163
NNAL (4-(methylnitrosamino)-1-(3-pyridyl)-
  1-butanol) 295, 305
NNK (4-(methylnitrosamino)-1-(3-pyridyl)-
  1-butanone) 240, 245, 258, 267
NO synthase 162
Non-steroidal antiinflammatory drugs (NSAIDs)
  190, 192, 206, 212, 214, 215, 259, 295,
  297, 306, 503–504, 510, 536
Noonon's syndrome 46
Noradrenaline (norepineprhine) 17, 97, 98,
  102, 103, 104, 108, 126, 241, 242, 487,
  491, 492
Norepinephrine see under noradrenaline
Norharman 167

Obesity 411
Octane-1,2-epoxide 468
Octopamine 103, 105, 111, 126
Oestradiol 166, 167, 174, 193, 198–199, 208,
  210, 288, 292, 371, 375, 376, 377, 382,
  383, 396, 526, 532, 535
Epoxides 469
Oestriol 375, 395, 397, 526
Oestrone 371, 375, 378, 391, 395, 397
Oltipraz 303, 341
Organic nitrates 162, 322
Organochlorine insecticides 248
Organophosphates 248
Organosulphates 2
Oxazepam 288, 304, 530
Oxidative deamination 7, 11
Oxidative dehalogenation 7,12
Oxidative stress 26, 78, 281, 303, 324
N-Oxides
  reduction 162
Oximes
  reduction 162–163
Oxipurinol 162, 167–168, 176
Oxprenalol 108, 109
Oxyphenbutazone 253
Ozone 213, 258

Pancreatitis 535
Paracetamol 3, 4, 11, 19, 23, 24, 195–196,
  201, 204, 293, 294, 302, 303, 306, 343,
  368, 385, 404, 406, 407, 410, 411, 412,
  449, 530
Parathion 238, 248–249, 251
Parathyroid hormone 210
Pargyline 95, 99, 102, 107, 116, 121, 128
Parkinson's disease 102, 119, 120, 257, 258,
  397–398, 489
PD 14616 261
Penciclovir 155-156, 174, 176
D-Penicillamine 494
Pentachlorophenol 400, 403–404, 449
Pentobarbitone 7, 8
Pentylamine 98
Peroxisomal proliferators 212, 360, 471, 475
Phase I metabolism 2-4, **5-9**, 16, 17, 281–282
Phase II metabolism 2–4, **9–17**, 16, 17,
  281–282, 303, 442, 521
  in bioactivation 24, 281, 442
Phase III metabolism 4–5, 281–282
Phenacetin 196
Phenanthridine 171
Phenelzine 122–123, 444
Phenethyl isothiocyanate 327
Phenethylamine 488
4-Phenetidine 196, 197
Phenidone 193, 253, 261
Pheniprazine 122, 123
Phenobarbital 120, 284, 301, 306, 340, 341,
  342, 398, 471, 556, 562
Phenothiazines 167, 239, 240, 252–254
Phenylalanine hydroxylase 47
Phenylbutazone 193, 198
Phenylenediamine 257
Phenylethanolamine 105, 106, 107, 126
2-Phenylethylamine 97, 98, 103, 105, 107,
  125, 128
o-Phenylphenol 215
2-Phenylpropanolamine 106, 107
Phenylramidol 448
Phenytoin 202, 253, 260, 267
PhIP (2-amino-1-methyl-6-phenylimidazo[4,
  5-b]pyridine 26, 294, 297
Phorbol 12-myristate 13-acetate 259
Phosgene 258
Phospholipase A_2 190, 214
Phthalazine 171
Pirimicarb 250
Piroxicam 214
Pistonia Amazonica 111
Plvalic acid 503
Polyamine oxidase 132, 134–136
Polycyclic aromatic hydrocarbons 1, 38, 44,
  242, 295, 302, 340
Primaquine 106, 107, 128
PPAR (peroxisome proliferator activator
  receptor) 37, 212, 328, 329, 474, 475

PPHP (5-phenyl-4-pentenyl hydroperoxide 204
Pregnanediol 294
Pregnenolone 371, 377
Pregnenolone-16-carbonitrile 562
Proadifen 107, 114, 166, 167, 259
Probenecid 357–358
Probucol 240
Procainamide 196, 201, 212, 443, 444, 445, 448
Procarbazide 127
Progesterone 102, 167, 210, 396, 536
Proguanil 444
Prolintane 159
Promazine 253, 254
Promethazine 167, 253, 254
Pronethalol 108
Propanil 9, 15
Propofol 293, 294
D-Propoxyphene 167
Propranolol 108, 109, 302, 399, 530
Prostacyclins 190, 208
Prostaglandins 190, 210, 320, 329, 330, 380, 507–508
Prostaglandin synthases **189–229**
    genetics 190–191, 215–216
    in xenobiotic metabolism and bioactivation 25, **191–203**, 243, 245, 246, 247, 248, 251, 444
    inhibition 54, 214
    isozymes 189
    mechanism **192–196**
    ontogenic expression 207–208
    physiological function 189, 232
    regulation by endogenous factors 208–210
    regulation by exogenous factors 210–214
    sex differences 208
    species differences 206–207
    subcellular localisation 191
    substrates 203–205
    tissue distribution 24, 205–206
Prostate cancer 411, 535
Protein kinases 360
Protocatechuic aldehyde
Putrescine 134
PXR (pregnane X receptor) 37
Pyrazine 152
Pyridazine 152
Pyridine 150, 152
Pyridoxal 159
2-Pyrimidinone 148, 152–153, 160
4-Pyrimidinone 153, 166
Pyrogallol 257

Quadricylane 44
Quercetin 292, 293, 303, 486, 488, 527

Quinacrine 167
Quinazoline 154
Quinidine 53, 152
Quinine 152
Quinoline 151
Quinone reductase see DT-diaphorase
Quinones 27, 29, 322, 325, 340, 559
Quinoxaline 154

Ranitidine 81, 82, 530
RAR (retinoic acid receptor) 37
Reactive intermediates
    fate **24–26**, 533
    generation **18–22**, 34
Reactive oxygen species (ROS) **26–28**, 159, 208, 213, 320, 325
Reduction 2, 5, 7, 13, 14, 160–165
    in bioactivation 20–21
Reductive dehalogenation 7, 9, 12, 164
Resveratrol 214–215
Retinal 159
Retinoic acids 292, 297, 323, 508
Retinol 241, 264, 363–364
Rifampicin 39, 399
Ritodrine 393–394, 407, 408
Rizatriptan 115
Ro 16–6491 99, 125
ROFA (residual oil fly ash) 211, 213
RXR (retinoid X receptor) 342
Ryanodine receptors 324, 344

Safrole 1, 366, 400, 402
    hydroxylated 362, 365, 402
(-)-Salbutamol 408
Salicylic acid 260, 511, 513
Salicylamide 406, 407, 411
Salmonella typhimurium 447
Sarcosine 129
Schistosoma mansoni 523
Scopolamine 364–366
Scopoletin 293
Selegiline see under deprenyl
Semicarbazide 95, 128, 132, 134
Semicarbazide-sensitive amine oxidase **125–132**
    effect of disease 126
    inhibitors 124
    species differences 127
    substrates 105, 111, 120, **127–132**
Serotonin see under 5–hydroxytryptamine
Sharpless chiral oxidation reagent 84
Silica 211, 213
Silymarin 214, 533, 536
SKF 525A see under proadifen

SN-38 (7-ethyl-10-hydroxycamptothecin) 293
Spermidine 134
Spermine 134
Spironolactone 397
Steroid sulphatase deficiency syndrome 526
cis-Stilbene oxide 473
trans-Stilbene oxide 334, 340, 341, 471, 473
Stiripentol 362, 401
Streptozotocin 258
Styrene 247, 468–469
Styrene 7,8-oxide 468, 473
Suicide inhibitors see under mechanism-based inhibitors
Sulindac 82, 163,193, 194, 204, 214
Sulphamethazine 444, 445, 447, 448, 449
Sulphamethoxazole 196, 197, 443
Sulphanilamide 4, 161
Sulphasalazine 7, 14, 444
Sulphatases **521–529**
    arylsulphatase A (ARSA) 522–524
    arylsulphatase B (ARSB) 524–525
    arylsulphatase C (ARSC) 525–527
    arylsulphatase D (ARSD) 527
    arylsulphatase E (ARSE) 527
    arylsulphatase F (ARSF) 527
    catalytic activity 521–522
    bacterial 529
    non-aryl sulphatases 527–528
Sulphinpyrazone 163
Sulphonamides 17, 201, 447–448
Sulphoraphane 342
Sulphotransferases **353–439**, 493
    assays **386–388**
    catalytic activity 354
    cellular distribution 392–393
    drug interactions 406
    effect of cell and tissue differentiation 395–396
    effect of disease 397–398, 399
    effect on nutrition 398
    function 357–366
        in dehydration reactions 363–366
        in endogenous metabolism 358–360
        in isomerisation reactions 362–363
        in xenobiotic bioactivation 361, 366, 395
    groups 355
    hormonal regulation 396–397
    in cells in culture 399–400
    induction 395, 398–399
    inhibition 373–374, 400–406
    nomenclature **366–382**
    ontogenic expression 390–391, **393–395**
    polymorphisms 407–410
        associations with disease 410, 411
    mechanism 382–385
    sex differences 376, 394, 395, 399, 410

species differences 389–390
strain differences 406
structure 381–382
subcellular localisation 393
tissue distribution, 368, 369, 370, 371, 372, 390–392
zonal distribution 392
Sulphoxides
    reduction 163
Sulphur oxidation 7, 10
Sumatriptan 114
Superoxide dismutase 26, 27
Superoxy anion 26
Suramin 523, 528
Synephrine 106, 111

Tamoxifen 82, 160, 161, 395, 402
TCCD (2,3,7,8-tetrachlorodibenzo-p-dioxin) 39, 211, 212–213, 302, 303
TER 286 (-glutamyl–amino–(2-ethyl-N,N,N,N-tetrakis[2-chloroethyl]phosphorodiamidate)-sulphonyl)propionyl-(R)-(-)phenylglycine) 326, 327, 328
Testosterone 102, 174, 290, 339
    6-hydroxylase 54
    conjugates 292, 358, 359, 371, 394
Tetrachloroethene 327
Tetrahydro–carboline 99
Tetramethiuram 250
1,2,4,5-Tetramethoxybenzene 4
Tetramethylbenzidine 257
Tetramethylphenyldiamine 204, 257
Thalidomide 202
Thiacetamide 399
Thiobenzamide 82, 237
6-Thioguanine 151, 157
Thioridazine 159
Thiotepa 325
Thromboxanes 190, 208, 213, 380
Thyroid hormones 294, 359, 379, 383, 395, 397, 522, 535
Tigogenin 293
Timolol 108
Toloxatone 99
TPA (tetradecanol phorbol acetate) 210–212, 258-259
N,O-Transacetylation see under AHAT
Tranylcypromine 107
Tresperimus 132, 133
Triamterene 361
1,1,1-Trichloro-2,3-propene oxide (TCPO) 471, 472
2,4,5-Trichlorophenoxyacetic acid (2,4,5-T) 510–511, 513
Trifluoperazine 253

Triflupromazine 253, 254
3,4,6-Trihydroxyphenylethylamine 95
Trimeprazine 253, 254
Trimethylamine 74–75, 81
Trimethylaminuria 71, 75–76, 85
Trimipramine 252, 253
Trimoprostil 505, 508
Triptans 114–115
Triton X-100 167
Troglitazone 328
Tryptamine 98, 103, 105, 111, 127
Tyramine 11, 76, 97, 98, 102, 103–105, 126,
    127
Tyrosinase 131
Tyrosine 329, 379, 380, 383, 405–406

UDP-Glucose dehydrogenase 300
Urethane 163
Uric acid 28, 149, 204

Valproic acid 297, 503, 507
Valpromide 468, 471–472
Vanillin 159–160
Vascular adhesion protein 126
Verapamil 73
Vinyl chloride 5–6
Vitamin A 264, 376, 535
Vitamin E (-tocopherol) 28, 240, 241, 264

Vitamin K epoxide reductase 469

Warfarin 522, 527
WR-1065 129, 131–131
WY-14,643 ([4-chloro-6-(2,3-xylidine)-2-
    pyrimidyl-thio]acetic acid) 211, 212

Xamoterol 412
Xanomeline 82
Xanthine 149, 155, 160, 171, 175
Xanthine dehydrogenase 147, 149, 171
Xanthine oxidase see under Molybdenum
    hydroxylases
Xanthinuria 177
XRE (xenobiotic response element) 37, 302,
    303, 340

Zaleplon 151, 154, 173
Zectran 250
Zidovudine 295, 304
Zileuton 261, 265
Zipper proteins 303
Zolmitriptan 114, 115
Zomepirac 298
Zonisamide 165